Sea Surface Sound

Natural Mechanisms of Surface Generated Noise in the Ocean

NATO ASI Series

Advanced Science Institutes Series

A Series presenting the results of activities sponsored by the NATO Science Committee, which aims at the dissemination of advanced scientific and technological knowledge, with a view to strengthening links between scientific communities.

The Series is published by an international board of publishers in conjunction with the NATO Scientific Affairs Division

A Life Sciences	Plenum Publishing Corporation
B Physics	London and New York
C Mathematical	Kluwer Academic Publishers
and Physical Sciences	Dordrecht, Boston and London
D Behavioural and Social Sciences	
E Applied Sciences	
F Computer and Systems Sciences	Springer-Verlag
G Ecological Sciences	Berlin, Heidelberg, New York, London,
H Cell Biology	Paris and Tokyo

Proceedings of the NATO Advanced Research Workshop on
Sea Surface Sound
Natural Mechanisms of Surface Generated Noise in the Ocean
Lerici, Italy
15–19 June 1987

Library of Congress Cataloging in Publication Data

NATO Advanced Research Workshop on Natural Mechanisms of Surface
  Generated Noise in the Ocean (1987 : Lerici, Italy)
    Sea surface sound : natural mechanisms of surface generated noise
  in the ocean / edited by B.R. Kerman.
        p.   cm. -- (NATO ASI series. Series C, Mathematical and
  physical sciences ; 238)
    "Proceedings of the NATO Advanced Research Workshop on Natural
  Mechanisms of Surface Generated Noise in the Ocean, held in Lerici,
  Italy, 15-19 June 1987"--T.p. verso.
    Includes index.
    ISBN-13: 978-94-010-7856-6
    1. Underwater acoustics--Congresses. 2. Ocean waves--Congresses.
  3. Noise--Congresses. 4. Hydrodynamics--Congresses.   I. Kerman, B.
  R. (Bryan R.) II. Title. III. Series: NATO ASI series. Series C,
  Mathematical and physical sciences ; no. 238.
  QC242.N34 1987
  534'.23--dc19   88-15570
                                                                 CIP

ISBN-13: 978-94-010-7856-6   e-ISBN-13: 978-94-009-3017-9
DOI: 10.1007/ 978-94-009-3017-9

Published by Kluwer Academic Publishers,
P.O. Box 17, 3300 AA Dordrecht, The Netherlands.

Kluwer Academic Publishers incorporates the publishing programmes of
D. Reidel, Martinus Nijhoff, Dr W. Junk, and MTP Press.

Sold and distributed in the U.S.A. and Canada
by Kluwer Academic Publishers,
101 Philip Drive, Norwell, MA 02061, U.S.A.

In all other countries, sold and distributed
by Kluwer Academic Publishers Group,
P.O. Box 322, 3300 AH Dordrecht, The Netherlands.

Sea Surface Sound
Natural Mechanisms of Surface Generated Noise in the Ocean

edited by

B. R. Kerman
Atmospheric Environment Service,
Toronto, Canada

Kluwer Academic Publishers

Dordrecht / Boston / London

Published in cooperation with NATO Scientific Affairs Division

CONTENTS

PROPAGATION EFFECTS

WIND AND WAVE NOISE

PANEL DISCUSSION REPORTS

PREFACE

In its relentless pursuit of further knowledge, science tends to compartmentalize. Over the years the pursuit of what might be called geophysical acoustics of the sea-surface has languished. This has occured even through there are well-developed and active research programs in underwater acoustics, ocean hydrodynamics, cloud and precipitation physics, and ice mechanics - to name a few - as well as a history of engineering expertise built on these scientific fields. It remained to create a convergence, a dialogue across disciplines, of mutual benefit.

The central theme of the Lerici workshop, perhaps overly simplified, was 'What are the mechanisms causing ambient noise at the upper surface of the ocean?' What could hydrodynamicists contribute to a better understanding of breaking wave dynamics, bubble production, ocean wave dynamics, or near-surface turbulence for the benefit of the underwater acoustics community? What further insights could fluid dynamicists gain by including acoustic measurements in their repertoire of instrumentation? While every attendee will have his or her perceptions of details, it was universally agreed that a valuable step had been taken to bring together two mature disciplines and that significant co-operative studies would undoubtedly follow.

The scope of the workshop was enlarged beyond its original intent to also include the question of ice-noise generation. The success of this decision can be seen in high quality of the presentations, the contribution of its disciples in the other workshop discussions and the heightened awareness and interest of we other novices.

I wish to salute all the participants, not only for their presentations and manuscript contributions, but for the stimulating atmosphere of inquiry that characterized the meeting. Among those who assured an open, challenging discussion were the invited speakers and panel chairmen, Owen Phillips, Michael Longuet-Higgins, Ira Dyer, John Ffowcs Williams, Larry Crum, Alex Kibblewhite, Bill Kuperman, David Farmer, Paul Crowther, Andrea Prosperetti, and panelists, Mike Banner, Norden Huang, Leif Bjorno, Doug Cato, Bob Mellen, Herman Medwin, Peter Wille and Ken Melville.

Of utmost importance to the initiation, organization and execution of the workshop were my fellow committee members, Leif Bjorno, Dick Heitmeyer, Reg Hollett and Andrea Prosperetti. The experience of Leif Bjorno in guiding us through the process is warmly appreciated. Reg Hollett and Dick Heitmeyer handled all the local details effectively with a cheery aplomb which we all appreciated.

I know that all others involved with the meeting would like to extend their heartfelt thanks to Ralph Goodman, Director, NATO ASW

Research Centre, La Spezia, for his encouragement and support that brought an idea to reality. Others who have contributed significantly are the late M. Di Lullo of the North Atlantic Treaty Organization, Vincenzo Damiani and Mario Astraldi of our host, the Energy and Environment Research Centre of Santa Teresa in Lerici, Skip Lackie of the Office of Naval Research, and James Young of the Atmospheric Environment Service.

Without the support of Anna Bizzari and Adolf Legner in Lerici, and of Evonna Mathis and Marg Stasyshyn in Toronto who handled all the myriad of important aspects that tied everything together, there would not have been a meeting and a book as an on-going record of everyone's efforts.

The financial support for the workshop was provided by the North Atlantic Treaty Organization through its Advanced Research Workshop program as well as by the United States Office of Naval Research and by the Atmospheric Environment Service of Canada.

The full transcript of the discussions of the panel discussions and of some individual papers is also available as an AES report at no cost. Those who would like a copy, should contact the editor.

To the reader, from all those whose work is bound up here, may you enjoy as much as we have.

Bryan R. Kerman

Atmospheric Environment Service
4905 Dufferin Street
Downsview, Ontario, M3H 5T4
Canada

December 31, 1987

NATO Advanced Research Workshop

Natural Mechanisms of Surface Generated
Noise in the Ocean

LERICI, ITALY
15-19 June 1987

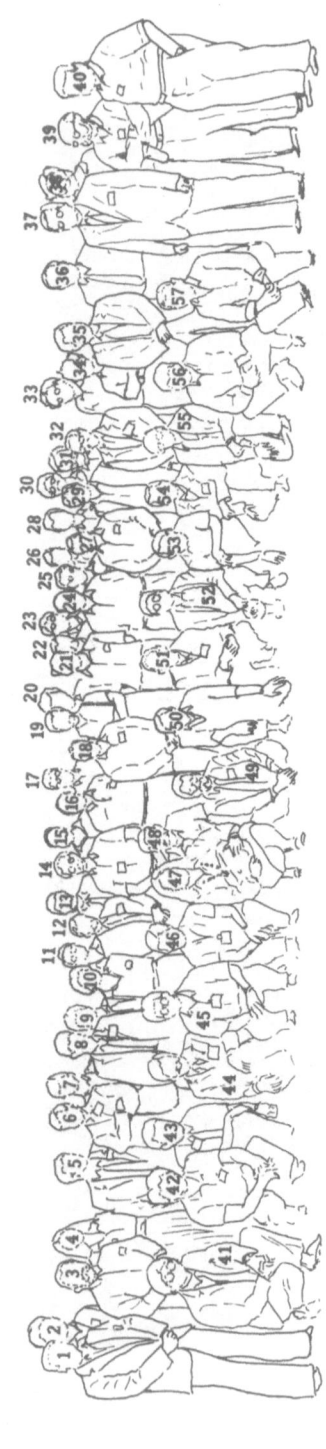

1	J. Esperandieu	12	W. Carey	23	R. Bannister	36	R. Goodman	47	A. Bizzarri
2	J. Roy	13	D. Cato	24	M. Longuet-Higgins	37	P. Wille	48	B. Kerman
3	P. Stein	14	M. Purshouse	25	H. Medwin	38	S. Thorpe	49	A. Prosperetti
4	R. Keenan	15	K. Melville	26	M. Buckingham	39	E. Sullivan	50	R. Heitmeyer
5	J. Ffowcs-Williams	16	W. Kuperman	27	F. Raichlen	40	R. Mellen	51	R. Hollett
6	L. Crum	17	T. Osborne	28	J. Nystuen	41	A. Lezzi	52	N. Huang
7	J. Lewis	18	I. Dyer	29	A. Plaisant	42	S. Mc Connell	53	S. Ling
8	H. Pumphrey	19	J. Papadimitrakis	30	M. Banner	43	F. Dias	54	S. Pao
9	W. Krichbaumer	20	B. Van Asselt	31	E. Monahan	44	E. Brumley	55	A. Kibblewhite
10	R. Hamson	21	O. Phillips	32	E. Yazgan	45	E. De Marinis	56	J. Wilson
11	M. Su	22	F. Brajou	33	D. Farmer	46	R. Marasco	57	A. Legner
				34	J. Charles				
				35	P. Crowther				

MECHANISMS OF WAVE BREAKING IN DEEP WATER

M. S. Longuet-Higgins
Department of Applied Mathematics and Theoretical Physics
Silver Street
Cambridge CB3 9EW
England

1. INTRODUCTION

Waves breaking at the sea surface are prime suspects as a source of underwater sound, but the precise way in which the sound is generated remains a matter for careful enquiry. In the present paper we shall review some of the known ways in which surface waves can break, with an eye to possible mechanisms of sound generation, and will describe in particular work in progress on the interactions between long and short waves.

2. LARGE-SCALE BREAKERS

Figures 1 - 3 illustrate some extreme conditions. Figure 1, taken from a sailing boat in the N. Atlantic in a wind of force 6, shows a typical spilling breaker, in which a whitecap appears near the crest of a steep wave and spreads more or less continuously down the forward face. It is accompanied by intense turbulence and bubble formation, particularly near the lower end, or toe. Each whitecap is active for at most a few wave periods, the wave eventually passing beneath it, so that a boat surfing on the wave crest is left behind in the following wave trough.

Figure 2 shows a plunging breaker, also in deep water. In this type, the crest has evolved into a distinct forwards jet, sometimes with lateral structure, which plunges forwards and downwards into the surface. The jet itself often disintegrates into droplets and spray even before impact.

Figure 3 shows a view of a North Sea oil rig during a severe storm. The quantity of spray shows how the air-sea boundary has become, in effect, a two-phase flow. To a lesser extent this is true at almost all wind speeds; see for example Wu (1973).

1

B. R. Kerman (ed.), Sea Surface Sound, 1–30.
© 1988 by Kluwer Academic Publishers.

2

Figure 1. Spilling breakers in the N. Atlantic, in wind force 6.
(from Coles 1967).

Figure 2. A deep-water plunging breaker in the N. Atlantic.
(from Coles 1967).

Figure 3. Breaking waves and spray in the North Sea in a strong gale.

3. LABORATORY STUDIES

Because the time-scale of overturning of the sea surface is short
compared to that of energy input by the wind, or of the slow nonlinear
interactions, it is very often appropriate to study wave breaking
induced by artificial means in the absence of wind. Three methods were
used by Longuet-Higgins (1974). In the first, a spilling breaker was
induced by sailing a small launch at a speed of 4 to 6 knots parallel
to a vertical-sided jetty (Figure 4). The reflection of the ship's wave
pattern from the vertical wall effectively doubled the wave amplitude,
causing the wave to break. Also the local increase in phase speed
turned the crest normal to the wall, producing a progressive, quasi-
steady breaker. The experiment was repeated in a ship's towing tank,
and the velocity field was measured with hot-wire instruments. Separation of the steady and turbulent components showed the strongest
turbulence occurring near the toe of the whitecap (Figure 5).

Secondly, spilling or plunging breakers were induced in relatively
deep water by a decelerating mechanical wavemaker. In this method the
wavemaker is first operated at a relatively high frequency, so that a
train of short waves propagates down the tank, the front travelling
with approximately the linear group-velocity. The wavemaker is then
gradually slowed, hence longer waves are generated which overtake the
shorter waves at some given observation point. The wavemaker can be
controlled so that the waves are highly reproducible. The method has
been used very effectively by Melville and Rapp (1985) some of whose
results are shown in Figure 6. Three effects can be seen:

(1) After striking the forwards face of the wave the jet partly
rebounds, producing further droplets and spray, somewhat like a
skipping stone. This process can be repeated several times.

(2) The part of the jet which penetrates the surface entrains with
it a large quantity of air, which immediately breaks up into bubbles.

(3) The volume of air trapped in the "tube" of the breakers is
forced downwards into the fluid, and again breaks up into a cloud of
bubbles. (In a three-dimensional situation some of this air would
of course escape sideways). The bubble clouds left after passage of the
breaker can be clearly seen in Figure 6d.

4

Figure 4. Breaking waves produced by reflection of a ship's wave pattern from a vertical wall. River Cam, October 1972.

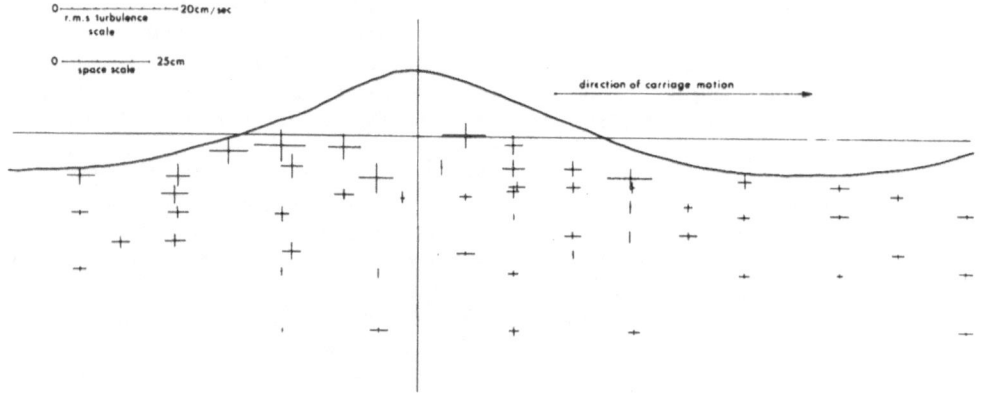

Figure 5. R.m.s. turbulent velocities in a spilling breaker. (from Longuet-Higgins 1974).

Figure 6. A deep-water plunging breaker produced by a decelerated wavemaker (from Melville and Rapp 1985).

Similar effects were first produced by Miller (1957) with a simple horizontal plunger. His sharp and elegant profiles of overturning jets have inspired several attempts to formulate analytic expressions for plunging breakers, as described in Section 7.

4. LEAP-FROGGING, INTERMITTENCY AND SHORT-CRESTEDNESS

In a normal sea state the dominant waves, i.e. those corresponding to the peak of the frequency spectrum occur in groups of high or low waves, the statistical properties of which have been derived theoretically by various methods (for a recent study see Longuet-Higgins 1984). In a narrow-band spectrum the wave groups correspond closely to the wave envelope, which advances in the direction of the dominant wavenumber with the theoretical group-velocity c_g. This is close to half the phase-speed c of the dominant waves. As a result, the dominant wave crests overtake their envelope and move forwards relative to the group with relative speed $\frac{1}{2}c$ (see Figure 7). In rough seas the waves

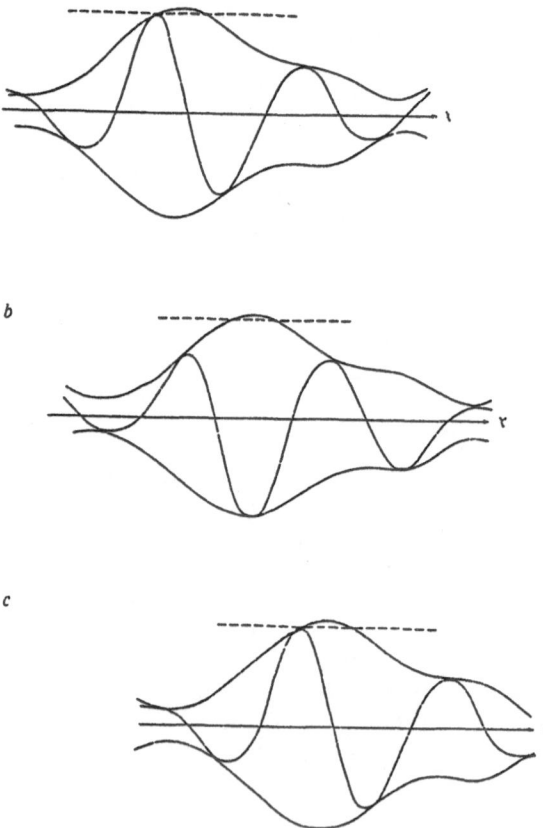

Figure 7. Intermittent whitecapping from high waves advancing through a wave group (from Donelan, Longuet-Higgins and Turner 1973).

7

at the centre of the group may exceed a critical steepness (usually somewhat less than the steepness ak = 0.443 of steady waves of limiting height) at which they will break and produce an active whitecap, until such time as the height of the envelope falls again below the critical level. Clearly, the crest of the wave next behind will tend to repeat the history of its predecessor at a time interval equal to the dominant wavelength L divided by the relative speed ½c, that is to say an interval of <u>twice</u> the wave period T.

This effect has been observed from the air (Donelan, Longuet-Higgins and Turner 1973) and can easily be demonstrated in the laboratory by the third of the methods described in Longuet-Higgins 1974). A wavemaker is started from rest at a steady rate, so that the waves in the resultant train, when well established, are just lower than the breaking steepness. Near the advancing wave front, however, the wave envelope (which is given approximately by a Fresnel integral) exceeds the steady amplitude by some 19%. So as each crest moves through the front it will break. Figure 8, taken from a film made in the 60 m wave channel at Wormley, illustrates the effect. Essentially it is because the frequency spectrum of the waves near the front itself has a small but finite bandwidth.

Figure 8. Waves breaking at the front of an advancing wave train.

Allied to this effect is the finite length of the wave crests and hence of the whitecaps. This is associated with the three-dimensionality of the wave envelope, which in turn is associated with the angular spread (in azimuth) of wave components in the spectral peak (for an analysis see Longuet-Higgins 1957a,b). Such a spread arises not only from a variability in the wind direction but also from certain three-dimensional instabilities (see below, Section 9). On the other hand, swells below a certain steepness may be highly two-dimensional, as is attested by ocean-going yachtsmen who report that after a severe storm the sea is covered with swells "marching in step from horizon to horizon".

8

Figure 9. A close-up view of the sea surface under a light wind

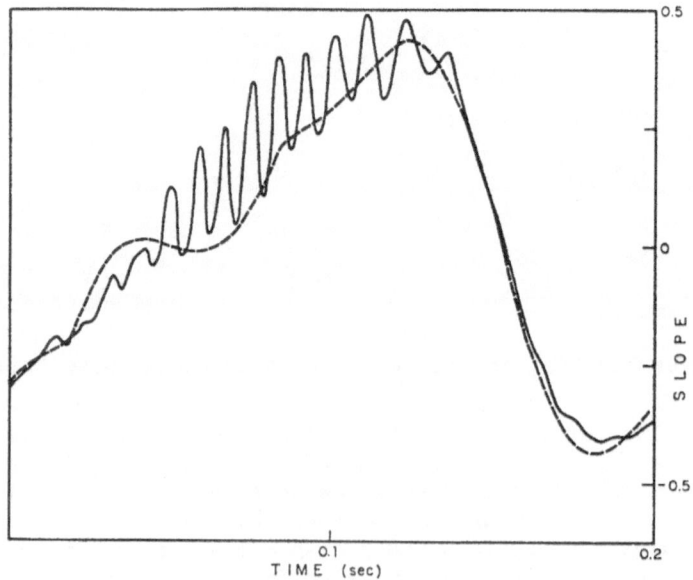

Figure 10. Surface slope as a function of time in steep gravity
waves, showing parasitic capillaries on the forward face. No wind.
(from Cox 1958).

We note that one of the principal differences between deep-water
breakers and those in shallow water is that in shallow water the group
velocity and phase velocity are nearly equal. Hence the breaking
are less intermittent, and whitecaps may persist on a given wave for
many wave periods, leading eventually to the almost total destruction
of the wave by breaking. In deep water, on the other hand, only a small
fraction of the wave energy may be lost in each whitecapping event.

As waves enter shallow water the directional spread of the wave
energy is also reduced by refraction, leading to an increase in the
relative length of the wave crests. A study of the changes in structure
of the wave groups was made by Longuet-Higgins (1956).

5.   SMALL-SCALE BREAKING

Figure 9 is a close view of the sea surface under a light wind. It can
be seen that there are roughnesses on all scales down to a few centi-
metres, or less. In particular one can pick out short capillary waves
riding on the forward slopes of steeper gravity waves, not far from the
gravity wave crests. This phenomenon was also observed by Cox (1958) in
a laboratory wave flume, even in a condition of zero wind-speed (see
Figure 10). These "parasitic capillaries" have been attributed to the
sharp curvature at the crest of the gravity-wave (Longuet-Higgins 1963;
Crapper 1970). As in the fish-line problem (Lamb 1932, §272) the
capillary waves appear upstream of the disturbance, viewed in a frame
of reference moving with the speed of the gravity wave. Such waves,
being strongly damped by viscosity, may extract a significant amount
of energy from the gravity waves, so delaying the onset of breaking
(see Longuet-Higgins 1963).

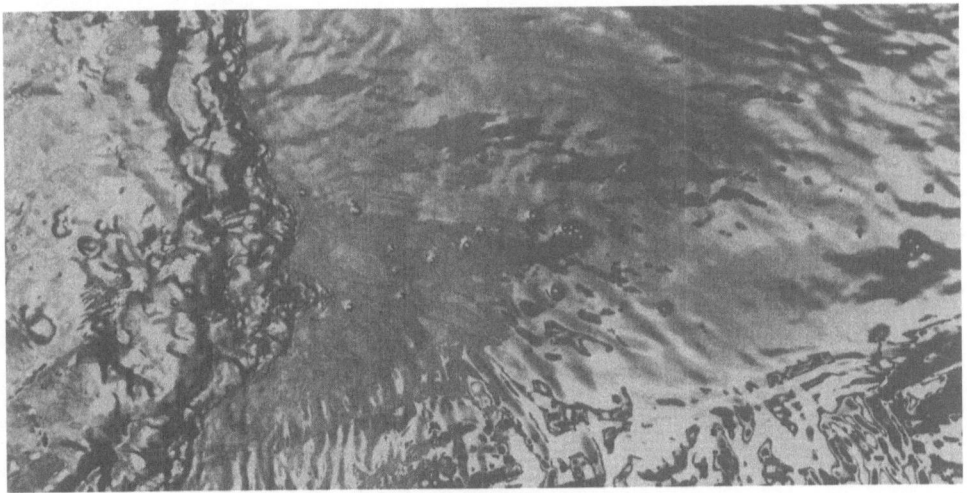

Figure 11. A small-scale spilling breaker on a steady current.
(from Banner and Phillips 1974).

In extreme cases the capillary wave steepness can be so great that a "block" forms on the forward face of the gravity wave, as seen in Figure 11. In this case the breaker was induced by a submerged obstacle in the flow.

Duncan (1981) has used a similar technique: he towed a subsurface aerofoil through still water at a steady speed, so as to induce a quasi-steady spilling breaker at the free surface. Study of the structure of such waves reveals strong turbulence, within the overriding roller, and a spread of turbulence into the lower part of the fluid, as was mentioned earlier for larger-scale spilling breakers.

6. EFFECTS OF VORTICITY

For relatively short waves, the shearing current near the surface induced by the wind stress and the momentum transfer from breaking short waves may have an appreciable effect on the limiting forms and speeds of short waves (Phillips and Banner 1974; Wright 1976). As a matter of fact it was Miche (1944) who first pointed out the effect of a finite vorticity on the shape of a sharp-cornered wave (see Figure 12). In an irrotational progressive wave the limiting form near the crest is the Stokes corner-flow, with a 120° interior angle (Stokes 1880, Michell 1893), as shown in Figure 12b. In such a flow the curvature vanishes everywhere at the free surface, excepting only at the crest itself. Now, in the presence of a locally uniform shear Ω the stream function ψ must satisfy $\nabla^2 \psi = \Omega$, and there exists a solution in the form

$$\psi = \frac{1}{4} \Omega r^2 + \frac{2}{3} g^{\frac{1}{2}} r^{\frac{3}{2}} \cos \frac{3}{2}\theta$$

where r and θ are polar coordinates with the origin at the crest and $\theta = 0$ vertically downwards. Applying the Bernoulli equation

$$p + \frac{1}{2} q^2 - gr \cos \theta + \Omega \psi = \text{constant}$$

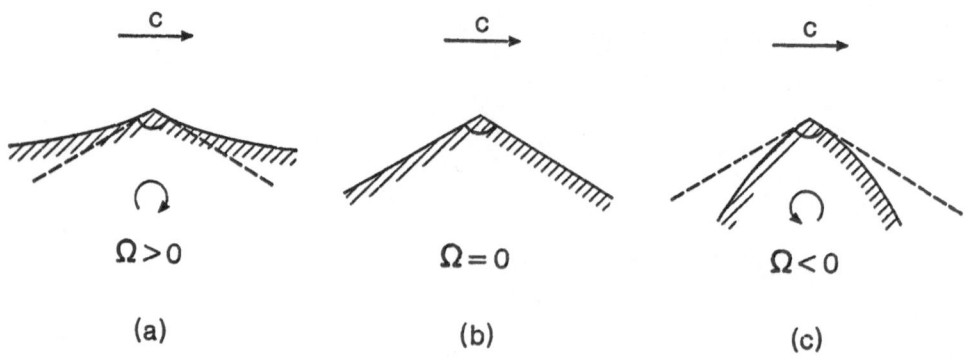

Figure 12. The effect of a finite vorticity ($-\Omega$) on the limiting form of gravity waves (a) $\Omega > 0$, (b) $\Omega = 0$, (c) $\Omega < 0$.

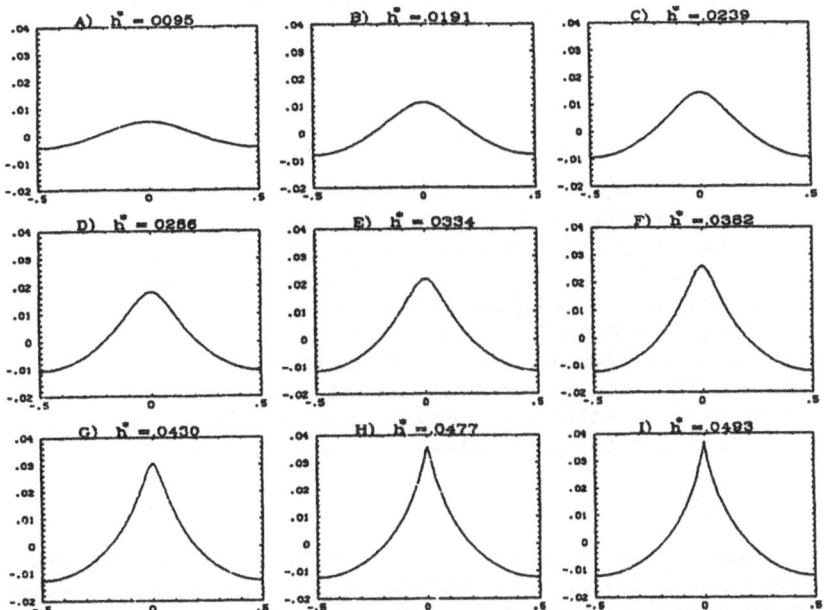

Figure 13. The profiles of surface gravity waves on a stream with uniform shear $\Omega^* = 1$ (The vertical scale is exaggerated). (from Simmen and Saffman 1985).

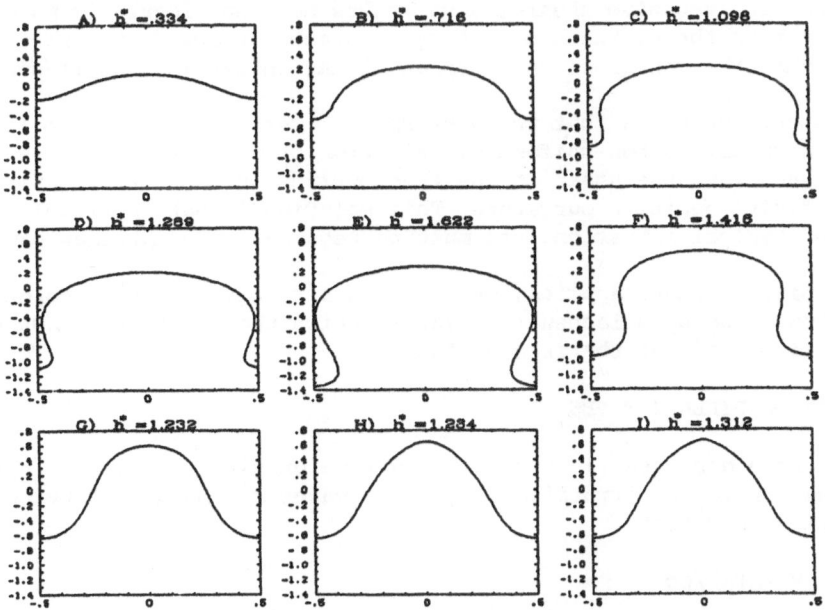

Figure 14. As Figure 13, but with $\Omega^* = -1.74$.

along the surface streamline, where $p = 0$ and $\psi = 0$, say, we easily find that as $r \to 0$,

$$|\theta| \sim \frac{\pi}{3} + \frac{\Omega}{4g^{\frac{1}{2}}} \, r^{\frac{1}{2}}.$$

So when $r \to 0$ the limiting curvature $d^2(r\theta)/dr^2$ becomes infinite, forming a double cusp, one on each face of the wave. When $\Omega > 0$ the free surface is concave upwards, as in Figure 12a; this is the case appropriate to a following wind. When $\Omega < 0$, the surface becomes convex, as in Figure 12c. Both Miche (1944) and Phillips and Banner (1974) point out that if $\Omega > 0$ the limiting wave amplitude will be reduced.

Complete wave profiles in waves with uniform shear, and neglecting viscosity and capillarity, have been calculated by Simmen and Saffman (1985). Some of their results are shown in Figures 13 and 14. It is convenient to define a dimensionless shear parameter

$$\Omega^* = \Omega/(gk)^{\frac{1}{2}}$$

where k is the wavenumber ($= 2\pi/\text{wavelength}$). Figure 13 shows the case $\Omega^* = 1$, giving sharper crests and a distinctly lower height/length ratio (note the vertical scale). Figure 14 shows an oppostie case $\Omega^* = -1.74$, when the crests are more rounded but the limiting height is greater. In extreme cases the surface can overhang, and even encloses a pocket of "air" between two adjacent "cheeks". Note that in such cases the limiting wave is neither the highest nor, probably, the most unstable.

Waves with negative shear may sometimes be seen clearly in water of finite depth on the backwash on a sloping beach. In deep water it is possible that they occur on the surface of the roller in a small-scale spilling breaker.

It should be noted that the very special Gerstner wave (Lamb 1932 C.9) has a specific, non-uniform distribution of negative shear. In the steepest such wave the shear at the free surface is negatively infinite, giving infinite negative curvature. This solution is not applicable to any known physical situation, and must be regarded as a mathematical fluke.

A similar comment applies, though less strongly, to the theoretical corner-flow found by Delachenal (1973) in which the vorticity tends to infinity like $\psi^{-1/3}$ at the free surface.

7. GRAVITY-CAPILLARY WAVES

So far we have discussed, in the main, surface gravity waves. The influence of surface tension on irrotational gravity waves may be gauged by the dimensionless parameter

$$K = Tk^2/\rho g$$

where T is the surface tension constant, k the wavenumber, ρ the density and g gravity. Thus $K = \infty$ corresponds to pure surface-tension waves, for

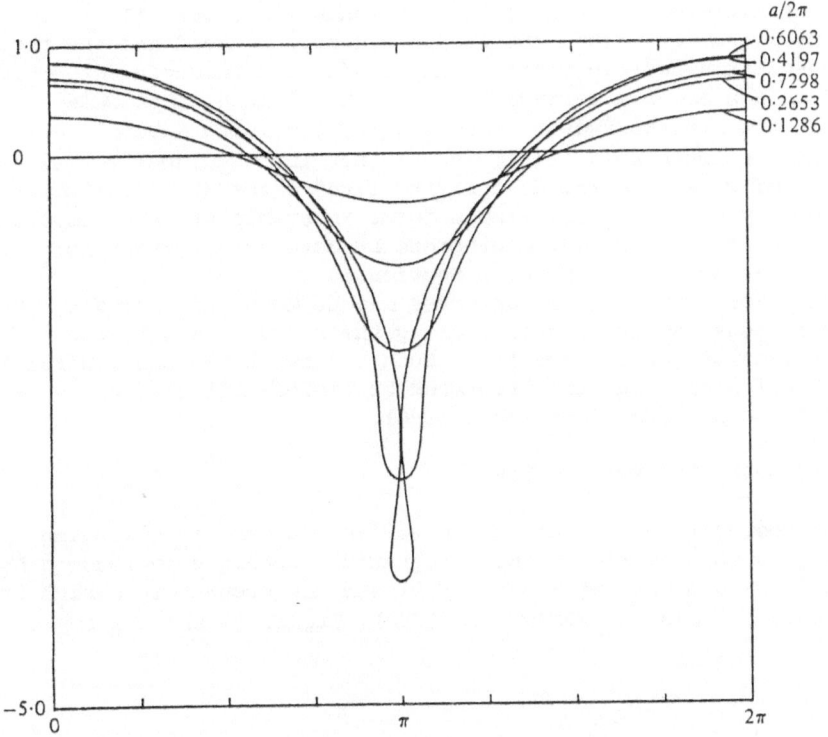

Figure 15. Surface profiles (or streamlines) of pure capillary waves: K = ∞.

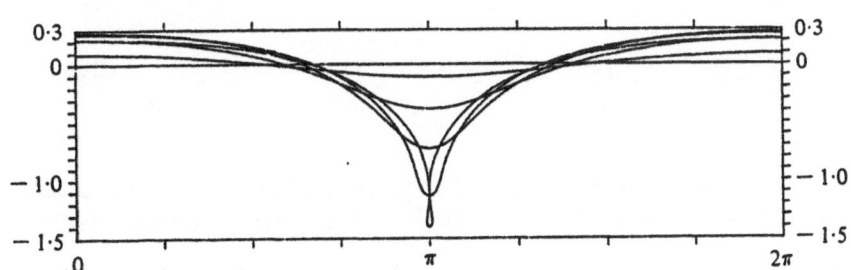

Figure 16. Surface profiles of capillary-gravity waves: K = 1.0.

14

which an exact solution was found by Crapper (1957). The sequence of profiles for different wave amplitudes is shown in Figure 15. In contrast to pure gravity waves, the crests are more rounded and the troughs more pinched. At a certain ratio of amplitude to wavelength, $a/L = 0.6063$, the free surface becomes vertical, and at the limiting amplitude $a/L = 0.7298$ the free surface touches itself, enclosing a bubble of "air". The same is true when $K = 1.0$ (Figure 16), but for the lower values of K two classes of waves are found: "gravity-like" waves and "capillary-like" waves. Probably none of these simple forms is stable at large amplitudes. The steeper gravity waves may degenerate by developing parasitic capillaries, as described above in Section 5.

In practice, surface tension waves may be much influenced by the variation in tension with distance along the surface, and by the relaxation time of the surface film. Such effects have been studied for waves of small slope, but the treatment of finite-amplitude waves is likely to be complicated (see Hogan 1986).

8. NUMERICAL AND ANALYTIC MODELS

A powerful tool for investigating the surface deformations leading up to breaking is the numerical time-integration technique introduced by Longuet-Higgins and Cokelet (1976, 1978) and now brought to a high state of efficiency by Dold and Peregrine (1984). Figure 17 shows a typical

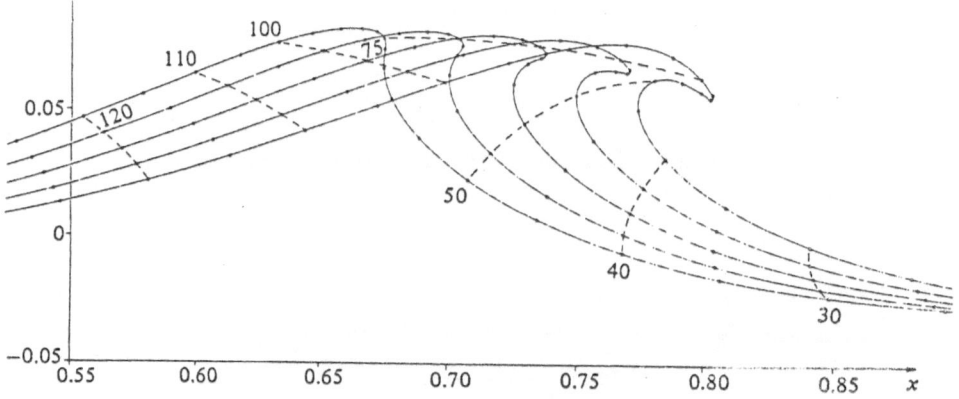

Figure 17. Successive profiles of a plunging breaker, calculated by numerical time-stepping (from New, McIver and Peregrine 1985).

computation for an energetic plunging breaker. The form of the overturning jet is remarkably universal, regardless of the initial conditions or the mean water detph. Even the initial overturning at the crest of a gentle spilling breaker appears to differ only in the local scale (see Longuet-Higgins and Cokelet 1978).

This has led to some attempts to find analytic expressions for the flow in the developing jet. A rotating hyperbolic flow (Longuet-Higgins 1980, 1983a,b) describes qualitatively several features of the tip of the jet. Greenhow (1983) has combined this with the rotating ellipse of New (1983) to produce a plausible model. These flows, however, do not include gravity; the picture of the flow is that seen by an observer in free-fall. A different analytic expression for the complex potential χ was suggested by Longuet-Higgins (1981). This consists of three terms

$$\chi = \frac{2}{3}ig^{\frac{1}{2}}z^{\frac{3}{2}} + Uz + 2Az^{\frac{1}{2}}$$

where U is a constant and A is a linear function of the time t. The first term corresponds to a Stokes 120° corner-flow (which includes gravity, of course). The second term represents a parabolic flow which is relatively important near the origin. The constants in A(t) are chosen so as to satisfy the condition of constant pressure at the free surface as closely as possible. A typical solution is illustrated in Figure 18. Solutions with sharp-pointed crests exist also.

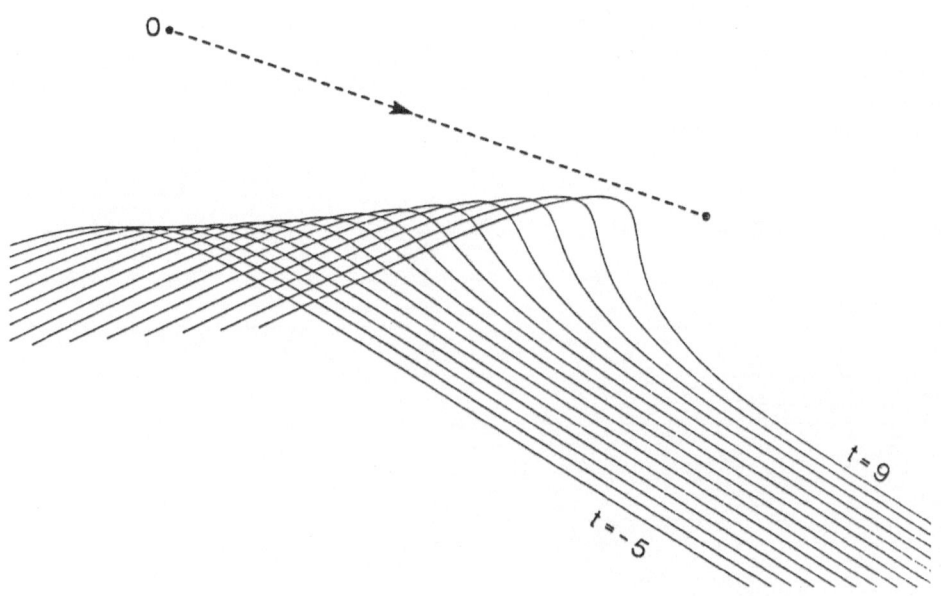

Figure 18. Successive profiles of a developing breaker, according to an approximate analytic expression.
(from Longuet-Higgins 1981).

9. WAVE INSTABILITIES

A particularly interesting set of initial conditions for the development
of breaking is a simple, uniform train of waves subject to a very small
perturbation in the form of a <u>normal mode</u>. This is a small perturbation
whose time-dependence is described by a factor eiot, in a reference
frame moving with the initial phase-speed. Normal modes of this type
were studied for two-dimensional waves by Longuet-Higgins (1978) and
for three-dimensional waves by McLean (1982). In some circumstances σ
can be complex; its imaginary part then corresponds to a growing or
decaying mode. Some two-dimensional growing modes were followed to
breaking by Longuet-Higgins and Cokelet (1978). The most recent computations by Dold and Peregrine (1966) show that while some initially
growing modes eventually break (see Figures 19 to 20), others fall back
eventually to almost zero, so that the steady wave train is restored -
an example of "recurrence". Figure 21 shows the boundary in parameter
space, between breaking and recurrence. The instabilities studied were
all subharmonic, of the Benjamin-Feir type.

The above results were for two-dimensional waves only. The calculations of McLean (1982) show that over an intermediate range of values
of the initial wave steepness ak, some of the three dimensional instabili
ties grow faster. Numerical time-stepping in three dimensions is however
very lengthy, and no results are yet available.

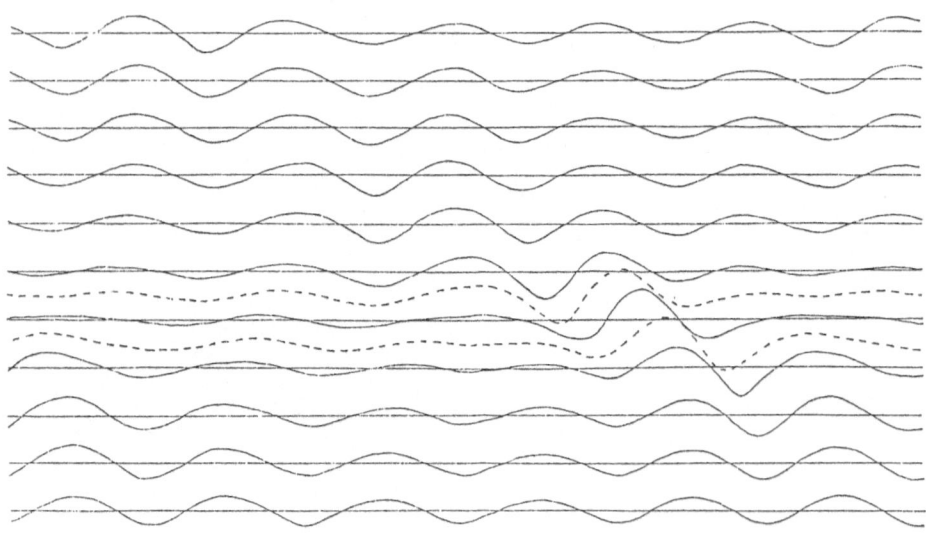

Figure 19. The development of an unstable normal mode, of length
5 waves, at intervals of 20 wave periods (full curves). The initial
wave steepness: ak = 0.11. Vertical exaggerations x 5.

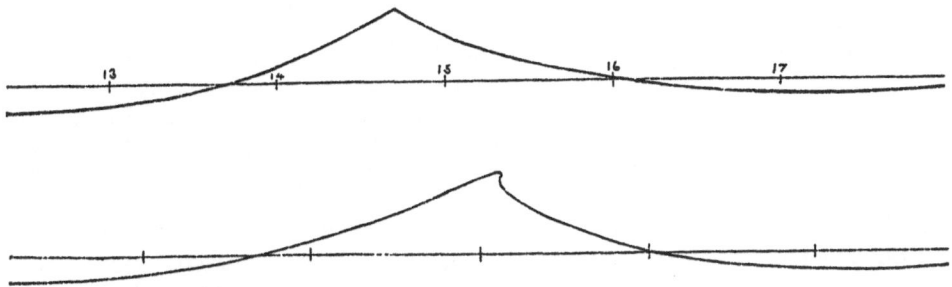

Figure 20. As Figure 17 but with ak = 0.12: the surface profile at breaking (a) with perturbation amplitude ε = 0.1, (b) ε = 0.05. No vertical exaggeration.

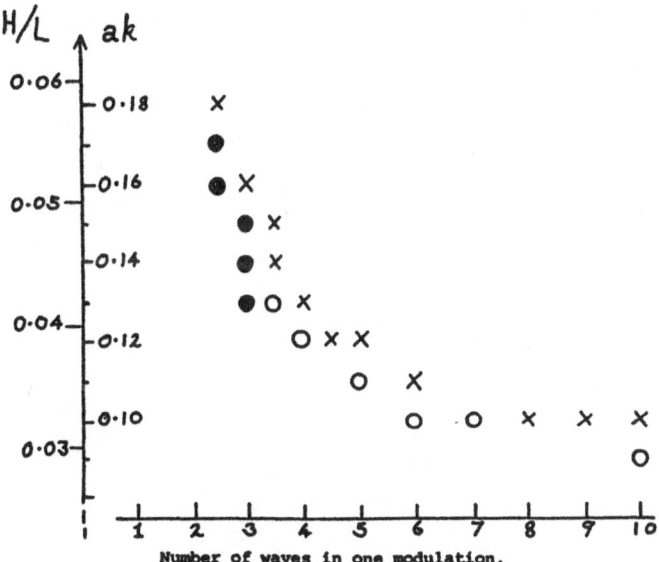

Figure 21. The boundary in parameter-space between conditions for breaking and recurrence. x Waves grew to breaking. 0 Waves recur. ● No growth of modulation.

10. LONG-WAVE - SHORT-WAVE INTERACTIONS

A very important phenomenon which indeed affects our imaging of the sea surface, visually or by radar, is the interaction between long- and short-scale surface waves. One basic mechanism is the stretching or compression of the short waves by the orbital motions in the long waves. The same horizontal convergence which raises the fluid to form the long-wave crests also contracts the free surface at a crest and similarly stretches it in a (long-wave) trough (the "concertina" effect. If a and k denote the amplitude and wavenumber of the short waves and A and K those of long waves, and if we assume ak << 1, AK << 1, k >> K, then it can be shown that the variation of a and k with the phase θ of the long waves is given by

$$k/\bar{k} = 1 + AK \cos \theta + 0 \, (AK)^2$$

$$a/\bar{a} = 1 + AK \cos \theta + 0 \, (AK)^2$$

(see Longuet-Higgins and Stewart 1962). Here $\bar{a}$ and $\bar{k}$ denote the values of a and k at the mean level, i.e. when $\theta \doteq \pm\pi/2$. The variation r in the short wave steepness ak is then given by

$$r = \frac{ak}{\overline{ak}} = 1 + 2AK \cos \theta + 0(AK)^2 .$$

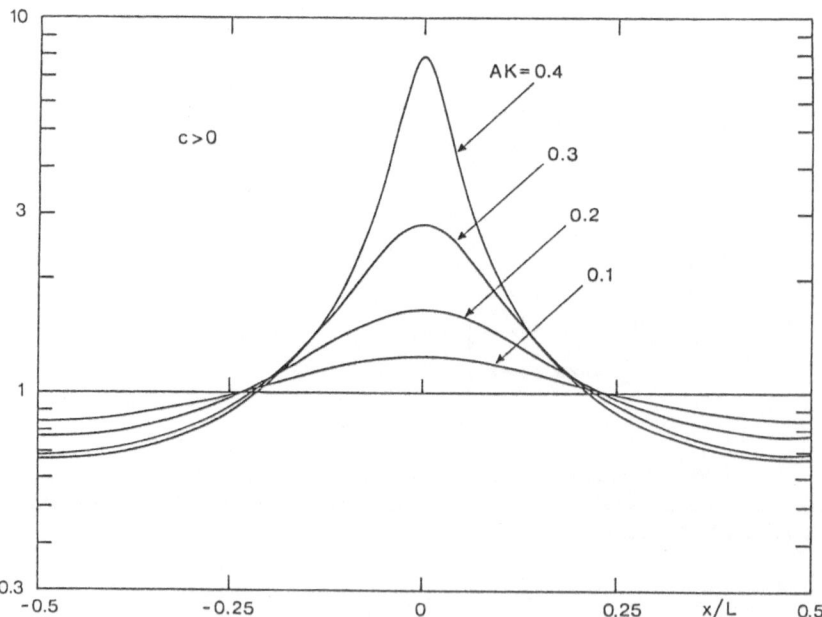

Figure 22. Ratio r of the steepness ak of the short waves relative to their steepness $\overline{ak}$ at the mean surface level.

For the maximum value 0.443 of AK this would indicate a relative
steepening of r at the wave crest ($\theta = 0$) by a factor less than 2.
However, more recent calculations for finite values of AK (Longuet-
Higgins 1987a) using the principles of phase and action conservation
for the short waves, indicate that when AK = 0.4, say, the steepness r
at the crest of the long waves is about 8 times that at the mean level
(see Figure 22). In the wave troughs r falls to about 0.6. The steepening is almost independent of the ratio $\bar{k}/K$ of the wave number. When
the phase-speeds are opposite the results are very similar.

To test these conditions, experiments are in progress using a
channel 16 m long 0.6 m wide and 40 cm deep at Cambridge. So as to
avoid interaction effects at the wavemaker itself, two separate wavemakers are used (called wavemakers A and B) for the long and the short
waves respectively. Wavemaker B is designed with a spring release,
enabling it to be quickly hoisted out of the water when desired. Thus,
to generate long and short waves travelling in the same sense, the
wavemakers are placed as in Figure 23, with wavemaker B in front
of wavemaker A. In operation, wavemaker B first generates a train of
short waves. It is then quickly hoisted out of the water (Figure 24).
Wavemaker A is immediately started, which generates a longer train of
waves overtaking the short waves near an observation window about half-
way down the tank. Both wave trains are dissipated on a sloping beach
at the far end.

To generate waves in opposite senses, wavemaker B is placed near
the far end of the channel, facing wavemaker A. First, wavemaker B
generates a train of short waves extending about half-way along the
channel. Wavemaker A is then started, and wavemaker B is hoisted out
the way to avoid reflection of the long waves, and to allow them to be
dissipated on the beach.

The latter arrangement has certain advantages, since it precludes
any breaking or other interactions between the wave trains before the
observation point is reached. A useful procedure is to accelerate
wavemaker A fairly slowly, so that at the observation point the long-
wave steepness gradually increases.

Figure 25 shows a short-wave train of (undisturbed) frequency
3.5 Hz interacting with a longer wave of frequency 0.8 to 1.0 Hz
travelling in the opposite sense. The relative steepening of the short
wave at the long wave crest can be seen, and in Figure 25d the short
wave is clearly breaking.

11. FIELD OBSERVATIONS

The results of Section 10 tend to explain some observations by
Holthuisen and Herbers (1986) who measured the height H and period T
of individual waves in a mixed sea off the Dutch coast. Simultaneously
they noted whether the crest of the wave was breaking. They found a
considerable overlap in values of H/gT^2 as between breaking and non-
breaking waves.

The situation has been clarified by some recent measurements of
breaking waves by Xu, Hwang and Wu (1986) in a wind-wave laboratory
channel. Figure 26 summarises their results. It shows the ratio H/gT^2
for individual waves observed to be breaking at the crest. There are

Figure 23. Long- and short-wave makers (A and B) in position to generate short waves.

Figure 24. As Figure 23, but with wavemaker B raised out of the water.

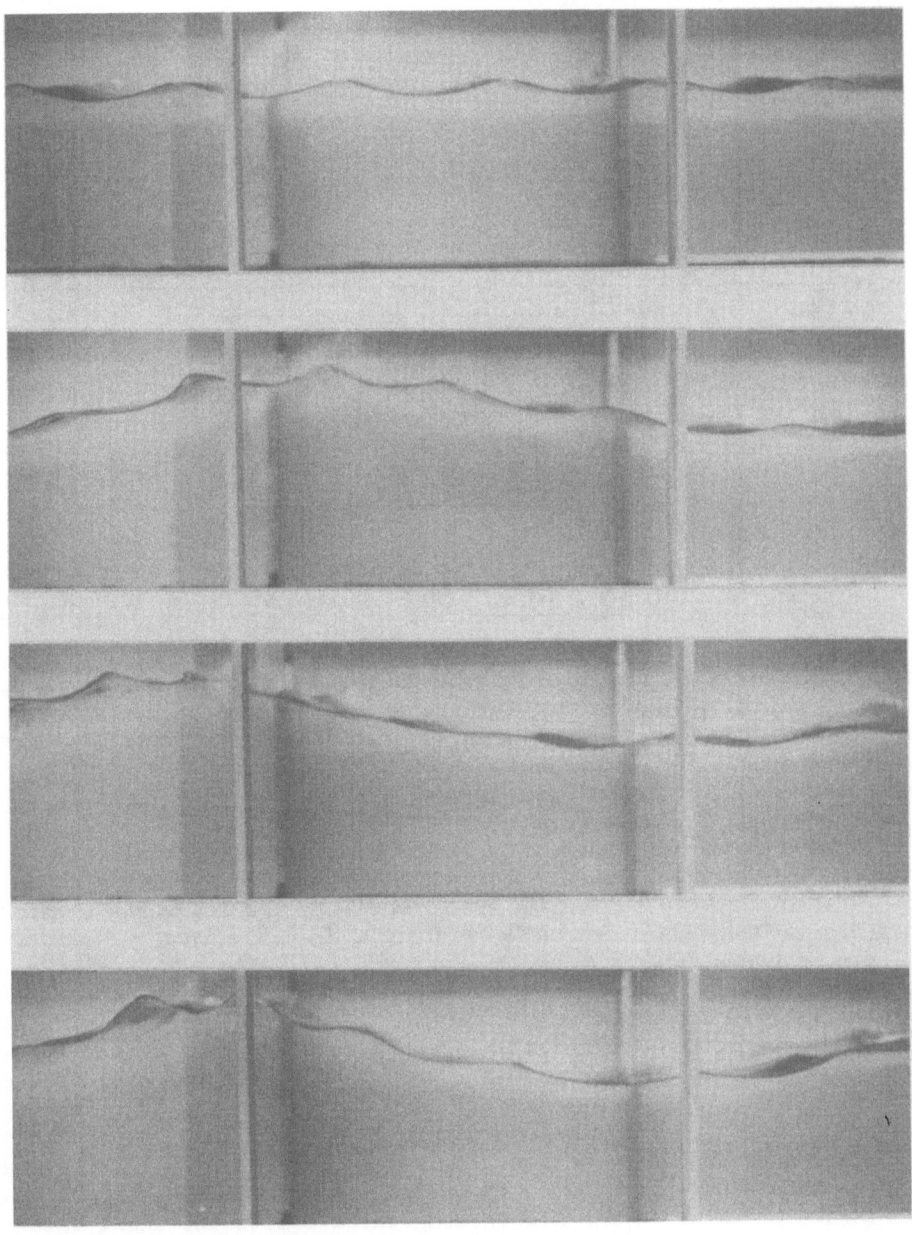

Figure 25. Interaction between long and short waves travelling in opposite senses. (a) AK = 0, (b) AK = 0.05, (c) AK = 0.10, (d) AK = 0.15.

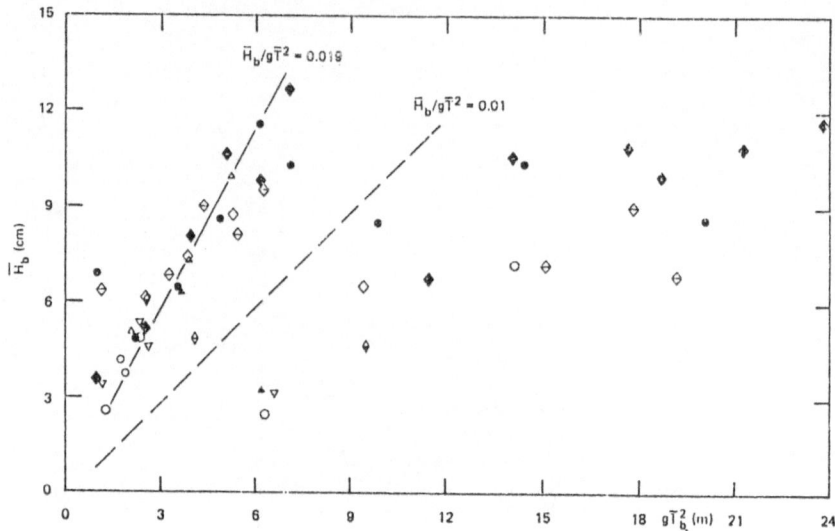

Figure 26. The ratio H/gT^2 for breaking waves in a wind-wave channel (from Xu, Hwang and Wu 1986).

apparently two distinct populations. On the one hand those with a high value of H/gT^2, which, as the authors suggest, seem to correspond to the dominant waves in the spectrum. On the other hand there is a more scattered population of breaking waves with a lower ratio H/gT^2 which appears to indicate lower waves with shorter waves breaking near the (longer) wave crests.

Thus a two-scale model of wave breaking gives qualitatively reasonable results.

We mention here that a two-scale model of the sea surface which includes (1) randomness in both the long and short waves, (2) the effects of breaking of the short waves and (3) regeneration of the short-wave energy by the wind, has recently been proposed (Longuet-Higgins 1987b). From this model one can calculate, for given wind and wave parameters, the probability of short-wave breaking at the long-wave crests. It is interesting that though this probability increases as a function of the wind-speed, it also decreases as a function of the r.m.s. amplitude of the long waves, at a given long wavelength. In other words, steepening of the long waves tends to reduce the probability of short-wave breaking on the average long wave. This is because short-wave breaking on the higher long waves limits the short-wave steepness on those waves; the subsequent straining reduces the short-wave steepness on the lower long waves.

This conclusion is in qualitative agreement with observations by Mitsuyasu (1966) in laboratory experiments where long, mechanically generated waves were superimposed on short, wind-generated waves, resulting in a reduction of the short-wave slopes. Previously the observation was explained by Phillips and Banner (1974) mainly as the result of the surface wind-drift.

12. STANDING AND PARTLY STANDING WAVES

In contrast to progressive waves, the breaking of standing waves has
been little studied. For the two-dimensional standing wave, Penney
and Price (1952) suggested a limiting crest angle of 90°; they assumed
that the pressure was not singular at the instant of maximum elevation.
Taylor (1953) doubted their argument but nevertheless confirmed by
experiment that the limiting angle was about 90°. The crest, however,
tended to become unstable to lateral disturbances. In spite of some
early attempts (Longuet-Higgins 1972, 1973) a rigorous solution for
the flow near the limiting configuration has yet to be found.

Just as in progressive waves, a breaking standing wave does not
necessarily pass through the limiting configuration. Numerical time-
stepping of overdriven standing waves suggests that a time-dependent jet
can be thrown vertically upwards (see Figure 27). Overturning of the
surface, as in a plunging breaker, can occur symmetrically on either
side. The form of the central jet itself is rather well described by a
"Dirichlet hyperbola" (see Longuet-Higgins 1972).

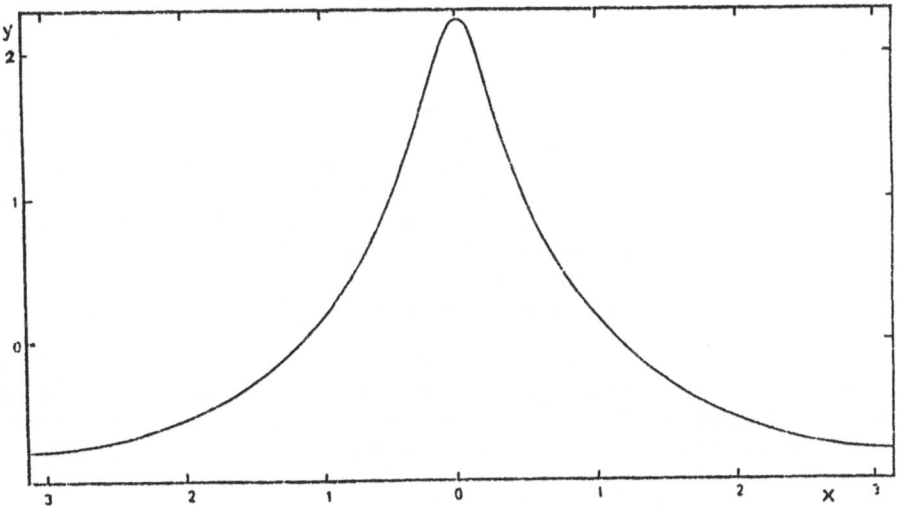

Figure 27. The computed form of a standing wave with initial
condition y=0, dy/dt=a cos x. (from McIver and Peregrine 1981).

In three-dimensional, axisymmetric standing waves, an argument
analogous to that of Penney and Price suggests a conical crest with a
limiting angle of 2 arctan √2, that is 109.5° (see Mack 1962). To test
this prediction the author (1983) generated standing axisymmetric
waves in a cylindrical flask of diameter 16.5cm and height 27cm, as

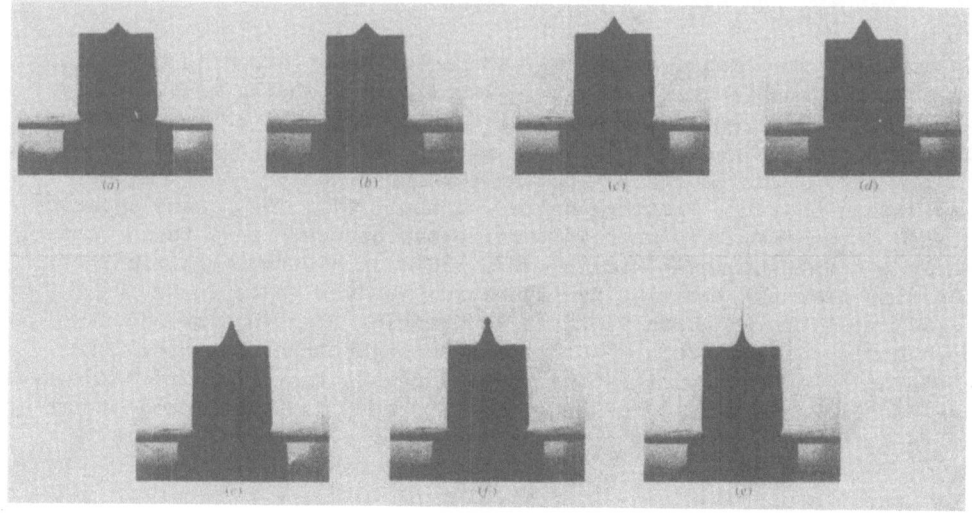

Figure 28. The forms of successive crests in an axisymmetric
standing wave, driven subharmonically (from Longuet-Higgins
1983c).

shown in Figure 28. The waves were generated subharmonically, by placing
the flask on a platform oscillating vertically at a frequency of 6.64 Hz
- slightly less than twice the natural frequency of the lowest axisymmetric standing wave in the flask. The pressure fluctuations on the
bottom of the flask thus have double the frequency of the standing waves.
This procedure is a reversal of the phenomenon whereby ocean waves with
opposite spectral components generate pressure fluctuations at double
their natural frequency in such a way as to cause microseisms (see
Longuet-Higgins 1953).

With the amplitude of the oscillator set at 0.3mm, say, at first
synchronous (6.64 Hz) waves appeared on the surface, but these soon gave
way to the desired subharmonic. A high-speed film (at 1000 frames/s)
recorded the growth. Figure 28 shows the form of successive standing
wave crests. The crest angle in Figures 28 (d) and (e), for example, is
clearly less than 109.5°. Finally the standing wave breaks (Figure
28 (e)) by a collapse of the wave trough, which emits a high-speed jet
sometimes rising as high as 2m above the surface of the flask. The
action is similar to a shaped charge (Bowen 1966) in producing a high-
speed jet; also to a collapsing bubble (MacIntyre 1968). An analysic
expression for the axisymmetric jet is given by Longuet-Higgins (1983).

Pure standing waves may occur only in rather special circumstances,
for example at the centre of a severe circular storm, or where waves
are reflected from a coatline or harbour wall. Partly standing waves,
however, which are formed by two oppositely travelling waves of the same
length but unequal amplitude, may be quite common. Such waves also

yield fluctuations in the potential energy, and hence unattenuated double-frequency pressures (Longuet-Higgins 1953). Such waves may be expected to break by emitting an asymmetric jet. Indeed the jet in a plunging breaker is quite similar to that thrown up by a standing wave, except that it falls over to one side.

13. MULTISCALE MODELS

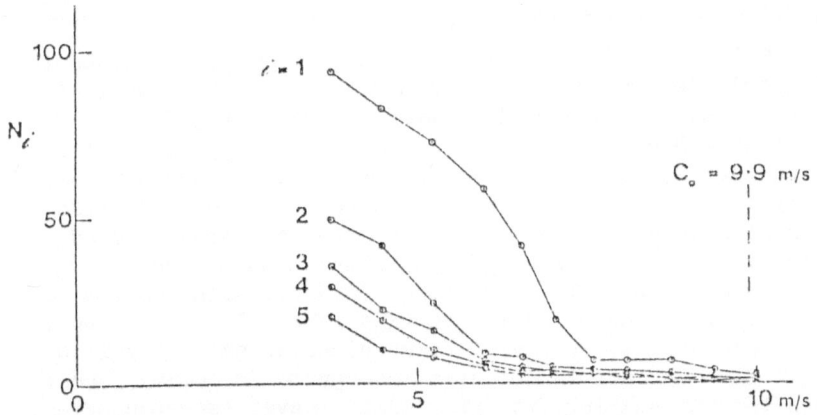

Figure 29. The number N_i of jumps having magnitude h between $(i - 1)\delta$ and $i\delta$, as a function of the critical rate of rise R (from Longuet-Higgins and Smith 1983).

Although two-scale models are useful for some purposes, wave breaking in a wind sea can in fact take place on many scales. A first attempt to define and quantify breaking waves on all scales was made by Longuet-Higgins and Smith (1983) using a vertical capacitance wire wave recorder. The signal was recorded and examined for "jumps", i.e. short intervals of time during which the rate of rise of the free surface exceeded a set limit R. The vertical extent of the "jump" was then measured automatically and a histogram of jump-heights constructed. Figure 29a shows the number of jump-heights lying between $(i - 1)$ and $i\delta$, where $\delta = 0.32$ cm, as a function of R, in a wind of 13.7 m/s. It will be seen that the small jumps are particularly numerous, especially at small values of R. Only at the larger values of R, where there are fewer jumps, do we encounter breaking of the dominant waves.

Such measurements suggest the representation of the sea surface as a fractal process, that is to say a surface elevation having a frequency spectrum with a negative power-law, as described for example by O.M. Phillips in the present conference proceedings. The vertial accelerations of the sea surface, which are closely associated with wave breaking, have also been measured (Ewing et al. 1987). We note that any spectral power law σ^{-n} for the surface elevation, where $n < 5$, will correspond to an acceleration field with infinite variance. The measured variance is therefore very sensitive to the high-frequency cut-off of the instrument.

14. WAVE BREAKING AND UNDERWATER SOUND

From the descriptions given above, the mostly likely way for breaking waves to generate underwater sound at high frequencies (or order 10^2 to 10^4 Hz) is by the sudden formation of air bubbles, either entrained by the whitecap or jet, or trapped by overturning of the free surface. A second source may be the splashing of jets and droplets of spray at the sea surface.

At intermediate frequencies (1 to 10 Hz) a possible source to be investigated is the turbulent, compressible bubble clouds which arise from both spilling and plunging breakers.

At low frequencies (10^{-1} to 1 Hz) the wavelength of the sound is such that it should be considered in conjunction with the propagation of elastic waves (especially Rayleigh waves) in the sea bed, as suggested by Longuet-Higgins (1950, 1953). Indeed the coupling between motions is such that the source or sources of the low-frequency noise are essentially the same as the source or sources of the microseisms. The generation of microseisms (and sound) by the mechanism studied by Longuet-Higgins (1950, 1953) involves the interaction of opposite components in the spectrum; its "signature" is a time-fluctuation of gravitational potential energy in the surface layer. In this process breaking is not necessarily involved. However, numerical integrations have shown that fluctuations in potential energy do accompany plunging breakers, even those resulting from progressive waves (see Longuet-Higgins and Cokelet, 1976).

REFERENCES

Banner, M.L. 1986 Oscillatory properties of spilling zones of quasi-steady breaking waves.
I.U.T.A.M. Symposium on Fluid Mechanics in the Spirit of G.I. Taylor, Cambridge, March 1986. (Abstract).

Banner, M.L. and Melville, W.K. 1976 On the separation of air flow over water waves.
J. Fluid Mech. 77, 825–842.

Banner, M.L. and Phillips, O.M. 1974 On the incipient breaking of small scale waves.
J. Fluid Mech. 65, 547–656.

Bowden, F.P. 1966 The formation of microjets in liquids under the influence of impact or shock.
Phil. Trans. R. Soc. Lond. A 260, 94–95.

Coles, K.A. 1967 Heavy Weather Sailing.
London, Adlard Coles Ltd. 303 pp.

Cox, C.S. 1958 Measurements of slopes of high-frequency wind waves.
J. Mar. Res. 16, 199–225.

Crapper, G.D. 1957 An exact solution for progressive capillary waves of arbitrary amplitude.
J. Fluid Mech. 2, 532-540.

Crapper, G.D. 1970 Nonlinear capillary waves generated by steep gravity waves.
J. Fluid Mech. 40. 149-159.

Delachanel, B. 1973 Existence d'ecoulement permanent de type coin pour un fluide parfait à surface libre.
C. R. Acad. Sci. Paris A 276, 1021-1024.

Dold, J.W. and Peregrine, D.H. 1984 Steep unsteady waves: an efficient computational scheme.
Proc. 19th Int. Conf. on Coastal Eng., Houston 1. 955-967.
New York, Am. Soc. Civil Eng.

Dold, J.W. and Peregrine, D.H. 1987 Water-wave modulation.
Proc. 20th Int. Conf. on Coastal Eng., Taipeh, Nov. 1986.
New York, Am. Soc. Civil Eng., 163-175.

Donelan, M., Longuet-Higgins, M.S. and Turner, J.S. 1972 Periodicity in whitecaps.
Nature. Lond. 239, 449-451.

Duncan, J.H. 1981 An experimental investigation of breaking waves produced by a towed hydrofoil.
Proc. R. Soc. Lond. A 377, 331-348.

Ewing, J.A., Longuet-Higgins, M.S. and Srokosz, M.A. 1987 Measurements of the vertical acceleration in wind waves.
J. Phys. Oceanogr. 17, 3-11.

Greenhow, M. 1983 Free-surface flows related to breaking waves.
J. Fluid Mech. 134, 259-275.

Hogan, J.H. 1980 Some effects of surface tension on steep water waves. Part 2.
J. Fluid Mech, 96, 417-445.

Hogan, J.H. 1981 Some effects of surface tension on steep water waves. Part 3.
J. Fluid Mech. 110, 381-410.

Hogan, S.J. 1986 Surface tension effects in nonlinear waves.
pp. 147-158 in Oceanic Whitecaps, eds. E.C. Monahan and G. Mac Niocaill. Dordrecht, D. Reidel Publ. Co., 294 pp.

Holthuisen, L.H. and Herbers, T.H.C. 1986 Statistics of breaking waves observed as whitecaps in the open sea.
J. Phys. Oceanogr. 16, 290-297.

Lamb, H. 1932 Hydrodynamics, 6th ed.
Cambridge Univ. Press, 738 pp.

Longuet-Higgins, M.S. 1950 A theory of the origin of microseisms.
Phil. Trans. R. Soc. Lond. A 243, 1-35.

Longuet-Higgins, M.S. 1953 Can sea waves cause microseisms?
Proc. Symp. on Microseisms, Harriman N.Y., Sept. 1952.
Washington D.C., U.S. Govt. Printing Off., NAS-NRC Publ. 302.

Longuet-Higgins, M.S. 1956 The refraction of sea waves in shallow
water.
J. Fluid Mech. 1, 163-176.

Longuet-Higgins, M.S. 1957a The statistical analysis of a random,
moving surface.
Phil. Trans. R. Soc. Lond. A 249, 321-387.

Longuet-Higgins, M.S. 1957b Statistical properties of an isotropic
random surface.
Phil. Trans. R. Soc. Lond. A 250, 157-174.

Longuet-Higgins, M.S. 1963 The generation of capillary waves by
steep gravity waves.
J. Fluid Mech. 16, 138-159.

Longuet-Higgins, M.S. 1972 A class of exact, time-dependent,
free-surface flows.
J. Fluid Mech. 55, 529-543.

Longuet-Higgins, M.S. 1973 On the form of the highest progressive
and standing waves in deep water.
Proc. R. Soc. Lond. A 331, 445-456.

Longuet-Higgins, M.S. 1974 Breaking waves - in deep or shallow
water.
Proc. 10th Symp. on Naval Hydrodynamics, Cambridge, Mass., June
1974, pp. 597-603. Arlington, Va., Office of Naval Research,
Publ. ACR-204. 792pp.

Longuet-Higgins, M.S. 1978 The instabilities of gravity waves of
finite amplitude in deep water.
Proc. R. Soc. Lond. A 360, 471-488 and 489-505.

Longuet-Higgins, M.S. 1980 On the forming of sharp corners of a free
surface.
Proc. R. Soc. Lond. A 371, 453-478.

Longuet-Higgins, M.S. 1981 On the overturning of gravity waves.
Proc. R. Soc. Lond. A 376, 377-400.

Longuet-Higgins, M.S. 1982 Parametric solutions for breaking waves.
J. Fluid Mech. 121, 403-424.

Longuet-Higgins, M.S. 1983a Rotating hyperbolic flow: particle
trajectories and parametric representation.
Q.J.Mech. Appl. Math. 36, 247-270.

Longuet-Higgins, M.S. 1983b Towards the analytic description of
overturning waves.
pp. 1-24 in Nonlinear Waves, ed. L. Debnath.
Cambridge Univ. Press 360 pp.

Longuet-Higgins, M.S. 1983c Bubbles, breaking waves and hyperbolic
    jets at a free surface.
    J. Fluid Mech. 127, 103-121.

Longuet-Higgins, M.S. 1984 Statistical properties of wave groups in
    a random sea state.
    Phil. Trans. R. Soc. Lond. A 312, 219-250.

Longuet-Higgins, M.S. 1987a The propagation of short surface waves
    on longer gravity waves.
    J. Fluid Mech. 177, 293-306.

Longuet-Higgins, M.S. 1987b A stochastic model of sea surface
    roughness. I. Wave crests.
    Proc. R. Soc. Lond. A 410, 19-34.

Longuet-Higgins, M.S. and Cokelet, E.D. 1976 The deformation of
    steep surface waves on water. I A numerical method of
    computation.
    Proc. R. Soc. Lond. A 350, 1-26.

Longuet-Higgins, M.S. and Cokelet, E.D. 1978 The deformation of steep
    surface waves on water. II Growth of normal mode instabilities.
    Proc. R. Soc. Lond. A 364, 1-28

Longuet-Higgins, M.S. and Smith, N.D. 1983 Measurements of breaking
    waves by a surface jump-meter.
    J. Geophys. Res. 88, 9823-9831.

MacIntyre, F. 1968 Bubbles: a boundary-layer "microtome" for micron-
    thick samples of a liquid surface.
    J. Phys. Chem. 72, 589-592.

McIver, P. and Peregrine, D.H. 1981 Comparison of numerical and
    analytical results for waves that are starting to break.
    Symp. on Hydrodynamics in Ocean Engineering, Preprints, 1,
    203-215. Trondheim, Norway. Nor. Hydro. Labs, 703pp.

Mack, L.R. 1962 Periodic, finite-amplitude, axisymmetric gravity
    waves.
    J. Geophys. Res. 67, 829-843.

McLean, J.W. 1962 Instabilities of finite-amplitude water waves.
    J. Fluid Mech. 114, 315-330.

Melville, W.K. and Rapp, R.J. 1985 Momentum flux in breaking waves.
    Nature, Lond. 317, 514-576.

Miche, M. 1944 Mouvements ondulatoires de la mer en profondeur
    constante ou décroissante. Forme limite de la houle lors de sa
    deferlement.....
    Annales des Ponts et Chaussées 2, 25-78, 132-164, 270-290 and
    369-406. See especially pp. 370-376.

Michell, J.H. 1893 The highest waves in water.
    Phil. Mag. (5) 36, 430-437

Miller, R.L. 1957 Role of vortices in surf zone prediction, sedimentation and wave forces. pp. 92-114 in Beach and Nearshore Sedimentation (eds. R.A. Davis and R.L. Ethington) Tulsa, Oklahoma: Soc. of Economic Palaeontologists and Mineralogists.

New, A.L. 1983 A class of elliptical free-surface flows. J. Fluid Mech, 130, 219-239.

New, A.L., McIver, P. and Peregrine, D.H. 1985 Computations of overturning waves. J. Fluid Mech. 150, 233-251.

Penney, W.G. and Price, A.T. 1952 Finite periodic stationary gravity waves in a perfect fluid. Phil. Trans. R. Soc. Lond. A 244, 254-284.

Phillips, O.M. 1981 The dispersion of short wavelets in the presence of a dominant long wave. J. Fluid Mech. 107, 465-485.

Phillips, O.M. and Banner, M.L. 1974 Wave breaking in the presence of wind drift and swell. J. Fluid Mech. 66, 625-640.

Simmen, J.A. and Saffman, P.G. 1985 Steady deep-water waves on a linear shear current. Studies in Appl. Math. 73, 35-57.

Smith, R. 1975 The reflection of short gravity waves on a non-uniform current. Math. Proc. Camb. Phil. Soc. 78, 517-525.

Stokes, G.G. 1880 On the theory of oscillatory waves. Appendix B: Considerations relative to the greatest height of oscillatory waves which can be propagated without change of form. Math. and Phys. Pap. 1, 225-228. Cambridge University Press.

Taylor, G.I. 1953 An experimental study of standing waves. Proc. R. Soc. Lond. A 218, 44-59.

Wright, J.W. 1976 The wind drift and wave breaking. J. Phys. Oceanogr. 6, 402-405.

Wu, J. 1973 Spray in the atmospheric surface layer: laboratory study. J. Geophys. Res. 78, 511-519.

Xu, D., Hwang, P.A. and Wu, J. 1986 Breaking of wind-generated waves. J. Phys. Oceanogr. 16, 2172-2178.

EQUILIBRIUM RANGE CHARACTERISTICS OF BREAKING WAVES

O. M. Phillips
Department of Earth and Planetary Sciences
The Johns Hopkins University
Baltimore, MD   21218   USA

ABSTRACT. This paper reviews and extends recent results from the dynamical theory of wave breaking at scales in the equilibrium range, that is, over wavelengths smaller than that of the dominant wave in a wind-generated sea. Expressions are given for the distributions of length of breaking front per unit area at any instant and for the frequency of occurrence of breakers per unit area as functions of friction velocity and breaking wave scale. The latter quantity is associated with the frequency of occurrence of 'sea spikes' as observed by radar back-scattering from the sea surface at large incidence angles. The results are also expressed as distributions with respect to rate of production of turbulent energy by breaking, a possible index of acoustic generation.

1. THE EQUILIBRIUM RANGE IN WIND-GENERATED GRAVITY WAVES

When wind blows over the ocean surface, waves are generated, first at small scales but then, as the fetch or duration of the wind action increases, the wavelength of the dominant wave continues to increase. For a given wind speed, however, the spectral density of the wave components that are significantly smaller than that of the spectral peak does not continue to grow--the spectrum broadens and its integral (proportional to the wave energy density) increases, but the spectral density of the shorter components attains an equilibrium that reflects the balance among the dynamical processes involved, namely the rates of energy input from the wind, transfer by wave-wave interactions and losses by wave breaking.

The precise nature of the balance has been a matter of considerable interest. Kitaigorodskii (1983) has proposed a Kolmogoroff-type hypothesis in which energy input is restricted to wave-numbers near the spectral peak, with spectral flux by wave-wave interactions to much smaller scales where dissipation occurs. The present author, on the other hand (1985), seeing no reason why wind energy input should be restricted to the wave-numbers near the spectral peak, and observing that wave breaking at sea occurs over a wide range of scales postulated

31

that in the equilibrium range all three processes remain important
throughout. The predictions of the two theories as regard to spectral
distributions in this range are not very different, but important differences do appear, particularly in the spectral distribution of energy
loss by breaking and the statistical characteristics of the breaking
events. The latter theory provides specific, testable predictions for
the distributions and characteristics of these breaking events and
these will be summarized and extended in this discussion.

For the present purpose, the most pertinent prediction from the
hypothesis that the dynamical balance in the equilibrium range involves
all three processes, is that the spectral rate of energy loss from the
wave field by breaking, or the spectral rate of energy input into near-
surface turbulence is

$$\varepsilon(k) = A\rho \left| \cos\theta \right|^{3/2} u_*^3 \, k^{-2}, \tag{1.1}$$

where $\underline{k}$ is the wave-number vector, of magnitude k, θ is the angle of
this wave-number to the wind, ρ the water density and u_* is the friction velocity of the wind. A is a numerical constant associated with
the scaling of the wave spectral density and the energy input from the
wind and its magnitude is of order 0.05. This result (1.1) provides
the pivotal connection between the dynamical inputs into the wave field
(characterized by u_*) and the statistical properties of the breaking
events themselves.

2. STATISTICAL CHARACTERISTICS OF BREAKING EVENTS

As the wind blows over the water surface, at any instant the fronts of
the breaking waves define a distribution of isolated line or arc
segments. The scales of the breaking waves may cover a very wide range
from very short gravity waves in which a moving convergent stagnation
point is marked by a group of capillary ripples, through intermediate
scales (15-30 cm or so) where the breaking is unsteady and turbulent,
but only a few bubbles are produced, to actual whitecaps in which the
breaking and the generation of turbulence is so vigorous that extensive
patches of foam are generated. There is clearly some association of
the breaking events with waves of different scales, but it is difficult
to make the association in an unambiguous way if we consider only the
surface configuration at one given instant. A breaking crest may
indeed be a local maximum in the instantaneous surface configuration,
but there is no guarantee that a local wavelength of the breaking wave
can be defined clearly. It seems more satisfactory to use the velocity
of the breaking front as a measure of the scale of the breaking, since
this is a well-defined quantity that might (conceptually at any rate)
be measured from movie images of the sea surface. In practice, it
could be obtained relatively easily for those breaking events that
generate whitecaps, though it may be difficult to discern the many
smaller scale, fugitive occurrences of breaking that do not generate
discernible bubble trains but which still turn over the water surface
as they advance.

Accordingly, let us define a distribution $\Lambda(c)$ such that $\Lambda(c)$ dc represents the average local length per unit surface area of breaking fronts that have velocities in the range c to c + dc. The total length of breaking fronts per unit area is then

$$L = \int \Lambda(c) \, dc \qquad (2.1)$$

In unit time, the fraction of sea-surface area traversed by breaking fronts with velocities between c and c + dc is c (c) dc, so that the fraction of total surface area turned over per unit time, the turnover rate is

$$R = \int c\Lambda(c) \, dc \qquad (2.2)$$

This quantity also expresses the total number of breaking waves of all scales passing a given point per unit time; the distribution specifies the expected number per unit time passing a fixed point with velocities in the interval c to c + dc.

What is the rate of energy loss from the waves to turbulence per unit length of front in these breaking events? This question has been examined by Duncan (1981) in a series of laboratory experiments: he showed that, in a continuing active breaker in deep water, the breaking zone extends down the forward face of the wave over a fixed fraction of its amplitude and that its shape is geometrically similar for waves of different scales. Furthermore, he found that the breaking waves themselves are geometrically similar, so that the cross-sectional area of the breaking zone is proportional to the square of the local wavelength, or to $(c^2/g)^2$. The weight of the breaking zone per unit length of the front then exerts a tangential force per unit length proportional to c^4/g that acts on the incoming stream, whose speed is approximately c. Consequently, the rate of energy loss per unit length of front is proportional (c^5/g), where the constant of proportionality is estimated by Duncan from his experiments as approximately 0.06. In terms of the wave-number of the breaking wave, the rate of energy loss per unit length of breaking front is proportional to $g^{3/2}k^{-5/2}$.

It is interesting to observe, even at this stage, that the rate of turbulent energy production increases very rapidly with speed of advance c of the breaking front or with the scale of the breaker. A few large-scale breaking events can produce as much energy loss from the wave field and input to the turbulence as many smaller ones. Nevertheless, the characteristic timescale for the duration of a breaking event, the ratio of the wave energy in one wavelength to the rate of loss by breaking, is proportional to the wave period, so that in this sense, both large and small-scale breaking events are equally transient.

The average rate of energy loss per unit area by breakers with speeds between c and c + dc is then

$$\varepsilon(c) \, dc = b \, g^{-1} \, c^5 \, \Lambda(c) \, dc. \qquad (2.3)$$

Let us now identify the scale of waves that are breaking by the speeds with which their fronts advance. For the larger-scale breaking events, i.e. those whose phase $c > (2\pi s) \, C$, where C is the phase speed of the dominant wave and s the significant slope, the associated wavenumber is simply $k = g/c^2$. The speed of advance of smaller or micro-scale breaking events is, however, influenced strongly by long-wave advection, and this introduces a substantial complication in the transformation. One might possibly argue that breaking occurs predominantly near the crests of the dominant waves where the orbital speed is u_0, so that $k \approx g/(c - u_* \cos\theta)^2$, where θ is the angle between the breaking front and the dominant wave. This matter will not be pursued further here, however; let us concentrate on the larger-scale breaking events.

If $\Lambda(k)$ dk represents the total length per unit surface area of breaking fronts associated with waves in the wave-number interval k, k + dk,

$$\Lambda(k) = \Lambda(c) \frac{\partial(c_1, c_2)}{\partial(k_1, k_2)}, \tag{2.4}$$

then the spectral rate of energy loss from the wave field by breaking, or the spectral rate of energy input into near-surface turbulence is

$$\varepsilon(k) \propto g^{3/2} k^{-5/2} \Lambda(k). \tag{2.5}$$

Equating this expression to (1.1) gives for the distribution with respect to wave-number of the expected length of breaking front per unit surface area

$$\Lambda(k) = \text{const} \left| \cos\theta \right|^{3/2} g^{-3/2} k^{1/2} u_*^3 \tag{2.6}$$

Since, when $c^2 = g/k$

$$\frac{\partial(c_1, c_2)}{\partial(k_1, k_2)} = \frac{c^6}{2g^2},$$

the corresponding distribution with respect to phase velocity is

$$\Lambda(c) \propto \left| \cos\theta \right|^{3/2} u_*^3 g^{1/2} k^{1/2} c^{-6} = \left| \cos\theta \right|^{3/2} u_*^3 g c^{-7} \tag{2.7}$$

These distributions are directionally considerably narrower than the spectral distribution itself, which is roughly proportional to $(\cos\theta)^{1/2}$—a preponderance of breaking wave events advance in directions close to that of the wind.

These expressions now enable us to estimate explicitly the expected number of breaking waves passing a given point with velocities in the interval c to c + dc, $c\Lambda(c)$ dc, or with speeds in the interval c to c + dc regardless of direction:

$$n(c) \, dc = \int_{-\pi/2}^{\pi/2} c\Lambda(c) \, c \, d\theta \, dc,$$

$$\propto u_*^3 g c^{-5} \, dc \tag{2.8}$$

The total density of breaking fronts (expected total length per unit surface area) that have speeds between c_0 and c_1 is therefore, from (2.7)

$$L(c_0, c_1) = \int_{-\pi/2}^{\pi/2} \int_{c_0}^{c_1} \Lambda(c) \, c \, dc \, d\theta,$$

$$\propto u_*^3 \, g(c_0^{-5} - c_1^{-5}), \tag{2.9}$$

and the expected number passing a given point per unit time with speeds in this range is

$$n(c_0, c_1) = \int_{c_0}^{c_1} n(c) \, dc$$

$$\propto u_*^3 \, g(c_0^{-4} - c_1^{-4}). \tag{2.10}$$

With the values given previously, the numerical values of the coefficients in (2.9) and (2.10) are both approximately 10^{-2}.

If we consider only those breaking fronts that generate a train of bubbles, the event being then identified as a whitecap, then one might postulate that only those breaking zones with a rate of energy release bc^5/g exceeding some threshold value, ε_T say, will contribute. The lower limit of integration in these expressions is then $c_0 = c_T$ where

$$\varepsilon_T = b \, c_T^5/g.$$

If, under light winds or at a short fetch, the longest waves that are breaking are shorter than those with speed c_T then virtually no whitecaps will be formed. Yet if they are even somewhat larger, then the terms c_1^{-5} and c_1^{-4} are much smaller than c_T^{-5} and c_T^{-4} respectively. Consequently, in this case, the expected length of whitecap fronts per unit area is almost independent of the speed of the longest waves in the field and is equal to

$$L_w \approx \text{const.} \ c_T^{-5} \, g u_*^3,$$

$$= \text{const.} \ \varepsilon_T^{-1} \, u_*^3, \tag{2.11}$$

while the expected number passing a given point per unit time

$$n_w \propto c_T^{-4} \, g u_*^3 \tag{2.12}$$

both increasing as the cube of the friction velocity.

Because of the geometrical similarity expected in wavebreaking, the average length of each breaking segment is proportional to k^{-1}. Consequently, the number of individual segments per unit area at any instant associated with wave-numbers in the range $\underline{k}$, $\underline{k} + d\underline{k}$, is proportional to $k_{\Lambda}(\underset{\sim}{k}) \, d\underset{\sim}{k}$. The lifetime of each event, is, for the same reason, proportional to (but a smallish fraction of) the wave period n^{-1}, so that the expected number of events appearing (and disappearing) per unit area per unit time in this wavenumber interval is this number density divided by the lifetime, or $nk_{\Lambda}(\underset{\sim}{k})d\underset{\sim}{k}$:

$$\nu(\underset{\sim}{k}) \, d\underset{\sim}{k} \propto nk_{\Lambda}(\underset{\sim}{k})d\underset{\sim}{k}. \tag{2.13}$$

Again, if we consider only the larger scale breakers, the advection by the orbital velocity of longer waves is of less significance, n $(gk)^{1/2}$ and (2.13) reduces to

$$\nu(\underset{\sim}{k}) \, d\underset{\sim}{k} \propto g^{1/2} \, k^{3/2} \, \Lambda(\underset{\sim}{k})d\underset{\sim}{k},$$

$$\propto |\cos\theta|^{3/2} \, g^{-1} \, k^2 \, u_*^3 \, dk, \tag{2.14}$$

from (2.6). The total number of breaking events per unit area per unit time associated with waves whose wave-numbers lie between k_o, that of the spectral peak, and k_1, some upper limit, is therefore approximately

$$\nu(k_o, k_1) \int_{-\pi/2}^{\pi/2} \int_{k_o}^{k_1} \nu(k) \, kdkd\theta \propto g^{-1} u_*^3 \, (k_1^4 - k_o^4). \tag{2.15}$$

Now individual breaking events have been shown by Kalmykov and Pustovoytenko (1976) to be identifiable as "sea spikes", particularly with short wavelength radar at glancing incidence. In such a radar measurement, suppose we count only those events whose intensity or whose duration exceeds some convenient threshold that enables them to be identified unambiguously. Since the threshold intensity and duration both increase monotonically with increase in wavelength, either choice corresponds to a threshold wave-number k_1 below which the associated sea spikes will be identified; if in a reasonably well-developed sea state, this wave-number is substantially larger than that of the spectral peak, k_1^4 is much greater than k_o^4 and (2.15) reduces to

$$\nu(k_1) \propto g^{-1} \, k_1^4 \, u_*^3, \tag{2.16}$$

giving the total number of breaking events per unit surface area associated with waves whose wave-numbers are less than k_1 or whose wavelengths are greater than $(2\pi/k_1)$.

3. DISTRIBUTION OF BREAKERS WITH RESPECT TO ENERGY RELEASE

In underwater acoustics, we may be interested not so much in the distributions of breaking with respect to scale or speed of advance, but with respect to the kinetic energy injected--the number of breaking events per unit surface area per unit time as a function of wind friction velocity, etc., that produce kinetic energy densities, $T = (1/2) u_t^2$ in the water in the range T, $T + dT$, or kinetic energy per unit surface area $E = \int (1/2) u_t^2 \, dz$, where u_t is the turbulent velocity fluctuation. The previous results can be re-expressed in this way.

For example, the rate of energy input to the turbulence by wave breaking per unit length of front is proportional to $g^{3/2}k^{-5/2}$ so that the kinetic energy left behind per unit surface area in the broken patch is:

$$E \propto c^{-1} g^{3/2} k^{-5/2} = gk^{-2}. \tag{3.1}$$

The frequency of occurrence of breaking events between wave-numbers k, $k + dk$ is from (2.14)

$$\nu(k)dk = \int_{-\pi/2}^{\pi/2} \nu(\underset{\sim}{k}) \, d\,\theta \cdot kdk \propto g^{-1} k^3 u_*^3 \, dk. \tag{3.2}$$

Consequently,

$$\nu dE = \nu \frac{dk}{dE} dE \propto (g^{-1} k^3 u_*^3) \frac{k^3}{g} \, dE,$$

$$\propto g \, u_*^3 \, E^{-3} \, dE, \tag{3.3}$$

from (3.1) and (3.2). The density of occurrence, number per unit area per unit time producing turbulent energy per unit surface area between E and $E + dE$ again varies as the cube of the friction velocity but also as E^{-3}; more energetic patches are much less frequent than less energetic ones. The variation is even stronger if we are concerned with the distribution of turbulent intensity or energy per unit volume $T = (1/2) u_t^2$. Because of the geometrical similarity

$$T \propto Ek \propto gk^{-1}$$

(the depth of each turbulent patch being proportional to the breaking wavelength), a similar calculation gives

$$\nu(T)dT \propto g^3 u_*^3 T^{-5} \, dT. \tag{3.4}$$

The frequency of occurrence of breakers producing a given turbulent intensity T thus decreases as T^{-5} as T increases.

Finally, the total turbulent energy in each patch is equal to E times the patch area, proportional to k^{-2}

$$E_T \propto gk^{-4},$$

so that

$$\nu(E_T)dE_T \propto u_*^3 E_T^{-2}, \tag{3.5}$$

by similar steps. The <u>average</u> rate of energy input into the surface
turbulence is then

$$\varepsilon_o = \quad E_T \nu (E_T) dE_T \propto u_*^3 \ell n \left(\frac{E_{To}}{E_{T1}} \right) ,$$

$$= u_*^3 \ln \left(\frac{k_1}{k_o} \right) .$$

where E_{T1} is the kinetic energy injected by the smallest breaking wave
in the equilibrium range at wave-number k_1 and E_{To} that by the largest,
at wave number k_o. This, of course, agrees with a result given in the
previous paper (Phillips, 1985).

4. ACKNOWLEDGMENT

I am happy to acknowledge the support for this research of the Office
of Naval Research, under contract N 00014-84-K-0080.

5. REFERENCES

Duncan, J. H., 1981   An experimental investigation of breaking waves
     produced by a towed hydrofoil. Proc. R. Soc. Lond., **A 377**, 331–348.
Kalmykov, A. I. and Pustovoytenko, V. V., 1976   On polarization
     features of radio signals scattered from the sea surface at small
     grazing angles. J. Geophys. Res., 81, 1960–64.
Kitaigorodskii, S. A., 1983   On the fluid dynamical theory of the
     equilibrium range in the spectrum of wind-generated gravity waves.
     J. Phys. Oceanogr., 13, 816–827.
Phillips, O. M., 1985   Statistical and spectral properties of the
     equilibrium range in wind-generated gravity waves. J. Fluid Mech.,
     156, 505–31.

EXPERIMENTS ON BREAKING WAVES

W. K. Melville & R. J. Rapp
Massachusetts Institute of Technology,
Cambridge MA 02139,
U.S.A.

ABSTRACT.   Laboratory experiments on controlled deep-water breaking are reviewed. Following Longuet-Higgins(1974), breaking events were produced in a modulated wave packet generated by a computer-controlled wavemaker. Measurements of surface displacement, mixing down of surface material, and the velocity field due to breaking are presented. The repeatability of the wave generation procedure permitted the collection of ensemble-averaged data and the separation of the turbulent and average fields. It was found that the breaking process may be separated temporally into two periods: a fast initial period followed by a slower asymptotic evolution. Power laws in time were found to describe the evolution of the primary variables describing the evolution of the flow.

1.   INTRODUCTION

Wave breaking is undoubtedly an important factor in the generation of noise at the surface of the ocean; however, we do not know enough about breaking nor about surface noise to directly quantify the relationship. Noise may be generated by the impact of the surface on itself, drop impact, and the generation, resonance and decay of bubbles resulting from breaking induced air entrainment. These separate mechanisms are the subject of other papers at this meeting. In this paper we wish to concentrate on describing laboratory experiments which we have undertaken to better understand the dynamics and kinematics of breaking. An improved understanding of surface noise may result from the incorporation of the results of studies of the separate noise generation processes into an improved understanding of the breaking process. Preliminary results of the work reported here were presented in Melville & Rapp(1985), with more complete measurements described in Rapp(1986) and Rapp & Melville(1987).

B R Kerman (ed.), Sea Surface Sound, 39–49.
© 1988 by Kluwer Academic Publishers.

Breaking at the ocean surface may result from the interaction of waves and currents, direct forcing by the wind (Phillips, 1977), intrinsic instabilities of the wave field (Melville,1981), and from the constructive interference of a number of Fourier components (Longuet-Higgins, 1974). We believe that the last mechanism may be the most important for waves near the peak of the spectrum. The groupiness of the wave field and observations of the periodicity in whitecaps(Donelan et al.,1972) offer some support for this view. We believe that the direct effects of wind forcing are likely to be of little significance near the peak of the spectrum but may dominate at higher frequencies. In the past the surface noise has usually been correlated with the atmospheric variables; but if breaking near the peak of the spectrum is a dominant source of noise, and if wind forcing is not a direct contributor to breaking, then it may be preferable to correlate the surface noise with the wave parameters. Thus the dependence of breaking on the wave parameters is of interest.

Breaking is a source of entrained air and turbulence at the ocean surface, each of which may contribute to the generation and propagation of sound. Thorpe (this meeting) has shown that the bubble clouds generated by breaking waves may be mixed down to depths of the order of 10 meters, and Farmer (this meeting) has shown that the bubble layer may significantly attenuate the surface noise at higher frequencies. As considered elsewhere in these proceedings, the bubbly layer influences the sound speed and therefore forms a waveguide whose properties may be of importance in propagation near the surface. Thus the turbulence and transport associated with breaking waves is of interest in problems of surface noise. Breaking then is of interest for both sound generation and propagation at the ocean surface.

Unfortunately it is very difficult to directly measure breaking in the field. It is an intermittent random process which is fast compared with the usual oceanographic variables. Much of the available theoretical modelling of breaking processes depends on hypothesizing a breaking criterion which applies at the instant of breaking. However, the probability of a single point measurement at the surface catching an event at the instant of its initiation is vanishingly small. Usually point measurements will catch breaking at some stage in its evolution. Since the event may last for a significant fraction of a wave period, during which the underlying wave variables may be changing rapidly, it is very difficult to directly test such theoretical models.Indeed, it may be that the noise generated by breaking may prove to be one of the more useful measures that can be used to identify and quantify breaking.

In this paper we shall review those results of our laboratory experiments that may prove useful in assessing the role of breaking in the generation and propagation of sound in the surface layer.

2. THE EXPERIMENTS

Single isolated packets of of waves were generated by superposition of Fourier components of equal amplitude a_0, in a radian frequency band, $(f_0 \pm \delta f/2)$. The initial phase for each component was chosen to give constructive interference of linear surface waves at a point x_0 down the channel. Dimensional analysis (Melville & Rapp,1985) shows that the dimensionless dependent variables will be a function of (xk_0, tf_0) ,with a parametric dependence on $(a_0 k_0, \delta f/f_0, x_0 k_0)$. The parameters are dimensionless measures of nonlinearity,bandwidth and phase.

A series of experiments were conducted over a range of these dimensionless parameters., and time series of the surface displacement measured at stations along the wave channel. The surface displacement measurements were then used to estimate the momentum flux and energy flux carried by the higher frequency waves down the channel.(It must be remembered that the modulation of the higher frequency carrier waves forces long lower frequency waves.) The change in these fluxes due to breaking are then evident in the data. In addition, we made a flow visualization study. In the absence of breaking, dye floated on the surface stays on the surface . Movies were taken of the mixing down of dye in the breaking events; the outline of the dye clouds were digitized and the evolution of the length, depth and area of the dye cloud was measured as a function of time from breaking. The dye studies also served to define the volume of fluid directly affected by breaking, and a series of experiments were undertaken to measure the along-channel and vertical velocity components in a vertical plane parallel to the side walls of the channel. These measurements were taken with a two-channel laser anemometer with between 10 and 20 repeats at each station to permit the reliable eduction of the ensemble average and turbulent contributions to the velocity field. The computer control and repeatability permitted this ensemble averaging, avoiding the disadvantages of linear filtration methods that have been used in the past to attempt separation of the wave and turbulent components of the velocity field. The experiment was run by two LSI/11 microcomputers which controlled the wavemaker and acquired and stored the data on tape for further processing.

42

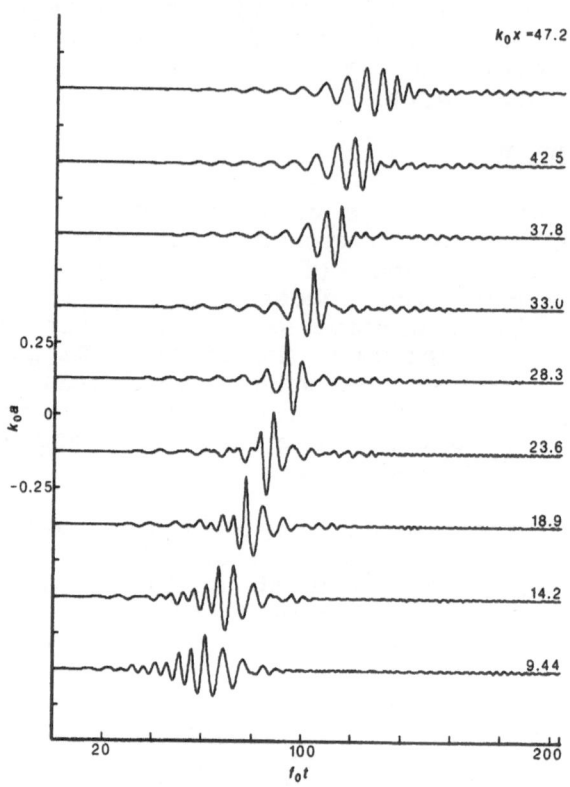

Figure 1. Time series of surface displacement down the channel for a spilling breaker: f_o=6.79 rad s^{-1}, $\delta f/f_o$=0.73, x_ok_o=27.4, a_ok_o=0.3. The time is normalized by f_o and the amplitude by k_o. Breaking occurs at x_ok_o=28, where xk_o is the normalized distance from the paddle (Melville & Rapp, 1985).

3. RESULTS

Figure 1 (Melville & Rapp, 1985) shows time series of the surface displacement in a wavepacket at different stations down the channel. Time series such as these can be used to infer the dynamical changes in the wave packet down the channel: most specifically those changes due to breaking.

Figure 2 (Melville & Rapp, 1985) shows examples of the breaking waves generated. It was found that the breaking was most sensitive to the amplitude parameter, a_ok_o. For fixed bandwidth and phase, small values of a_ok_o did not produce breaking, just a modulation- demodulation cycle of the wave packet. Above a threshold value of a_ok_o, breaking

began with a single very weak (incipient) breaking event. An
increase in $a_o k_o$ gave weak multiple breaking events until a
single spilling event occurred at a particular value of
$a_o k_o$. A subsequent increase in $a_o k_o$ led to multiple breaking
events until a single plunging event was attained. A further
increase in $a_o k_o$ led again to multiple events.

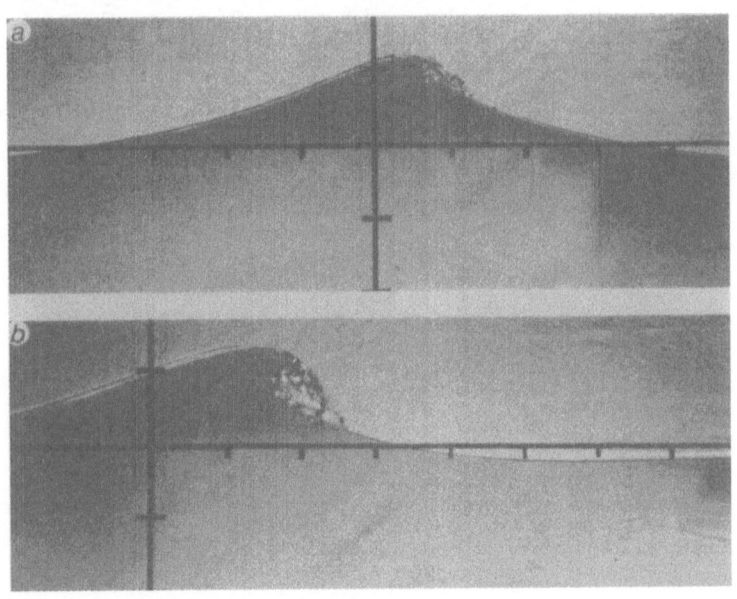

Figure 2. Examples of (a), single spilling; and (b), single
plunging waves (Melville & Rapp, 1985).

The surface displacements were used to estimate the
excess momentum flux down the channel and the change across
the breaking region. An example of the change in excess
momentum flux is shown in figure 3 (Melville & Rapp, 1985).
This is also an estimate of the dissipation due to breaking
(Rapp, 1986; Rapp & Melville, 1987) and the values plotted
have been normalized over the total packet upstream of the
breaking region. These results suggest that up to
approximately 25% of these fluxes may be lost in a single
breaking event.

Analysis of the mixing experiments produced the results
shown in figures 4-6. Figure 4 represents the evolution of
the length of the dye cloud for plunging breaking events in
which the center frequency varied from 0.88 to 1.28Hz. This
figure shows the two-time-scale structure that was found in
all the flow visualization; with a fast increase in length
followed by a much slower increase for large times. The fast

increase lasted for 1-2 wave periods. This figure also
contains the only clear departure from scaling based on the
ideal-fluid wave variables. In particular, we draw attention
to the fact that the longest waves show a significantly
larger dimensionless length of the mixing region. The
photographic evidence suggests that this is due to much more
spray being generated at the surface-surface impact in this
case, and the effects of surface tension being neglected in
this scaling used in the figure. Figure 5 and 6 show
corresponding results for the depth and area of the dye
cloud. Again the fast initial and slow asymptotic regions
are evident, with the depth increasing like $t^{1/4}$ and the
area like $t^{1/2}$.

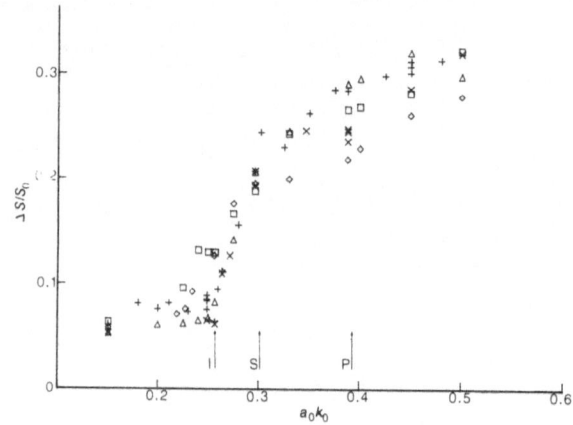

Figure 3. Excess momentum flux lost from the wave field due
to breaking, plotted as a function of $a_0 k_0$ for $\delta f/f_0$ in the
range [0.4-1.4], $x_0 k_0 = 27.4$. The arrows show the amplitudes for
incipient breaking (I), single spilling (S), and single
plunging waves (P), for $\delta f/f_0 = 0.73$ (Melville & Rapp, 1985)

Two-component velocity measurements were made and
analyzed to give the ensemble averaged and turbulent
contributions. Figure 7 shows an example of the ensemble
averaged velocity field evolving in time from one to
approximately sixteen wave periods after breaking. The
obvious feature is the eddy approximately one wavelength
long with local speeds in the range 0.01-0.1 C initially, (C
is the characteristic phase speed) decaying to 0.001-0.01C
after 16 wave periods .The total kinetic energy in the
residual motion was also estimated from the velocity
measurements and found to follow a t^{-1} power law.

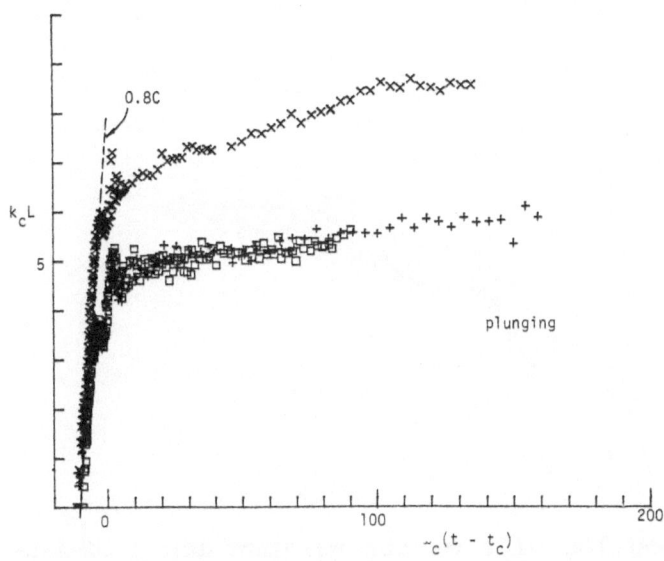

Figure 4. Length of dye boundary vs time for plunging waves with center frequencies of 0.88(x), 1.08(+) and 1.28([])Hz; $\delta f/f_c$=0.73. (Rapp,1986; Rapp & Melville, 1987) (Note that the subscripts "o" and "c" are used interchangeably in this paper, where figures have been taken from different sources.)

4. DISCUSSION

The results summarized here and presented in more detail in Rapp & Melville(1987) and in the oral presentation at the meeting suggest that despite the randomness of the post-breaking kinematics, the main dynamical and kinematic features of breaking do scale with the pre-breaking wave variables. The only exception to this observation being the processes associated with enhanced spray generation in the largest scale events observed. The breaking process divides temporally into an initial period which lasts for 1-2 wave periods and in which the velocities are comparable to the characteristic phase speed. This is followed by an asymptotic decay which we observed for O(10-100) wave periods.(See below.)

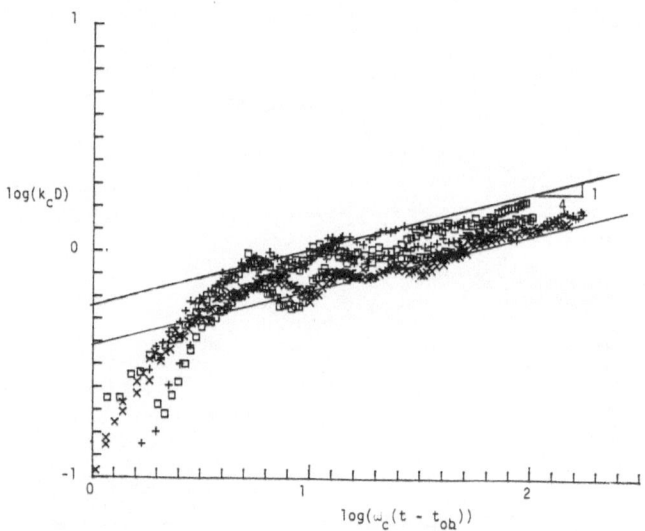

Figure 5. Log-log plot of the maximum depth of the dye cloud vs time. Same parameters as figure 4. Plot shows fit of the data to a $t^{1/4}$ power law. (Rapp, 1986; Rapp & Melville, 1987)

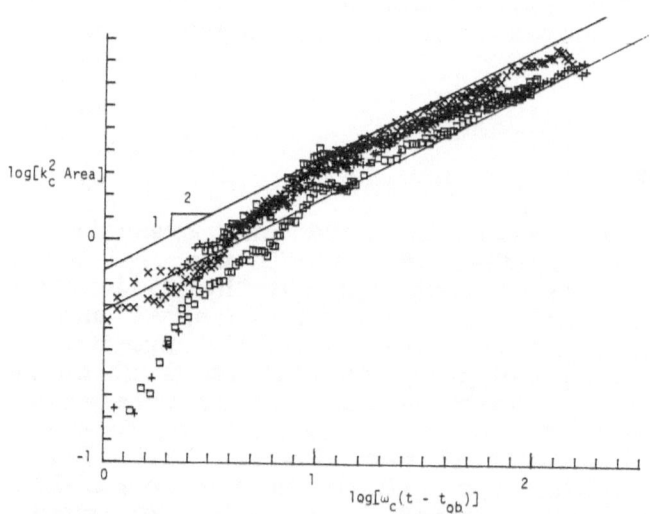

Figure 6. Log-log plot of the dye cloud area vs time. Same parameters as figure 5. (Rapp, 1986; Rapp & Melville, 1987).

The measurements of the losses of excess momentum flux and energy from the wave field varied smoothly independently of whether there were single or multiple events. The measurements of the ensemble averaged velocity field remaining after the waves had left the breaking region showed an average vorticity of $O(0.01)f_o$, where f_o is the characteristic radian frequency. The formation of this eddy involves the forced long waves as well as the higher frequency carrier waves (Rapp,1986; Rapp & Melville, 1987). These results support the view that it is not sufficient to consider the breaking of individual waves in isolation, but that the modulation of the wave field and the associated forced waves are important throughout the breaking process.

Analysis of the mixing data and the velocity measurements show that in the asymptotic decay period the turbulence scales vary as follows:

length, $l \propto t^{1/4}$,

velocity, $v \propto t^{-3/4}$.

If time is transformed to a distance variable, these power laws correspond to the intermediate asymptotic regime for the 2-D momentumless wake(Tennekes & Lumley,1972) . This contrasts with the experiments in which a hydrofoil has been used to generate breaking(Battjes & Sakai, 1981), which display the plane wake scaling:$l \propto x^{1/2}, v \propto x^{-1/2}$. These differences point to the need for care in using quasi-steady models to describe unsteady breaking.

Finally, it is instructive to consider the implications of these measurements for the field. If we assume that these two-dimensional results can be used to estimate the three-dimensional breaking in the field,a breaking wave of characteristic period 6 seconds (length 60 m,phase speed 10 ms^{-1}) will mix to a depth of 10-20 meters in 6-60seconds, and will produce a surface velocity in the direction of wave propagation which will decrease from 30-5 cms^{-1} over 6 minutes, with typical turbulent velocity scales decreasing from approximately 20 cms^{-1}. Available field data show that at a point the number of breaking waves per wave may be in the range $O(0.01-0.1)$(Thorpe & Humphries, 1980). Thus the measured decay times may be comparable to the recurrence time of breaking.

48

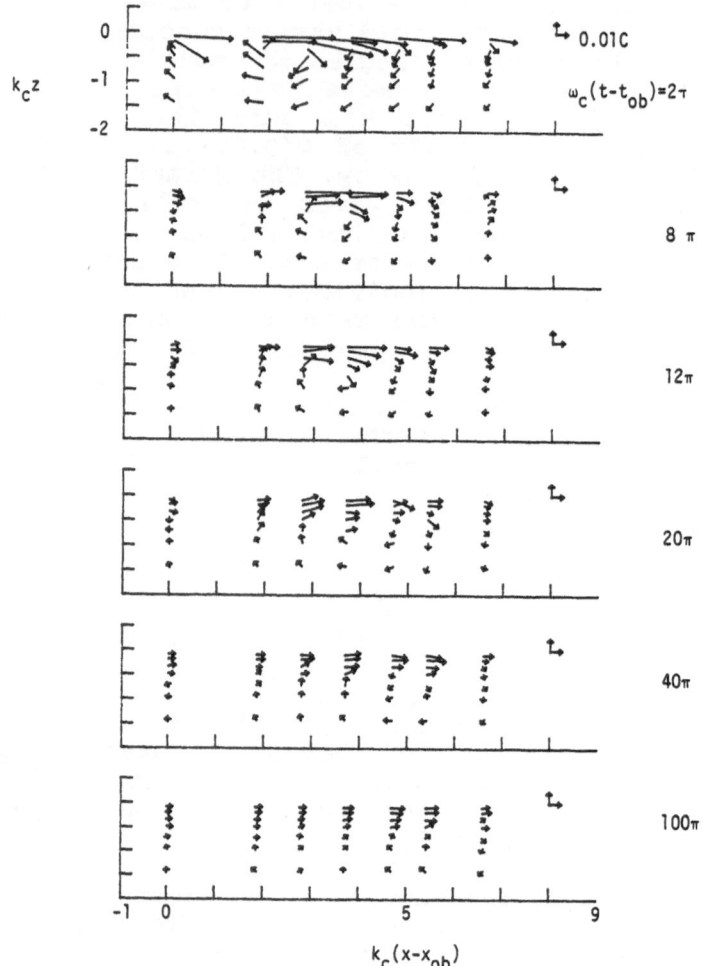

Figure 7. Ensemble averaged velocity field as a function of time after a plunging break; $f_C=1.28Hz$, $a_0k_0=0.42$. (Rapp, 1986).

5.   ACKNOWLEDGEMENTS

This research was supported National Science Foundation Grants, 8214746-OCE and 8614889-OCE.

6. REFERENCES

Battjes, J.A. & Sakai, T. 1981, 'Velocity field in a steady breaker", J. Fluid Mech., **111**,421-437.

Donelan, M.A., Longuet-Higgins, M.S. & Turner, J.S. 1972, 'Periodicity in whitecaps', Nature, **239**, 449-451.

Longuet-Higgins, M.S. 1974, ' Breaking waves in deep or shallow water', Proc. 10th Conf. on Naval Hydrodynamics,597-605 (U.S. Gov. Printing Office, 1976).

Melville, W.K. 1981,'The instability and breaking of deep-water waves', J. Fluid Mech., **115**, 165-185.

Melville, W.K. & Rapp, R.J. 1985, ' Momentum flux in breaking waves', Nature, **317**, 514-516.

Phillips, O.M. 1977, <u>The Dynamics of the Upper Ocean</u>, Cambridge University Press, Cambridge.

Rapp, R.J. 1986,'Laboratory measurements of deep-water breaking waves', Doctoral Thesis, Ocean Engineering, Massachusetts Institute of Technology.

Rapp, R.J. & Melville, W.K. 1987, 'Laboratory measurements of deep-water breaking waves', Phil. Trans Roy. Soc. Lond., sub judice.

Tennekes, H. & Lumley, J.L. 1972, <u>A First Course in Turbulence</u>, MIT Press, Cambridge, Massachusetts.

Thorpe, S.A. & Humphries, P.N. 1980, ' Bubbles and breaking waves', Nature, **283**, 463-465.

WIND WAVES AS A COUPLING PROCESS BETWEEN AIR AND WATER TURBULENT BOUNDARY LAYERS

Y. Toba, H. Kawamura and I. Yoshikawa
Department of Geophysics
Faculty of Science
Tohoku University
Sendai 980
Japan

ABSTRACT. Experimental evidence indicates that the boundary layers above and below wind waves have a common structure, which is similar to that of the turbulent boundary layer over a rough solid wall. Overall constraints for the coupling of the two-sided air and water turbulent boundary layers, combined with the 3/2-power law of wind waves, require that all the characteristic velocities related to the air-water boundary processes are proportional to one another. A physical interpretation of this state of the local equilibrium is presented with a concept of the "breaking adjustment of wind waves". The above proportionality of the characteristic velocities provides a physical basis for the observed friction-velocity scaling of the ocean mixed layer.

1. INTRODUCTION

A particular feature of the wind wave is that it is generated as a coupling process between the two adjoining turbulent boundary layers of air and water. As well as the element of surface wave motion, the local wind drift and turbulence in the air and water boundary layers are also important elements in the wind wave. This paper summarizes our recent studies on wind-wave breaking and turbulence at the air-water interface, with reference to our two manuscripts: "Turbulent structure in water under laboratory wind waves" by I. Yoshikawa, H. Kawamura, K. Okuda and Y. Toba, and "Similarity laws of the wind wave and the coupling process of the air and water turbulent boundary layers" by Y. Toba.

2. TURBULENT BOUNDARY LAYER BELOW THE WIND WAVE

We first describe results of an experimental study in a laboratory wind-wave tank on turbulence underneath wind waves, by using a combination of a high-sensitivity thermister array with a sonic flowmeter. The temperature fluctuations have been used to detect movements of water parcels, with temperature as a passive quantity.

As shown in Figure 1, the turbulent energy is dominant in the

B. R. Kerman (ed.), Sea Surface Sound, 51–62.
© 1988 by Kluwer Academic Publishers.

52

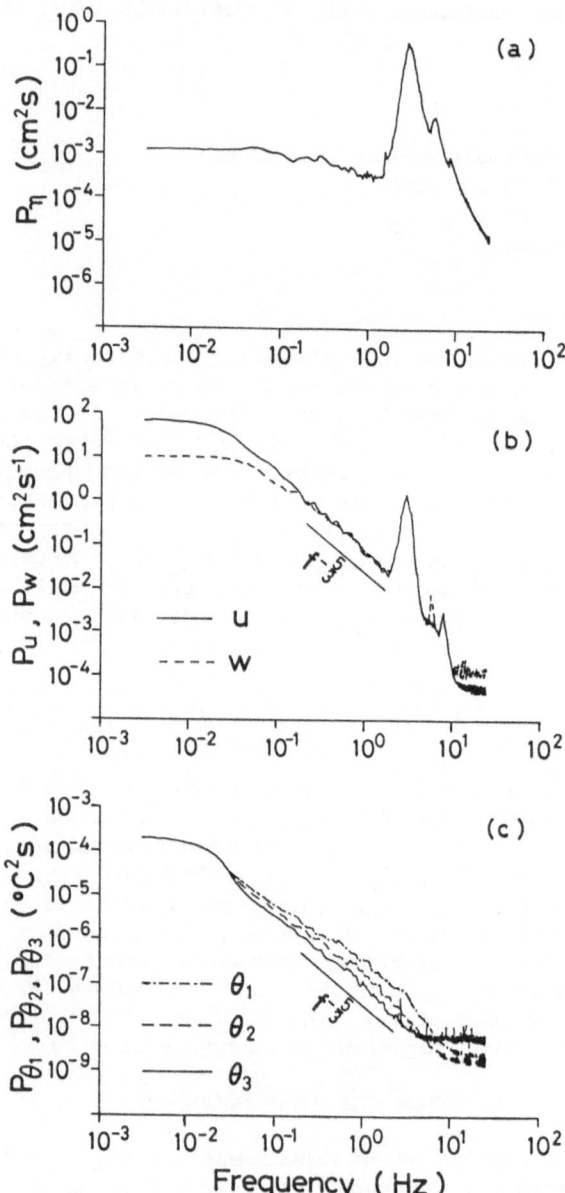

Figure 1. Power spectra of surface displacement (a), water velocities
u and w (b) and temperature fluctuation at three depths (c), under
laboratory wind waves. The conditions were $u_* = 0.36$ m s^{-1} and
fetch = 8.2 m. (Cited from Yoshikawa et al., 1987.)

frequency range of 0.01-0.1 Hz, which is much smaller than the wind-wave frequencies of 2-5 Hz. In the former range the turbulence is anisotropic and transfers of momentum and heat proceed. By using a variable-interval time-averaging technique (VITA) for analyses of the time series, it has been found that conspicuous events in this main turbulence energy range consist of downward bursting from near the air-water interface. There is a frequency range 0.2-2 Hz for velocity, and 0.2-5 Hz for temperature fluctuation, where the turbulence is isotropic and has the -5/3 slope. After the energy for the wind-wave frequencies is eliminated by using a filter, the profiles of the turbulent intensities, $(\widehat{u^2})^{1/2}$ for the horizontal direction, $(\widehat{w^2})^{1/2}$ for the vertical direction, and the Reynolds stress $(-\widehat{uw})^{1/2}$, are shown in the left panels of Figure 2. A layer of constant momentum flux is seen in the profiles of the Reynolds stress. In Figure 3 an example of the structure of the turbulent boundary layer above laboratory wind waves is shown for a single wind condition. In this turbulent boundary layer above wind waves, ordered motions have been clearly observed (Kawamura and Toba, 1987). It is noted that the boundary layers above and below wind waves have a common structure which is similar to that of the turbulent boundary layers over rough solid walls.

However, at the same time, the turbulent intensities below wind waves can just as well be scaled by properties of the wind wave, i.e., the root mean squared water-level fluctuation $(\overline{\eta^2})^{1/2}$ and the wave peak frequency σ_p, as seen in the right panels of Figure 2. From this we deduce that, the turbulence underneath wind waves develops to keep a close relationship to the wind waves. It is noted that the distributions of points in the right panels of Figure 2 correspond closely to those of Figure 3 when the vertical axis is reversed.

The fact demonstrated by the right panels of Figure 2 can be interpreted if we consider the existence of the 3/2-power law (Toba, 1972) between the nondimensional wave height and period together with overall constraints for the coupling between the air and water turbulent boundary layers, as discussed in the following section.

3. THE 3/2-POWER LAW OF WIND WAVES AND COUPLING CONDITIONS BETWEEN AIR AND WATER TURBULENT BOUNDARY LAYERS

A similarity law was proposed by Toba (1972) for growing wind waves. It is expressed by a power law,

$$H^* = BT^{*3/2}, \quad B = 0.062, \tag{1}$$

for the nondimensional wave height $H^* = gH_{1/3}/u_*^2$ and nondimensional period $T^* = gT_{1/3}/u_*$, where g denotes the acceleration of gravity, u_* the friction velocity of the air, and $H_{1/3}$ and $T_{1/3}$ the significant wave height and period of these waves, respectively.

There are several characteristic velocities which are concerned with

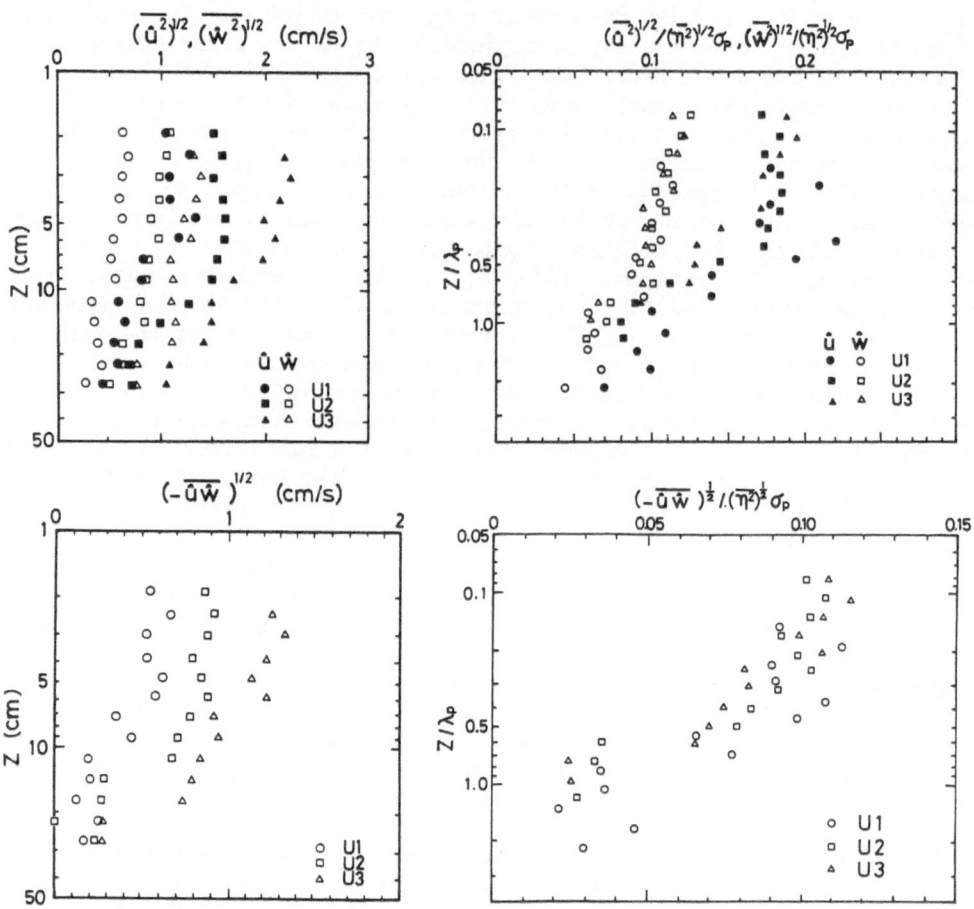

Figure 2. Measured profiles, at three different wind conditions, of square root of turbulent intensities and Reynolds stress below laboratory wind waves (left), and their scaling by wind-wave properties (right). (Cited from Yoshikawa et al., 1987.)

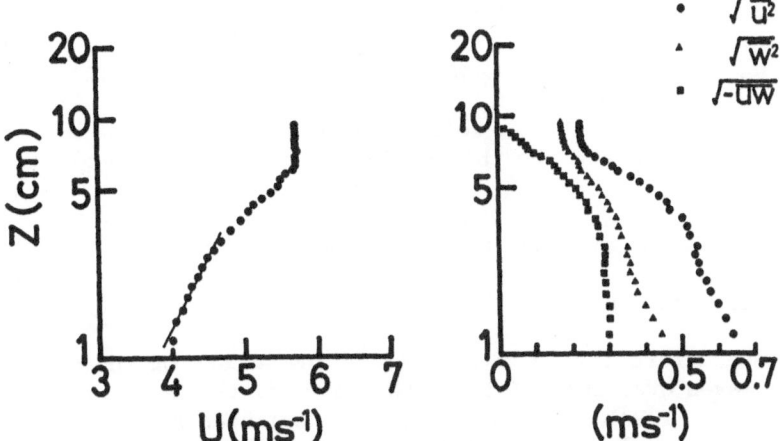

Figure 3. Measured profile of wind speed (left), square root of turbulent intensities and of Reynolds stress above laboratory wind waves (right). The conditions were $u_* = 0.29$ m s^{-1} and fetch = 6 m. (Cited from Kawamura and Toba, 1987.)

the air-water boundary processes (cf. Figure 4):

(a) u_*: the friction velocity of the air flow defined by $u_* = (\tau/\rho)^{1/2}$, where τ is the total wind stress and ρ the density of the air. Since the boundary layer above the wind waves is a conventional turbulent boundary layer, other turbulent intensities are proportional to u_*^2.

(b) $\overline{u_d}$: the average of the Lagrangian velocity of water on the air-water interface.

(c) u_0: the Stokes-drift velocity of the irrotational component of the wind wave. We assume that it is mostly due to waves of the peak frequency σ_p for simplicity. For laboratory experimental data, u_0 can

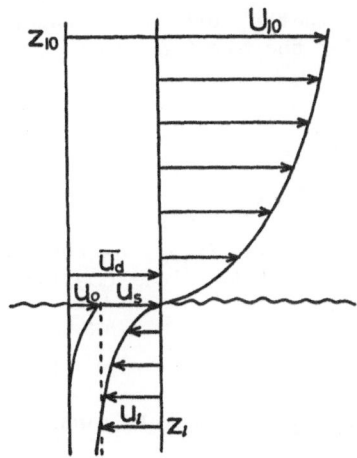

Figure 4. Schematic representation of the characteristic velocities related to the air-water boundary processes.

be expressed in terms of the average wave height H and the average wave period $\bar{T}$ ($\simeq 2\pi/\sigma_p$), as follows:

$$u_0 = (\pi/2)\, \delta\, \bar{H}\, \sigma_p = 2\,\pi^3 \bar{H}^2/g\bar{T}^3, \qquad (2)$$

where δ is the wave steepness defined by $\delta = \bar{H}/\bar{\lambda}$, $\bar{\lambda}$ the wave length of waves of period $\bar{T}$, and the third-order terms have been neglected in Eq. (2).

(d) u_s: the effective wind-drift velocity defined by

$$u_s = \bar{u_d} - u_0 . \qquad (3)$$

This speed may alternatively be defined as the excess of the phase speed of the wind waves over that of the water waves of the same wave number (Tokuda and Toba, 1982). Note that u_s is an Eulerian mean velocity while u_0 is a Lagrangian velocity. However, the average Lagrangian velocity $\bar{u_d}$ of water on the air-water interface comprises the two components.

(e) u_{*w}: the friction velocity of water in the turbulent boundary layer underneath the wind waves. This is equal to the square-root Reynolds stress $(-\overline{\hat{u}\hat{w}})^{1/2}$. The turbulent intensities, $(\overline{\hat{u}^2})^{1/2}$ and $(\overline{\hat{w}^2})^{1/2}$ are found proportional to u_{*w}^2, as seen in Figure 2.

Now we consider the continuity of momentum flux between the two-sided turbulent boundary layers above and below the air-water interface. The friction velocity of water, u_{*w}, can be related to u_* by

$$u_{*w} = (\rho/\rho_w)^{1/2}(1-G)^{1/2}u_*, \qquad (4)$$

where G is the fraction of momentum transfer retained in the wave growth and ρ_w the density of water. The value G is at most about six percent (Toba, 1978), and we may replace Eq.(4) by an approximate form:

$$u_{*w} = (\rho/\rho_w)^{1/2} u_* . \qquad (5)$$

From the logarithmic velocity law in the turbulent boundary layer, the following expression can be used for the air flow above the wind waves:

$$\frac{U_{10} - \bar{u_d}}{u_*} = \frac{1}{\kappa} \ln \frac{z_{10}}{z_0} = C_D^{-1/2} , \qquad (6)$$

where U_{10} is the wind speed at the 10-m height (z_{10}), z_0 the roughness length, κ the von Karman constant, and C_D is the drag coefficient defined by

$$\tau = \rho\, C_D U_{10}^2 , \qquad (7)$$

where $\bar{u_d}$ has been neglected compared to U_{10}. Similarly, when the mean depth is measured by z, the following equation can be used for the turbulent boundary layer underneath the wind waves:

$$\frac{u_1}{u_{*w}} = \frac{1}{\kappa} \ln \frac{z_1}{z_{0w}} = C_{Dw}^{-1/2} , \tag{8}$$

where u_1 is the representative speed of flow to be measured backward from a coordinate system moving at a speed of $\overline{u_d}$, at the depth z_1 (see Figure 4). The depth z_1 might be taken as z_{10}. However, in wind-wave tunnels, there are differences usually in the measured absolute values of $\overline{u_d}$ because of the existence of a return flow in the lower part of the tank; the value of $\overline{u_d}$ depends upon difference in dimensions and other conditions of tanks. For the moment, let us choose a value of z_1 $\lambda_p/2$ (cf. Figure 2), where λ_p is the wavelength of the waves of frequency σ_p. Since we cannot determine the flow below this depth, we assume for simplicity that $u_1 = u_s$. The z_{0w} is the roughness parameter, C_{Dw} the drag coefficient for the water boundary layer, and the latter is defined by

$$\tau_w = \rho_w C_{Dw} u_1^2 . \tag{9}$$

If we assume

$$C_{Dw} = b \, C_D \tag{10}$$

for the drag coefficients, where b is a constant, we obtain, using Eqs. (5), (6) and (8), with $\overline{u_d}$ neglected in Eq. (6),

$$\frac{u_{10}}{u_*} = b^{1/2} \frac{u_s}{u_{*w}} \tag{11}$$

and then it follows

$$\frac{u_s}{u_*} = (\frac{\rho}{\rho_w})^{1/2} C_D^{-1/2} b^{-1/2} . \tag{12}$$

Since C_D has incidentally a similar value with ρ/ρ_w, we get

$$\frac{u_s}{u_*} \simeq b^{-1/2} . \tag{13}$$

If we assume alternatively that $z_{0w} = a z_0$ with a as a constant, a somewhat similar result is obtained.

It has been deduced from the above reasoning that the condition of the continuity of the momentum flux between the two turbulent boundary layers requires the proportionality $u_s \propto u_*$, at least within the order of approximation of C_D = constant. We can express this situation in another way, which is that, once the drag coefficients are given, the nature of the coupled boundary layers gives a fixed value of u_s for a fixed value of U_{10}.

Next we consider the consequence of the 3/2-power law Eq. (1). A combination of Eq. (1) and the expression for Stokes drift Eq. (2) requires the proportionality of the Stokes-drift velocity u_0 with the friction velocity u_*, as already indicated in Toba (1972):

$$u_0 \, / \, u_* = 2\,\pi^3 B^2 \, . \tag{14}$$

Equation (3) for the surface drift combined with Eq. (14) then requires

$$\frac{\overline{u_d}}{u_*} = \frac{u_s}{u_*} + 2\,\pi^3 B^2 \, . \tag{15}$$

By virtue of Eqs. (5), (13), (22), (14) and (15), the following relationship holds approximately:

$$\overline{u_d} \propto u_0 \propto u_s \propto u_* \propto u_{*w} \propto (\,\overline{u^2}\,)^{1/2} \, . \tag{16}$$

Namely, all the characteristic velocities which are concerned with the air-water boundary processes are proportional to one another, and to u_*.

Now we examine the experimental evidence for this relationship (16). Tokuda and Toba (1982) gave the values of $u_s/u_* = 0.21$ and $u_0/u_* = 0.11$. Wu (1983) reported that $u_0/\overline{u_d} = 0.13$ with $u_s/u_* = 0.53$, resulting in $u_0/u_* = 0.069$. The difference of the observed values may be due to the difference of the methods of measurement, together with the small possible difference of the influence of the return flows at the lower layers of the tanks. Ebuchi, Kawamura and Toba (1987) gave support to Tokuda and Toba value of u_0/u_*, from the measurement of spacing of capillary waves on the forward face of larger waves, using a longer wind-wave tank. From wind-wave tank data by Hsu et al. (1982) (their Table 1), we can estimate a definite value of $\overline{u_d}/u_* = 0.55$. The statistical proportionalities among values u_*, $\overline{u_d}$, u_0, and u_s are thus considered supported experimentally.

From the form of the expression for u_0/u_* in (14), the value of B should have a definite order of magnitude. The above mentioned values of $u_0/u_* = 0.11$ by Tokuda and Toba (1982) and 0.069 by Wu (1983) give the value of B of 0.043 and 0.033, respectively. The former value is coincident with the value of B given by Tokuda and Toba (1982) for the 3/2-power law for average wave measures ($\overline{H}$ and $\overline{T}$) in their experiment. The value of u_0/u_* and B are thus consistent with each other as Tokuda and Toba reported already.

In the right panel of Figure 2 the value of the normalized square-root Reynolds stress is approximately expressed as

$$(-\overline{\hat{u}\hat{w}})^{1/2} \Big/ (\overline{\eta^2})^{1/2} \sigma_p = 0.105 \tag{17}$$

in the constant momentum flux layer of depth $z/\lambda_p < 0.25$. Using the values of $(\overline{\eta^2})^{1/2} = H_{1/3}/4 = \overline{H}/2.5$ $(H_{1/3} = 1.6\ \overline{H})$ and the value of wave steepness $\delta = 0.10$ together with Eq. (2) for the Stokes drift, the above relation is converted to

$$(-\overline{\hat{u}\hat{w}})^{1/2}/u_0 = u_{*w}/u_0 = 0.27. \tag{18}$$

This then gives, using the ratio of the air and water friction velocities Eq. (5), $u_0/u_* = 0.12$. This value of u_0/u_* is coincident with the Tokuda and Toba (1982) value, though the experiment by Yoshikawa et al. (1987) of Figure 2 was performed in a wind-wave tunnel which is different from that of Tokuda and Toba. Thus experimental values are regarded as consistent.

A further physical interpretation of the interrelationship of velocities in Eq. (16) will be presented in the next section.

4. BREAKING ADJUSTMENT OF WIND WAVES: A PHYSICAL INTERPRETATION

In order to investigate the physical situation underlying the relationship (16) among the characteristic velocities in the air-water boundary processes, it is necessary to examine local phenomena occuring at the air-water interface. There exists a thin wind-drift shear layer along the air-water interface. Banner and Phillips (1974) and Phillips and Banner (1974) assumed such a kind of thin sheet of vorticity parallel to the surface that is strained and convected by underlying larger scale velocity field, in their study of the breaking-wave conditions. However, the actual situation is transient and inhomogeneous, since the layer is driven by the shear stress which is much higher at the windward face of the wind-wave crests and low or sometimes negative in the lee face due to the sporadic flow separation (Okuda, Kawai and Toba, 1977; Okuda, 1982; Kawai, 1982).

For those wind-wave crests where the air entrainment is occuring, it is usual that the speed of water particles at the crest is faster than the phase speed of the waves (Koga, 1981). Even when air entrainment is not occuring, recirculation of water exists irregularly at the crest, accompanied by small scale breaking (Okuda, 1982). However, if we observe the conditions in the average, a kind of statistical equilibrium should exist where, for a given average wind and wave conditions, we can define the average values such as $\overline{u_d}$, u_0 and u_s.

Let us consider a microscopic picture of the fluctuation of these values. If for example the wave height is increased by the wave modulation, the Stokes-drift velocity increases, and at the same time the local wind-drift increases. At the leeward face of the crest air flow separation occurs (Kawai, 1982), which causes a larger wind-drift at the windward face and a smaller wind-drift at the leeward face (Okuda et al., 1977). The accelerated volume of water in the windward face must go into a deeper layer, since the thickness of the wind-drift shear layer is confined to be thin because the layer below this thin layer is the turbulent boundary layer. This is nothing but the wave breaking, even if there is no air entrainment.

The breaking of wind waves is a phenomenon corresponding to the flow separation in the air side, and it sometimes causes downward bursting. The excess momentum accumulated locally in the surface shear layer is transferred into a deeper layer by the breaking of waves and the downward bursting. As a result, the wave height is decreased (Koga, 1984), the Stokes drift velocity, the effective wind-drift and the surface water velocity become small, and the local wind stress will also become small, as the air-flow separation ceases. Since a large energy dissipation due to the wave breaking ceases, the wave again experiences the wave growth, to repeat these cycles.

Thus the values of $\overline{u_d}$, u_s and u_0 are average values of the locally fluctuating quantities, and the average state or the local equilibrium state is maintained by the breaking of wind waves, and by the transfer of the excess momentum at the interface into the deeper layer at the site of the wave breaking. Consequently, it is very natural, in this statistical equilibrium condition, that the turbulent velocities such as ($\widehat{u}^2)^{1/2}$ in the water underneath the wind waves are proportional to the characteristic velocities at the interface, such as $\overline{u_d}$, u_s and u_0. The relationship (16) between the characteristic velocities is interpreted as the manifestation of this state of equilibrium.

In other words, the relationship (16) is maintained by the ever-breaking situation of the wind waves. We may call this situation the "breaking adjustment of wind waves". We thus consider that the proportionality of the characteristic velocities, and the 3/2-power law Eq. (1) are all manifestation of the breaking adjustment of the wind wave.

Mitsuyasu (1985) discussed the budget of the momentum transfer from the wind to the wind wave, synthesizing recent experimental results of growth of the wind wave and the pressure measurements above the wind wave, also taking account of the value of G as given by Eq. (3). He concluded that, an amount of about 50% of the total momentum flux from the wind to the water enters once into the waves, but a major part of the momentum is lost immediatly by the wave breaking and only a small fraction goes into the wave growth. Melville and Rapp (1985) supported his concept by measurement of the momentum loss due to wave breaking. Such a situation of the momentum budget is consistent with the vision of the breaking adjustment in the present article.

The magnitude of the vertical turbulent eddy diffusion coefficient

K_z is one of important issues, in relation to the distribution of bubbles or gases in the upper ocean. Many studies (e.g., Jones and Kenny, 1977) support the form $K_z = \mathcal{K} u_{*_w} z$, though recent bubble diffusion measurements by Thorpe (1986) suggest higher values of u_{*_w} than can be estimated from u_* by using Eq. (5). The scaling of the ocean mixed layer, which has been observed to depend on u_*, follows from the proportionality of all the characteristic velocities Eq. (16), and does not need to include wave breaking as an additional independent variable.

Acknowledgments

The authors express their thanks to Profs. T. Tatsumi, K. Kajiura, H. Mitsuyasu, Drs. I.S.F. Jones, A. Masuda, K. Hanawa, and Mr. N. Ebuchi, for their valuable discussion. This study was partially supported by Grant-in-Aid for Scientific Research by the Japanese Ministry of Education, Science and Culture, No. 61540294, and was also performed as a part of the Ocean Mixed Layer Experiment (OMLET), a Japanese program for WCRP.

References

Banner, M.L. and O.M. Phillips, 1974: 'On the incipient breaking of small scale waves'. *J. Fluid Mech.*, **65**, 647-656.

Ebuchi, N., H. Kawamura and Y. Toba, 1987: 'Fine structure of laboratory wind-wave surfaces studied by using an optical method'. *Boundary-layer Met.*, **39**, 133-151.

Hsu, C.T., H.W. Wu, E.Y. Hsu and R.L. Street, 1982: 'Momentum and energy transfer in wind generation of waves'. *J. Phys. Oceanogr.*, **12**, 929-951.

Jones, I.S.F. and B.C. Kenney, 1977: 'The scaling of velocity fluctuations in the surface mixed layer'. *J. Geophys. Res.*, **82**, 1392-1396.

Kawai, S., 1982: 'Structure of air flow over wind wave crests'. *Boundary-Layer Met.*, **23**, 503-521.

Kawamura, H. and Y. Toba, 1987: 'Ordered motion in the turbulent boundary layer over wind waves'. (Submitted to *J. Fluid Mech.*)

Koga, M., 1981: 'Direct production of droplets from breaking wind waves -- its obsevation by a multi-colored overlapping exposure photographing technique'. *Tellus*, **33**, 552-563.

Koga, M., 1984: 'Characteristics of a breaking wind-wave field in the light of the individual wind-wave concept'. *J. Oceanogr. Soc. Japan*, **40**, 105-114.

Melville, W.K. and R.J. Rapp, 1985: 'Momentum flux in breaking waves'. *Nature*, **317**, 514-516

Mitsuyasu, H., 1985: 'A note on momentum transfer from wind to waves'. *J. Geophys. Res.*, **90**, 3343-3345.

Okuda, K., 1982: 'Internal flow structure of short wind waves. I. On the internal vorticity structure'. *J. Oceanogr. Soc. Japan*, **38**, 28-42.

Okuda, K., S. Kawai and Y. Toba, 1977: 'Measurement of skin friction distribution along the surface of wind waves'. *J. Oceanogr. Soc. Japan,* **33**, 190-198.

Phillips, O.M. and M.L. Banner, 1974: 'Wave breaking in the presence of wind drift and swell'. *J. Fluid Mech.,* **66**, 625-640.

Thorpe, S.A., 1986: 'Measurements with an automatically recording inverted echo sounder; ARIES and the bubble clouds'. *J. Phys. Oceanogr.,* **16**, 1462-1478.

Toba, Y., 1972: 'Local balance in the air-sea boundary processes, I. On the growth process of wind waves'. *J. Oceanogr. Soc. Japan,* **28**, 109-120.

Toba, Y., 1978: 'Stochastic form of the growth of wind waves in a single parameter representation with physical implications'. *J. Phys. Oceanogr.,* **8**, 494-507.

Toba, Y., 1987: 'Similarity laws of the wind wave and the coupling process of the air and water turbulent boundary layers'. In press in *Fluid Dynamics Research.*

Tokuda, M. and Y. Toba, 1982: 'Statistical characteristics of individual waves in laboratory wind waves. II. Self-consistent similarity regime'. *J. Oceanogr. Soc. Japan,* **38**, 8-14.

Wu, J., 1983: 'Sea-surface drift currents induced by wind and waves'. *J. Phys. Oceanogr.,* **13**, 1441-1451.

Yoshikawa, I., H. Kawamura, K. Okuda and Y. Toba, 1987: 'Turbulent structure in water under laboratory wind-waves'. (Submitted to *J. Oceanogr. Soc. Japan*)

ON THE MECHANICS OF SPILLING ZONES OF QUASI-STEADY BREAKING WAVES

Michael L. Banner
School of Mathematics
University of New South Wales
P.O. Box 1, Kensington
N.S.W. 2033. AUSTRALIA

ABSTRACT. Wave breaking is a widespread transient phenomenon occurring on the wind-driven sea surface, yet our understanding of the attendant hydrodynamics is incomplete. This contribution reports on the basic flow characteristics associated with quasi-steady spilling breakers via a detailed laboratory study of their typical local geometric, kinematic and dynamic characteristics. It is believed that many of these local characteristics are shared by transient spilling breakers for which such detailed observations are far more elusive. Results of an extension of the study to examine unsteady phenomena arising from the particular local characteristics of the spilling zone are also reported.

1. INTRODUCTION

Wave breaking is a commonly observed feature on the wind-driven ocean surface occurring over a wide range of length scales and arising from a variety of contributing mechanisms. These include excessive hydrodynamical straining by longer wave components, local reinforcement by focussing of neighbouring directional wave components as well as aerodynamic forcing by the wind stress. The most conspicuous manifestation of wave breaking at sea is 'white-capping', which results from concomitant air entrainment, but breaking of very short waves occurs on a far more widespread basis without air entrainment as 'microscale' breaking. These are both forms of spilling breaking, typically a deep-water breaking mode less violent than the plunging breakers often associated with waves incident on sloping shorelines.

Despite the significant role of oceanic wave breaking in several processes of fundamental and technological importance, including underwater ambient noise generation, our detailed knowledge of the phenomenon is fragmentary at present, both observationally and theoretically. Several contributions on various aspects of wave breaking have been published in the open literature as well as in recent conference proceedings (eg. Phillips and Hasselmann (1986),Toba and Mitsuyasu (1985)). These contributions include studies of the instabilities which appear to precede breaking, assessments of

63

B. R. Kerman (ed.), Sea Surface Sound, 63–70.
© 1988 by Kluwer Academic Publishers.

momentum transferred from the wave field resulting from breaking and of
the resulting subsurface currents and turbulence as well as criteria
for the onset of wave breaking at sea.   Indeed, these serve to
highlight the myriad of questions surrounding the phenomenon of wave
breaking at sea.

Not unexpectedly, direct observation of the local details of an
intermittent, turbulent free surface phenomenon such as wave breaking is
a particularly challenging task, but the local hydrodynamics is of
primary concern in understanding and quantifying the physical
contributions from these events. To simplify this task initially, it
was decided to investigate the typical properties of quasi-steady spilling breaking waves in a laboratory flume, with the aim of revealing
detailed local geometric, kinematic and dynamical properties, many of
which are likely to be shared by transient spilling breakers for which
these local characteristics would be much more difficult (if not
impossible) to observe directly. With this approach this investigation
extends aspects of earlier studies by Battjes and Sakai (1981) and
Duncan (1981) by providing a detailed probing of the spilling zone and
some of its key local mean and unsteady properties. This approach has
been adopted with success in previous studies aimed at isolating and
probing related breaking wave phenomena (eg. Banner and Melville (1976)
Banner and Fooks (1985)). From previous experience, local phenomena
identified using quasi-steady wave observations have been confirmed
under transient breaking conditions, despite differences in the factors
contributing to the onset of breaking.

2. EXPERIMENTAL CONFIGURATION

The quasi-steady breaking wave was realized in the laboratory
water tunnel shown schematically in Figure 1.

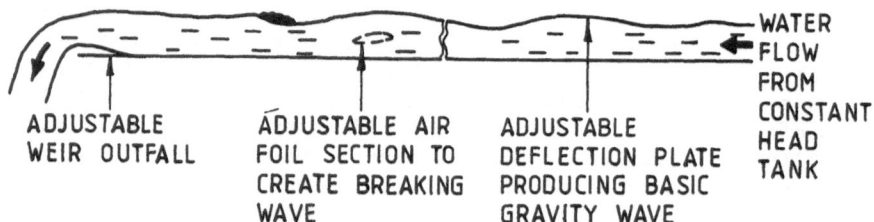

Figure 1. Experimental wave flume configuration: working length 7.3 m ;
channel width 0.225 m ; water depth 0.22 m. The measurement site was
halfway along the channel.

The wavelength and water depth were set by the volume flow and outflow
level controls. The hydrofoil section (96 mm × 21 mm), located at
mid-water depth was orientated to provide a given strength of breaking.
Measurements of the surface elevation were made with a standard
capacitance wire (0.1 mm dia) gauge. The upslope and downslope motion
of the spilling zone was determined with a specially configured
capacitance gauge. The velocity data was measured with a single channel

laser anemometer in the forward scatter mode, with a frequency shifting Bragg cell to accommodate the near zero mean velocities within and near the spilling region.

   In this initial study, a gently spilling breaking wave with a wavelength of ~ 0.4 m was established in the working section. With a very slight adjustment of the hydrofoil inclination, the strength of the breaking was increased. Figure 2 shows photographs of the quasi-steady breaking waves studied and for comparison, a propagating spilling

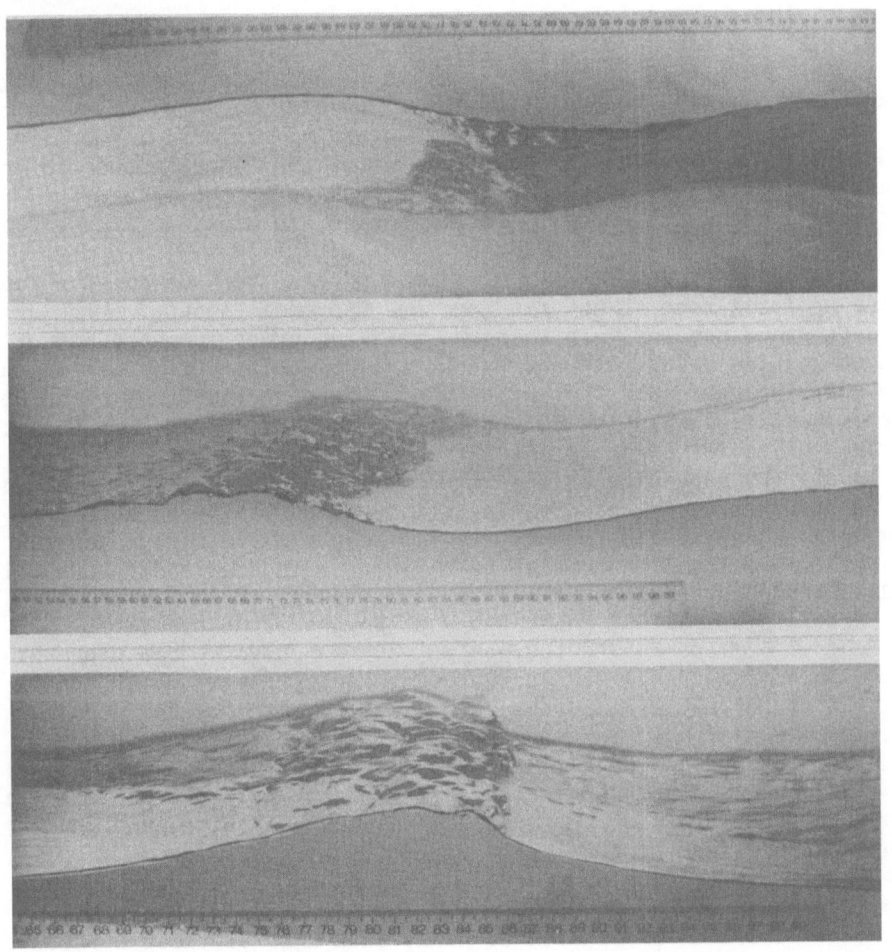

Figure 2. Typical laboratory realizations of short breaking waves.
Upper photograph : 'lightly' breaking quasi-steady spilling breaker (wavelength ~ 0.4 m)
Middle photograph : 'moderately' breaking quasi-steady spilling breaker (wavelength ~ 0.4 m)
Lower photograph : propagating spilling breaker induced by bottom steepening and 5 m/s wind (wavelength ~ 0.6 m).

breaker of comparable proportions which was generated by a wave paddle,
locally steepened by gentle bottom topography and induced to break by a
moderate (5 m/s) wind, photographed at one phase (not necessarily the
strongest breaking phase) of its transient breaking cycle.

3. MEAN GEOMETRIC AND KINEMATIC STRUCTURE

The mean surface profile was determined for each of the two
breaking conditions — 'light' and 'moderate'. The vertical distribution of the near-surface tangential velocity was measured at several
sites, with the aim of defining (a) the geometry of the spilling region
and (b) the mean velocity profiles at the leading and trailing troughs,
at the breaking crest and half-way down the rear face (c) measures of
the vertical distribution of r.m.s. fluctuating tangential velocity at
these sites. The velocities have been normalized with respect to the
linear wave speed $c_\ell = 0.85$ m/sec appropriate to the observed
upstream wavelength of 0.45 m. These results are shown in Figure 3 below.

Reference to the mean velocity data gives a very detailed picture
of the spilling zone structure. For the hydrofoil-induced breaking,
the data shows conclusively that the breaking crest location is not a
mean stagnation point. If one adopts the objective definition of the
mean spilling region as the region where the mean fluid motion is at
rest or downslope relative to the quasi-stationary waveform, then it is
evident that this region is less extensive than had been assumed
previously (Longuet-Higgins and Turner (1973), Duncan (1981)). It does
not extend back up to the breaking crest, which had been tacitly
assumed in these earlier studies. This result is of importance in
establishing a local tangential force balance (Duncan 1981)) and will
be discussed further in section 4.

Within the spilling zone, the peak downslope mean velocity was
measured to be less than a few percent of the phase speed, with this
region dominated by r.m.s. fluctuating velocity levels of around
twenty-five percent of the phase speed, as revealed by the vertical
distribution of r.m.s. turbulent velocity shown in Figure 3. From
these plots the local turbulent intensity can be readily calculated.

The combination of geometric and kinematic data provide a detailed
picture of the nature of the spilling region. The attendant underlying mean shear is exceptionally high, reaching as high as
$180 \; sec^{-1}$ (0.9 m/s over 5 mm) at the leading edge. The consequences of
this high shear region on the fluctuating structure of the spilling
region is described in section 5. This data also provides a basis for
a simple mathematical model of the response of the spilling zone to
perturbations from equilibrium. Progress toward this goal is also
reported in section 5.

4. MEAN DYNAMIC STRUCTURE

A significant goal of the present approach is a direct determination of the effective tangential and normal forces induced by the
spilling region on the underlying wave motion. This approach was

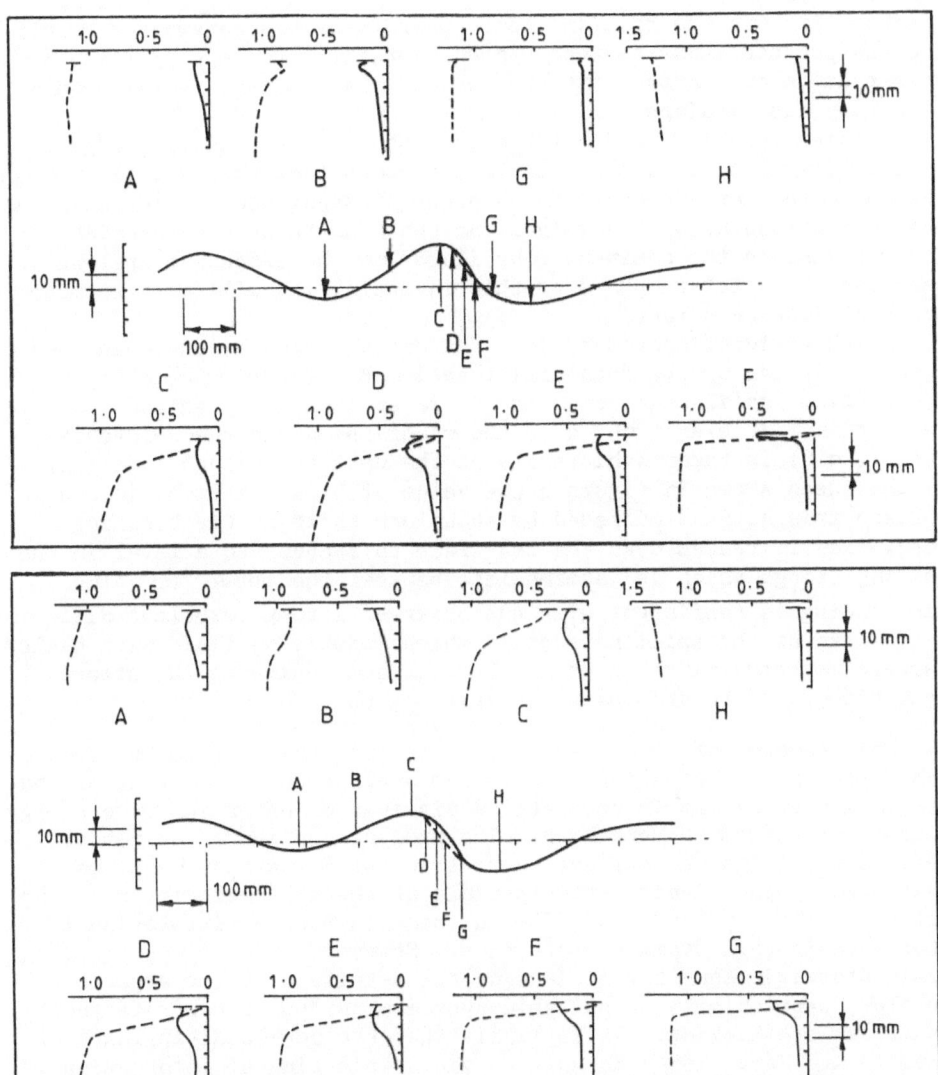

Figure 3. Vertical profiles of the normalised mean velocity (----) and r.m.s. turbulent velocity (——) parallel to the local mean water surface at the various sites indicated along the breaking wave profile (wavelength ~ 0.4 m) Upper figure corresponds to 'light' breaking. Lower figure corresponds to 'moderate' breaking.

explored by Duncan (1981) who studied the properties of quasi-steady breaking waves in the lee of a hydrofoil towed at a steady speed. Duncan proposed a static balance in which the upslope shear force supported the downslope weight component of the spilling region fluid with the normal force balancing the normal component of its weight.

He also attempted to relate the increase in the momentum flux of the downstream mean current to the rate of conversion of wave-coherent momentum flux via the spilling zone shear force. A noteworthy result from the present study is that Duncan over-estimated the spilling zone cross-section area A and that the use of more realistic values would strengthen his findings.

Despite the high value of the mean shear at the leading edge of the spilling zone, a simple calculation shows that the mean shear force based on molecular viscosity is an order of magnitude smaller than the downslope component of the weight and that the turbulent Reynolds stresses must be the dominant contributor to the effective upslope shear force. A future phase of this investigation will seek to measure this contribution directly.

Another significant dynamical feature revealed by the measurements is the variation of the Bernoulli function $B = (\rho g z + \frac{1}{2}\rho q^2)$ along the free surface profile, where ρ is the water density, g is the gravitational acceleration, q is the water speed and the reference level (z = 0) is taken arbitrarily at the upstream trough location. For the cases shown in Figure 3 the value of B is a maximum in the approach trough, is diminished by about one third at the breaking crest, then increases down the rear face to recover to a level in the trailing trough which is intermediate between the above two values. This finding is consistent with the observed strong turbulent diffusion in the wake of the spilling zone in which underlying fluid with higher rearward momentum mixes up to the surface and indicates the strong local effects of rotationality in this region.

The transfer of excess wave-coherent momentum flux to the surface layer arising from breaking is a fundamental process occurring at the air-sea interface, as is the rate of dissipation of excess wave energy which results from the breaking. The present configuration reveals the presence of an induced surface current in the downstream wake, as manifested by the significant departure of the mean velocity near the surface from the velocity profile in the trough of an irrotational water wave in this frame of reference. However, it does not provide a straightforward means for estimating the details of these aspects of the flow, particularly as suitable upstream and downstream reference conditions are lacking. It is likely that the non-local approach of Melville and Rapp (1984) is able to yield this kind of information more readily.

5. PERTURBATIONS OF THE QUASI-STEADY SPILLING ZONE

5.1. Surface Disturbances Short Compared with the Breaking Wavelength

Referring to figure 2 it is evident that accompanying the spilling zones are short free surface disturbances which persist downstream of the breaking crest. The nature of these disturbances was investigated in detail by Banner and Fooks (1985) in a study of the strong local microwave reflectivity of small-scale breaking waves. Depending on the breaking wave length and on the strength of the breaking, detailed observations revealed that the disturbances generated by quasi-steady breaking waves had a fairly sharply tuned frequency spectrum rather than a more diffuse spectrum which might have been anticipated. Their longitudinal coherence was found to be a significant fraction of the breaking wavelength, but the lateral coherence of the disturbances was very short. A two-probe technique showed the disturbances to be predominantly rearward-travelling.

An attempt to explain these findings was made by Miklavcic (1984) who studied the linear inviscid stability characteristics of the mean velocity profile associated with a quasi-steady spilling zone. He investigated piecewise linear and truncated hyperbolic tangent profile approximations to the observed profiles with an appropriately located free upper surface. He found an intermediate band of unstable wave number components, corresponding to rearward travelling surface shear wave modes, with apparent frequencies in close correspondence with the observed frequencies. This initially encouraging agreement has motivated an ongoing extension of this work with the aim of examining the nonlinear stability characteristics of such highly sheared free surface flows.

This class of short (on the scale of the breaking wavelength) disturbances emanating from spilling breakers is believed to be a universal property which can be observed visually during the transient whitecapping process at sea. It has been shown to be of potential importance in microwave backscatter from the ocean surface : its role in the generation of underwater noise is an interesting question which awaits further investigation.

5.2. Surging of the Spilling Zone.

In his study of quasi-steady breaking waves induced by a steadily towed hydrofoil, Duncan (1981) observed low frequency (relative to the wave frequency) upslope/downslope surging of the extent of the spilling zone on a timescale of the order of four to five wave periods and whose amplitude decayed in time. Duncan associated these oscillations with wave components produced when the hydrofoil motion was initiated from rest.

Based on the mean flow observations reported in section 3, a simple mathematical model has been developed with the view of providing an approximate equation of motion for spilling zone. At this stage the model has been formulated to describe departures from a quasi-steady equilibrium in which the downslope spilling region weight is balanced by an effective upslope shear force. This provides a form for the effective viscosity, which is a function of the local spilling zone geometry and characteristic shear. An approximate equation of motion can be constructed, for the bulk spilling zone upslope/downslope motion, assuming that during departures from equilibrium, the spilling zone fluid has a common mean upslope/downslope velocity. This leads to a second order nonlinear ordinary differential equation for the centroid of the spilling zone. To find solutions, it is necessary to specify the variation of the underlying characteristic free stream speed. In the absence of any more detailed information, an initial calculation has been done assuming quasi-static variation of the underlying fluid speed based on Bernoulli's equation and solving for the return to equilibrium from an initial upslope/downslope displacement from equilibrium. These initial calculations showed encouraging agreement with measurements of the relaxation timescales of the spilling zone back to equilibrium after impulsive perturbation to the crest region. The form of the motion is a strongly damped oscillation with a typical relaxation time of 4-5 times the wave period, not unlike the observations reported by Duncan (1981). The model is presently being refined and offers the useful possibility of estimating energy dissipation rates associated with spilling breaking in terms of the (observable) time evolution of this region.

CONCLUSIONS.

This ongoing detailed probing of quasi-steady breaking waves has revealed interesting fundamental aspects of the steady and unsteady structure of the breaking zones of short spilling breakers. There is scope for further extension of this work, both experimentally and through mathematical modelling in the quest for a better understanding of the breaking process and its implications for upper ocean dynamics and related physical processes which occur concomitantly, including contributions to the ambient underwater noise spectrum.

REFERENCES

Banner, M.L. and Melville, W.K. (1976) *On the separation of air flow over water waves.* J. Fluid Mech., 77, 825-42.

Banner, M.L. and Fooks, E.H. (1985) *On the microwave reflectivity of small-scale breaking water waves.* Proc. Roy. Soc. A, 399, 93-109.

Battjes, J.A. and Sakai, T. (1981) *Velocity field in a steady breaker.* J.Fluid Mech., 111, 421-37.

Duncan, J.H. (1981) *An experimental investigation of breaking waves produced by a towed hydrofoil.* Proc. Roy. Soc. A, 377, 331-348.

Longuet-Higgins, M.S. and Turner, J.S. (1974) *An 'entraining plume' model of a spilling breaker.* J. Fluid Mech., 63, 1-20.

Melville, W.K. and Rapp, R.J. (1985) *Momentum flux in breaking waves.* Nature, 317, 6037, 514-516.

Miklavcic, S.J. (1984) *Instability of a shear current with a free surface.* Honours Thesis, Dept. App. Math., Univ. of N.S.W., 61 pp.

Phillips, O.M. and Hasselmann, K. (1986) <u>Wave Dynamics and Radio Probing of the Ocean Surface</u>. Plenum Press, N.Y., 694 pp.

Toba, Y. and Mitsuyasu, H. (1985) <u>The Ocean Surface</u>. D.Reidel, Dordrecht, 586 pp.

AN ESTIMATE OF WAVE BREAKING PROBABILITY FOR DEEP WATER WAVES

Y. A. Papadimitrakis, N. E. Huang
NASA/GSFC Greenbelt, MD 20771-USA
L. F. Bliven, and S. R. Long
NASA/GSFC Wallops Flight Facility
Wallops Island, VA 23337-USA

ABSTRACT.   Analytical expressions for the probabilities of both wave breaking and of the fractional energy dissipation losses generated by wave breaking, are derived on the basis of a joint distribution for wave frequencies and amplitudes.   Direct effects of wind forcing on wave breaking are explicitly considered in the form of Phillips and Banner (1974) mechanism. The results are found to depend on the wave spectrum band width parameter θ and the characteristic frequency σ_0 (definitions follow), and are not restricted to narrow-band spectrum cases. When the computed total energy dissipation losses (per average wave period) are used together with appropriate wave growth models to predict the fraction of wind momentum and energy fluxes transferred to water waves, the results appear to be in good agreement with their experimental counterparts.   These expressions are also compared to other recent work on wave breaking and their similarities and differences are discussed.

INTRODUCTION

Wave breaking in the ocean is caused by non-linear modulations and distortions of the various frequency components, associated with the wave motions and the wind forcing. The effects of non-linearity on a single deterministic wave train have been extensively studied, but the results so obtained, though elegant, have limited applicability to the real ocean.   As far as the dissipation of wave energy and/or whitecapping is concerned, most of the available information is based on empirical correlations (rather than on physical arguments) which express, for example, whitecapping as a function of wind speed. Typical examples are the correlations proposed by Monahan and O'Muircheartaigh (1980), namely

$$\tilde{\omega} = 2.95 \times 10^{-6} U^{3.52} \qquad ; \qquad \tilde{\omega} = 3.84 \times 10^{-6} U^{3.41} \qquad (1)$$

where U is the wind speed in m/sec (measured at about 10 m above the mean water level-MWL), and $\tilde{\omega}$ is the fraction of the sea surface covered by whitecaps. Although there is no denying that, such expressions can probably give a reasonable approximation of the whitecap coverage under

71

B. R. Kerman (ed.), Sea Surface Sound, 71–83.
© 1988 by Kluwer Academic Publishers.

certain wind and sea conditions, it is important to realize that none of them has general validity, let alone that they all are dimensionally wrong.

Alternatively, Ochi and Tsai (1983), and Snyder et al. (1983), considered the dynamics of the wave field and proposed a breaking threshold mechanism which, essentially, in both cases involves the vertical acceleration. Then this threshold mechanism was applied to the probability density function of the wave field in order to compute whitecap coverage. Their results appear to be reasonable, but their general validity may be questionable again, mainly because the direct effects of wind forcing on the wave breaking were completely ignored by them.

Recently, Huang et al. (1986) proposed another whitecap coverage model which was based on the breaking threshold mechanism originated by Longuet-Higgins (1969), but with the wind effect included in the form of Phillips and Banner (1974) mechanism. The resulting expression depends on various wind and wave parameters and their mutual coupling. Comparison of model predictions with the limited data of Snyder et al. (1983) showed reasonable agreement. On detailed scrutiny, however, one can still identify an artificial limitation in that model, that is, the requirement of a narrow-band width spectrum assumption. In this paper, we have removed this restriction and present a more general wave energy dissipation model applicable also to a finite-band width ocean spectrum.

THE STATISTICAL MODEL

The joint distribution of amplitude and frequency (or period), as well as some other pertinent distributions, play an important role in the investigation of many statistical properties of sea waves. Yet, most of these distributions (see, for example, Longuet-Higgins 1975, 1983; and Tayfun 1980, 1981), except those derived by Yuan (1982), Huang et al. (1983, 1984) and a few othes, are based on linear wave theories and are restricted to narrow-band width spectra. Yuan (1982) derived a relationship among the zero crossing intervals, the sea surface elevation, η, and its second derivative (or vertical acceleration), η'', at the extreme points below and above the MWL, by analyzing the geometry of random waves. This relationship was then applied to linear, breaking, and non-linear waves. For non-linear waves the Stokes model was used, as done by Tayfun (1980) and later by Huang et al (1983,1984). In the breaking wave model, waves break whenever the vertical acceleration at any point on the surface reaches $g/2$ (g being the gravitational acceleration). Then, the acceleration is limited to $g/2$ at that point, and the surface elevation is reduced according to the ratio of $g/2$ and the acceleration of the original wave, that is:

$$\eta_b = \eta \left\{ \frac{g}{2|\eta''|} \cdot \text{He}(|\eta''| - \frac{g}{2}) + \text{He}(\frac{g}{2} - |\eta''|) \right\} \tag{2}$$

$$\eta_b'' = \frac{g}{2} \text{sign}(\eta'') \cdot \text{He}(|\eta''| - \frac{g}{2}) + \eta'' \cdot \text{He}(\frac{g}{2} - |\eta''|) \tag{3}$$

Here prime indicates differentiation with respect to time, the subscript b refers to breaking conditions, and He(.) denotes the Heaviside unit step function.   Our calculations are based on the joint amplitude-frequency density function, p(h,σ), proposed by Yuan (1982), namely:

$$p(h,\sigma) = a_y h^2 \exp(-b_y h^2) \tag{4}$$

where

$$a_y = \frac{4\sigma^3}{\alpha^2 (2\pi)^{\frac{1}{2}} m_0^{3/2} \sigma_0^4} \times \frac{1}{(\theta^2 - 1)^{\frac{1}{2}}(\theta + 1)} \tag{5}$$

$$b_y = \frac{1}{2m_0} \left[1 + \frac{\left(1 - \alpha^{-1}\left(\frac{\sigma}{\sigma_0}\right)^2\right)^2}{(\theta^2 - 1)} \right] \tag{6}$$

$$\alpha = \left(\frac{\sigma_p}{\sigma_0}\right)^2 \times \frac{7}{[2 + [25 + 21(\theta^2 - 1)]^{\frac{1}{2}}} \tag{7}$$

$$\theta = (m_0 m_4 / m_2^2)^{\frac{1}{2}} \quad ; \quad \sigma_0 = (m_2/m_0)^{\frac{1}{2}} \tag{8a,b}$$

Here h is the wave amplitude, σ is the intrinsic frequency, and m_i is the ith moment of the spectrum.   The derivation of Eq. (4) is based on Longuet-Higgins (1957) original joint distribution of η, η' and η'', and a suitable definition of σ, namely:

$$p(\eta,\eta',\eta'') = \frac{1}{(2\pi)^{3/2} m_2^{\frac{1}{2}} \Delta^{\frac{1}{2}}} \exp\left\{ -\frac{\eta'^2}{2m_2} - \frac{1}{2\Delta}(m_4\eta^2 + 2m_2\eta\eta'' + m_0\eta''^2) \right\} \tag{9}$$

$$\sigma^2 = -\alpha\eta''(t_0)/\eta(t_0) \quad ; \quad \Delta = m_0 m_4 - m_2^2 \tag{10a,b}$$

where t_0 indicates the time of occurrence of the sea surface extrema; θ is the ratio of the expected number of extrema to that of zero crossings per unit time.   It is also a measure of the spectrum band width, and is related to the similar parameter, ε, introduced by Cartwright and Longuet-Higgins (1956), since $\theta^2 = 1/(1 - \epsilon^2)$.   For a narrow band width case, θ = 1.0.   Eq. (9) involves no limitation on θ, but it does assume that η, η', η'' have to be jointly Gaussian.
        Longuet- Higgins (1983) has also proposed another form for the joint amplitude-period density function which is similar to Yuan's (1982) model but with different functionals a_y and b_y, namely:

$$a_y = \frac{L(\nu)}{(\pi)^{\frac{1}{2}} \nu (2m_0)^{\frac{1}{2}} m_1} \tag{11}$$

$$b_y = \frac{1}{2m_0} \left[1 + \left(1 - \frac{\sigma m_0}{m_1}\right)^2 / \nu^2 \right] \tag{12}$$

where

$$\nu = [m_0 m_2 / m_1^2 - 1]^{\frac{1}{2}} \quad ; \quad L^{-1}(\nu) = \frac{1}{2}[1 + (1 + \nu^2)^{-\frac{1}{2}}] \tag{13a,b}$$

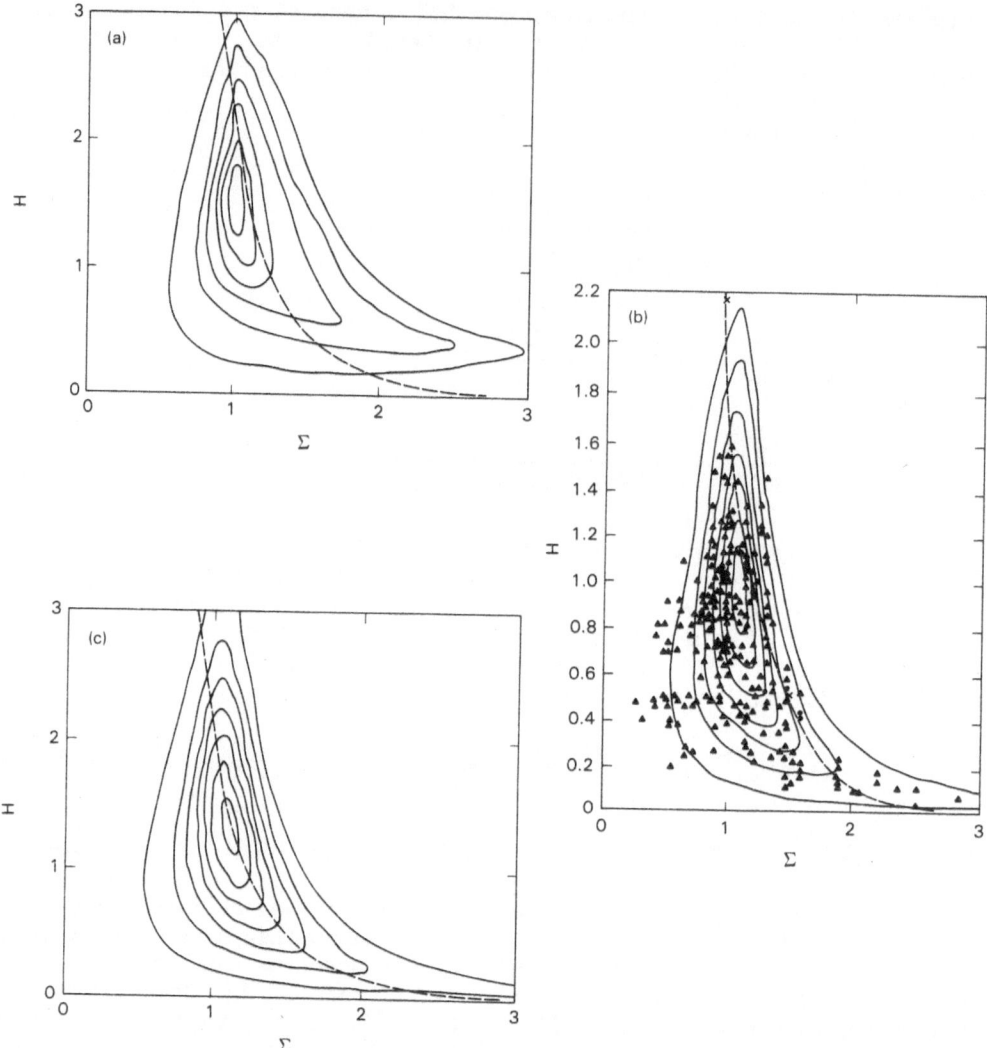

Figure 1: Contours of the normalized joint distibution of amplitude and frequency, $p(H,\Sigma)/p_{max}$ (a): laboratory data; $p(H,\Sigma)/p_{max}$ takes the values 1.10, 0.90, 0.75, 0.60, 0.40 and 0.2 from the center contour outwards (b): ▲; ocean data from Yellow sea (see Yuan 1982); $p(H,\Sigma)/p_{max}$ takes the values 0.65, 0.55, 0.45, 0.30, 0.25 and 0.15 from the center contour outwards (c): theoretical results; $p(H,\Sigma)/p_{max}$ takes the values 1.15, 1.10, 0.85, 0.65, 0.45, 0.25 and 0.15 from the center contour outwards. The dashed line indicates Phillips and Banner (1974) breaking limit.

This model, however, requires the narrow-band width spectrum assumption, but depends only on the three lowest moments m_0, m_1, m_2 of the spectral density function.

Both Longuet-Higgins (1983) and Srokosz (1986) have pointed out that m_4 may depend rather critically on the behavior of the wave spectrum at high frequencies. Although Phillips (1985) suggested a σ^{-4} variation of the spectrum tail, the more recent work of Lin (1986), based on wave surface elevation and wave slope data, favors a σ^{-5} variation of the tail. In either case, there are certainly difficulties in determining m_4. Yet, since m_4 represents the variance of the vertical acceleration it has to be finite. We will, therefore, assume that m_4 exists, and that it can be determined either directly from a measured wave spectrum or indirectly from a spectral model by applying a suitable cutoff frequency.

Next, we have adopted Phillips and Banner (1974) breaking criterion defined by the limiting wave amplitude, h_0, at frequency σ, namely:

$$h_0 = \frac{c^2}{2\alpha_1 g}\left(1 - \alpha_2\frac{u_*}{c}\right)^2 \tag{14}$$

where u_* is the wind friction velocity, c is the phase velocity of the water wave, and α_2 is a coefficient of $O(1)$. A comment or two are here in order. The maximum real acceleration at the wave surface is certainly not exactly $g/2$ and in fact is smaller than this value (Longuet-Higgins 1986). For this reason we have introduced the coefficient α_1, of $O(1)$, such that, in the absence of drift currents, the limiting acceleration at the wave surface reaches the value of $g/2\alpha_1$; α_1 can be determined as described in Longuet-Higgins (1986).

RESULTS - DISCUSSION

Figure 1 shows typical contours of $p(H,\Sigma)/p_{max}$, where $H = h/(2m_0)^{\frac{1}{2}}$, $\Sigma = \sigma/(\alpha^{\frac{1}{2}}\sigma_0)$, and p_{max} indicates the maximum value of $p(H,\Sigma)$; p_{max} can be found from the condition that $\partial p/\partial H = \partial p/\partial \Sigma = 0$. Hence we find:

$$\Sigma_{max} = [-1 + (3\theta^2 + 1)^{\frac{1}{2}}]^{\frac{1}{2}} \quad ; \quad H_{max} = \frac{1}{2}\left\{\frac{\theta^2 - 1}{\theta^2 - \Sigma^2_{max}}\right\}^{\frac{1}{2}} \tag{15a,b}$$

and the value of $p(H,\Sigma)$ at this point is given by

$$p_{max} = \frac{2\exp(-1)(\theta^2 - 1)^{\frac{1}{2}}\Sigma^3_{max}}{\pi^{\frac{1}{2}}(\theta + 1)(\theta^2 - \Sigma^2_{max})} \tag{16}$$

It is clearly seen that the joint density distribution shows some asymmetry with respect to Σ, in general, but in the neighborhood of $\Sigma = 1$ it becomes symmetric about the mean wave frequency, independently of H. Table 1 lists some values of α, H_{max}, Σ_{max}, and p_{max} for representative values of the parameter θ between 1.0 and 2.0. Clearly broadening of the spectrum reduces and enhances, respectively, the 'most probable' joint values of the wave amplitude and frequency and also reduces their maximum probability density. For very large values of Σ and H the normalized contours become asymptotically tangent to the Σ and H axes.

Table 1: Parameters of $p(H,\Sigma)$

θ	α	Σ_{max}	H_{max}	p_{max}
1.1	0.60	1.07	0.95	1.92
1.2	0.57	1.14	0.91	1.40
1.3	0.54	1.21	0.87	1.17
1.4	0.51	1.27	0.84	1.04
1.5	0.49	1.34	0.82	0.95
1.6	0.46	1.40	0.80	0.88
1.7	0.44	1.45	0.78	0.83
1.8	0.42	1.51	0.76	0.79
1.9	0.41	1.56	0.75	0.73
2.0	0.39	1.61	0.73	0.72

These values correspond to: $m_0 = 0.022472$ cm^2, and $\sigma_0 = 5.34$ Hz.

The density of the wave amplitude H is now obtained by integrating $p(H,\Sigma)$ with respect to Σ over all positive frequencies, namely:

$$p(H) = \frac{8}{(\pi)^{\frac{1}{2}}(\theta^2 - 1)^{\frac{1}{2}}(\theta + 1)} H^2 \exp(-H^2) \int_0^\infty \Sigma^3 \exp\left\{-H^2 \frac{(\Sigma^2-1)^2}{\theta^2-1}\right\} d\Sigma$$

$$= \frac{2}{(\theta + 1)} H \exp(-H^2) F(B) \tag{17}$$

where

$$F(B) = 2 - \text{erf}(1/B) + \frac{B}{\pi^{\frac{1}{2}}} \exp\left(-\frac{1}{B^2}\right) ; \quad B = \frac{(\theta^2-1)^{\frac{1}{2}}}{H} \tag{18a,b}$$

and erf(.) denotes the standard error function. Equation (17) states that the density of H has a Rayleigh like distribution, but must be corrected by a factor proportional to $F(B)$. For large values of H (0(1) or greater), the correction is small. However, when H is of order of $(\theta^2 - 1)^{\frac{1}{2}}$, the correction becomes significant.

The total density of Σ is obtained by integrating $p(H,\Sigma)$ with respect to H over $0 < H < \infty$, namely:

$$p(\Sigma) = 2(\theta - 1)\left\{\frac{\Sigma}{[\theta^2 - 1 + (\Sigma^2 - 1)^2]^{\frac{1}{2}}}\right\}^3 \tag{19}$$

Figure 2 shows a typical $p'(\Sigma) = p(\Sigma)/(\alpha^{\frac{1}{2}}\sigma_0)$ distribution. Then we write for the breaking probability, p_B:

$$p_B = \int_{h_0}^\infty p(h, \sigma) dh \tag{20}$$

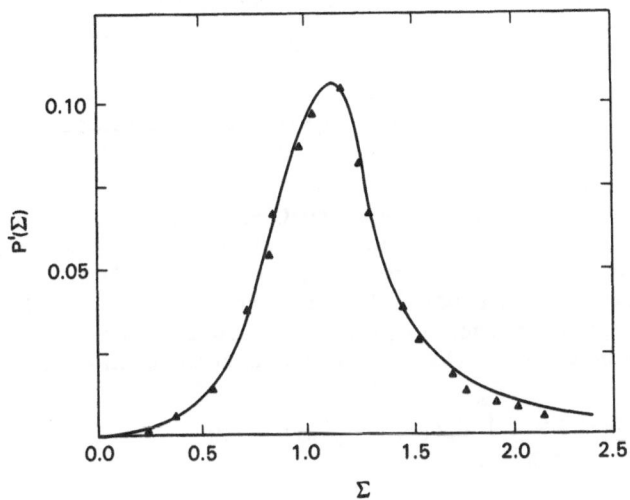

Figure 2: Typical normalized frequency probability density; ▲ : laboratory data ($\theta = 1.13$, $\sigma_0 = 5.35$ Hz, and $\alpha = 0.62$).

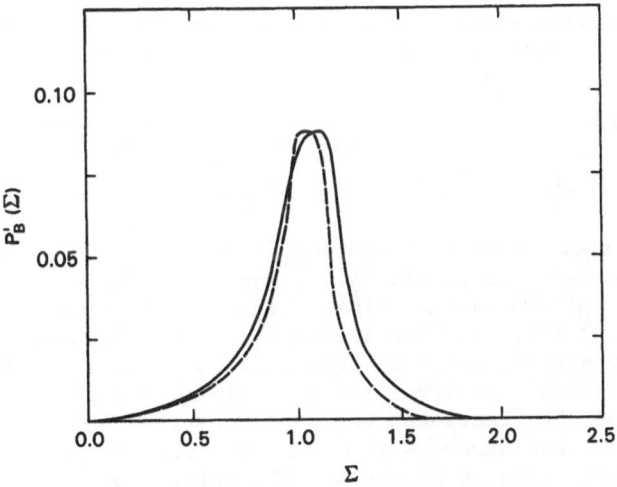

Figure 3: Typical breaking probability density. The dashed line indicates the corresponding density according to Longuet-Higgins (1983) model ($\nu = 0.21$, $\bar{\sigma} = 5.35$ Hz).

Integration of Eq.(20) with respect to wave height, h, yields:

$$p_B(\sigma) = [a_y/(2b_y^{3/2})]\Gamma(3/2,h_0^2) \tag{21}$$

$$p_B(\Sigma) = 4\pi^{-\frac{1}{2}}(\theta - 1)\Gamma(3/2,H_0^2)\left\{\frac{\Sigma}{[\theta^2 - 1 + (\Sigma^2 - 1)^2]^{\frac{1}{2}}}\right\}^3 \tag{22}$$

Here $\Gamma(,)$ represents the incomplete Gamma function and $H_0 = h_0/(2m_0)^{\frac{1}{2}}$. Since the present model makes no assumption on the spectrum band width, it is expected to cover a wider range of sea states. An extension of this model gives the probability of fractional energy losses per unit time due to wave breaking, $p_L(\sigma)$. This is obtained by integrating the wave energy density over all amplitudes exceeding h_0, that is:

$$p_L(\sigma) = \frac{1}{\rho g \overline{h^2}}\int_{h_0}^{\infty} \rho g(h^2 - h_0^2)p(h,\sigma)dh \tag{23}$$

where $\overline{h^2} = 2m_0$. Eq. (23) yields

$$p_L(\sigma) = \frac{a_y}{4m_0 b_y^{5/2}}\left[\Gamma(5/2,h_0^2) - b_y h_0^2\Gamma(3/2,h_0^2)\right] \tag{24}$$

When this function is expressed in terms of frequency, it yields an analytic expression which is proportional to the rate of energy dissipation, often used in wave prediction models, namely:

$$p_L(\Sigma) = 4\pi^{-\frac{1}{2}}(\theta - 1)\cdot\left\{\frac{\Sigma^3}{[\theta^2 - 1 + (\Sigma^2 - 1)^2]^{5/2}}\right\}\times$$
$$\times\left\{(\theta^2 - 1)\Gamma(5/2,H_0^2) - [\theta^2 - 1 + (\Sigma^2 - 1)^2]H_0^2\Gamma(3/2,H_0^2)\right\} \tag{25}$$

Figures 3, 4 show typical $p'_B(\Sigma) = p_B(\Sigma)/(\alpha^{\frac{1}{2}}\sigma_0)$ and $p'_L(\Sigma) = p_L(\Sigma)/(\alpha^{\frac{1}{2}}\sigma_0)$ distributions, respectively. For comparison their counterpart distributions $p_B(\Sigma)/\bar{\sigma}$ and $p_L(\Sigma)/\bar{\sigma}$, obtained from Longuet-Higgins (1983) model (Eq. 11,12,13), are also included in these Figures. The agreement is quite satisfactory, since in this particular case we are dealing with a rather narrow-band spectrum, provided that $\theta \cong 1.13$ and $\nu = 0.21$. The distribution of the fractional energy losses (per unit time) obtained by properly converting Hasselmann's et al. (1985) spectral dissipation function is also shown in Figure 4. The latter is given by:

$$p_{LH}(\sigma) = \frac{3.2\pi\hat{a}^2}{m_0}\left(\frac{\sigma}{\bar{\sigma}}\right)^2\Phi(\sigma) \tag{26}$$

where

$$\hat{a} = \frac{m_0\bar{\sigma}^4}{g^2} \quad ; \quad \bar{\sigma} = \frac{m_1}{m_0} \tag{27a,b}$$

and $\Phi(\sigma)$ represents the spectral density. In this work we have adopted a modified Wallops spectral representation (see Huang et al. 1981),

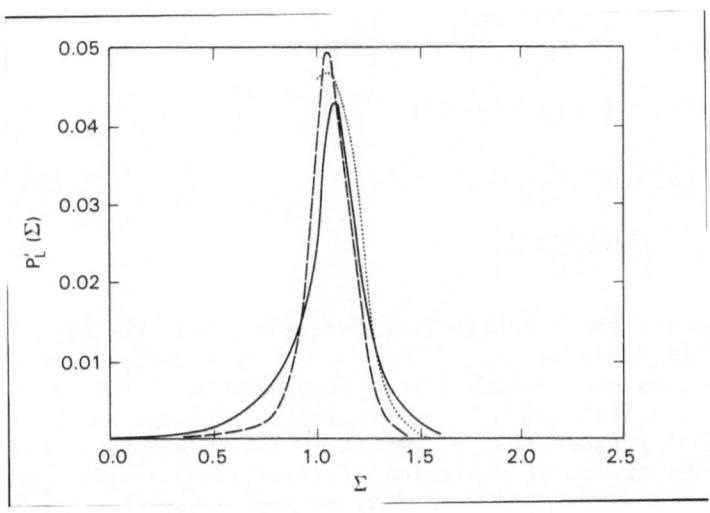

Figure 4: Typical probability density of the fractional energy energy losses (per unit time). The dashed and dotted lines indicate the corresponding losses according to Longuet-Higgins (1983) and Hasselmann et al. (1985) models.

namely:

$$\Phi(\sigma) = 0 \qquad\qquad\qquad\qquad\qquad\qquad\qquad 0 < \sigma < \sigma_p$$

$$\Phi(\sigma) = \beta g^2 \sigma_p^{m-5} \sigma^{-m} \left(1 - \alpha_2 \frac{u_*}{g} \sigma\right)^4 \qquad\qquad \sigma_p \leq \sigma \leq A_1 \sigma_p$$

$$\Phi(\sigma) = \beta_1 g u_* \sigma^{-4} = \beta g^2 A_1^{-(m-4)} \sigma_p^{-1} \sigma^{-4} \left(1 - \alpha_2 \frac{u_*}{g} \sigma\right)^4 \quad A_1 \Sigma_p < \Sigma \leq A_2 \Sigma_p$$

$$\Phi(\sigma) = 0 \qquad\qquad\qquad\qquad\qquad\qquad\qquad A_2 \sigma_p < \sigma$$

$$m = \left| \frac{\log(\sqrt{2}\pi\S)^2}{\log 2} \right| \qquad\qquad\qquad\qquad\qquad 28(a,b,c,d,e)$$

Here σ_p denotes the spectral peak frequency, $\S$ is the significant slope as defined by Huang et al. (1981), and β is a coefficient which can be determined from the condition that $\int\Phi(\sigma)d\sigma = m_0$. The coefficients A_1, A_2 are taken as 2.0 and 3.0, respectively, although the data of Donelan et al. (1985) suggest that A_1 may be as low as about 1.5 and A_2 as high as 3.5. The effect of variation of these coefficients (in the range: $1.5 \leq A_1 \leq 2.0$ and $3.0 \leq A_2 \leq 3.5$) on $p_{LH}(\sigma)$ has also been explored in this study. As seen in Figure 4, Hasselmann's et al. (1985) distribution is also in good agreement with its other two counterparts.

Finally the fraction of total wave energy losses, $\hat{\omega}$, is obtained by integrating Eq.(24) over all (positive) frequencies, σ, namely:

$$\hat{\omega} = \int_0^\infty p_L(\sigma) d\sigma \qquad\qquad\qquad\qquad\qquad (29)$$

The computed $\hat{\omega}$ values according to Yuan's (1982) model (Eqs. 5-7) are smaller, by a factor of about 2, than their counterparts computed according to Longuet-Higgins model (Eqs. 11-12). They are, however, both comparable to Hasselmann's et al. (1985) predictions. These values are also different from those obtained by Huang (1986) where he used the original Longuet-Higgins (1969) threshold criterion without considering the effect of wind forcing on wave breaking.

The total energy losses, $\bar{W}$, per average wave period $\bar{T}$ can then be calculated as:

$$\bar{W} = \frac{\hat{\omega}\rho g m_0}{\bar{T}} \quad ; \quad \bar{T} = \frac{2\pi}{\sigma_0} \quad \left(\text{or} = \frac{2\pi}{\bar{\sigma}}\right) \qquad\qquad (30a,b,c)$$

Although the energy loss part lacks independent comparison, our predictions can be useful in studying the dynamics of the upper ocean as discussed by Huang et al. (1986). In this respect we have used the predicted total energy losses together with suitable wave growth models (see, for example, Huang et al. 1981; Hsu et al. 1982; and Mitsuyasu and Honda 1982) to estimate the fraction of the wind momentum (γ_M) and energy (γ_E) fluxes which are transmitted to water waves, under a variety of wind and sea state conditions. The theoretical formulation of the problem is described elsewhere (see Papadimitrakis et al. 1987), and will not be repeated here for lack of space. Briefly, the steady state form of the radiation-transport equation proposed by Hasselmann et al. (1976) has been integrated with respect to frequency and the wave propagation direction, and used to describe the evolution of the wave field. From the total energy balance among wind input, wave growth, and energy dissipation, and invoking the fact that the mean momentum transferred to water waves can be approximated by the ratio $\bar{E}_W/(\bar{c} - \bar{q})$, where $\bar{E}_W$ represents the wave-supported energy and $\bar{c}$, $\bar{q}$ are a characteristic phase velocity (say $= g/\sigma_0$, or $g/\bar{\sigma}$) and a mean value of the total surface drift current, one obtains the following expressions for γ_M and γ_E:

$$\gamma_M = \frac{A}{\left\{A + 2\left(\dfrac{\bar{c}}{u_*} - \dfrac{\bar{q}}{u_*}\right)\right\}} \quad ; \quad \gamma_E = \frac{\left(\dfrac{\bar{c}}{u_*} - \dfrac{\bar{q}}{u_*}\right)}{\left\{\dfrac{\bar{q}}{u_*} + \left(1 + \dfrac{2\bar{q}}{Au_*}\right)\left(\dfrac{\bar{c}}{u_*} - \dfrac{\bar{q}}{u_*}\right)\right\}}$$

$$A = \kappa\left(\frac{\rho}{\rho_a}\right)\alpha_3\left(\alpha_4\frac{c_p}{u_*} + 2\alpha_2\right) + 2\pi\alpha_4^{-1}\S^2\tilde{\omega}\left(\frac{\rho}{\rho_a}\right)\left(\frac{c_p}{u_*}\right)^3 \qquad (31a,b,c)$$

Here $\kappa = 1.6 \times 10^{-4}$, c_p is the phase velocity at the spectral peak frequency, and the coefficients α_3, and α_4 are of O(1). The predicted γ_M and γ_E values have been compared with the laboratory measurements of Hsu et al.(1982), and Papadimitrakis et al. (1986). Table 2 shows such a comparison. It lists γ_M, γ_E along with their experimental counterparts γ_{Mm} and γ_{Em} as a function of either c_p/u_*, or the non-dimensional fetch $\tilde{x} = xg/u_*^2$, where x represents fetch. The agreement is considered to be satisfactory (within 15-25%). Comparisons (not shown here) of the predicted γ_M and γ_E values based on Longuet-Higgins (1983) and Hasselmann's et al. (1985) models, described previously, show that our model produces results somewhat closer (by as much as 10-15%) to the experimental data.

Table 2: Momentum and energy partition between waves and currents

c_p/u_*	1.10	1.12	1.24	1.33	1.37	1.43	1.44	1.64	1.77	1.85	10.0	13.0	17.6
$\tilde{x}$	207	290	424	373	407	644	559	875	1096	1233	5240	9006	16468
γ_{Mm}	0.60	0.48	0.38	0.59	0.54	0.53	0.64	0.58	0.59	0.64	0.62	0.43	0.26
γ_M	0.46	0.41	0.35	0.49	0.49	0.45	0.49	0.45	0.46	0.45	0.60	0.45	0.24
γ_{Em}	0.17	0.17	0.17	0.33	0.29	0.34	0.13	0.34	0.32	0.46	0.42	0.76	0.77
γ_E	0.22	0.22	0.23	0.31	0.31	0.32	0.34	0.37	0.40	0.43	0.41	0.72	0.74

CONCLUSIONS

We have derived analytical expressions for the probabilities of both wave breaking, and of the fractional energy losses (per unit time) generated by wave breaking. The derivation is based on a joint distribution of wave frequencies and amplitudes which applies to narrow-band, as well as to finite-band width ocean wave spectra. These expressions also include the direct effects of wind forcing on the wave breaking in the form of Phillips and Banner (1974) mechanism.

For purely wind-generated (laboratory) waves, the computed fraction of the total wave energy losses, $\hat{\omega}$, according to Yuan's (1982) model compares favorably with the results obtained from both Longuet-Higgins (1983) and Hasselmann's et al. (1985) models.

Predictions of the fraction of wind momentum and energy fluxes which are transferred to water waves, by utilizing appropriate wave growth models and the total energy losses computed from (our) fractional energy loss results, agree reasonably well with available laboratory observations taken under a variety of wind and wave conditions.

Detailed comparisons of γ_M and γ_E values with the open ocean data of Snyder et al. (1981), using all of the above models and different spectral representations, $\Phi(\Sigma)$, are currently under way and will be reported somewhere else.

REFERENCES

Cartwright, D. E., and M. S. Longuet-Higgins, 1956: 'The statistical distribution of the maxima of a random wave function,' Proc. Roy. Soc., **A237**, 212-232.

Donelan, M. A., J. Hamilton, and W. H. Hui, 1985: 'Directional spectra of wind-generated waves,' Phil. Trans. R. Soc. Lond., **A315**, 509-562.

Hasselmann, K., D. B. Ross, P. Muller, and W. Sell, 1976: 'A parametric wave prediction model,' J. Phys. Oceanogr., **6**, 200-228.

Hasselmann, S., K. Hasselmann, J. H. Allender, and J. P. Barnett, 1985: 'Computations and parameterizations of the nonlinear energy transfer in a gravity-wave spectrum. Part II. Parameterization of the nonlinear energy transfer for application in wave prediction models,' J. Phys. Oceanogr., **15**, 1378-1389.

Hsu, C. T., H. Y. Wu, E. Y. Hsu, and R. L. Street, 1982: 'Momentum and energy transfer in wind generation of waves,' J. Phys. Oceanogr., **12**, 929-951.

Huang, N. E., S. R. Long, and L. F. Bliven, 1981: 'On the importance of the significant slope in empirical wind wave studies,' J. Phys. Oceanogr., **11**, 569-573.

Huang, N. E., S. R. Long, and L. F. Bliven, 1983: 'A non-Gaussian statistical model for surface elevation of nonlinear random wave fields,' J. Geophys. Res., **88**, C12, 7597-7606.

Huang, N. E., S. R. Long, L. F. Bliven, and C. C. Tung, 1984: 'The non-Gaussian joint probability density function of slope and elevation for a non linear gravity wave field,' J. Geophys. Res., **89**, C2, 1961-1972.

Huang, N. E., 1986: 'An estimate of the influence of breaking waves on the dynamics of the upper ocean,' **Wave Dynamics and Radio Probing of the Ocean Surface**, 295-313, Eds. O. M. Phillips, and K. Hasselmann, Plenum Press.

Huang, N. E., L. F. Bliven, S. R. Long, and C. C. Tung, 1986: 'An analytic model for oceanic whitecap coverage,' J. Phys. Oceanogr., **16**, 1597-1604.

Lin, J. T., 1986: 'Empirical prediction of wave spectrum for wind-generated gravity waves,' 20th Int. Coastal Engrg. Conf., Taipei, Taiwan, Nov. 9-14.

Longuet-Higgins, M. S., 1957: 'The statistical analysis of a random waving surface,' Phil. Trans. R. Soc. Lond., A249, 321-387.

Longuet-Higgins, M. S., 1969: 'On wave breaking and equilibrium spectrum of wind-generated waves,' Proc. R. Soc. Lond., A310, 151-159.

Longuet-Higgins, M. S., 1975: 'On the joint distribution of the periods and amplitudes of sea waves,' J. Geophys. Res., **80**, C18, 2688-2694.

Longuet-Higgins, M. S., 1983: 'On the joint distribution of wave periods and amplitudes in a random wave field,' Proc. R. Soc. London, A389, 241-258.

Longuet-Higgins, M. S., 1986: 'Eulerian and Lagrangian aspects of surface waves,' J. Fluid Mech., **173**, 683-707.

Mitsuyasu, H., and T. Honda, 1982: 'Wind-induced growth of water waves,' J. Fluid Mech., **123**, 425-442.

Monahan, E. C. and I. O'Muircheartaigh, 1980: 'Optimal power law description of oceanic whitecap coverage dependence on wind speed,' J. Phys. Oceanogr., **10**, 2094-2099.

Ochi, M. K. and C. H. Tsai, 1983: 'Prediction of occurrence of breaking waves in deep water,' J. Phys. Oceanogr., **13**, 2008-2019.

Papadimitrakis, Y. A, E. Y. Hsu, and R. L. Street, 1986: 'The role of wave-induced pressure fluctuations in the transfer processes across an air-water interface,' J. Fluid Mech., **170**, 113-137.

Papadimitrakis, Y. A., N. E. Huang, S. R. Long, and L. F. Bliven, 1987: 'On the parameterization of momentum and energy fluxes and their partition across the air-sea interface,' J. Geophys. Res., Submitted.

Phillips, O. M. and M. L. Banner, 1974: 'Wave breaking in the presence of wind drift and swell,' J. Fluid Mech., **66**, 625-640.

Phillips, O. M., 1985: 'Spectral and statistical properties of the equilibrium range in wind-generated gravity waves,' J.Fluid Mech., **156**, 505-531.

Snyder, R. L., L. Smith, and R. M. Kennedy, 1983: 'On the formation of white caps by a threshold mechanism. Part III: Field experiment and comparison with theory,' J. Phys. Oceanogr., **13**, 1505-1518.

Snyder, R. L., F. W. Dobson, J. A. Elliott, and R. L. Long, 1981: 'Array measurements of atmospheric pressure fluctuations above surface gravity waves,' J. Fluid Mech., **102**, 1-59.

Srokosz, M. A., 1986: 'On the probability of wave breaking in deep water,' J. Phys. Oceanogr., **16**, 382-385.

Tayfun, M. A., 1980: 'Narrow-band non-linear sea waves,' J. Geophys. Res., **85**, C3, 1548-1522.

Tayfun, M. A., 1981: 'Breaking limited wave heights,' J. Waterways Harbors and Coastal Eng. Div., ASCE, **107**, WW2, 59-69.

Yuan, Y. L., 1982: 'On the statistical properties of sea waves,' Ph.D. Thesis, North Carolina State University, Raleigh, N. C.

WHITECAP COVERAGE AS A REMOTELY MONITORABLE INDICATION OF THE RATE OF BUBBLE INJECTION INTO THE OCEANIC MIXED LAYER

Edward C. Monahan
Marine Sciences Institute
University of Connecticut
Avery Point, Groton
Connecticut 06340, U.S.A.

ABSTRACT: The observed cubic dependence on friction velocity of near-surface bubble concentration is independently deduced from a simple geometrical model for the whitecap bubble plume. This exponential plume model likewise implies that the bubble injection rate is proportional to instantaneous whitecap coverage. Explicit quantitative expressions for surface bubble flux, and for near-surface bubble concentration, are derived as functions of bubble radius and sea-surface whitecap coverage. Calculated bubble spectra agree well with optically measured spectra for equivalent conditions.

1. INTRODUCTION

It has recently been shown by Farmer and Lemon (1984) that over a range of relatively high wind speeds, the bubble density in the mixed surface layer of the ocean varies with U_*^3; i.e., with the cube of the friction velocity. Detailed analyses of extensive sets of whitecap photographs taken from shipboard (Monahan and O'Muircheartaigh, 1980; 1986) show that the fraction of the ocean surface covered by whitecaps, W, also manifests an essentially cubic dependence on U_*. Herein a model will be set forth that explains why bubble density in ·the mixed layer and sea surface whitecap coverage show the same dependence on U_*.

2. BUBBLE PLUME MODEL

The U_*^3 dependence of whitecap coverage was postulated by Wu (1979), and could have been inferred for the case of dynamic equilibrium from Cardone's (1969) finding that on the North American Great Lakes, whitecap coverage (Monahan, 1969) was proportional to the rate at which energy was being dissipated out of the high frequency end of the wave spectrum. Now W is proportional to the rate of new whitecap area production, $\dot{W}$, the coefficient of proportionality being τ, the time constant characterizing the exponential decay of the individual whitecap (Monahan, 1971).

$$W = \tau \dot{W} \qquad (1)$$

B. R. Kerman (ed.), Sea Surface Sound, 85–96.
© 1988 by Kluwer Academic Publishers.

In turn, $\dot{W}$ is proportional to $\dot{V}$, the rate at which aerated plume, i.e., new bubble cloud, volume is being formed, the coefficient of proportionality in this case being D, the scale depth defining the exponential attenuation with depth of the horizontal cross-sectional area of the initial aerated plume (Monahan, 1986), the surface manifestation of which is the whitecap. In the upper panel of Figure 1

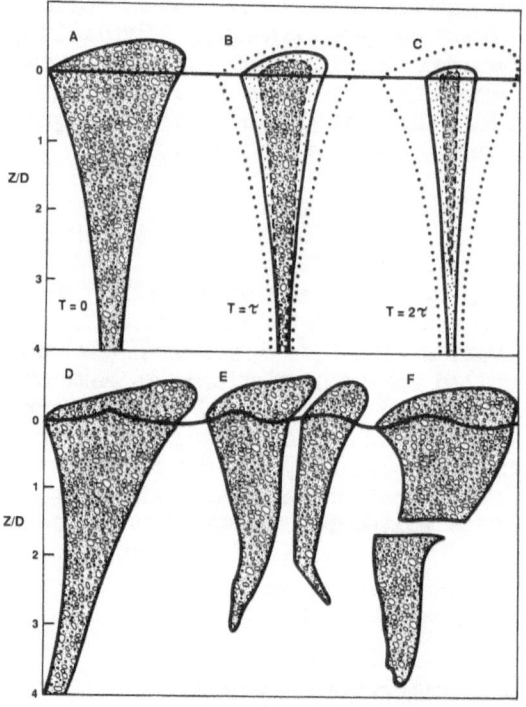

FIGURE 1. Vertical sections, and partial plan views, of the bubble plumes produced by breaking waves. All views reflect significant vertical exaggeration. A: idealized plume immediately after formation by breaking wave. B: same plume after time τ, note large bubbles constrained to narrower cone than the smaller bubbles, as a consequence of the large bubble field being displaced upward faster than the assemblage of small bubbles. C: same plume after time 2τ, note that area of surface whitecap is only 14% (e^{-2}) of original area outlined by dotted line. D,E,F: more realistic plumes subjected to current shears immediately after formation.

is depicted the exponentially attenuated column of aerated water that corresponds to this idealized model of a bubble plume. If $A_o(0)$ represents the initial horizontal cross-sectional area of the plume at zero depth, i.e., the initial area of the associated whitecap, then the volume of the just-formed bubble plume, V_o, is given by

$$V_o = \int_0^\infty A_o(0) e^{-z/D} \, dz = DA_o(0) \tag{2}$$

Likewise, the horizontal cross-sectional area of the bubble plume at depth Z and at time t after formation is simply,

$$A = A_o(0) \ e^{-z/D} \ e^{-t/\tau} \tag{3}$$

this geometrical model of a newly formed bubble cloud is consistent

$$\dot{V} = D \dot{W} \tag{4}$$

with the exponential decrease with depth of the scattering cross-section of bubble clouds reported by Thorpe (1982, 1986a, 1986b). This model ceases to be meaningful when a sufficient time t is allowed to pass, subsequent to a breaking wave event, to permit wholesale, significant, turbulent dispersion of the slow-rising smaller bubbles, and extensive dissolution of the individual bubbles. In the presence of large-scale shearing, as reflected in the sketches incorporated in the lower panel of Figure 1, this model can be applied as long as the aggregate cross-sectional area of the several resulting plume elements can still be identified at each depth.

It follows from this model that $\dot{V}$ should be proportional to U_*^3.

$$\dot{V} \sim \dot{W} \sim W \sim U_*^3 \tag{5}$$

Recognizing that the W and τ values obtained from photographic and video records are in practice a function of the optical resolution of the camera employed, and indeed are based on the detection of only the large bubbles and bubble clusters at the sea surface, it can nonetheless be argued that $W\tau^{-1}$; i.e., $\dot{W}$, is not a function of camera resolution, and hence is a valid measure of the rate of new bubble plume volume formation. If the bubble spectrum, and the bubble concentration within the newly formed plume, resulting from breaking waves do not themselves show a strong U_* dependence, then $\dot{V}$ should be simply proportional to the rate of bubble production, $\dot{C}$, and this source function in turn should, if the bubble spectrum is essentially invarient with changes in wind speed, be proportional to the resident bubble density, C. Hence C should be proportional to U_*^3, as indeed

$$C \sim \dot{C} \sim \dot{V} \sim \dot{W} \sim W \sim U_*^3 \tag{6}$$

was found by Farmer and Lemon (1984).

3. THE PRACTICAL SIGNIFICANCE OF WHITECAP COVERAGE

The use of W as an indication of the rate at which bubbles are being injected into the oceanic mixed layer is to be preferred to the use of the 10m-elevation wind speed U_{10}, or even of the friction velocity U_* for several reasons. Whitecap coverage is more directly related to bubble injection rate than is either U_{10} or U_*, as has been shown above. In order to estimate injection rate from U_*, one must in essence take into account the influence of wind duration and fetch, in addition to the effect of U_*, on W. If U_{10} is selected as the measured quantity to be used in the estimation of bubble injection rate or W, then the thermal stability of the lower atmosphere is a further factor that must be considered (Monahan and O'Muirchear-taigh, 1986). Whitecap coverage, due to its effect on sea surface emissivity, and hence on sea surface apparent brightness temperature, can be detected by satellite-borne passive microwave radiometers such as those that comprise the SMMR. Likewise, whitecaps have an active microwave (radar) signature. Finally, the selection of whitecap coverage as the indicator of bubble injection rate makes it possible to readily identify those low wind circumstances when, because of the total absence of whitecapping, the injection rate can be confidently assumed to be nil.

4. MODELING SEA SURFACE BUBBLE FLUX

It has been shown that the rate of marine aerosol production, per unit area of the sea surface, per unit radius increment, $\partial F_o/\partial r$, is likewise proportional to V, W, and W (Monahan, 1986). Furthermore, an explicit quantitative expression for $\partial F_o/\partial r$, as a function of aerosol droplet size at 80% relative humidity, r, and of W, was determined from the results of detailed experiments carried out in a whitecap simulation tank (Monahan, et al., 1982; 1983; Monahan, 1986). It should be possible to "work backwards" from this expres-

$$\partial F_o/\partial r = f(r,w) = 3.57 \times 10^5 r^{-3} (1+0.057\ r^{1.05}) \times$$

$$\times\ 10^{1.19e^{-B^2}}\ W, \tag{7}$$

$$B = (0.380 - \text{Log } r)\ /0.650$$

sion for the aerosol flux to arrive at the associated bubble flux.

If G represents the bubble flux to the surface (which is equal to C for steady state conditions), and if R is the bubble radius, then, in the case where all the aerosol particles are taken to be jet drop-lets, and where each bursting bubble is assumed to produce a single such droplet, an expression for the rate of bubble bursting, per unit area of sea surface, per unit increment of bubble radius, can be readily written.

$$\partial G/\partial R = \partial C/\partial R = \partial F_o/\partial r\ \frac{\partial r}{\partial R} \tag{8}$$

Since $\partial r/\partial R$ can be evaluated from the data on daughter jet-droplet and parent bubble sizes given in Blanchard (1963), as was done by Monahan (1965),

$$r = aR + b = 8.77 \times 10^{-2}R + 0.98 \qquad (9)$$

and as $\partial F_o/\partial r$ is already known, an explicit expression for $\partial G/\partial R$ can be set down.

$$\partial G/\partial R = \partial \dot{C}/\partial R = a\; f(aR + b, W) = \qquad (10)$$

$$3.13 \times 10^4 (8.77 \times 10^{-2}R + 0.98)^{-3} \times$$

$$\times\; [1 + 0.057(8.77 \times 10^{-2}R + 0.98)^{1.05}] \times$$

$$\times\; 10^{1.19e^{-B^2}}\; W,$$

$$B = [0.380 - \text{Log}(8.77 \times 10^{-2}R + 0.98)]/0.650$$

Recognizing that not all the marine aerosol particles produced are jet droplets (indeed most of the droplets with radii less than 2 µm produced during the decay of a whitecap are film droplets), the recent results of Woolf, et al (1987) on film droplet/jet droplet partition can be used to modify the above expression as follows:

$$\partial G/\partial R = \partial \dot{C}/\partial R = aP(aR + b)f(aR + b, W) \qquad (11)$$

where P(r) is the jet-droplet fraction of the bubble generated aerosol,

$$P = 1 - e^{-0.343r} = 1 - 0.715e^{-0.030R} \qquad (12)$$

A further refinement to the bubble injection/decay model can be introduced by taking into account the observation that typically more than one jet droplet is produced as each bubble bursts (Hayami and Toba, 1958; Blanchard, 1963), and by letting J represent the jet-droplet production per bubble.

$$\partial G/\partial R = \partial \dot{C}/\partial R = aJ^{-1}P(aR + b)\; f\; (aR + b, W) \qquad (13)$$

The sea surface bubble flux spectrum for a 100% whitecap covered sea that results from the substitution of unity for W in the basic $\partial G/\partial R$ expression (Equation 10) is illustrated by curve A on Figure 2. Inclusion of the jet droplet partition factor, i.e., use of Equation 11, yields the surface bubble flux spectrum for an entirely foam covered sea represented by curve B on this same figure. Assuming, as did Cipriano and Blanchard (1981), that each bubble produces five jet drops, and evaluating Equation 13 for J equals 5 and W equals 1, gives the bubble flux spectrum, for the unrealistic case of an ocean entirely covered by "white water," or more pertinently, for the surface of an active, albeit isolated, whitecap, represented by curve

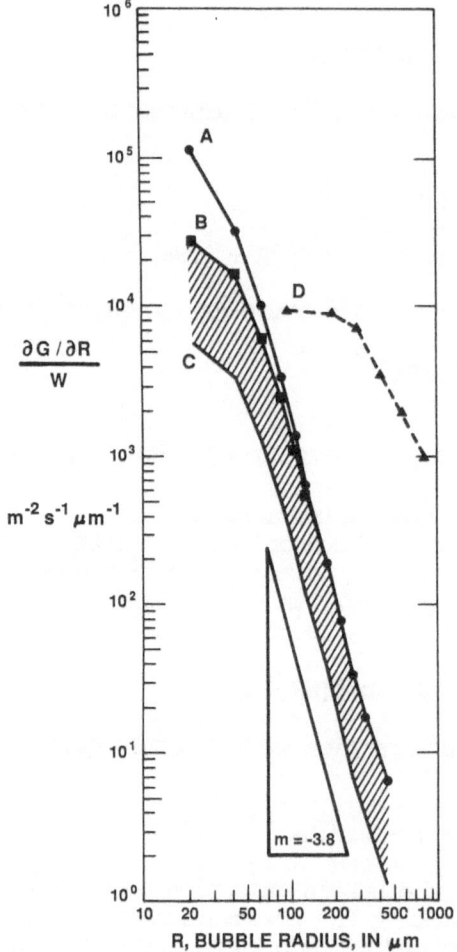

FIGURE 2. Sea surface bubble flux spectra, normalized for whitecap coverage, consistent with the sea surface aerosol flux spectrum obtained from the Whitecap Simulation Tank measurements of Monahan, et al. (1982, 1983). A: basic bubble flux spectrum as given by Equation 10. B: spectrum after account has been taken, via Equation 11, of Jet-Droplet/Film-Droplet partition. C: spectrum based on Equation 12, where 5 jet droplets are assumed to be produced per bubble. Note that slope of these three bubble flux spectra, on this log-log plot, is close to −3.8 for R greater than 70 μm. D: bubble flux spectrum derived from results of Cipriano's (1979) Weir-waterfall experiments.

C on Figure 2. (Curve D on this figure was obtained by dividing the number of bubbles per second in each size range that Cipriano (1979) determined were reaching the surface of the water in his weir-waterfall experiments by the effective surface area, 0.020 m^2, of the resulting perpetual whitecap.)

5. INFERRED SPECTRUM OF NEAR-SURFACE BUBBLE POPULATION

Surface bubble flux expressions, such as those given by equations 11 and 13, when divided by v(R), the terminal rise velocities for bubbles, e.g., those summarized for hydrodynamically clean and dirty bubbles in Figure 16 of Thorpe (1982), yield estimates of the near-surface bubble population, $\partial C/\partial R$,

$$\partial C/\partial R = v^{-1}\partial G/\partial R = v^{-1}\partial \dot{C}/\partial R \qquad (14)$$

$$= av^{-1}J^{-1}P(aR + b)f(aR + b,W)$$

The resident bubble spectrum for 100% foam coverage inferred using Equation 11 and the dirty bubble v(R) curve from Thorpe (1982) is given by Curve B_D on Figure 3. Using the same $\partial G/\partial R$ expression, and again taking W equal to 1.00, but making use of the clean bubble v(R) information from Thorpe, results in bubble population spectrum B_C on Figure 3. The near-surface resident bubble spectra deduced from the surface bubble flux spectrum given by Equation 13, for 100% foam cover, J equals 5, and dirty and clean bubbles, are depicted by curves C_D and C_C respectively. In light of the foregoing, it can be concluded that the near-surface bubble population spectrum $\partial C/\partial R$ to be expected immediately under a whitecap falls somewhere between curves B_D and C_C on Figure 3.

6. DISCUSSION

Confidence in the validity of the approach to modeling sea surface aerosol generation embodied in Equation 7 was reinforced when the resident aerosol spectrum inferred from this surface droplet flux model was shown to be similar to the aerosol spectra measured for comparable conditions in the Marine Atmospheric Boundary Layer (Monahan, 1986). It is appropriate, as an initial test of the sea surface bubble flux model described in this present work, to now compare published near-surface resident bubble spectra with the bubble populations deduced from Equation 14 for the appropriate values of whitecap coverage.

Blanchard and Woodcock (1957) have described the band of bubble spectra they obtained just off-shore using a small, manual, bubble trap at a depth of 0.10m several seconds after the passage of a breaking wave. Since this procedure effectively corresponds to carrying out measurements immediately under an active whitecap, their results have been plotted as Region E on Figure 3, the illustration

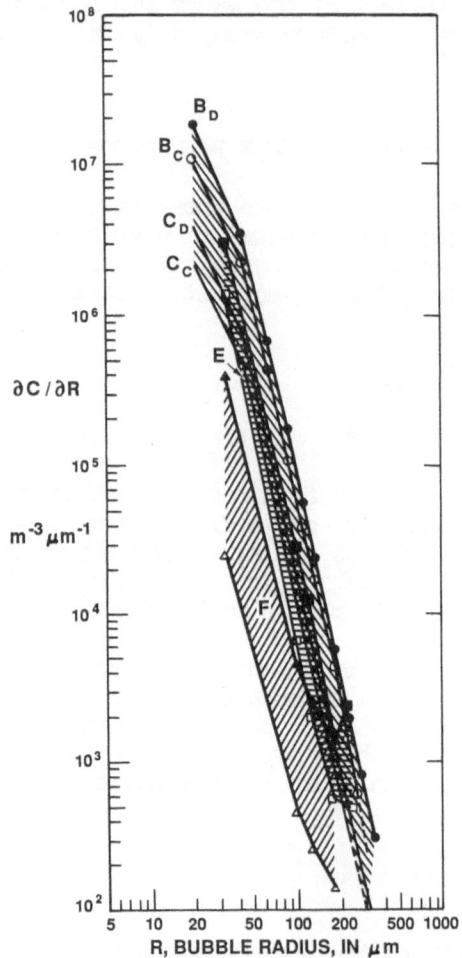

FIGURE 3. Near-surface resident bubble spectra. Curves B_D, B_C, C_D, and C_C from Equation 14 for W equals 1. B_D, J equals 1, dirty bubbles. B_C, J equals 1, clean bubbles. C_D, J equals 5, dirty bubbles. C_C, J equals 5, clean bubbles. Region E, band of bubble spectra measured just after passage of shallow water breaking waves (Blanchard and Woodcock, 1957). Region F, band of bubble spectra observed in lee of wave inundated rock (Blanchard and Woodcock, 1957).

which also contains the family of resident bubble spectra predicted by Equation 14 for W equals 1.00. The band of bubble spectra observed by these same authors in the lee of a wave inundated rock is also included, as Region F, on Figure 3. The extensive overlap between Region B_D-C_C and Region E is a noticeable feature of Figure 4, and suggests that Equation 14 can be used quite successfully to predict the resident bubble population in the new plume of aerated water beneath an active whitecap.

Near-surface bubble spectra associated with winds of 11-13 ms^{-1} have been measured by several investigators. Kolovayev (1976) measured bubble spectra in the Atlantic Ocean using a semi-automatic cylindrical bubble trap equipped with a camera and three lights to illuminate the bubbles, a system based on one designed by Neuimin (Glotov, et al, 1962). Using a camera in a waterproof housing, equipped with three external strobe lamps, Johnson and Cooke (1979) measured the in situ bubble population in the coastal waters of Nova Scotia. An additional near-surface resident bubble spectrum, in this case identified with a wind speed of 13 ms^{-1}, was obtained by Baldy and Bourguel (1985) using a sophisticated optical bubble probe (Avellan and Resch, 1983; Resch, 1986) that utilizes the scattering by individual bubbles of light from a laser source. Unlike the other bubble measurements, which were made at sea, these measurements were obtained with the probe just below the fresh-water surface at a fetch of 26.2 m in the large wind-wave flume at the Institut de Mecanique Statistique de la Turbulence in Marseille.

At a wind speed of 13 ms^{-1}, a typical value for whitecap coverage is 0.0241. This was determined using the following expression for W in terms of the 10m-elevation wind speed U, expressed in ms^{-1} (Monahan and O'Muircheartaigh, 1980; Monahan, 1986),

$$W = 3.84 \times 10^{-6} \, U^{3.41} \tag{15}$$

Introducing the above value of W into the f(aR + b,W) term of Equation 14, and evaluating this equation while allowing J to range from 1 to 5, and assuming alternatively the dirty-bubble and clean-bubble terminal rise velocities, leads to the band of near-surface resident bubble spectra which can appropriately be compared to the bubble spectra of Kolovayev, of Johnson and Cooke, and of Baldy and Bourguel. When this comparison is effected (see Chapter 2, Monahan and Woolf, 1987) it transpires that for all bubble radii, R, greater than 120 μm Kolovayev's spectrum falls effectively within the band of spectra predicted by Equation 14 for these conditions, while Johnson and Cooke's spectrum agrees well with the model predictions for all R greater than 50 μm. The discrepancy in the 50 to 120 m radius interval between Kolovayev's spectrum on the one hand, and Johnson and Cooke's spectrum and the spectra obtained using Equation 14 on the other hand, may, as was suggested as a possibility by Johnson and Cooke (1979), be due to some dissolution, and perhaps also coalescence, of the bubbles in this size range while they are within the Neuimin bubble trap. The entire Baldy and Bourguel spectrum, with the exception of one point, falls within the spectral band predicted

94

by the model for this wind speed. This agreement extends even to the small radius (32 μm) end of the Baldy and Bourguel spectrum.

It would be desirable to test the bubble spectra predicted by the model presented in this paper against additional spectra collected using the various optical techniques, and to also have recourse to acoustically derived spectra, but given the variety of fates that Thorpe (1982) has documented await small bubbles in the sea, it may be unreasonable to expect close agreement between the small bubble end of the resident spectrum predicted by this model, a model which by the nature of its derivation is sensitive only to those bubbles that reach the sea surface, burst, and produce spray droplets, and in situ observations of the small bubble fraction.

The agreement between the wind dependence of bubble concentration found by Farmer and Lemon (1984), and the wind dependence implicit in the simple whitecap dependence of bubble concentration which is a central feature of the model presented in this paper, coupled with the demonstrated ability of this model to predict the approximate shape and amplitude of the large radius portion of the in situ bubble spectrum, give support to the contention that the modeling approach set forth in this paper is a valid one, and thus lend weight to the suggestion that this model merits further development and refinement.

ACKNOWLEDGMENTS

We thank Mr. D. K. Woolf for his beneficial comments during the preparation of this paper, and are grateful for the assistance provided by Mrs. E. Minik, who typed this paper, by Ms. C. Venti-Rashan, who inked the illustrations, and by the other staff of the Sea Grant Office, Marine Sciences Institute, University of Connecticut.

Our research on whitecaps, bubbles, and sea spray droplets has been supported throughout by the Office of Naval Research, currently via Contracts N00014-87-K-0185, N00014-87-K-0069, and N00014-86-K-0276.

REFERENCES

Avellan, F., and F. Resch. 1983. 'A Scattering Light Probe for the Measurement of Oceanic Air Bubble Sizes.' Int. Journal of Multiphase Flow, 9, 649-663.

Baldy, S., and M. Bourguel. 1985. 'Measurements of Bubbles in a Stationary Field of Breaking Waves by a Laser-Based Single-Particle Scattering Technique.' J. Geophys. Res., 90, 1037-1047.

Blanchard, D.C. 1963, 'The electrification of the atmosphere by particles from bubbles in the sea.' Prog. Oceanog., 1, 71-202.

Blanchard, D.C., and A.H. Woodcock. 1957. 'Bubble formation and modification in the sea and its meteorological significance.' Tellus, 9, 145-158.

Cardone, V.J. 1969. 'Specification of the wind disruption in the marine boundary layer for wave forecasting.' Tech. Rep. GSL-69-1, New York University, 1-131.

Cipriano, R.J. 1979. 'Bubble and Aerosol Spectra Produced by a Laboratory Simulation of a Breaking Wave.' Ph.D. Thesis, State University of New York at Albany, 1-265.

Cipriano, R.J., and D.C. Blanchard. 1981. 'Bubble and Aerosol Spectra Produced by a Laboratory "Breaking Wave."' J. Geophys. Res., 86, 8085-8092.

Farmer, D.M., and D.D. Lemon. 1984. 'The influence of bubbles on ambient noise in the ocean at high wind speeds.' J. Phys. Oceanogr., 14, 1762-1778.

Glotov, V.P., P.A. Kolobaev, and G.G. Neuimin. 1962. 'Investigation of the Scattering of Sound by Bubbles Generated by an Artificial Wind in Sea Water and the Statistical Distribution of Bubble Sizes.' Sov. Phys., Acoustics,7, 341-345.

Hayami, S., and Y. Toba. 1958. 'Drop production by bursting of air bubbles on the sea surface (1) experiments at still sea water surface.' J. Ocean. Soc. Japan, 14, 145-150.

Johnson, B., and R.C. Cooke. 1979. 'Bubble populations and spectra in coastal waters: a photographic approach.' J. Geophys. Res., 84, 3761-3766.

Kolovayev, P.A. 1976. 'Investigation of the concentration and statistical size distribution of wind-produced bubbles in the near-surface ocean layer.' Oceanology, 15, 659-661.

Monahan, E.C. 1965. 'A Field Study of Sea Spray and its Relationship to Low Elevation Wind Speed - Preliminary Results.' Tech. Rep., Woods Hole Oceanographic Institution, 1-147.

Monahan, E.C. 1969. 'Fresh water whitecaps.' J. Atmos. Sci., 26, 1026-1029.

Monahan, E.C. 1971. 'Oceanic Whitecaps.' J. Phys. Oceanogr., 1, 139-144.

Monahan, E.C. 1986. 'The Ocean as a Source of Atmospheric Particles,' in The Role of Air-Sea Exchange in Geochemical Cycling, P. Buat-Menard, Ed., D. Reidel Pub. Co., Dordrecht, Holland, 129-163.

96

Monahan, E.C., K.L. Davidson, and D.E. Spiel. 1982. 'Whitecap aerosol productivity deduced from simulation tank measurements.' J. Geophys. Res., 87, 8898-8904.

Monahan, E.C., and I. O'Muircheartaigh. 1980. 'Optimal power-law description of oceanic whitecap coverage dependence on wind speed.' J. Phys. Oceanogr., 10, 2094-2099.

Monahan, E.C., and I.G. O'Muircheartaigh. 1986. 'Whitecaps and the Passive Remote Sensing of the Ocean Surface.' Int. J. Remote Sensing, 7, 627-642.

Monahan, E.C., D.E. Spiel, and K.L. Davidson. 1983. 'Model of marine aerosol generation via whitecaps and wave disruption.' Ninth Conference on Aerospace and Aeronautical Meteorology, 6-9 June, 1983, Omaha, Nebraska, American Meteorological Society, Preprint Volume, 147-158.

Monahan, E.C., and D.K. Woolf. 1987. Oceanic Whitecaps and Sub-surface Bubble Clouds; Their Influence on the Marine Microlayer and the Marine Atmospheric Boundary Layer. Whitecap Report No. 3, to ONR from MSI, UConn (in press).

Resch, F. 1986. 'Oceanic Air Bubbles as Generators of Marine Aerosols,' in Oceanic Whitecaps and Their Role in Air-Sea Exchange Processes, E.C. Monahan and G. MacNiocaill, Eds., D. Reidel Pub. Co., Dordrecht, Holland, 101-112.

Thorpe, S.A. 1982. 'On the clouds of bubbles formed by breaking windwaves in deep water, and their role in air-sea gas transfer.' Phil. Trans. R. Soc. London, A304, 155-210.

Thorpe, S.A. 1986a. 'Measurements with an automatically recording inverted echo sounder; ARIES and the bubble clouds.' J. Phys. Oceanogr., 16, 1462-1478.

Thorpe, S.A. 1986b. 'Bubble clouds: A review of their detection by sonar, of related models, and of how K_v may be determined, in Oceanic Whitecaps and Their Role in Air-Sea Exchange Processes, E.C. Monahan and G. MacNiocaill, Eds., D. Reidel Pub. Co., Dordrecht, Holland, 57-68.

Woolf, D.K., P.A. Bowyer, and E.C. Monahan. 1987. 'Discriminating Between the Film Drops and Jet Drops Produced by a Simulated Whitecap.' J. Geophys. Res., 92. (in press).

Wu, J. 1979. 'Oceanic whitecaps and sea state.' J. Phys. Oceanogr., 9, 1064-1068.

WAVE AND BUBBLE CHARACTERISTICS IN THE SURF ZONE

P. Papanicolaou
W. M. Keck Laboratory of
Hydraulics & Water Resources
California Institute of Technology
Pasadena, California
U.S.A.

F. Raichlen
W. M. Keck Laboratory of
Hydraulics & Water
Resources, California
Institute of Technology
Pasadena, California

ABSTRACT. Attention is given to the shape of the bubble cloud produced during breaking of shallow water solitary waves in a sloping tank. The growth and collapse of the bubble cloud on the front face of the breaking wave is evaluated. The noise generated by these bubble clouds after breaking is investigated in an exploratory manner.

1. INTRODUCTION

In recent years, attention has been devoted by the profession to different aspects of wave breaking in both shallow water and deep water. Studies have dealt with the kinematics of waves at and just after breaking and, to some extent, certain acoustical properties of these waves. Two representative examples of experimental studies of wave kinematics are those of Nadaoka (1986) and Skjelbreia (1987), both of whom have applied laser-Doppler velocimetry (LDV) to evaluate the mean velocity field and the turbulence intensity in the breaking region. These studies both have dealt with shallow water wave breaking and the wave modifications which result in the surf zone after breaking. Svendsen (1984), in one approach to describing the process analytically, has associated the energy dissipation and the wave height decay with the "roller" formed after wave breaking. Dally, Dean, and Dalrymple (1985) also proposed a quantitative description of the energy dissipation associated with breaking waves, considering the change in the local energy flux with distance after breaking. Thus, an important facet of the energy dissipation taking place during wave breaking for shoaling waves appears to be the generation of the bubble plume (or roller) which forms on the front face of the breaking wave. One descriptor of this process may be the associated noise which is produced. A summary of several aspects of the underwater sound occurring with breaking waves has been presented by Kerman (1984).

In this discussion the results of visual observations and limited acoustic measurements describing aspects of the process of the breaking of solitary waves on a sloping bottom are presented. The solitary wave, which is a nonlinear shallow water wave, is a useful wave to describe

97

B. R. Kerman (ed.), Sea Surface Sound, 97–109.
© *1988 by Kluwer Academic Publishers.*

periodic breaking waves in shallow water. In addition, it is particularly useful in the laboratory for investigating such nearshore
phenomena, since the first wave (and the only wave) is well formed and
permanent, and after the wave is generated the only sound being produced in the wave tank is that of the propagating wave. Thus, it becomes
an ideal shallow water wave to investigate both visually and acoustically where various breaking characteristics can be isolated for study.
Since the solitary wave responds to decreasing depths, waves which
produce spilling or plunging breakers can be generated easily in the
laboratory. The processes which take place some distance after
breaking for shallow water and deep water waves are quite different,
since shallow water waves propagate to shore and collapse forming
bores which then run up on the shore where deep water waves continue
to propagate with a reduced height. However, the initial characteristics of the breaking wave just at the beginning of formation of the
bubble plume on the front face of the wave must be similar for shallow
water and deep water waves both for their plunging and spilling breaker
counterparts. Thus, although this study simplifies the wave system by
using a solitary wave, a great deal can be learned about breaking
processes in general through the close observation of the characteristics of such waves breaking in the laboratory.

2.   EXPERIMENTAL EQUIPMENT AND PROCEDURES

A tilting wave tank was used for all experiments; the tank is 39.6 m
long, 1.1 m wide, and 0.61 m deep, and is constructed with glass sidewalls throughout and a stainless steel bottom which is plane to about
±1.0 mm. The tank is supported at a central hinge point with two
motorized jacks upstream and two downstream, geared to operate together
thereby permitting a continuous adjustment of slope from horizontal to
1:50. A bulkhead wave generator located at one end of the tank is an
integral part of the wave tank support structure that tilts with the
tank; in this manner waves can be produced on a slope and caused to
break at predetermined locations. The generator is driven by an
electro-hydraulic-servo-system with voltage time history determined so
the velocity of the wave plate, at a given time, matches the water
particle velocity of the wave as it propagates away from the plate.
An iterative procedure can be used to compensate for the dynamics of
the hydraulic-mechanical system. A photograph of a portion of the tank
is presented in Figure 1; the wave machine can be seen to the left at
the end of the tank. In the foreground the carriage used to support a
dual beam forward scattering laser-Doppler velocimeter (LDV) is shown.
(Although the LDV was developed and used by Skjelbreia (1987) in
breaking wave studies, it was not used in this study.)
    All measurements dealing with the waves before, during and after
breaking have been obtained from film using a high speed motion picture
camera (Redlake Corporation, Model LOCAM 51-003) mounted on a movable
carriage which can travel smoothly on the instrument rails extending
the length of the wave tank and seen in Figure 1. The camera can
operate at film speeds of up to 500 frames per second with a shutter

speed of 1/4000 sec. For the experiments described, generally the film speed used was 300 frames per second with a shutter speed of approximately 1/1200 sec. Floodlights were mounted to the carriage to provide the required illumination. Motion pictures of the wave breaking process have been taken through the tank sidewalls by moving the camera with the wave. Frame-by-frame observations were made to define the variation of wave height during the breaking process and the development, growth, and collapse of the bubble plume associated with breaking. Various characteristics of the wave at and after breaking were obtained from the films using 1/100 sec timing marks automatically recorded on the film and fiduciary marks on the tank sidewalls.

Acoustic measurements were made using a Celesco LC-32 hydrophone (approximately 2 cm in diameter with a sensitive length of approximately 6 cm) in conjunction with the laboratory data acquisition system. The transducer was mounted, with its long axis vertical, to a strut located on

Figure 1 Photograph of Tilting Wave Tank, Wave Generator, and LDV Carriage.

the center line of the tank with the center of its acoustic field approximately at mid-depth. According to the manufacturer's specification, the response of the hydrophone is flat from 1 Hz to 1 kHz and begins to decrease somewhat at 6 kHz falling by about 2 db at 20 kHz. The arrangement of the hydrophone in the tank will be described later.

3. RESULTS AND DISCUSSION OF RESULTS

The main objective of this paper is to report on observations which were conducted to describe the generation and collapse of the bubble plume on the front face of a breaking wave along with exploratory measurements of the associated noise. The solitary wave is a unique shallow water wave whose volume is completely above the still water level with the shape and breaking characteristics defined by the relative height of the wave at generation and the depth variation as it propagates shoreward. This is demonstrated in Figure 2 where the variation of the relative wave height at breaking is presented as a function of the slope of the wave tank. The relative breaking wave height, defined as the ratio of the wave height at breaking to the local depth at breaking, increases from approximately 0.9 at the smallest slope to nearly 1.25 at the maximum wave tank slope. Interesting in this figure are the data at a slope of 0.0126 corresponding to three waves with different relative initial heights; the

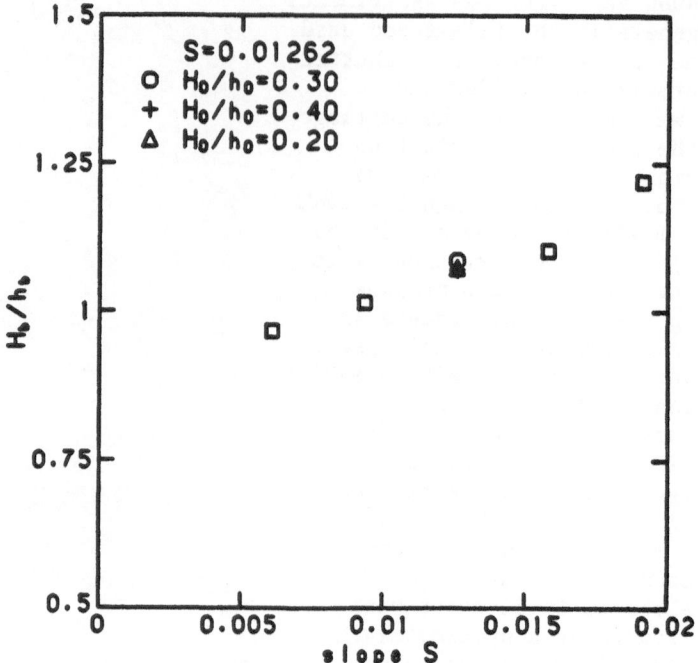

Figure 2   Variation of Relative Wave Height at Breaking With Bottom
Slope.

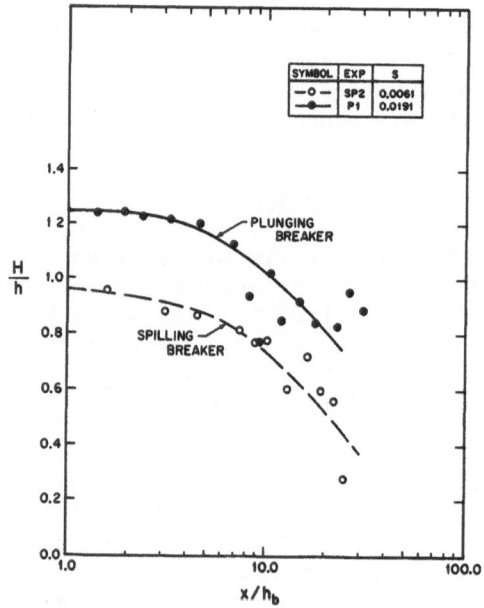

Figure 3   Variation of Local Relative Wave Height With Relative Distance
from Breaking.

relative initial wave height is defined as the ratio of the wave height to the depth at the wave generator. These waves exhibit similar breaking conditions for the same slope which should be expected and is due to the careful means of generation used in this study.

The conditions for the experiments to be discussed are presented in Table I. These have resulted in waves which range from spilling waves (Experiment SP2) to plunging waves (Experiment P1).

Table I Experimental Condition for Experiments

Experiment	S m/m	h_o (cm)	H_o/h_o	h_b (cm)	H_b (cm)
P1	0.0191	43.25	0.20	13.50	16.50
P2	0.0158	39.05	0.25	15.42	17.00
PS2	0.0126	35.45	0.20	13.04	14.00
PS	0.0126	35.45	0.30	16.57	18.00
PS1	0.0126	35.45	0.40	18.86	20.20
SP1	0.0094	31.55	0.35	17.15	17.40
SP2	0.0061	27.65	0.40	17.06	16.50

The variation of the relative wave height, H/h, with relative distance from breaking, x/h_b, is shown in Figure 3 for two waves: a spilling breaker (SP2) and a plunging breaker (P1). As these waves propagate shoreward, initially there is little change in relative wave height for about five breaking depths shoreward from breaking and then there is a rapid decrease in height which continues to about twenty depths from breaking (the limit of the data).

To help understand this change in wave height and the breaking process, a sequence of photographs taken from frames of high speed motion pictures are presented in Figure 4a for a plunging wave (P1) and Figure 4b for a spilling breaking wave (SP2) as each shoals and propagates shoreward. The frames were chosen to better depict several features of the breaking process. The camera film speed was 300 frames per sec with a shutter speed of 1/1200 sec, and since the celerity of the unbroken and broken waves varied from between 100 cm/sec to 150 cm/sec, the temporal resolution of the movies was considered to be adequate.

The uppermost photograph in Figure 4a shows the wave at incipient breaking. The near vertical face at the crest leads to the over-turning process (seen in the second photograph) that entrains air and generates the bubble plume in the third photograph of the sequence. The plume which develops is both a source of the energy dissipation and a measure of that dissipation. After the initial breaking, the reduced wave appears to catch up and break again creating a second roller parallel to and co-rotating with the first, and so on, leading to the configurations in the last two photographs in the sequence. In the fourth photograph down, several large bubble clouds relatively evenly spaced are left behind as the wave propagates through the plume toward shore. These bubbles, although formed at the surface, appear to be driven toward the bottom of the tank by vorticity generated

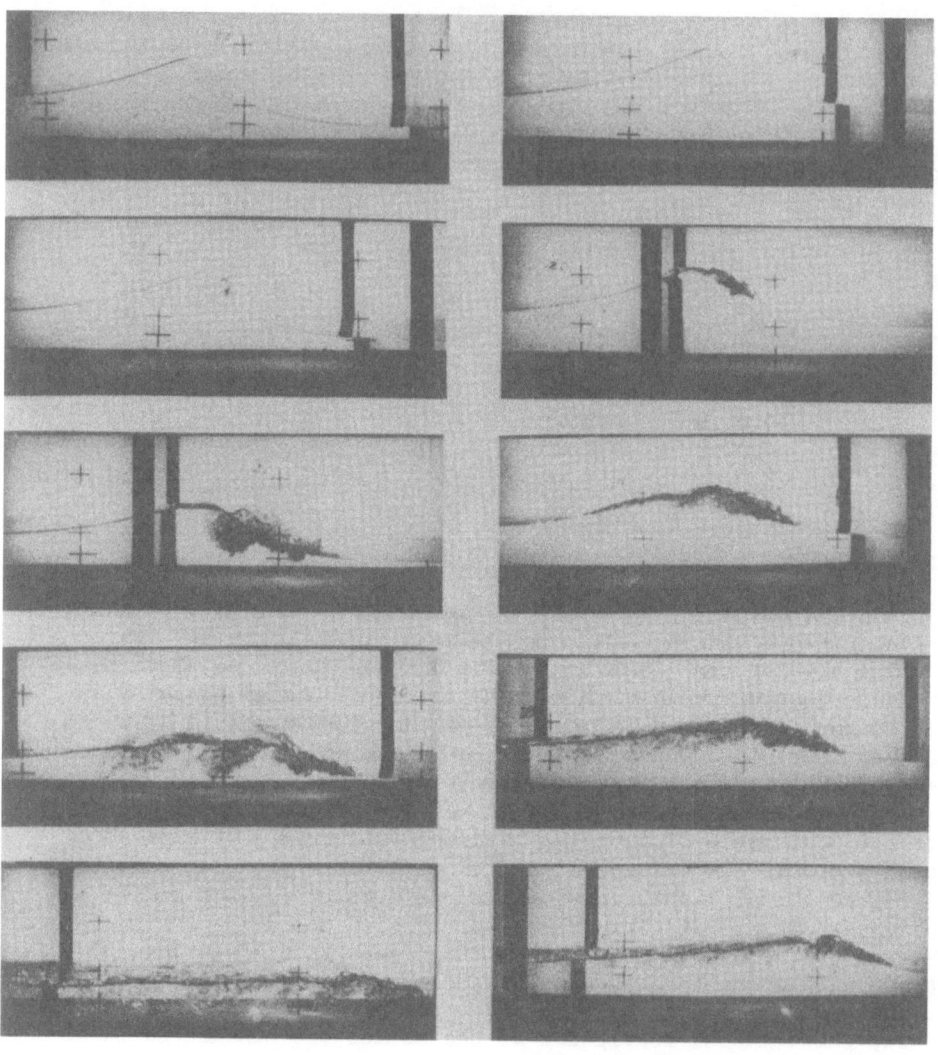

(a)   (b)

Figure 4   Sequence of Photographs Showing Breaking Process:
(a) Plunging Wave, Exp P1, S = 0.0191
(b) Spilling Wave, Exp SP2, S = 0.0061

during the breaking process. A "borelike" wave propagating shoreward is seen in the last photograph of the sequence with bubbles distributed throughout the depth which decrease in concentration by rising to the surface and bursting. It was observed that this last process took place rapidly; the vorticity mentioned could explain the delay in the bubbles rising to the surface.

The sequence denoted as Figure 4b shows the breaking process of a spilling wave. In the uppermost photograph the wave at incipient breaking also has a near vertical front face at the crest similar to the plunging wave but smaller in extent. Looking closely at the photographs the shapes of the bubble plume for the spilling and plunging waves have some similarity especially near the interface between the wave and the bubble mass. As the wave propagates shoreward, as seen in the fourth photograph, air bubbles that are left behind are more uniformly distributed through the upper part of the depth than for the plunging wave. This suggests a smaller intensity of vorticity through the depth for the spilling breaker compared to the plunging one. Finally, in the last photograph of the sequence a borelike wave can be seen propagating shoreward; one major difference between the bores generated by the two waves is the spilling wave produces a finite volume bore which is more wavelike than that resulting from a plunging breaker.

In Figure 5 water surface profiles are shown for the plunging breaker corresponding to Experiment P1 which have been traced from frames of the high speed motion picture, examples of which were presented in Figure 4a. Indicated on this figure are shaded areas which generally correspond to the bubble masses observed on the film. Indicated in each frame is the relative distance of that frame from breaking. The most striking feature is the initially small decrease in wave height for about four depths from breaking followed by a general collapse of the wave and the associated significant change in wave height. This process of wave height reduction is the reason for the two distinct decay regions observed earlier in Figure 3.

In Figure 6 the growth and collapse of the bubble plume is shown with distance from breaking; the ordinate is the ratio of the area of the bubble mass on the front face of the wave as seen on the sidewall of the wave tank divided by the square of the breaking wave height, and the abscissa is the relative distance of the observation shoreward of breaking. The bubbles seen in Figure 4 in the fourth sequence both for the plunging and spilling wave which occur on the rear face of the wave are not included in the area estimate. All seven experiments conducted, with the experimental conditions shown in Table I, are presented here. It is interesting, considering the difficulty in accurately defining the plume cross-section area as seen on the side of the tank, that all data for the seven different types of breaking waves demonstrate the same variation of growth and collapse of the bubble area with distance. It should be realized that such a view from the side of the tank does not describe fully the observed bubble mass, since the bubble concentration is unknown and perhaps appears greater than it really is due to the opacity of the cloud when viewed from the side. Initially, the bubble plume cross-section area grows at a rate somewhat

104

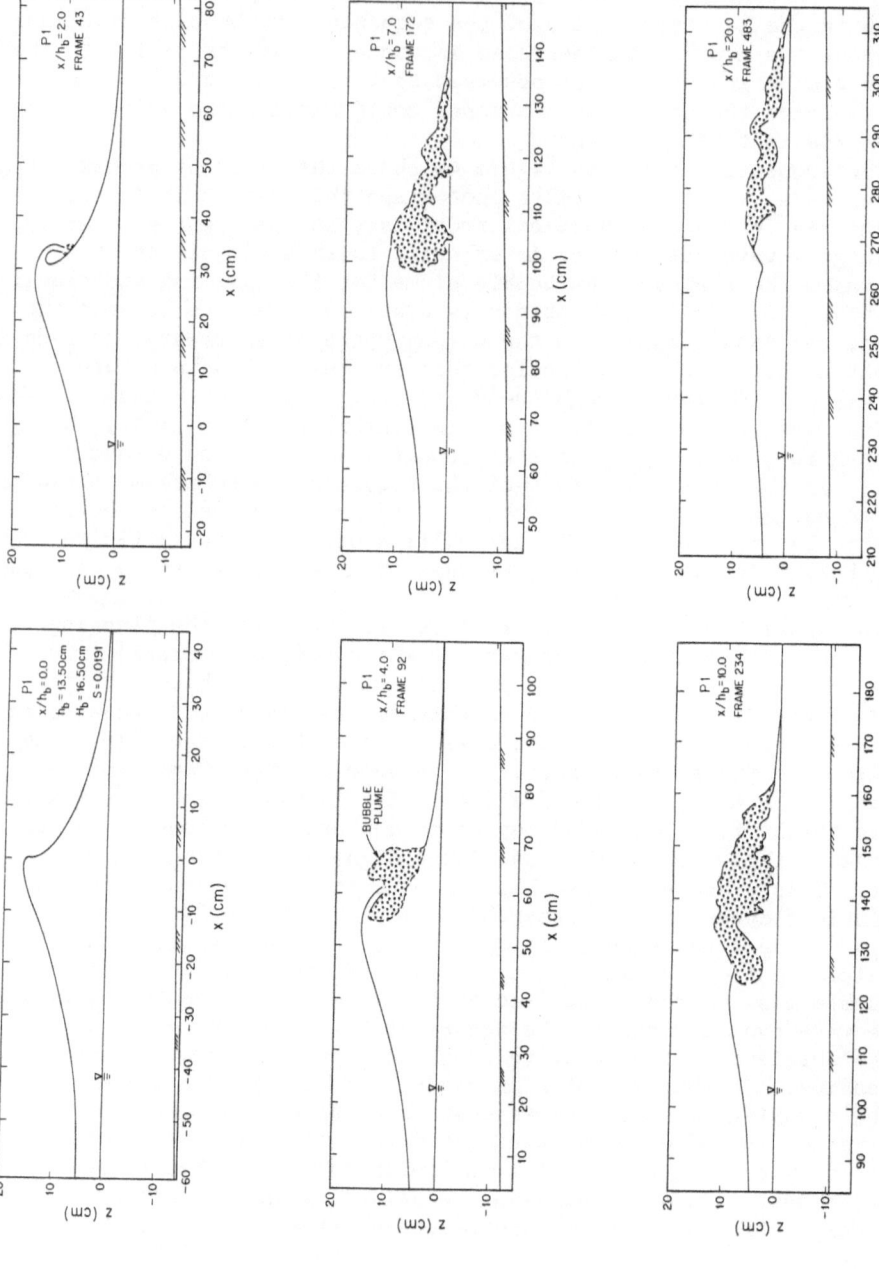

Figure 5 Water Surface Profiles at and After Breaking; Exp P1, S = 0.0191.

greater than the square of the distance from breaking and then the
growth rate decreases between about seven and ten breaking depths and
the relative area reaches a maximum. The relative area then decreases
at a rate which increases with distance from breaking. Thus, in a
sense, Figure 6 shows the growth of newly formed bubbles and then the
persistance of the bubbles on the front face of the wave as it propagates. These data imply that trends in the rate of growth of the
bubble region and its collapse are relatively independent of the type
of breaking wave. This suggests that even though the early stages of
breaking for the plunging wave appear different from those of the
spilling wave, the mechanism of bubble formation must be similar to
result in similar growth and collapse characteristics.

The same data are presented in Figure 7 where the relative bubble
cross-section area is plotted as a function of normalized time measured
from breaking. The data for the seven different experiments show the
trend of the relative bubble area growing to a maximum and then
decreasing to a value of about 0.2 at a normalized time of about twenty.

A schematic drawing is presented in Figure 8 of the arrangement
used for the measurement of noise associated with these breaking solitary waves. The hydrophone (Celesco LC-32) was mounted vertically with
the center of its acoustic field (indicated as 3.96 cm from one end of
its sensitive length) located approximately 6.4 cm from the bottom of
the tank. The measurements of noise were made for the condition where
the normalized time of the wave shoreward of breaking, τ, was equal to
four which corresponds to a distance Δx of the transducer from the wave
of between 1.4 m and 1.75 m with the former corresponding to a spilling
wave and the latter to a plunging wave (see Table II). Since these

Table II Conditions for Noise Measurement at $\tau = 4$

Exp	S (m/m)	h_b (cm)	x_h (m)	x (m)	Δx (m)	$\Delta x/h_b$
P1	0.0191	13.5	2.25	0.50	1.75	12.96
SP2	0.0061	17.06	2.00	0.60	1.40	8.21

waves are shallow water waves, to avoid the effect of the flow
velocity on the noise measured by the transducer, only those measurements where the wave did not influence the hydrophone were considered.
Because the wave itself acts as a traveling noise source a slight
Doppler shift in frequency is introduced. This is essentially negligible since the velocity of the sound source (the wave) is only about
1 m per sec. More serious is the fact that the sound radiated by the
wave will reverberate underwater in the region which is bounded at one
end by the bulkhead wave generator, on the top by the free surface, and
on the bottom by the sloping bottom of the tank. Many acoustical modes
of oscillation are induced by the sound from the traveling wave in this
triangular shaped region.

The normalized noise spectrum for the unbroken wave which became
a spilling breaking wave is presented in Figure 9a and the spectrum for
the unbroken wave which became a plunging breaking wave is presented in

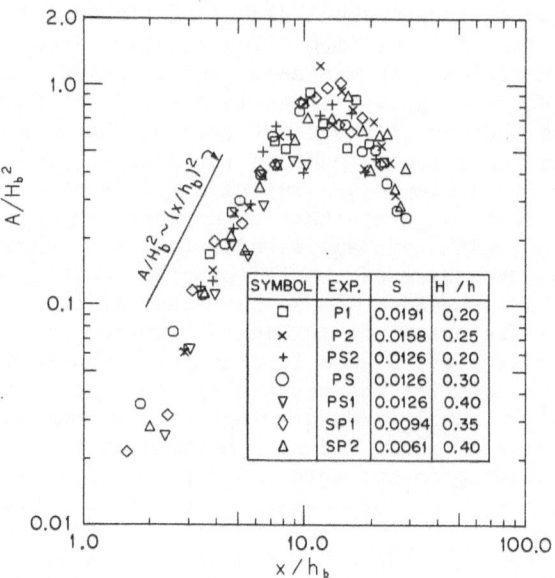

Figure 6   Relative Bubble Plume Area as a Function of Distance
from Breaking.

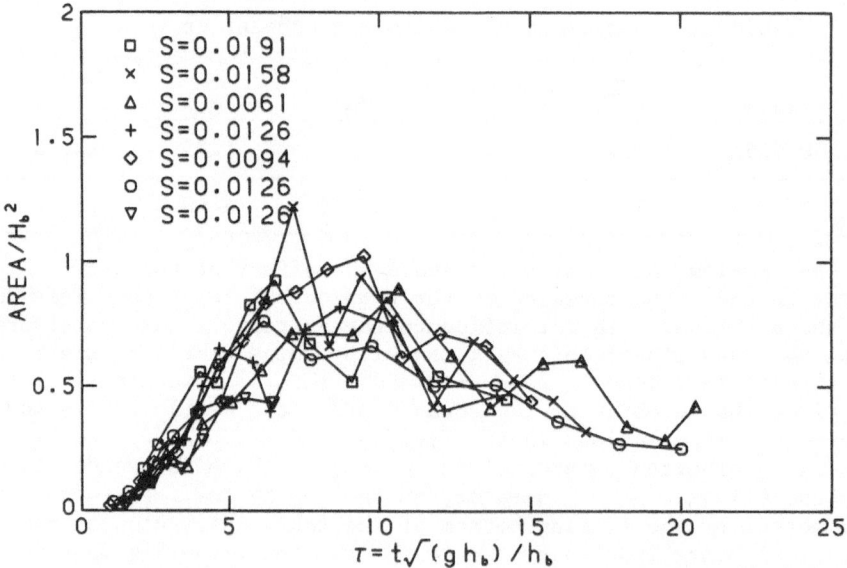

Figure 7   Relative Bubble Plume Area as a Function of Time After
Breaking.

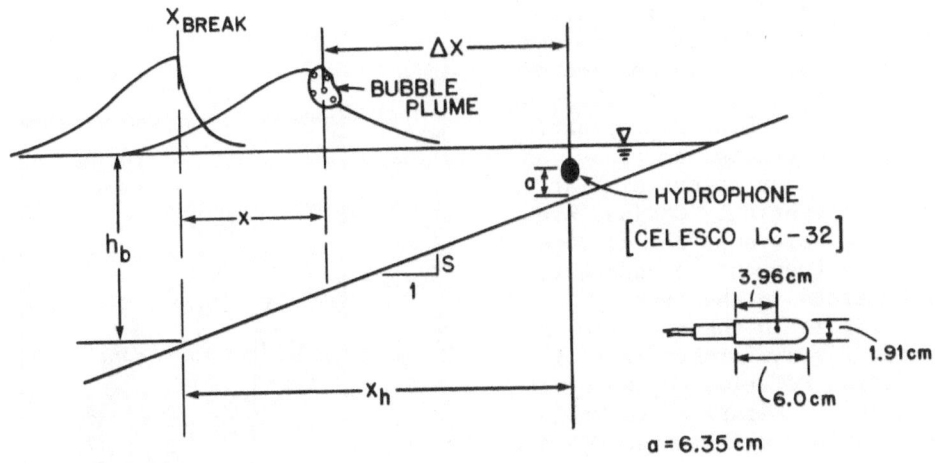

Figure 8 Experimental Arrangement for Noise Measurements (see
Table II)

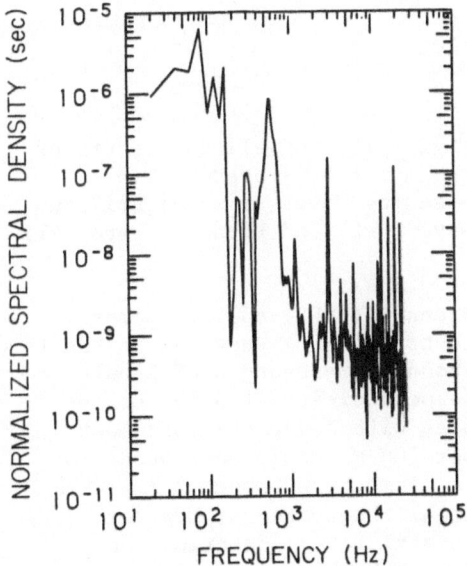

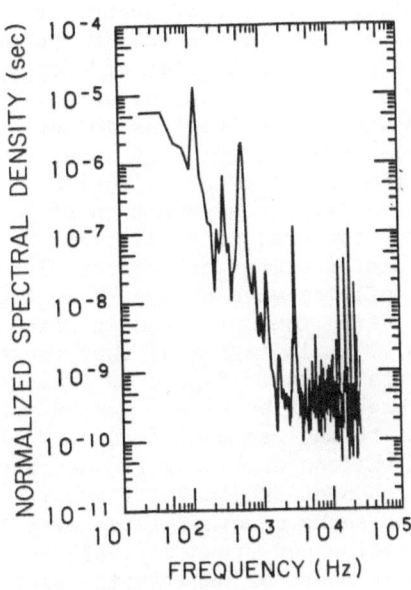

Figure 9a Normalized Spectrum of
Wave Before Breaking; Spilling
Breaker (SP2).

Figure 9b Normalized Spectrum
of Wave Before Breaking;
Plunging Breaker (P1).

108

Figure 9b. The spectral estimates have been normalized by the areas under the respective spectrum, and these are plotted as a function of frequency. Realizing that these are the spectra of unbroken propagating waves, when one compares these, on the average, they appear similar in energy distribution. Obvious are various periodic spectral components which may correspond to underwater reverberations in the tank.

In Figure 10 the spectrum is presented for the broken wave with its location relative to the hydrophone shown in Figure 8 and Table II. Spectra for the spilling wave and the plunging wave are presented together, and there are several important features which are apparent in comparison. First, there appears to be a concentration of energy for both the spilling wave and the plunging wave at a frequency of about 5 kHz. Each spectrum appears to have a concentration of energy near 500 Hz which may be associated with reverberations in the tank. At a frequency of about 200 Hz, the plunging wave spectrum shows a large concentration of energy compared to a relative absence of energy associated with the spilling wave at that frequency.

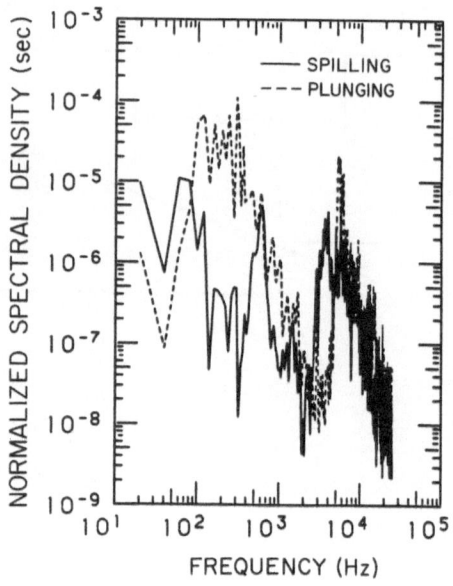

Figure 10 Normalized Spectra of the Noise Generated by Two Breaking Waves at $\tau = 4$; Spilling Wave (SP2) and Plunging Wave (P1).

Using the work of d'Agostino and Brennen (1983) and Prosperetti (1987), it is suggested that these concentrations of energies in the spectra shown in Figure 10 could correspond to resonances of bubbles and clouds of bubbles. For example, the periodicity at 5 kHz corresponds to the resonant frequency of bubbles with a diameter of about 2 mm whereas the lower frequency of about 500 Hz would correspond to the resonant frequency of bubbles with a diameter of about 2 cm. The latter seems somewhat large from the visual observations; however, d'Agostino and Brennen (1983) and Prosperetti (1987) describe the lowered modes of oscillation associated with bubble clouds due to the interaction of large numbers of smaller bubbles. Thus, the lower frequency could indeed be due to the bubble cloud which is associated with the mass of bubbles apparent, for example, in the third photograph in the sequence presented in Figure 4a.

4.   CONCLUSIONS

The following major conclusions are drawn from this study of shallow
water solitary waves shoaling on relatively gentle slopes:

(1)   When viewing the breaking process for solitary waves,
in terms of the distance from breaking, two distinct
regions of wave decay appear to exist.

(2)   The observed bubble plume on the front face generated
during breaking grows and collapses and a maximum
cross-section area is reached somewhere near where
the rate of decay of the wave height increases.

(3)   With regard to the noise generated after breaking,
it appears that higher frequency noise is similar
both for plunging and breaking waves where the
frequencies correspond to those which one would
expect from the resonance of bubbles which are
approximately 2 mm in diameter, and a low frequency
peak of energy in the spectrum for the plunging
breaker may correspond to the resonant oscillations
of the bubble cloud produced after breaking.

5.   ACKNOWLEDGMENTS

This research was supported by the National Science Foundation under
Grant MSM-8311374. Photographic equipment was obtained using funds
supplied by the Miriam and Omar J. Lillevang fund at Caltech.   The
assistance of John Lee in obtaining the noise measurements reported
herein is greatly appreciated.

6.   REFERENCES

d'Agostino, L. and Brennen, C.E., 'On the Acoustical Dynamics of
Bubble Clouds,' ASME Cavitation and Multiphase Flow Forum, 1983.

Dally, W.R., Dean, R.G., and Dalrymple, R.A., 'Wave Height Variation
Across Beaches of Arbitrary Profile,' Jour. Geophys. Res., 90(C6),
1985.

Kerman, B.R., 'Underwater Sound Generation by Breaking Wind Waves,'
Jour. Acoustic Soc. Am., 75(1), Jan. 1984.

Nadaoka, K., 'A Fundamental Study on Shoaling and Velocity Field
Structure of Water Waves in the Nearshore Zone,' Tech. Report No.
36, Dept. of Civil Engr., Tokyo Inst. of Tech., 1986.

Prosperetti, A., 'Bubble Dynamics in Oceanic Ambient Noise,' Proc. NATO
Advanced Research Workshop, Lerici, Italy, June 1987.

Skjelbreia, J.E., 'Observations of Breaking Waves on Beaches by Use of
Laser-Doppler Velocimetry,' Ph.D. thesis, Calif. Inst. of Tech.,
1987.

Svendsen, I.A., 'Wave Heights and Set-up in a Surf Zone," Coastal
Engr. 8, 1984.

LDA AS A NEW TOOL TO DETECT AIR-SEA INTERACTION MECHANISMS

Ch. Werner and W.A. Krichbaumer
DFVLR- Institute for Optoelectronics
D-8031 Wessling
Fed. Rep. of Germany

ABSTRACT. The laser Doppler anemometer (LDA) is an active measuring system for indirect determination of wind characteristics. It can be used to measure wind profiles within a short time. In November 1984, the system was flown to a research platform in the North Sea to investigate the possibilities of the LDA in the field of air-sea interaction. It was shown that the LDA is capable of simultaneous detection of the interaction between wind and water.

1. INTRODUCTION

Why is a large, complicated remote sensing system like the laser Doppler anemometer (LDA) flown to a research platform in the North Sea for measuring the wind profile? On one hand, the wind profile over sea surfaces has already been measured quite a few times and on the other hand, there are standard wind measuring instruments which measure a 24-hour profile from buoys or platforms.

Some of the decisive reasons for this experiment were:
- it was intended that the LDA, as a non-contacting remote sensing system, should supply a wind profile which is unaffected by the measuring instrument itself,
- the signatures were to be compared for future global measurement of the wind above oceans by means of microwave scatterometers in satellites,
- it was intended to determine the influence of the platform on the wind field,
- and, finally, the "logarithmic wind profile" [1] is only a working hypothesis and there is room for advances in the knowledge regarding the interaction between wind and ocean. According to Kinsman [1], "... a marked advance in the study of oceanic turbulence and turbulent diffusion waits only on a sufficiently sensitive velocity meter that will operate realiably at sea."

B. R. Kerman (ed.), Sea Surface Sound, 111–121.
© *1988 by Kluwer Academic Publishers.*

2. MEASURING PRINCIPLE AND EXPERITMENTAL SET-UP

2.1. Laser Doppler wind measurements

The laser Doppler anemometer is an active measuring system for indirect determination of the wind characteristics [2]. The radiation source used is a CO_2 continous-wave laser, the output power of which is focussed through a telescope on to the volume to be measured at distance R (Figure 1). A part of the radiation is backscattered by the aerosol particles moving with the wind through the volume to be measured. The scattered radiation is Doppler shifted with respect to the incident radiation:

(1) $\Delta f_0 = 2 \, (V_{LOS} \, / \, c) \, x \, f_0$

where f_0 = frequency of the CO_2 laser (approx. 28.3×10^{12} Hz at a
            wavelength of 10.6 μm)

    c = velocity of light

    V_{LOS} = component of the wind vector in the direction of measuring

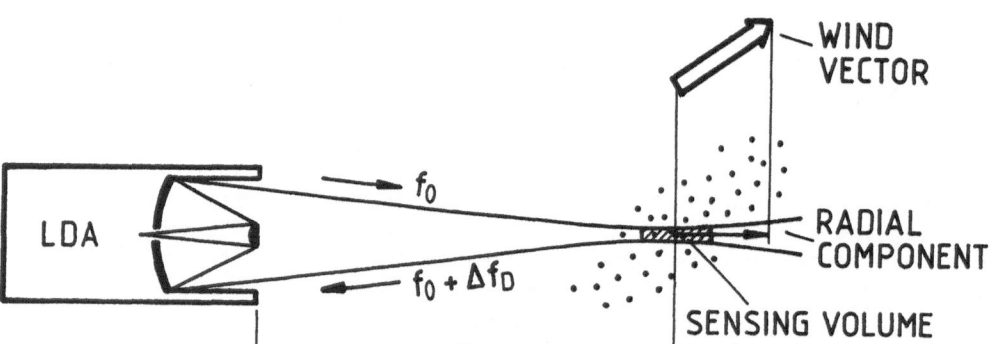

Figure 1. Measuring principle of the Laser Doppler anemometer.

A wind velocity V_{LOS} of, for example, 1 m/s causes a frequency shift of 189 kHz. The backscattered and frequency-shifted radiation is col-lected again by the telescope.

The system uses a 30 cm transceiver aperture diameter and a two-mirror scanning system. For the range scan the doublet lens is moved. The range resolution is ± 6 m at 100 m focussed range and ± 15 m at 175 m focussed range. The resolution decreases with range. The length of the volume is measured at half-width value of the sensitivity curve.
Figure 2 shows a typical spectrum of the spectrum analyser. The scales are signal voltage versus line-of-sight velocity. Two corresponding scales show the digitized values. Specifically, the peak positions in the spectrum correspond to the magnitude of the radial wind components in and outside the sensing volume. Fog and cloud components can have their origin from outside the sensing volume [3, 4].

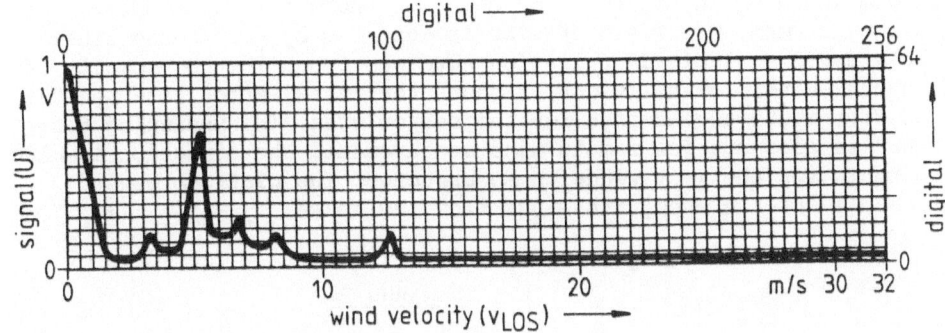

Figure 2. Doppler spectrum in a digital frame.

To reduce the amount of data to be transfered, only the peak values of each spectrum are used. The amplitude of the wind peaks depends on the amount of scatterers in relation to the focussed range. Using the LDA as a homodyne system, one cannot determine the sign (plus or minus) of the radial velocity. The first peak in figure 2 is the strong 0-frequency spectral peak.

By scanning the azimuth angle θ step by step from 0° to 360°, a sinusoidal representation of the radial components is obtained:
(1) $V_{LOS} = | u \cdot r \cdot \sin \theta + v \cdot r \cdot \cos \theta + w \cdot \sin \phi |$

u, v, w   wind components in the x, y, and z direction
u   is the east-west wind component
v   is the north-south wind component
w   is the vertical component
θ   is the scan angle clockwise from north
ϕ   is the angle of elevation
r   radius of the cone.

Since the field of view of an LDA system is frequently restricted for a full VAD scan by buildings, trees and other obstacles, only a certain sector can be used for sine-wave fitting. Investigations into how far this sector can be restricted without losses in accuracy have shown that an optimum is obtained between 45° and 90° [4, 5]. The sine-wave fitting is carried out for a predetermined sector by a large computer.

2.2. Preparation for using the LDA experiment on the "Nordsee" research
     platform (FPN)

The aim of the test experiment was the remote sensing of the wind profile and the detection of the interaction between wind and waves without interfering influence. The possible applications of the LDA system had to be co-ordinated with the situation prevailing at the "Nordsee" research platform (FPN ⊨Forschungsplattform "Nordsee"). The FPN research platform is located at 7° 10' E and 54° 42' N. This is approximately 70 nautical miles north west of Helgoland on the latitude of the southern tip of the island of Sylt. The top view of the working platform is shown in Figure 3. On one side there is a helicopter landing platform. The field of view is obscured by cranes and other structures. A suitable installation site was found in the vicinity of the north-western crane. Figure 3 shows the installation site and the range of azimuth angles free for the sector scan. This scanning range of 106° is adequate for remote wind measurement. The direction is also favourable for the west winds to be expected. The platform caused noticeable interference with winds from the south.

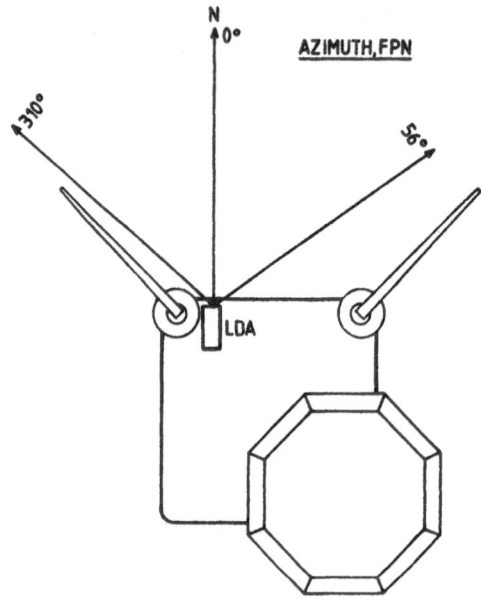

Figure 3. Installation site of the LDA container on FPN with limiting azimuth angles.

3.   RESULTS

The experiment was carried out during November 1984 [6].

3.1. Influence of the platform

If the wind blows from south, the platform influence can be seen at the leeward side with a scan around north.

The wind shadow behind an object against which the wind is blowing is called the wake. The velocities in the wind shadow are lower than in the outer flow. With increasing distance from the platform, the wind shadow zone becomes wider and the differences in velocity with respect to the outer flow become smaller. This is according to Schlichting's theory [7].

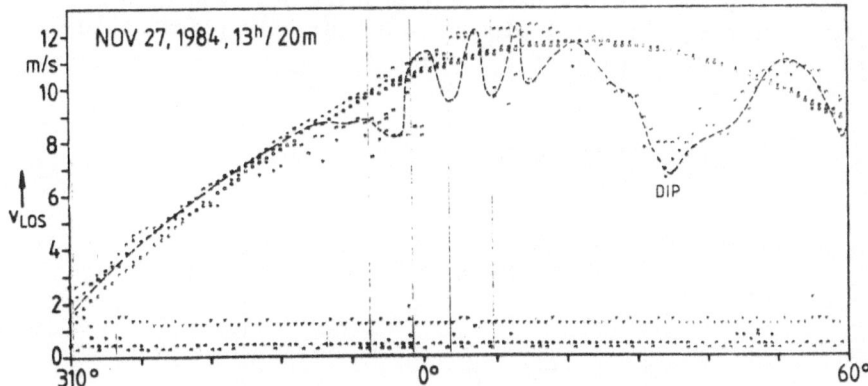

Figure 4. Computer printout of the section from the VAD sector scan .
          Date: Nov 27th 1984, 13.00 h CET, height 20 m.

Figure 4 shows a section of data recorded on tape. The azimuth angle
varied from 310° to 60° at a range of 175 m, 20 m above sea surface.
Each vertical line corresponds to one spectrum (see figure 2).

The points represent peak signals. A sine fitting is included. There
are deviations from the sine-wave fitting between 20° and 45° with a
maximum "dip" at approximately 35°. On that day the wind came from 219°
(219° - 180° = 39°).

From the available results it can be said that the influence is still
noticeable even at a range of 250 m and can be qualitatively described
with simplifying assumptions. At a height of 50 m and above, no further
influence was measurable. The influence of the platform on the remote
wind measurement is obvious and can be by-passed by restricting oneself
to appropriate measuring directions. The angular range to be left open
is ± 20° around the main wind direction. Even reducing the range of
analysis to 50° is adequate for the calculation.

3.2. Wind Profile

The following heights to match the known logarithmic profiles [1] were
selected: 0, 2, 5, 10, 20, 50, 100 m above sea-level at standard sea-
level.

According to [1], the mean velocity profile U(z) with a logarthmic
height variation is:
(2) $U = U_0 + U_1 \ln (z/z_0)$

where z_0 is the length of roughness
      U_0 is the lower limit for the logarithmic profile, and
      U_1 is the reference velocity.

Figure 5 shows one representative example (out of about 150 profiles)
for the unstable temperature structure [6].

116

The measurement shows almost constant wind conditions in the lower
100 m above the sea. The hypothetical profile has been determined in
accordance with equation (2) with z_0 = 0.0001 (fit point at 10 m
height).

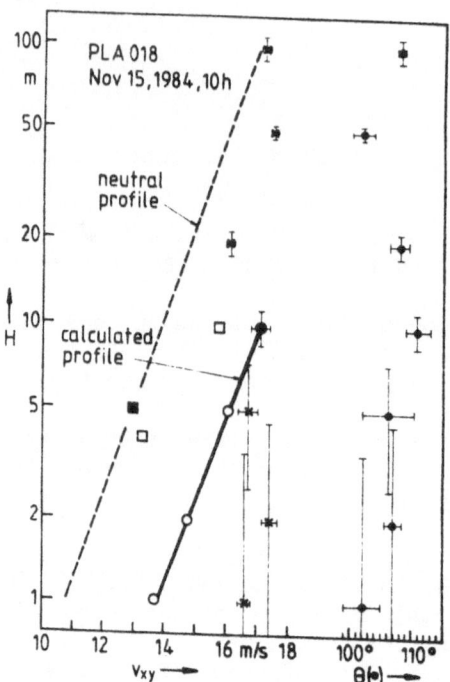

Figure 5. Vertical wind profile of the horizontal wind velocity v_{xy} and
wind direction θ. A neutral profile (Smith [8]) is drawn
starting at 100 m, another profile is included starting at
10 m with z_0 = 0.0001. The squares at 8.4 and 3 meters are
20-minutes-average values.

To explain the deviation from the expected logarithmic wind profile in
the lowest 5 meters, the influence of the water surface was studied.

3.3. Water influence

Figure 6 shows the measuring geometry at a focused range of 175 m.

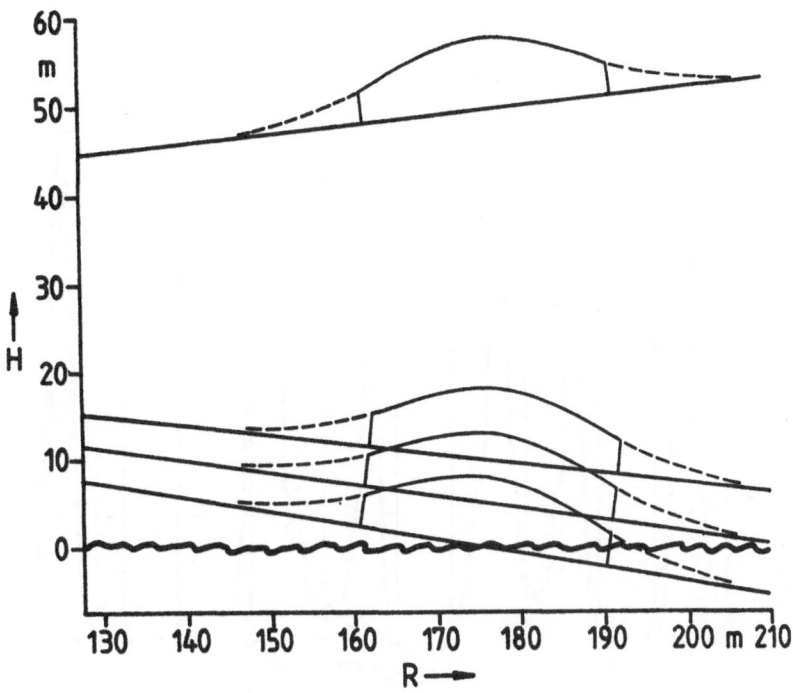

Figure 6. Measuring geometry at 175 m focused range, altitudes 0 m,
5 m, 10 m, and 50 m.

For each altitude, the line of sight (LOS) is drawn together with the
sensitivity curve of the LDA system for this range. The length of the
volume measured at the half-width value of the snsitivity curve is 30 m.
At 0 m altitude one half of this volume is in the water and the other
out of it. At 2 m, parts of the measured volume are still in the water,
and at 5 m only the fringes are still in the water. At 10 m height, the
water could supply a contribution only if its reflection is much
stronger than that of the aerosol particles. This is a similar case as
is known from the determination of the wind profile in the presence of
clouds [3] .
Therefore, what we expect to measure is an increasing contribution from
the water surface to the signal with decreasing altitude. As an example
consider the following measurements from Nov. 26th, 1984. For each
frequency or wind channel the signals were continuously recorded (for
about 3 minutes). After summation these continuous recordings give the
probabilities of occurrence for each wind velocity (Figure 7).

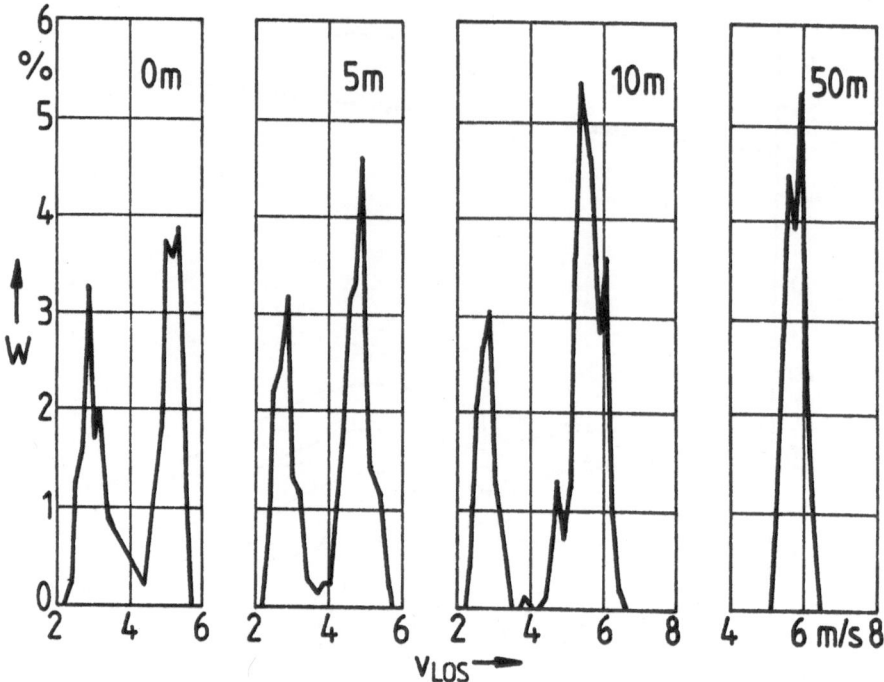

Figure 7. Probability of the occurrence of wind components per
measuring channel versus wind velocity for 4 altitudes.

At low wind velocities, the wind signal is clearly separated from the
water signal. At 0 m, 5 m, and 10 m altitude the water signal is recorded simultaneously with the wind signal. As follows from fig. 6,
in these cases the laser aims in the direction of the water. This is
not so at 50 m altitude, where no water signal can be measured. The
decrease of the water signal together with an increase of the wind
signal is a clear proof of the operation of the LDA. At larger wind
velocities the separation of the two peaks disappears even at sea
level, because the strong perturbation of the sea increases the number
of different velocities seen by the LDA.

3.4. Breaking waves and the explanation of the water influence

Figure 8 shows a section from a computer printout of the measured velocity V_{LOS} during a continuous observation of the sea surface .

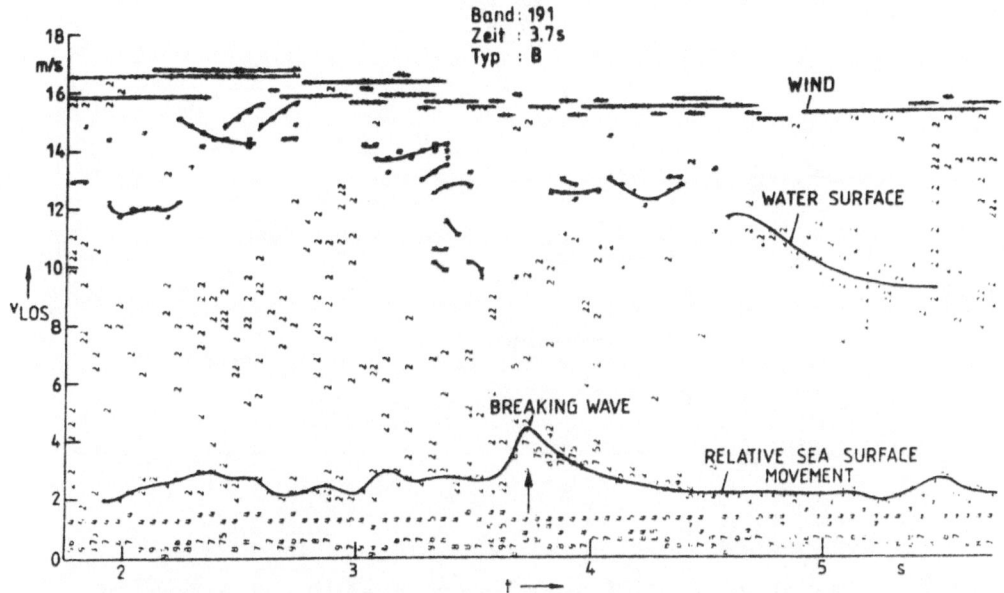

Figure 8. Computer printout of a continuous recording at a range of
175 m and water surface height (against the wind).

First, this section shows hydrosols from the breaking waves. These
particles, having various sizes (and velocities), are driven through
the measuring volume and provide a significant signal structure. This
is repeated at intervals corresponding to the spacing between breakers.
In addition, the wind signal is split into two components which indicates that a contribution is supplied both by the wind and by the water
surface velocity. But the water velocity is much larger than expected.
From the data of the platform, the wave height was between 2 and 4
meters for wind velocities larger than 10 m/s. The wave period was 5
to 7 seconds.

There are four kinds of velocity:
1. wind velocity
2. water surface velocity
3. surface movement towards the laser
4. hydrosols from the breaking events.
There should be a difference in the velocities 3 and 4 by looking in
both directions.Figure 9 is the comparison to figure 8.

120

Band 68
Zeit 931s
Typ A

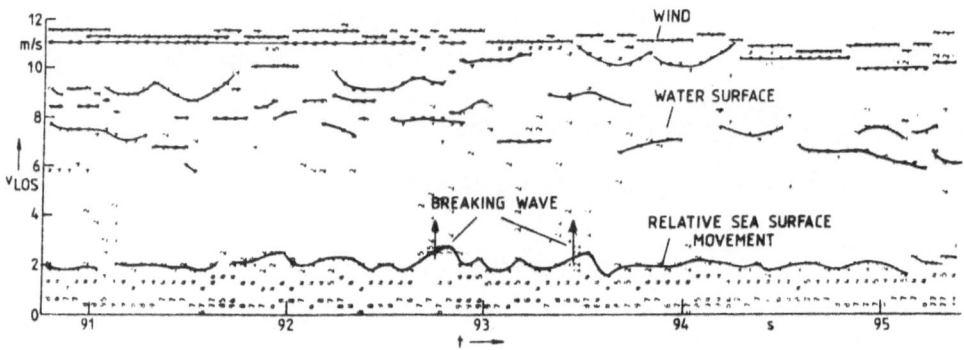

Figure 9.   Computer printout of a continuous recording at a range of
            175 m and water surface height (with the wind).

4.   Summary

The preparation and performance of the laser Doppler wind measurements
over sea surfaces has shown that the LDA system can be used as a modern
remote wind sensing system for questions involving the interaction
between atmosphere and ocean.

It permits wind profiles to be measured within a short time. The wind
profiles do not show any logarithmic relationship to height. Both the
profile measurements and the "continuous" recordings display almost
constant wind velocities with height.

The influence of the water on the measurements at 0 m height is obvious.
The platform and its wake dip also cause interference during measurements in the wind direction.

5.   References

[1]   Kinsman B., "Wind waves", New York: Dover Publications (1984).
[2]   Köpp F., Herrmann H., Werner Ch., Schwiesow R.L., Bachstein F.,
      "Erstellung und Erprobung des Laser- Doppler-Anemometers zur
      Fernmessung des Windes. (Creating and testing the laser Doppler
      anemometer for remote sensing of the Wind)", DFVLR-FB 83-11 (1983).

[3] Werner Ch., Köpp F., Schwiesow R.L., "Influence of clouds and fog on LDA wind measurements", Appl. Opt. 23, 2482-2484 (1984).

[4] Werner Ch., "Fast sector scan and pattern recognition for a cw laser Doppler anemometer", Appl. Opt. 24, 3557-3564 (1985).

[5] Schwiesow R.L., Köpp F., Werner Ch., "Comparison of lidar-measured wind values obtained by full conical scan, conical sector scan and two point techniques", J. Atmosph. Oceanic Techn. 2, 3 (1985).

[6] Werner Ch., Köpp F., Biselli E., Biselli E., "Laser-Doppler-Windprofil über See", DFVLR FB 85-48 (1985) or ESA-TT 971(1986).

[7] Schlichting H., "Grenzschichttheorie (Boundary layer theory)", Karlsruhe: G. Braun (1985).

[8] Smith S., "Private Communic. Report", Bedford Inst. of Oceanography, BI-R 81-3 (1981).

FEASIBILITY STUDY OF AT-SEA MEASUREMENT
OF OCEAN-AIR INTERFACE PARAMETERS NEEDED TO EVALUATE
OCEAN SURFACE GENERATED NOISE

Dr. James H. Wilson
Wilson's Arctic Research Inc.
16333 Keeler Drive
Granada Hills, CA   91344
U.S.A.

ABSTRACT. The feasibility to collect at-sea measured acoustic and
supporting environmental data to both describe the ocean surface
source level density function theoretically and to model ambient noise
accurately is demonstrated in this paper. An exercise involving high
sea state capable data collection buoys is presented so that the risk
of data retrieval and exercise cost are minimized.

1.   BACKGROUND

Theoretical investigations of ocean surface generated noise
measurements are at a critical stage. Experimental measurements of
basic ocean-air interface parameters are badly needed to provide
surface source level density levels needed to model ocean surface-
generated noise and to evaluate the importance of certain natural
mechanisms of ocean surface noise. Since it is difficult to simulate
a state six sea with large fetch in the laboratory, and, since at-sea
measurement capabilities have improved significantly in recent times
by using satellite linked drifting buoys capable of measuring wave
spectra and other environmental parameters, the time for planning an
at-sea measurement effort may have arrived.
   A primary objective of performing an at-sea measurement of
simultaneous ambient noise levels and supporting environmental sea-air
interface parameters is to determine the source level density of ocean
surface noise as a function of frequency, local wind speed, storm
fetch and duration, and source vertical directionality. A careful
measurement of source level density is a required input into an
ambient noise model of ocean surface generated noise. An ambient
noise model must include a reliable, range dependent propagation loss
model capable of predicting propagation for surface sources in
changing sound speed, ocean bottom and surface roughness environments.
Propagation models such as the Parabolic Equation (PE) or ASTRAL seem
fully capable of predicting propagation for this type of ambient noise
model.
   Once the source level density function is determined more

B. R. Kerman (ed.), Sea Surface Sound, 123–129.
© 1988 by Kluwer Academic Publishers.

precisely, the dominant physical mechanism(s) of ocean surface generated noise can be analyzed from the measurements. For a given frequency interval, mechanisms that produce the correct spectral shape, wind speed dependence, source directivity and absolute level for the measured source level density can be evaluated. The ocean surface source level density can be deduced from the ambient noise measurements, the ocean surface environmental measurements, and the propagation loss model as others have done in the past. These measurements are vital in determining which physical mechanism is dominant in a given frequency interval.

Current theories in this area are in need of at-sea environmental measurements so that accurate calculations can be made. Turbulence just above and below the ocean surface, the process of ocean wave generation, growth and breaking and the vertical profile of wind speed within 10 meters of the ocean surface are a few examples of near surface processes that are important in surface generated noise theories. It is currently feasible to measure some of the parameters that characterize these processes using a drifting buoy system recently developed and successfully tested by the National Data Buoy Center (NDBC). During long term sea tests, both surface and subsurface environmental data were transmitted via satellite to a ground station. Design of a subsystem for measuring low frequency ambient noise has also been completed and a prototype system is partially finished. Recent advances in data compaction using the Fourier Series Method (FSM) and other preprocessing approaches make it possible to send a significant amount of data via satellites with limited data throughput capability. Naturally, one obvious advantage of this measurement technique is that the risk of not recovering the data is eliminated. A tape recording system can also be added so that in cases where buoys can be recovered the data time series can be compared to the data received earlier via satellite.

The specific measurement capabilities of this type of system are addressed in this paper in detail. One version of this buoy is called the Air Sea Interaction Drifter (ASID) and it has been successfully deployed in the North Atlantic and North Pacific Ocean and other high sea state areas. ASID has successfully transmitted data before, during and after a hurricane and during prolonged sea state six conditions (A wave spectrum capability has now been added - this version is called WASID). The life of the buoy batteries is about 450 days with the current data measurements made (air and sea temperature, barometric pressure, wind speed and wave data). It is hoped that a list of critical environmental/acoustic measurements needed to understand ocean surface noise mechanisms can be developed at this workshop so that their at-sea measurement feasibility can be assessed with the improved measurement capabilities available.

For the sake of discussion, an experiment is outlined in this paper that would meet the measurement objectives discussed above. It is proposed that 10 to 15 WASID buoys be distributed over the Northeastern Pacific Ocean, as shown in Figure 1, starting in the fall of the year. Actual buoy placement will depend on normal storm patterns and predicted acoustic propagation during a one year period.

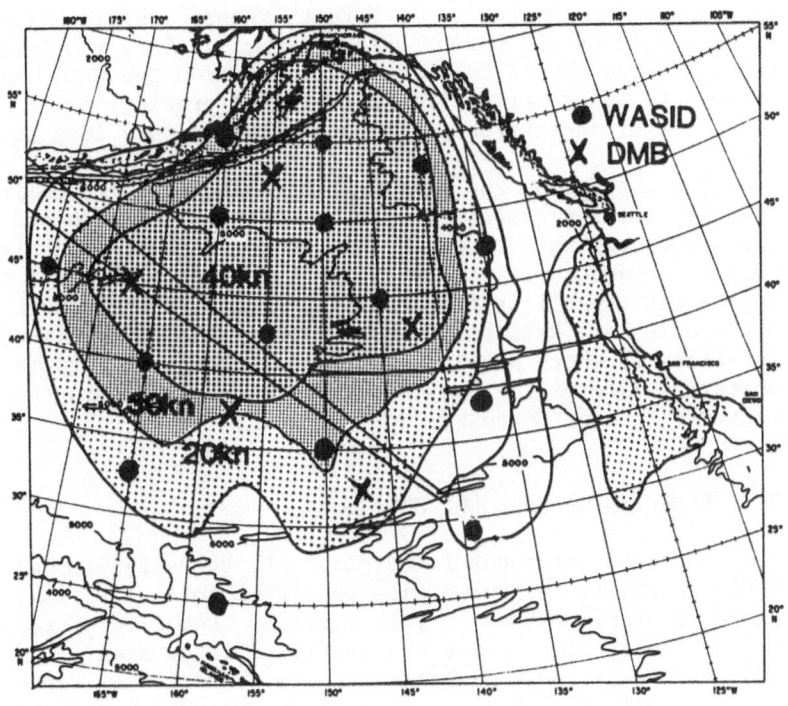

FIGURE 1. PROPOSED EXERCISE SCENARIO

Each buoy will have a vertical array (whose element spacing is not yet specified) capable of omnidirectional measurements from 5Hz to 10kHz and capable of beamforming 3 sub-frequency intervals. Each WASID buoy will be equipped with an environmental measurement subsystem capable of measuring the following parameters:

- o Wind speed vertical profile (3 different heights)
- o Low frequency wave spectra
- o Barometric pressure
- o Air and sea temperature
- o Subsurface bubble density*
- o Ocean Turbulence*
- o High frequency wave spectra*

* System currently not capable of measuring
  this parameter.

In addition to the WASID system, five or more deep moored buoys with bottom mounted sensors could be deployed to collect acoustic (and environmental data) near the bottom. These buoys will be placed in areas where the acoustic properties of the bottom are poor and, if possible, where the effects of distant noise sources – such as shipping – are shielded. In effect, these acoustic measurements will be dominated by the ocean surface directly above the measurement site. Based on wind speed/sea state estimates from the WASID buoys and state of the art atmospheric circulation model predictions, ocean surface source level density data will be collected on a long term statistical basis.

The motivation for this type of exercise is to collect a large quantity of high quality data at relatively low risk and expense. Since oceanographic research ships are extremely expensive, they will only be needed to initially deploy the buoys. In fact, an air deployment capability from rear door cargo type aircraft for WASID is now available. Reliable transmission of the data collected via satellite to shore stations provides recovery of data for the full operating life of the buoy even if the buoy is lost precluding data recovery from an on board tape recorder.

2. EXERCISE MEASUREMENT ASSETS

The most important features of the buoys (WASID and DMB) and the ARGOS satellite data transmission system for a low risk, at sea exercise are discussed in this section. Detailed system descriptions of the measurement assets are specified elsewhere.[1,2,3]. The section addresses only those measurement/data transmission capabilities that are currently feasible or can be added to existing systems with some additional development effort.

A concept of the exercise proposed here is to collect data over a long (90 day to 1 year) time period and transmit the data to shore via satellite. This method avoids both the risk of loss of data if the buoy is lost and the large expense inherent to using oceanographic research ships. However, the limited satellite throughput of ARGOS limits the data that can be retrieved.

Each WASID or DMB data buoy can be assigned 1 to 3 ARGOS identification (ID) numbers. For each ID number, 256 bits of data can be transmitted per data transmission and multiple transmissions are received during each satellite pass. There are six ARGOS passes per day at the equator and 28 passes per day at the Poles. The buoys currently transmit "in the blind" with approximately 60 second intervals between transmissions. Barometric pressure is currently transmitted with 10 bit precision with all other information transmitted with 8 bit precision. Buoy latitude/longitude information comes "free" because ARGOS determines this by analysing the transmission doppler information.

Detailed data throughput analyses will not be made in this paper. However, in summary it can be stated that it is feasible to transmit all the environmental parameters listed in the last section at a sampling period on the order of 1 to 3 hours, as well as transmitting the acoustic data from a small (6 to 18 element) vertical array. A data compaction technique[4] developed by the author makes it possible to transmit the covariance matrices at several frequencies. In this manner, both single phone data vs. depth and beam data using conventional beamforming (or any type of high resolution processing) will be available to the user analyzing the data on shore.

A nested "3-sub array" sparse filled vertical array is proposed for a preliminary array design. The 3-sub arrays would be spaced so that the 3 array design frequencies would be 10Hz, 50Hz and 500Hz.

The important points concerning the surface buoy, WASID, and the deep moored buoy, DMB, is that they have passed the time of time at sea during high (SS6 and hurricanes) sea state conditions. In addition, there is a version of DMB which drifts on the surface which has received extensive development attention. Thus, the cost to develop a platform to withstand high sea states <u>and</u> collect and transmit high quality environmental and acoustic data should be relatively low. A schematic of the current DMB system concept and WASID can be found in References 2 and 3.

3. EXERCISE TECHNICAL OBJECTIVES

The objective of this exercise is to provide a measured data base to help determine the source level density function for surface generated noise in frequency bands from 5Hz to 10KHz. Recently there have been several[5,6,7] attempts to empirically determine the surface-generated noise source level density function by use of models and measured data. Kewley and others[7] have recently refined Wilson's estimate of source level density function by using the so called DUNES model to account for all the propagation paths involved. These estimates provide a very important input into ambient noise models, but some very fundamental questions remain unanswered.

- o Does the source level density depend directly on only surface wind speed or are there other factors, such as subsurface bubble density, or sea turbulence, or storm fetch, that also impact the source level density function?
- o What are the underlying physical mechanism(s) that dominate surface generated noise from 5Hz to 10KHz?
- o What is the source directivity pattern and how close does the "rocking dipole" model it?
- o What is the relationship between the physical processes that cause waves to grow and become a fully developed sea?

There are many more questions that could be added to this list, but the need for measured data to test many of the theories of surface generated noise is clear. If two physical mechanisms of surface generated noise result in the <u>same</u> source level density function, then there is no need to study these theories separately. It is not the academic beauty of a particular theory that is important. It is the fact that the current theories predict different characteristics for the source level density function that is so important. Unfortunately, the theories all suffer from a lack of high quality, at-sea data on which to calculate the source level density function. Many of the theories use data from wind tunnels, laboratory experiments and even data over land.

There are two basic questions that guide this experimental design. First, what is the <u>right</u> set of data to measure in order to evaluate the theories of physical mechanisms? Second, is it feasible to measure these data with current technology and the measurement/data transmission assets available using the exercise concept proposed here? A list of environmental parameters that can now be measured or could easily be measured to support interpretation of vertical array acoustic data was given in Section 1.0. It is hoped that the participants of the workshop will refine and/or expand this list as critical factors are evaluated for each theory. With this list completed, the author can then address the feasibility of collecting this data.

4. AN EXERCISE SCENARIO

For purposes of discussion, an exercise scenario is discussed for the Northeastern Pacific Ocean. Figure 1 shows a typical winter storm in the Pacific and several initial buoy locations are shown. Each of the WASID buoys will provide a sample every hour of the environmental parameters listed in Section 1.0, as well as simultaneous vertical array data. With these points of "ground truth", large scale atmospheric models can be used to determine the surface distribution of most of these environmental parameters. This distribution of WASID buoys will provide a sampling of the sea surface environment within the storm fetch (north and south of the oceanographic front near 45°N) and far south of the main path of the storm fetch. For the first time, simultaneous measurements of wind speed/sea state and acoustic data will be made both inside and outside of the storm fetch. For the WASID buoys south of the storm, the noise level and noise vertical directionality will indicate the importance of distant storm noise.

The DMB provide a complementary set of measurements to the WASID data. The bottom vertical arrays will provide a long term sample of a single ocean location with the hope that the surface source level density function can be deduced from these measurements. Only then can the theoretical estimates of different physical mechanisms be compared to measured data so that the dominant mechanism can be determined. Also, the microseismic activity of the bottom can be measured as a function of wind speed and noise level.

In conclusion, the need for at-sea measured data to both model ambient noise and describe the ocean surface source level density function theoretically has been established in this paper.

5. REFERENCES

1. J. Anderson, 'Remote Controlled/Waves Air-Sea Interaction Drifting (RC/WASID) Buoy', PRL Technical Report in preparation.

2. J. Anderson, et al, 'Deep Ocean Ambient Noise Measurements from the Slack Moored BREAM Buoy', PRL Technical Report #59, February 1985

3. ARGOS Location and Data Collection Satellite System – User's Guide, ARGOS-NASA/NOAA/CNES

4. J. Wilson, 'Data Compaction using the Fourier Series Method', PRL Technical Report in preparation.

5. J. Wilson, 'Wind-Generated Noise Modeling', J. Acous. Soc. Am. 73(1), January 1983

6. J. Wilson, 'Distant Storm Noise Versus Local Wind Noise at 165Hz in the Northeastern Pacific Ocean', J. Acous. Soc. Am. 74(5), November 1983

7. D. J. Kewley, et al., 'Directional Underwater Noise Estimation', paper presented at 112th meeting of Acoustical Society of America, Anaheim, CA, 10 December 1986.

BUBBLE NOISE CREATION MECHANISMS

P A Crowther
Marconi Underwater Systems Limited
Blackmoor Lane
Watford, Hertfordshire
WD1 8YR, England

ABSTRACT. The contribution of noise created by bubbles at the instant
of their formation is considered as a major contribution to sea noise
over 1 - 60 kHz. Estimates of the noise from this mechanism have been
made by combining laboratory measurements of the average noise made
per bubble with the estimated rate of creation of bubbles at sea, both
in relation to bubble size. Appeal is made to dimensional similarity
arguments for the bubble output, confirmed by experiment, and for the
rate of creation of bubbles, consistent with experiment. Frequency
dependence predicted, at $f^{-1.5}$, is close to that observed.   The
estimated absolute level and windspeed dependence tend to exceed those
observed; it is argued that this may indicate a modification in the
bubble output, through a local modification in the medium in the
vicinity of the breaking wave. Experimental evidence on the spatial
concentration of sea noise is shown.

1.   INTRODUCTION

The musical sound associated with any liquid surface when broken,
whether from a babbling brook or a dripping tap, is universally
familiar. It is also observed at sea that there is a particular
region of great significance to sonar, say from 1 - 100 kHz and
sometimes referred to as the Knudsen region, where the measured
ambient noise spectrum level appears frequently to be dominated by
surface agitation. The object of this paper is to explore the idea
that the Knudsen spectrum is generated by noise emissions arising from
the creation of bubbles at the sea surface, from the breaking of wave
caps. Two pieces of information are needed: on the one hand we need
to know the average amount of sound produced by a bubble at creation,
and on the other, we need to know the rate of creation of bubbles per
area, in a given sea state.

The author has studied bubble creation noise emission in the
laboratory, the results of which are presented and analysed in Section
3, after a brief review of bubble acoustics in Section 2. It is found
that bubble noise emissions can be related to the energy required to
create the bubble, in such a way that the average noise emission may

B. R. Kerman (ed.), Sea Surface Sound, 131–150.
© 1988 by Kluwer Academic Publishers.

be estimated as a function of frequency to within a numerically determined pure scaling factor. The rate of creation of bubbles has to be estimated rather indirectly from scanty data on their distribution, and this question has been approached in combination with a dimensional analysis on the expected windspeed and bubble radius dependence of the bubble creation rate, in Section 4.

When the two pieces of information are combined, an order of magnitude estimate of the bubble contribution to sea noise is possible, which broadly supports the bubble noise hypothesis, but tends to estimate a level rather higher than observed, especially in high windspeeds. This is further discussed in Section 5, in the context of spatial variability of noise. Experimental data on noise measured in a narrow beam are shown, which verify that noise is generated principally in the breaking of the waves alone, and is therefore highly inhomogeneous spatially. Finally, in Section 6, the interaction of a bubble with the surface is studied theoretically, and it is concluded that this helps to explain why the experimental bubble noise output is somewhat greater than might be expected on a non-interacting assumption.

GLOSSARY OF PRINCIPAL SYMBOLS

a	=	bubble radius, $\quad a_o$ = outer scale of $a \approx 5.10^{-3}$ m
c	=	speed of sound
d	=	bubble centre depth
D	=	bubble dipole strength
E_s	=	bubble static energy, E = bubble acoustic excitation energy
f	=	frequency, $\quad f_o$ = inner scale of $f\left(\approx 700 \text{ Hz}\right)$
g	=	gravity = 9.81 ms^{-2}
$N(f)$	=	noise pressure power spectrum (Pa^2/Hz)
p	=	acoustic pressure, p_a = atmospheric pressure
p_o	=	initial acoustic pressure excess inside the bubble.
P	=	rate of working of wind wave dissipation (Watt/m^2)
r	=	$|\underline{r}|$ = radius from bubble centre
t	=	time
T	=	integrated source level = $\underset{r \to \infty}{\text{Lt}} \int p^2(t) \, r^2 dt$
T_{expt}	=	experimental T
Ts, Td	=	static and dynamic reference scales for T
U_{10}	=	windspeed at 10 m
u	=	bubble rise speed
v	=	initial bubble translation speed
X	=	bubble creation rate per area per radius (m^{-4} s^{-1})
σ	=	surface tension
$\delta_{rad}, \delta_{th}$	=	radiation, thermal terms in bubble loss tangent
Δ	=	bubble emission bandwidth
θ	=	angle with respect to normal
$\mu_1 \, \mu_2 \, \mu_3 \, \mu_{12} \, \mu_{13}$	=	scaling factors
ρ	=	water density = 1000 kg/m^3

ρ_a = air density
σ = surface tension = 74.10^{-3} J/m^2

2. THE ACOUSTICS OF BUBBLES

It is necessary first to summarize some of the acoustic properties of bubbles. A full account of the acoustic properties of bubbles in the free field may be found in Devin, [1] from which we may summarize the following, for bubbles resonant between 0.5 and 60 kHz. The resonant frequency is approximately that of the adiabatic stiffness of the gas:

$$f = (3 \gamma p_a/\rho)^{1/2}/2 \pi a = 3.26/a . \quad \text{(MKS)} \tag{1}$$

The loss tangent is dominated by thermal and free field radiation terms:

$$\delta = \delta_{rad,f} + \delta_{th} , \quad (f \lesssim 60 \text{ kHz}) \tag{2}$$

where:
$$\delta_{rad,f} = 2 \pi af/c = (3 \gamma p_a/\rho)^{1/2}/c = 0.014 \tag{3}$$

$$\delta_{th} = 4.4.10^{-4} f^{1/2} / (1 + f/2.510^5) , \quad \text{(MKS)} \tag{4}$$

the last equation being a numerical fit to Devin's result at $p_a = 10^5$ Pascal, accurate to within a few per cent over $f \lesssim 60$ kHz.

We must deal with bubbles radiating when close to the sea surface. Under such conditions, the radiation is dramatically altered. To the lowest order, we retain (1) and (4) as working approximations, but we reconsider the radiated energy. If the bubble depth at the instant of excitation is d, then if $d \ll c/f$, the bubble will radiate as a dipole, in association with its image in the surface. The radiated pressure wave becomes, from simple image theory:

$$p(t) \doteq D \frac{\cos\theta}{r} (\cos (w)/kr - \sin (w)) \exp (-\pi f \delta t),$$

$$(r \gg a) \tag{5}$$

where:
D = $4 \pi fda \, p_o / c$ = dipole strength
k = $2 \pi f/c$
w = $2 \pi f (t - r/c)$
θ = polar angle w.r.t. normal.

The integrated source level signature is:

$$T \stackrel{def}{=} \underset{r \to \infty}{Lt} r^2 \int p(t)^2 dt \doteq (D \cos \theta)^2/(4 \pi f \delta_{th}) \tag{6}$$

whilst the total radiated power, if we allow for integrating the $\cos^2\theta$ radiation pattern over only 2π solid angle, is:

134

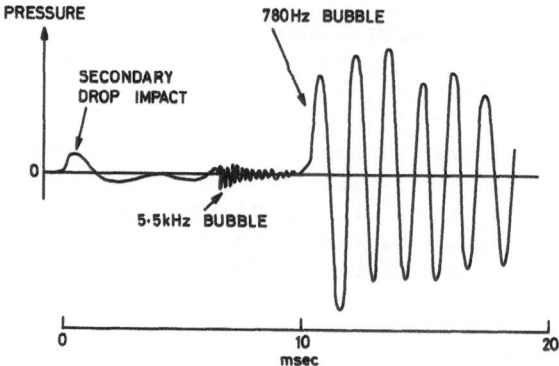

Fig 1a.   Example of double bubble emission.

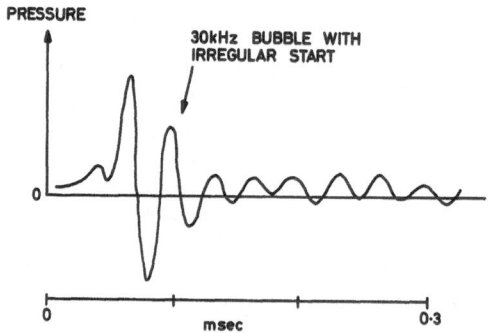

Fig 1b.   Example of start of an ultrasonic bubble emission, with some drop pulse over-pressure.

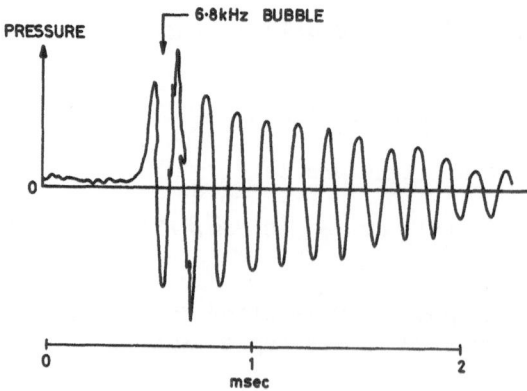

Fig 1c.   Typical Exponentially modulated pulse.

$$W = \frac{2\pi}{3} \quad T \,(\theta = 0)\, /\rho c = \frac{2}{3}\,(2\pi f\, d/c)^2\, W_{free}\, . \qquad (7)$$

This means that the radiation loss tangent will be reduced from the free field radiation loss to:

$$\delta_{rad} = \frac{2}{3}\,(2\pi f\, d/c)^2 \cdot \delta_{rad'f}$$

$$= 1.7\ 10^{-6}\ (d/a)^2 \qquad (8)$$

Since we are considering bubbles formed at depths of order of the bubble radius, or not much more, in practice $\delta_{rad} \ll \delta_{th}$, and so can be ignored in the expected loss tangents $\delta \approx \delta_{th}$.

A word about non-sphericity: since the bubbles of present concern are in a state of just having been formed from surface closure, it seems likely that they will frequently be non-spherical. Non-sphericity does not seriously affect the resonant frequency if we relate to the spherical bubble of equal volume, [2,3] but it will increase the loss tangent δ_{th}, since the thermal area/volume ratio is increased. This appears to be consistent with observations.

3.   BUBBLE SIGNATURES

3.1   Experiment

In order to estimate the amount of noise made in bubble creation, some laboratory experiments were made with drop and splash-induced bubbles. The equipment was similar to that used by Franz for drop noise measurements, [4]. Drops of various sizes from 2 - 10 mm diameter were allowed to fall into the water in a test tank from heights of 15 - 50 cm, thus generating impact speeds of from 1.5 - 3.0 m/s. Directly below the impact point, a small probe hydrophone was mounted, at a depth of typically 6 cm. It was found that with this arrangement, bubble noise could be generated at frequencies from about 0.5 kHz to 50 kHz or more. The aim of the present work has been to study the sound created by individual bubbles, and relate this to the bubble size, rather than to study the average spectrum level of noise in relation to the number of drops, as was done by Franz. Drops and splashes were used thus as a convenient source of energy, but the emphasis is on the bubbles produced.

Hydrophone signals after amplification were displayed and analysed. In early work, previously reported, use was also made of a Spectral Dynamics Type 335 Spectrum analyser, in transient capture or peak capture mode, for spectrum analysis [5]. In more recent work, a Data Precision type 6000 transient analyser was used. Some specimen time domain waveforms obtained from the latter are shown in fig (1). Bubble noises were commonly found between 30 - 200 msec after the initial drop impact, sometimes following a secondary drop impact pulse. An example of this occurs in figure (1a), where the drop impact, and two bubble pulses are all evident. Whilst the principal drop impact waveform is found to be very closely reproduced from one pulse to another, the bubble pulses were found to vary greatly in both frequency and amplitude from drop to drop in a random manner, although a general tendency was noted that larger more energetic drops were needed to excite the lower frequency bubbles. Sometimes single bubbles are found, and sometimes two frequencies are seen to be generated simultaneously. Sometimes an approximately exponentially falling envelope is found, whereas in other cases this is not so. Bubble noise greatly exceeded impact noise on average.

Bubble pulses have been analysed for spectral content, total energy and width of individual lines in the spectrum. By correcting for the near field effect of the hydrophone, it was possible to obtain the total integrated source level of each bubble pulse, as follows:

$$T_{expt} = h^2 (1 + (c / 2 \pi f h)^2)^{-1} \int p^2 (t) \, dt \qquad (9)$$

where h = the hydrophone depth. The measured half power total linewidth, Δ , after correction for the instrument resolution, was also extracted for each bubble. We are therefore in a position to compare experimental results with the theoretical figure for an ideal spherical bubble in the vicinity of the surface, which would be expected, from Fourier transformation of (5), to be $\Delta \sim f . \delta_{th}$, for an exponentially decaying pulse. The distribution of total bubble signatures, T_{expt}, is shown in the form of a scatter diagram against frequency in figure (2). A very large range is evident, but it is also evident that there is some systematic dependence of mean level against frequency.

3.2 Theoretical Ideas

In order to make further progress, we must now consider a scaling law for bubble pulse energies. It appears probable that it is the surplus centripetal kinetic energy in the water at the instant of surface closure which is converted into potential pressure energy, that thereafter may radiate some of the energy as sound. We suppose that if bubble formation over some range of bubble radii is a scale model phenomenon, it would be necessary for the excess energy in the bubble to be distributed in a universal law in relation to the bubble's static energy. For a sufficiently small bubble close to the surface, the static energy is mostly due to surface tension. For a large bubble, some gravitational energy will also be involved; thus for a bubble just below the surface, we take the static energy as:

$$E_s = 4 \pi \sigma a^2 + 4 \pi \rho g a^4 /3. \qquad (10)$$

If we now suppose that the mean initial excess energy of the bubble scales to the static energy, we should have:

$$\langle E \rangle = \mu_1 E_s \qquad , \qquad (11)$$

and it follows that the initial excess pressure corresponding to E scales according to:

$$\langle p_o^2 \rangle = \mu_1 (6\sigma/a + 2\rho g a) \gamma p_a \quad , \tag{12}$$

where μ_1 is a pure number, of unknown value. From the physical argument, we might anticipate that the numerical value of μ_1 might be of order unity.

We must now consider what happens to the bubble after closure. There are two possibilities. The bubble may either (a) remain at the same point during its acoustically active life time of order $1/(2\pi\delta_{th} f)$ – the static model; or (b) it may be supposed to move over this period – the dynamic model.

In the static model, the pulse envelope will be exponential, as already discussed. Retaining the scale model idea, we must suppose that the effective depth of excitation will scale to the bubble radius, in the static model:

$$\langle d^2 \rangle = \mu_2 a^2 \tag{13}$$

where μ_2 is another pure number of order ~ 1. Applying (12, 13) in (6), we would expect a law for the total source level in the form:

$$\langle T \rangle_{\theta=o} = \mu_{12} \cdot T_s \tag{14}$$

where T_s = the static model scale, $\mu_{12} = \mu_1 \cdot \mu_2$, and

$$T_s = 24\pi f \gamma p_a \sigma a^3 [1 + (a/a_o)^2] /c^2 \delta_{th}, \tag{15}$$

in which $a_o = 3\sigma/\rho g = 4.8 \ 10^{-3}$ m.

If we interpret the fa relationship as if for the free field [eq (1)], the integrated source level scale goes to:

$$T_s = \sqrt{243} \ \pi^{-2} \sigma c^{-2} \delta_{th}^{-1} (\gamma p_a)^{5/2} \rho^{-3/2} f^{-2}.$$
$$\cdot [1 + (f_o/f)^2], \tag{16}$$

where $f_o = (\gamma p_a g/\sigma)^{1/2} / (2\pi) = 690$ Hz $\tag{17}$

Next, let us consider the dynamic model. Suppose that the bubble, in addition to its centripetal kinetic energy, has some downward kinetic energy. A simple scale for the translational kinetic energy is $\sim \frac{1}{2}\rho v^2 \cdot 4\pi a^3/3$, where v is the bubble speed. If we suppose that the average translational energy also scales to the static energy of creation, E_s, then we may deduce that:

$$\langle v^2 \rangle = \mu_3 \ 6\sigma a^{-1} \rho^{-1} [1 + (a/a_o)^2] \stackrel{def}{\simeq} \mu_3 v_e^2 \tag{18}$$

where μ_3 is the energy scaling ratio. If the bubble starts at an acoustically effective depth of d = o, and moves at speed v away from the surface, the envelope of the pulse would no longer be exponential, as for a free bubble; but because the dipole strength now increases with time, (5) would now be modified to have a 'Rayleigh' envelope:

$$p(t) = \frac{\cos\theta}{r} \; 4\pi a p_o \; \frac{v}{c} \cdot t(\cos(w)/kr - \sin(w)) \exp(-\pi f \delta_{th} t) \quad (19)$$

From this, we may further deduce a mean integrated source level of:

$$\langle T \rangle_{\theta=0} \;=\; \mu_{13} \; T_d, \quad (20)$$

where $\mu_{13} = \mu_1 \cdot \mu_3$, and T_d is the dynamic scale signature

$$T_d \;=\; 72\sigma^2\gamma p_a \; (\pi\rho c^2 \delta_{th}^3 f)^{-1} \; [1 + (f_o/f)^2]^2 \quad (21)$$

We finally consider the distribution of T_{expt}. Suppose that the conjecture of energy scaling is valid, and further that if the creation of a bubble is a random event, then the excess energy available might be roughly Maxwell distributed, with a cummulative distribution of

$$F(E) \;=\; 1 - \exp(-E/\mu_1 E_s), \quad (22)$$

for pressure excitation energy E. This would imply a distribution for T_{expt} on the static model, ie.

$$F(T_{expt}) \;=\; 1 - \exp(-T_{expt}/\mu_{13} T_s). \quad (23)$$

For the dynamic model, however, T is the product of two energy-like terms $T \propto \langle v^2 \rangle \cdot E$, both of which might be exponentially distributed, in which case it is possible to show that the resulting distribution for T_{expt} will be a modified Bessel:

$$F(y) \;=\; 1 - 2y^{\frac{1}{2}} K_1 (2y^{\frac{1}{2}}) \quad (24)$$

where K_1 is the modified Bessel function of order one, and

$$y = T_{expt} / \mu_{13} T_d \quad . \quad (25)$$

To test the distributions, the cumulative distribution of $y_s = T_{expt}/T_s$ and $y_d = T_{expt}/T_d$ respectively were tested for maximum likelihood fit to the generalisation of (24), namely:

$$F(y) \;=\; 1 - z K_1 (z) \quad (26)$$

where $z \;=\; 2 \left(\Gamma(1 + \tfrac{1}{2}b)^2 \; y/m \right)^b \quad (27)$

y being either of the ratios y_s, y_d, and m being the mean values of the ratios $- \mu_{12}$, μ_{13} respectively. Clearly, for the dynamic model with a Maxwellian energy distribution we should expect $b = 0.5$.

TABLE 1

Analysis of bubble energy/theoretical scale ratio distribution, by maximum likelihood fit to $F = 1 - z\,K_1(z)$, according to Eqs (26,27)

Group		1	2	3	4	5	6	7
Frequency band: kHz		0.5–1.27	1.27–2.0	2.0–3.1	3.1–4.5	4.5–6.7	6.7–9.0	9.0–18.2
T_{expt}/T_s	$m = \mu_{12}$	108	140	49	19	24.5	54	24
	90% Confidence	62–237	56–550	27–113	10.9–43	12.3–65	27–147	12.8–57
	b	0.46	0.30	0.43	0.45	0.38	0.38	0.41
	90% Confidence	0.35–0.58	0.23–0.38	0.33–0.53	0.34–0.57	0.29–0.47	0.29–0.48	0.32–0.52
T_{expt}/T_d	$m = \mu_1\,\mu_3$	18.2	34	13.0	5.4	6.9	15.2	6.8
	90% Confidence	10.3–40	13.6–135	7.1–30	3.0–11.9	3.5–18.1	7.6–42	3.6–16.2
	b	0.45	0.30	0.42	0.45	0.38	0.38	0.41
	90% Confidence	0.35–0.57	0.23–0.38	0.34–0.53	0.34–0.57	0.29–0.47	0.29–0.48	0.32–0.52

3.3   Results

Distributions of pulse energy have been plotted in seven bubble frequency bands, and maximum likelihood fits to the distribution of genus (26,27) was made. Results for the values of m and b are tabulated in Table 1. No difference in distribution was found between frequency groups (4 – 7) – frequencies 3.1 – 18.1 kHz – significant at the 5% level; and over 2.0 – 18.1 kHz, differences were significant only just at the 5% level. The main anomaly occurs in the 1.27 – 2.0 kHz band, since it is found that the distribution for the 0.5 – 1.27 kHz band is not statistically different at the 5% level from that for the 3.1 – 18.1 kHz band for the dynamic model, although it is for the static model. The reason for the anomaly over 1.27 – 2.0 kHz may be that this is a region of transition from capillary to gravity energy control, so that scale free assumptions break down.

    The values of b found are quite close to the guessed figure 0.5 for the dynamic assumption. Other facts which appear to favour the dynamic model are   that   the maximum   likelihood   estimate   of m, at $m = T_{expt}/T_d = \mu_1 \cdot \mu_3 \approx 11$, with b = 0.41 for the dynamic model, against $m = \mu_1 \cdot \mu_2 \approx 47$, with b = 0.41 for the static model, over all frequencies other than 1.27 – 2 kHz. If we assume equipartition between the centripetal and translational kinetic energies, $\mu_1 = \mu_3 = 3.4$ on the dynamic model, which is indeed of order unity, as expected.   The mean   energy   estimate at   this   scale   ratio   is

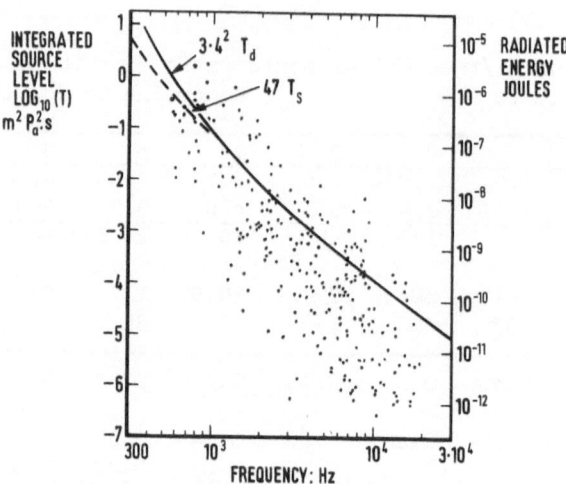

Fig 2.   Scatter of bubble emission energies, with the maximum
likelihood mean energy fits, on static (T_s) and dynamic (T_d) models.

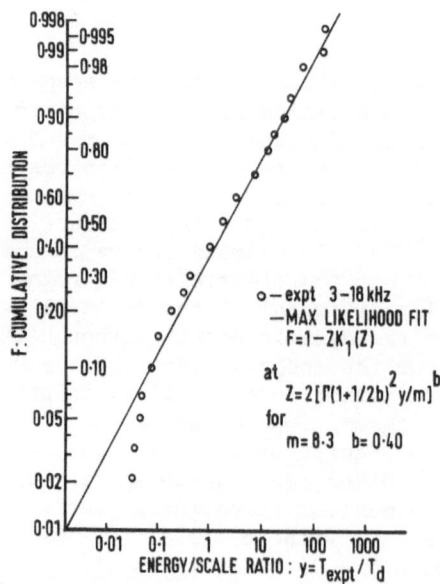

Fig 3.   Illustrating maximum likelihood fit to bubble energy
emission distribution.

shown in fig (2). Fig (3) shows the fit to experimental T/T_d distribution over 3-18 kHz.

The measured bandwidth, Δ , was also analysed for its distribution in relation to the half height bandwidth of the ideal static bubble, δ_{th} f. The ratio $\Delta/(\delta_{th} f)$ was found to be randomly distributed, but with frequency dependence in the median values, which were 4.1 for 0.5 - 1 kHz, 2.3 for 1 - 2 kHz, 1.6 for 2 - 5 kHz, 1.0 for 5 - 10 kHz, 0.6 for 10 - 20 kHz. Thus, at low frequencies, most bubbles are far more damped than would an ideal static spherical bubble be, whereas at high frequencies, the damping is somewhat less. Non sphericity could be expected to be more severe for large low frequency bubbles, because of the increased gravity/surface tension energy ratio, and perhaps it is this which increases the damping.

Another possible cause of increase bandwidth, not due to damping, is some frequency modulation of the resonance through surface interaction, if the effective depth of the bubble changes during its emission. This is discussed in section 6.

4.   SEA NOISE CONTRIBUTION

We must now turn from the laboratory to the sea, using the results from the laboratory experiments to estimate how much noise the creation of bubbles in breaking waves can produce. If we have a hydrophone at depth h in the sea, then the mean noise spectrum level will be, with neglect of absorption loss, just

$$N(f) = \int S(\theta, f) \frac{1}{\cos\theta} \, d\Omega = 2\pi \int_o S(\theta,f) \frac{1}{\cos\theta} \, d\cos\theta , \qquad (28)$$

where $S(\theta,f)$ = Power spectrum radiated per area per solid angle per frequency band, and $d\Omega$ = solid angle. Now the bubble contribution to this source function S will be:

$$S(\theta,f) \, df = X(a, U_{10}) da. \, \cos^2\theta \, \mu_1 \, \mu_3 \, T_d \, (f), \qquad (29)$$

X(a) being the rate of creation of bubbles per area per radius. If we now use (1), and substitute into (28), we deduce a noise spectrum level:

$$N(f) = 0.5 \, (3 \gamma \, p_a/\rho)^{\frac{1}{2}} \, \mu_1 \, \mu_3 \, f^{-2} \, T_d \, (f) \, X \, (a,U_{10})$$
$$= 12 \pi^{-1} \, (3\gamma p_a/\rho)^{3/2} \, \mu_1 \, \mu_3 \, f^{-3} \, c^{-2} \, \sigma^2 \, \delta_{th}^{-3} \, f^{-3}.$$
$$. \, [1 + (f/f_o)^2]^2 \, X \, (a,U_{10}) \qquad (30)$$

Now we must complete the argument by attempting to estimate X(a). Sources of data for this are not very extensive, so we must first consider dimensional arguments for how the rate of creation of bubbles depends on windspeed and radius. We do this by again appealing to the idea that bubble creation is a scale model phenomenon as regards sizes, at least within limits, and that there is sufficient energy locally in the breaking waves to suppose that the shape of $X(a, U_{10})$ against a is independent of windspeed, the latter only affecting the level. Furthermore, if we regard wave breaking as the major source of dissipation of wave energy, and we assume that the rate of creation of bubbles will be proportional to the rate of dissipation of energy per area, P, then for those scale-free bubbles governed by surface tension, we assume a dependence of X on a, σ, P, which by dimensional form must be:

$$X \propto a^{-3} \sigma^{-1} P \tag{31}$$

We must now consider how the rate of working per area, P, depends on windspeed. This is best approached by assuming that dissipation nearly matches wave growth in a 'fully developed' sea. The rate of input of wave energy from the wind per area from the wind forcing (modified Miles theory) is:

$$\frac{d}{dt} E_w = \beta E_w \tag{32}$$

where β is the growth coefficient, E_w = wave energy per area. DeLeonibus and Simpson have shown that for waves of frequency from the dominant frequency upwards, β is approximately proportional to wave frequency, and is numerically of order

$$\beta \sim 2.10^{-3} f_w \sim 1.5 (\rho_a/\rho) f_w \tag{33}$$

where f_w = wave frequency [7]. To obtain the dimensional scaling law on wave energy, we simply consider the dominant wave frequency and total energy, and therefore anticipate

$$P \sim 1.5 (\rho_a/\rho) f_w E_w \quad . \tag{34}$$

For the Pierson-Moskowitz sea wave spectrum, for example, we have

$$E_w = 2.2 \ 10^{-3} \rho g^{-1} U_{10}^4 \quad , \tag{35}$$

and the modal point on frequency is

$$f_w = 0.15 \ g/U_{10} \quad . \tag{36}$$

Substitution into (31) gives the final dependence as:

$$X(a) \propto a^{-3} \rho_a \sigma^{-1} U_{10}^3 \quad (68\mu m \lesssim a \lesssim a_o) \tag{37}$$

In order to fit a numerical figure to (37), it is necessary to appeal to experiment. The approach is to estimate the mean rate at which bubbles return to the sea surface, and then assume approximate dynamic equilibrium between creation and return, to estimate:

$$X(a) \simeq B(a,o) \cdot u(a) \quad , \tag{38}$$

where $B(a,z)$ = bubbles per volume per radius at depth z .

A reasonable approximation to the rise speed is, from Reference [6]:

$$u(a) = \text{Min} \left[(0.15 \, a^{-1/2} + 4.6 \, 10^{-7} \, a^{-2})^{-1}, \; 0.25 \right] \text{ (MKS)} \qquad (39)$$

Sources of data for $B(a,o)$ are scanty. Photographic measurements in situ by Johnson and Cooke [8] and via a bubble trap by Kolovayev [9] have yielded some information over the bubble size ranges $17 \lesssim \mu m \leqslant a \leqslant 250 \, \mu m$, corresponding to resonant frequencies 13 - 200 kHz, although they are only in mutual agreement to within about an order of magnitude. Acoustic measurement by Thorpe tend to confirm the U_{10}^3 law quite well, but do not resolve radius dependence, since only one frequency was used [10]. We have therefore assumed the U_{10}^3 law, and attempted to fit to the results of Johnson and Cooke to a form of the kind:

$$B(a,d) \varpropto e^{-d/d_o} \; U_{10}^3 \; a^{-3}/u(a) \qquad (40)$$

The resulting fit, for 2 windspeeds only ($U_{1c} \approx 11.3$ m/s and 15.7 m/s respectively) was estimated quantitively at:

$$X(a) = 4.10^{-9} \, U_{10}^3 \, a^{-3} \text{ (MKS) } (\doteqdot 6.10^{-7} \, a^{-3} \, \sigma^{-1} \text{ P)}, \qquad (41)$$

$$(68 \, \mu m \leqslant a \leqslant 250 \, \mu m)$$

which would imply order of 10^{-6} of wave dissipation energy going into the creation of bubbles. This law would appear to break down for bubbles of radius $\leqslant 68 \, \mu m$, where some form of cut-off in the rate of creation appears to arise, the mechanism for which is not clear. The fit to Johnson and Cooke's data extrapolated to zero depth, allowing a slight correction for bubble compressibility with depth, is shown in figure (4).

If we now substitute the estimate (41) into (30) with $\mu_1 = \mu_3 = 3.3$ we find an estimated noise spectrum in the sea due to bubble creation of:

$$N(f) = 10^{-3} \, U_{10}^3 \, f^{-1.5} \left[1 + (690/f)^2 \right]^2 . \left[1 + f / 2.5 \; 10^5 \right]^3$$

$$(\text{Pa}^2 /\text{Hz}) \quad (10^3 \lesssim f \lesssim 6.10^4 \text{ Hz ?)} \qquad (42)$$

This estimate is about ten times higher than a previous estimate, [5] largely because of the re-estimation of the bubble creation rate $X(a)$.

In comparing (42) with the measured sea noise spectrum, we note that although of similar order, (42) tends to exceed the measured total surface noise at most frequencies and windspeeds. The predicted frequency dependence $f^{-1.5}$ is approximately as measured $-f^{-1.7}$ empirically, but the windspeed dependence at U_{10}^3, is rather stronger than found experimentally - usually closer to U_{10}^2. Eq (42) is compared with the Knudsen/Wenz spectrum in figure (5).

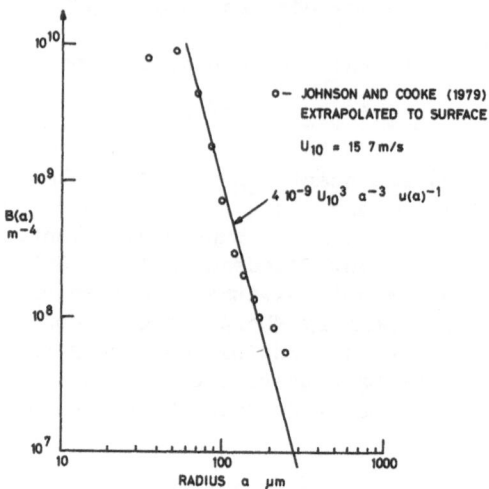

Fig 4.   Bubble density from Johnson and Cooke (1979) Extrapolated to zero depth, compared with Eq (40).

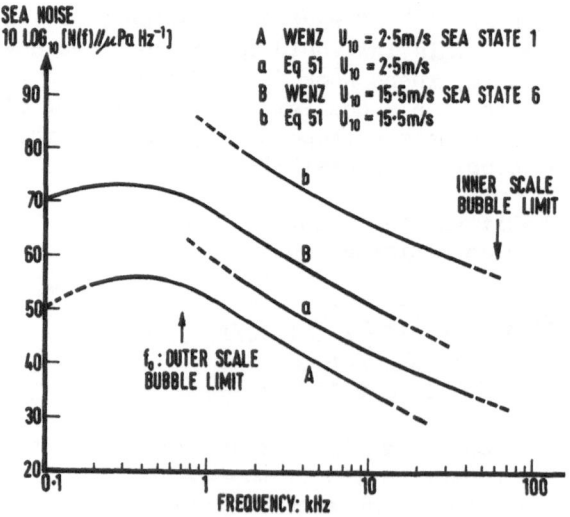

Fig 5.   Estimated bubble creation noise at sea, compared with observed spectrum [G M Wenz J Acoust Soc Am **34**, 1936 (1962) and J Acoust Soc Am **48**, 362 (1970)]

The fact that (42) matches or exceeds the measured noise spectrum is evidence for the importance of the bubble creation noise mechanism as a principal, possibly dominant, source of noise in the higher sonic and ultrasonic frequency bands; but the discrepancies need to be discussed. Regarding absolute level, it is possible that bubbles created at sea do not radiate as much energy as do the laboratory bubbles, and it is also possible that the bubble creation rates are lower than has been assumed. Certainly the photographic data does not cover larger bubbles responsible for noise at over say 1 - 10 kHz, although there does not seem much likelihood or error in the extrapolation of X(a) from smaller a, if we believe the a^{-3} law. Regarding the windspeed dependence, despite the general $U_{10}{}^2$ law, it has been pointed out that there is some evidence for a $U_{10}{}^3$ law at low windspeeds [11]. One possible explanation of both the windspeed law and absolute level suggests itself: that the water in the immediate vicinity of the bubble creation is so strongly modified as to reduce the radiation from the bubble. If wave breaking is confined to a very small area, so many bubbles may be created that they mutually mask one another. Evidence for this will be considered in what follows.

5.   SPATIAL VARIABILITY

Experiments at sea are being carried out by the author and co-workers at present, aimed at clarifying a number of the physical features of sea noise in the 1 - 100 kHz band. The equipment is sea-bed mounted, and has a set of receive beams, with one transmit beam, capable of angular steer. In addition to the study of noise, reverberation is also being measured, and by the use of narrow beams with 150 - 300 μs transmitted pulses, bubble layers are being studied over the frequency range 30 - 200 kHz.

Evidence is emerging on the extreme spatial variability in the noise radiated from the sea surface, and of the rate of creation of bubble layers. An example is shown in figure (6), which is a time history of a section of record of the noise power, filtered over a number of 2 kHz wide bands, and integrated after square law detection to smooth over 0.4 s. It is evident that where resolved spatially, the sea does not behave as a uniform source, but consists of a number of local patches, typically occupying $\sim$ 0.05 of the total surface at windspeed $\sim$ 18 m/s, for example. This is consistent with visual observations of whitecap coverage, estimated as of order $1.2 \ 10^{-5} \ U_{10}{}^{3.3}$ as a fraction of the sea surface [12]. Since the 'active' part of the whitecap, in which bubbles are being created, will be only a fraction of the total whitecap foam area, it seems reasonable to suppose that the local rate of creation of bubbles in the breaking zones will typically be at least 100 times the mean level.

At such a concentration, there may well be scope for reduction in the noise output per bubble, because of the aeration of the surrounding water at the instant of creation. If the degree of this perturbation increases progressively with the violence of the wave

146

breaking, and hence with the windspeed, we might explain the reduction
of the U^3 law to a lower power.

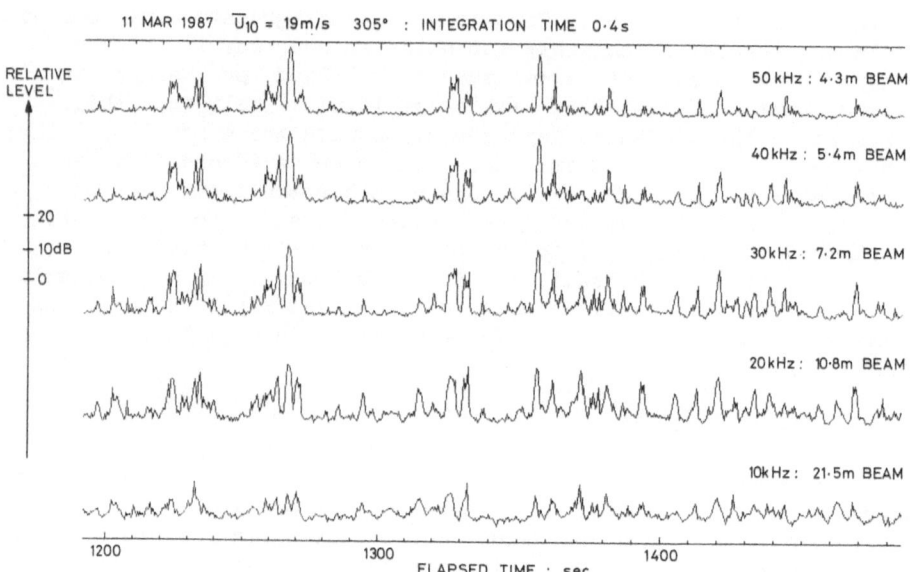

11 MAR 1987 $\overline{U}_{10}$ = 19 m/s   305° : INTEGRATION TIME 0·4 s

RELATIVE
LEVEL

50 kHz : 4·3 m BEAM

40 kHz : 5·4 m BEAM

30 kHz : 7·2 m BEAM

20 kHz : 10·8 m BEAM

10 kHz : 21·5 m BEAM

1200   1300   1400

ELAPSED TIME : sec

Fig 6   Illustrating inhomogeneity of noise over 2 kHz bandwidths,
integrated over 0.4 s . Observed temporal fluctuations of up to 20 dB
are due to passage of breaking waves over the vertical receive beam,
and are thus indicative of spatial inhomogeneity.
Water depth is 70 m.

6.   HIGHER ORDER EFFECTS

In the treatment of bubble signatures so far, we have ignored the
effect of proximity of the surface or of other bubbles, on the
frequency and radiation from an excited bubble, so that the scale
levels for bubble radiation output as given in (16,21) assume that
bubble frequency and dipole strength are just as if the free field
output of the bubble were simply reflected from the surface - a
non-interacting assumption. We now consider interactions.
    The method of computing this interaction is a combination of wave
series expansion and image technique. Briefly, for a bubble at
centre-depth d, the acoustic pressure may be expanded, with neglect of
the time factor, as:

$$p\ (\underline{r}) = \sum_{s=0}^{\infty} b_s \left[\Phi_s\ (\underline{r}) - \Phi_s\ (\underline{r} - 2\underline{d}) \right] \tag{44}$$

where $\underline{r}$ = space position re bubble centre, $\underline{d}$ is a vector of length d, directed towards the surface, b_s are coefficients, $\Phi_p(\underline{r})$ the appropriate wave components - zonal harmonics times outward going spherical Hankel functions. We may then compute the mean velocity and mean potential on the surface $|\underline{r}|$ = a (spherically symmetric components), and match to the known compressibility and loss of the air in the bubble and required pressure amplitude, as a function of bubble radius and frequency. By iterating the real and imaginary parts of the frequency until convergence, it is then possible to obtain frequency, loss tangent, and the b_s coefficients, up to some sufficiently high finite order of expression. Furthermore, it is possible then to obtain the far field source level of the bubble and image.

Results for this type of calculation are illustrated in Figure (7), which shows the resonant frequency, loss tangent and dipole to bubble pressure ratio D_i/p_o, as a function of the bubble centre depth, for a 1 mm bubble. Also shown, for interest, is the resonant frequency of a system in which the bubble is reflected in a rigid, rather than pressure release, boundary. By symmetry, the two reflection situations are of course equivalent to the two possible oscillation modes of a pair of bubbles in an infinite liquid, and with centres separated by 2d. The symmetric (rigid boundary) oscillation would normally dominate in such a circumstance, however, since it preserves a monopole output, and hence a far greater radiation efficiency.

Returning to the pressure-release boundary interpretation, we note from Figure (7) that the proximity to the surface raises the resonant frequency. Not much happens to the loss tangent, but a very considerable modification in the relative dipole strength appears to occur, for instead of falling monotonically as we approach the surface, as might be expected, it appears to reach a minimum when the top of the bubble is about half a bubble radius from the surface, and for closer spacing to increase with the increasing frequency. The resulting dipole strength is thus always greater than that for a non-interacting source also shown in Figure (7) as asymptote, (D/p_o = 4 , d 3.26/1500 MKS), and very considerably greater than the value taken in defining the nominal static scale of reference for bubble signatures - D/p_o = 2.710^{-5} m., for d = a = 1 mm. Taken in conjunction with the elevation in resonant frequency, this means that the actual scale for the integrated source level of a static bubble will be considerably greater than that given by (16).

From the general formula (6), we see that if we put T_{si} as the scale of integrated source level for the bubble at depth d, with inclusion of surface interaction, we can show that the comparison with the nominal scale emission T_s in (16), of the same frequency, with the same p_o^2 (a) scaling for a = 1mm, is in ratio:

$$T_{si}(f)/T_s(f) = 411\ (D_i/p_o)^2\ f\ a^{-1}\ \delta_{th}(f)/\delta_i \quad \text{(MKS)} \tag{45}$$

148

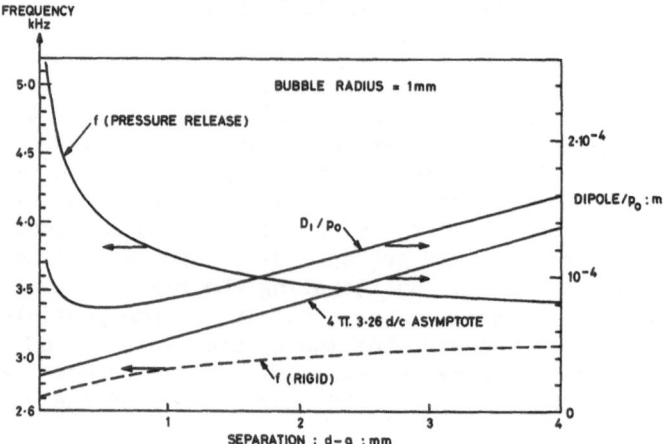

Fig 7.   Frequency and dipole/internal pressure ratio for a spherical
bubble close to a flat surface, with bubble centre at depth d.
Resonant frequency increases as the bubble approached the surface,
whilst dipole strength deviates strongly from the asymptotic non-
interacting theory.   The effect of a right surface is shown for
comparison.

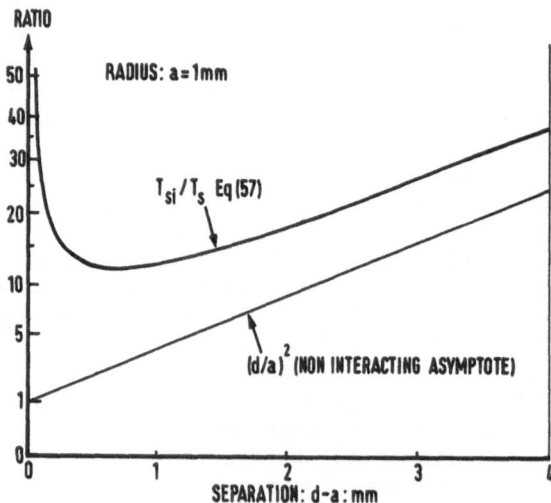

Fig 8.   Comparison of energy output scale of an interacting bubble
with the nominal scale, T_s, for the same pressure energy endownment
rule.   Interaction increases output by $\sim$ 12.

Using the computed f, (D_i/p_o) and δ_i, loss tangent, the ratio
of effective scale to nominal scale was calculated as a function of
depth. The result is plotted in Figure (8), where it can be seen that
the effect of the surface interaction is quite pronounced. Instead of
reducing to the nominal value T_s of (16) as the bubble approaches
contact, T_{si} has a minimum of $\sim 12\, T_s$ at about a bubble surface
separation of about two-thirds radius, and then increases again as the
bubble come closer to the surface, reaching 50 at $(d - a)/a = 0.05$.

This rather unexpected result – that the bubble radiated energy
scale should be at least an order of magnitude greater than the simple
nominal figure hitherto assumed – may help explain the rather high
values of T_{exp}/T_s found in the laboratory, in that with allowance for
this surface interaction, the required average energy excitation of
the bubble at creation would now be much closer to the static bubble
energy, Es, even on the static model, than had been supposed according
to the simpler interaction-free supposition.

One final point: the increasing slope of the resonant frequency
versus separation curve in Figure (7) could indicate another mechanism
for excessive bandwidth in relation to the ideal, which was found for
some bubble emissions, in that the emission will become frequency
modulated if the separation is changing over the acoustic lifetime of
the bubble, as the dynamic model suggests will happen.

7.   CONCLUSIONS

A combination of dimensional similarity argument, experimental
measurement and theoretical study of bubble surface interaction all
appear to support one another, and lead to the conclusion that
although bubble noise emission is highly variable from pulse to pulse,
it tends to dominate direct impact noise for low impact speed, and the
average level can be predicted in relation to bubble resonant
frequency. The dimensional scaling law is confirmed. The idea of a
bubble as something dominated by surface tension puts a limit on the
frequency range for which bubbles are acoustically significant. The
energy argument suggests that a frequency of ~ 700 Hz would be the
lower limit of surface tension dominated bubble action, and it is of
interest to note, and in general support of the hypothesis, that the
'Knudsen' spectrum appears to cut off below just such a frequency. The
upper limit to bubble action is unclear, although photographic data on
bubble distributions suggests that bubble creation may become
ineffective for radii $\lesssim 50\ \mu m$, equivalent to about 60 kHz. Within the
1 - 60 kHz band, however, the bubble noise hypothesis, when worked out
by combining laboratory bubble emission scaling with creation rates
estimated from photography produces a calculated noise spectrum very
nearly correct in frequency dependence, slightly stronger in windspeed
dependence – U^3 against experimental U^2 – and about an order of
magnitude high in absolute level, compared with experiment.

It is possible that these discrepancies are due to the fact that
bubble creation is confined to a small fraction of the surface area at
any instant; so that local bubble creation rates may be 100 times
greater than the average, and that this could progressively modify the

150

bubble output with increasing windspeed by very local attenuation within the whitecap region. Experimental results verify that surface noise output is also highly spatially inhomogeneous. There is scope for further experimental work to study this question, perhaps by combining active detection of bubble generation with passive listening to the bubble emissions.

REFERENCES

1. C Devin, 'Survey of Thermal Radiation and Viscous Damping of Pulstating Air Bubbles in Water' J Acoust Soc Am 31, 1654-1667 (1959)
2. M Strasberg 'The Pulsation Frequency of Non spherical Gas Bubbles in Liquids' J Acoust Soc Am 25, 536 (1953)
3. M Strasberg 'Gas Bubbles as Sources of Sound in Liquids' J Acoust Soc Am 28, 20 - 26 (1956)
4. G J Franz 'Splashes as Sources of Sound in Liquids' J Acoust Soc Am 31, 1080 - 96 (1959)
5. P A Crowther 'Near Surface Bubble Excitation and Noise in the Ocean', in Advances in Underwater Acoustics, Proc Inst. of Acoustics Conf. Portland (1981)
6. P A Crowther 'Modelling of Sea Surface Scattering', in Scattering Phenomeman in Underwater Acoustics, Proc Inst. Acoustics (UK) 7 (3), 49 - 57 (1985)
7. P S De Leobinus and L S Simpson, 'Case Study of Duration - Limited Wave Spectra' J Geophys Res 77, 4555 - 4569 (1972)
8. B D Johnson and R C Cooke, 'Bubble Populations and Spectra in Coastal Waters' J Geophys Res 84, 3761 - 3766 (1979)
9. D A Kolovayev 'Investigation of the Concentration and Statistical Size Distribution of Wind-Produced Bubbles in the Near Surface Ocean, Oceanology (Eng Transl) 15, 659 - 661, (1976)
10. S A Thorpe 'On the Clouds of Bubbles Formed by Breaking Wind-Waves in Deep Water' Phil. Trans Roy Soc A304, 155 - 210 (1982)
11. B R Kerman et al 'Wind Dependence of Underwater Ambient Noise', Boundary-Layer Meteorology 26, 105 - 113 (1983)
12. E C Monahan 'Ocean Whitecaps', J Phys Oceanogr, 1, 139-144 (1971)

BUBBLE DYNAMICS IN OCEANIC AMBIENT NOISE

Andrea Prosperetti
Department of Mechanical Engineering
The Johns Hopkins University
BALTIMORE MD 21210
U.S.A.

Abstract. A concise review of the physical processes involving bubbles which appear to have a bearing on oceanic ambient noise is presented. In the second part of the paper elements of a theory of bubble-related noise in the ocean are outlined.

1 Introduction

There is ample experimental evidence that the upper layers of oceans and other water bodies contain a large number of gas bubbles [1-5]. While some of these bubbles are the product of biological activity, the vast majority is due to entrainment of air into the water by a number of different processes such as breaking waves, surface turbulence, and water drop impact. In his classic review paper Wenz [6] mentions several mechanisms by which these bubbles can affect ambient noise, and many authors have developed models to establish a quantitative connection between these mechanisms and overall noise levels (see e.g. [7-11]). In the present paper we shall review several aspects of bubble dynamics relevant to the emission and absorption of sound, and we shall sketch possible connections with oceanic ambient noise.

2 Equilibrium Radius

Small bubbles are spherical because of surface tension. Their equilibrium radius R_0 is determined by the competition between the forces that tend to collapse the bubble, which are the ambient pressure p_∞ and the excess pressure $2\sigma/R_0$ due to surface tension, and the forces which tend to expand the bubble, i.e. the liquid vapor pressure p_v and the pressure p_g of the gas in the bubble. Since from the equation of state $p_g = K/R_0^3$, where K is a constant, the equation which determines the equilibrium radius is

B. R. Kerman (ed.), Sea Surface Sound, 151–171.
© *1988 by Kluwer Academic Publishers.*

$$p_v + \frac{K}{R_0^3} = p_\infty + \frac{2\sigma}{R_0} \tag{2.1}$$

It is readily shown that this equation has one real positive root for $p_\infty > p_v$, while it has two real positive roots in the opposite case. In this latter case the smaller root is stable while the larger one is unstable, with the bubble radius growing indefinitely or shrinking spontaneously towards the smaller, stable value. For $p_\infty < p_{0cr}$ where $p_{0cr} = p_v - (32\sigma^3/27K)^{1/2}$ corresponding to a radius

$$R_{0cr} = \frac{4}{3} \frac{\sigma}{p_v - p_{0cr}} \tag{2.2}$$

no physically acceptable solution of (2.1) exists, the expanding forces dominate, and the bubble grows indefinitely. The vapor pressure of pure water at $15^\circ C$ is 0.017 bars and can therefore safely be neglected for the present purposes. The condition for the appearance of the second unstable branch of solutions of Eq. (2.1) is therefore that the pressure in the neighborhood of the bubble becomes negative, $p_\infty < 0$. It is hard to imagine that such large pressure fluctuations occur at all in the upper layers of the ocean, but if they do they must be quite exceptional events. The unstable branch of solutions of Eq. (2.1) seems therefore to be quite irrelevant for the ambient noise problem and it is surprising that an equation like (2.2) sometimes finds its way into ambient noise models [7].

Hydrostatic effects tend to perturb the spherical shape of a bubble in equilibrium. An estimate of their importance may be obtained by balancing the pressure difference over the bubble due to gravity, $2\rho g R_0$, with the pressure difference $2\sigma/R_0$ due to surface tension, with the result

$$R_0 \simeq (\sigma/\rho g)^{1/2} \tag{2.3}$$

We thus find that the spherical shape of bubbles begins to be appreciably distorted by gravity only when their radius exceeds 2 mm approximately. Since a considerable fraction of the bubble population is smaller than this limit [1-5], we shall assume a spherical shape for most of the following.

3 Radial Dynamics

The radial dynamics of a spherical bubble in an incompressible liquid is governed by the Rayleigh-Plesset equation

$$R\ddot{R} + \frac{3}{2}\dot{R}^2 = \frac{1}{\rho}\{p_B - p_\infty[1 + F(t)]\}, \tag{3.1}$$

where the dimensionless function $F(t)$ describes the fluctuations of the ambient pressure about its static level, and p_B denotes the pressure on the wet surface of the bubble surface. This quantity is related to the bubble internal pressure p, by conservation of momentum across the bubble interface

$$p_i = p_B + \frac{2\sigma}{R} + 4\mu\frac{\dot{R}}{R} \tag{3.2}$$

where μ is the liquid viscosity. The internal pressure is the result of a complex interplay of physical processes, but for the time being we make the simple approximation that it is adequately described by an isothermal law

$$p_i = p_{i0}(R_0/R)^3 \tag{3.3}$$

where p_{i0} is the internal pressure corresponding to the equilibrium radius R_0 and p_v has been neglected. An expression for p_{i0} can be obtained noting that, for equilibrium, $F(t) = 0, \dot{R} = \ddot{R} = 0$ so that, from (3.1), $p_B = p_\infty$. Equation (3.2) gives then

$$p_{i0} = p_\infty + \frac{2\sigma}{R_0} \tag{3.4}$$

The dynamical behavior of a bubble in a variable pressure field is essentially that of a non-linear oscillator the inertia of which is due to the displaced liquid and the restoring force to the compressed gas. This leads one to the expectation, which is indeed borne out by detailed numerical solutions [12], that dynamical effects, represented by the left-hand side of Eq. (3.1), become appreciable only for forcings having a time scale comparable with the natural frequency of the system. As will be shown below, for the range of bubble radii of interest, this time scale is far shorter than that associated with wave motion or turbulent fluctuations in the water and therefore, in describing the bubble response to these agents, it is possible to make use of the quasi-static relation

$$p_{i0}(R_0/R)^3 - p_\infty[1 + F(t)] - 2\sigma/R = 0, \tag{3.5}$$

obtained by equating the right-hand side of (3.1) to 0 and using (3.3). A graph of the solutions of this equation for different values of F is presented in Fig. 1 for several bubble radii and $\sigma = 72$ erg/cm^2. It is evident from this figure that, for realistic pressure fluctuations in the ocean, one cannot expect changes in the radius of more than 10-20%. The violent phenomena associated with cavitation as conventionally understood are therefore completely ruled out.

154

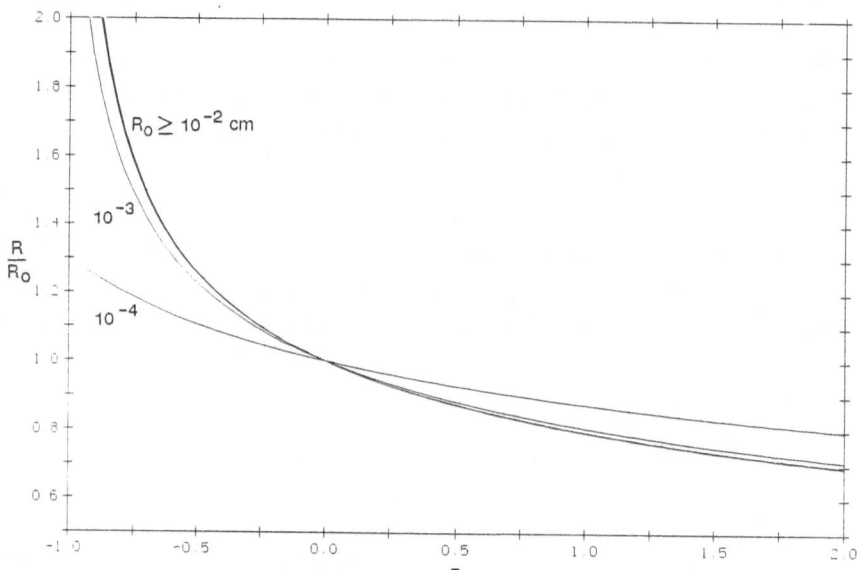

Figure 1: Bubble compression and expansion in response to changes in the dimensionless ambient pressure according to Eq. (3.5).

4 Radial Oscillations

The preceding considerations indicate that most processes by which gas bubbles affect oceanic ambient noise involve small-amplitude oscillations about the equilibrium radius. Larger-amplitude motion may also be possible, for example at the moment of creation of a bubble by a breaking wave or by a droplet impinging on the water surface, but the essence of the phenomena may be captured by considering a linear model. It may also be noted that oscillations in which the spherical shape of the bubble is distorted may also occur. However motions in which the volume of the bubble changes are by far more important as far as sound generation is concerned, so that again we are led to the consideration of radial pulsations.

Equation (3.1) is inadequate to study sound radiation because it treats the liquid as incompressible. Various forms of the radial equation including acoustic effects exist, which are all equivalent to first order in the liquid compressibility [13, 14]. The most convenient one for our purposes is [14]

$$\left(1 - 6\frac{\dot{R}}{c}\right) R\ddot{R} + \frac{3}{2}\left(1 - \frac{4}{3}\frac{\dot{R}}{c}\right)\dot{R}^2 - \frac{1}{c}R^2\,\dddot{R} \tag{4.1}$$

$$= \frac{1}{\rho}\{p_i - p_\infty[1 + F(t)] - \frac{2\sigma}{R} - 4\mu\frac{\dot{R}}{R}\}$$

where c is the speed of sound in the liquid. To study small-amplitude oscillations about the equilibrium radius R we set

$$R = R_0[1 + X(t)] \quad , p_i = p_{io}[1 + P(t)] \tag{4.2}$$

and linearize with the result

$$\ddot{X} - \frac{R_0}{c}\dddot{X} + 4\cdot\frac{\mu}{\rho R_0^2}\dot{X} - \frac{2\sigma}{\rho R_0^3}X = \frac{p_{io}}{\rho R_0^2}[P - (1-w)F] \tag{4.3}$$

where use has been made of (3.4) and w is given by

$$w = \frac{2\sigma}{R_0 p_{io}} = \frac{2\sigma}{R_0 p_\infty + 2\sigma}. \tag{4.4}$$

For the more detailed analysis which we are now pursuing the simple relation (3.3) for the internal pressure is no longer suitable. This quantity must be obtained by solving the complete set of conservation equations in the bubble. This is feasible for linear harmonic motion for which

$$F \propto \exp(-i\omega t) \tag{4.5}$$

and one finds a result that may be written as [15,16]

$$P = -X \operatorname{Re}\phi + \frac{1}{\omega}\dot{X} \operatorname{Im}\phi \tag{4.6}$$

where ϕ is a complex-valued function of the variable

$$\theta = R_0 \left(\frac{2\omega}{\chi}\right)^{1/2} \tag{4.7}$$

in which χ is the gas thermal diffusivity.

Even though we are now concerned only with solutions depending on time through the exponential factor $\exp(-i\omega t)$ it is interesting to rewrite (4.3) using (4.6) and the relation $\dddot{X} = -\omega^2\dot{X}$ in the form

$$\ddot{X} + 2\left(\frac{\omega^2 R_0}{2c} + \frac{2\mu}{\rho R_0^2} + \frac{p_{io}}{2\rho R_0^2\omega}\operatorname{Im}\phi\right)\dot{X} + \frac{p_{io}}{\rho R_0^2}(\operatorname{Re}\phi - w)X = -\frac{p_\infty}{\rho R_0^2}F(t) \tag{4.8}$$

which enables one to determine at a glance the effective stiffness of the bubble (the coefficient of X) and the effective damping (the coefficient of $2\dot{X}$). It is clear from (4.7) that both these quantities depend on the oscillation frequency ω. In particular, the frequency of oscillation for free motion must be determined by solving the implicit equation

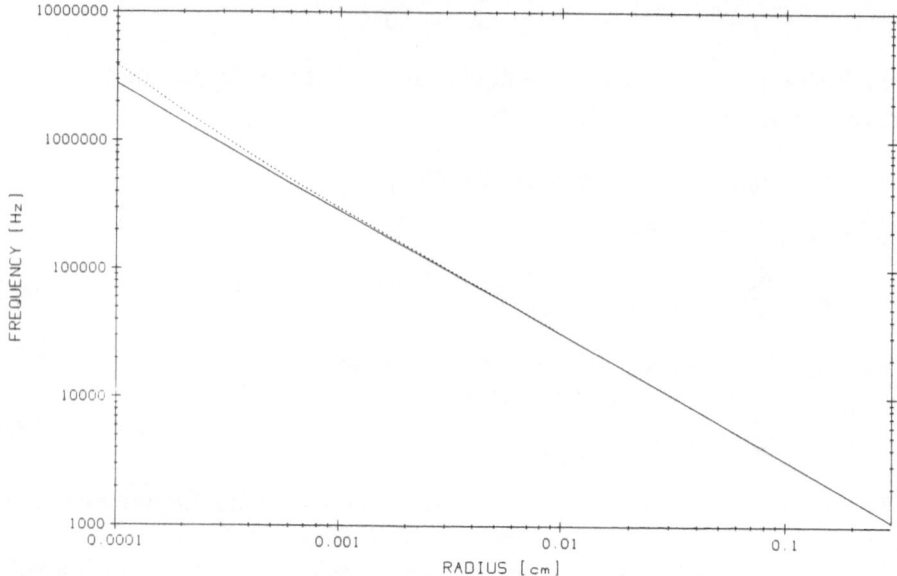

Figure 2: Natural frequency for radial pulsations of air bubbles in water at 1 atm, Eq.(4.9). Dottes line: zero surface tension

$$^{\bullet}\omega_0^2 = \frac{p_{io}}{\rho R_0^2}[\mathrm{Re}\phi(\theta_0) - w], \tag{4.9}$$

in which θ_0 is given by (4.7) with ω_0 in place of ω . (In this procedure we neglect the small frequency shift caused by damping). We show in Fig. 2 a graph of $f_0 = \omega_0/2\pi$ where ω_0 is the solution of this equation for an ambient pressure p_∞ of 1 atm and $\sigma=72.5$ cm^2/sec (continuous line) and $\sigma=0$ (dashed line). It may be noted that for small values of θ (i.e., nearly isothermal bubble) $\phi \to 3$ so that (4.9) gives

$$\omega_0^2 = \frac{p_{io}}{\rho R_0^2}(3 - w) \tag{4.10}$$

This result is applicable for R_0 smaller than 10 μm, approximately. For large values of θ (i.e., nearly adiabatic bubble) one has instead $\phi \to 3\gamma$ so that (4.9) becomes

$$\omega_0^2 = \frac{p_{io}}{\rho R_0^2}(3\gamma - w) \tag{4.11}$$

This result is applicable for R_0 greater than 1 mm approximately. For both estimates we have considered air bubbles in water at 1 atm.

The coefficient of $\dot{X}$ in Eq. (4.8) is twice the oscillations. We may distinguish three contributions to this quantity, the thermal damping, the acoustic damping and the viscous damping given by

$$eq : 10\beta_{th} = \frac{p_{io}}{2\rho R_0^2 \omega} \mathrm{Im}\phi, \quad \beta_{ac} = \frac{\omega^2 R_0}{2c}, \quad \beta_v = \frac{2\mu}{\rho R_0^2}, \qquad (4.12\mathrm{a,b,c})$$

respectively. All but the last one depend on the frequency ω. We show in Fig.3 their values calculated at the resonance frequency (4.9) as a function of the bubble radius, and in Fig. 4 the ratio of each contribution to the total damping. This figure is particularly interesting and shows that for air bubbles in water at 1 atm thermal damping is dominant for radii between 3 μm and 3 mm, approximately. Acoustic energy loss becomes significant for radii greater than about 200 μm, while viscous dissipation is only important below this value.

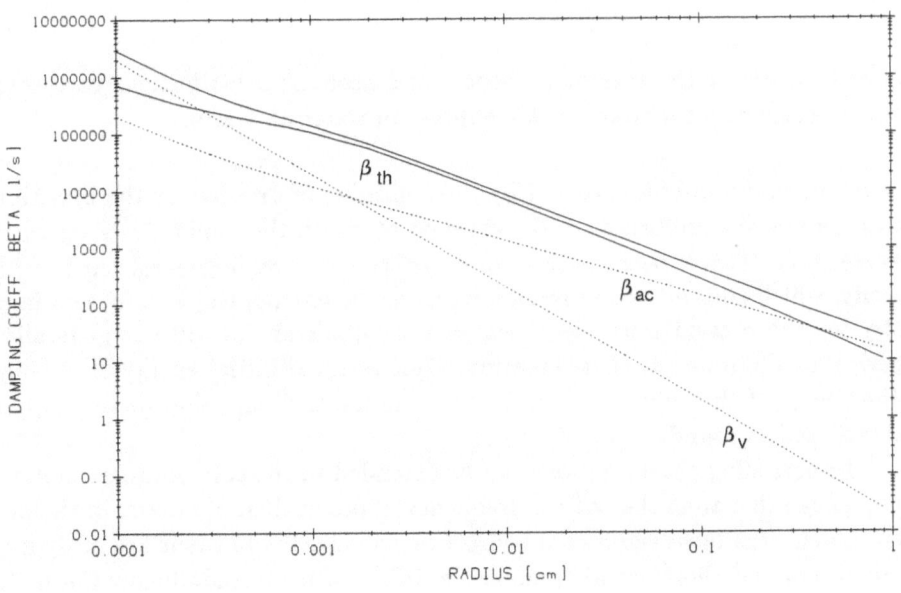

Figure 3: Damping coefficient at resonance for air bubbles in water at 1 atm. The thermal, acoustic, and viscous contributioins are shown separately.

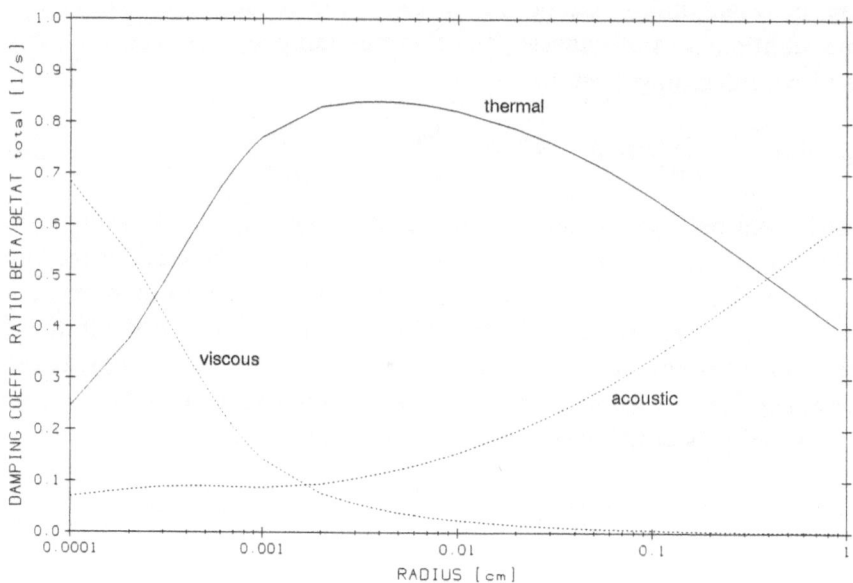

Figure 4: Ratio of the thermal, viscous, and acoustic contributions to the total damping coefficient for resonant air bubbles in water at 1 atm.

In the ocean bubbles created by the splashing of droplets or the breaking of waves are not in equilibrium at the moment at which the liquid closes up entrapping some air. This is because the liquid surface will have in general some residual velocity, which goes into compression work on the entrapped gas and, less importantly for the present purposes, because the bubble shape will not generally be spherical at the moment of its creation. This residual initial energy is dissipated in the course of the subsequent oscillatory motion and only the fraction β_{ac}/β of it is radiated as sound.

The preceding considerations can be extended to the case of forced motion at a frequency other than the natural frequency of oscillation. However, in the ocean environment, the most significant sources of pressure fluctuations are of hydrodynamic origin and therefore at frequencies orders of magnitude below the natural frequency. For these conditions, the simple approach of section 3 is adequate.

5 Shape Oscillations

In the case of small amplitude all the oscillation modes superpose linearly and, to study shape oscillations, we may consider a bubble having a fixed equilibrium radius R_0 with the free surface given by

$$r = R_0 + a_n(t)P_n(\cos\theta) \tag{5.1}$$

where P_n is a Legendre polynomial and $a_n(t)$ is the time- varying amplitude of the n-th mode. Assuming a time dependence for a_n proportional to $\exp(-i\bar{w}_n t)$, with $\bar{w}_n$ the complex frequency, we may write the equation of motion for a_n in a form which enables us to determine at a glance the damping effects, namely [17-19]

$$\ddot{a}_n + 2\left[(n+2)(2n+1)\frac{\mu}{\rho R_0^2}\right.$$

$$\left. + \frac{R_0}{2c}\frac{\omega_n^2}{n+1}\frac{H_{n-1/2}^{(1)}(\bar{w}_n R_0/c)}{iH_{n+1/2}^{(1)}(\bar{w}_n R_0/c)}\right]\dot{a}_n + \omega_n^2 a_n = 0 \tag{5.2}$$

where the $H^{(1)}$'s are Hankel functions and ω_n is given by

$$\omega_n^2 = (n-1)(n+1)(n+2)\sigma/\rho R_0^3. \tag{5.3}$$

If the damping effects are negligible, this quantity is the well-known frequency of the shape oscillations in an incompressible, inviscid fluid [20]. For small damping ω_n can be substituted for $\bar{w}_n$ in the coefficient of $\dot{a}_n$ and one finds the following expressions for the viscous and acoustic damping constants of the shape oscillations

$$\beta_{n,v} = (n+2)(2n+1)\frac{\mu}{\rho R_0^2}, \quad \beta_{n,ac} = \frac{R_0}{2c}\frac{\omega_n^2}{n+1}\frac{H_{n-1/2}^{(1)}(\omega_n R_0/c)}{iH_{n+1/2}^{(1)}(\omega_n R_0/c)}. \tag{5.4}$$

In all cases of interest $\omega_n R_0/c \ll 1$ so that the arguments of the Bessel functions can be approximated to find, for the n-th mode, a value of the radiation damping of the order of $\omega_n \times (\omega_n R_0/c)^{2n+1}$, and hence very small. Thus we conclude that a bubble executing shape oscillations is a very inefficient acoustic radiator.

6   Acoustic Radiation

The pressure perturbation caused by a bubble executing linear radial oscillations at a frequency ω is given in the far field (i.e., more than a wavelength away from the bubble), by [17]

$$p_0 = -\rho\omega^2 R_0^2 X_0\left(\frac{R_0}{r}\right)\exp\{i[k(r-R_0)-\omega t]\} \tag{6.1}$$

In writing this formula we have assumed $\omega R_0/c \ll 1$, r is the distance from the bubble center, and X_0 is the (gradually decreasing because of damping) amplitude of the oscillations [cf. (4.2)]. An alternative form in which this equation may be written is

$$p_0 = \frac{\rho}{4\pi r}\left[\frac{d^2 V}{dt^2}\right]_{ret} \tag{6.2}$$

where $V(t)$ is the bubble volume and the subscript ret indicates the retarded time $t - (r - R_0)/c$. This form, which coincides with the standard law of monopole radiation, is approximately valid also in the case of non-spherical bubbles since in this case V differs from the volume of a sphere having the same mean radius only at second order in the quantities a_n defined in the previous section.

The far-field expression for the pressure radiated by shape oscillations at frequency $\bar{\omega}_n$ is instead [17]

$$p_n = -\frac{2^n n! i^{-n}}{(n+1)(2n)!}\omega_n^2 \rho R_0 \left(\frac{\bar{\omega}_n R_0}{c}\right)^n \left(\frac{R_0}{r}\right) \\ \exp\{i[k(r - R_0) - \bar{\omega}_n t]\}P_n a_n. \tag{6.3}$$

Neglecting phase factors and angular dependence the ratio of the pressure field due to the n-th non-spherical mode to that due to radial oscillations is

$$\frac{p_n}{p_0} \sim \frac{2^n n!}{(n+1)(2n)!}\left(\frac{\bar{\omega}_n}{\omega}\right)^2 \left(\frac{\bar{\omega}_n R_0}{c}\right)^n \frac{a_n/R_0}{X_0}. \tag{6.4}$$

The factor $(\bar{\omega}_n R_0/c)^n$ renders this ratio very small. For example, for a 1 mm bubble in water (which, at 1 bar, has a natural frequency ω_0 of 2.78 kHz), taking $\bar{\omega}_n = \omega_0, n = 2, a_2/R_0 = X_0$ we find $p_2/p_0 = 1.4 \times 10^{-7}$. This ratio decreases even further with the radius and with increasing n, which justifies the neglect of the acoustic radiation due to shape oscillations. A very enlighting discussion of several aspects of sound emission by oscillating bubbles can be found in the classic paper by Strasberg [17].

7 Mass Diffusion Effects

The equilibrium radius of a bubble is determined by the mass m of gas contained in it according to the perfect gas law

$$R_0 = \left(\frac{3}{4\pi}\frac{m}{M}\frac{R_G T}{p_{io}}\right)^{1/3} \tag{7.1}$$

where M is the gas molecular weight, T is the temperature, and R_G is the universal gas constant. Diffusion processes across the bubble interface affect m and cause therefore changes in R_0. In considering these phenomena it must be stressed that they occur on time scales orders of magnitude longer than those for oscillations and amount therefore to changes in the parameter R_0 free of any dynamical effect.

If c denotes the gas concentration in the liquid (expressed as a density), the variation in m is described by

$$\frac{dm}{dt} = 4\pi R^2 D \left.\frac{\partial c}{\partial r}\right|_R \tag{7.2}$$

where D is the diffusion coefficient. For the case of mass diffusion in a constant pressure field, the relevant length scale for the gradient in (7.2) is R_0 and we may rewrite the equation approximately as

$$\frac{dm}{dt} = \frac{3D}{R_0^2} \frac{\Delta c}{\rho_G} m, \tag{7.3}$$

where Δc is the difference in gas concentrations at the bubble surface and far away and ρ_G is the gas density in the bubble. From this equation we can read off directly the time scale t_D for this process

$$t_D = \frac{R_0^2}{3D} \frac{\rho_G}{\Delta c} \tag{7.4}$$

Typically $D = 2 \times 10^{-5}$ cm^2 /sec, $\rho_G = 10^{-3}$ g/cm^3, and $\Delta c = 2 \times 10^{-5}$ g/cm^3 so that, for $R = 1, 0.1$, and $0.01mm$ we find $t_D = 2.5 \times 10^4$, 250, and 2.5 sec respectively. While these estimates are very crude, they are a clear indication of the slowness of mass diffusion processes in a constant pressure field, which is in agreement both with experiment and theory [21,22].

Another mass diffusion process peculiar to bubbles is the so-called rectified mass-diffusion. In this a bubble executing oscillations about its equilibrium radius can very slowly change the amount of gas it contains due to the fact that the mass of gas which diffuses into it during expansion is slightly greater than that which diffuses out during contraction. In this way the gas content of the bubble averaged over one oscillation undergoes a slow (i.e., on time scales measured in hundreds or thousands of oscillation periods) variation in time. This process can change the mass contained in the bubble by orders of magnitude over a period of time (of the order of tens of minutes), but each oscillation has a small amplitude about the instantaneous value of the equilibrium radius, much in the same way of a pendulum the length of which is a slowly varying function of time. Therefore, as far as the radiated sound is concerned, a bubble growing by rectified diffusion differs from a non-growing one only in the fact that the frequency of sound emission during free oscillations changes very slowly with time [cf. Eq. (4.9)]

Mass gain by rectified diffusion occurs only for pressure amplitudes greater than a well-defined threshold. This threshold is completely unrelated to that associated with the unstable radius (2.2) and with the threshold for transient cavitation (which, it might be added, is a very ill-defined quantity). Confusion among this three completely unrelated concepts has led in the past to totally fallacious ambient noise modelling efforts [7].

Strict rectified diffusion presupposes a long-time coherence of the driving pressure, which is perhaps unrealistic to expect for a bubble entrained in a turbulent field. However it is certain that the gas contained in oceanic bubbles varies in response to ambient pressure changes caused by hydrostatic variations (e. g., due to entrainement at greater depths by turbulence or Langmuir circulation), wave motion, and turbulence. Presumably the mode in which this change is effected is intermediate between the two outlined above. Whatever the precise mechanism, however, it is important to keep in mind the relative slowness of the process.

8   Collective Phenomena

The volume of air entrained in a spilling breaker can be in the range 10-20% and therefore interactions among bubbles may become important. Furthermore, even at much lower bubble concentrations, there may be a large number of bubbles present in one sound wavelength which may not interfere with each other hydrodynamically, but can have a large integrated effect on acoustic waves. Despite some recent advances (see e.g. [25,26]) the theory of bubbly liquids is still in a very rudimentary state and we shall be content here with the derivation of an expression for the complex celerity of pressure waves by a simple argument which bypasses the complexities associated with bubble-bubble interactions. Let α denote the gas volume fraction in a bubbly liquid, i.e. the volume occupied by the gas in a unit volume of mixture. Since the air density is so much lower than that of the water we may approximate the mixture density ρ_m by

$$\rho_m = (1 - \alpha)\rho \tag{8.1}$$

We suppose for simplicity that all bubbles have the same radius R_0 and have a number density n, so that $\alpha = (4/3)\pi R^3 n$. The speed of sound c_m in the mixture is given by $c_m^{-2} = d\rho_m/dp$ and, differentiating (8.1) taking n=constant (i.e., the bubbles move with the liquid) and ρ=constant (justified by the far greater compressibility of the gas with respect to the liquid) we find

$$\frac{1}{c_m^2} = -4\pi \rho R^2 n \frac{dR}{dp}. \tag{8.2}$$

The derivative dR/dp can be calculated in the linear approximation as $dR/dp = (R_0/p_\infty)X/F$ and, from (4.8), we find

$$\frac{dR}{dp} = -\frac{1}{\rho R_0}(\omega_0^2 - \omega^2 + 2i\beta\omega)^{-1}, \tag{8.3}$$

from which

$$\frac{1}{c_m^2} = \frac{3\alpha}{R_0^2(\omega_0^2 - \omega^2 + 2i\beta\omega)} \tag{8.4}$$

with β the total damping constant given by the sum of the three contributions (4.12). The imaginary part of c_m is responsible for the attenuation of sound waves, which can be very dramatic. We show in Fig. 5 a comparison with experiment of the attenuation coefficient in dB/cm calculated (essentially) from (8.4) using the theory given in section 4 for the damping and the resonance frequency. The bubbles have a radius of 2.6 mm and a natural frequency $\omega_0/2\pi = 1.25$ kHz approximately. The volume fraction is 1%. The large attenuation over a very broad band which extends from the resonance frequency up to about 20 kHz is particularly worthy of notice. Lowering the gas volume fraction has the effect of reducing this high absorption band. If a distribution of bubble sizes is present, the absorption band can become very broad even at low volume fraction.

For oceanic ambient noise the particular form of (8.4) which obtains at frequencies far below ω_0 is important. In this limit, neglecting surface tension, we may use (4.10) to find

$$c_m^2 = \frac{p_\infty}{\rho \alpha},\tag{8.5}$$

independent of the bubble radius. For $p_\infty = 1$ atm and $\rho = 1$ g/cm we find $c_m = 318$ and 101 m/sec for $\alpha = 0.1$ and 1% respectively. This shows that even a minute quantity of bubbles has a large effect on the speed of sound which, in pure water, is of the order of 1500 m/sec.

As an application of this result consider, in a liquid, a cloud of bubbles having a linear dimension of order L. This cloud, regarded as a system of coupled oscillators, has a number of normal modes the lowest one of which can be estimated to be at a frequency given by

$$\omega_c \simeq \frac{c_m}{L}.\tag{8.6}$$

It will be seen that, when the cloud contains a relatively large number N of bubbles, ω_c is much smaller than the frequency of a single bubble and therefore the approximation (8.5) can be used for c_m . In this way, from (4.10) we find the following estimate of the ratio of the lowest normal mode of the cloud to that of one of its constituent bubbles (all assumed equal for simplicity)

$$\frac{\omega_c}{\omega_0} = \frac{1}{\alpha^{1/6} N^{1/3}}.\tag{8.7}$$

The volume fraction enters in this expression raised to such a small power as to contribute a number very close to 1. The ratio of the two frequencies is therefore seen to vary as the inverse cube root of the number of bubbles present in the cloud.

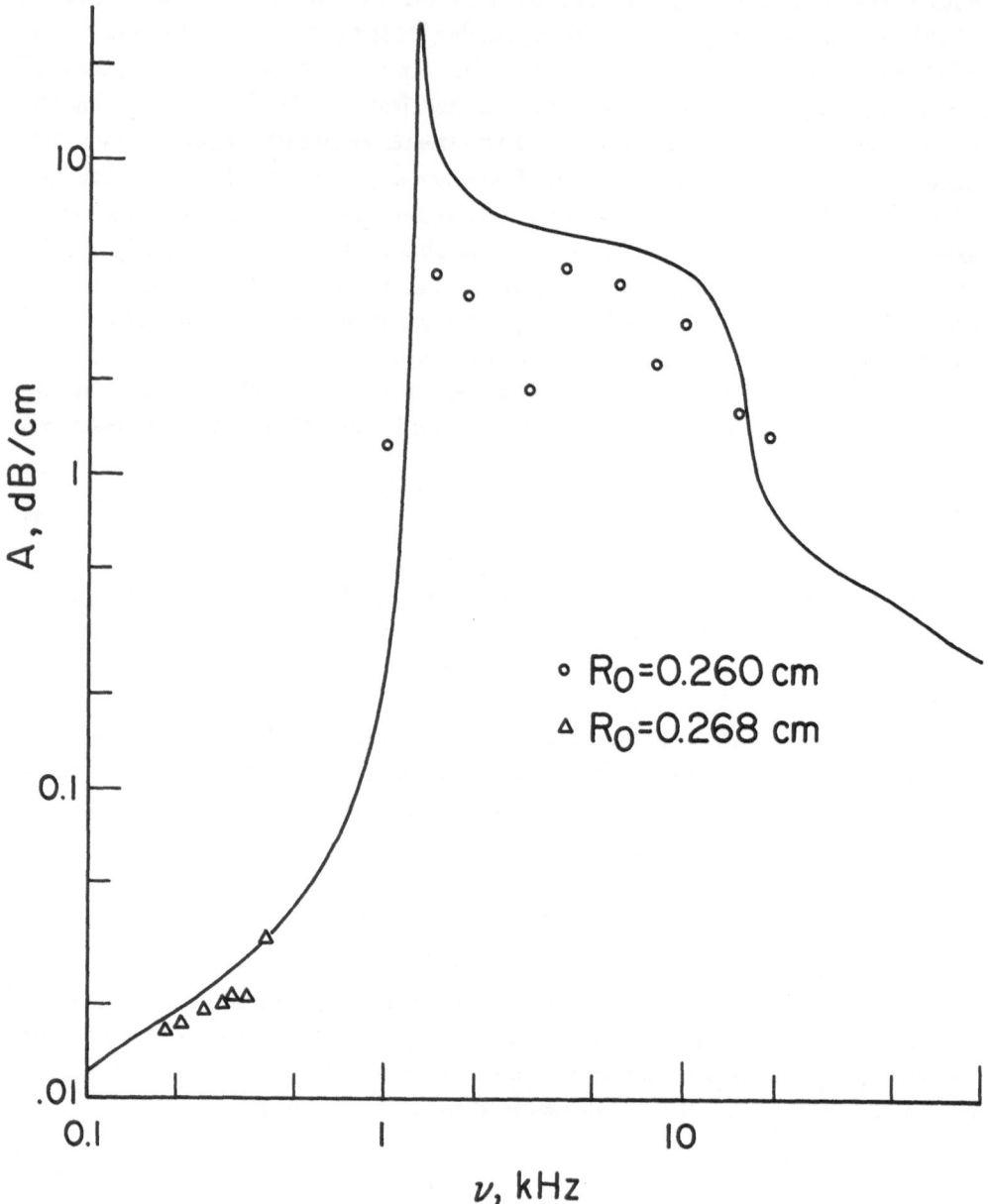

Figure 5: Comparison between theory and experiment for the attenuation coefficient of pressure waves in bubbly water at 1% gas volume fraction as a function of frequency.

9   Low Frequency: Amplification of Turbulence Noise

We present in this and in the following sections some elements for a theory of bubble-related ambient noise in the ocean. The emphasis will be on a qualitative description, and the framework for a quantitative theory will be only mentioned. More details will be found in Ref. [11].

It is well known that, acoustically, turbulence behaves as a quadrupole source [27, 28]. As a consequence turbulence is a weak acoustic source since the intensity radiated by a quadrupole is a factor $(L/\lambda)^4$ smaller than that of a monopole of the same strength. Here L is the linear dimension of the source and λ is the sound wavelength. The previous result presupposes $L < \lambda$, which is typical of the application to ambient noise at low frequency, and therefore $(L/\lambda)^4 << 1$.

The presence of bubbles in the neighborhood of the turbulent eddy increases the radiation efficiency dramatically [29]. The bubble responds to the turbulent pressure fluctuations by contracting and expanding in the manner characteristic of a simple acoustic source or monopole. The acoustic output of a turbulent eddy containing bubbles is therefore a factor $(\lambda/L)^4$ greater than that of the same eddy without bubbles. The increase can be estimated in several tens of dB's.

The bubbles most strongly affected by the mechanism described are those near the turbulent eddy and therefore they partake of the eddy's motion. For this reason the frequency spectrum of the acoustic emission is not the Eulerian frequency spectrum of the turbulence, but rather the Lagrangian one. The understanding of the Lagrangian aspects of turbulence is much less developed than that of its Eulerian characteristics, but certainly the Lagrangian frequency content extends to much lower frequencies, which can be estimated to be several tens of Hz, perhaps 100 Hz in strongly turbulent conditions. Studies in air turbulence have shown that typical Lagrangian frequencies are about one order of magnitude smaller than Eulerian ones [30].

A very rough estimate of the acoustic intensity that can be expected from this mechanism can be obtained by a crude procedure which has however the merit of highlighting its physical aspects. If the volume occupied by the bubbles in a unit control volume decreases or increases by $\Delta\alpha$, the same amount of liquid is expelled or attracted into the volume. The associated variation of liquid mass per unit volume is $\rho \, \Delta\alpha$. The apparent rate of liquid mass creation or destruction is therefore

$$q = \rho \, \frac{\partial \alpha}{\partial t}, \qquad (9.1)$$

where q is the monopole source intensity. Since $\rho\Delta\alpha$ equals very nearly the variation of the bubbly mixture density, we may alternatively write

$$q = \frac{1}{c_m^2} \frac{\partial p}{\partial t}, \qquad (9.2)$$

where c_m is the speed of sound in the mixture given, at low frequency, by (8.5). The acoustic pressure field in the liquid does not respond to q but to $\partial q/\partial t$. Furthermore, the free surface introduces a negative image source the effect of which can be roughly accounted for by the factor L/λ, where L is the depth of the acoustically active region. With these elements we deduce the following expression for the far-field pressure perturbation p' at a distance x from the source

$$p' \sim \frac{1}{c_m^2 x}\frac{L}{\lambda}\frac{\partial^2}{\partial t^2}\int pdV, \tag{9.3}$$

where the integral is over the source volume of linear dimension L. The instantaneous acoustic intensity is $p'^2/\rho c$. If an acoustic event (e.g., a turbulent spot) has the duration θ and is observed for a time T, however, only an intensity of the order (θ/T) times the instantaneous value is recorded on the average. Furthermore, what is important is the ensemble-averaged intensity $< I >$ over many such events. We have then

$$< I >= \frac{\theta}{T\rho c}\left(\frac{L}{\lambda c_m^2 x}\right)^2 < \left(\frac{\partial^2}{\partial t^2}\int p(\mathbf{y},t)d^3y\right)\left(\frac{\partial^2}{\partial t^2}\int p(\mathbf{y'},t)d^3y'\right) > \tag{9.4}$$

In the frequency domain $\partial^2/\partial t^2 \to \omega^2$ and the spectral density $\hat{I}(\omega)$ must be approximately given by $\hat{I}(\omega) \sim< I > /\omega_{max}$ if the turbulence spectrum is taken to be essentially flat up to a maximum frequency ω_{max}. This maximum frequency must be of the order of the inverse minimum Lagrangian correlation time scale τ_0. From (9.4) we then have

$$\hat{I}(\omega) \sim \frac{\theta L^2 \omega^6 \tau_0}{T\rho c^3 c_m^4 x^2}\int d^3y \int d^3\xi < p'(\mathbf{y},t)p'(\mathbf{y}+\xi,t) > \tag{9.5}$$

As a function of ξ the integral vanishes at distances exceeding a typical correlation length ℓ. For distances less than ℓ the turbulent pressure fluctuations have the order $\rho u'^2$, where u' is the turbulent velocity fluctuation. We then finally find

$$\hat{I}(\omega) \sim (\rho\overline{u'^2})^2\frac{\theta L^5 \ell^3 \omega^6 \tau_0}{T\rho c^3 c_m^4 x^2}, \tag{9.6}$$

a result which is borne out by the more detailed calculation of [11].

In order to obtain a numerical estimate of the noise level to which this result can give rise, we introduce an effective surface source density $\hat{S}(\omega)$ by setting

$$\hat{I}(\omega) = \frac{L^2}{x^2}\hat{S}(\omega). \tag{9.7}$$

From (9.5) the explicit expresson of $\hat{S}$ is

$$\hat{S} \sim (\rho \overline{u'^2})^2 \frac{\theta L^2 \ell^3 \omega^6 \tau_0}{T \rho c^3 c_m^4}. \tag{9.8}$$

We take the following numerical values: $\ell = 10cm$, $L = 1m$, $\theta = 20sec$, $T = 1sec$, $\overline{u'^2} = (10cm/sec)^2$, $p_0 = 1bar$, $\rho = 1g/cm^3$, $\omega = 2\pi \times 10Hz$, $c = 1500m/sec$, $\alpha = 5\%$, $\tau_0 = 0.1sec$ to find $\hat{S} \sim 63$ dB re 1 $\mu Pa/Hz$, which is of the order of the estimated surface source intensity at 10 Hz for a wind speed of 10 m/sec. A more realistic estimate should weigh this result with the space-time density of turbulent events and is therefore likely to result in a somewhat lower figure. Nonetheless it may be concluded that the proposed mechanism can give a very significant contribution to ambient noise in this frequency range where additional mechanisms (wave-wave interactions, air turbulence, and others) may be active as well.

A very convenient quantitative framework for the treatment of the preceding physical concept can be found in Lighthill's theory of aerodynamic noise [27-29] which is used for this purpose in Ref. [11]

10 High Frequency: Single-Bubble Oscillations

If a bubble is created with non-zero initial energy (for instance, in a compressed state) it executes an oscillatory motion during which this energy is dissipated and sound is radiated. In the ocean, in addition to drop impact, favorable conditions for this process occur especially during wave breaking. Indeed there is ample evidence of large bubble clouds formed by wave breaking and reaching a depth of several meters [4,5]. The details of air entrainment are poorly understood but it is known that very large gas volume fractions, 10-20%, are encountered in the "white water" topping a breaking wave [31].

Again it is instructive to give, in the same spirit as in the previous section, a rough estimate of the noise levels that can be expected owing to this mechanism.

We suppose that bubbles are created at the front of the mass of water spilling down the breaking wave. Let ℓ be the thickness of this "active" region, δ its depth, and L its lateral extent perpendicular to the direction of propogation of the wave. If the duration of the process is θ and the characteristic frequency of the oscillations ω_0, there will be a phase difference $\theta \omega_0$ between the oscillatory motion of the first and last bubbles to be created, and this phase difference grows linearly with time. The situatin is therefore similar to that of Fraunhofer diffraction from a slit, where the amplitude radiated by the slit must be multiplied by a factor $\sin (\gamma/2)/(\gamma/2)$, if γ is the phase difference associated with the different distances of the edges of the slit from the observer. As before, the image sources can roughly be accounted for by introducing a factor δ/λ. From the theory of elementary monopole radiation, using as a source strength q defined in (9.1), we then have

$$p' \sim \frac{1}{x} \rho \frac{\delta}{\lambda} \frac{\sin \frac{1}{2}\omega_0\theta}{\frac{1}{2}\omega_0\theta} \frac{\partial^2}{\partial t^2} \int \alpha dV \tag{10.1}$$

or, substituting ω_0 for $\partial/\partial t$,

$$p' \sim \frac{1}{x} \rho \frac{\delta}{\lambda} \omega_0^2 \Delta\alpha (L\ell\delta) \frac{\sin \frac{1}{2}\omega_0\theta}{\frac{1}{2}\omega_0\theta}. \tag{10.2}$$

Let X_0 be the oscillation amplitude of a bubble relative to its undisturbed radius R_0. Then $\Delta\alpha \sim nR_0^3 X_0$, where n is the number of bubbles per unit volume. Since the oscillations are damped with a damping constant β, the spectral distribution of p' is $\hat{p}' \sim p'/\beta$ around ω_0, and zero elsewhere. For the spectral density $\hat{I} \sim \hat{p}'^2/\rho c$ averaged over a time T we then have

$$\hat{I}(\omega) \sim \frac{\rho}{Tc^3x^2} \left(\frac{n\omega_0^3 L\ell\delta^2 R_0^3 X_0}{\beta} \frac{\sin \frac{1}{2}\omega_0\theta}{\frac{1}{2}\omega_0\theta} \right)^2. \tag{10.3}$$

Similarly to (9.7) a source density can be defined by

$$\hat{I} = \frac{L\ell}{x^2} \hat{S}(\omega), \tag{10.4}$$

to find

$$\hat{S} \sim \frac{\rho L\ell}{Tc^3} \left(\frac{n\omega_0^3 \delta^2 R_0^3 X_0}{\beta} \frac{\sin \frac{1}{2}\omega_0\theta}{\frac{1}{2}\omega_0\theta} \right)^2 \tag{10.5}$$

or, since with the neglect of surface tension $\omega_0^2 R^2 \sim p/\rho$,

$$\hat{S} \sim \frac{L\ell p^3}{Tc^3\rho^2} \left(\frac{n\delta^2 X_0}{\beta} \frac{\sin \frac{1}{2}\omega_0\theta}{\frac{1}{2}\omega_0\theta} \right)^2 \tag{10.6}$$

where p is the static ambient pressure acting on the bubble. With the numerical values $L = 1m$, $\ell = 10cm$, $\delta = 10cm$, $p_0 = 1bar$, $X_0 = 0.2$, $n = 10cm^{-3}$, $T = 1sec$, $c = 1500m/sec$, $\rho = 1g/cm$, $\beta = 500Hz$, $\omega_0 = 2\pi \times 3 \ kHz$, $\theta = 20sec$ we find $S \sim 67.2 \ dB$ re $1\mu Pa/Hz$ which is of the order of the measured values for winds blowing with a velocity of 15 m/sec. As in the previous estimate, this figure must be reduced to account for the fact that the process described does not occur 100% of the time nor occupies 100% of the ocean surface. However it is clear that the noise level which can be expected is consistent with measured values and that this process can be a viable candidate as a source mechanism.

To proceed beyond the preceding rough estimates an approach based again on Lighthill's theory can be developed [11]. However, since at high frequency the mixture speed of sound has a rather complex structure, it is more useful to phrase the theory in terms of the effective mass source (9.1) in a manner similar to that of Ref. [29].

11 Intermediate Frequencies: Collective Oscillations

Most ambient noise data show a broad maximum, at least relative, in the neighborhood of 500 Hz [6, 9]. Difficulties arise in trying to ascribe the origin of this part of the spectrum to single bubble oscillations because, as shown in Figure 2, a bubble resonant in this frequency region would have a diameter of the order of one centimeter. It does not seem possible to imagine a mechanism capable of giving rise to a sufficient number of such large bubbles in practically any condition as is implied by the data. We believe that one possible mechanism for this emission is the collective oscillations of bubble clouds. As was remarked above, bubbles are created in a large number during wave breaking. They find themselves in the near field (in a hydrodynamic sense) of each other and behave therefore as a system of coupled oscillators with a whole range of collective normal modes, and it is the frequencies corresponding to these normal modes which would be radiated as the bubbles in the cloud oscillate. The highest of these modes lie close to the natural frequency of single bubbles. The lowest one has been estimated above in section 8.

A rough estimate of the acoustic levels that can be expected can be obtained from the result (10.6) derived earlier simply by substituting ω_N for ω_0. By use of (8.7) and (4.10) this step leads to

$$\hat{S} \sim \frac{L\ell p^3}{Tc^3\rho^2} \left(\frac{n\delta^2 X_0}{\alpha^{\frac{1}{2}}N^{\frac{1}{2}}\beta} \frac{\sin\frac{1}{2}\omega_N\theta}{\frac{1}{2}\omega_N\theta} \right)^2 \tag{11.1}$$

With the same numerical values as before and $\alpha = 10\%$, $N = 200$, $\beta = 500Hz$, $\omega_N = 2\pi \times 0.5kHz$ we find $\hat{S} \sim 69.7\ dB$ re $1\mu Pa/Hz$ which, again, is of the right order.

Also in this case an adaptation of aerodynamic noise theory can be developed for more quantitative estimates [11].

12 Conclusions

In the preceding sections we have presented an account of several aspects of bubble dynamics which appear to be pertinent to the question of ambient noise in the ocean, and we have tried to indicate a possible connection between bubble behavior and observed noise. For a fuller exposition of this second part of the work the reader is referred to [11]. We would like to conclude by briefly mentioning other possible mechanisms by which bubbles might contribute to oceanic noise.

The bursting or "popping" noise of bubbles at the ocean surface has recently been suggested as an acoustic source [10]. On the basis of some simple estimates it appears that, in order to match observed levels, a number of bursting bubbles per square centimeter per second in the order of hundreds of millions or more is required. This conclusion casts some doubts on the physical realism of this mechanism.

The wake of bubbles ascending in a non-rectlinear motion has also been indicated as a possible acoustic source [9]. The acoustic emission would have a quadrupole nature, and would therefore be rather weak. However, volume pulsations of the bubbles would be generated in these conditions, with correspondingly higher acoustic levels. The frequency of such emissions, which are forced by the hydrodynamics of the bubble motion, would be limited to a few tens of Hz at the most.

In conclusion it may be noted that acoustic emissions from bubbles created by the impact of drops on the ocean surface have long been known [32], as well as those arising from the break-up of a large bubble in a turbulent field [17]. In both cases one deals with free oscillations similar to those described in sections 4 and 10.

This study has been supported by the Underwater Acoustics Division of the Office of Naval Research.

References

1. H. Medwin, J. Geophys. Res. 82, 971-976, 1977.
2. B.D.Johnson and R.C.Cooke, J. Geophys. Res. 84, 3761, 1979.
3. J. Wu, J. Geophys. Res. 86, 457, 1981.
4. S.A. Thorpe, Phil. Trans. R. Soc. Lond. A304, 155, 1982.
5. S.A. Thorpe and A.J. Hall, Cont. Shelf Res. 1, 353, 1983.
6. G.M. Wenz, J. Acoust Soc. Am. 34, 1936, 1962.
7. A.V. Furduev, Atmos. Oceanic Phys. 2, 314, 1966.
8. P.A. Crowther, presented at the Underwater Acoustics Meeting of the Institute of Acoustics, Portland, 1980.
9. B.R. Kerman, J. Acoust. Soc. Am. 75, 149, 1984.
10. E.C. Shang and V.C. Anderson, J. Acoust. Soc. Am. 79, 964, 1986.
11. A. Prosperetti, to be submitted to J. Acoust. Soc. Am.
12. W. Lauterborn, J. Acoust. Soc. Am. 59, 283, 1976.
13. A. Prosperetti and A. Lezzi, J. Fluid Mech. 168, 457, 1987.
14. A. Prosperetti, submitted to The Physics of Fluids.
15. A. Prosperetti, Ultrasonics 22, 69, 1984.
16. A. Prosperetti, L.A. Crum, K.W. Commander, Acoust. Soc. Am., in press.
17. M. Strasberg, J. Acoust. Soc. Am. 28, 20, 1956.
18. A. Prosperetti, Q. Appl. Math. 35, 339, 1977.
19. A. Prosperetti, Ultrasonics 22, 115, 1984.
20. H. Lamb, Hydrodynamics, 6th Edition, Cambridge University Press, 1932.

21. P. Epstein and M.S. Plesset, J. Chem. Phys. **18**, 1505, 1950.
22. I.M. Krieger, G.W. Mulholland, and C.S. Dickey, J. Phys. Chem. **71**, 1123, 1967.
23. L.A. Crum, J. Acoust. Soc. Am. **68**, 203, 1980.
24. L.A. Crum, Ultrasonics **22**, 215, 1984.
25. R.E. Caflisch, M.J. Miksis, G.C. Papanicolaou, and L. Ting, J. Fluid Mech. **153**, 259, 1985.
26. M.J. Miksis and L. Ting, Phys. Fluids **29**, 603, 1986.
27. M.J. Lighthill, Proc. Roy. Soc. Lond. **A211**, 564, 1952.
28. D.G. Crighton, Prog. Aerospace Sci. **16**, 31, 1975.
29. D.G. Crighton and J.E. Ffowcs-Williams, J. Fluid. Mech. **36**, 585, 1969.
30. D.J. Shlien and S. Corrsin, J. Fluid Mech. **62**, 255, 1974.
31. M.S. Longuet-Higgins and J.S. Turner, J. Fluid Mech. **63**, 1, 1974.
32. G.J. Franz, J. Acoust. Soc. Am. **31**, 1080, 1958.

THE HORIZONTAL STRUCTURE AND DISTRIBUTION OF BUBBLE CLOUDS

S.A.Thorpe
Department of Oceanography
The University
Southampton
SO9 5NH, U.K.

ABSTRACT. We review and extend the knowledge of the shape of bubble clouds and describe the factors which lead to their uneven distribution in space and time in local region of the ocean. The information is based on observations using upward-pointing moored, or bottom-mounted, sonars and towed sonars. Side-scan sonar in particular provides much information about the movement of bubble clouds, their persistence, and the processes which affect their distribution.

1. INTRODUCTION

Surface waves breaking in deep water produce clouds of bubbles which persist below the sea surface as identifiable acoustic targets for periods of minutes. They are carried and dispersed by the near-surface turbulence, and serve to identify fluid regions directly affected by the breaking waves which produce the bubble clouds. The subsequent distribution of bubbles in a cloud and the cloud's shape are functions of the properties of the bubbles themselves as well of the turbulence. The principal objective of our research has been, through sonar observations of the bubble clouds and models of the bubble dynamics, to describe the turbulent motion in the upper ocean boundary layer (Thorpe, 1985) and to estimate the importance of bubble clouds in the exchange of gases between the atmosphere and the ocean (Thorpe, 1982, 1984a; Merlivat and Mémery, 1983).

We here focus attention on the horizontal pattern of bubble clouds. Information about their vertical distribution is given elsewhere (e.g. Thorpe, 1986b; Crawford and Farmer, 1987).

2. SONAR OBSERVATIONS

Observations of the near-surface bubble field using upward-pointing sonars seem to have been used by submarines in World

B. R. Kerman (ed.), Sea Surface Sound, 173–183.

174

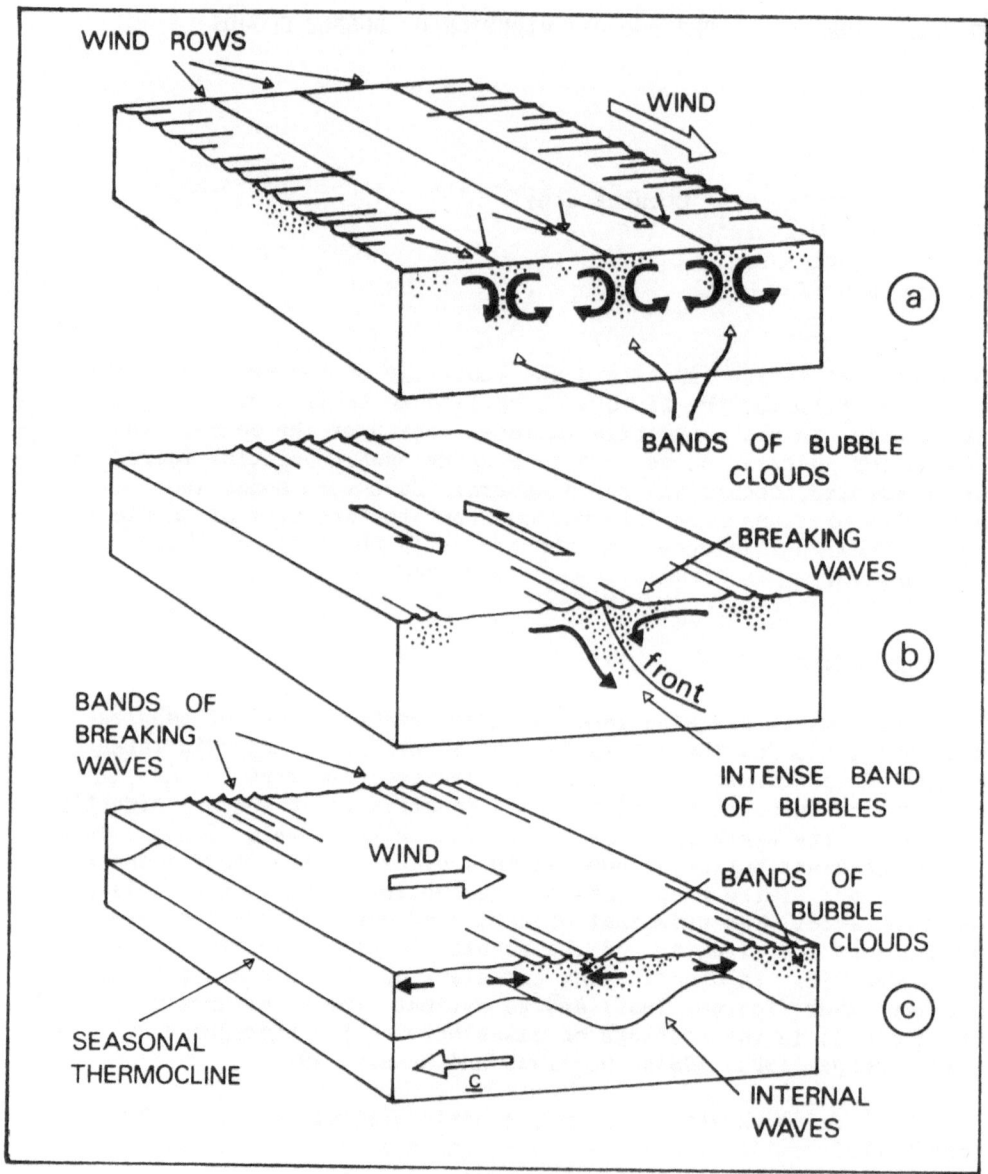

Figure 1. Processes leading to the formation of aggregations of bubble clouds. (a) Langmuir circulation; zones of intense bubble clouds form in convergence regions beneath wind rows. (b) Fronts: these are zones of convergence where bubbles may accumulate, as below wind rows; the associated currents also modify the surface waves, sometimes promoting breaking and more intense bands of bubbles parallel to the front. (c) Internal waves: these modulate the surface currents and thereby the surface waves, promoting bands of breaking waves ('walls of white water') and bubble clouds parallel to the internal wave crests.

War II as a means to locate the wakes of surface vessels, but the earliest reference I know to the use of inverted sonars in the scientific study of bubble clouds is by Aleksandrov and Vaindruk (1974). They describe a bottom-mounted upward-pointing sonar 1 km off-shore in a water depth of 20m. They noticed an increase in the depth of the bubble clouds as wind speed increased and, with considerable foresight, recognised the potential of the technique as a means of studying near-surface turbulence.

Our observations were first made in a fresh water lake (Loch Ness: Thorpe and Stubbs, 1979; Thorpe and Humphries, 1980; Thorpe, 1982) using a narrow-beam upward-pointing sonar operating at 248 KHz at a fixed location. The data, the intensities of the pulsed signal returns, were displayed on a sonagraph and digitized in a series of range bins. In winds exceeding about 2.5 m s^{-1}, bubble clouds were observed to form intermittent patches. At higher wind speeds, exceeding about 7 m s^{-1}, the patches are sufficiently numerous and persistent to produce a 'stratus' below the surface beyond which individual, deeper-penetrating, clouds can still be seen. The sonar was calibrated to provide estimates of the acoustic scattering cross section per unit volume of the bubble clouds to determine its vertical distribution with time and variation with wind speed. Measurements were subsequently made at a coastal site in Western Scotland using the same narrow upward-pointing beam (Thorpe, 1982), but also adopting a telesounder (multiple finger-beam side-scan) technique (Thorpe et al, 1982) and a more conventional fixed 2-beam side-scan system (Thorpe and Hall, 1983) which had the advantage of being able to detect the motion of bubble clouds which, partly, resolved their space-time properties. This was a significant advance, for it showed that bubble clouds persist at detectable levels for periods of several minutes, being advected by the near-surface currents. Their distribution in the horizontal depends upon their random formation by individual waves, or by the pattern of waves breaking in a wave group, as well as on their duration.

The plan-form shape of individual bubble clouds appears roughly elliptical, on average slightly elongated in the down-wind direction with aspect ratio of about 1.4, but considerable variations occur. The horizontal dimensions appear to decrease with depth.

The horizontal distribution, or aggregation, of bubble clouds is sometimes ordered by the presence of Langmuir circulation (Figure 1a), that is by a process operating on the bubble clouds once formed (loc.cit., Figure 13). Other observations have found that waves do not appear to break preferentially, or with greater effect, in one part of the pattern of the circulating Langmuir cells than in another (Thorpe and Hall, 1980; although this is a result which, having important consequences in the use of drifting wave buoys, deserves further careful examination). Langmuir circulation is also effective in producing ordered bands of subsurface bubbles produced by heavy rain (Thorpe and Hall, 1983, Figure 12) The convergence at fronts

176

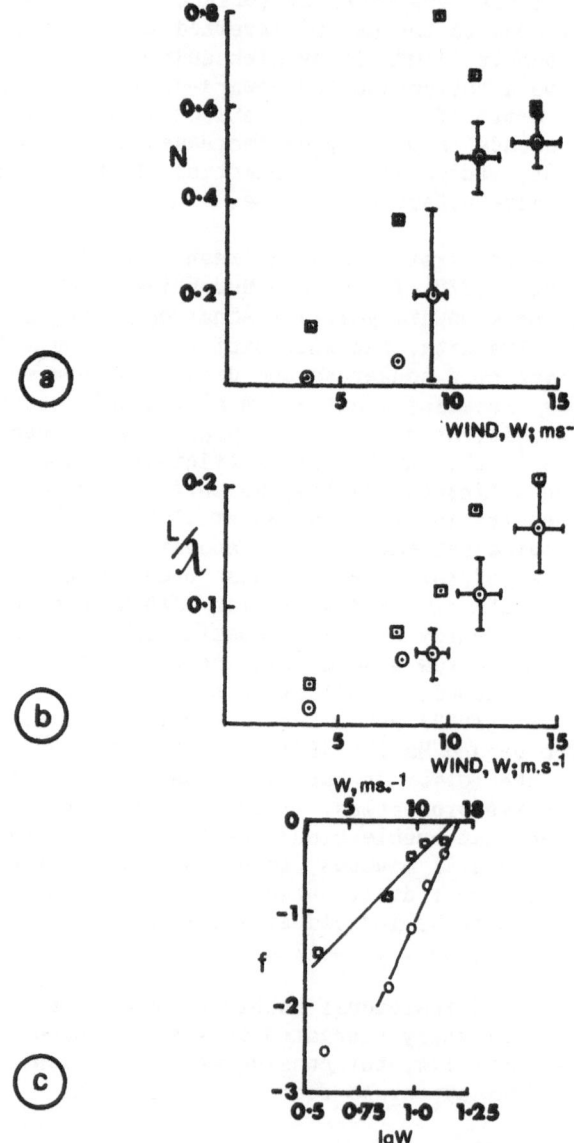

FIGURE 2

Caption for Figure 2.

(a) The number of distinct bubble clouds per wavelength, N, as a function of wind speed, W (squares and circles),

(b) The length of bubble clouds, L, divided by the dominant wavelength λ as a function of W (squares and circles).

(c) The fraction, f, of the sonar path occupied by bubbles as a function of wind speed, plotted on log scales.

Data was selected from tows made within 30 deg of the wind direction using the sonar returns from 3.38m ahead of the spar and at approximately 3.8m depth. A zero and a noise level for the sonar has been established by plotting the mean recorded intensity data values and their standard deviations against wind speed to provide estimates at zero wind speed. The zero wind mean plus n standard deviations (the 'noise threshold') have been subtracted from other records to establish 'signals', where n=2 (squares) or 6 (circles). Comparison with earlier work shows that the noise threshold n=2 corresponds to a scattering cross-section for bubble clouds, M_v of $(1 \pm 0.6) \times 10^{-3}$ m^{-1} at 248 KHz whilst that at n=6 corresponds to about $M_v = 3 \times 10^{-3}$ m^{-1} at 248 KHz. The 'signal' has been recognised only when it exceeds these noise thresholds. The vertical and horizontal bars are error estimates. These are particularly large in the variable conditions found at wind speeds of $8-10 ms^{-1}$.

may similarly lead to greater local concentrations of bubbles in
bands following the alignment of the front (Figure 1b, see also e.g.
loc.cit., Figure 6).

Structured bands of bubble clouds are also produced by the effect
of surface wave modulation and preferential breaking caused by
surface currents induced by internal gravity waves travelling on a
shallow thermocline, that is by a process which localises the source
of the bubble clouds (figure 1c; see also e.g. loc.cit., figures
10,11). Fronts may act in a similar way by promoting surface wave
breaking in regions of varying surface currents (Figure 1b; see also
e.g. loc.cit., Figure 8).

Observations of the variation at a fixed point, the temporal
variability, have also been made using an automatically recording
narrow beam sonar, ARIES, moored near a data buoy measuring winds,
temperatures and currents on the edge of the Continental Shelf S.W.
of the U.K., (Thorpe, 1986b). ARIES provided further evidence of the
clouds' advection and measurements of their vertical structure.

Measurements of the spatial distribution of bubbles have been
made using sonar mounted on a submarine (Crawford and Farmer, 1987).
These observations appear generally consistent with those of our
work, finding similar cloud shapes and distributions.

3. RECENT STUDIES OF BUBBLE CLOUDS AND TEMPERATURE VARIATION

Recent experiments have used an array of thermistors mounted on
a spar hanging beneath a catamaran towed by a ship (Thorpe and Hall,
1987). The spar carries at 1 MHz forward-pointing narrow beam sonar
to detect bubbles and to measure the horizontal scale and frequency
of bubble clouds. The intense sonar reflections from the bubble
clouds have, on average, associated temperature signals, with maxima
in the temperature field being found in regions of intense scattering
(i.e. in the bubble clouds). The temperature 'ramps' in the mixing
layer (Thorpe and Hall, 1980; Thorpe, 1985) have companion sonar
ramps. These 'large-scale coherent structures' in the temperature
and bubble field of the turbulent upper ocean layer are thus related.

The bubble clouds become more numerous and intense as the wind
increases. Figure 2a, b, shows that the number of bubble clouds per
wave and the length of the clouds per wavelength, both increase with
wind speed. In consequence the fraction of record in which bubble
clouds are detected by the sonar increases rapidly with wind speed,
approaching a total 'stratus' of bubbles at the depth of the sonar
target scattering volume, approximately 3.8m, at wind speeds of about
15 m s^{-1} (Figure 2c; see also section 2). This stratus scatters
sound and, in particular, reduces the amount of the high-frequency
sound produced by breaking waves which propagates down into the
underlying ocean (Farmer and Lemmon, 1984).

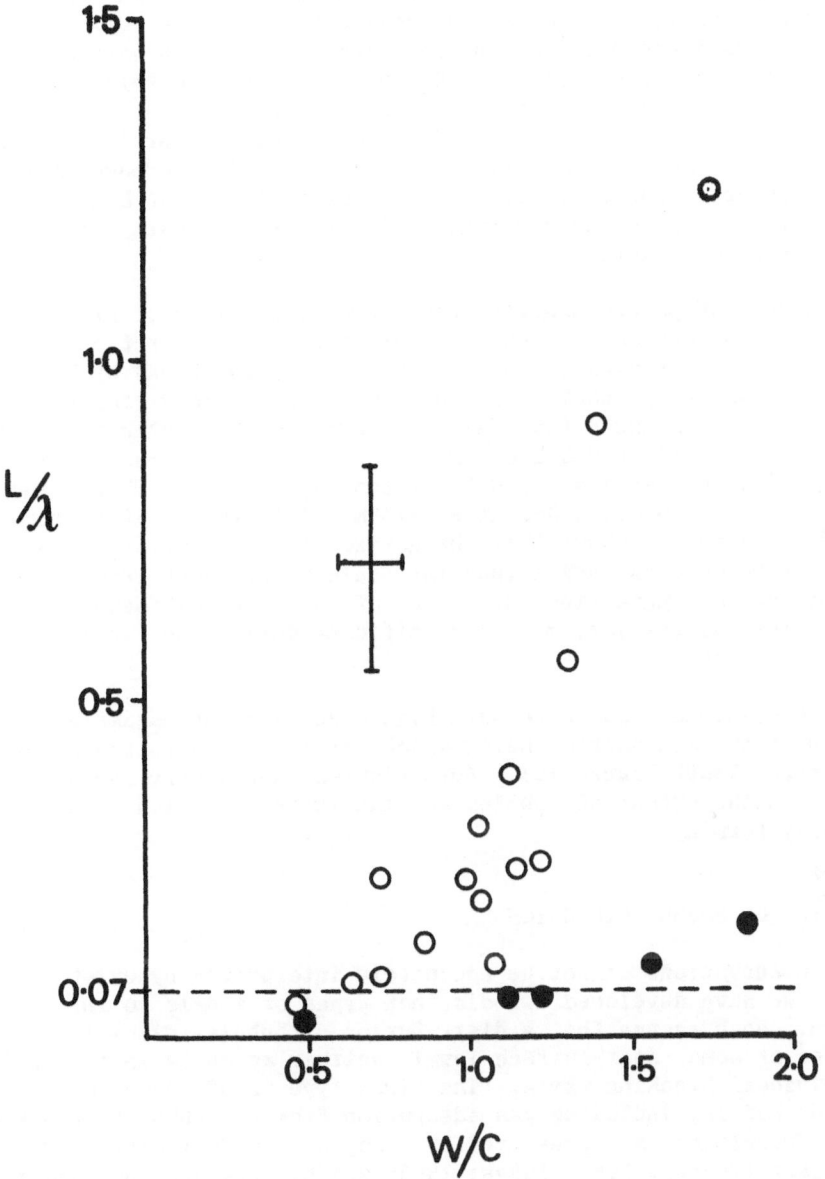

Figure 3. Estimates of the mean length of bubble clouds, L, as a fraction of wavelength, λ, plotted against W/c. The circles (to which the error bars apply) are derived from ARIES data (see text). The dots are from the towed thermistor spar (for error bars in L/λ see 2b). The two data sets differ as explained in the text.

Figure 3 shows the ratio of the average bubble cloud length in
the upwind direction divided by the dominant wavelength, L/λ, as a
function of wind speed divided by wave speed, W/c, with data derived
from the towed sonar (dots) at 3.8m and from ARIES (circles) at
10.3m. The towed sonar (1MHz) used a signal to noise threshold of
3×10^{-3} m^{-1} in the acoustic scattering cross section (corrected to 248
KHz) whilst for ARIES (248K Hz) it is 1.7×10^{-5} m^{-1}. Moreover ARIES
is a stationary, moored, instrument and the estimates of L are
derived indirectly. Because of these differences the data sets are
not strictly comparable.

Both sets of points however show increasing values of L/λ as W/c
increases. The dashed line at $L/\lambda=0.07$ represents the ratio of the
mean period of wave breaking divided by wave period found by Xu et al
(1986) in laboratory studies in a wind wave tank. They compared
their results with those (how few there are!) of Longuet-Higgins and
Smith (1983): 0.013 at $W=6$ m s^{-1} and large fetch, Weissman et al
(1984) : 0.086 at $W=6$ m s^{-1} and 8km fetch, and those of Thorpe and
Humphries (1980): 0.025-0.065 at $W=5-13$ m s^{-1} in the limited fetch of
Loch Ness. Figure 3 shows that the horizontal scale of the bubble
clouds may be more extensive than the scale of regions in which wave
breaking occurs, suggesting the effect of turbulent diffusion,
perhaps shear dispersion, may be significant during the lifetime of
the bubble clouds.

It is possible also to compare Figure 2c with the estimated
fraction of the sea surface below which, at 10.3m, there are bubbles
(see Thorpe, 1986b Figure 15). Both figures show a very rapid
increase in the extent of bubbles as wind increases, with a 'status'
eventually forming.

4. MODELS OF BUBBLE DIFFUSION

The observations cannot be adequately interpreted by existing
models. We have developed two distinct types of models to interpret
the data. Both assume that a distribution of bubbles sizes is
specified at some near-surface level. Neither accounts specifically
for individual breaking waves. The first type is of advection-
diffusion models, including gas adsorption from the bubbles into the
water. Turbulence is represented by a depth-dependent diffusion
coefficient (Thorpe, 1982; 1984b; 1986a and b; Crawford and Farmer,
1987), the value of which is determined by comparing the model with
the observations. This methodology offers an inverse technique to
measure turbulence using sonar observations of bubble clouds. The
second type is of random-walk or Monte-Carlo models (Thorpe, 1984c)
which follow individual bubbles. This is particularly useful in
examining the gas flux associated with different components (e.g. O_2,
N_2). Both model types have been extended to include Langmuir
circulation (Thorpe, 1984d).

The future development of models requires information about the specific source of bubbles, the breaking waves. Whilst some progress has been made in developing a model relating the bubble cloud distributions to the breaking waves (Thorpe, 1986b), information is required about both the turbulence generated by individual breaking waves and their local injection of bubbles. This would allow progress towards a model which specifically and more realistically represents the surface wave field. There is a need here for studies of the ocean surface in the location of bubble injection (i.e. in breaking waves) and its relation to the background wave and wind fields, for studies of the local generation of turbulence and the transfer of momentum (Mitsuyasu, 1985; Melville and Rapp, 1985), and for studies of the bubble size distribution in space and time. Significant and rapid progress could be made with the collaborative use of existing instruments, provided the facilities for their deployment (and these may be sophisticated and expensive) are made available or developed. (Plans exist for the study of turbulence in bubble clouds using a submarine).

5. SUMMARY

We have reviewed some of the information presently available about the horizontal distribution of bubble clouds and the factors which effect them. Processes leading to uneven distributions of bubble clouds are summarised in Figure 1. Further study is needed to improve the models. In particular:-

(a) The size distribution of bubbles, its variation with depth, and its time history in bubble clouds are poorly known. The mean size distribution is presently estimated in some models, and further observations at sea would provide a test of their validity.

(b) The effect of surface-active films on the bubbles' surfaces may retard gas adsorption, reducing the gas transfer rates and extending the bubbles' lifetime. The importance of these effects is unknown, but might usefully be studied in the laboratory.

(c) Specific representation of breaking waves in the models; how should this best be achieved?

Further sea-going experiments are required, in particular simultaneous observations might be made (see section 4) of bubble clouds (including size distribution of bubbles), surface waves, wind speed, temperatures and in situ turbulence,

(i)   to assess the rate of production of turbulence by breaking waves in comparison with that produced by shear in the upper ocean;

(ii)   to understand how turbulence produced by breaking waves decays within the ambient field of turbulence;

(iii) to test the predictions and validity of the models and of the inverse acoustic technique for 'measuring' turbulence;

(iv) to assess the effect of breaking waves on momentum, heat and gas transfer.

The value of the information obtained would not only help in the improvement of models of bubble clouds and their related transfer of gases and particles to and from the ocean, but would also better the understanding of ocean turbulence and diffusion in the 'mixed layer' of the upper ocean, and develop our capability to predict the effect of surface waves on the underlying ocean.

Much of this work would usefully be augmented by studies of ambient noise, especially that associated with breaking waves, bubble coalescence (more prevalent in fresh water), bubble bursting and rain.

REFERENCES

Aleksandrov, A.P. and Vaindruk, E.S., 1974 In 'The investiga-tion of the variability of hydrophysical fields in the ocean' (ed. R.Ozmidov) pp122-128, Moscow: Nauka Publishing Office.

Crawford, G.B. and Farmer, D.M., 1987 On the spatial distribution of bubbles generated by breaking waves (submitted to J.Geophys. Res.).

Farmer, D.M. and Lemmon, D.D., 1984 J.Phys.Oceanogr. 14, 1762-1778.

Longuet-Higgins, M.S. and Smith, N.D., 1983 Measurements of breaking waves by surface jump meter. J.Geophys. Res. 88, 9823-9831.

Melville, W.K. and Rapp, R.J., 1985 Momentum flux in breaking waves. Nature, 317, 514-516.

Merlivat, L. and Mémery, L., 1983 Gas exchange across the air-water interface: experimental results and modelling of bubble contribution to transfer. J.Geophys. Res. 88, 707-724.

Mitsuyasu, H., 1985 A note on the momentum transfer from wind to waves. J.Geophys. Res. 90, 3343-3345.

Thorpe, S.A., 1982 On the clouds of bubbles produced by breaking wind-waves in deep water, and their role in air-sea gas transfer. Phil. Trans. Roy. Soc. London A, 304, 155-210.

Thorpe, S.A., 1984a The role of bubbles produced by breaking waves in super-saturating the near-surface ocean mixing layer with oxygen. Annales Geophysicae 2, 53-56.

Thorpe, S.A., 1984b On the determination of K_v in the near-surface ocean from acoustic measurements of bubbles. J.Phys. Oceanog. 14, 855-863.

Thorpe, S.A., 1984c A model of the turbulent diffusion of bubbles below the sea surface. J.Phys. Oceanog. 14, 841-854.

Thorpe, S.A., 1984d The effect of Langmuir circulation on the distribution of submerged bubbles caused by breaking wind waves. J. Fluid Mech. 142, 151-170.

Thorpe, S.A., 1985 Small-scale processes in the upper ocean boundary layer. Nature 318, 519-522.

Thorpe, S.A., 1986a Bubble Clouds : A review of their detection by sonar, of related models, and of how K_v may be determined, in 'Ocean Whitecaps'. E.C. Monohan & G. MacNiocaull. D. Reidel Publishing Co., 57-68.

Thorpe, S.A., 1986b Measurements with an automatically recording inverted echo sounder; ARIES and the bubble clouds. J.Phys. Oceanog 16, 1462-1478.

Thorpe, S.A. and Hall, A.J., 1980 The mixing layer of Loch Ness. J.Fluid Mech. 101, 687-703.

Thorpe, S.A. & Hall, A.J., 1983 The characteristics of breaking waves, bubble clouds and near-surface currents observed using side-scan sonar. Cont. Shelf Res. 1, 353-384.

Thorpe, S.A. and Hall, A.J., 1987 Bubble clouds and temperature anomalies in the upper ocean. Nature, 328, 48-51.

Thorpe, S.A. & Humphries, P.N., 1980 Bubbles and breaking waves. Nature, 283, 463-465.

Thorpe S.A. & Stubbs, A.R., 1979 Bubbles in a freshwater lake. Nature, 279, 403-405.

Thorpe, S.A., Stubbs, A.R., Hall, A.J. & Turner, R.J., 1982 Wave produced bubbles observed by side-scan sonar. Nature, 296, 636-638.

Weissman, M.S., Dean, J.R., Marra, R., Price, J.F., Francis, E.A., and Broadman, D.C., 1984 Detection of breaking events in a wind-generated wave field. J.Phys. Oceangr. 14, 1608-1619.

Xu, D., Hwang, T.A. and Wu, J. 1986 Breaking of wind-generated waves. J.Phys.Oceanogr. 16, 2172-2178.

ON THE DISTRIBUTION OF BUBBLES NEAR THE OCEAN SURFACE*

Bryan R. Kerman
Atmospheric Environment Service
4905 Dufferin Street
Downsview, Ontario, M3H 5T4
Canada

ABSTRACT. A model is presented for the size distribution and vertical number profile of bubbles near the ocean surface in the presence of breaking waves. A source of bubbles is postulated at the surface which arises primarily from the shattering of large cavities near an outer scale determined jointly by gravity and surface tension. The fragmentation rate varies as the area of the individual bubbles down to an inner scale where no further shattering is possible. The resulting Weibull distribution of bubble sizes adequately fits the bulk of observed bubble distributions except in the limits of small and large sizes. The vertical profile of bubble density is modelled on the premise of a depth-independent but scale dependent eddy diffusivity. Other aspects of the characteristics of bubble fields as presently understood are adequately reproduced by the model.

1. INTRODUCTION

Bubbles in the uppermost layer of the ocean play a crucial role in several important air-sea interactive processes. The entrainment of air associated with breaking waves, its buoyant rise and its final expulsion contribute to the exchange of gas as well as the production of sea salt aerosols, an electrical charge exchange, and chemical fractionation in addition to a net upward flux of organics and bacteria. Bubbles near the surface are also significant in applied studies of the generation, scattering and propagation of sound underwater.

Actual observations of bubble fields in the ocean based on different experimental techniques, have all demonstrated that the size distribution of bubbles, after normalization by depth and wind dependence, follow an approximately common geometric shape. In each case a power law dependency in radius has been suggested. Estimates of

---

*This paper represents a general condensation, with an elaboration of key points, of a paper by the author appearing in 'Atmosphere-Ocean, Vol. 24, 169-188, 1986.

B. R. Kerman (ed.), Sea Surface Sound, 185–196.
© *1988 by Kluwer Academic Publishers.*

the slope with increasing bubble radius vary from -3.5 based on Kolovayev's bubble trap method to -5 based on Johnson and Cooke's bubble photography method. The observations of Bortkovskiy and Timanovskiy uniquely represent the actual bubble field in an actively breaking wave.

Both Kolovayev and Johnson and Cooke's data indicate that the number of bubblesdecreases with depth, z, approximately as exp - $z\lambda^{-1}$, where λ is some scaling length. Wu has combined both these data sets to demonstrate that the total number of bubbles per volume, increases very rapidly with wind speed, U_{10} (measured at 10m height) perhaps as rapidly as U_{10}^5. However, Wu's estimate of the wind speed sensitivity is preliminary and tentative. In fact, some recent acoustic backscattering estimates of bubble densities extending to significantly higher wind speeds indicate a power law dependence on wind speed of about 2.

Crowther in a somewhat abbreviated and tentative discussion has proposed a very interesting model. Crowther argued that the bubble distribution, which he pictured as arising from the combined effects of breaking waves, spray and capillary waves, with the bubbles sub-sequently carried down by the turbulence, was described by the equation

$$\frac{dn}{dt} = \frac{\partial n}{\partial t} \quad -K \frac{\partial^2 n}{\partial z^2} \quad -v_t \frac{\partial n}{\partial z} \quad = \frac{\partial s}{\partial z} \ \delta(z-\epsilon) \tag{1}$$

where n(r,z) is the number of bubbles per volume per radius interval, s(r,z) is the number of bubbles produced or destroyed per surface area per radius interval per time interval at a depth ϵ, and K and v_t are the turbulent diffusivity of bubbles and the terminal velocity respectively. z is positive downwards. Crowther has quoted a steady state solution to Eq. 1, as $\epsilon \rightarrow o$, given by

$$n(r,z) = \frac{s(r,o)}{v_t(r)} \ exp - z \ \lambda^{-1} \tag{2}$$

where

$$\lambda = K \ v_t^{-1} \tag{3}$$

Equation 2 has the interesting property of demonstrating a negative exponential form, similar to observations, with a scale length, λ, given explicitly in terms of the turbulent diffusivity and terminal velocity. The key variable in Eq. 2 is the source rate of bubbles, s(r,o), identically at the surface.

Crowther attempted to close his model by proposing a dimensional analysis argument for s(r,o) in terms of certain of the dynamically significant variables of the problem. He proposed that the rate of work, W, done by the atmosphere on the ocean by the wind, the kinematic surface tension, γ, and the bubble's radius formed a sufficient set to describe the number production rate per unit area per unit radius increment at the surface, that is

$$s(r,o) \sim W \ \gamma^{-1} \ r^{-3} \tag{4}$$

Because the dimensional analysis is underspecified (more dimensions appear than relations) an arbitrary exponent of unity was specified for W_0.

Several objections can be made to Crowther's arguments. The first is that nowhere is the gravitational acceleration, g, which is associated with a critical vertical acceleration of waves leading to breaking introduced into the development leading to (4). In fact, for an equilibrium wind-wave field consisting of energy input through wind stress on the surface and waves themselves and dissipation by breaking waves at all scales, Phillips has argued that the mean square wave amplitude at a given wavenumber is solely a function of g. Phillips has recently refined his arguments such that the surface elevation is a joint function of W_0, g and wavenumber. Likewise, g and essentially W_0 are the two variables assumed by Charnock in making his classical estimate of the roughness height of the wavy surface for airflow. It is therefore difficult to see why g should not enter into deliberations leading to the source term for the production of bubbles.

Crowther's result implies that there is no accumulation of bubble volume in the size range of interest. Such a situation could arise either if there were a uniform probability for the formation of fragmentable cavities, or shattering was combined to a narrow range of shatterable cavities with formation probabilities zero elsewhere. While there is reason to expect a minimum size for a fragmentable cavity according to a critical Weber number criterion based on an rms turbulence velocity, there is no reason a priori to expect equal shattering probabilities because of the distribution of turbulent energy with scale size, or to expect the fragmentation to be limited to a given scale. Instead, it is more realistic to expect that the probability of a large cavity shattering is related positively to scale, that is, large cavities will shatter more readily than smaller ones and that the production of bubbles occurs over a finite range of scales.

2. MODEL

2.1 Diffusivity

Thorpe and Crowther have provided solutions to the bubble production equation for various postulated variations of turbulent diffusivity with depth. The motivation for their models stems in part from the lack of sensitivity of the location of the maximum bubble density in Johnson and Cooke's data with increasing depth. As a point of departure it is assumed here that K is depth-invariant but scale-dependent. We represent the unknown dependency of the diffusivity of bubbles by

$$K = \frac{u_* \ell}{\alpha} G(\frac{v_t}{u_*})$$ (5)

where any scale dependence is carried in the terminal velocity contribution to the argument of the similarity function, G.

188

With increasing r the buoyant transport increases in importance relative to the turbulent transport to the point where sufficiently large cavities are not transported below the surface. Accordingly no matter what the turbulent transport velocity, the buoyancy, as represented by the terminal velocity, controls the process. For large r it is expected that

$$G = \beta \frac{v_t}{u_*} \tag{6}$$

so that the diffusivity has a form independent of direct reference to the turbulent mixing intensity, that is,

$$K = v_t \ell \alpha^{-1}\beta \tag{7}$$

Equation (7) represents the same solution for the vertical mixing length of bubbles quoted by others (Kv_t^{-1}). For sufficiently small scales, it is reasonable to assume that turbulent transport dominates buoyant transport and from the previous discussion of time scales, that the scale dependency of K, and hence G, disappears so that , without loss of generality,

$$G = 1 \tag{8}$$

and

$$K = u_* \ell \beta^{-1} \tag{9}$$

A corollary of this development is the expectation that very weakly buoyant bubbles behave as if they were buoyantly neutral.

2.2 Shattering Process

The model development can be reduced to separable problems – estimating the total number of bubbles, N_0^*, and deducing their source distribution f(r). In order to motivate the derivation of an expression for N_0^* let us consider the following properties of entrainment, breaking waves and fragmentation.

Bubbles break up when the internal circulation induced by the (turbulent) flow near them produces sufficient kinetic energy of radial velocity, v_r, components to exceed the surface energy. The condition is often written in terms of the Weber number, $W(=rv_r^2\gamma^{-1})$, that is

$$W > W_c \tag{10}$$

where W_c is the critical value of W at which bubbles shatter (≈ 1.3 according to Sevik and Park).

It has been argued by the author that separation, and possibly incipient small wave breaking, will occur when the phase speed, c, of the slowest (gravity-capillary) waves is exceeded by the friction velocity, u_*, or at least some multiple of u_* which is of the order of unity. This phase velocity, c_{min}, alternatively defined as u_{*c}, is given by

$$c_{min} = u_{*_c} = (4\ g\gamma)^{1/4} \qquad (11)$$

It is argued here that at incipient breaking the entrained air will shatter to smaller bubbles when its characteristic dimension, r, exceeds a characteristic length scale, r_m, given by

$$r_m = W_c \gamma\ u_{*_c}^{-2} \qquad (12)$$

Accordingly there can never be cavities of larger size and r_m forms on outer scale to the problem. At higher wind speeds ($u_* > u_{*c}$) all entrained air cavities (large bubbles) will continue to shatter until

$$r < \frac{\gamma}{u_*^2}\ W_c \qquad (13)$$

which defines an inner scale to the problem.

On this basis it is hypothesized that the essential physical variables associated with the volume entrained and shattered are u_*, g and γ. No reference is made to scale-dependent parameters either directly in terms of r or indirectly in terms of v_t because N_o^* is an integral property. Dimensional analysis leads to

$$r_m^3\ N_o^* \sim (\frac{u_*}{u_{*_c}})^{3\varepsilon} \qquad (14)$$

where ε is an undefined coefficient due to the underspecification of the problem. Eq. 14 represents an expression for the void fraction of air in water. It can intuitively be understood by expecting that most of the void fraction will be related to the size range near r_m, and that the number of bubbles outside this range will be proportional to the number scattering in this region.

Let us now turn to the second aspect of the problem – the estimation of the size distribution of the fragmenting source-by considering some results taken from the statistical theory of failure (Miller and Freud). The process of failure is here associated with a bubble which shatters. The failure rate at size r, $b(r)$, defined as the probability of fragmentation given survival to that scale in a size increment, dr, at r, is related to the pdf by

$$f(r) = b(r)\ exp - \int_o^r b(x)\ dx \qquad (15)$$

If b is constant in r, the resulting prediction of f(r) is the well-known Poisson process. If b is a power law in r, the resulting expression for f(r) is referred to as a Weibull distribution. In particular if

$$b(r) = \zeta \; \mu r^{\mu-1} \tag{16}$$

then

$$f(r) = \zeta \; \mu r^{\mu-1} \; \exp{-\zeta} \; r^{\mu} \tag{17}$$

Intuitively the survival factor is related to local bubble size relatlive to r_m so that the implied length scale of Eq. 16 ($\zeta^{-1/3}$) is proportional to r_m.

A theoretical formulation of the size distribution of dispersions including bubbles has been reported in the chemical engineering literature but apparently overlooked by the geophysical community. Angelidou et al. have examined the problem of fragmentation of large bubbles such that surface energy is negligible compared to the internal compressional energy. Their basic hypothesis, based on concepts used in statistical mechanics, is that the gas is partitioned into bubbles in a completely random way with respect to the bubble's internal energy. The result is a Weibull distribution with $\mu=3$.

In terms of the above formulation in terms of failure the model of Angelidou et al. ($\mu=3$) implies that b(r) varies as r^2, that is the larger the bubble area the more probable it is to shatter. While the result is intuitively reasonable, there is no obvious a priori argument that leads to this result if the scale dependence of the turbulence surrounding the bubble is also considered. In their model and the subsequent discussions the scaling volume of the problem, ζ^{-1}, is not related directly to either the maximum shattering rate near r_m, or the entrained volume at scales near r_m, that is those cavities which are ultimately shattered to the observed bubble size range. Here the inherent scaling of the shattering rate is $(r/r_m)^2$ implying less shattering progressively down to the inner scale.

A sample of the ambient distribution as observed experimentally relates in some unknown way to previous events. It is assumed here that the sampled distribution evolved from either single or multiple breaking wave events in such a way that the observed mass weighted scale, $\zeta^{1/3}$, is a reflection of the original mass weighted scale, r_m, of the active fragmentation phase. Accordingly the aging process is represented by

$$\zeta^{-1/3}(t) = r_m \; \Psi(t, \; u*) \tag{18}$$

where t=0 refers to the time of whitecap initiation. It is expected that the similarity function, Ψ is constant throughout the life of the breaking wave and decreases rapidly after the entrainment is complete. Such assumptions are equivalent to hypothesizing that the observed ambient distribution maintains geometric similarity with the source distribution.

3. EXPERIMENTAL COMPARISON

Bubbles counted directly by Johnson and Cooke and Kolovayev, at a
common depth of 1.5m, and some bubble densities implied from an
acoustic resonance technique of Lovik for 38 kHz at 8m depth are
presented in Fig. 1. Clearly the best fit from bubble photography and
bubble trap data indicates a power law slope for u_* greater than 3
while the acoustic data indicates a slope less than 3. While no
precise conclusion can be drawn, it is noted that, overall a value of
$\varepsilon \approx 1$ is predicted from the results.

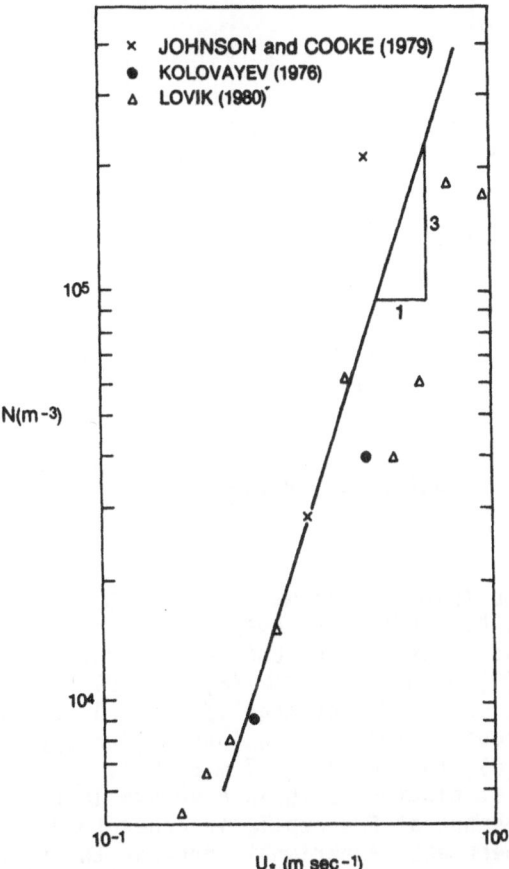

Fig. 1 Number of bubbles per volume near the surface as a function of
the friction velocity.

Angelidou et al. tested their predicted size distribution against
several data sets by examining the similarity in shape with reported
histograms. However in order to examine the size dependence of the
modelled distribution more exactly, it is useful to utilize the cpf,
F, instead of the pdf, f.
Several data sources were assembled for the comparison. The data
sources according to their assigned index on Figure 2, are from

192

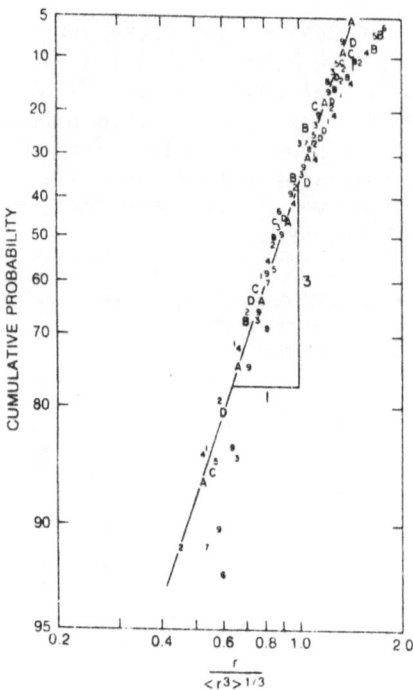

Fig. 2 Cumulative distribution of bubble sizes. See text for data sources.

Kolovayev for the wind speed, U, between 11 and 13 m/sec at depths of (1) 1.5m, (2) 4m, (3) 8m and for (4) U=6 to 8 m/sec at a depth of 1.5m; from Johnson and Cooke for (6) U=8 to 10 m/sec at 1.5m and for U=11 to 13 m/sec at (5) 0.7m, (7) 1.8m, (8) 4m; (9) from Glotov et al. from (A) a laboratory experiment of Angelidou et al. and from Bortkovskiy and Timanovskiy for observations of the upper surface of breaking waves (B), (C), (D) for about 7 m sec $^{-1}$.

The data were first plotted as ln ln F versus ln r, then the data were shifted to correspond at F = exp-1, in order to eliminate mean size differences between each experiment. Because the size at x=1 corresponds to $\zeta^{-1/\mu}$, the bubble size was further normalized by $<r>=\zeta^{-1/\mu}$. For convenience the double logarithm scale has been labelled in cumulative probability converted to percentage. The results are presented in Figure 2. A slope of $\mu = 3$ is established through much of the data. Deviations from linearity are confined to the uppermost and lowermost 15% of the data which is the region where experimental methods and procedures are most demanding. It is concluded that a Weibull distribution with $\mu= 3$ satisfactorily represents both the active source and the aged ambient bubble field. Apparently the assumption of shape similarity with time expressed in Eq. 18 is not seriously violated.

4. DISCUSSION

Breakup of bubbles in a turbulent water flow occurs when the total local shear stress imposed by the water acts to deform the bubble and ultimately overcome the restoring forces of surface tension and viscous stresses inside the bubble. The force actually imparted to a bubble will be proportional to the product of the stress and the bubble's area. If the stress is relatively constant in the size region determined by the critical Weber number then the shattering process of each bubble is dominated by its area as specified by the model. Such an assumption is reasonable if the efective size range of fragmenting bubbles is small enough such that the scale variations of stress are negligible.

If Crowther's model is generalized somewhat to include explicitly g and the scale of volume containing region, $r_v = \zeta^{-1/3}$, an expression for the source mass production rate per unit area and radius increment, $m(r,0)$, is

$$m(r,o) = \rho_a \frac{u_*^3}{g\, r_v^2} \Sigma\, (\frac{r}{r_v}) \tag{19}$$

There are now no disposable exponents as the problem is specified to within a similarity function involving the relationship between r and r_v. If the mass flux is not a function of r, then the similarity function is itself independent of r and reduces to a constant. From our major assumption that the volume containing size is proportion to r_m, $(\sim(\gamma g^{-1})^{1/2})$, that is to the critical bubble size for incipient breaking and entrainment, the resulting expression for $m(r,0)$ leads directly to Crowther's model, (Eq.2). The assumption of independence of the production rate on scale is compatible with assuming that the source of bubbles is at another scale, r_m far removed from the local scale, r. It is noteworthy that although explicit reference is then lost to g, implicit reference is contained in the definition and physical significance of r_m. It is concluded that Crowther's power law formulation is limited to a size range where the mass flux density is independent of scale. The Weibull distribution on the other hand postulates a failure rate at all scales albeit dominate near r_m where there is a maximum which rapidly decreases towards smaller bubble sizes as r^2. The failure rate is of the order of unity near the outer scale decreasing to a negligible value at the inner scale. Alternatively the failure rate is negligible everywhere except in a range of scales near r_m.

In summary the model proposed for bubble density in the upper ocean under active breaking wave conditions is

$$n(r,z)dr = 3\, \hat{\xi} \frac{\hat{u}_*^3}{r_m^3} \hat{r}^2 \exp{-\hat{r}^3} \exp{-z/\lambda}\, \hat{dr} \tag{20}$$

where $r = r\zeta^{1/3}$. The similarity coefficient ξ for an ambient distribution was calculated from Johnson and Cooke's data for U=12 m/sec by extrapolating their data for $N^*(z)$ to the surface and substituting the derived estimate, $N_o^*=1$ 10^6 m^{-3} in the relationship

$$N_o^* = \xi \frac{\hat{u}_*^3}{r_m^3} \tag{21}$$

derived from integrating Eq. 20 over r. The value of $u*$ (=0.48 m/sec) was estimated using a drag coefficient formulation. The value of r_m is about 1.8 mm based on an estimation of $W_c=1.3$. The estimate of ξ by this method is 0.72 10^{-3}. A similar calculation using Koloyayev's data for U=12 m/sec results in an estimate of $\xi=0.14$ 10^{-3}. Observations of Blanchard and Woodcock in the wake of a breaking wave and of Cipriano and Blanchard near a laboratory waterfall provide estimates of N_o^* at least 10^2 greater than the ambient bubble concentration observed by Johnson and Cooke.

Another estimate of N_o^* follows from the data of Bortkovskiy and Timanovskiy. For 3 breaking waves observed about 15 cm from the surface, they observed about 140 bubbles in a 140 cm^2 field of view. Since their observations were made at the threshold of breaking (U $\simeq$ 7 m sec^{-1}) it is reasonable to estimate $\hat{u}*\simeq1$. Further if the effective depth of field of the sample is about r_m so that $N_o^* \simeq 140/(140 r_m)$ (cm^{-3}), then $\xi \simeq 1 \cdot r_m^2$. For the above estimate of r_m ($\simeq.18$ cm), $\xi \simeq 3.2$ 10^{-2}, or a factor of about 50 greater than Johnson and Cooke's ambient density and comparable with an estimate based on Blanchard and Woodcock's relative increase (X100) for bubble density in a whitecap's wake.

5. SUMMARY

A model of bubble densities in the near-surface layer of the ocean in the presence of a breaking wave field has been presented. Several aspects drawn from diverse fields have been incorporated into the model. It is argued that the total number of bubbles present in a given volume close to the surface is related to the work done to create breaking waves and the characteristic size of fragmenting cavities. The modelled wind dependence is compatible with observations. The similarity coefficient for the interrelationship of the whitecap bubble density and the controlling physical parameters is poorly defined but probably within an order of magnitude of unity.

The depth variation of the bubble population is modelled here by a depth invariant but size dependent turbulent diffusivity which accounts for the variation in effective mixing rms velocity depending on the bubble's buoyancy. The formulation of bubble mixing length tends to become scale independent in the limits of large and small bubble radius. A function of simple geometric form has been proposed to match the two asymptotic regions. The resulting formulation for mixing length is compatible with observations which indicate that the most prevalent bubble size does not vary appreciably with depth although the overall density varies as exp-z.

A theoretical development based on kinematical arguments of statistical mechanics proposed by Angelidou et al. has been evaluated. It is shown that indeed the Weibull distribution associated with the failure, or fragmentation, of cavities in a source region adequately fits a number of observed source and ambient bubble populations. It is argued that an appropriate physical model to fit the observations is one in which the failure rate is proportional to the area of a bubble. Necessarily the scale variation of turbulent shear stresses over the range of fragmenting cavities is assumed to be negligible.

In conclusion the model as presented conforms to much of our present understanding of the ambient bubble field in the presence of a breaking wave field. While the model requires confirmation in several areas and will accordingly be modified as more is learned of the hydrodynamics of the bubble field, it is expected that the model will provide a reasonable and useful approximation in the diverse fields mentioned in the introduction.

6. REFERENCES

Angelidou, C., M. Psimopoulos, and G.J. Jameson, 1979: 'Size distribution functions of dispersions'. Chem. Eng. Sc., 34, 671-676.

Blanchard, D.C., and A.H. Woodcock, 1957: 'Bubble formation and modification in the sea and its meteorological significance'. Tellus, 9, 148-158.

Bortkovskiy, R.S., and D.F. Timanovskiy, 1982: 'On the microstructure of the breaking crests of wind waves'. Atmos. Oceanic Physics. 18, 225-256.

Charnock, H., 1955: 'Wind stress on a water surface'. Quart. J. Roy. Meteor. Soc., 81, 639-640.

Cipriano, R.J.,. and D.C. Blanchard, 1981: 'Bubble and aerosol spectra produced by a laboratory "breaking wave"'. J. Geophys. Res., 86, 8085-8092.

Crowther, P.A., 1980: 'Acoustical scattering from near-surface bubble layers'. In Cavitation and Inhomogeneities in Underwater Acoustics (ed. W. Lauterborn). Springer-Verlag, 194-204.

Crowther, P.A., 1985: 'Modelling of sea surface scattering'. Proc. Inst. Acoustics. 7 (3), 49-57

Glotov, V.P., P.A. Kolobaev, and G.G. Neuimin, 1962: 'Investigation of scattering of sound by bubbles generated by an artificial wind in sea water and the statistical distribution of bubble sizes'. Sov. Phys. Acoust., 7, 341-345.

Johnson, B.D. and R.C. Cooke, 1979: 'Bubble population and spectra in coastal waters: A photographic approach'. J. Geophys. Res., 84, 3761-3766.

Kerman, B.R., 1984: 'A model of interfacial gas transfer for a well-roughened sea'. J. Geophy. Res., 89, 1439-1446.

Kolovayev, D.A., 1976: 'Investigation of the concentration and statistical size distribution of wind-produced bubbles in the near-surface ocean'. Oceanology, 15, 659-661.

Lovik, A., 1980: 'Acoustic measurements of the gas bubble spectrum in water'. In <u>Cavitation and Inhomogeneities in Underwater Acoustics</u>. (ed. W. Lauterborn). Springer-Verlag, 211-218.

Miller, I., and J.E. Freud, 1965: <u>Probability and Statistics for Engineers</u>. Prentice-Hall.

Phillips, O.M., 1958: 'The equilibrium range in the spectrum of wind-generated waves'. J. Fluid Mech., 4, 526-434.

Phillips, O.M., 1985: 'Spectral and statistical properties of the equilibrium range in wind-generated gravity waves'. J. Fluid Mech., 156, 505-531.

Sevik, M., and S.H. Park, 1973: 'The splitting of drops and bubbles by turbulent fluid flow'. J. Fluids. Eng., 95, 53-60.

Thorpe, S.A., 1984: 'On the determination of K in the near-surface ocean from acoustic measurements of bubbles'. J. Phys. Oceanog., 14, 855-863.

Wu, J., 1981: 'Bubble populations and spectra in near-surface ocean: summary and review of field measurements'. J. Geophys. Res., 86, 457-463.

STUDY OF MICRO-BUBBLES IN THE NORTH SEA

S. C. Ling
The Catholic University of America
Washington, D.C.   20064, U.S.A.

H. P. Pao
Johns Hopkins University Applied Physics Laboratory
Laurel, Maryland   20707, U.S.A.

ABSTRACT. This paper presents the study of micro-bubbles in the North Sea and other natural waters. Detailed measurements of both the spatial variation and the time variation of micro-bubbles in the sea were performed by using an optical technique.   Two major sources of micro-bubbles were identified: the persistent population of micro-bubbles associated with the activities of zooplankton and the transient population associated with breaking waves.

INTRODUCTION

It is known that both the transmission of sound for underwater acoustic and the inception of cavitation for marine equipment are strongly affected by the number density of micro-bubbles in the ocean.   The objective of this study is to find out the variability of bubbles in the sea due to the natural environment of both biological activities and wind. In general, bubbles may be generated as a volumetric source by biological organisms, breaking waves, upwelling of bottom water, and by a host of other natural processes such as cosmic rays, dust, pollen, rain, etc. As a first step to resolve this problem, we have conducted extensive studies in laboratories, lakes, rivers, seas, and oceans. This report presents the key results of these studies.

The bubble detector system used in this work is based on the technique originally developed by Ling(1982) for the study of micro-bubbles in water tunnels. The micro-bubbles are detected by using a darkfield specular-reflection technique which also differentiates a bubble from a non-bubble. A detailed discussion of this technique is presented in this paper.

Due to the limited fetch of a small lake, the production of micro-bubbles under high wind condition is relatively insignificant as compared to that in the open seas and oceans. This is because small breaking waves can produce only relatively large bubbles with a very short life span. By comparing the measured results of waters having very low concentration of organic matter to natural waters with high organic concentration, one may conclude that the persistent micro-bubbles less than 100 microns in diameter are closely associated with the living micro-organisms; while the transitory micro-bubbles above   200 microns

B. R. Kerman (ed.), Sea Surface Sound, 197–210.

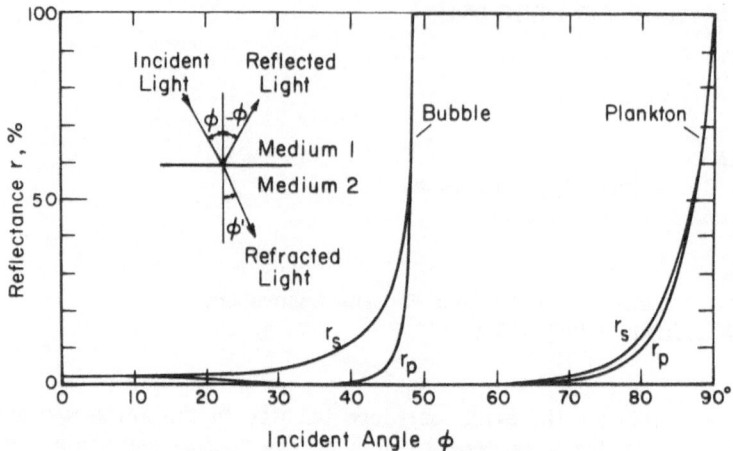

Fig. 1. Reflectance on the surface of bubble under different incident angles of light Ø.

are associated with the breaking waves. Under varying sea states, the small bubbles exhibit a very long time constant for both their increase and decrease in concentration with time, while the opposite is true for the large bubbles.

MEASUREMENT TECHNIQUES

The detector used in these experiments is based on the critical reflection of a uniform white light at the water-gas interface of a bubble, see Fig. 1. In the figure, the reflectance of light r is given for different incident light angles Ø, where Ø is the angle of incident light with respect to the normal of the interface between the two mediums, Jenkins and White(1951). One notes that at the

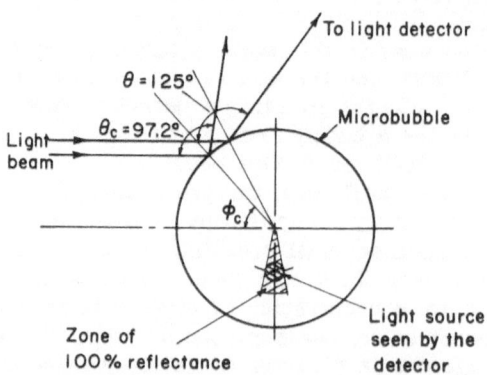

Fig. 2. Rays of light reflecting specularly on the surface of a gas bubble. The lower part of the figure shows the rotated top-view of the zone of 100% reflectance seen by the detector.

critical angle of reflection $\emptyset_c \geq 48.6$ deg, both polarized components of the reflected light r_s and r_p are equal with 100% reflectance. The subscripts s and p indicate the components of polarized light which are perpendicular and parallel to the plane of the incident light, respectively. The ray of incident light with $\emptyset = 62.5 \pm 5$ deg is selected for detection, see Fig. 2. Hence, the reflected ray with a forward-scattering angle of $\theta = 2\emptyset = 2 \times 62.5$ deg $= 125 \pm 10$ deg is used by the light detector to differentiate a bubble from a non-bubble. For a non-bubble, such as a plankton, the surface reflectance is almost negligible at the specific angle of detection $\emptyset = 62.5$ deg, see Fig. 1. This is due to the difference in the index of refraction for the water-plankton interface. Furthermore, most light scattering objects other than micro-bubbles are not spherical in shape, hence the probability of a surface reflection at the specific detection angle is rare. The effective light reflecting-surface of a bubble as seen by the light detector is shown in Fig. 2. This surface reflects an intense spot of light approximately 1/10 of the diameter of the bubble, in contrast to a generally diffused, weak, multiple external or internal scattering of light of a non-bubble. This characteristic also enables one to differentiate a bubble from a non-bubble through the signal pulse height to pulse width ratio. In general, living organisms less than 500 microns in size can be differentiated from bubbles larger than 17 microns by the difference in signal pulse amplitudes, while organisms larger than 500 microns can be differentiated by their larger signal pulse width to pulse height ratios. Figure 3 shows the general isometric view of the bubble detector. Two parallel white-light beams, with high uniformity in light intensity, are produced from a Koehler-lamp light source. Gas bubbles crossing the sampling area A_1 and A_2 are detected by the photomultiplier tube 1 and 2, respectively. The width of the sampling area is defined by the length of the

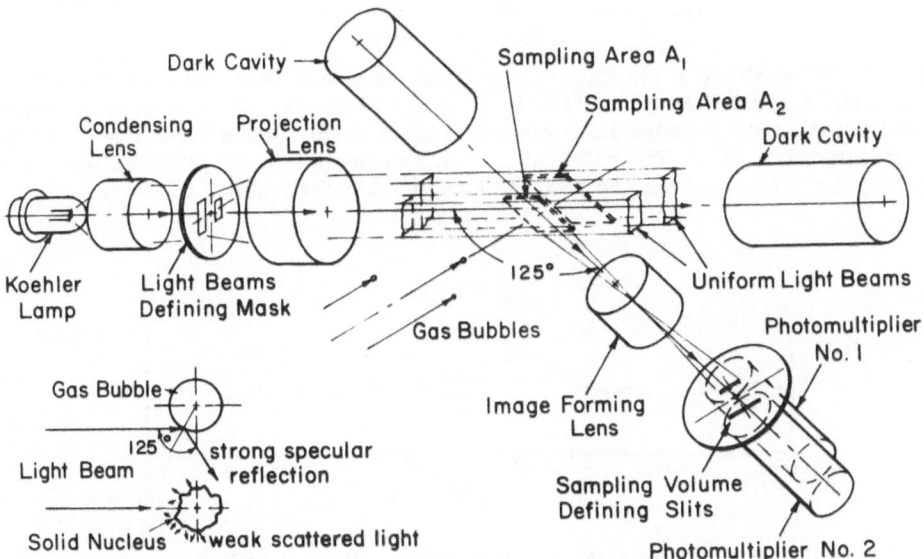

Fig. 3. Detection of micro-bubbles by the darkfield specular-reflection technique.

light masking-slit placed over the light detector. Similarly the thickness of the detecting volume is defined by the width of the slit, see Fig. 3. The detector has a sampling area $A_1 = 0.30$ cm^2 and a sampling volume $V = 0.012$ cm^3. Hence, the probability of having two bubbles in the same detecting volume V for a maximum bubble concentration of $C = 10/$cm^3 is

$$P = 100(C\ V)^2 = 1.4\ \% \ . \tag{1}$$

This error is greatly reduced for lower values of C. The bubble-size band-spread error due to bubbles crossing the perimeter of A_1 and detected as half-size bubbles can be expressed for a maximum diameter d = 400 microns bubble as

$$C_{d/2}/C_d = 100d/5A_1^{0.5} = 1.5\ \% \ . \tag{2}$$

This error also is reduced for smaller bubbles. The dark cavities shown in Fig. 3 are used to provide a dark field for the detector. The time it takes for the bubble to cross the sampling areas A_1 and A_2 gives the speed U of the bubble. Hence, the concentration of bubbles C may be obtained as

$$C = q/A_1U \ , \tag{3}$$

where A_1 is the sampling cross-sectional area, and q the number of bubbles N/sec crossing the area A_1. The value of the mean velocity U is obtained by performing the cross correlation of A(t) and B(t) time series signals from the light detector 1 and 2, respectively as

$$R(\tau) = \frac{1}{T}\int_0^T A(t-\tau)B(t)dt \ , \tag{4}$$

where τ is the delay time, and T the total sampling time. For the present experiments, a typical sampling time of T = 1000 sec was used to obtain bubble statistics with good repeatability and confidence level; i.e., a total bubble count of approximately 10^6. The value of τ corresponding to the peak of the $R(\tau)$ function, see Fig. 4, gives the mean delay time $\Delta\tau$, from which the mean

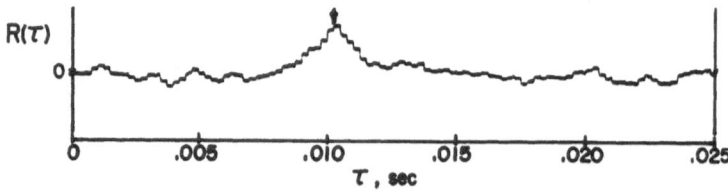

Fig. 4. Cross correlation $R(\tau)$ of the A(t) and B(t) time series for different delay time τ. The value of $R(\tau)$ is shown in arbitrary units. The peak of $R(\tau)$ indicates the mean delay time $\Delta\tau$ of B(t) with respect to A(t).

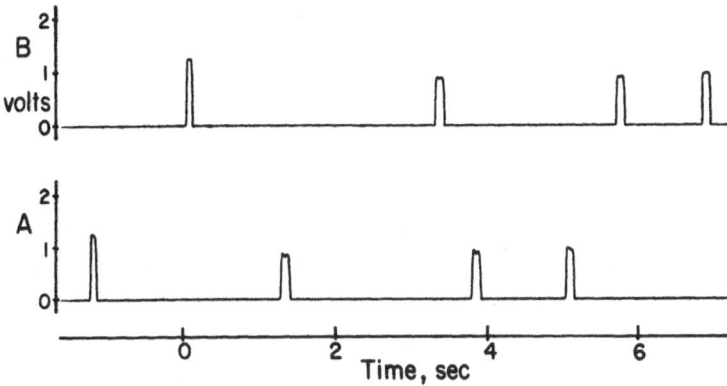

Fig. 5. Calibration of micro-bubbles in water containing a
high concentration of swimming zooplankton.

flow velocity can be expressed as

$$U = \Delta x / \Delta \tau ,$$
(5)

where $\Delta x = 0.70$ cm is the spacing between the sampling areas A_1 and A_2.
   Because the detector can measure bubble velocity, it is capable of self-
calibration by measuring the terminal velocity of individual hydrogen bubbles
rising vertically through the detector. Hence, the bubble size can be found
through the known size-velocity relationship. Since the signal intensity of the
light detector is proportional to the surface area of the bubble, the square-root
of the signal is taken to provide a signal which is proportional to the diameter
of the bubble. The bubble detector shown in Fig. 3 is placed in a position so
that the sampling area A_2 is vertically on top of A_1. An artificially
generated hydrogen bubble ascending under its terminal velocity will produce a
pair of time delayed signal pulses. A typical recording of the hydrogen bubble
calibration-signals is shown in Fig. 5. The first bubble signal in Fig. 5 shows a
delay of t = 1.25 sec. Since the distance between A_1 and A_2 is $\Delta x = 0.70$
cm, this indicates a terminal velocity of U = 0.55 cm/sec. From the well known
Stokes-Oseen equation, this corresponds to a bubble with d = 100 microns. Thus
the peak signal-pulse of 1.25 volts corresponds to a 100 microns bubble. The
preceding calibration was performed with water containing a high concentration
of zooplankton. One notes that the recording is free of noise produced by the
plankton. Figure 6 shows the non-square-rooted noise signal due to the
swimming of zooplankton in and out of the detecting volume. In the square-rooted
signal, noise signals less than 0.05 volt are cut off due to the non-ideal
analogue square-rooting process at input voltage below 0.05 volt.   Hence, non-
bubbles produce no noise signal in the recording shown in Fig. 5.   The resulting
micro-bubble calibration curve is shown in Fig. 7. One notes that the signal
output is linearly related to the bubble diameter d, and there is a cutoff effect

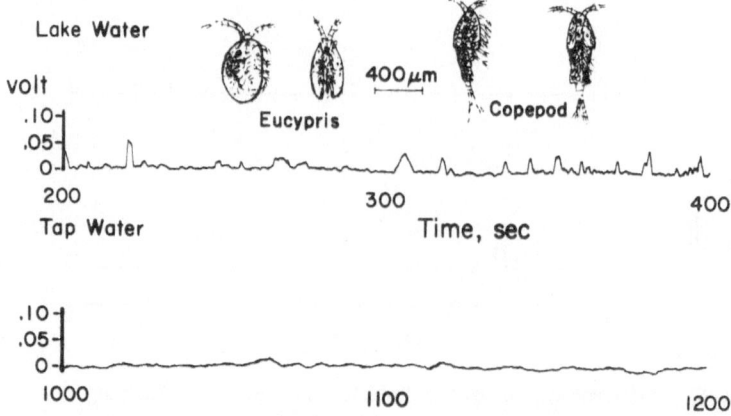

Fig. 6. Noise signals produced by plankton swimming in and out of the bubble detector. This lake water sample contains the typical high concentration of plankton. The long lapse time ensures all mechanically introduced bubbles have left the volume. Also shown is the background noise of clean tap water.

at d less than 17 microns. In Fig. 8 is a typical sample of bubble signal recording from field measurements. The four signal pulses shown in the figure represent a 35, a 70, a 50, and a 48 microns diameter bubble, respectively. The typical signal pulse width is noted to be approximately 0.00063 sec. This is associated with the thickness of the sampling volume W = 0.04 cm and the mean bubble velocity U. From Fig. 4, one notes that the mean delay time $\Delta \tau$ is - 0.0104 sec, which gives a mean bubble velocity of $U = \Delta x / \Delta \tau = 67$ cm/sec. Hence the pulse width should be $\approx (W+d)/U = 0.00065$ sec. The pulse width to pulse

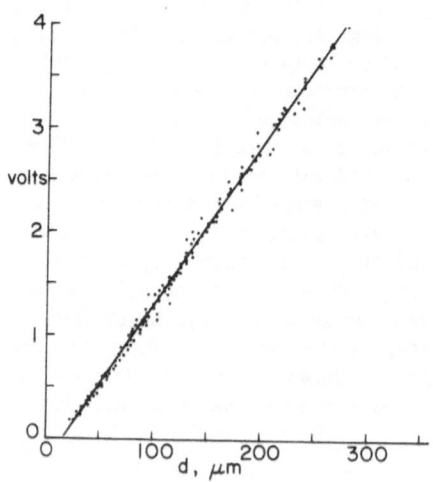

Fig. 7. Calibration curve for the bubble detector.

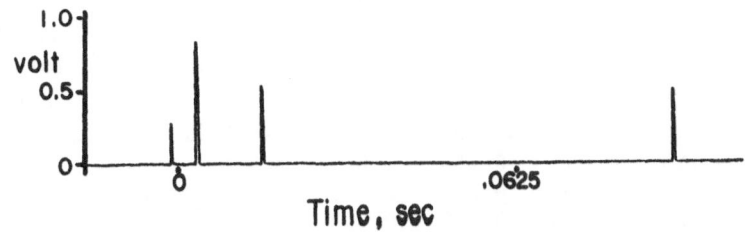

Fig. 8. A typical recording of micro-bubble signals.

height ratio is noted to fit the specification of a micro-bubble.

As an operating system, the bubble detector shown in Fig. 3 is housed inside a high pressure and streamlined casing which is suitable for mounting on either a tow body or a moving platform. In this mode of operation, the water sample is continuously guided through the detector by the towing motion. However, for studies presented in this paper, the detector is deployed as a heavily weighted, non-moving, object suspended at the end of a steel cable. In this mode of operation, a pump is used to gently draw the water sample through the detector. To minimize the recirculation of the sampled water, the water is exhausted more than 4 m away from the intake of the detector.

STUDY IN A CONSTANTLY FILTERED WATER

The study conducted in a constantly filtered and chemically treated water with minimum living organisms, such as in a large swimming pool, shows remarkably low concentration of micro-bubbles. A typical bubble size-concentration spectrum is shown in Fig. 9. The bubble concentration is plotted as the number of bubbles

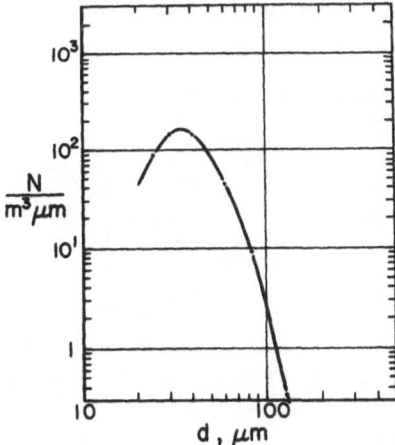

Fig. 9. Micro-bubble concentration spectrum for filtered and chemically treated water.

Table I. Average Micro-bubble Concentration in the Lake

Depth m	Micro-bubble concentration C, N/cm^3, for different ranges of bubble diameters, μm							Total C N/cm^3
	20-30	30-40	40-60	60-80	80-100	100-150	>150	
5	0.054	0.115	0.271	0.080	0.029	0.030	0.010	0.59
10	0.075	0.152	0.384	0.103	0.031	0.029	0.016	0.79
25	0.102	0.221	0.753	0.279	0.067	0.025	0.015	1.46
50	0.056	0.126	0.477	0.298	0.117	0.041	0.006	1.12
100	0.038	0.083	0.322	0.192	0.079	0.039	0.004	0.76

N per cubic meter of water per micron of bubble-diameter bandwidth N/m^3 μm versus the bubble diameter d in microns. One notes that the spectrum has a very steep slope; i.e., bubbles greater than 200 microns in diameter are almost non-existent. The total number of micro-bubbles from 25 to 200 microns is 0.0045 per cubic centimeter. This is more than 100 times less than those found in natural waters. One may conclude from this evidence that the persistence of micro-bubbles in the natural environments, a subject which is to be discussed in the following sections, is largely associated with biological activities.

STUDY IN A SMALL LAKE

The study of micro-bubbles was conducted in a small glacial lake in the State of Idaho, Ling(1987). The lake is approximately 30 km in length, 8 km in width, and 300 m in depth. The measured micro-bubble concentration at each given depth was found to be uniform over a wide area of the lake. The average concentrations for different bubble-size ranges and depths of the lake are tabulated in Table I. One notes that most bubbles are 50 microns in diameter with a maximum concentration located at the depth of 25 m. This corresponds to the lower boundary of the mixed layer of the lake. The water temperature in the

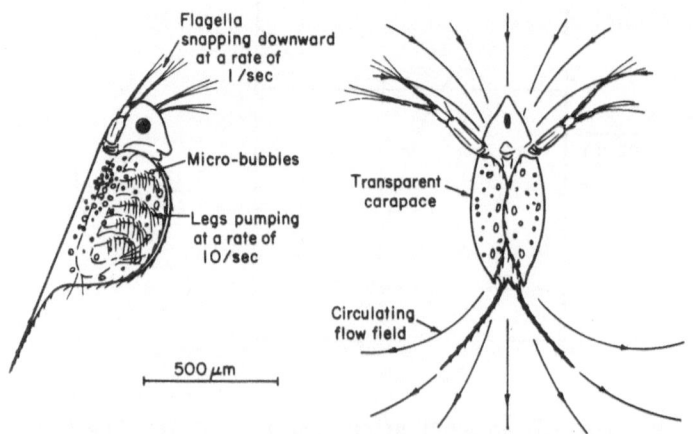

Fig. 10. Side and ventral views of a zooplankton(Daphnia).

mixed layer is typically 15°C, while below this layer is the large body of very cold 4°C water. High concentrations of zooplankton during daylight hours are known to exist at the thermocline zone, Raymont(1963).

Study of many species of zooplankton under a microscope has revealed a substantial collection of micro-bubbles under the inside surface of the carapace of the crustacean. Bubbles less than 20 microns in diameter are spherical in form while larger bubbles are in a flattened shape wet to the carapace. Since the carapace is made of rather dense material having a specific gravity of 1.4, micro-bubbles evidently are formed as part of the normal respiration cycle of the animal to provide for the necessary neutral buoyancy effect. Figure 10 shows the pumping of fluid through the carapace by the sweeping action of the legs hidden inside the closed bivalve. Water pumped out of the tail opening of the bivalve is recirculated back into the neck opening, creating a large circulating flow field. This provides an effective means for the animal to collect food from a space more than 8,000 times its own body volume. In addition, the zooplankton swims constantly with a jerk-like motion by flapping its pair of flagella. In relationship to the body mass, zooplankton are one of the most active and prodigious feeding animals in the living world. It consumes approximately 0.5 mm^3 of oxygen per hour, Raymont(1963). Hence, the production of CO_2 must also be substantial. Under constant observation, the animal is noted to occasionally discharge a cloud of mucous substance containing digested food particles and micro-bubbles. Thus, during the daily up and down migration of feeding depths, a significant number of micro-bubbles are released by each animal. This could very well be a major source of micro-bubbles observed in the natural waters. These bubbles, typically in the size of 30 microns, are noted to be coated with mucopolysaccharides and have a long life span even in water with a low dissolved gas content.

Near the surface of the lake, the concentration of micro-bubbles is noted to be at a minimum. This also can be related to the fact that during daylight hours, when the bubble measurements were performed, most zooplankton are at the bottom of the mixed layer. Here, local concentration of plankton can be as high as one per cubic centimeter. Hence, it is believed that the permanency of micro-bubbles in natural waters is largely associated with the zooplankton.

Under 10 to 15 m/sec high wind conditions, only small breaking waves, approximately 0.6 m in height were observed in the lake. This is due to the very limited fetch of the small lake. Measurements of bubbles 30 to 100 microns in size , taken one hour after high wind condition and at a depth of 5 m, show a small increase of 6% in number above the ambient concentration under quiescent condition. Bubbles below the mixed layer are relatively stable and unaffected by the weather.

STUDY IN THE NORTH SEA

The two preceding studies give a general background knowledge concerning the basic sources that sustain the persistent field of micro-bubbles in natural waters. The study in the North Sea provides additional information concerning wind generated micro-bubbles under high sea states. The experiment was conducted in the North Sea at the Nordsee Platform of Germany, located at 7°10'E and 54°42'N. The bubble detector was deployed by a crane on the south-west corner of the platform, and was placed at a distance of 20 m from the nearest

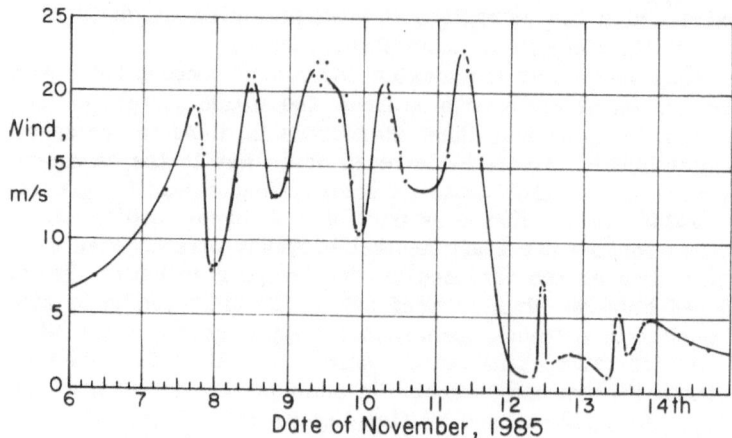

Fig. 11. Recording of 40m wind during the North Sea experiment.

support of the platform. During the experiment, the wind was mostly from the S, SW, and NW directions, and the tidal flow was from either SE or NW directions. Hence, the detector was free from contamination by bubbles generated by waves on the platform supports.

The experiment was performed in the month of November when stormy seas prevail. The North Sea is relatively shallow with a depth of approximately 31 m. Unlike the deep ocean, the whole depth of the sea is a mixed layer with a temperature uniform to within ±0.1°C. Figure 11 shows a record of the 40 m

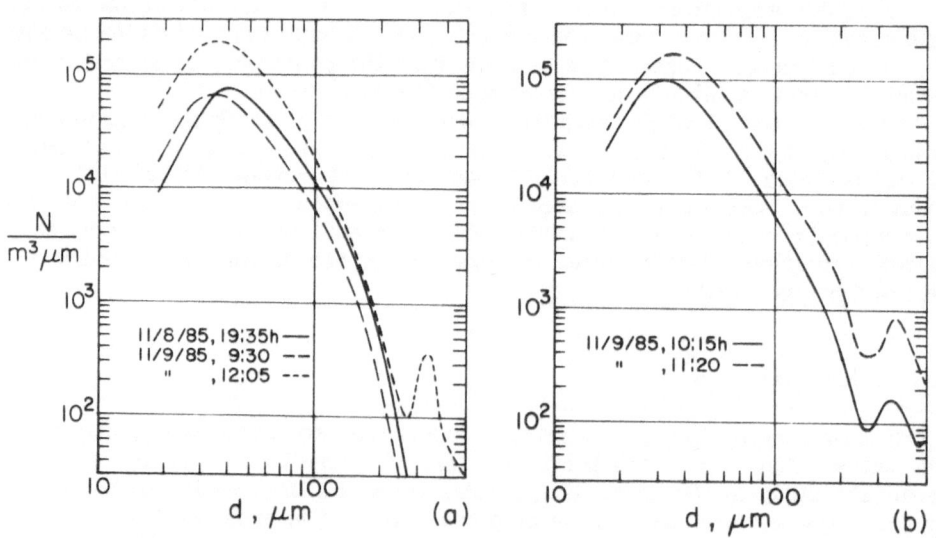

Fig. 12. a) Micro-bubble spectra at the 6m depth and
b) at the 3m depth.

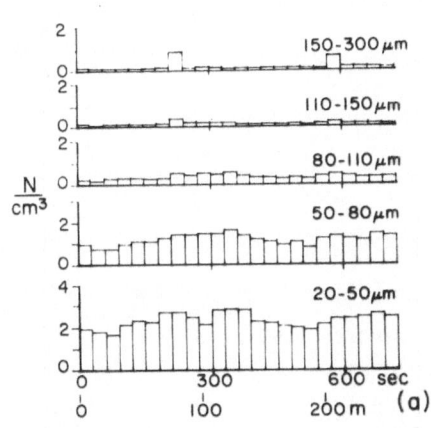 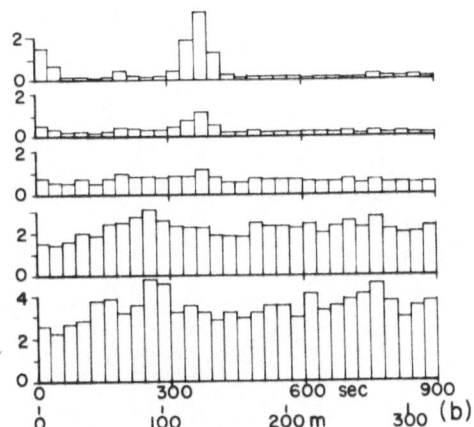

Fig. 13. Local time and spatial variability of micro-bubbles corresponding to bubble spectra shown in Fig. 12b.

level wind starting from the 7th to 14th of November, 1985. During the period the experiment was performed, there was an extended time of very high wind velocity averaging 18 m/sec for a duration of 4 days. Shortly after noontime on the 12th, the wind subsided suddenly to a complete calm sea-state for a duration of 1.5 days. It was followed with an increased light wind for 2 more days. This sudden drop-off of extended high wind gives one an excellent opportunity to study the nature of micro-bubble decaying characteristics. It was unfortunate that the deployment system for the bubble detector broke-down at the height of the storm on the evening of the 9th, and it was not until the morning of the 12th when the storm had subsided that the system could be recovered for further operation. Fortunately, the collected data have covered a sufficient period of this episode to give one an insight into the generation and decaying processes of micro-bubbles produced by intense breaking waves.

Figure 12a shows the micro-bubble spectra taken at a depth of 6 m on the 8th at 19:35h and on the 9th at 9:30h and 12:05h. Referring to the wind record in Fig. 11, one notes that there is a delay of several hours for the concentration of large bubbles >50 microns to follow the fast up, down, and up cycle of the wind, as depicted in the figure, and that the smaller bubbles <50 microns continue to increase following the long trend of continuing high wind. This would indicate that the response time for the increasing concentration of small bubbles is much slower than the larger bubbles. Figure 12b shows the bubble spectra for the depth of 3 m taken on the 9th at 10:15h and 11:20h. One notes that the bubble concentration continued to rise during the one hour interval when the data were collected. Note that the wind was at its peak of 22 m/sec, see Fig. 11.

Figure 13 shows the time series of the local variation of bubble concentration for the bubble spectra shown in Fig. 12b. The abscissa represents the sampling time and the ordinate the concentrations of various size range of micro-bubbles. One notes that for bubbles >100 microns in diameter there are

208

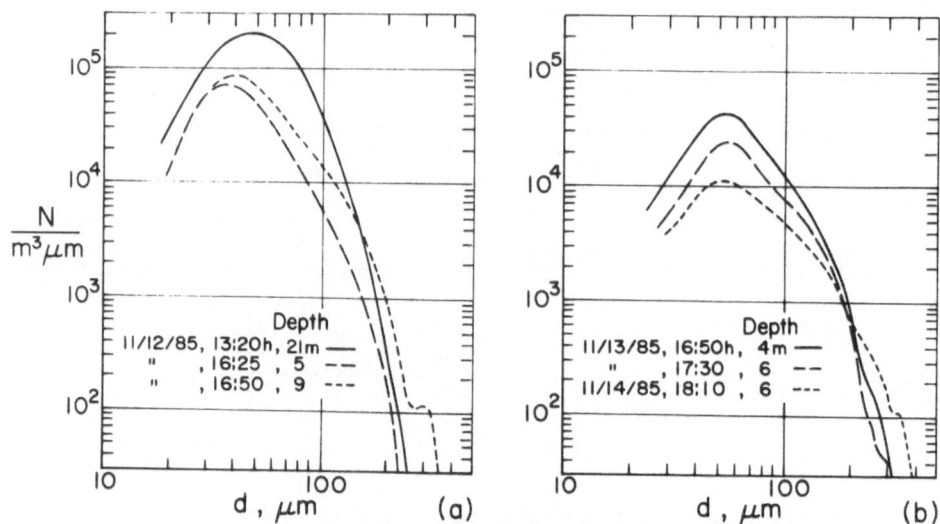

Fig. 14. a) Micro-bubble spectra taken on the 12th and
b) taken on the 13th and the 14th of November, 1985.

definite correlations in the peaking of concentrations with respect to time.
These are bubbles produced by local breaking waves, and they are being convected
by the tidal current towards the stationary detector. Note that the period
between the occurrence of breaking waves is approximately 400 sec. Since the
spatial distribution of micro-bubbles is being convected by a tidal speed of 0.36
m/sec, the corresponding time scale can be converted into a distance scale as
shown in the figure. It indicates a typical spacing of 130 m between the
breaking waves along the direction of tidal flow. One further notes that the
peaking of large bubbles does not correlate with the distribution of bubbles <80
microns. This clearly indicates that breaking waves do not directly produce
bubbles <100 microns. However, the aging of some of the large bubbles do
contribute slowly towards the population of smaller bubbles.

It is worthy to note that within the area of a local breaking wave, the
concentration of large bubbles is a major percentage of the total bubble
population. This effect is greatly reduced when the bubble concentration is
averaged over a larger sea surface where there is no breaking wave nearby, as
indicated by the relatively small concentration of the large bubbles in the
corresponding mean bubble-spectra shown in Fig. 12b.

As the wind subsided suddenly, the source for large bubbles ceased. This
causes the large bubble population to decay quickly, while the population of
smaller bubbles remains stable due to contribution from the decaying process of
large bubbles, as is seen in the bubble spectra shown in Fig. 14a. As time
progresses the small bubbles begin to decay significantly due to the decreasing
source of large bubbles, see Fig. 14b. These bubble spectra represent data taken
1 to 3 days after the cessation of high wind conditions. The corresponding local
variability of bubble concentrations is shown in Fig. 15. One notes that there

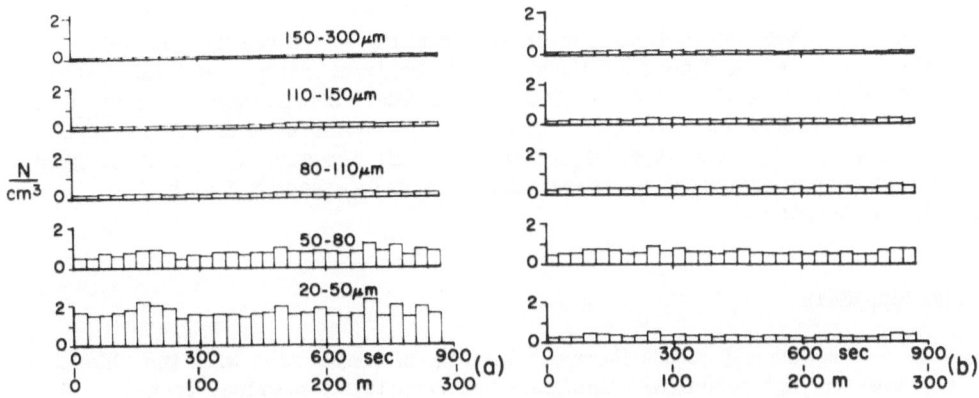

Fig. 15. Local time and spatial variability of micro-bubbles
corresponding to bubble spectra shown in a) Fig. 14a at the
16:25h and b) Fig. 14b at the 17:30h.

is no local peaking of large bubbles in Fig. 15a, because the sea was glass
smooth on the 12th. On the 13th, Fig. 15b, there is a small increase of wind to
4.5 m/sec. The small increase in the large bubble concentration is due to small
breaking waves. However, the small bubbles are noted to continue to decrease due
to the lack of a sustaining bubble source from the small breaking waves. Thus
without a large scale turbulence associated with large breaking waves the large
bubbles have very little chance of being deeply entrained into the sea to
contribute towards the population of the smaller bubbles. A tabulation of bubble
concentrations at the 6 m depth is given in Table II.

Bubble measurements taken at 21 m depth shows consistently higher
concentration of bubbles 50 microns in diameter, see Fig. 14a. This could be
attributed to the larger population of zooplankton at this level as found in the
case study of the lake.

Table II. North Sea Bubble-Concentration at the 6m Depth.

Date Time	Wind speed m/s	\multicolumn{7}{c}{Micro-bubble concentration C, N/cm^3, for different ranges of bubble diameters, µm}							Total C N/cm^3
		20-30	30-40	40-60	60-80	80-100	100-150	>150	
9th 12:05	20.5	1.10	1.80	2.40	1.10	0.52	0.36	0.07	7.4
12th 16.25	2.5	0.32	0.70	0.86	0.35	0.13	0.08	0.01	2.5
13th 17:30	4.5	0.05	0.08	0.38	0.26	0.16	0.21	0.03	1.2

CONCLUDING REMARKS

One may briefly conclude that high intensity breaking waves mainly produce
bubbles >200 microns in diameter. When these bubbles are deeply entrained into
the sea, they can contribute substantially to the small bubble population through
diffusive aging processes. The time constant for such increase and decrease of
micro-bubbles is very long, typically of the order of one day. On the other hand,
the more permanent source of micro-bubbles is largely associated with the
activities of zooplankton.

ACKNOWLEDGEMENTS

This work was conducted at the Nordsee Platform in conjunction with the NOREX
'85 experiment. The technical assistance and hospitality provided by the
Forschungsantalt der Bundeswehr fur Wasserschall und Geophysik is gratefully
acknowledged.

REFERENCES

Jenkins, F.A. and White, H.E. 1957 Fundamentals of Optics, p. 509-517, McGraw-
Hill, New York, N.Y.

Ling, S.C., Gowing, S. and Shen, Y.T. 1982 Role of micro-bubbles on the inception
of cavitation over headforms, Proc. 14th Symposium on Naval Hydrodynamics, p.
547-582, Office of Naval Research, Arlington, Va.

Ling, S.C. 1987 Study of micro-bubbles in Lake Pend Oreille, David Taylor Naval
Ship Research and Development Center Report No. DTNSRDC/SPD-1237-01, Bethesda,
Maryland.

Raymont, J.E. 1963 Plankton and Productivity in the Oceans, p. 120-466,
MacMillan, New York, N.Y.

M. -Y. Su[1], S. -C. Ling[2] and J. Cartmill[3]

1 Naval Ocean Research and Development Activity, NSTL, MS 39529 U.S.A.

2 Catholic University of America, Washington, D.C. 20064 U.S.A.

3 Planning Systems Incorporated, Slidell, LA 70458 U.S.A.

ABSTRACT. Microbubbles from 2 m below the wave trough to the largest depth of 21 m, and within the diameter ranges of 20 to 400 μm were measured near the NORDSEE platform in the North Sea by means of an optical device based on the dark-field specular reflection. The most striking results are (1) a close similarity in shape of bubble concentrations under a wide range of wind/wave conditions and over the depth range of the measurements, and (2) a distinctive concentration maximum appearing between the bubble diameter of 40 to 50 μm, (3) a power law dependence of bubble density on wind speed, and (4) a lack of a clear depth dependence in the bubble concentrations.

1. INTRODUCTION

Breaking of wind-generated ocean waves generates bubble plumes near the ocean surface as evidenced by direct visual observations, underwater photography (Su et.al., 1984), and acoustical scattering techniques (Medwin 1977 and Thorpe 1982). These air bubbles play important roles in many air-sea interaction processes (see the recent review article by Wu, 1981). Our main interest lies in effects of air bubbles, in addition to high-frequency surface waves, on high-frequency acoustical scattering, particularly at low grazing angles. Moreover, we are most interested in obtaining in-situ measurement of bubble concentrations and distributions from which acoustical scattering may be computed theoretically. This specific interest leads us to employ optical, rather than acoustical, in-situ techniques for bubble detection.

The purpose of this paper is to report and discuss some results of bubble concentrations from a joint experiment among three U.S. and F.R.G. agencies in the North Sea during November-December 1985 (NOREX 1985). In this experiment, measurements of acoustical scattering and surface waves, were made, which are reported elsewhere (Nutzel, et. al., 1986). The bubble data discussed in the paper by Ling and Pao (1988) in this volume is also from the same series (Records No. 1-20) of the bubble measurement made during NOREX 85.

211

2.   MEASUREMENT AND ANALYSIS OF MICROBUBBLES

Microbubbles were measured near the ocean surface from 2 m below the
wave trough to the largest depth of 21 m.   A diameter range of 20 μm
to 400 μm was measured in-situ by means of an optical device designed
by S.C. Ling,   Catholic University of America, under U.S. Navy
sponsorship.   The basic principle of bubble detection is based on the
dark-field specular reflection from a single bubble residing in a
small sampling volume.   Hypothetically, in the mean, there is less
than .01 probability for more than one bubble to appear in the sample
volume when the maximum bubble density is 10^7 bubbles/m^3.   Figure
1 shows the main components of the detection system.   The overall
dimensions of the system with the casing is 32 x 9 x 4 in.   It is
noted in Figure 1 that there are actually two scattering volumes,
separated by a distance of 7 mm together with two separate
photomultipliers for optical signal detection.   The purposes for the
two scattering volumes are (1) for determining the mean velocity of
passing bubbles in order to facilitate the estimation of the bubble
concentration versus different bubble diameters and, (2) for use in
calibrating the system by means of the well-known relationship
between the bubble diameter and its terminal velocity (so called
Stokes law).   A pump is used to suck water containing the bubbles
through the two sampling volumes with a mean velocity of about 0.62
m/sec.   Refer to Ling and Pao's paper for a more detailed description
and discussions of the optical bubble sensor.

During NOREX 85, the optical bubble detector, a pump, and counter
weights were suspended from the tip of either the 20 m crane or the
26 m crane on the deck of the NORDSEE Research platform, that is
located approximated 60 km west of the German and Danish coasts at
the mean water depth of 30 m.   In order to ensure that the detector
was always submerged during high sea states, the sensor depth closest
to the sea surface was chosen to be 2 m below the mean wave trough.
Five other depths of 4, 6, 10, 15, and 21 m were also chosen.   A
14-minute long record was made at each depth for a given wind
condition.   A total of 121 records were obtained under conditions
from relatively calm sea with the wind speed $\leq$ 2 m/s to very rough
sea with the wind speed close to 18 m/s.   These records consist of 20
depth profiles, each under a similar sea state.

3.   RESULTS AND DISCUSSIONS

Figures 2, 3, and 4 show the bubble concentration (number of bubbles
per m^3 per μm of bubble diameter) for a mean wind speed range of
9.5 m/s to 13 m/s on three separate dates.   Superimposed on these
figures is the bubble concentration at 0.7 m, 1.8 m, and 4.0 m, as
obtained by Johnson and Cooke (J-C) (1979) using a photographic
method under similar wind speed of 11-13 m/s at a Canadian coastal
location in 30 m water depth during winter.   To the best of our
knowledge, the J-C represents the best in-situ measurements of near

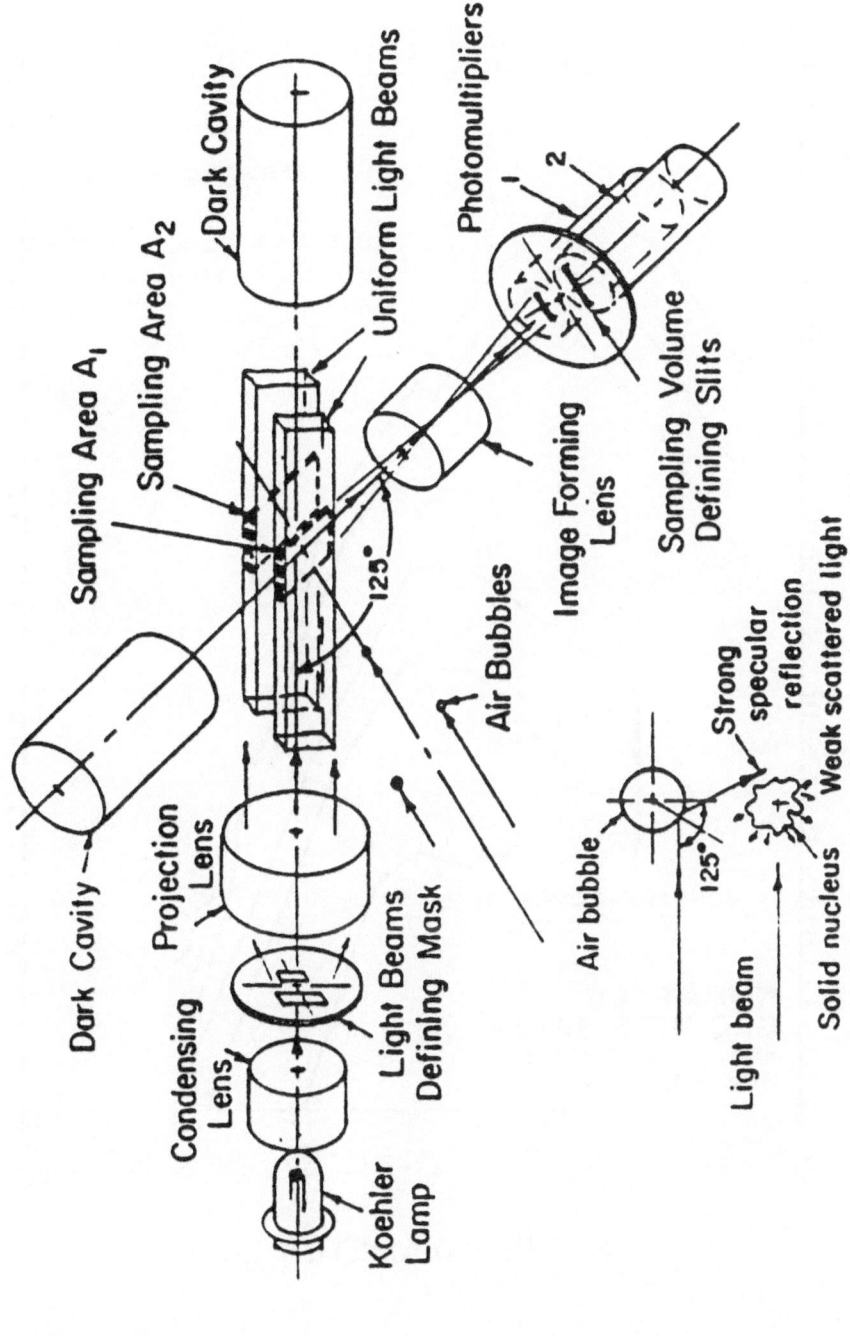

Figure 1. A sketch of detection of microbubbles by dark-field specular reflection

214

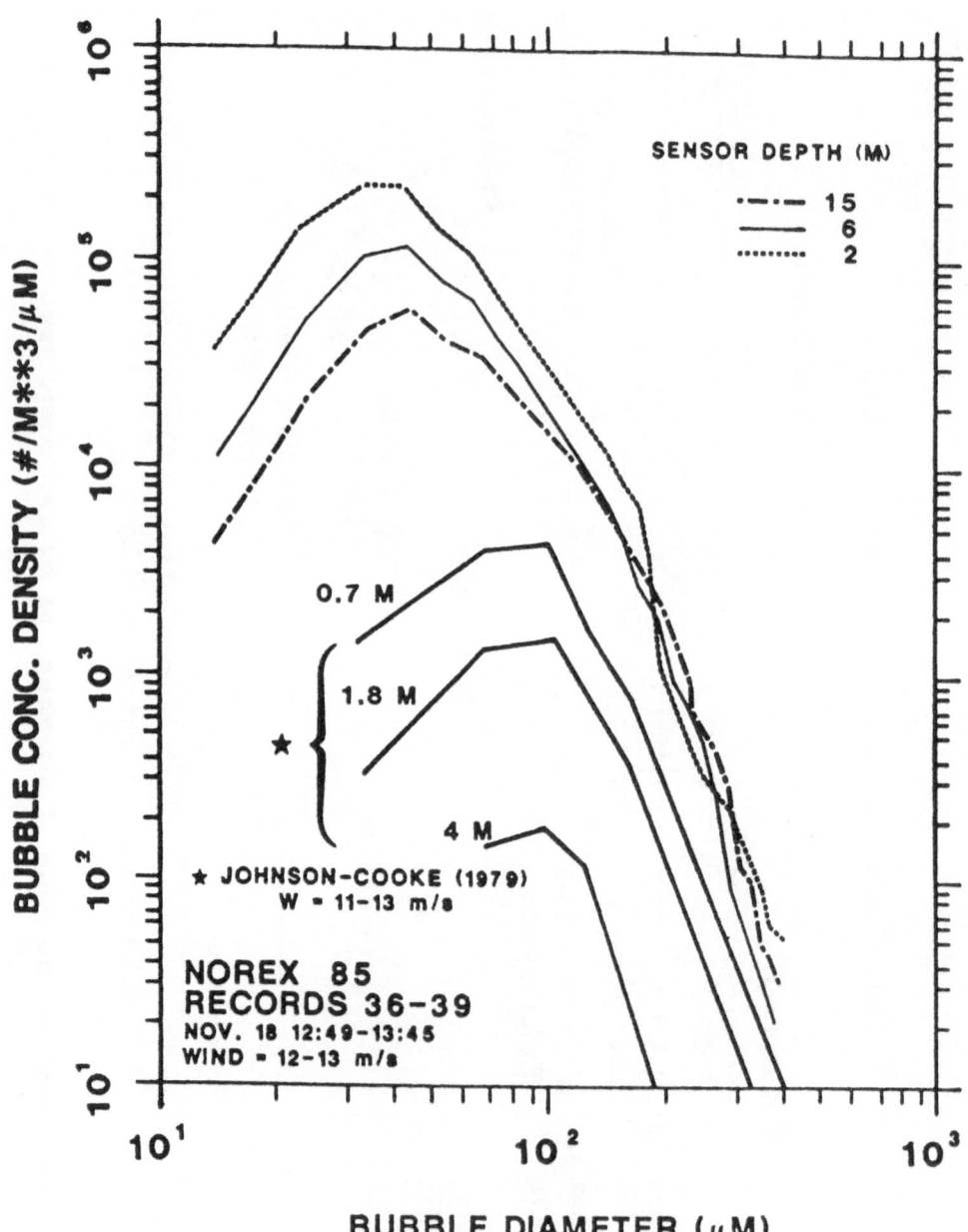

Figure 2. Bubble concentration for Record No. 36-39
with a comparison with Johnson and
Cooke's (1979)

ocean surface bubbles to date without obvious limitations in measurement techniques or affected by particle contamination (see the review by Wu, 1981), and is thus the only independent data used for comparison with the data from NOREX 85. Obviously, as can be noted in the figures, there exists considerable differences between these two sets of bubble concentrations; these differences will be briefly discussed.

The NOREX 85 measurements show a dominant peak in bubble concentrations at diameters (d) between 40 and 50 μm, at all depths, while J–C peaks near 90 μm, a factor of two larger than the NOREX 85 data. Private communication with Johnson (1986) indicates that his photographic technique becomes increasingly less accurate for d < 100 μm. Therefore, the decrease in the J–C bubble density for d < 90 μm is very likely due to the technique used. Kolovayev (1976) used a bubble trap, originally developed by Glotov et al (1962), to measure oceanic bubbles, which show a dominant peak in bubble concentrations between 140 and 160 μm in diameter for similar wind speeds. The longer delay time necessary for the smaller bubbles to rise to the transparent upper end before photographing raises serious questions about dissolution, growth and coalescence. According to Stokes' law, a bubble 50 μm in diameter requires 8 min to traverse the 60 cm tube length of Kolovayev's trap, but the same bubble, if clean and without surfactant coating, requires only 3 min to dissolve at 10 cm depth in air-saturated water (Blanchard and Woodcock, 1957). Therefore, the bubble trap method is not reliable for measuring bubbles with d < 100 μm and gives a false peak near d = 150 μm in Glotov and Kolovayev's measurements. Therefore, the peak bubble diameter between 40 and 50 μm from NOREX 85 is deemed to be more accurate.

The depth dependence of the bubble concentration is not clear in the NOREX 85 data, while it is very noticable in the J–C data. Assuming that near-surface bubbles are mainly generated by wave breaking and considering the acoustical scattering measurement by Thorpe (1982), one would expect that the type of rapid decrease as shown by the J–C data is more reasonable. In general, the bubble concentrations from NOREX 85 are much higher than J–C; however, near d = 350 μm, the NOREX 85 data at 2.0 m is fairly close to J–C at 1.8 m for two of three cases shown in Figures 3 and 4.

The possibility of contamination of the bubble measurements due to wave-platform interactions and the existence of a subsurface coolant exhaust located 10 m below the sea surface cannot be totally dismissed at this time. However, the bubble sensor was a minimum distance of 20 m from the N–W or S–W leg of the platform, and the coolant exhaust was located at the N–E leg of the platform. The main tidal direction is from N–W to S–E. The wind direction is from the east during the measurement of Record No. 36–39 (Figure 2), from the west during Record No. 112–116 (Figure 3), and from the south-west during Record No. 117–121 (Figure 4). A definitive resolution of this issue will be determined with planned future measurements removed from the platform using different deployment methods.

216

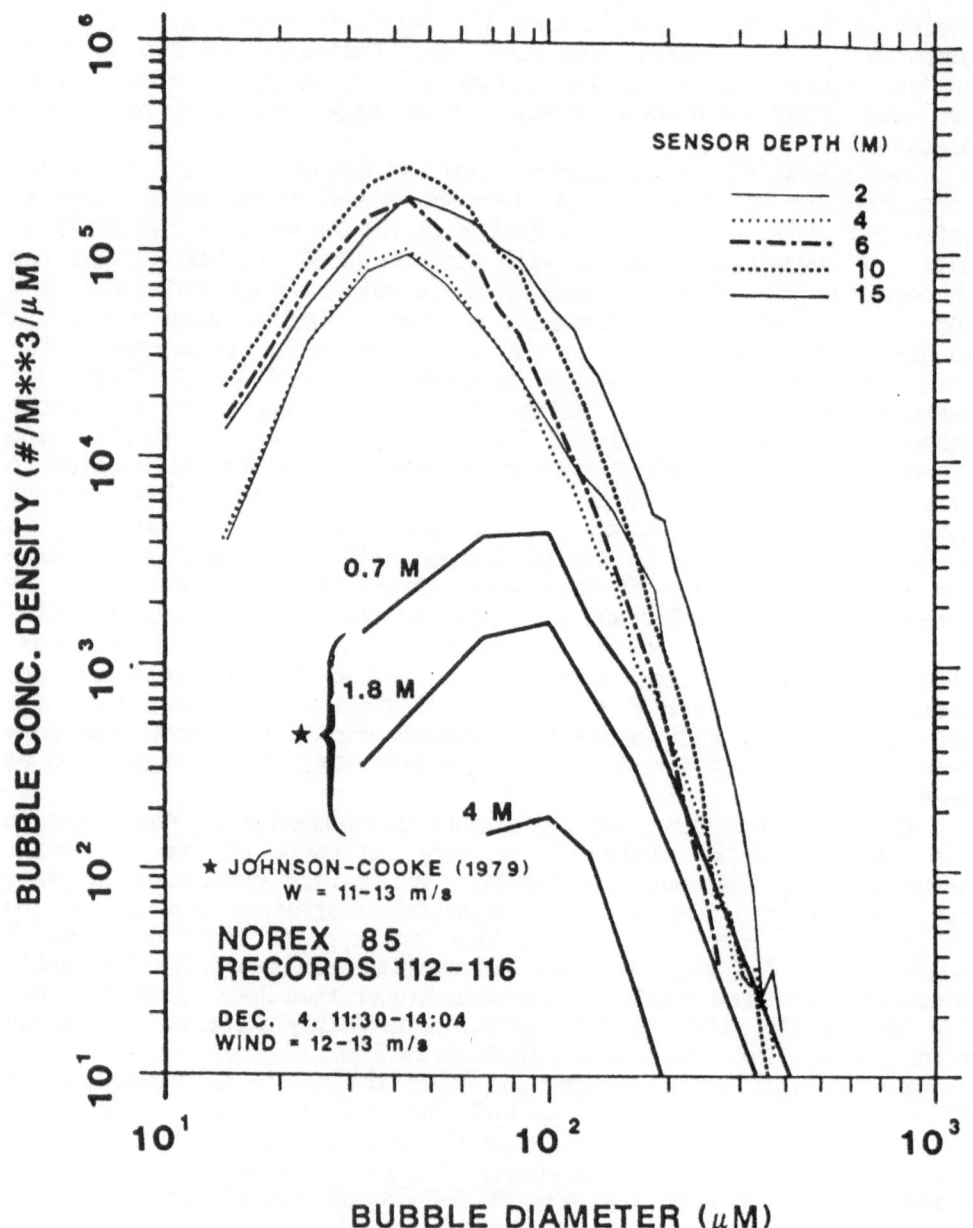

Figure 3. Bubble concentration for Record No. 112-116
with a comparison with Johnson and
Cooke's (1979)

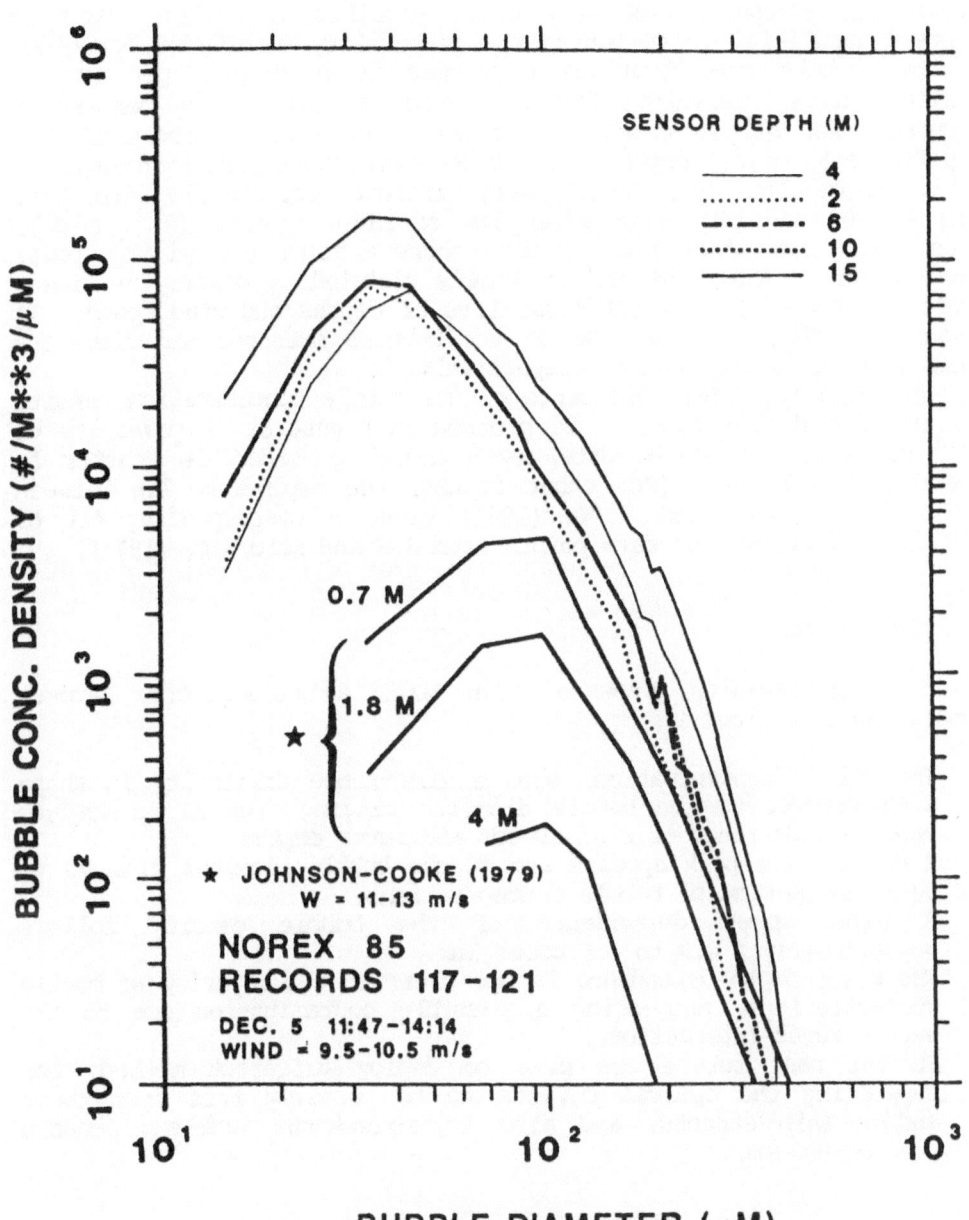

Figure 4. Bubble concentration for Record No. 117-121
with a comparison with Johnson and
Cooke's (1979)

The similarity in the shapes of the bubble concentrations from d = 20 to 400 μm is nevertheless very striking, as can be noted in Figures 2, 3 and 4 (and many other profiles as well). This is further quantified by computing the probability density of occurence of the bubble concentrations expressed as a percent per 10 μm diameter class intervals. Figure 5 shows the probability density of occurence for all 20 cases of 2 m depth for Record Numbers 21-121, and Figure 6 for all depths. For bubble diameters less than 30 μm, a cubic power law (d^3) fits best, whereas for bubble diameters between 200 and 400 μm a power law of minus 5 or 6 (d^{-5} or d^{-6}) fits well. Note that the J-C data shows a minus 4.5 ($d^{-4.5}$) power law. Figure 7 shows the average bubble probability density in linear scale for Record No. 21-121 regardless of depths and wind speed. It shows that about 82% and 99% of the bubbles measured had diameters less than 100 μm and 200 μm, respectively.

The wind speed (W) dependence of the bubble concentration density for D = 2 m during NOREX 85 is plotted in Figure 8. A power law of $W^{4.3}$ seems to follow the data, even though considerable scatter in the data can be seen (for other depths, the dependence lie between 4.0 and 4.6 power law). Wu (1981) gives a corresponding fit of $W^{4.5}$ based on only four data points from J-C and Kolovayev (1976).

4. CONCLUSIONS

Based on the results presented from NOREX 85 we may draw several conclusion as follows:

a. The bubble concentrations show a remarkable similarity in shape with respect to the bubble diameter ranging from 20 to 400 μm under a wide range of wind speeds and water depths.
b. A distinctive peak appears around the bubble diameter from 40 to 50 μm among most of bubble concentrations.
c. A wind speed dependence of the bubble density follows approximately a 4.0 to 4.6 power law.
d. The water depth dependence is not clear in the majority of bubble concentrations, suggesting a possible contamination due to the wave - tower interactions.
e. In the near future, we plan to deploy different methods for supporting the optical devices to be possibly free from tower and/or ship effects, and also to extend the bubble diameters beyond 1000 μm.

ACKNOWLEDGEMENTS

The authors would like to thank the assistance and hospitability provided by Dr. Heinz Herwig and his collegues at Defense Research Institute for Underwater Sound and Geophysics, Kiel, F.R.G. during the NOREX 85.

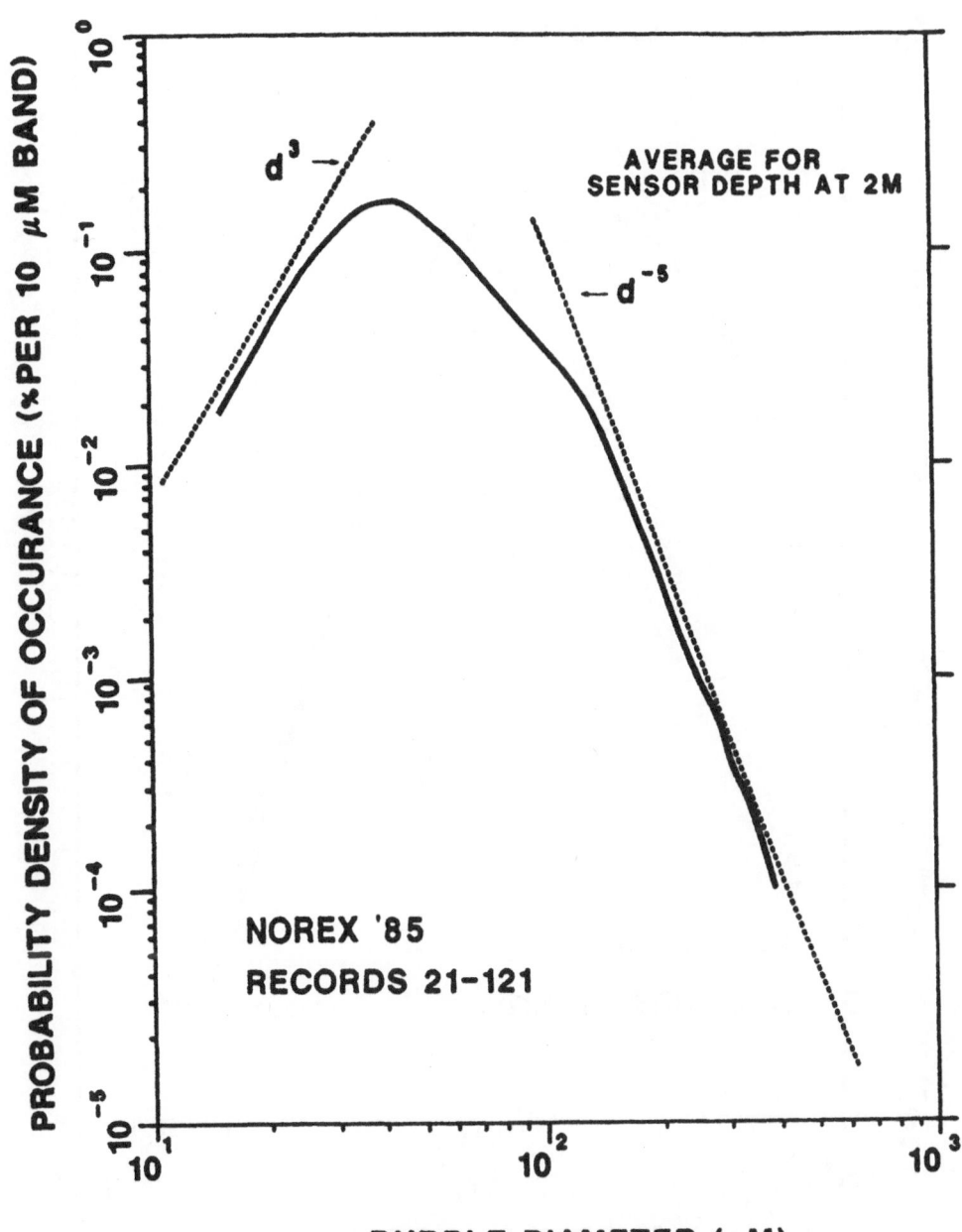

Figure 5. The average probability density of bubble occurrence for all the records with depth=2 m during NOREX 85

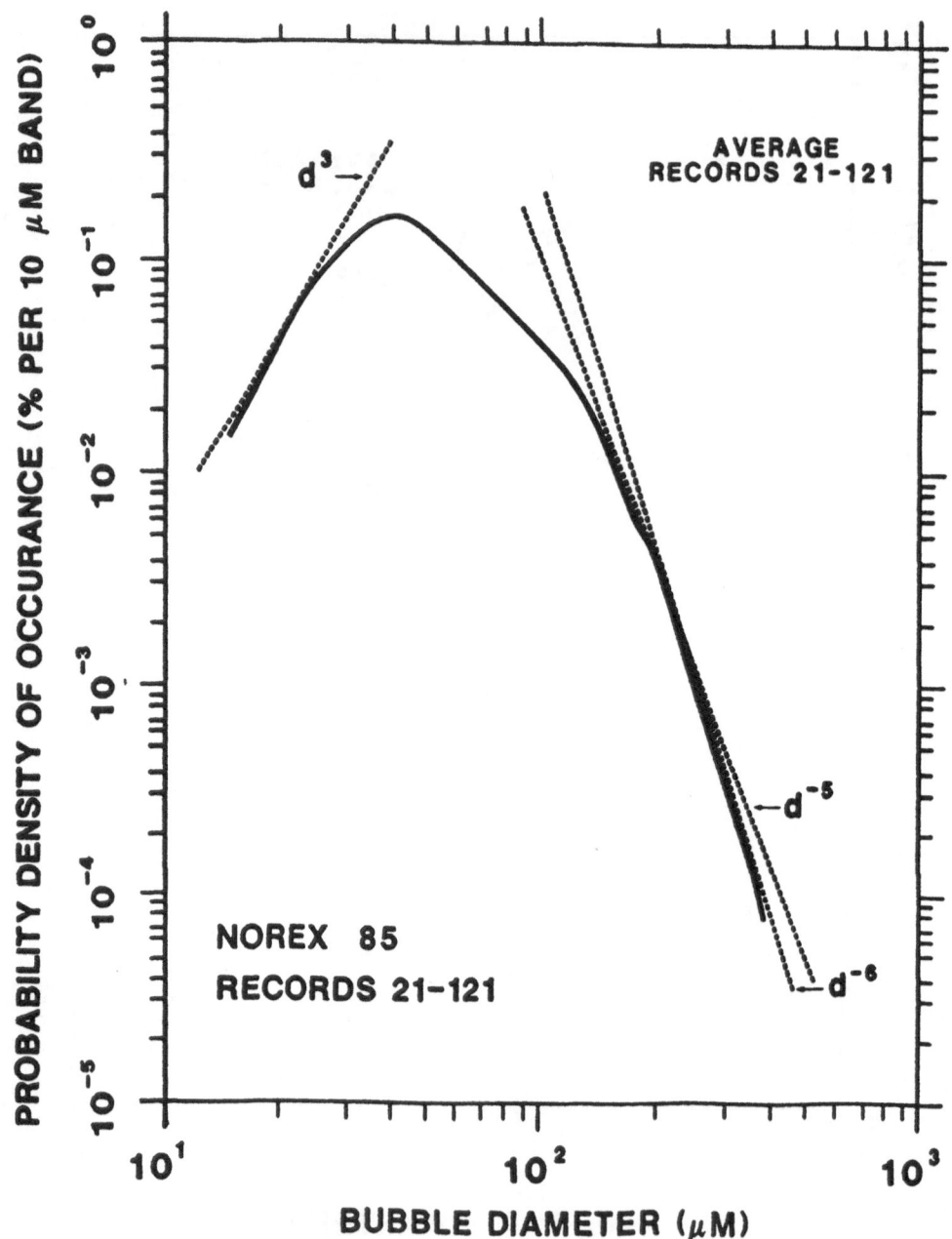

Figure 6. The average probability density of bubble
occurrence for NOREX 85 Record No. 21
through 121 regardless of depths

221

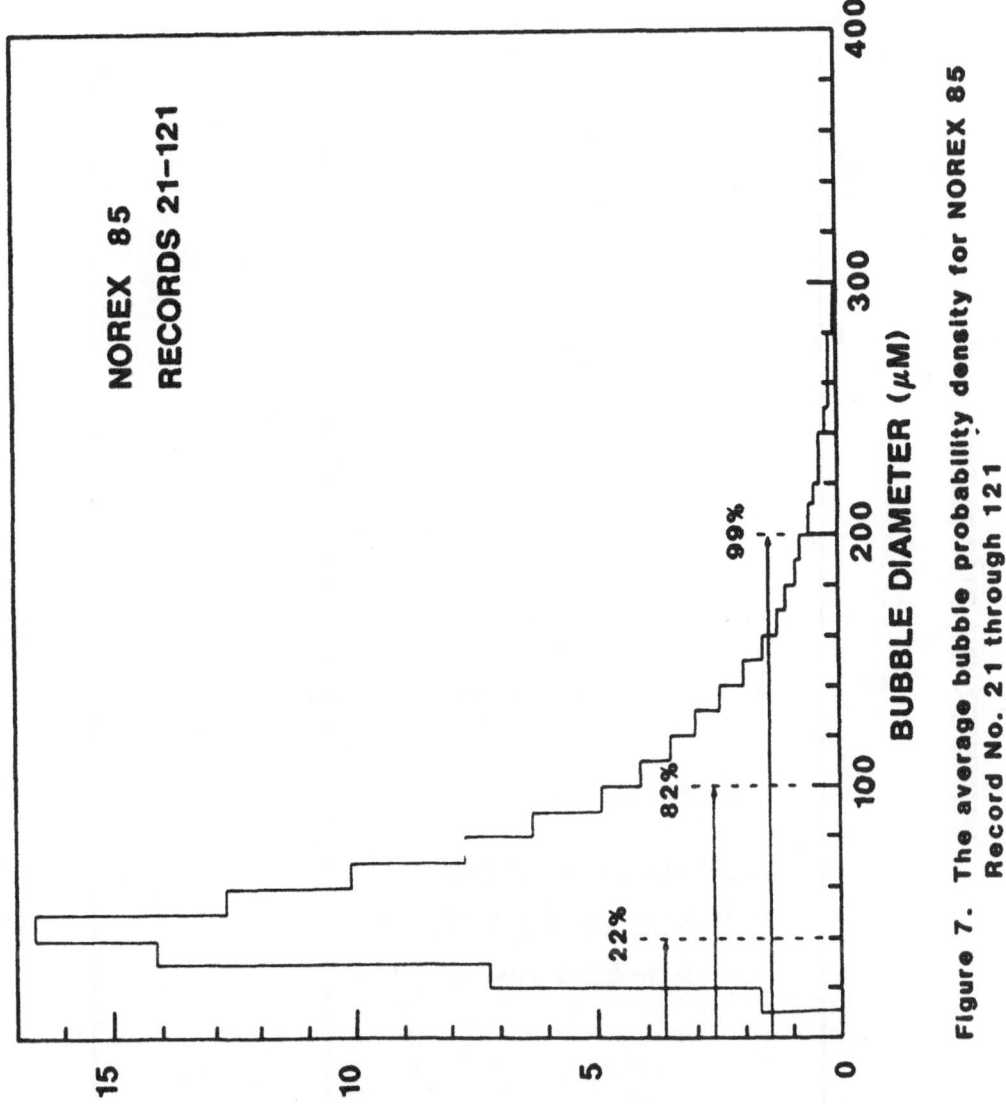

Figure 7. The average bubble probability density for NOREX 85
Record No. 21 through 121

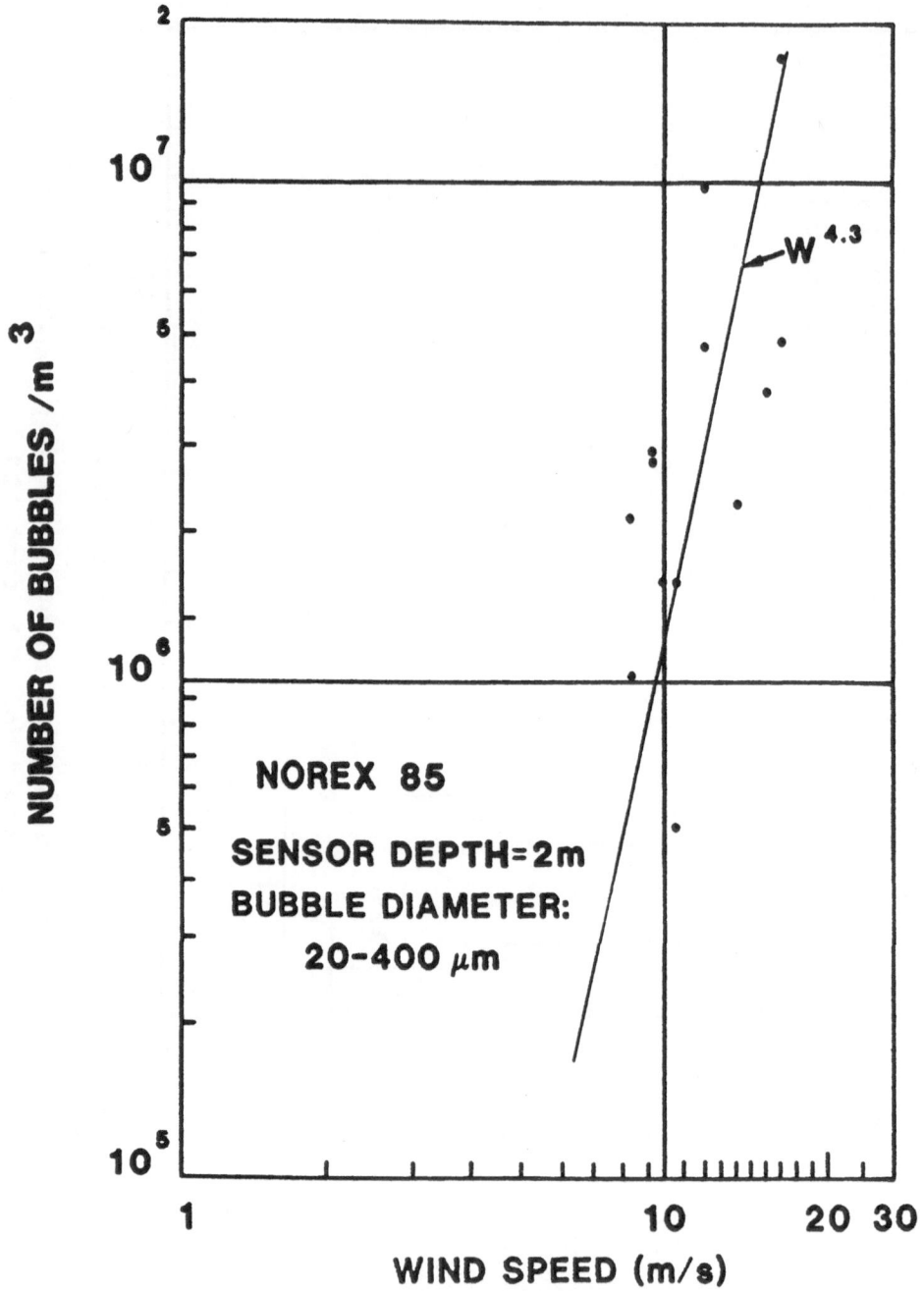

Figure 8. Dependence of number of bubbles per m³
on the wind speed for sensor depth=2 m

REFERENCES

Blanchard, D.C. and A.H. Woodcock (1957). "Bubble Formation and Modification in the Sea and its Meteorological Significance". Tellus, 9, pp. 145-158.

Glotov, V.P., P.A. Kolobaev, and G.G. Neuimin (1962). "Investigation of Scattering Of Sound by Bubbles Generated by An Artificial Wind in Sea Water and the Statistical Distribution of Bubble Size". Sov. Phys. Acoust. Engl. Transl. 7, pp. 341-345.

Johnson, B.D. and R.C. Cooke (1979). "Bubble Populations and Spectra in Coastal Waters: A Photographic Approach", Jour. Geop. Res., Vol. 84, C7, pp. 3761-3766.

Kolovayev, D.A. (1976). "Investigation of the Concentration and Statistical Size Distribution of Wind-Produced Bubbles in the Near-Surface Ocean", Oceanology, Eng. Transl. 15, pp. 659-661.

Ling, S.C. and F. Pao (1988). "Study of Micro-bubbles in the North Sea". In this volume.

Medwin, H. (1977). "In-situ Acoustic Measurements of Microbubbles at Sea". Jour. Geop. Res., 82, pp. 971-987.

Nutzel, B., H. Herwig, J.M. Monti, and P.D. Koenigs (1986). "A Further Investigation of Acoustical Scattering from the Sea Surface". NUSC Tech. Doc. 7685. Naval Underwater System Center.

Su, M.Y., A.W. Green and M.T. Bergin, (1984). "Experimental Studies of Surface Wave Breaking and Air Entrainment". In: Gas Transfer at Water Sufaces, pp. 211-219, eds. Brutsaert and Jirka, D. Reidel.

Thorpe, S.A. (1982). "On the Clouds of Bubbles Formed by Breaking-waves in Deep Water, and their Role in Air-Sea Gas Transfer", Phil. Trans. R. Soc. London, A 304. pp. 155-210.

Wu, J. (1981). "Bubble Populations and Spectra in Near-Surface Ocean: Summary and Review of Field Measurements", Jour. Geop. Res., Vol. 86, C1, pp. 457-463.

ACOUSTICAL ESTIMATES OF SUBSURFACE BUBBLE DENSITIES IN THE OPEN OCEAN
AND COASTAL WATERS

Suzanne T. McDaniel
Applied Research Laboratory
The Pennsylvania State University
University Park, PA 16804

ABSTRACT:

High-frequency acoustic backscattering data are inverted to determine the dependence of subsurface bubble populations on wind speed,
resonant bubble radius, surficial water temperature, and the proximity
of land. Relationships of bubble densities with resonant radius and
wind speed are found to compare favorably with the results of previous
studies. No dependence on surface water temperature is apparent for
temperatures ranging from 12 to 25° C. However, a strong dependence on
the proximity to land masses is evident, with bubble densities in
coastal waters being almost an order of magnitude greater than those in
the open ocean at elevated wind speeds.

I.   INTRODUCTION

Subsurface bubbles play an important role in many oceanic processes including the generation and propagation of high-frequency ambient
noise. Forced bubble oscillations are a source of noise [Kerman
(1984)] and large numbers of resonant bubbles act to absorb the acoustic energy produced by other ocean surface phenomena [Farmer and Lemon
(1984)]. A number of experimental studies [Medwin (1970), Johnson and
Cooke (1979), Kolovayev (1975)] have been performed to measure subsurface bubble populations. The techniques used for this purpose
ranged from deriving bubble densities from acoustic absorption and
backscatter measurements to direct optical observations. All of these
techniques have certain basic limitations, leading to a very wide
spread in the bubble populations reported by researchers using
different techniques.

Medwin (1970) inferred subsurface bubble densities from
measurements of acoustic absorption. Because the bubble densities
measured by Medwin were almost independent of depth, he concluded that
they were due to biological processes. Kolovayev (1976) used a bubble
trap to measure bubble densities in temperate parts (14° C) of the
Atlantic Ocean. Johnson and Cooke (1979) conjectured that the

225

B. R. Kerman (ed.), Sea Surface Sound, 225–236.
© 1988 by Kluwer Academic Publishers.

disparity between their measurements and Kolovayev's may have been due to the dissolution and coalescence of the smaller bubbles on rising to the top of the trap. They used a photographic approach to measure bubble densities in the frigid (2° C) coastal waters of Nova Scotia. The bubble densities observed in these two measurements are attributed to wind-wave action because of their rapid decay with depth below the surface.

Wu (1981) reviewed these latter two measurements and despite their differences was able to derive functional relationships for dependences on depth, wind speed and bubble radius. He found, that in the upper 3 m of the ocean, the bubble population decayed exponentially with depth. For bubbles greater than 50 μ in diameter, the bubble density spectrum F obeyed a power law dependence on bubble radius a_r: $F = a_r^{-s}$, where $3.5 < s < 5$. Wu found that the wind speed dependence also followed a power law: the bubble density increased with the 4.5th power of the wind speed. From his review of these data, Wu also proposed that the physical mechanisms governing subsurface bubble production may be dependent on the proximity of land masses and on surficial water temperature, as well as wind speed, since bubble populations in coastal waters may be affected by the presence of contaminants which alter the life-time of the bubbles and the increase of the viscosity of water with decreasing temperature leads to slower bubble rise times.

More recently, Farmer and Lemon (1984) applied an elegant technique to derive subsurface bubble densities from wind generated ambient noise data. Their experiments were performed in Queen Charlotte Sound north of Vancouver Island where the surficial water temperature was 8.5° C. By assuming that the linear frequency relationship that exists between ambient noise levels at wind speeds below 10 m/s holds at higher wind speeds, departures from this relationship were used to infer bubble densities for wind speeds of 10 to 20 m/s. This technique permitted an estimation of depth integrated bubble densities which compared quite favorably with Johnson and Cooke's results.

In this paper, acoustic backscatter data are analyzed to determine bubble populations in the open ocean and in coastal waters. This method allows only the determination of depth integrated bubble densities and thus provides no information on their distribution in depth. In addition, the acoustic data on which it is based may be contaminated with scatter from marine life or ambient noise, leading to overestimates of bubble densities, particularly at low wind speeds. The major advantage of this method is that a vast amount of acoustic backscatter data is available, thus permitting studies of the dependence of bubble densities on the proximity of land masses and on water temperature as well as relationships with wind speed and bubble radius.

In Section II of this paper, the basic equations used to invert the acoustic backscattering data are presented. In Section III, the acoustic backscatter data are briefly reviewed and in Section IV inverted to obtain estimates of depth integrated bubble densities. The dependence of bubble densities on wind speed and bubble radius is found to be in accord with the results of previous studies. For the range of surficial water temperatures 11° to 25° C for which acoustic data are available, no significant dependence on temperature is obtained. The data, however, strongly support Wu's conjecture that the proximity of land masses influences subsurface bubble concentrations. In Section V these findings are discussed and summarized.

II   THEORY

The basic theory used follows closely that of Crowther (1980), the significant difference being that all four paths by which an ensonified bubble may backscatter energy to the source are considered. These four paths are shown in Fig. 1. At very low wind speeds when few bubbles are present, all four paths contribute equally. However, because bubbles absorb as well as scatter acoustic energy, at high wind speeds the relative energy scattered via paths b, c, and d of Fig. 1 is greatly attenuated compared to that scattered along path a. For paths b, c, and d, the losses incurred on reflection from the surface are negligible for bubbles less than 2 m in depth.

To proceed, it will be assumed that scatter from different bubbles and scatter along disjoint paths from the same bubble add incoherently. The scattered pressure field P for a source of unit strength and frequency f then obeys

$$\langle PP^* \rangle = \sum_{n,m=1}^{2} \iint S^2(a) \, dV \, N(z,a)(a_n^2 a_m^2 / r_m^2 r_n^2) da, \tag{1}$$

where S is the scattering cross section of a scatterer of radius a, $N(z,a)$ is the number of bubbles per unit volume per unit bubble radius at depth z, and a_n and a_m are the attenuation over the respective paths 1 and 2. The major contribution to the integral over a in Eq. (1) comes from a narrow region about resonance, hence, the attenuation factors may be approximated by their value at resonance. Furthermore, Medwin (1977) has shown that the integral over a of S^2 may be

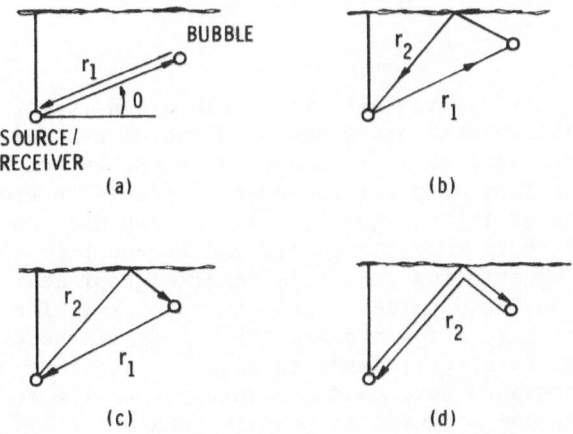

Figure 1. Paths by which backscatter can occur.

approximated by $S^2 = 3.4 \pi a_r^3/4d$, where a_r is the radius at resonance, $a_r = 3.2/f$, and d is a damping factor $d = .0136 + 3.8 \times 10^{-4} f^{1/2}$.

Following Crowther (1980), the absorption factors along each of the paths take the form

$$a_n^2 a_m^2 = \begin{cases} \exp[-2q\ F(z)]; \text{ path a} \\ \exp[-2q]; \text{ paths b and c} \\ \exp[-4q+2qF(z)]; \text{ path d} \end{cases} \qquad (2)$$

where $q = 3.4\ \pi^2 a_r^3\ N/[d_r \sin(\theta)]$, with d_r being the reradiation damping factor, $d_r = 0.136$, θ the nominal grazing angle as shown in Fig. 1, and N the depth integrated bubble density $N = \int dz\ N(z)$. With the function $F(z)$ given by $dF(z)/dz = N(z)/N$, the integration over z in Eq. (1) may be performed, to obtain

$$\langle PP^* \rangle = (d_r A \sin(\theta)/8\pi d\ r_1^4)[1 + 4q\ \exp(-2q) - \exp(-4q)], \quad (3)$$

where $A = \int dx\ dy$ is the illuminated area and it has been assumed that the paths r_1 and r_2 are sufficiently close in length that they can be taken as equal. The backscattering strength s for a source of unit strength is defined as

$$s = 10\ \log_{10}(PP^* r_1^4/A). \qquad (4)$$

Equations (1) through (4) provide relationships between backscattering strength and depth integrated bubble density that can be inverted to obtain the density given the backscattering strength.

In the limit of high bubble densities when the attenuation becomes significantly great, Eq. (3) reduces to

$$\langle PP^* \rangle = d_r A \sin(\theta)/(8\pi dr_1^4), \qquad (5)$$

which depends only on the grazing angle and the ratio of the two damping constants. At sufficiently high wind speeds, the backscattering strength is independent of bubble density, and thus the inversion of backscatter data is limited to bubble densities less than this saturation limit.

III ACOUSTIC DATA

Most of the acoustic data that will be inverted to obtain sub-surface bubble densities were described by McDaniel and Gorman (1982). In addition, data acquired at a frequency of 60 kHz by Garrison, Potter and Murphy (1966) in Dabob Bay are included to provide a comparison with open ocean data at this frequency. All of the data considered were acquired using short acoustic pulses and narrow beam sources and receivers. The backscattering strengths reported represent an average over 100 to 200 transmitted pulses, with the total time for one measurement being roughly a few minutes. For most of the experimental geometries employed, relatively large surface areas were ensonified. Environmental measurements were conducted along with the acoustic data aquisition. Wind speeds measured at heights ranging from 2 m to 25 m above the mean sea surface have been translated to the wind speed at a standard height of 10 m, assuming a neutrally stable atmosphere.

Prior to proceeding with the inversion, let us examine some of the features of these data. Lilly and McConnell (1978) have shown that in the open ocean, the acoustic backscattering strength is almost independent of frequency at fixed wind speed. This lack of frequency dependence will be used to group data in the frequency range of 20 to 30 kHz together. That is, all of the data over this frequency range will be inverted using parameters for a nominal bubble radius of 130 μ corresponding to the mean radius for this frequency range.

It is of interest to examine the data in the saturation limit to determine if the basic theory developed in Section II is correct and also to verify the bubble parameters used. Figure 2 compares high wind speed backscatter data in the 18 to 25 kHz range with the predictions of Section II for a frequency of 22.5 kHz. The 18 kHz data shown in this example were recently acquired from a platform in the North Sea by Koenigs et. al. (1986). The theoretical result is the saturated limit given by Eq. (5). The results are generally within 2 dB of the predictions which is roughly the accuracy of the acoustic measurements.

IV BUBBLE POPULATIONS

Equation (3) was used to invert backscattering strength data to obtain depth integrated bubble densities. To perform this inversion, contributions arising from the sea surface roughness were first estimated to insure that the scatter was not from this source. Then Eq. (3) was solved iteratively for N and a final check was performed to eliminate data in the saturated regime.

Figure 3 shows depth integrated bubble densities as a function of wind speed for a bubble radius of 53 μ. The first thing that strikes

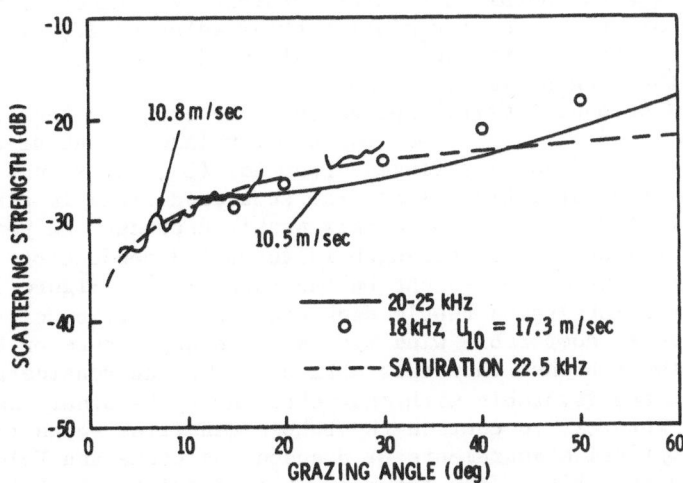

Figure 2. Measured and predicted backscattering strengths in the saturation limit.

one on examining this figure is the wide spread in the data. The accuracy of the acoustic measurements of about ± 2 dB can account for only a fraction of the scatter in the data. Other factors contributing to the scatter are the variation in wind speed during the course of a measurement and the inaccuracies in measuring this parameter. In addition, the ordinate in this figure is U_{10}, the wind speed at standard height rather than surface friction velocity u* which is a more relevant environmental parameter. The relationship between U_{10} and u* depends on atmospheric stability. The existence of non-stable atmospheric conditions during these measurements is another possible source of the scatter in the data.

Data are shown in Fig. 3 for near surface water temperatures differing by about 10°C. In contrast to what one would expect, the bubble densities in the warmer water are somewhat higher than those in the cooler water, although given the spread in the data it is doubtful if this difference is statistically significant. A least squares fit to the data is shown by the solid line in this figure, with the standard deviation shown by the dashed lines. The dependence on wind speed follows a 3.6 power law which is somewhat lower than the 4.5 dependence determined by Wu from optical bubble measurements. Figure 4 shows bubble densities derived from data acquired in Dabob Bay by Garrison, Potter and Murphy (1966). The somewhat larger spread in the data in this case can be attributed to refractive effects which were not removed from the data. Comparing Figs. 3 and 4, it is evident that at the higher wind speeds there are significantly greater bubble densities in coastal waters.

Open ocean depth integrated bubble densities for bubbles having a nominal radius of 130 μ are shown in Fig. 5 for two water temperatures. Again, the bubble densities for the warmer water appear to be slightly, but not significantly, higher than those for the cooler water. This data set has the highest wind speed dependence of all the cases examined. Figure 6 shows bubble densities obtained from acoustic measurements in coastal waters for a nominal bubble radius of 130 μ. Bubble densities at the higher wind speeds are almost an order of magnitude greater than the open ocean densities.

Lilly and McConnell (1978) showed that the acoustic backscattering strength at a fixed wind speed was essentially independent of frequency over a 10 to 60 kHz frequency range. From Eq. (3), for values of q well away from the saturation limit, the scattered field is directly proportional to Na^3/d. Since d is only weakly dependent on resonant frequency, if the scattering strength is to be independent of frequency then N must be inversely dependent on the cube of a_r. Figure 7 shows open ocean depth integrated bubble densities as a function of resonant radius measured at comparable wind speeds. The dependence on bubble radius is somewhat higher than that obtained from the considerations above and compares favorably with that obtained by by other researchers.

It is of interest to compare the bubble densities shown in Figs. 3-6 with the optical measurements of Johnson and Cooke and Kolovayev. To effect this comparison, the optical bubble densities must be depth integrated. This integration was performed for the higher wind speed data reported by matching the data with an exponential in depth N(z) =

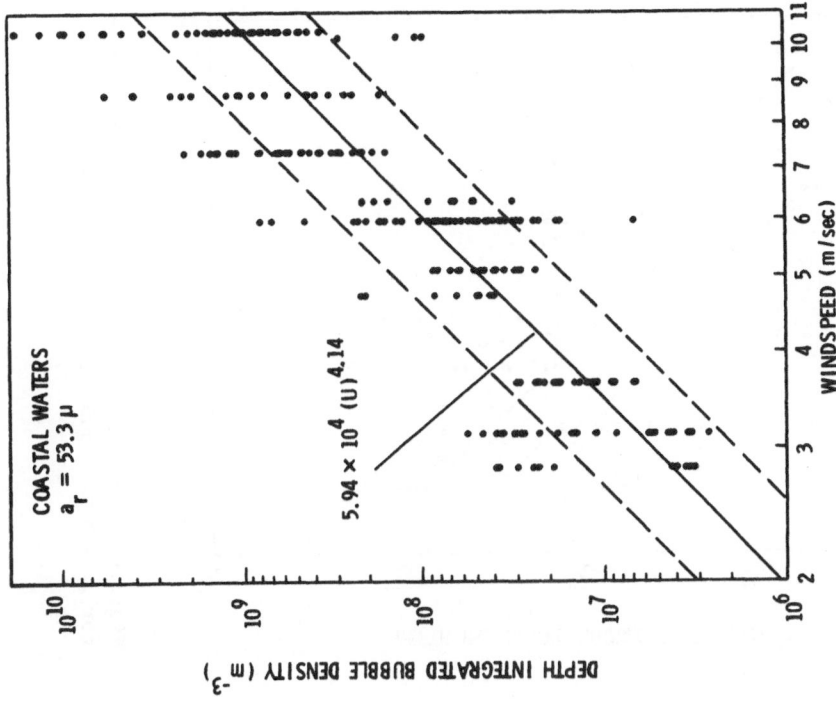

Figure 4. Depth integrated bubble densities in Dabob Bay for a bubble radius of 53 μ.

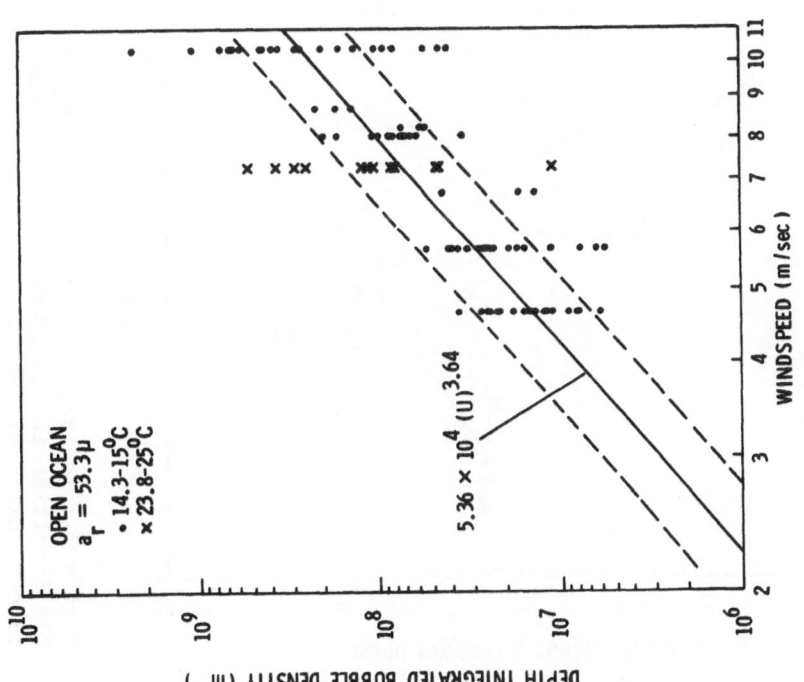

Figure 3. Estimates of open ocean depth integrated bubble densities for a bubble radius of 53 μ.

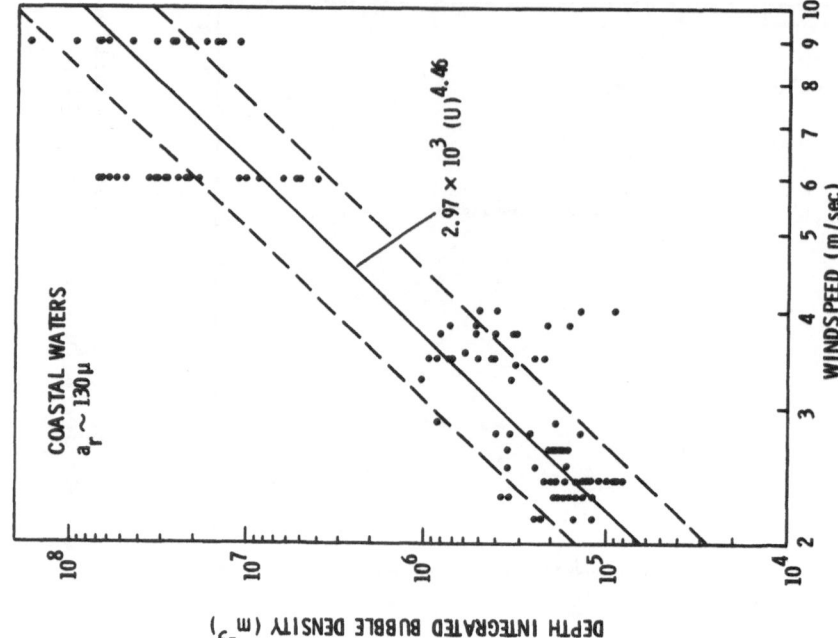

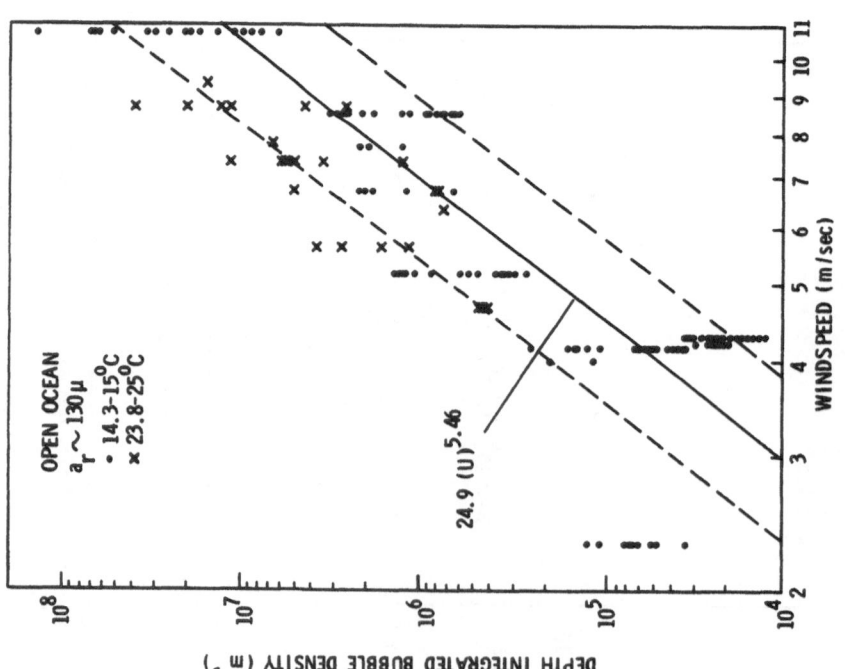

Figure 6. Bubble densities in coastal waters for a nominal bubble radius of 130 μ.

Figure 5. Estimates of depth integrated bubble densities in the open ocean for a nominal bubble radius of 130 μ.

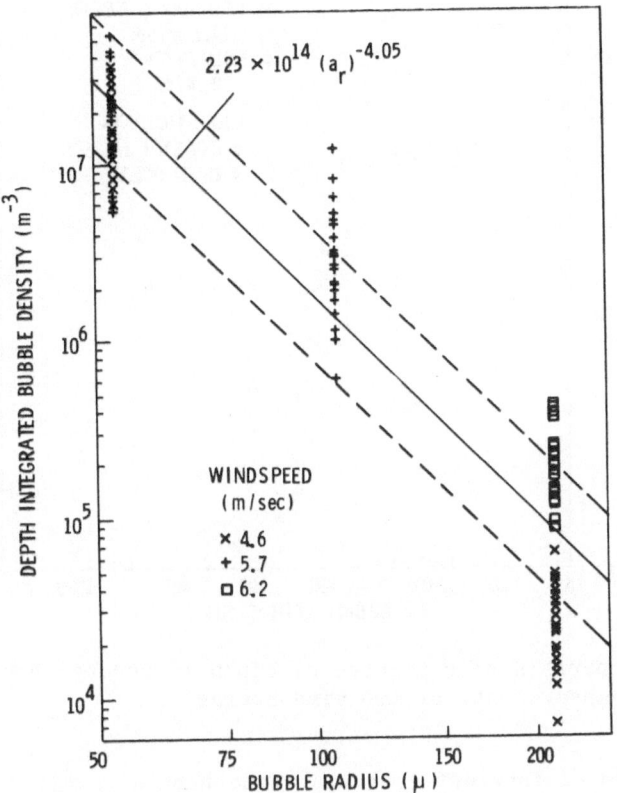

Figure 7. Dependence of bubble density on radius.

$N_0 \exp(-z/L)$. Only the first two depth points were used for the Kolovayev data, whereas all three depth points were used, where available, for the Johnson and Cooke data. The depth integrated bubble density N is then given simply by $N = N_0 L$.

Figure 8 shows depth integrated optical bubble densities where the wind speed for the Johnson and Cooke data has been translated to standard height. Also shown in this figure are estimates of depth integrated bubble densities obtained by extrapolating the least squares curves of Figs. 3-6 to wind speeds of 12 and 14.7 m/s. For bubbles having a radius of 53 μ, the Johnson and Cooke data are significantly higher than either the open ocean or coastal densities at the higher wind speed. Kolavayev's results agree with the coastal densities for 12 m/s winds but exceed the open ocean density for this wind speed. For bubbles of larger radius, the coastal estimates of bubble density agree quite favorably with the Johnson and Cooke data. Open ocean density estimates for this bubble radius are roughly an order of magnitude lower at the two wind speeds and hence fall significantly below the optical data.

234

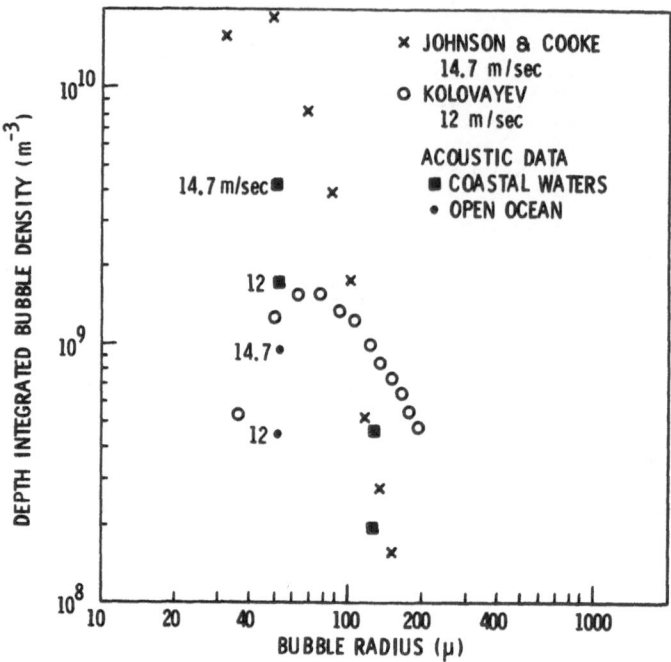

Figure 8. Comparison of estimates of depth integrated bubble densities with optical measurements at two wind speeds.

It is also of interest to compare the bubble densities inferred from acoustic backscatter with those of Farmer and Lemon (1984) which rely on the absorption rather than the scattering of acoustic energy by bubbles. In Fig. 9 the results obtained for coastal waters are compared with Farmer and Lemon's results for a nominal bubble radius of 130 μ. To perform this comparison, the relationship $U_{10} = .65 \, u_*^{.74}$ with u_* in cm/s was used based on stable atmospheric conditions for this range of wind speeds. For wind speeds up to 15 m/s, $\log(u_*/100) = 2$, where such an extrapolation is reasonable, the agreement is good. As in the case of the optical comparisons, the open ocean bubble densities are an order of magnitude low.

V. CONCLUSIONS

In this paper, acoustic backscatter data were inverted to establish relationships between subsurface bubble densities and the following parameters: wind speed, bubble radius, surficial water temperature and the proximity of land. The wind speed dependence of the bubble densities obtained was found to follow a power law dependence, with extreme variations in the exponent of 3.6 to 5.5. However, most of the results yielded a power law of between 4 and 5 for both coastal and open ocean conditions. An extrapolation of the densities obtained to

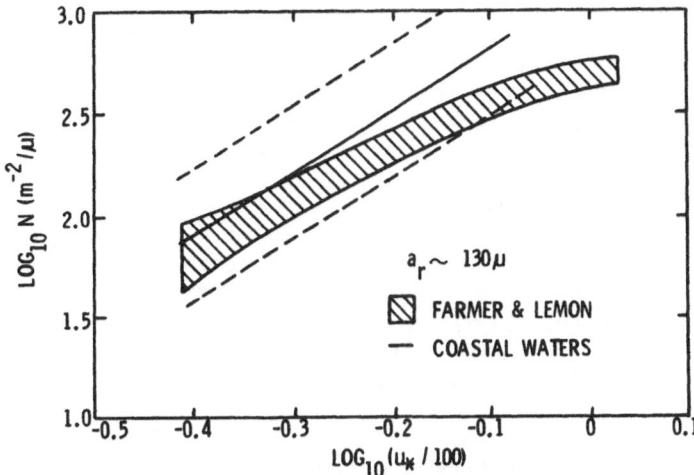

Figure 9. Comparisons of depth integrated bubble densities inferred from acoustic scattering and absorption.

higher wind speeds yielded depth integrated bubble densities that were generally lower than those derived from optically measured bubble populations. An exception to this were the coastal bubble densities for a nominal bubble radius of 130 μ which agreed well with both the Johnson and Cooke, and Farmer and Lemon data acquired in coastal waters. The dependence of bubble density on radius was also found to follow a power law dependence. An exponent of -4. was obtained for bubble radii between 50 and 200 microns in agreement with the results of many other studies.

Little dependence on surficial water temperature was obtained for the range of water temperatures at which acoustic data were acquired. Slightly but not significantly higher densities were found for water temperatures of 25° C than those observed for temperatures of 15° C, in contrast to what one would expect on the basis of the higher viscosity of the cooler water. It is quite possible that the effects of higher viscosity may be counterbalanced by the greater soluability of gases in colder waters.

Bubble densities obtained from acoustic backscatter data exhibited a strong dependence on the proximity of land, with estimated densities being almost an order of magnitude higher in coastal areas than in the open ocean at the higher wind speeds. Because the densities differ little at lower wind speeds, it appears unlikely that dissolved aerosols are the source of this difference. A more likely cause is the presence of contaminants which may act either as surfactants that increase bubble lifetimes or as nuclei for bubble formation.

236

ACKNOWLEDGEMENT

This research was performed at the Applied Research Laboratory of the Pennsylvania State University under the support of the Naval Ocean Research and Development Activity.

REFERENCES

Crowther, P. A., 'Acoustic scattering from near-surface bubble layers,' in Cavitation and Inhomogeneities in Underwater Acoustics, edited by W. Lauterborn, Springer, New York (1980).

Farmer, D. M. and D. D. Lemon, 'The influence of bubbles on ambient noise in the ocean at high wind speeds,' J. Phys. Oceanog., 14, 1762-1778 (1984).

Garrison, G. R., S. R. Murphy and D. S. Potter, 'Measurements of backscattering of underwater sound from the sea surface,' J. Acoust. Soc. Am., 32, 104-110 (1960).

Johnson, B. D. and R. C. Cooke, 'Bubble populations and spectra in coastal waters: A photographic approach,' J. Geophys. Res., 84, 3761-3766 (1979).

Kerman, B. R., 'Underwater sound generation by breaking wind waves,' J. Acoust. Soc. Am., 75, 149-165 (1984).

Koenigs, P. D., J. M. Monti, B. Nutzel, and H. Herwig, 'The influence of bubbles on sea surface backscatter measurements,' NUSC Technical Document 7811, Oct. (1986), Naval Underwater Systems Center, New London, CT 06320.

Kolovayev, P. A., 'Investigation of the concentration and statistical size distribution of wind-produced bubbles in the near-surface ocean layer,' Okeanologiya, 15, 1013-1017 (1975).

Lilly, J. G. and S. O. McConnell, 'Surface reverberation measurements in Dabob Bay and the open ocean,' J. Acoust. Soc. Am., 63, S24 (1978).

McDaniel, S. T. and A. D. Gorman, 'Acoustic and radar sea surface backscatter,' J. Geophys. Res., 87, 4127-4136 (1982).

Medwin, H., 'In situ acoustic measurements of bubble populations in coastal ocean waters,' J. Geophys. Res., 75, 599-611 (1970).

Medwin, H., 'Counting bubbles acoustically: A review,' Ultrasonics, 15, 7-13 (1977).

Wu, J., 'Bubble populations and spectra in near-surface ocean: Summary and review of field measurements,' J. Geophys. Res., 86, 457-463 (1981).

ACOUSTIC MEASUREMENTS OF BUBBLE DENSITIES AT 15-50 kHz

S. O. McConnell
Applied Physics Laboratory
College of Ocean and Fishery Sciences
University of Washington
1013 N.E. 40th Street
Seattle, WA 98105

ABSTRACT. Acoustic measurements of near-surface bubbles generated by
breaking waves were made off Whidbey Island, Puget Sound, Washington. A
limited amount of data was also gathered in the open ocean off the coast
of Washington. The intent of these measurements was to delineate the
role of near-surface bubbles in surface backscattering, forward loss,
and ambient noise at high frequencies. High resolution vertical
incidence backscatter was the principal method used for determining the
near-surface volume scattering strength profile attributable to bubbles.
For one of the data sets low grazing angle forward loss measurements
were interspersed with the vertical incidence measurements. It was
clear from the vertical incidence measurements that bubbles are
acoustically observable at wind speeds as low as 3 m/s and that for wind
speeds greater than about 5-6 m/s the surface forward loss can become
quite large (>10 dB).

1. INTRODUCTION

     The research presented at this meeting is focused on the role of
near-surface bubbles generated by breaking waves in acoustic scattering
and noise processes. Whether surface roughness or a near-surface bubble
layer is the dominant source of high-frequency surface backscatter at
low grazing angles is a controversial issue[1,2,3] that the present
measurements were intended to address. Perhaps more controversial is
whether observed high surface forward losses can be attributed to these
near-surface bubbles.[4,5] Thirdly, an issue directly related to the
purpose of this meeting is the role of bubbles in ambient noise
generation and propagation. Presently, the ambient noise effect of
greatest importance to us is the absorbing effect of bubbles that serves
to reduce the noise levels generated at or very near the air-sea
interface because the noise must propagate through the bubble layer to
be detected. Recent work has clearly implicated bubbles as the cause of
the dramatic reduction in the noise level at 25 kHz at wind speeds above
12 m/s.[6] Presumably this effect would be even more pronounced at higher

B. R. Kerman (ed.), Sea Surface Sound, 237–252.
© 1988 by Kluwer Academic Publishers.

frequencies since bubble density measurements have shown increasing numbers of acoustically resonant bubbles up to frequencies of 60 kHz and beyond.[7,8] The expected dipole radiation pattern for ambient noise[9] should be affected as well since the path length through the bubbles varies with grazing angle as $1/\sin\theta$. For noise generation, knowledge of the bubble population is essential to determination of the relative strength of certain contributory mechanisms such as bubble oscillations excited by near-surface turbulence and bubbles bursting at the air-sea interface.

The principal type of measurement was narrow-beam, short pulse length vertical incidence backscatter measurements at 15-50 kHz that permitted determination of the near-surface volume scattering strength profile attributable to bubbles. From this profile it is possible to specify the scattering strength for any arbitrary bistatic geometry. It is also possible to infer the surface forward loss. To ascertain the accuracy of this inference, low grazing forward loss measurements were taken in conjunction with vertical incidence measurements. Measurements of low angle backscatter and ambient noise (contaminated by other noise interference) were also made but are not reported here. The data reported here represent only a small portion (10%) of the total data set but nearly all of the data digitized in the field. The remainder was recorded on analog tape after the digitizing computer was damaged by several power outages resulting from storms (the very storms we hoped to observe acoustically!).

2. EXPERIMENTAL DESCRIPTION

Figure 1 shows the measurement locations off Whidbey Island, Puget Sound, Washington, and off the coast of Washington. The Whidbey Island location was chosen such that fairly open exposure to passing storms was common at varying fetches up to 50 km (winds from the west). Mile long cables connected the acoustic and environmental instrumentation to a shore site trailer containing the electronic systems, computer, and recorders. For the open ocean measurements a 500 m steel-armored cable connected the acoustic transducers to the surface support vessel placed in a 1-point moor. The water depth was about 50 m in the open ocean and 35 m at Whidbey Island.

The measurement geometry at Whidbey Island is shown in Figure 2. The principal acoustic measurement system was a suite of transducers placed atop a 5 m tripod tower on the sea bed. For the vertical incidence measurements two linear arrays 1.3 m in length and arranged perpendicular to each other were used. The acoustic pulses were transmitted on one of the arrays and received on the adjacent array, thus producing a very narrow combined beam with a 3 dB beamwidth ranging from 1.2° at 50 kHz to 4.1° at 15 kHz. An omnidirectional broadband hydrophone was also placed next to the line arrays. The frame on which all the transducers were mounted could be varied in both elevation (0° to 90°) and azimuth (300° total). In contrast, the open ocean vertical incidence measurements were gathered with a 20° wide transmitter and an omnidirectional receiver. The system in that case was placed in a three-point moor 25 m below the sea surface.

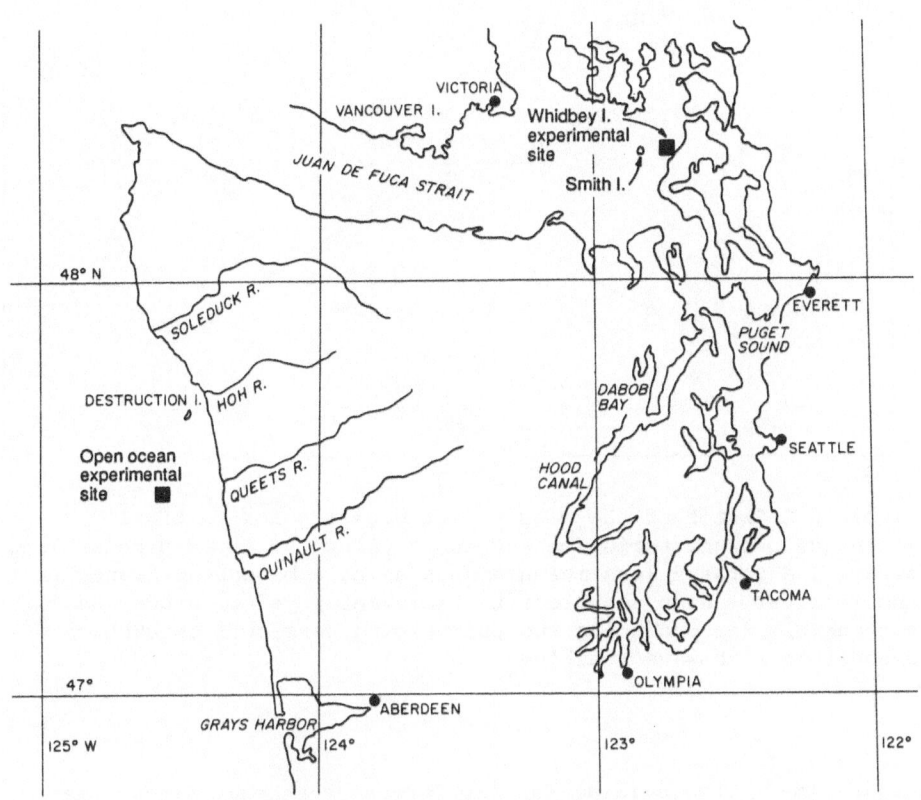

Figure 1. Map showing locations of the Whidbey Island and open ocean experimental sites.

For the forward scattering measurements a nearly omnidirectional transmitter was placed 5 m above the bottom at a range of 500 m as depicted in Figure 2. The acoustic transducers were trained in the direction of this transmitter and data gathered on the sea surface forward loss. Because of the 35 m water depth, a number of multipath arrivals were detected at the receiver including the single surface bounce path that is of primary interest. The dominant arrival involving the surface is this single bounce path since paths involving the bottom suffer a high loss because the bottom is soft mud with an approximate porosity of 0.7.

Both the vertical incidence backscatter and forward scattering measurements used a sequence of 100 short pulses for each data set. Pure tone pulses with lengths of 0.3 ms and 1.0 ms were interleaved and transmitted at a rate of one every 2 s giving a total averaging time of 200 s for 50 pings of the same pulse length. To obtain the average scattered intensity a thresholding technique was devised to align the

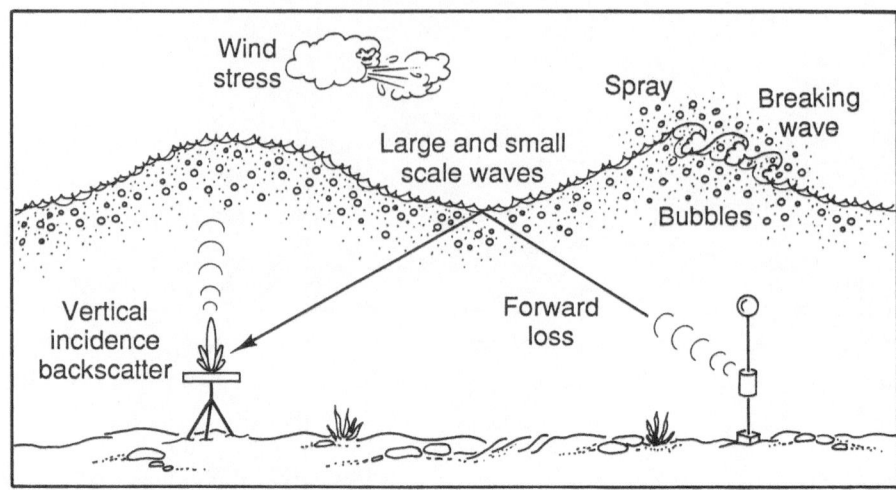

Figure 2. Sketch of the measurement geometry for vertical
incidence backscattering measurements using two perpendicular line
arrays and forward loss measurements using a broadbeam transmitter
and receiver. Also shown are the relevant physical processes
responsible for producing and influencing near-surface bubbles
associated with wave breaking.

pings in time. Thresholding for the forward scattered signal was
performed on the combined direct and single bottom bounce paths (time
difference less than 0.1 ms between paths) and this removed travel time
variations due to motion of the transmitter (see Figure 2) in the tidal
currents. The strong air-sea interface return was used for thresholding
the vertical incidence returns permitting the scattering strength
profile to be determined as a function of depth below the instantaneous
sea surface, rather than the mean sea surface. For cases where the
bubble density and receiver gain were high (thus saturating the air-sea
interface return), implementation of this thresholding technique on the
vertical incidence signal became quite difficult. The difficulty arose
because the bubble signal level became comparable to the interface
signal level, and the bubble signal, rather than the interface signal,
often triggered the threshold. In these cases a large proportion of
individual pings needed to be carefully examined to determine the proper
temporal shift.

Environmental measurements included CTD casts, wind speed, and wave
height. CTD casts were made on a few occasions and showed the sound
speed profile to be very nearly constant with a value of 1460 m/s for
the December through February time frame of the Whidbey Island
measurements. For the wind speed and wave height measurements (and
air-sea temperature difference) a spar buoy with the appropriate sensors
was constructed and deployed near the tripod tower. Unfortunately, the

buoy was apparently sabotaged a few days before the acoustic measurements began in earnest and was not repaired and redeployed until several weeks later. Thus, wind speed was measured with a cup anemometer at the shore site, and weather reports issued every 3 hours from Smith Island (about 5 miles from shore and 4 miles from the experimental site) were recorded routinely. Wave height was determined from the travel time fluctuations produced by the thresholding technique on the air-sea interface return of the vertical incidence backscatter signal.

3. EXPERIMENTAL RESULTS

The vertical incidence backscatter results will be presented first, followed by the surface forward loss measurements. Table 1 summarizes the data and the corresponding environmental conditions for all the data presented herein.

TABLE 1. Acoustic data parameters and environmental conditions.

Run No.	Frequency (kHz)	σ_I (dB)	σ_0, σ_1 (dB re m^{-1})	L_0, L_1 (m)	z_{knee} (m)	SBL (dB)	Wind Speed (m s^{-1})	Wind Direction	rms Wave Height (m)
43	30	-28.1	-28.2	0.61	---	---	12	W	0.34
44	30	-30.7	-36.0,-39.1	6.3,0.61	4.27	---	12	W	0.40
47	30	-23.6^a	---	---	---	23.1	12	W	---
48	30	-25.1	-25.0,-38.0,-40.5	0.61,6.3,0.61	2.05,6.2	---	12	W	0.50
49	15	-16.1	-13.1	0.54	---	---	17	W	0.63
51	15	-23.3^a	---	---	---	19.5	17	W	---
53	15	-23.5^a	---	---	---	18.9	17	W	---
55	15	-20.4	-23.5	1.55	---	---	17	W	0.62
57	25	-22.3	-25.9,-34.6	2.49,1.26	5.1	---	13	W	0.66
60	25	-25.2^a	---	---	---	15.0	13	W	---
62	25	-25.3	-24.4	0.82	---	---	13	W	0.51
67	20	-34.7	-31.3	0.45	---	---	10	W	0.44
68	20	-26.0^a	---	---	---	11.5	10	W	---
72	20	-37.6	-38.4	1.05	---	---	10	SW	0.38
75	30	-37.6	-38.3,-40.5	2.67,0.39	1.32	---	12	SW	0.36
78	40	-33.6	-33.2,-44.8	1.00,0.38	2.67	---	12	SW	0.47
83	20	-52.4	-48.7	0.36	---	---	7	SE	0.10
84	30	-42.7	-40.8	0.66	---	---	7	SE	0.10
88	40	-40.6	-40.9,-44.7	1.79,0.42	1.48	---	7	SE	0.10
89	50	-37.0	-35.1	0.57	---	---	7	SE	0.10
91	50	---	---	---	---	0.5	4	SE	0.02
134	50	-26.1	-30.9,-41.4	3.10,0.57	5.82	---	11.0	SE	0.17
135	20	-31.0	-29.8	0.79	---	---	11.0	SE	0.17
136	40	-26.1	-27.8,-32.9	1.91,0.52	2.35	---	11.0	SE	0.17
137	30	-31.6	-30.3	0.67	---	---	11.0	SE	0.17
138	25	-25.8	-22.5	0.46	---	---	11.0	SE	0.17
139	50	-27.9	-28.3,-40.6	-1.21,0.61	3.16	---	11.0	SE	0.17
2101^b	25	---	---	---	---	---	~1	---	0.57
2301^b	25	-23.0	-17.0	0.25	---	---	2.8	W	0.60
2302^b	25	-22.8	-20.1	0.54	---	---	9.7	SW	0.64

^{a} Inferred σ_I using surface bubble loss
^{b} Open ocean data

For the vertical incidence backscatter measurements there are three characteristics of the signal that are of primary interest. First is the integrated volume scattering strength, which is a measure of the total number of resonant bubbles corresponding to a given frequency. From this integrated quantity the surface forward loss can be estimated directly using scattering and absorption theory[8,10] for "clean" resonant bubbles. The second quantity of basic interest was the shape of the volume scattering profile. This quantity is of interest for determining the depth dependence of the bubble density. Thirdly, the frequency dependence of these two quantities just discussed is of interest since this is related to the bubble size spectrum.

3.1 Whidbey Island Vertical Incidence Backscatter

The average volume scattering strengths at 50 kHz and 20 kHz and two wind speed conditions are shown in Figure 3 as a function of travel time. This travel time is equivalent to depth from the instantaneous

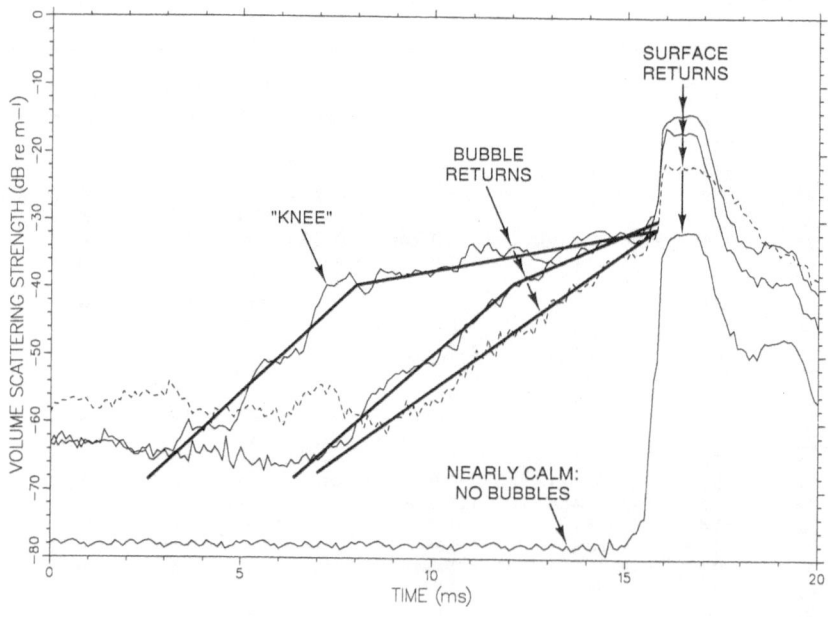

Figure 3. Volume scattering strength averaged over 50 pings (200 s) versus time into the ping cycle. The individual pings are aligned in time with respect to the instantaneous sea surface by thresholding on the surface returns. Bubbles are clearly visible for the data taken at 11 m/s and do not appear for nearly calm conditions (2 m/s). Three data runs at 50 kHz are shown as solid lines and one at 20 kHz as a dashed line. Straight line "eyeball" fits to the data are shown as well.

sea surface when referenced to the air-sea interface return. Note first that when the conditions are nearly calm no discernible bubble return is observed, and the leading edge of the air-sea interface return is very sharp, rising 40 dB in about one-third millisecond. This clearly indicates that bubbles produced by mechanisms other than wave-breaking[10] are negligible, thus simplifying the data analysis and allowing us to focus solely on bubbles created by wave-breaking and their acoustical effects. At a relatively high wind speed of 11 m/s the bubble returns are plainly visible above the background signal level at both 20 kHz and 50 kHz. The expected exponential rise in scattering level can be seen at 20 kHz as represented by the straight line "eyeball" fit to the data, but both 50 kHz data sets display an unexpected "knee" in the profile. The "knee" is quite noticeable for the upper curve and there is a dramatic change in slope where the exponential depth constant decreased from 3.10 to 0.57 m (see Table 1 for depth constant values for other data sets). At 50 kHz, then, the differential volume scattering coefficient profile could be described as follows,

$$
\sigma_V = \begin{cases} \sigma_0 e^{-z/L_0} & , \quad 0 < z < z_{knee} \\ \sigma_1 e^{-z/L_1} & , \quad z_{knee} < z < \infty , \end{cases}
$$

where z is positive downward from the sea surface. Most often the data evidenced an exponential profile as can be noted from Table 1 (entries with a single depth constant, L_0), but two-segment profiles like those that are shown in Figure 3 were relatively common, and in one case a three-segment profile was observed. The deviation from an exponential profile tended to occur more often at higher wind speeds and frequencies.

Besides the behavior of the average scattering strength profile, examination of individual pings is also helpful to elucidate the dynamics of bubble plumes. Figure 4 shows a waterfall plot of a sequence of 12 pings separated by 4 s between pings. The data set shown is the same as that shown in the upper curve of Figure 3. The ping creating the distinctive sharpness of the "knee" in Figure 3 is visible as the very large feature occurring early in ping 4. With a 4 s repetition period it is difficult to follow plume evolution, but it appears that this distinctive feature begins at ping 2 and is still noticeable at ping 7. Inspection of many similar features shows that they tend to be approximately 1 pulse length in duration. When short 0.3 ms pulses (22 cm resolution) were employed the strong returns tended to appear as a number of sharp "fingers" protruding above the background. The sharpness of these features indicates that bubbles are carried away from the surface where they are generated as a series of thin sheets, rather than as extended billows.[11]

A convenient and informative method for condensing the data for each run to a single number is to compute the integrated volume scattering strength. This dimensionless quantity is found by integrating under the bubble return portion of the average profile as

244

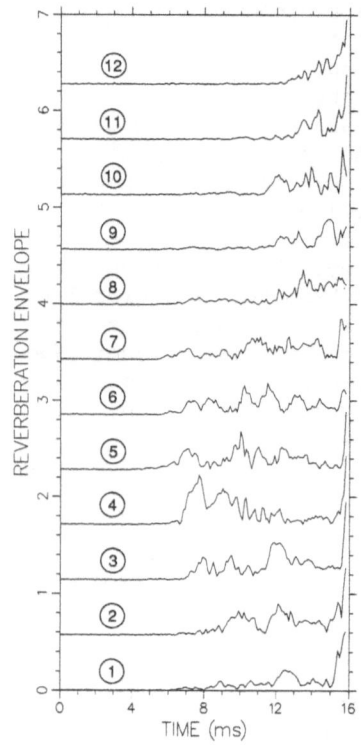

Figure 4. A sequence of individual
pings for the data corresponding to the
upper curve in Figure 3. The source of
the dramatic "knee" in Figure 3 is
visible as the first large feature in
ping 4.

shown in Figure 3. To this integral is added a small portion extending
the measured profile by a distance of $c\tau/4$ (c = speed of sound and τ =
pulse length) beyond the air-sea interface return. This accounts for
the unmeasurable part of the profile dominated by the interface return.
This integral was also calculated using the straight line fits to the
scattering strength profile, and the resultant values were all within
± 1 dB of the directly computed values thus supporting the accuracy of
these "eyeball" fits. The integrated volume scattering strength for
two sets of runs covering the frequency range 20-50 kHz is shown in
Figure 5. For the upper set of data the mean wind speed was 11.0 m/s,
and for the lower set the wind speed was about 7 m/s (from Smith
Island). Other processed data taken at wind speeds less than 3 m/s
showed no measurable bubbles. Hence, the expected increase in
scattering level with wind speed is clearly evident in this data. From
this small amount of data a wind speed power law coefficient for σ_I
cannot be extracted, but it is certainly in the range of 3 to 6.5
implied by other data and analyses.[5,6,12]

The frequency dependence of the integrated volume scattering
strength is also of interest. Shown for comparison with the data in
Figure 5 is a frequency dependence ($\sigma_I \propto f$) for σ_I based on a fourth
power law fall-off in bubble density with radius for radii between 65 μm

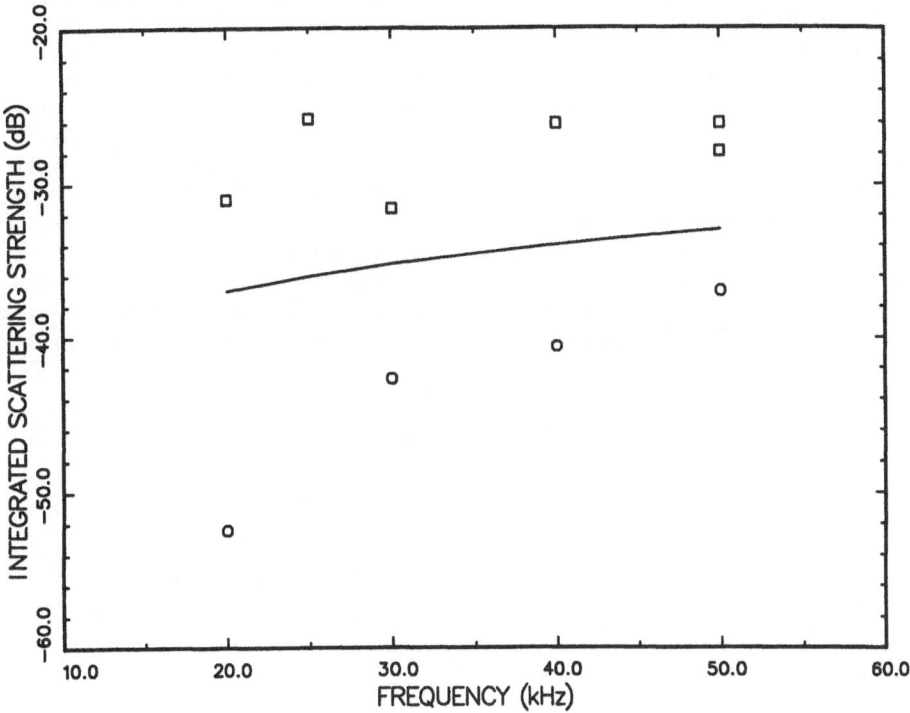

Figure 5. The integrated volume scattering strength attributable to bubbles versus transmit frequency. The upper set of data (squares) was taken at a wind speed of 11 m/s (see run nos. 135 - 139 in Table 1), and the lower set was taken at 7 m/s (run nos. 83-89). Also shown is an f^{+1} curve for comparison.

(resonant radius at 50 kHz at z = 0) and 163 μm (resonant radius at 20 kHz).[4,7] Both sets of data are consistent with this frequency dependence but with an uncertainty of ± 1 in the exponent. The upper set of data shows variations as large as 3 dB away from an f^{+1} dependence, and the lower data set shows a marked drop at 20 kHz. For this latter data a continuous wind speed recording was not available, and thus it is not possible to say whether the drop was associated with a lower speed. A continuous recording at the shore site was available for the upper set of data, and the trends in the data roughly followed the trends in the wind speed (averaged over 100 s) such as for the difference between 25 and 30 kHz (two runs taken back-to-back) and the two 50 kHz runs made at the beginning and end of the frequency dependent series (total time period covered is 35 min). Limited averaging time for the 50-ping runs can also lead to variations in the average σ_I values, particularly since the spatial sampling size was relatively small (surface patch size of order 1 m diameter). To examine this concern regarding averaging time, the integrated scattering strength was computed for individual pings and displayed as a function of time. Figure 6 shows this time dependence for selected frequencies for the two sets of data. The interesting characteristic of the time variation of σ_I is that at higher wind speeds there appear to be longer time scales of variation. This certainly suggests that a longer time averaging period may be necessary, at least at the higher wind speeds, if the fluctuations in the final average are to be reduced below ± 2 dB, for example. Extension of the time for each run made at a single frequency cannot be done, however, without risking significant changes in the environmental conditions. Fortunately, most of the subsequent data recorded on analog tape were taken in a multifrequency mode with 0.1 s between frequencies. This will give a nearly instantaneous frequency-to-frequency comparison for individual pings.

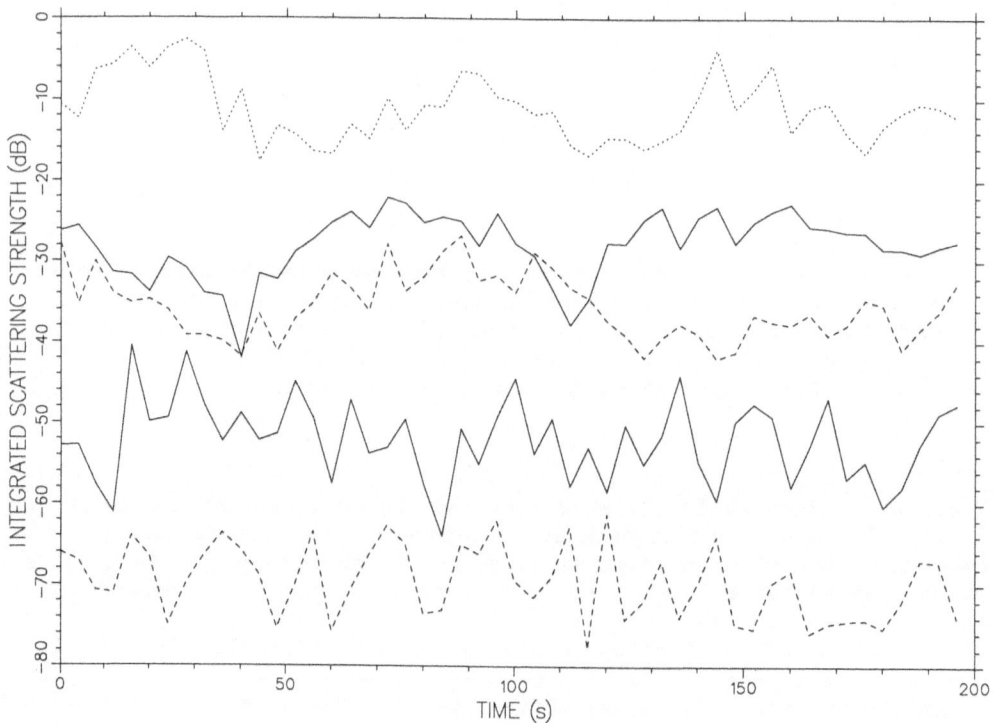

Figure 6. Time dependence of the individual (single ping) integrated volume scattering strengths. The 50 kHz data are shown as solid lines (lower curve is adjusted down 10 dB), the 30 kHz data as the dashed lines (lower curve is adjusted down 25 dB), and the 25 kHz data as the dotted line (adjusted up 10 dB).

3.2 Open Ocean Vertical Incidence Backscatter

A small amount of vertical incidence data was also gathered in the open ocean off the coast of Washington. Because of equipment problems and other conflicting experiments (this was a multi-experiment, multi-lab expedition) only 3 data runs could be made at a single intermediate frequency of 25 kHz. In addition, forward loss data from 15-40 kHz were gathered on separate days and have been reported elsewhere.[5] The fortunate aspect of this data was that the wave height (a large proportion of which was swell) remained the same and only the wind speed changed. Figure 7 shows the averaged volume scattering strength profile for the three different wind speeds. Again, it is clear that the volume scattering level associated with bubbles increases with wind speed and that bubbles are not evident for calm conditions. Since the surface wavelengths in the open ocean are markedly longer (about 100 m) and the returns from crests vis-a-vis troughs were quite discernible, it was

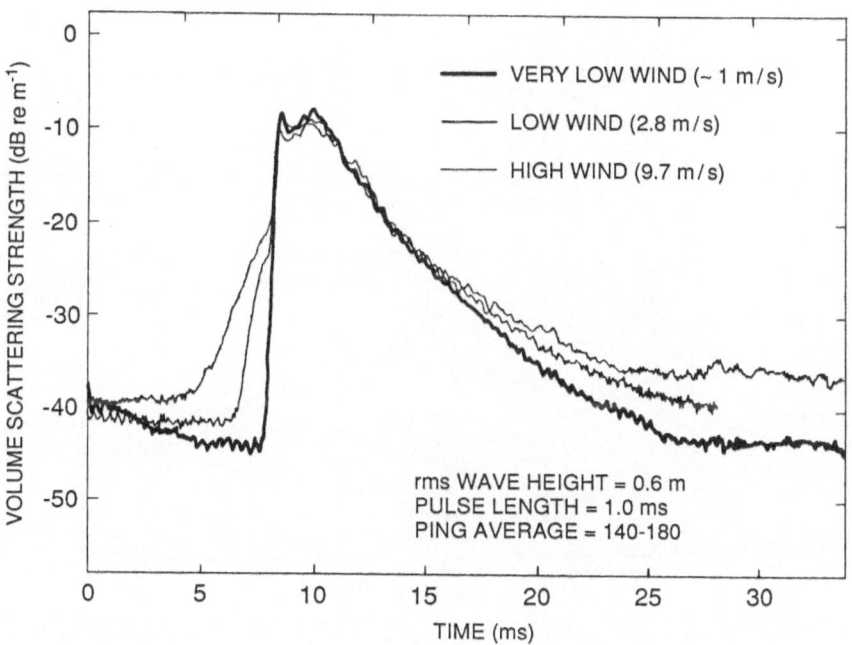

Figure 7. Open ocean vertical incidence backscattering strength data at 25 kHz. The wave height remained essentially the same at 0.6 m rms and only the wind speed changed. Like the Whidbey Island data, no bubbles are discernible for calm conditions, and the bubble layer thickness increases with wind speed. The data taken at 9.7 m/s were obtained 5 min after an abrupt increase in wind speed from 3 m/s.

248

possible to conditionally average the data for these two locations along the dominant wave profile. These conditionally averaged data, when compared to the overall average, showed virtually no difference in depth dependence, and the integrated scattering strengths differed by less than 1 dB. From a modeling viewpoint this implies that it may be appropriate to use the instantaneous rather than the mean sea surface[3] as the reference depth for z = 0 for the volume scattering strength profile. Eventually, however, the effect of plume structure needs to be incorporated in the models, especially with regard to fluctuation statistics and very low grazing angle scattering and noise levels.

3.3 Whidbey Island Forward Scattering Loss

For one series of data runs (50 pings each), low angle forward loss measurements were taken between two vertical incidence measurements. Figure 8 shows two examples of the forward scattered signal, one for a low wind/wave condition and one for a relatively high wind/wave

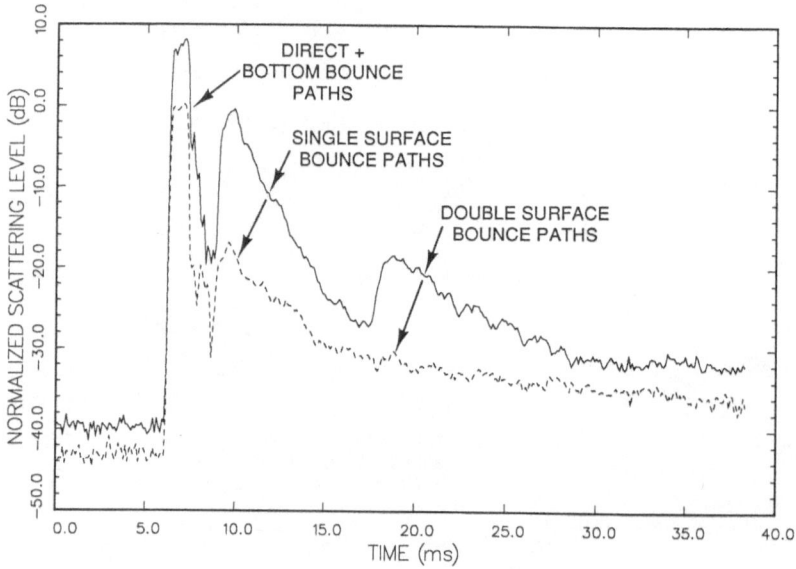

Figure 8. Normalized forward scattering level averaged over 50 pings (200 s) versus time into the ping cycle. The normalization is such that a single direct path should appear as a transmit pulse replica with a level of 0 dB if spherical spreading plus chemical absorption describes the path loss. The upper set of data was taken at 50 kHz with a wind speed of 4 m/s (run no. 91 in Table 1), and the lower set at 20 kHz and wind speed of 10 m/s (run no. 68).

condition. The combined direct and bottom bounce path signals are
evident as the early, near-replica arrival. This arrival is followed by
the surface bounce and other multipath arrivals. The first surface
bounce arrivals show a rapid increase to a peak over a one pulse length
period and then a gradual decay beyond. For the low wind/wave condition
another peak occurs at 12 ms following the direct path arrival, and
these are the multipaths involving two surface bounces. A comparison of
the curves for the two wind/wave conditions shows that the higher
wind/wave case is about 10 dB lower in level and displays a much slower
decay in level beyond the first peak (also, no second peak is visible).
The surface bounce loss was computed by integrating the scattered
intensity from the leading edge of the single-surface bounce arrivals to
the leading edge of the double-surface bounce arrivals. From the
decibel equivalent of this integral, 1.5 dB is subtracted to account for
the inclusion of arrivals involving bottom bounces. For the low
wind/wave case this integral differs from the expected level for no loss
at the surface by 0.5 dB (run no. 91). For the high wind/wave case the
loss is computed to be 11.5 dB (see Table 1 for run no. 68).

Figure 9 presents a comparison of the integrated volume scattering
strength computed from the vertical incidence backscatter measurements
and this same quantity inferred from the forward scatter measurements.
The inferred σ_I is calculated from the surface bounce loss (attributed
to bubbles) as,[4,8]

$$\sigma_I \, [dimensionless] = \frac{\sin\theta}{109.2} \frac{\delta_R}{\delta} \, SBL \, [dB]$$

where

$\delta = 2.55 \times 10^{-3} \, f^{1/3}$ (fit to data in Reference 13)
   (total damping coefficient)

$\delta_R = 0.0136$
   (radiation damping coefficient)

$\theta = 6.9°$
   (grazing angle for single surface bounce).

From Figure 9 it can be seen that the integrated scattering strength
inferred from the forward scattering losses are similar over the
frequency range 15 - 30 kHz, but the vertical incidence backscatter
results exhibit significant scatter. The largest differences between
the forward scattering and vertical incidence backscattering results are
seen at 20 kHz where the difference is 9 dB at best. These large
variations are not well understood at present, but two contributing
causes can be readily identified. One is simply statistical
fluctuations in the average σ_I values. This cause of variation was
discussed earlier and seems to be worse at higher wind speeds (say >
10 m/s) where longer time scales of variation in individual (single
ping) σ_I values are observed (see Figure 6). The data shown in

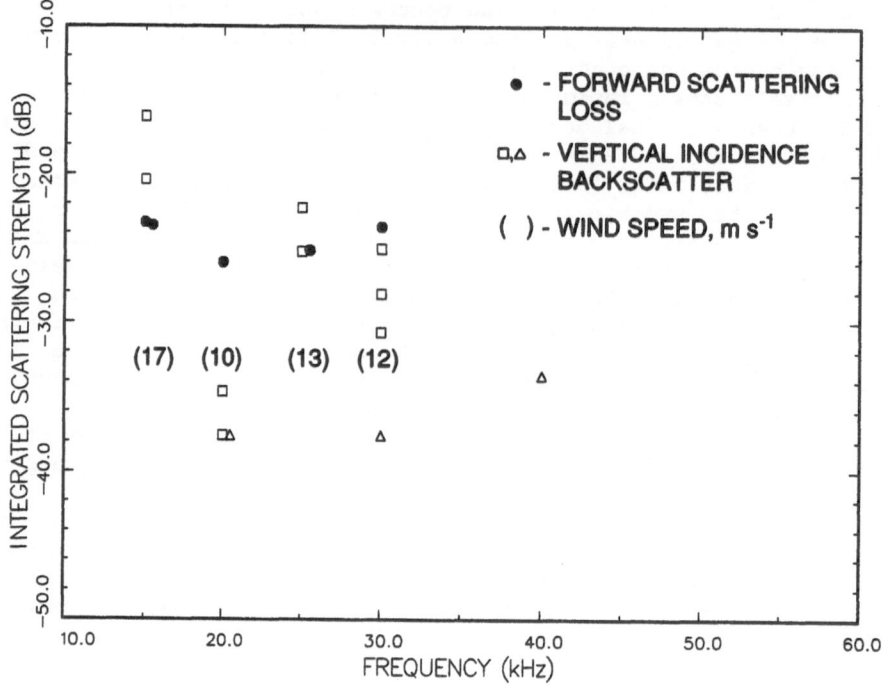

Figure 9. A comparison of the integrated volume scattering strength obtained directly from vertical incidence backscattering measurements with those inferred from forward loss measurements. The wind speeds obtained from Smith Island for each of the four data sets at different frequencies are shown as the numbers in parentheses. The triangles represent the last three data runs (nos. 72, 75, and 78 in Table 1), and the 20 kHz data (no. 72) are doubly represented as both a square and a triangle.

Figure 5 and the data runs shown in Figure 9 that were closely spaced in time (e.g., 3 triangle points taken at the conclusion of this series of data runs) indicate that fluctuations of ± 3 dB might be attributable to these statistical fluctuations. The other easily identifiable reason for fluctuations in averaged σ_I values is non-stationarity in the wind field. This mechanism for producing fluctuations in averaged σ_I values is enhanced by the wind speed power law coefficient for σ_I that other studies have indicated to be quite large, between 3 and 6.5.[5,6,12] If an intermediate value of 5 is chosen, a variation in σ_I of 7.6 dB is produced for a change in the mean wind speed from 12 to 17 m/s (see Figure 9 for nominal wind speeds for each frequency obtained from Smith Island). Other not so easily identifiable effects (e.g., surfactants[10]) might also play a role in enhancing fluctuations and producing differences between backscattering and absorption from bubbles. However, the two mechanisms just discussed appear to be amply capable of explaining the results shown in Figure 9.

4. SUMMARY

The principal results of the data taken to date can be summarized as follows:

1) The integrated volume scattering strength owing to near-surface bubbles generated by breaking waves increased with wind speed in a manner consistent with other published work $(\sigma_I \propto U^q,\ 3 \leq q \leq 6.5)$.

2) The frequency dependence of the integrated volume scattering strength is also consistent with previous published work on bubble densities $(\sigma_I \propto f^{+1})$.

3) The depth dependence of the near-surface volume scattering strength follows the expected exponential behavior for the majority of cases, but significant departures occur in many other cases where a two-part exponential behavior is observed.

4) Although the surface bounce loss is zero (within experimental error) when no bubbles are observed in the vertical incidence backscatter measurements, the correlation of the measured loss with the loss inferred from vertical incidence backscatter displays a wide range of variation about the expected trend.

Processing of the remainder of the Whidbey Island data should lead to a better quantification of these trends and hopefully clarify the result noted in 4).

ACKNOWLEDGEMENTS

I would like to thank most Eric Thorsos who led the Whidbey Island experiment and who provided very helpful discussion during the course of the data analysis. This experiment would not have been a success without the indefatigable efforts of the support staff, and I would especially like to note the efforts of Herb Hiegel, Marv Strenge, Fred Karig, Bob Doerr, Mel Bohleen, and Kate Bader. This work is supported by the NORDA High Frequency Acoustics Block program managed by Code 243 (R.W. Farwell).

REFERENCES

1. P. D. Koenigs, J. M. Monti, B. Nutzel and H. Herwig, "The Influence of Bubbles on Sea Surface Backscatter Measurements," NUSC TD 7811 (also FWG-Bericht 1986-9), Naval Underwater Systems Center, New London, 26 October 1986.

2. D. Middleton, "Acoustic Scattering from Composite Wind-Wave Surfaces II: Backscatter Cross Sections and Doppler Effects at High Frequencies and Small Angles for "Bubble-Free" Regimes," NUSC

252

TD 7635, Naval Underwater Systems Center, New London, 22 July 1986.

3. S. T. McDaniel and A. D. Gorman, "Acoustic and Radar Sea Surface Backscatter," J. Geophys. Res. **87**, 4127-4136 (1982).

4. E. I. Thorsos, "Surface Forward Scattering and Reflection," APL-UW 7-83, Applied Physics Laboratory, University of Washington, May 1984.

5. E. I. Thorsos, "High Frequency Surface Forward Scattering Measurements," presented at the 108th meeting of the ASA, 8-12 October, 1984 (J. Acoust. Soc. Am. Suppl. 1, **76**, S55(1984)).

6. D. M. Farmer and D. D. Lemon, "The Influence of Bubbles on Ambient Noise in the Ocean at High Wind Speeds," J. Phys. Ocean. **14**, 1762-1778 (1984).

7. J. Wu, "Bubble Populations and Spectra in Near-Surface Ocean: Summary and Review of Field Measurements," J. Geophys. Res. **86**, 457-463 (1981).

8. H. Medwin, "In-Situ Acoustic Measurements of Bubble Populations in Coastal Waters," J. Geophys. Res. **75**, 599-611 (1970).

9. S. O. McConnell, "Remote Sensing of the Air-Sea Interface Using Microwave Acoustics," Proceedings of OCEANS '83, 85-92, 29 August-1 September (1983).

10. F. MacIntyre, "On reconciling optical and acoustical bubble spectra in the mixed layer," 75-94, in <u>Oceanic Whitecaps</u> (E. C. Monahan and G. MacNiocaill (eds.)), D. Reidel Publishing Co., 1986.

11. S. A. Thorpe, "On the Clouds of Bubbles Formed by Breaking Wind-Waves in Deep Water, and Their Role in Air-Sea Gas Transfer," Phil. Trans. Roy. Soc. Lond. A **304**, 155-210 (1982).

12. B. R. Kerman, "Distribution of Bubbles Near the Ocean Surface," Atmos.-Ocean **24**, 169-188 (1986).

13. C. Devin, "Survey of Thermal, Radiation, and Viscous Damping of Pulsating Air Bubbles in Water," J. Acoust. Soc. Am. **31**, 1654-1667 (1959).

PROPAGATION EFFECTS ASSOCIATED WITH AMBIENT NOISE

W. A. Kuperman
Naval Research Laboratory
Washington, DC 20375

ABSTRACT. The spatial distribution of ocean surface noise generated by natural mechanisms depends on the sound propagation characteristics of the ocean. Here we review the ocean acoustic environment, the physics of sound propagation in the ocean and then couple these concepts to the existence of natural sources of surface noise in order to explain some of the basic features of how sound is distributed in the ocean.

1. INTRODUCTION

In underwater acoustics, noise [1-19] becomes an issue when it masks a signal of interest. Ocean acoustic signal processing [20-23] is essentially a procedure for extracting a signal embedded in noise. The noise is irrelevant if the signal is very strong. However, the more interesting case is the marginal situation of low signal to noise ratio ($SNR < 1$). Here we would like to exploit the difference in the physical properties of the signal of interest and the noise to be "rejected." Simple examples of this procedure are: omnidirectional noise can be reduced by a directional receiver with narrow "look directions" (beams); directional noise can be avoided by not looking in the direction of the noise. The more general case of achieving "noise gain" i.e.,enhancement of SNR, is to somehow factor into the design of a receiving system knowledge of the general spatial distribution of the ambient noise including its coherence properties. Here we will be concerned with these spatial properties.

The levels and distribution of ambient noise are determined by both the source properties and the acoustic propagation conditions of the ocean environment. Though the topic of natural source mechanisms is the main theme of this workshop, understanding the ultimate distribution of the noise requires a fundamental knowledge of how sound propagates in the ocean [24-30]; it is this latter subject that will be reviewed here. The topics we will review are:

1. the ocean acoustic environment;

2. the various modes of sound propagation in the ocean;

3. models of sound propagation;

4. the coupling of surface generated noise to the water column;

5. propagation effects that determine the level and distribution of surface generated noise in the ocean and seabed.

Before continuing, it is important to note here that the alternative to noise generated from natural sources is noise from man-made sources which we can call "shipping." Figure 1 is a conceptual summary of the origin of noise spectra in the ocean. No matter what the source is, though, the

B. R. Kerman (ed.), Sea Surface Sound, 253–272.

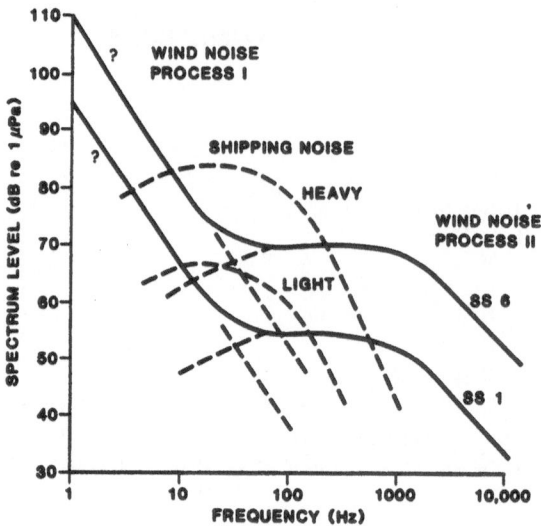

Figure 1: Conceptual summary of the spectra and origin of noise sources in the ocean (taken from [3]). SS is sea state and the question marks refer to the specific mechanisms of the natural noise.

physics of the sound propagation that determines the spatial distribution of the noise is all the same. Of course, in the absence of man-made sources it can be said that all the noise, no matter what the mechanism, is from natural sources.

2. THE OCEAN ENVIRONMENT

In underwater acoustics, the ocean is an acoustic waveguide and the speed of sound plays the same role as the index of refraction does in optics. The sound speed in the ocean is an increasing function of temperature, salinity and pressure. Seasonal and diurnal changes affect these oceanographic parameters in the upper ocean. In addition, all of these parameters are a function of geography. Figure 2 illustrates a typical set of sound speed profiles indicating greatest variability near the surface as function of season and time of day. In a warmer season (or warmer part of the day), the temperature increases near the surface and hence the sound speed increases with decreasing depth. In nonpolar regions, the oceanographic properties of the water near the surface result from mixing activity originating from the air-sea interface. This near surface "mixed layer" has a constant temperature (except in calm, warm surface conditions as described above). Hence, in this isothermal mixed layer we can have a sound speed profile which increases with depth because of the pressure gradient effect. This is the "surface duct" region, and its existence depends on the near surface oceanographic conditions. Note, the more agitated the upper layer is the deeper the mixed layer and the less likely there will be any departure from the mixed layer part of the profile depicted in Fig. 2. Hence, an atmospheric storm passing over a region where a surface duct exists is likely to mix the surface waters and cause the duct to change.

Below the mixed layer is the thermocline where the temperature decreases with depth and therefore the sound speed also decreases with depth. Below the thermocline, the temperature is constant (about 4°C–a thermodynamic property of salt water at high pressure) and the sound speed increases because of increasing pressure. Therefore, between the deep isothermal region and

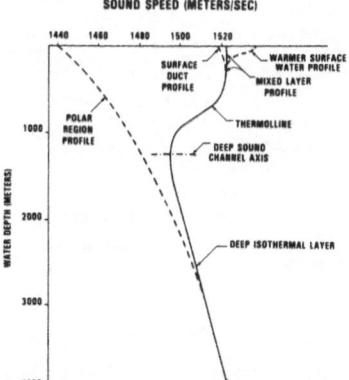

Figure 2: Generic ocean sound speed profiles.

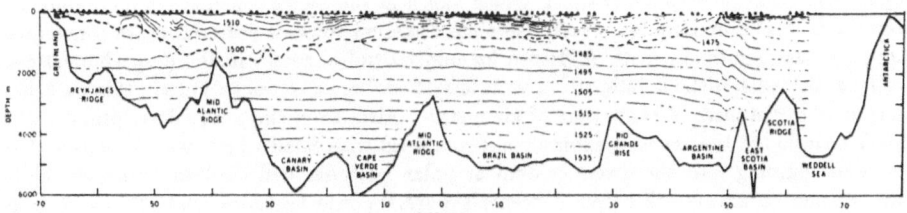

Figure 3: Sound speed contours [31] taken from the North and South Atlantic along the $30.50°W$ Meridian. Dashed line indicates axis of the deep sound channel (also referred to as the SOFAR channel axis).

the mixed layer, we must have a minimum sound speed which is often referred to as the axis of the "deep sound channel." However, in polar regions, the water is coldest near the surface and hence the minimum sound speed is at the ocean/air(or ice) interface as indicated in Fig. 2. In continental shelf regions (shallow water) with water depth of the order of a few hundred meters, only the upper region of the sound speed profile in Fig. 2, which is dependent on season and time of day, affects sound propagation in the water column.

Figure 3 is a contour display of the sound speed structure of the North and South Atlantic [31] with the deep sound channel indicated by the heavy dashed line. Note the geographic(and climatic) variability of the upper ocean sound speed structure and the stability of this structure in the deep isothermal layer. For example, as explained above, the axis of the deep sound channel becomes shallower toward both poles, eventually going to the surface.

3. SOUND PROPAGATION PATHS IN THE OCEAN

Figure 4 is a schematic of the basic types of propagation in the ocean resulting from the sound speed profiles (indicated by the dashed lines) discussed in the last section. These sound paths can be understood from a simplified statement of Snell's Law: sound bends locally toward regions of low sound speed (or sound is "trapped" in regions of low sound speed). Simplest to explain are

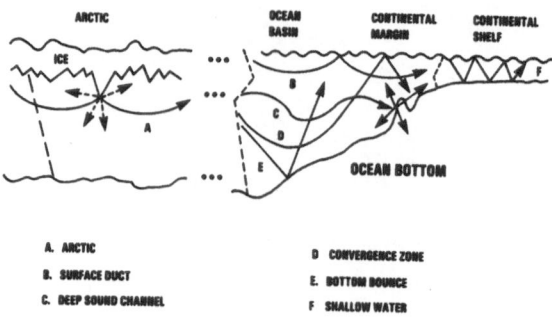

A. ARCTIC   D   CONVERGENCE ZONE

B. SURFACE DUCT   E   BOTTOM BOUNCE

C. DEEP SOUND CHANNEL   F   SHALLOW WATER

Figure 4: Schematic of various types of propagation in the ocean.

the paths about local sound speed minima: A,B, and C. Paths A and B correspond to surface duct propagation where the minimum sound speed is at the ocean surface (or at the bottom of the ice cover for the Arctic case). Path C, depicted by a ray leaving a deeper source at a shallow horizontal angle propagates in the "deep sound channel" whose axis is at the shown sound speed minimum. As stated above, this local minimum tends to become more shallow towards polar latitudes converging to the Arctic surface minimum, path A. Hence for mid latitudes, sound in the deep channel can propagate long distances without interacting with lossy boundaries; propagation via this path has been observed over distances of thousands of kilometers. Also, from the above description of the geographical variation of the acoustic environment combined with Snell's Law, we can expect that shallow sources coupling into the water column at polar latitudes will tend to propagate more horizontally around an axis which becomes deeper toward the mid latitudes. Path D, which is at slightly steeper angles than those associated with path C, is "convergence zone" propagation, a spatially periodic ($\sim 35 - 65$ km) refocusing phenomenon producing zones of high intensity near the surface because of the upward refracting nature of the deep sound speed profile. Referring back to Fig. 2, there may be a depth in the deep isothermal layer at which the sound speed is the same as it is at the surface. This depth is called the "critical depth" and, in effect is the lower limit of the deep sound channel. A receiver below this depth will only receive sound from distant shallow sources by interacting with the ocean bottom. A "positive critical depth" specifies that the environment supports long distance propagation without bottom interaction, whereas "negative" implies that the bottom ocean boundary *is* the lower boundary of the deep sound channel. The bottom bounce path, E, which interacts with the ocean bottom is also a periodic phenomenon but with a shorter cycle distance and a shorter total propagation distance because of losses when sound is reflected from the ocean bottom. Finally, the right hand side of Fig. 4 depicts propagation in a shallow water region such as a continental shelf. Here sound is channeled in a waveguide bounded above by the ocean surface and below by the ocean bottom. Because of the latter, negative critical depth environments exhibit much of the sound propagation physics descriptive of shallow water environments.

The modeling of sound propagation in the ocean is further complicated because the environment varies laterally ("range dependent") and all environmental effects on sound propagation are dependent on acoustic frequency in a rather complicated way which often makes the ray type schematic of Fig. 4 misleading, particularly at low frequencies. Finally, a quantitative understanding of acoustic loss mechanisms in the ocean is required for modeling sound propagation. These losses are, aside from geometric spreading: volume attenuation, bottom loss, and surface, volume (including fish) and bottom scattering loss.

Volume attenuation increases with increasing frequency. Returning to Fig. 4, the losses associated with path C, deep channel propagation, will in general be volume attenuation and scattering

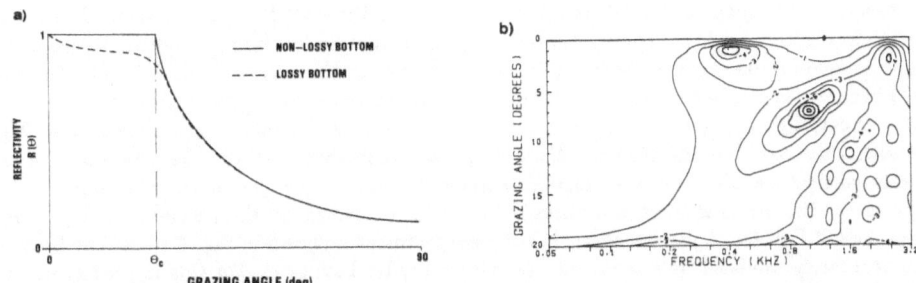

Figure 5: Bottom reflection loss. (a) Generic bottom reflectivity vs grazing angle. (b) Contours of bottom reflection loss in dB vs grazing angle and frequency.

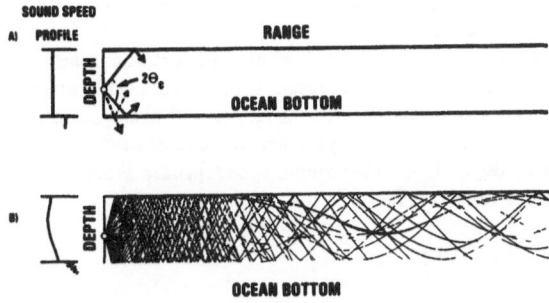

Figure 6: Ocean waveguide propagation. The curved paths in B) result from the sound speed profile varying with depth.

since this path does not involve interaction with the boundaries. The volume scattering in this case comes from the environmental condition that the upper boundary of the deep sound channel is typically the region of greatest internal wave activity in the ocean. Both of these effects are small for low frequencies and hence, deep sound propagation has been observed to distances of thousands of kilometers. This same internal wave region is also on the lower boundary of the surface duct allowing scattering out of the surface duct thereby constituting a loss mechanism for the surface duct. This mechanism also "leaks" sound into the deep sound channel, a region which without scattering would be a "shadow zone" for a surface duct source. This type of volume scattering from internal waves is also thought to be a major source of fluctuations.

When sound interacts with the ocean bottom, the structure of the ocean bottom becomes important. Figure 5a depicts a simple bottom loss curve with unit reflectivity indicating perfect reflection. For loss in dB, 0 dB is perfect reflecting, 6 dB loss is an amplitude factor of one-half, 12 dB loss is one-fourth, etc. For a "nonlossy" bottom we still get severe loss above a certain critical angle (with respect to the horizontal) in the water column due to transmission into the bottom. However, for the "lossy" (more realistic) bottom we never get perfect reflection even though the curves look similar. Path E in Fig. 4, the bottom bounce path, often involves paths which corresponds to angles near or above the critical angle; therefore, after a few bounces, the sound level will be highly attenuated. On the other hand, for shallow angles, many more bounces are possible. Hence, in shallow water, path F, most of the energy that propagates is close to the horizontal and this type of propagation is most analogous to waveguide propagation. In fact, as shown in Fig. 6, there exists a small cone from which energy propagates long distances (θ_c is typically $10° - 20°$). Energy outside the cone is referred to as the near field (or "continuous modes") and is rapidly lost into the bottom whereas the propagating field originating from within the cone is referred to the normal mode field (or "discrete modes") because there will be a set of angles which corresond to discrete paths which constructively interfere, essentially making up the normal (natural) modes of the shallow water environment. The modes within the cone are discrete since the near perfect reflectivity permits the existence of a set of discrete vertical standing waves analogous to those of a vibrating string or an organ pipe.

Very often, the ocean is more complicated than that described by the frequency independent reflection loss of Fig. 5a. A more realistic bottom environment is shown in Fig. 8 where the bottom is more accurately modeled by a sediment overlying a harder basement. For this layered structure the reflectivity will be a complicated function of frequency and incident grazing angle with respect to the horizontal. Figure 5b displays some model results of loss contours of such a layered bottom. The more familiar reflectivity versus grazing angle curve such as depicted in Fig. 5a is obtained from a vertical cut through the loss contours and noting that the loss contours are exressed in decibels. For example, at low frequencies, Fig. 5b indicates a critical angle of about $18°$ whereas at high frequencies we do not see a critical angle.

Finally, surface and bottom scattering are loss and fluctuation mechanisms. Ocean surface and bottom roughness attenuate the mean acoustic field. Again, this attenuation increases with increasing frequency. Because the ocean surface moves, it will also generate acoustic fluctuations. Bottom roughness can also generate fluctuations when the sound source and/or receiver is moving. The effect of the roughness depends on the sound speed profile which determines the degree of interaction of sound with the rough surface.

4. SOUND PROPAGATION MODELS

Sound propagation in the ocean is mathematically described by the wave equation, whose parameters and boundary conditions are descriptive of the ocean environment. There are essentially four types of models (computer solutions to the wave equation) to describe sound propagation in the sea: ray theory, fast field program (FFP), normal mode (NM) and parabolic eqaution (PE). All of these models permit the ocean environment to vary with depth. A model that also permits horizontal variations in the environment, i.e., sloping bottom or spatially variable oceanography, is termed "range dependent." For high frequencies (few kilohertz or above), ray theory, the infinite frequency approximation, is still the most practical whereas the other three model types become more and more applicable and useable below, say, a kilohertz.

The wave equation for an acoustic field of angular frequency ω is

$$\nabla^2 G(\mathbf{r}, z) + K^2(\mathbf{r}, z)G(\mathbf{r}, z) = -\delta^2(\mathbf{r} - \mathbf{r}_s)\delta(z - z_s); \quad K^2(\mathbf{r}, z) = \frac{\omega^2}{c^2(\mathbf{r}, z)}, \tag{1}$$

where the subscript "s" denotes the source coordinates. The range dependent environment manifests itself as a coefficient, $K^2(\mathbf{r}, z)$, of the partial differential equation for the sound speed profile and the range dependent bottom type and topography appears as both coefficients (elasticity effects are an added complication) and complicated boundary conditions.

4.1 Ray Theory [24]

Ray theory, a geometrical approximation, in its simplest form is just an application of Snell's Law. It is an infinite frequency approximation to Eq. 1 and assumes a solution of the form

$$G(\mathbf{R}) = A(\mathbf{R}) \exp[iS(\mathbf{R})], \tag{2}$$

where the exponential term allows for rapid variations as a function of range and $A(\mathbf{R})$ is a more slowly varying "envelope" which incorporates both geometrical spreading and loss mechanisms. Substituting Eq. 2 into Eq. 1 in a source free space, we obtain from real and imaginary parts,

$$\frac{1}{A}\nabla^2 A - (\nabla S)^2 + K^2 = 0; \quad \nabla A \cdot \nabla S + \frac{1}{2}A\nabla^2 S = 0. \tag{3}$$

The geometrical approximation is that the amplitude varies slowly with range,i.e., $(1/A)\nabla^2 A \ll K^2$, so that the first part of Eq. 3 goes to

$$(\nabla S)^2 = K^2, \tag{4}$$

which is the eikonal equation. The ray trajectories are perpendicular to surfaces of constant phase (wavefronts), S, and may be expressed mathematically as follows:

$$\frac{d}{dl}\left[K\frac{d\mathbf{R}}{dl}\right] = \nabla K, \tag{5}$$

where l is the arc length along the direction of the ray and $\mathbf{R}$ is the displacement vector. One can determine that the direction of average flux (energy) follows that of the trajectories and the amplitude of the field at any point can be obtained from the density of rays. Once S is obtained, The second part of Eq. 3 yields the amplitude. We mention here, also, that "corrected" ray theory assumes that A can be expanded in powers of inverse frequency–the leading term is the infinite-frequency result with the additional terms being frequency corrections.

The ray theory method is computationally rapid, extends to range dependent problems and the ray traces give a very physical picture of the acoustic paths. It is helpful in describing how noise redistributes itself when propagating long distances over paths that include shallow and deep environments and/or mid latitude to polar regions. The disadvantage of ray theory is that it does not include diffraction and such effects that describe the low frequency dependence ("degree of trapping") of ducted propagation.

4.2 Fast Field Program (FFP) [32,33,34,35]

Range independent wave theory solves the wave equation exactly when the ocean environment does not change in range. One of the possible derivations of the solution technique is to Fourier decompose the acoustic field into an infinite set of horizontal waves,

$$G(\mathbf{r}, z) = \frac{1}{2\pi}\int_{-\infty}^{\infty} d^2\mathbf{k}\, g(\mathbf{k}, z, z_s)\exp[i\mathbf{k}\cdot(\mathbf{r} - \mathbf{r}_s)], \tag{6}$$

and from Eq. 1, the depth dependent Green's function, $g(\mathbf{k}, z, z_s)$, satisfies

$$\frac{d^2 g}{dz^2} + (K^2(z) - k^2)g = -\frac{1}{2\pi}\delta(z - z_s). \tag{7}$$

Assuming azimuthal symmetry, we can integrate Eq. 6 over the angular variable to Hankel functions and their asymptotic form reduces Eq. 6 to (for simplicity, we take $\mathbf{r}_s = 0$)

$$G(\mathbf{r}, z) = \frac{\exp(-i\pi/4)}{(2\pi r)^{1/2}}\int_{-\infty}^{\infty} dk\, (k)^{1/2}g(\mathbf{k}, z, z_s)\exp(i\mathbf{k}\cdot r). \tag{8}$$

Note that the factor $r^{-1/2}$ arises from cylindrical spreading. We now discretize the above integral and transform to a form amenable to the FFT technique by setting $k_m = k_0 + m\Delta k$; $r_n = r_0 + n\Delta r$

where $n, m = 0, 1, ... N - 1$ with the additional condition $\Delta r \Delta k = 2\pi/N$ and N is an integral power of two. The discretization scheme limits the solution to outgoing waves and Eq. 7 becomes

$$G(r_n, z) = \frac{\Delta k \exp[i(k_0 r_n - \pi/4)]}{(2\pi r)^{1/2}} \sum_{m=0}^{N-1} X_m \exp(2\pi i m n/N), \tag{9}$$

$$X_m = (k_m)^{1/2} g(k_m, z, z_s) \exp(i m r_0 \Delta k).$$

The above equation is now easily evaluated using the FFT algorithm with the bulk of the effort going into evaluating g by solving Eq. 7. Although the method is labeled "fast field" it is fairly slow because of the time required to calculate the g's. However, it has advantages when one wishes to calculate the "near field" region or to include shear wave effects in elastic media. Because of this latter capability, it can be used as the propagation component of a description of (micro)seismic noise. The FFP method is often used as a benchmark for other less exact techniques. One such technique, not applicable to the near field but exact for a large class of range independent far-field problems is the computationally faster normal mode method.

4.3 Normal Mode Model (NM) [26,27,29]

Rather than evaluate Eq. 7 for each g for the complete set of k's (typically solving Eq. 7 1024 to 8196 times), one can utilize a normal mode expansion of the form

$$u(\mathbf{k}, z) = \sum a_n(\mathbf{k}) u_n(z), \tag{10}$$

where the quantities u_n are eigenfunctions of the following eigenvalue problem:

$$\frac{d^2 u_n}{dz^2} + \left[K^2(z) - k_n^2 \right] u_n(z) = 0. \tag{11}$$

The eigenfunctions, u_n, are zero at $z = 0$, satisfy the local boundary conditions descriptive of the ocean bottom properties and satisfy a radiation condition for $z \to \infty$. They form an orthonormal set in a Hilbert space with weighting function $\rho(z)$, the local density. The range of discrete ("standing waves" discussed in Section 3) eigenvalues is given by the condition

$$\min[K(z)] < k_n < \max[K(z)]. \tag{12}$$

The eigenvalues typically have a small imaginary part which serve as modal attenuation coefficients representative of the losses in the ocean environment.

We now substitute the normal mode expansion of g given by Eq. 10 into Eq. 6 (we take the source at the origin, for simplicity) and multiply both sides by $\rho(z) u_n(z)$ and integrate over all z space to obtain with the Hilbert space orthonormality conditions,

$$G(r, z) = \frac{\rho(z_s)}{(2\pi)^2} \int_{-\infty}^{\infty} dk_x \int_{-\infty}^{\infty} dk_y \sum_n \frac{u_n(z_s) u_n(z)}{k^2 - k_n^2} \exp(i\mathbf{k} \cdot \mathbf{r}). \tag{13}$$

We choose a contour about the poles which leads to a solution with outgoing waves from the point source. Each integral in Eq. 13 is proportional to the two-dimensional plane wave representation of the zero-order Hankel function of the first kind, and thereofre, G can be expressed as

$$G(r, z) = \frac{i}{4} \rho(z_s) \sum_n u_n(z_s) u_n(z) H_0^1(k_n r). \tag{14}$$

The asymptotic form of the Hankel function can be used in the above equation to obtain the well known normal mode representation of a cylindrical (axis is depth) waveguide:

$$G(r,z) = \frac{i\rho(z_s)}{(8\pi r)^{1/2}} \exp(-i\pi/4) \sum_n \frac{u_n(z_s)u_n(z)}{k_n^{1/2}} \exp(ik_n r). \tag{15}$$

Equation 15 is a "far field" solution of the wave equation and neglects the "continuous spectrum" $(k_n < \min[K(z)]$ of Eq. 12) of modes; therefore, we are considering only the discrete modes. For purposes of illustrating the various portions of the acoustic field, we note that k_n is a horizontal wave number so that a "ray angle" associated with a mode with respect to the horizontal can be taken to be $\theta = cos^{-1}[k_n/K(z)]$. For a simple waveguide the maximum sound speed is the bottom sound speed which therefore corresponds to $\min[K(z)]$. At this value of $K(z)$, we have from Snell's law $\theta = \theta_c$, the bottom critical angle. In effect, if we look at a ray picture of the modes, the continuous portion of the mode spectrum corresponds to rays with grazing angles greater than the bottom critical angle, θ_c, of Fig. 5 and therefore outside the cone of Fig 6. This portion, as explained in Section 3 undergoes severe loss. Hence,we note that the continuous spectrum is the near (vertical) field and the discrete spectrum is the (more horizontal, profile dependent) far field (falling within the cone in Fig. 6). Note also, in describing the acoustic paths in these two regimes we use the result from Snell's law that for a medium with a vertical variability in index of refraction like the "horizontally stratified" ocean, rays that start close to the horizontal undergo "more bending" than the more vertical rays. Hence we say that the continuous, more vertical portion of the sound field, is less profile dependent than the discrete, more horizontal, part.

The advantages of the NM procedure are: that once the eigenvalue problem is solved one has the solution for all source and receiver configurations, and, that it is easily extended to moderate range dependent conditions using the adiabatic approximation.

4.4 Adiabatic Mode Theory [36]

We mention this model because all of the range independent normal mode "machinery" developed for environmental ocean acoustic modeling applications can be adapted to mildly range dependent conditions using adiabatic mode theory. Very briefly, the underlying assumption is that individual propagating normal modes adapt (but do not scatter or "couple" into each other) to the local environment. The coefficients of the mode expansion, a in Eq. 10 now become mild functions of range, i.e., $a_n(k) \rightarrow a_n(k,r)$.This modifies the Eq. 15 as follows:

$$G(r,z) = \frac{i\rho(z_s)}{(8\pi r)^{1/2}} \exp(-i\pi/4) \sum_n \frac{u_n(z_s)v_n(z)}{\overline{k_n}^{1/2}} \exp(i\overline{k_n}r). \tag{16}$$

where the range-averaged wave number (eigenvalue) is

$$\overline{k_n} = \frac{1}{r} \int_0^r k_n(r)\, dr, \tag{17}$$

and the $k_n(r)$'s are obtained at each range segment from the eignevalue Eq. 11 evaluated at the environment at that particular range along the path, that is for $K(z)$ and the relevant boundary conditions at r. The quantities u_n and v_n are the sets of modes at the source and the field positions, respectively.

Simply stated, the adiabatic mode theory, as formulated by Eqs. 16 and 17 leads to a description of sound propagation such that the acoustic field is a function of the modal structure at both the source and the receiver and some average propagation conditions between the two. Thus, for example, when sound emanates from a shallow region where only two discrete modes exist and propagates into a deeper region with the same bottom (same critical angle) , the two modes from the shallow region adapt to the form of the first two modes in the deep region. However, the deep region can support many more modes; intuitively, we therefore expect the resulting two modes in the deep region will take up a smaller more horizontal part of the cone of Fig. 6 than they take up in the shallow region. This means that sound rays going from shallow to deep tend to become more horizontal.

Recently, a fully coupled mode theory for range dependent environments has been developed [37]; the computations, however, are still quite costly in time.

4.5 Parabolic Equation Model (PE) [38,39,40,41]

The PE method is presently the most practical and encompassing wave-theoretic *range – dependent* propagation model. In its simplest form, it is a far-field narrow-angle ($\sim \pm 20°$ with respect to the horizontal—adequate for most underwater propagation problems) approximation to the wave equation. Assuming azimuthal symmetry about a source, we express the solution of Eq. 1 in cylindrical coordinates in a source free region in the form

$$G(r, z) = \psi(r, z) \cdot J(r), \tag{18}$$

and we define $K^2(r, z) \equiv K_0^2 n^2$, n therefore being an "index of refraction" c_0/c, where c_0 is a reference sound speed. Substituting Eq. 18 into Eq. 1 in a source free region and taking K_0^2 as the separation constant , J and ψ satisfy the following two equations:

$$\frac{d^2 J}{dr^2} + \frac{1}{r}\frac{dJ}{dr} + K_0^2 J = 0; \tag{19}$$

$$\frac{\partial^2 \psi}{\partial r^2} + \frac{\partial^2 \psi}{\partial z^2} + \left(\frac{1}{r} + \frac{2}{J}\frac{\partial J}{\partial r}\right)\frac{\partial \psi}{\partial r} + K_0^2 n^2 \psi - K_0^2 \psi = 0. \tag{20}$$

Equation 19 is a Bessel equation and we take the outgoing solution, a Hankel function, $H_0^1(K_0 r)$, and we take its asymptotic form (see Eqs. 14 and 15) to substitute into Eq. 20, together with the "paraxial" (narrow angle) approximation,

$$\frac{\partial^2 \psi}{\partial r^2} \ll 2K_0 \frac{\partial \psi}{\partial r}, \tag{21}$$

to obtain the parabolic equation (in r),

$$\frac{\partial^2 \psi}{\partial z^2} + 2iK_0 \frac{\partial \psi}{\partial r} + K_0^2 (n^2 - 1)\psi = 0, \tag{22}$$

where we note that n is a function of range and depth. We use a marching solution to solve the parabolic equation. There has been an assortment of numerical solutions [40,41] but the one that still remains the most robust is the so called "split step algorithm" [38], which we briefly describe below.

For this derivation, we take n to be a constant; it is shown elsewhere [38,39] that the error this introduces can be made arbitrarily small by the appropriate numerical gridding. The Fourier transform of ψ with respect to z can then be written as

$$\psi(r, s) = \frac{1}{2\pi}\int_{-\infty}^{\infty} \psi(r, z)\exp(-isz)\,dz \tag{23}$$

which together with Eq. 22 gives

$$-s^2 \psi + 2iK_0 \frac{\partial \psi}{\partial r} + K_0^2 (n^2 - 1)\psi = 0. \tag{24}$$

The solution of Eq. 24 is simply

$$\psi(r, s) = \psi(r_0, s)\exp\left[-\frac{K_0^2(n^2 - 1) - s^2}{2iK_0}(r - r_0)\right], \tag{25}$$

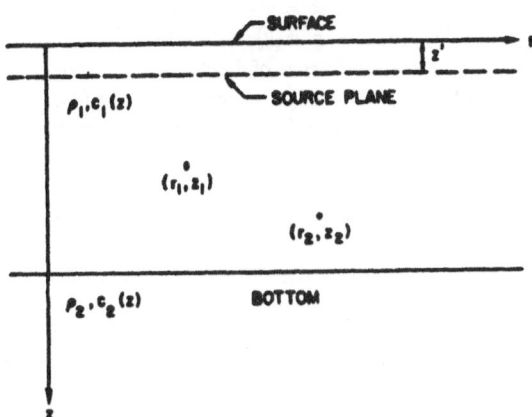

Figure 7: The model geometry showing the noise source plane, at depth z', below the surface, and the two field points. The local environment is simplified to sound speed c and density ρ.

where the initial condition at r_0 must be specified. The inverse transform then gives the field as a function of depth,

$$\psi(r,z) = \int_{-\infty}^{\infty} \psi(r_0,s) \exp\left[\frac{iK_0}{2}(n^2-1)\Delta r\right] \exp\left[-\frac{i\Delta r}{2K_0}s^2\right] \exp(isz)\, ds, \tag{26}$$

where $\Delta r = r - r_0$. Introducing the symbol $\mathcal{F}$ for the Fourier transform operation from the z-domain (as performed in Eq. 23) and $\mathcal{F}^{-1}$ as the inverse transform, Eq. 26 can be summarized by the range-stepping algorithm,

$$\psi(r+\Delta r,z) = \exp\left[\frac{iK_0}{2}(n^2-1)\Delta r\right] \mathcal{F}^{-1}\left[\left(\exp(-\frac{i\Delta r}{2K_0}s^2)\right)\cdot \mathcal{F}\left[\psi(r,z)\right]\right], \tag{27}$$

which is often referred to as the "split step" marching solution to the PE. The Fourier transforms are performed using an FFT. It is the solution for n constant, but the error introduced when n (profile or bathymetry) varies with range and depth can be made arbitrarily small by increasing the transform size and decreasing the range-step size.

5. MODELING OCEAN NOISE FROM SURFACE SOURCES

The distribution of sound from surface noise sources can be calculated by coupling these noise sources into the water column using any of the above mentioned propagation models together with Green's theorem. We then would include both the physics of the noise sources and the environmental propagation effects. Figure 7 shows the geometry for such a model. The distribution of the field is characterized by the "cross-spectral density," which reduces to the local sound intensity when the two field points are at the same position. Otherwise, this quantity is representative of the directional/spatial properties of the noise field which would be measured by a distributed acoustic receiving array. The cross-spectral density is ($<>$ is the "expectation" value symbol)

$$<\phi(\mathbf{r}_1,z_1)\phi^*(\mathbf{r}_2,z_2)> = \int\int d^2\mathbf{r}'d^2\mathbf{r}'' <S(\mathbf{r}')S^*(\mathbf{r}'')> G(\mathbf{r}_1,\mathbf{r}',z_1,z')G^*(\mathbf{r}_2,\mathbf{r}'',z_1,z''), \tag{28}$$

where S is the surface noise source function and z'' is at the same depth as z'. The correlation function of the surface noise sources which is the expectation quantity in the integrand of Eq. 2 is taken to be spatially dependent only on the separation ($s \equiv r' - r''$) between sources. We denote this correlation function as $<S(r')S^*(r'')> \equiv q^2 N(s)$, with q being surface source strength. We also express the noise field points through the difference of displacement vectors, $R \equiv r_1 - r_2$, and denote the cross-spectral density given by Eq. 28 for angular frequency ω as $C_\omega(R, z_1, z_2)$. Substituting in the full field representation of G given by Eqs. 6-7 into Eq. 28 and performing the integration over the azimuth angle associated with k, gives

$$C_\omega(R, z_1, z_2) = 2\pi q^2 \int ds N(s) \int_0^\infty k\, dk g(k, z_1, z')g^*(k, z_2, z')J_0(k|R - s|), \qquad (29)$$

where J_0 is the Bessel function of zero order. An alternative form to Eq. 29 which is particularly useful since it can directly call upon the algorithms associated with the FFP approach presented above involves expressing the correlation function of the surface noise sources in terms of its spatial spectral representation, $P(k)$. Then, if we substitute into Eq. 28 for N, its Fourier transform

$$N(s) = \frac{1}{2\pi} \int dk\, P(K) \exp(ik \cdot s), \qquad (30)$$

together with the other substitutions used to get Eq. 29, we obtain

$$C_\omega(R, z_1, z_2) = 2\pi q^2 \int dk P(k) \exp(ik \cdot R)g(k, z_1, z')g^*(k, z_2, z'). \qquad (31)$$

Depending on the type of numerics being applied to a specific problem, Eqs. 29 and 31 are suitable for model computations. Note that the noise cross-spectral density evaluated at the same field points, $C_\omega(0, z, z)$, is the intensity of the noise at depth z for this range-independent formulation of the distribution of noise.

It has been shown [8] that the correlation function of surface sources which produces a far field radiation (of multipole order m) pattern $cos^m\theta$ with respect to the vertical is given by

$$N(s) = \begin{cases} \frac{2\delta(K(z')s)}{[K^2(z')s]} & m = 1 \\ 2^m m! \frac{J_m[K(z')s]}{[K(z')s]^m} & m > 1 \end{cases} \qquad (32)$$

In simple general terms, we know from the literature that much of the noise sources can be categorized as and are associated with [2]:

1. monopole: mass addition, heat addition, volume change, compression—e.g., bubbles, rain droplet impact, etc.

2. dipole: force, translation, acceleration (sloshing)—e.g., vibration of unbaffled rigid bodies;

3. quadrupole: moment, shear, distortion, rotation, turbulence.

These noise sources are then taken to be on the source plane in Fig. 7 where the surface is pressure release so that the radiation couples into the water column as a higher order pole radiator; for example, monopole sources near the surface couple as dipoles just by using an ocean sound propagation Green's function, G because G satisfies the pressure release condition. For this case we then use $m = 1$ in Eq. 32, a dipole sheet at the surface corresponding to monopole plane below the surface, which gives the $cos^m\theta = cos\theta$ radiation pattern with respect to the vertical. Then, the noise distribution of sound in the ocean from surface sources can be constructed from the above expressions using the propagation models described in Section 2 for G and the noise source distributions discussed throughout this conference and expressed in the multipole form of Eq. 32.

Of interest is a particular simplification when the noise can be modeled as dipole-coupled into a range independent environment and an expression for the distribution of noise is sought in terms of normal modes. Then, that part of the noise that corresponds to discrete propagation from mainly

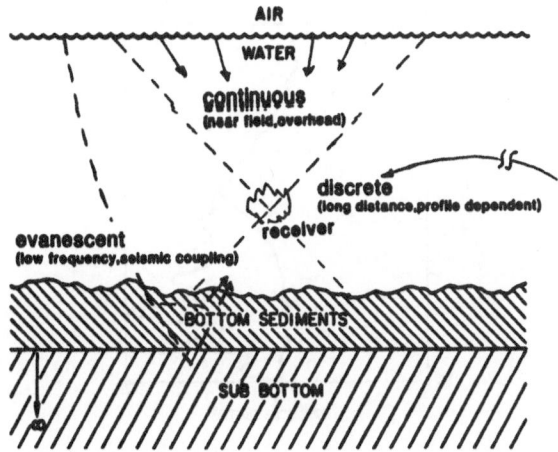

Figure 8: Surface generated noise propagation paths to an acoustic receiver.

far away sources and corresponds to the inner cone of Fig. 6 and the inequality of Eq. 12 can be derived from Eqs. 29 and 15. The simple result can be obtained for this important case, with $m = 1$ as given by Eq. 32. After a complex integration, we get the result for the discrete portion of the noise field (suppressing the subscript ω),

$$C_{disc}(\mathbf{R}, z_1, z_2) \approx \frac{\pi q^2 \rho^2(z')}{2K^2(z')} \sum_n \frac{u_n^2(z') u_n(z_1) u_n(z_2)}{\alpha_n k_n} J_0(k_n R), \tag{33}$$

where α_n is the attenuation coefficient of the $n - th$ mode and as described in Section 4 is the imaginary part of the eigenvalues resulting from the existence of loss in the system. It is clear from the physics of this problem that contributions from noise sources long distances away in a waveguide with only cylindrical spreading will begin to diverge if there was no loss in the system; hence, we have the attenuation coefficients in the denominator of Eq. 33.

6. SPATIAL PROPERTIES OF SURFACE GENERATED NOISE

As discussed in Section 3, the ocean is a waveguide and we can therefore draw some general conclusions about how sound sources at the upper boundary of a waveguide will couple into the waveguide. Quantitaive calculations as discussed in Section 4 confirm these insights. As mentioned earlier, the ray and mode pictures provide the most straightforward physical insight into the actual coupling process.

First, for reference, let us consider the noise field in an unbounded semi-infinite isovelocity medium with constant distribution of monopole sources radiating into the medium from the upper boundary. Summing all the sources over an infinite area and noting that in this situation we have spherical spreading, it is easy to show that the resulting sound field will be finite and its directional properties will be isotropic in the upward direction (the only direction from which sound is coming) and the intensity of the sound will be independent of position. All these characteristics come from three simple considerations:

1. the sources are spherically symmetric;

2. contributing source area goes up as the square of the radius;

3. the field of the sources decay by spherical spreading.

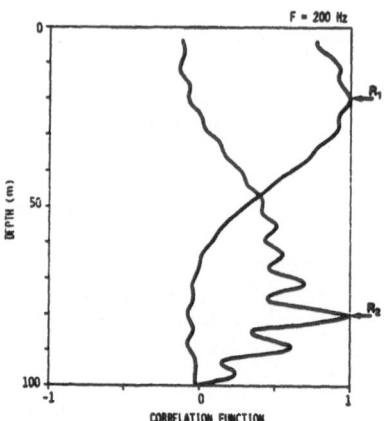

Figure 9: Normalized spatial correlation in a waveguide environment of noise for two different receiver depths.

However, in the ocean, which is a waveguide, we have cylindrical spreading, meaning that if sound can propagate long distances, large areas from very long distances will contribute to local noise. As a matter of fact, mathematically, a rigorous model must include some kind of attenuation when we have the contribution to noise field increasing with the area of the contributing noise sources and the propagation loss decaying only by cylindrical spreading; the real ocean, of course is lossy (see discussion below Eq. 33). In a waveguide it makes sense that noise coming directly from above looks like the semi-infinite unbounded case discussed in the last paragraph whereas noise from long distances has the ocean channel properties discussed in Section 3.

With this background of some knowledge of sound propagation in the ocean and some intuition of how sources on a surface plane couple either into a semi-infinite space (spherical spreading) or into a waveguide (cylindrical spreading) we can understand much of how ambient noise is distributed in the ocean. Figure 8 depicts the possible propagation regimes by which surface generated noise can arrive at a receiver near the ocean bottom (so as to include the possibility of seismic paths). The continuous is the vertical structured noise not too affected by the sound speed profile. The discrete (or channelled) corresponds to horizontally arriving noise from, typically, distant sources. We already have reviewed the mathematical tools to represent this structure. We can write the continuous portion of the noise field as the right hand side (rhs) of Eq 31 where the limit on the integration is $|\mathbf{k}| < min[K(z)]$ (see discussion below Eq. 15); we define this quantity to be C_{cont}. Then, the total cross-spectral density of the noise field which is also referred to as the spatial correlation or spatial coherence of the noise field is given by

$$C_w(\mathbf{R}, z_1, z_2) = C_{cont}(\mathbf{R}, z_1, z_2) + C_{disc}(\mathbf{R}, z_1, z_2).$$ (34)

Equation 34 indicates that: the noise field varies with depth; the cross-spectral density of the noise is not only a function of the mean depth of the field points but also the absolute depth location of the field points, i.e., noise is not homogeneous in depth (homogeneous means only a function of separation distance). The cross-spectral density of the noise is homogeneous in range for a range-independent environment since there is no reference range. These conclusions are particularly true for horizontally stratified continental shelf areas in which waveguide theory is most applicable.

Figure 9, taken from [42], indicates the above mentioned vertical inhomgeneity in a waveguide

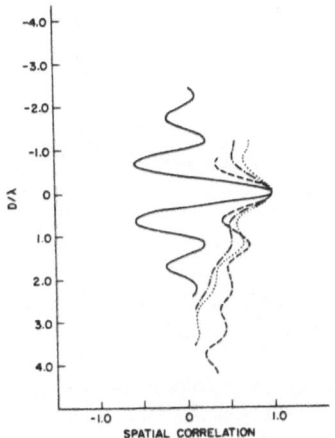

Figure 10: Normalized vertical spatial correlation functions for an upward refracting environment with one receiver fixed at 10 m. Three frequencies are: - - - 200 Hz; ·· 400 Hz; - ·- 800 Hz. The solid line is the semi-infinite homogeneous result [12].

type environment by plotting the cross-spectral density of the noise related to receivers located at two different depths. The noise at receiver R_1 is correlated with the noise at all other field points of the water column and the same is true for receiver R_2. We see that the two correlation functions appear different, indicating the spatial inhomogeneity.

Also contrary to simplified noise models [7], the cross-spectral density of noise in a stratified waveguide should have a different structure than noise in a homogeneous semi-infinite medium. Figure 10, taken from [12] shows the modeled vertical spatial correlation (cross-spectral density for a receiver at the same horizontal location but different vertical positions) between a receiver located at 10 m depth and another receiver separated by distances D in units of wavelength, λ. The particular environment for this case was a 50 m water depth and an upward refracting sound speed profile. Here we see that the vertical correlation of the noise vs separation in units of wavelength is a function of frequency, not agreeing with the simpler [12] for a semi-inifinite medium where the spatial correlation scales with wavelength. This different result arises from the additional scales that are introduced when considering the stratified environment. Note also, the noise is more correlated over the water column than the simple case because a few lower order modes dominate the propagation environment due to upward refraction and therefore there is very little interaction with a lossy bottom for the horizontal long distance contribution as opposed to the more vertical lossy propagation. That is, the noise tends to come from the horizontal rather than from all directions thereby increasing the spatial correlation.

Figure 11, also taken from [12] shows the results for the horizontal correlation for the same kind of environment as for the case in Fig. 10 except that the sound speed is constant in the water column. Here, we plot the discrete and continuous contributions separately. The continuous (vertical) part, is more coherent for small receiver separations but quickly becomes less coherent (and looks like the simpler semi-infinite case) than the discrete part. Again, the physics of the propagation environment indicates the relative importance of the discrete and continuous parts. For low loss, the discrete modes dominate as they can be propagated very large distances from a very large area; see discussion below Eq. 33. In high loss cases the continuous modes tend to dominate since they are important near the surface while the long range contributions of the discrete

268

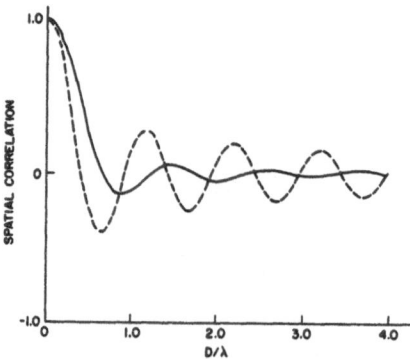

Figure 11: Horizontal spatial correlation of noise for an isovelocity waveguide environment showing the discrete (- - -) and continuous (—) contributions, both normalized.

are severely attenuated; hence, we expect a more vertical structure for noise in a lossy environment. As a matter of fact, the structure of the noise can be used as a diagnostic tool for estimating the acoustic properties of such an environment [43].

In Sections 2 and 3 we discussed the impact of the deep water sound speed structure on propagation. In particular, we mentioned in Section 3 the existence of a "critical depth" in the deep isothermal layer if the water is deep enough. For such a case we expect from the above discussion that when there is a "positive critical depth" that sound from surface sources can travel long distances without interacting with the ocean bottom but that a receiver below this critical depth should sense less surface surface noise because any propagation from distant shallow sources to these depths must involve interaction with the lossy bottom. Indeed, Fig. 12 illustrates this effect using data from the Northeast Pacific [3,44] The noise in this figure comes from a combination of natural sources and shipping but the propagation effects for these surface sources are the same. We should note here also, that higher frequency noise (especially the multi-khz region and above) diminishes with depth simply because volume attenuation increases with frequency.

The vertical directionality of deep water noise, should also vary with depth as discussed above. Figure 13 shows some depth dependence of the directionality of low frequency noise in the Pacific [3,45]. The shallower depth is at the axis of the deep sound channel while the other is at the critical depth. We see, as expected, that the pattern is narrower at the critical depth where the sound paths tend to be horizontal since the rays are turning around at the lower boundary of the deep sound channel.

In deep water, sound from surface sources should arrive, on the average, off the horizontal ($\sim \pm 12°$ to $\pm 15°$) if the ocean did not have a range varying environment. But, as explained in Section 4.4, noise from continental shelf areas tend to have paths which become more horizontal as the sound propagates to deeper water; we also have an analogous effect that polar latitudes have shallow sound channels so that sound propagating from polar latitudes will go deep and horizontal in the mid latitude regions. Hence, the accurate modeling of this horizontal component of the noise [46] requires the use of range dependent models.

Finally, we mentioned seismic possibilities. Very low frequency sound will couple directly into the ocean bottom. This "microseismic" noise [13,18] also obeys the basic physics of waveguide propagation we review in this paper. Briefly, when the ocean surface sources are "acoustically" close to the ocean bottom, their evanescent field will couple into seismic waves. This coupling occurs below the acoustic "cutoff" frequency of the waveguide [47].

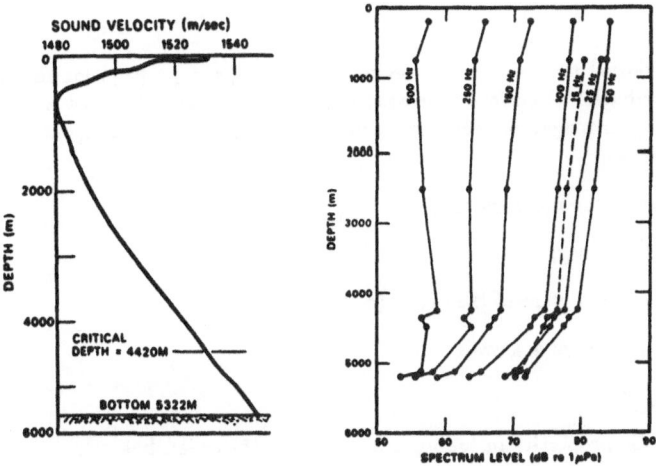

Figure 12: Sound speed profile and noise level as a function of depth in the Pacific [44].

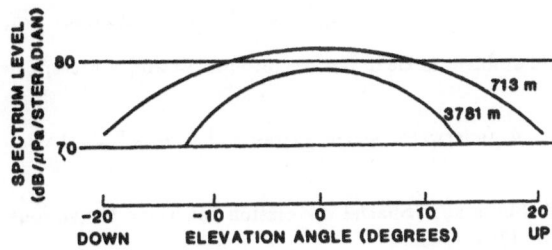

Figure 13: Vertical directionality of noise at the axis of the deep sound channel and at the critical depth [45].

7. SUMMARY

This paper has reviewed the basic ocean environmental acoustic propagation factors that directly impact on the distribution of surface generated noise in the ocean. The most important aspects of the physics of sound propagation that impact on the distribution of noise generated from natural surface mechanisms are:

1. the directivity of the noise sources;

2. the acoustic frequency;

3. the sound speed structure of the ocean to long ranges;

4. acoustic attenuation of sea water

5. the water depth;

6. the geo-acoustic properties including attenuation of the ocean bottom.

8. REFERENCES

[1] This brief reference list [2-30] is representative of the ocean noise, signal processing and propagation literature but for lack of space, is very incomplete. The bibliographies contained within the papers cited below are, taken together, much more encompassing.

[2] D. Ross, *Mechanics of Underwater Noise*, Pergamon Press, New York, [1976].

[3] R. J. Urick, *Ambient Sea Noise in the Ocean*, published by The Undersea Warfare Technology Office, Naval Sea Systems Command, Department of the Navy, Washington, D.C. 20362, [1984].

[4] R. J. Urick, *Principles of Underwater Sound*, 3rd ed., McGraw-Hill Book Co., New York, [1983].

[5] G. M. Wenz, "Acoustic ambient noise in the ocean: spectra and sources," *J. Acoust. Soc. Am*, 34, p. 1936, [1962].

[6] G. M. Wenz, "Review of underwater acoustics research: noise," *J. Acoust. Soc. Am*, 51, p. 1010, [1972].

[7] B. F. Cron and C. H. Sherman, "Spatial correlation functions for various noise models," *J. Acoust. Soc. Am*, 34, p. 1732, [1962].

[8] W. S. Liggett and M. J. Jacobsen, "Covariance of surface generated noise in a deep ocean," *J. Acoust. Soc. Am*, 38, p. 302, [1965]

[9] E. M. Arase and T. Arase, "Correlation of ambient sea noise," *J. Acoust. Soc. Am*, 40, p. 205, [1966]

[10] L. M. Brekhovskikh, "Underwater sound waves generated by surface waves in the ocean," Izv, Atmos.Ocean.Phys., 2, p. 582, [1966].

[11] M. A. Isakovich and B. F. Kuryanov, "Theory of low frequency noise in the ocean," Sov. Phys. (Acoust.), 16, p. 49, [1970].

[12] W. A. Kuperman and F. Ingenito, "Spatial correlation of surface generated noise in a stratified ocean," *J. Acoust. Soc. Am*, 67, p. 1988, [1980].

[13] A. C. Kibblewhite and K. C. Ewans, "Wave-wave interaction,microseisms, and infrasonic ambient noise in the ocean," *J. Acoust. Soc. Am*, 78, p. 981, [1985].

[14] J. H. Wilson, "Wind-generated noise modeling," *J. Acoust. Soc. Am*, 73, p. 211, [1983].

[15] R. H. Nichols, "Infrasonic ambient noise measurements: Eleuthera," *J. Acoust. Soc. Am*, 69, p.974, [1981].

[16] P. C. Wille and D. Geyer, "Measurements on the origin of the wind-dependent noise variability in shallow water," *J. Acoust. Soc. Am*, 75, p. 173, [1984].

[17] R. Dashen and W. Munk, "Three models of global ocean noise," *J. Acoust. Soc. Am*,76, p. 540, [1984].

[18] R. G. Adair amd J. A. Orcutt, "Low-frequency noise observation in the deep ocean," *J. Acoust. Soc. Am*, 80, p. 633, [1986].

[19] R. A. Wagstaff, "A Comprehensive Ambient Noise Bibliography," TP 333, [1973].

[20] H. Cox, "Resolving power and sensitivity for mismatch of optimum array processors,", *J. Acoust. Soc. Am*, 54, p. 771, [1973].

[21] D. H. Johnson, "The application of spectral estimation methods to bearing estimation problems,: *Proceedings of the I.E.E.E.*, 70, p. 1018, [1982].

[22] R. M. Hamson, "The theoretical response of vertical and horizontal vertical line arrays in wind-induced noise in shallow water," *J. Acoust. Soc. Am*, 78, p. 1702, [1985].

[23] A. B. Baggeroer, W. A. Kuperman and H. Schmidt, "Matched field processing: source localization in correlated noise as an optimum parameter estimation problem," submitted to *J. Acoust. Soc. Am*, [1987].

[24] J. B. Keller and J. S. Papadakis, (eds.), *Wave Propagation and Underwater Acoustics*, Springer-Verlag, Berlin, [1977].

[25] C. S. Clay and H. Medwin, *Acoustical Oceanography*, Wiley-Interscience, New York, [1977].

[26] W. A. Kuperman and F. B. Jensen, "Deterministic propagation modeling I, Fundamental Principles," in *Underwater Acoustics and Signal Processing*, ed. L. Bjorno, Reidel Publishing Co., Dordrecht, Holland, [1980].

[27] F. B. Jensen amd W. A. Kuperman, "Consistency test of acoustic propagation models," Appendix, Report Sm-157,SACLANT ASW Research Centre, La Spezia, Italy, [1982].

[28] W. A. Kuperman, "Models of sound propagation in the ocean," *Naval Research Reviews*, 3, [1985].

[29] F. B. Jensen, "Numerical models in underwater acoustics," *Proceedings of NATO Advanced Study Workshop on Hybrid Formulation of Wave Propagation and Scattering*, ed. L. B. Felsen, IAFE, Castel Gandolfo, Italy, Martinus Nijhoff Publishers, Dordrecht, [1984].

[30] W. A. Kuperman and B. E. McDonald, "Linear and Nonlinear Ocean Acoustic Propagation Models," *Ocean Seismo-Acoustics*, eds. T. Akal and J. M. Berkson, Plenum Press, New York, [1986].

[31] J. Northrup and J. G. Colborn, "Sofar Channel Axial Sound Speed and Depth in the Atlantic Ocean," *J. Geophys. Res.*, 79, p. 5633, [1974].

[32] F. R. DiNapoli and R. L. Deavenport, "Theoretical and numerical Green's function field solution in plane multilayered media," *J. Acoust. Soc. Am*, 67, p. 92, [1980].

[33] H. W. Kutschale, "Rapid computation by wave theory of propagation loss in the Arctic Ocean," Rpt. CU-8-73, Columbia University, Palisades, New York, [1973].

[34] H. Schmidt and F. B. Jensen, "A full wave solution for propagation in multilayered media with application to gaussian beam reflections at fluid interfaces," *J. Acoust. Soc. Am*, 77, p. 813, [1985].

[35] H. Schmidt, "SAFARI. Seismo-acoustic Fast Field Algorithm for Range Independent Environments,"

[36] L. M. Brekhovskikh and Yu. Lysanov, *Fundamentals of Ocean Acoustics*, Springer-Verlag, Berlin, [1982].

[37] R. B. Evans, "A coupled mode solution for acoustic propagation in a waveguide with stepwise depth variations of a penetrable bottom," *J. Acoust. Soc. Am*, 74, p. 188, [1983].

[38] F. D. Tappert, "The Parabolic Approximation Method," in [24], p. 224.

[39] F. B. Jensen and K. R. Krol, "The use of the parabolic equation method in sound propagation modelling," Rpt SM-72, SACLANT ASW Research Centre, La Spezia, Italy, [1975].

[40] J. A. Davis, D. White and R. C. Cavanagh, "NORDA parabolic equation workshop," Rpt. TN-153, Naval Ocean Research and Development Activity, NSTL Stn., MS, [1982].

[41] D. Lee and K. E. Gilbert., "Recent progress in modeling bottom interacting sound propagation with parabolic equations," in *Ocean 82*, MTS-IEEE, p. 172, Washington, DC, [1982].

[42] F. B. Jensen and W. A. Kuperman, "Environmental Acoustic Modeling at SACLANTCEN," Report Sr-34, SACLANT ASW Research Centre, La Spezia, Italy, [1979].

[43] M. J. Buckingham and S. A. S. Jones, "A new shallow-ocean technique for determining the critical angle of the seabed from the vertical directionality of the ambient noise in the water column," *J. Acoust. Soc. Am*, 81, p. 938, [1987].

[44] G. B. Morris, "Depth dependence of ambient noise in the Northeastern Pacific Ocean," *J. Acoust. Soc. Am*, 64, p. 581, [1978].

[45] V. C, Anderson, "Variations of the vertical directivity of noise with depth in the North Pacific," *J. Acoust. Soc. Am*, 66, p. 1446, [1979].

[46] R. A. Wagstaff, "Low-frequency ambient noise in the deep sound channel–The missing component," *J. Acoust. Soc. Am*, 69, p. 1009, [1981].

[47] T. Akal, H. Schmidt, W. A. Kuperman, "Low frequency ambient noise in shallow water," This volume.

LOW FREQUENCY WIND GENERATED AMBIENT NOISE IN SHALLOW WATER

Henrik Schmidt and Tuncay Akal
SACLANT ASW Research Centre
I-19026 La Spezia, Italy

W.A. Kuperman
Naval Research Laboratory
Washington, DC 20375

ABSTRACT. The waveguide nature of a shallow water environment bounded
below by a viscoelastic stratified bottom permits wind generated ambient
noise to propagate as both waterborne sound and seismic waves. Geophone
and hydrophone measurements have shown that below a threshold frequency
of about 10 Hz in 100 m of water, there is a significant increase in the
ambient noise recorded by the geophones in particular. A full wavefield
solution for distributed noise in a stratified viscoelastic medium has
been used to explain this behavior, which is due to coupling into
seismic interface waves at low frequencies. The excellent qualitative
agreement obtained between the experimental and theoretical results has
yielded the possibility of directly estimating the actual noise source
levels and calculating these to the observed wind speeds and sea states.
The obtained source levels are in close agreement with published noise
levels obtained in deep water.

1. INTRODUCTION

The ambient noise recorded by sensors in the ocean or on the seabed is
governed by the type and spectral level of the generating sources, but
also by the environmental conditions under which the sound propagates.
The waveguide nature of the ocean, bounded above by a pressure release
surface and below by a stratified viscoelastic bottom, allows the
acoustic energy to propagate along many different paths [1].

Above the cut-off frequency for the waveguide it is well known,
that for distant sources, the measured acoustic field is dominated by
the normal modes of the waveguide, the so-called discrete wavenumber
spectrum.

For sources at short range, however, the continuous part of the
wavenumber spectrum, corresponding to propagation at angles larger than
critical, becomes important. Further, the viscoelastic bottom allows
energy to propagate as seismic interface waves. This propagation regime
becomes particularly important below the acoustic cut-off frequency.

B. R. Kerman (ed.), Sea Surface Sound, 273–280.
© *1988 by Kluwer Academic Publishers.*

In the case of wideband ambient noise produced by distributed sources, such as the surface generated noise due to wind and non-linear surface wave interaction, the measured field will obviously reflect these different propagation mechanisms. In order to extract the true spectral distribution of the source level, it is therefore necessary to remove propagation effects from the measured data. This is particularly important for low frequency noise in shallow water where bottom interaction is significant. It is therefore not surprising that the shallow water noise levels reported by Akal et al [2] do not agree well with the deep water data presented by Kibblewhite et al [3].

We will here use a recently developed model for surface generated noise in a stratified viscoelastic medium [4] to predict the surface generated noise in a shallow water environment where experimental data [2] were obtained. Analysis of the wavenumber spectra in different spectral regimes, enables propagation effects on the measured noise field to be related to the spectral features observed in the experimental data. Then the synthetic noise levels for a white-noise source spectrum are subtracted from the real data recorded at different sea states, to obtain the true source spectra. These are then compared to the deep water data of Ref. [3].

2. EXPERIMENTAL RESULTS

The setup for the experiment reported in [2] is illustrated in Fig. 1. The ambient noise was recorded by means of a 3-component geophone station (OBS) and a hydrophone close to the seabed in a relatively flat area of the Ligurian shelf having a 100 m water depth. The sea state was measured by means of a wave rider buoy and the data were continuously transferred to a shore station via radio link.

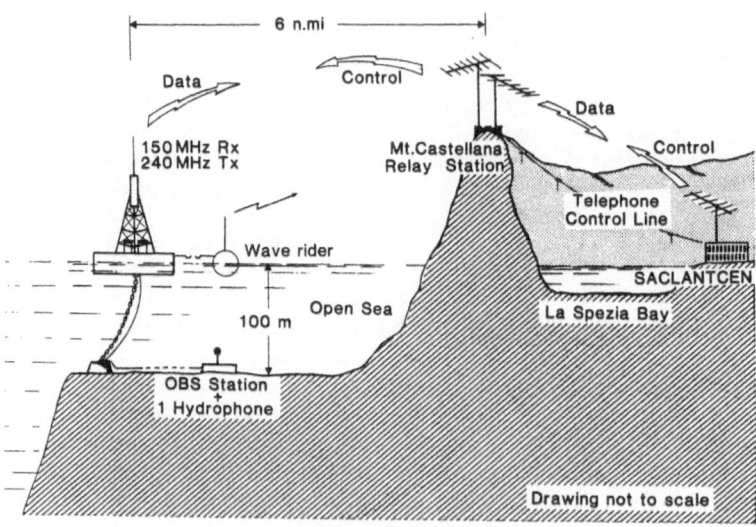

Fig. 1: Experimental setup for measuring ambient noise on the Ligurian shelf.

Data were recorded during a period of 105 days with acquisition in daily 5 minute periods. This setup allowed registration of ambient noise under many different conditions, both with and without nearby shipping.

Fig. 2 shows a characteristic example of the noise spectrum obtained during a period having a prevalent sea state of around 2 (30-60 cm significant wave height) and low amount of shipping. The solid line indicates power spectrum in $dB//(10^{-6}$ Pa$)^2$/Hz for the hydrophone whereas the dashed and dotted curves represent the power spectra for a horizontal and vertical geophone, respectively, given in $dB//(10^{-12}$ m/s$)^2$/Hz. Only one horizontal component is shown, but it was a general feature in periods of no nearby shipping that the two horizontal components registered almost identical spectra, indicating an isotropic noise field under these conditions.

3. THEORETICAL RESULTS

The mathematical model of Kuperman and Ingenito [5] for surface generated ambient noise has recently been combined with the SAFARI propagation model [6] to allow simulation of surface generated noise in general stratified viscoelastic media [4]. In addition to simulating the pressure field detected by hydrophones, this model also has the capability of calculating the corresponding geophone signals, which makes it well suited for modelling the experiments described above. The fact that the sensors were placed in a relatively flat area of the shelf makes this experiment well suited to the range-independent assumption of the model. The fact that the measured noise field was horizontally isotropic supports this assumption.

Based on the environmental data taken during the experiment, a water depth of 100 m was assumed and the bottom was represented by a 10 m silt layer overlaying a sand halfspace. Surface noise monopole sources (coupling as dipoles) of source level of 70 dB (defined as the pressure level obtained in an infinitely deep, isovelocity ocean) in the frequency range 0 to 100 Hz was assumed. The simulated noise spectra are shown in Fig. 3 for a hydrophone (solid curve), a horizontal geophone (dashed curve) and a vertical geophone (dotted curve). The qualitative agreement with the experimental data in Fig. 2 is evident. Below 20 Hz the hydrophone level for the experimental data decreases more rapidly than predicted by the theory, but this is mainly due to the low-frequency cutoff of the hydrophone response.

By examining the wavenumber spectra of the synthetic results it can be demonstrated that above the cutoff frequency of the waveguide (8 Hz) the oscillations in the spectra are related to the cutoff of the various lower order modes. Further, the steep increase in the level of the theoretical results at very low frequencies is due to a significant excitation of seismic interface waves.

To remove these environmental effects from the experimental data and hence enable determination of the true source spectrum (defined as the acoustic pressure that would be obtained in an infinitely deep ocean from the same source distribution), the normalised synthetic spectra are

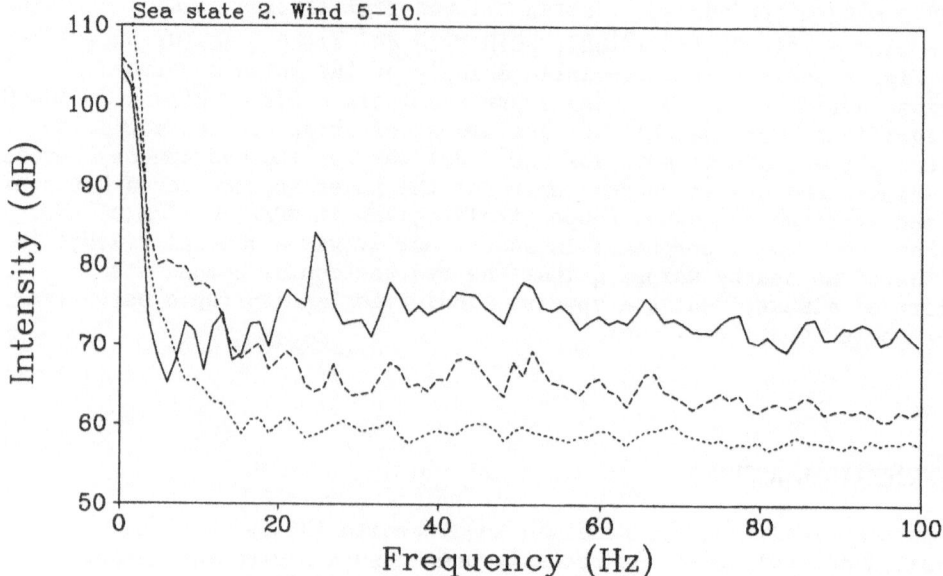

Fig. 2: Power spectra for ambient noise at sea state 2 as measured by hydrophone (solid curve), horizontal geophone (dashed curve) and vertical geophone (dotted curve).

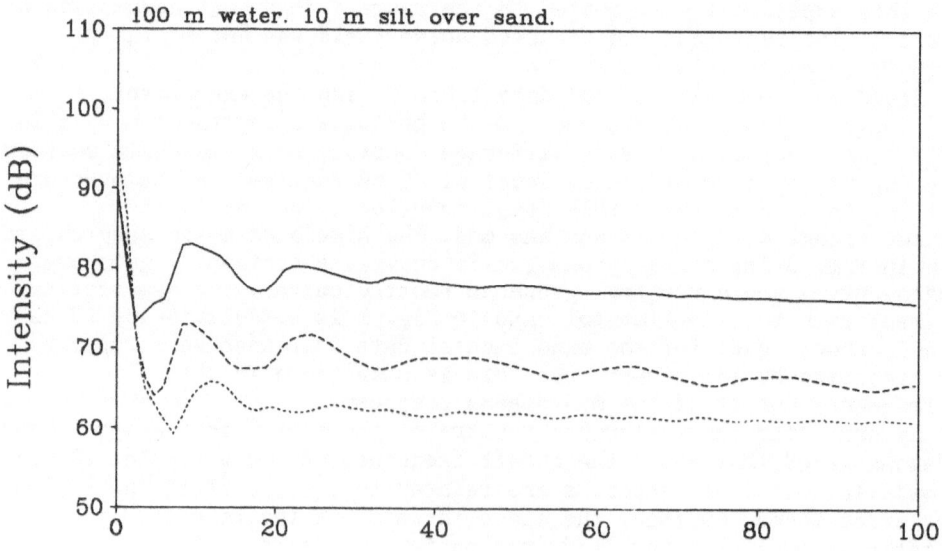

Fig. 3: Power spectra predicted by noise model for source level 70 dB as detected by a hydrophone (solid curve) a horizontal geophone (dashed curve) and a vertical geophone (dotted curve).

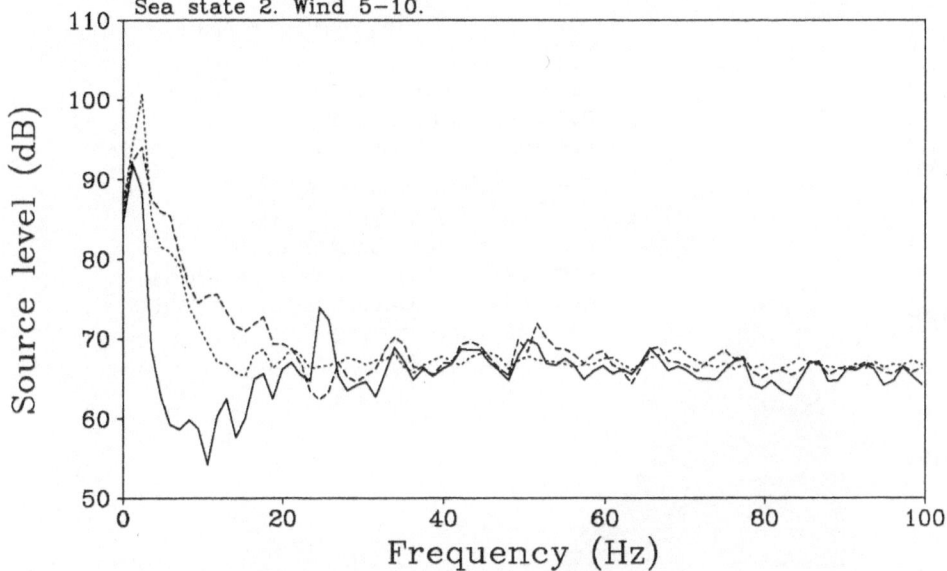

Fig. 4: Source spectra at sea state 2 obtained by eliminating propagation effects from hydrophone data (solid curve), horizontal geophone (dashed curve) and vertical geophone (dotted curve).

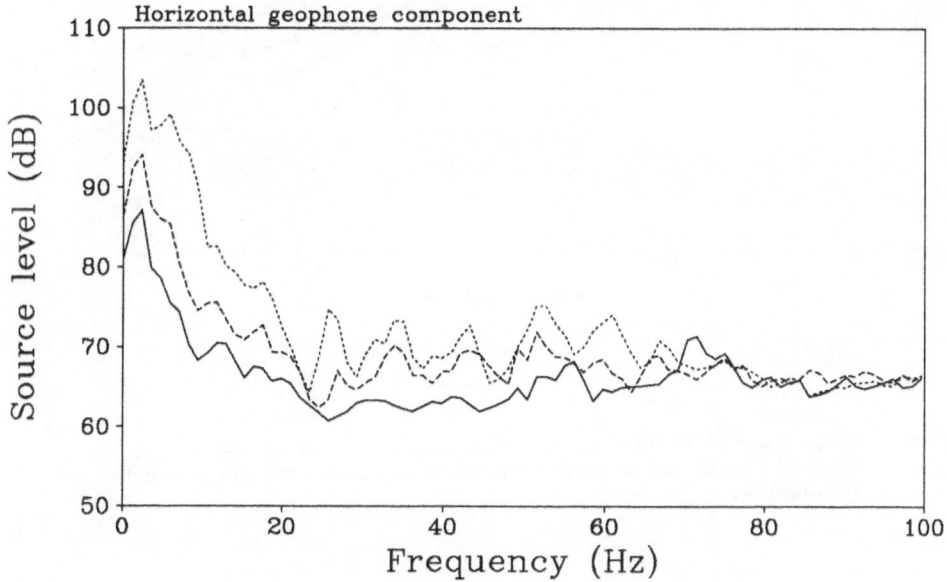

Fig. 5: Source spectra obtained from horizontal geophone ~ ta for sea levels 0 (solid curve), 2 (dashed curve) and 6 (dotted curve).

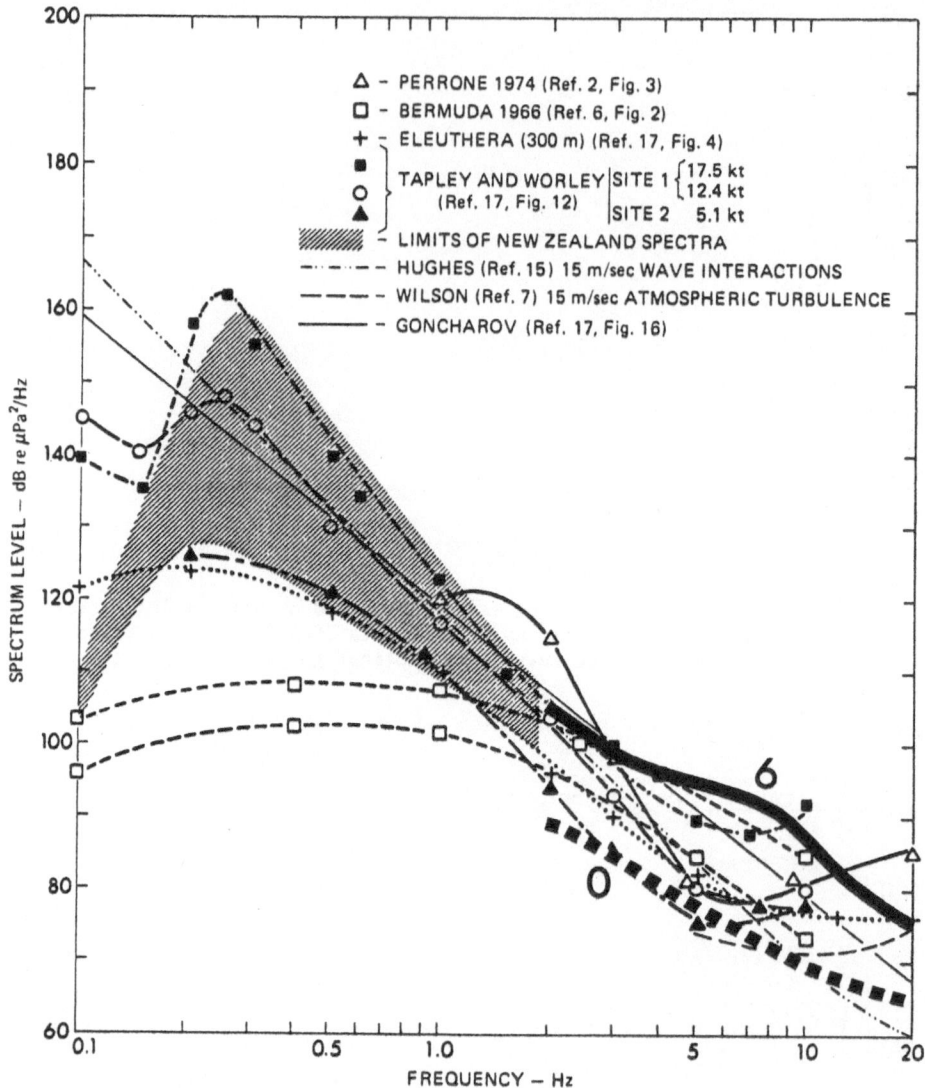

Fig. 6: Comparison of shallow water noise source levels at sea states 0 (dashed curve) and 6 (solid curve) with noise data presented by Kibblewhite [3].

subtracted from the experimental spectra with the result shown in Fig
Here the solid curve represents the source spectrum obtained from the
hydrophone data, whereas the dashed and dotted curves indicate the
spectra obtained from the horizontal and vertical geophone,
respectively. Except for the hydrophone below 20 Hz, the agreement
between the three source spectra is excellent in the entire frequency
range. This, in turn, indicates that the noise propagation model is
capable of accurately estimating the noise spectra on the different
sensors and also that the environmental model assumed is a good
approximation to the one existing in reality.

Fig. 5 shows the source levels obtained from a horizontal geophone
at three different sea states: 0 (solid curve), 2 (dashed curve) and 6
(dotted curve). The dependence on the sea state is evident, however with
decreasing sea state dependence with increasing frequency. Above 80 Hz
the sea state effect is negligible, probably because the noise field at
these frequencies is dominated by man-made noise. Unfortunately the sea
state spectrum is not available above a few Hz, so direct correlation
could not be determined.

The source spectra shown in Fig. 5 represent the fields that would
be obtained in an infinitely deep ocean with the same source
distributions. These spectra can therefore be directly compared to
measured data from deep water, where the most dramatic propagation
effects - those due to the excitation of seismic waves - only become
significant at extremely low frequencies. In fact the source levels
obtained by removing the propagation effects from the shallow water data
are in excellent agreement with the results presented by Kibblewhite et
al [3] as demonstrated in Fig. 6. Here the source levels obtained in the
frequency band 2 to 20 Hz at sea states 0 and 6 have been plotted on the
original illustration from [3] as thick dashed and solid curves,
respectively.

4. CONCLUSIONS

It has been demonstrated the the generalisation of the Kuperman-Ingenito
model to viscoelastic environments is capable of predicting the
hydrophone and geophone detected ambient noise in shallow water,
yielding the possibility of extracting the actual source spectra from
experimental data obtained in such an environment. The excellent
agreement between the obtained source spectra and deep water data
enables prediction of the low-frequency ambient noise in other
environments, based on measurement of sea states only.

REFERENCES

[1]   W.A. Kuperman, 'Propagation effects associated with ambient
       noise,' This volume.

[2]   T. Akal, A. Barbagelata, G. Guidi and M. Snoek, 'Time dependence
       of ambient seafloor noise on a continental shelf,' In: Ocean
       Seismo Acoustics, Eds. T. Akal and J.M. Berkson, Plenum Press
       (1986): 767-778.

[3]   A.C. Kibblewhite and K.C. Ewans, 'A study of Ocean and Seismic
       Noise,' In: Ocean Seismo Acoustics, Eds. T. Akal and J.M. Berkson,
       Plenum Press (1986): 731-741.

[4]   H. Schmidt and W.A. Kuperman, 'Estimation of surface noise source
       level from low-frequency seismo-acoustic ambient noise
       measurements,' J. Acoust. Soc. Am. (1987): submitted.

[5]   W.A. Kuperman and F. Ingenito, 'Spacial correlation of surface
       generated noise in a stratified ocean,' J. Acoust. Soc. Am. 67
       (1980): 1988-1996.

[6]   H. Schmidt, 'SAFARI. Seismo-Acoustic Fast field Algorithm for
       Range Independent environments,' SACLANT ASW Research Centre, La
       Spezia, Italy, Rep. SR-113 (1987).

THE EFFECT OF PROPAGATION CONDITIONS ON WIND-GENERATED NOISE
AT REAL SHALLOW WATER SITES

Rachel M. Hamson
SACLANT ASW Research Centre
I-19026 La Spezia
Italy

ABSTRACT. A theoretical model for the propagation in shallow
water of noise generated by an infinite surface layer of
sources is used to investigate the variability of wind noise
measurements at two shallow water sites in the Mediterranean
and the Baltic. The noise at a receiver is described by two
components; the discrete normal modes of the particular
shallow water channel, and the continuous spectrum. The
extent of the mode contribution is controlled by the
environmental conditions (bottom type, sound speed profile),
and in hard bottom conditions these can dominate the noise.
The model can account for the differences in measured levels
between the two sites for frequencies between 400 - 3200 Hz
and wind speeds from 5 - 20 kn, provided certain assumptions
are made regarding bottom types. The model is further used to
predict the directional response of a vertical array of
hydrophones at a third site where measurements have recently
been carried out. Reasonable model/measurement comparisons
are obtained if the sources are assumed to radiate with 'cos
α' directionality.

1. INTRODUCTION

      In the absence of a well-established physical model for
the generation of wind noise at the ocean surface, acoustic
modelling studies of the noise at a deep receiver have
usually been based on the postulate that noise generation can
be represented by an infinite sheet of point sources located
at or near the surface. The way in which these sources
radiate sound can be described either by a directionality
function or by a source correlation function. Use of such a
source model with an acoustic propagation model leads to the
prediction of the underwater noise (relative to a unit source
level) and its characteristics (depth dependence,
directionality) at a receiver in a specified position. Thus
the propagation model provides the link between the source

B. R. Kerman (ed.), Sea Surface Sound, 281–293.

field and actual noise measurements in the ocean. Combination of careful measurements (as a function of frequency and wind conditions) and known propagation effects can result in estimates of the unknown source parameters, i.e. the source levels and directivities, which in turn may be utilized in identifying a particular physical mechanism as being the main source of sound.

This source sheet model has been used in the past by Cron and Sherman [1] to derive horizontal and vertical correlation coefficients between deep receivers in an idealised deep water environment. More recently Kuperman and Ingenito [2] incorporated the source model into a wave theory derivation of the acoustic field in a shallow water channel. The present paper makes use of this latter work to investigate the effects of propagation conditions (sound speed profile, bottom type) on noise level and vertical array responses, and to account for the variability of wind noise measured at two shallow water sites for frequencies between a few hundred Hertz and a few thousand Hertz.

2. MODEL DESCRIPTION

An infinite layer of monopole sources is located just below the ocean surface, as indicated in Fig 1, the sources having specified correlation functions and unit strength. It has been shown [Liggett and Jacobsen,3] that this is an equivalent model to considering a layer of directional sources coinciding with the ocean surface, when the source correlation functions and source directionalities have particular relationships. We consider here source pressure directionalities of the form $\cos^m \alpha$ where α is measured from the downward vertical (see Fig 1) and $m \geq 1$. The m=1 case corresponds to a dipole at the surface.

The noise field generated in the water column is determined by the solution of the wave equation for specified environmental conditions. The full solution for a single source is composed of a set of discrete normal modes, which arise from interactions with the bottom at grazing angles less than the critical angle, and the continuous modal spectrum caused by direct paths and bottom interaction at steeper angles than critical. Both of these field components are important in modelling the noise in shallow water as we shall see later. (In an idealised infinitely deep ocean the continuous component provides the total noise as no discrete modes exist.)

A full mathematical treatment for the derivation of the complex spatial correlation function between two points in the water column is given in references [2,4] and a computer model based on this theory is in use at SACLANTCEN. The discrete mode computation is provided by the Centre's normal mode propagation model, SNAP, and the continuous field by a Fast Field Program (FFP). Both these models contain a three-layer environment : the water column, the sediment layer and the remaining sea bed extending to infinity. The

SPATIAL CORRELATION MODEL FOR STRATIFIED OCEAN
(KUPERMAN/INGENITO)

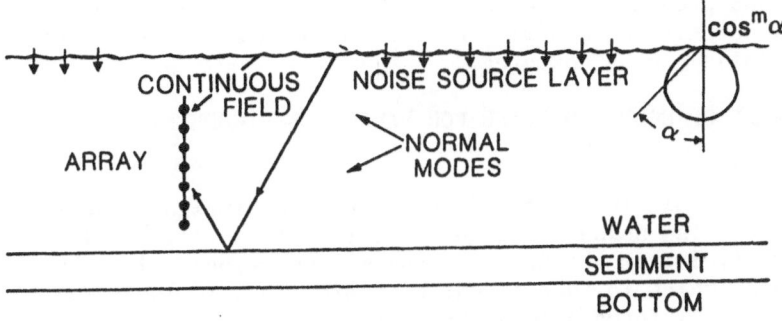

Fig.1. The shallow water wind noise model.

DISCRETE MODE COMPONENT – ISOVELOCITY/m=1

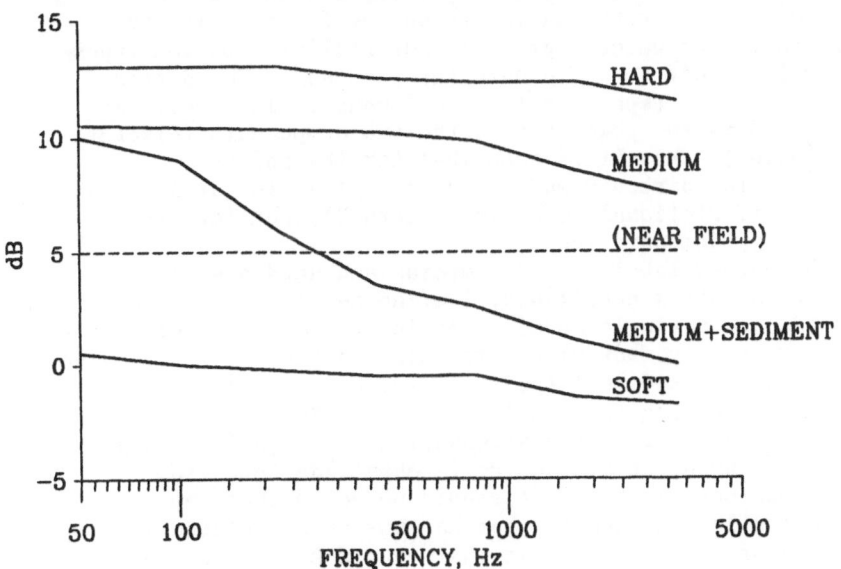

Fig.2. Discrete mode component vs. frequency for
various bottom types.

speed of sound can vary with depth in the first two layers.
Appropriate densities and attenuation coefficients are
required for the two bottom layers and shear losses can be
included in the sub - bottom. The noise model uses the inputs
from the two propagation models, forms the spatial
correlation function integrated over the infinite surface of
sources and produces noise level vs depth, horizontal and
vertical correlation coefficients, from which array responses
to wind noise can be computed.

3. EXAMPLES OF MODELLED NOISE FOR VARIOUS ENVIRONMENTAL CONDITIONS.

In an ideal, infinitely deep ocean with straight line
propagation, the noise level relative to unit source level
can be derived from the Cron and Sherman equations [1] to be
simply π/m, which is independent of depth. Thus 10 log (π/m)
(= 5dB for m=1) forms a lower bound for the noise expected in
shallow water. Studies with the model have shown that the
continuous field in shallow water is only weakly dependent on
the particular conditions and is generally 1 or 2dB above the
10 log (π/m) level. The extra noise in shallow water is
provided by the discrete modes, the contribution of these
being controlled by the bottom type, sound speed profile and
the parameter m. In hard bottom conditions this component can
produce up to 13 dB of noise. Fig 2 shows a plot of the
discrete mode contribution vs. frequency for isovelocity
conditions, 100m water depth,m=1, for realistic bottom types
varying from soft (mud) to hard (coarse sand). The dotted
line is the 5 dB (approximate) continuous field level. The
total field is the power sum of the two components (discrete
+ continuous). Thus it is seen that for the softest
conditions the discrete modes may be negligible (at least as
far as omnidirectional level is concerned), but for hard
bottoms they dominate the noise.
    The curves labelled soft, medium and hard are all for
homogeneous bottom conditions, i.e. no sediment layer is
included, and a single sound speed in the bottom is used.
These curves only show slight frequency dependence. Inclusion
of an absorbent sediment (medium + sediment),6m thick in this
case, corresponding to a real area in the Mediterranean, has
the effect of considerably attenuating the high frequency
noise, e.g the noise at 1000 Hz is about 7dB below the
homogeneous bottom level.(It should be noted that one
propagation loss mechanism that has not been included in the
computations of this paper is that caused by surface and
bottom roughness. In high wind conditions surface roughness
almost certainly has some effect, but the present model does
not handle fully such scattering losses in a broadband source
situation. These effects require further study.)
    The discrete mode component of the field also causes the
depth dependence of the noise if any exists. In some
conditions e.g.a winter sound speed profile, this effect is
not large, maybe 1-2 dB variation over the water column, but

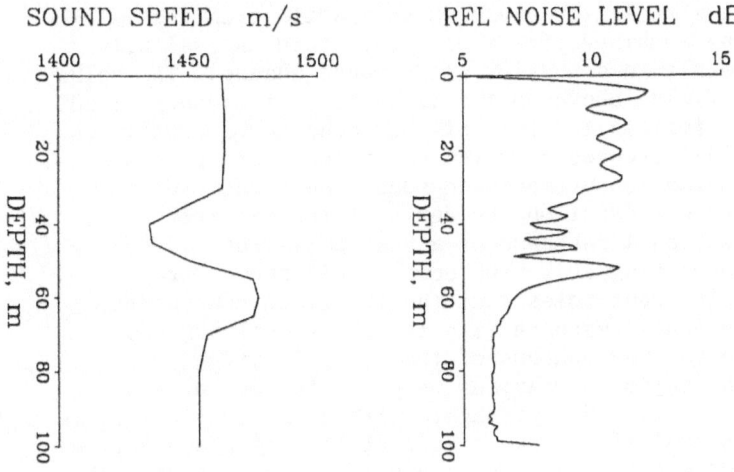

Fig.3. A sound speed profile from a site in the Baltic sea on the left, and the corresponding 800 Hz modelled noise level (rel. unit source level) vs. depth, on the right.

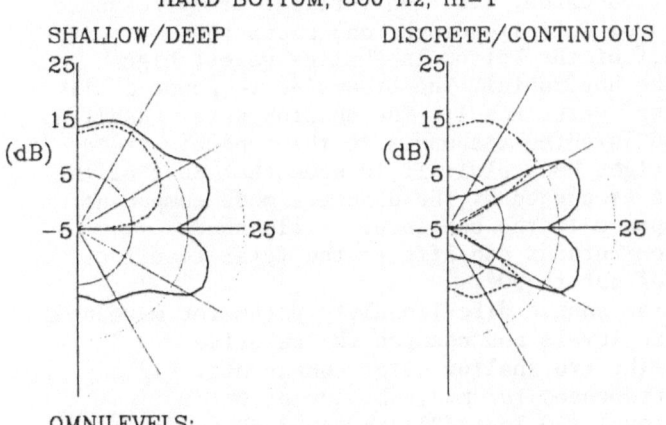

OMNILEVELS:

SHALLOW = 13.1 dB

DEEP = 5 dB

DISCRETE = 12.1 dB

CONTINUOUS = 5.8 dB

Fig.4. Modelled vertical array responses for the parameter m=1 plotted relative to the shallow water omnidirectional level (13.1 dB).

in extreme profiles it can be significant. Fig.3 shows an
example using a realistic sound speed profile from the Baltic
sea, where the combined effect of temperature and salinity
variations result in a shallow water sound channel. The noise
level (total field) shown on the right for a frequency of 800
Hz (with m=1) follows the form of the sound speed profile on
the left and differences over the water depth of up to 6 dB
occur. Wind noise measurements in such conditions are
therefore very dependent on the depth of the receiver.

The directional response of a vertical array of
hydrophones operating in a wind noise field shows more
clearly the different roles that the two field components
play. A directional response is obtained by applying phase
weightings to the hydrophones of the array in order to
'steer' the direction of maximum response in the vertical
plane. The discrete modes propagate within the critical angle
defined by the relative sound speeds of the water and bottom
(typically 20° - 40°); the continuous component contains the
steeper surface/bottom interactions and arrives at the array
from near vertical directions. Fig 4 shows an example for a
31 element array in isovelocity/hard bottom conditions for
the parameter m = 1 and a frequency of 800 Hz. The two plots
of Fig.4 are of vertical array output in dB vs. steering
angle from θ = -90° looking downwards to the bottom, through
broadside (θ = 0°) to θ = 90° looking upwards to the surface.
The array output has been normalised to the omnidirectional
level for a unit source level. The left hand plot shows the
shallow water result (solid curve) with the Cron and Sherman
result as the dotted curve, i.e. the vertical array response
in an ideal ocean of infinite depth, and shows the
significant effect of the bottom in shallow water; high
response below the horizontal. The broadside response is 11
dB above the 'deep' water result. The shallow water result
can be decomposed into the responses to the separate fields
as shown on the right hand plot. It is seen that most of the
increase in noise is caused by the discrete mode component,
although the response to the continuous field shows
significant bottom returns and affects the total result for
steering angles of 35° to 90°.

Increasing the source directionality parameter m reduces
the received noise levels and changes the relative
contributions of the two shallow water components. Fig 5
shows the array responses for m=2, the 'deep' water
omnidirectional level ($10 \log(\pi/2)$) is now 2 dB and the
shallow water level 4.7 dB. This reduced noise in shallow
water is mainly caused by high losses in the mode component.
The lower order modes which propagate close to the horizontal
are particularly attenuated (see [4]), hence the deep null at
the horizontal.

Use of a vertical array then provides a method of
identifying the two components and measuring their separate
contributions. The array response pattern contains a lot more
information than a noise level, features such as depth of
broadside null, mode peaks and surface/bottom responses are

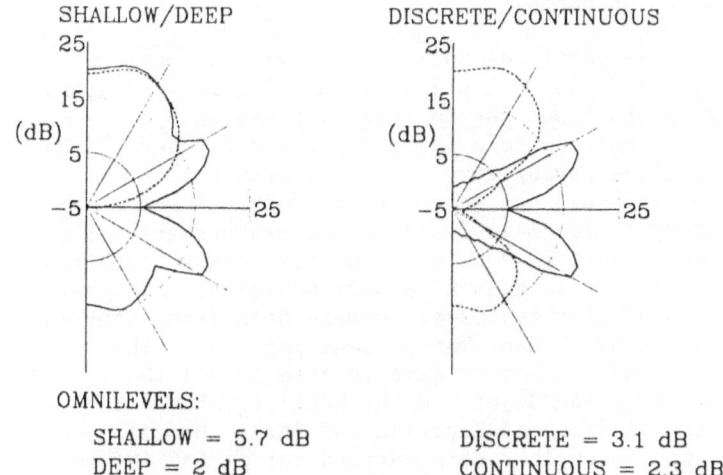

HARD BOTTOM, 800 Hz, m=2

SHALLOW/DEEP   DISCRETE/CONTINUOUS

OMNILEVELS:

SHALLOW = 5.7 dB   DISCRETE = 3.1 dB
DEEP = 2 dB   CONTINUOUS = 2.3 dB

Fig.5. Modelled vertical array responses for the parameter m=2 plotted relative to the shallow water omnidirectional level (5.7 dB).

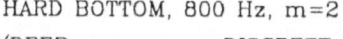

MEASURED NOISE LEVELS

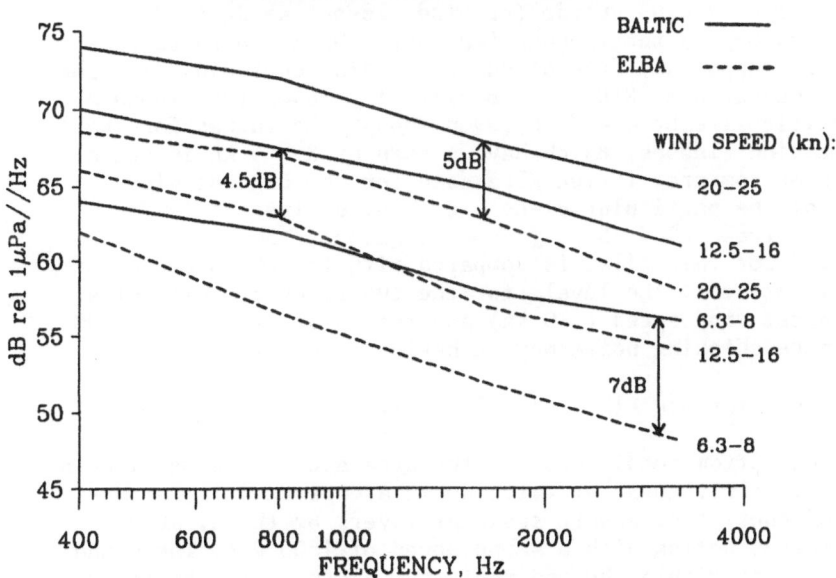

Fig.6. Measured wind noise spectra from a site in the Baltic sea and a site in the Mediterranean (near Elba) for three wind speed classes.

affected by both the environmental and the source parameters.

4. WIND NOISE MEASUREMENTS.

Reported measurements of noise due to wind at shallow water sites (water depth $\leq$ 200m) as a function of wind speed and frequency can be found in the following papers: Piggott[5], Kuperman and Ferla [6] and Wille and Geyer [7]. These refer to sites having very different propagation conditions (bottom types and sound speed profiles) and a variability of up to 10 dB is found in the measurements. The Piggott data were taken at a hard bottom site off the Scotian shelf, the Kuperman-Ferla data at a soft bottom site in the Mediterranean, and Wille and Geyer compare data from sites in the Baltic and the North Sea. The present paper uses the model described above to investigate in more detail the Baltic data of Wille and Geyer and the Mediterrenean data (near the island of Elba) of Kuperman and Ferla. Modelling studies of transmission loss data carried out at SACLANTCEN for these two areas have established sets of bottom parameters, which, together with sound speed profiles taken during the noise measurements, enable us to run the model (for a particular m) and compare the differences in the results for the two sites with differences exhibited in the measured data.

Fig 6 shows the measured data for these two sites for three selected wind speeds (or wind classes as defined by Wille and Geyer) for frequencies above 400 Hz where both data sets are supposedly free of shipping noise contamination. The levels measured at Elba are consistently lower than those at the Baltic site by 4 - 7 dB (even though the values for the Baltic wind classes, which have a span of 3 - 5 kn in speed, have been compared to the Elba data for the highest wind speed of the particular class - e.g. wind class 15 is for speeds between 20 - 25 kn grouped together, the Baltic noise measured for this class is compared with the 25 kn curve for Elba.) Note that the levels for the two sites are closest at the lowest wind speed (10 kn) and the lowest frequency, 400 Hz, where shipping noise may be having some effect.

4.1 Models for the Elba and Baltic sites.

The bottom conditions for the area south of Elba along a track of fairly constant water depth are given in reference [8] and consist of a soft sediment layer, 6m thick, above a homogeneous bottom with a sound speed of 1600 m/s. The sound speed varies within the sediment, but at the water-bottom interface it is very low (1470 m/s), significantly lower than the speed in the water at the bottom (1502 m/s). The sound speed profile taken during the wind noise measurements in November 1980 is shown in Fig.7. The receiving positon was in a sloping area about 15 km from the mainland. Therefore isotropic sound propagation and wind fields would not be expected. A string of hydrophones between depths of 20m and

E.BALTIC (OCT) / S.ELBA (NOV)

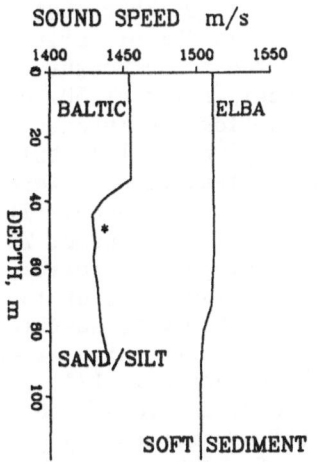

Fig.7. Sound speed profiles used in the model for the Baltic and Elba sites.

MODELLED NOISE LEVELS REL. UNIT S.L. (m=1)

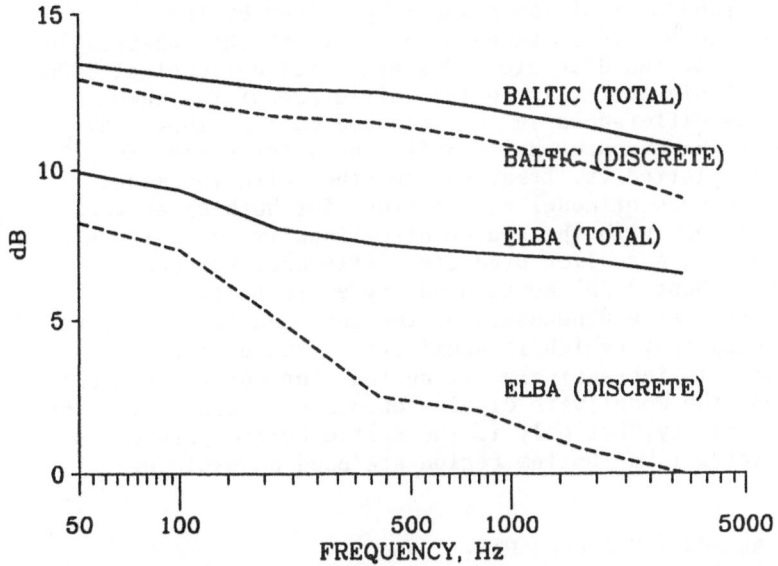

Fig.8. Modelled noise spectra (rel. unit source level) for the two sites.

the bottom were used and the average noise level reported.
Little depth dependence was found in the data or predicted by
the model.

The measurements in the Baltic in October 1981 were at a
site in the Eastern area of the sea in 90m of water [7]. The
sound speed profiles taken during the noise measurements are
given in Fig.7, and the single receiver was at 50m depth,
below the thermocline. Transmission loss measurements along a
track going west from the site are given in [7], but no
bottom data directly related to the measurements were
available. Instead bottom parameters from a model fit to
transmission loss data about 40 km east were used. This fit
in a sloping area used a 1600 m/s homogeneous bottom, which,
together with the low sound speed in the water of 1440 m/s,
results in fairly high reflectivity. However, near the site
the bottom type changes, giving rise to the high attenuation
in the western direction which is reported in [7].

Thus both the Elba and Baltic sites are subject to
azimuthal changes in environmental parameters and possibly in
source distribution, the effects of which are not included in
the present stratified noise model. The above sets of
parameters were used to run the noise model for the two sites
and obtain noise levels (relative to unit source level) vs.
frequency at the appropriate receiver depths. The results are
plotted for the total noise and the discrete mode components
in Fig.8 using the parameter m=1. Comparing the Baltic
(total) curve with the Elba (total) curve we see a difference
in predicted noise levels of about 5 dB over a wide frequency
range. For Elba most of the noise is provided by the
continuous component for frequencies above 500 Hz, whereas in
the Baltic case the discrete modes are still dominant at 3000
Hz. The 5 dB higher levels in the Baltic are of the same
order as the differences in the measured data as shown in
Fig.9. Here the measured noise differences for seven wind
classes are plotted vs. frequency together with the model
results. The sets of model computations for both sites were
also carried out with the source directionalty parameter m=2.
These result in a smaller predicted difference between the
two sites (about 2 dB) as seen on Fig.9. In fact, as m
increases, the site dependence of the noise decreases, since
the mode component (which is sensitive to the particular
environment) is increasingly attenuated. For our two present
sites then, the model with dipole sources can account for the
noise variability, but only if the Baltic bottom parameters
for the 'reflective' bottom region are used as explained
above.

5. VERTICAL ARRAY MEASUREMENTS.

Measurements of wind noise with a vertical array were
carried out by SACLANTCEN in October 1985 at a shallow water
site in the Mediterranean off the west coast of Sardinia. No
propagation measurements or modelling has previously been
done in this area and only very general information was

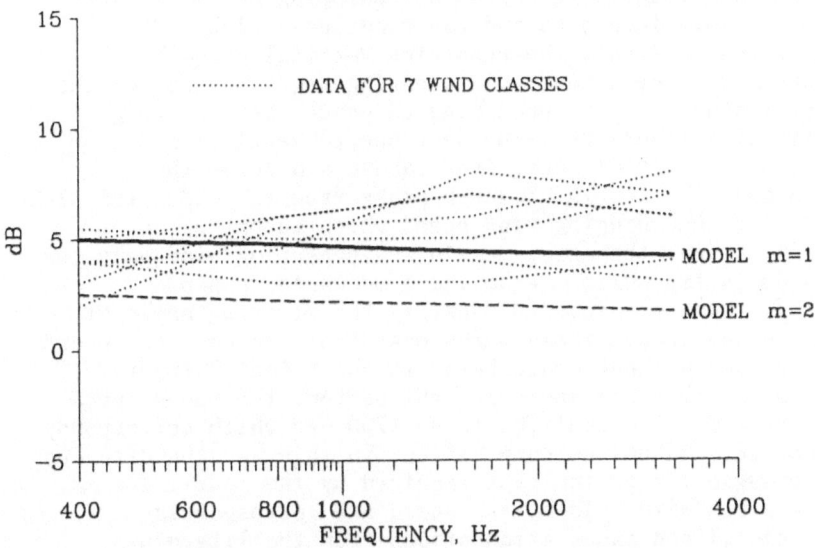

Fig.9. Differences in noise spectra (Baltic-Elba) from the measurements and the model.

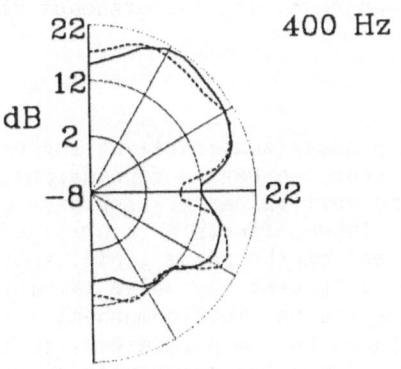

Fig.10. Vertical array response to wind noise at a site near Sardinia; solid curve from measurements, dotted curve from the model.

available on the bottom characteristics. Thus no attempt has
been made to model the noise differences between this and the
previous two sites. An array of 32 hydrophones measured the
noise up to a frequency of 1500 Hz. Shipping noise was very
low in the immediate area and the wind speed high and
variable (up to 22kn). The resulting vertical array
responses, i.e. beam power vs. steering angle, exhibited the
general features of the predicted response pattern (e.g. Fig
4), with high levels from the near-upward vertical and from
the two discrete mode directions (above and below the
horizontal). The levels of these peaks from the mode arrivals
varied with the changing wind speed which supports the
supposition that the array was indeed measuring wind only (or
wind - dominated noise). From these measured response
patterns it was possible to identify the steering angle at
which the continuous field takes over from the mode component
in being the dominant contributor to the noise. Using this
angle as the critical angle for the bottom, the sound speed
in the bottom was calculated to be 1700 m/s which corresponds
to a coarse sand bottom composition. In this way the data was
used to suggest a bottom type required by the model. The
model was run with this bottom speed and corresponding
compressional and shear attenuations from the literature,
together with the sound speed profile measured at the site.
The resulting array response prediction is plotted with the
measured pattern in Fig.10, and we see that the two match
reasonably well not only in critical angle, but in other
features such as relative levels of modal peaks, responses
from vertically up and down, and the presence of a null at
broadside.

6. CONCLUSIONS

In conclusion, propagation modelling suggests that
environmental conditions can have a considerable influence on
wind noise levels and vertical array responses in shallow
water. The extent of these effects is controlled by the
discrete mode component of the noise field, which can cause
an increase of up to 8 dB over the noise level expected in an
ideal, infinitely deep ocean. This component is very
sensitive to the assumed bottom parameters, therefore these
need to be well-established for noise modelling at real
sites. The differences between measured wind noise spectra at
two shallow water sites in the Mediterranean and the Baltic
have been investigated by a model which includes both the
discrete mode component and the continuous field. Although
the model with dipole sources can account for these
differences, its assumptions of a stratified environment and
an isotropic source distribution may not be justified at
these sites. The noise at both sites is probably influenced
by azimuthal changes in bottom type and water depth and by
non-isotropic wind fields. Therefore a conclusive
verification of noise dependence on propagation effects would
require measurements in isotropic wind conditions at sites

with known propagation conditions in all directions.

Vertical array data taken at a third site in high wind/low shipping conditions indicates the presence of mode arrivals at the array. The directional response agrees reasonably well with that obtained by modelling using estimates of the parameters required to describe the bottom type.

The implications of this study on the investigation of the wind noise source mechanism are that measurement sites with well-known environmental conditions should be chosen so that propagation effects can be eliminated by modelling. Also vertical array measurements are considerably more valuable than omnidirectional levels as they allow the separation of the two components of the noise field, and the use of both to identify source parameters.

REFERENCES

1. B.F.Cron and C.H.Sherman,'Spatial correlation functions for various noise models',J.Acoust.Soc.Am.,34, 1732-1736(1962).
2. W.A.Kuperman and F.Ingenito,'Spatial correlation of surface generated noise in a stratified ocean', J.Acoust.Soc.Am.,67,1988-1996(1980).
3. W.S.Liggett and M.J.Jacobsen,'Covariance of surface generated noise in a deep ocean', J.Acoust.Soc.Am.,38,303-312(1965).
4. R.M.Hamson,'The theoretical responses of vertical and horizontal line arrays to wind-induced noise in shallow water', J.Acoust.Soc.Am.,78,1702-1712(1985).
5. C.L.Piggott,'Ambient sea noise at low frequencies in shallow water of the Scotian shelf', J.Acoust.Soc.Am.,36,2152-2163(1965).
6. W.A.Kuperman and M.C.Ferla,'A shallow water experiment to determine the source spectrum level of wind-generated noise', J.Acoust.Soc.Am.,77,2067-2073(1985).
7. P.C.Wille and D.Geyer,'Measurements on the origin of wind-dependent ambient noise variability in shallow water', J.Acoust.Soc.Am.,75,173-185(1984).
8. F.B.Jensen,'Comparison of transmission loss data for different shallow water areas with theoretical results provided by a three-fluid normal mode propagation model',La Spezia,Italy,SACLANT ASW Research Centre, SACLANTCEN CP 14(1974).

SIMULTANEOUS MEASUREMENTS OF SURFACE GENERATED NOISE AND ATTENUATION AT THE FIXED ACOUSTIC SHALLOW WATER RANGE "NORDSEE"

P.C. Wille and D. Geyer
Forschungsanstalt der Bundeswehr
für Wasserschall- und Geophysik
Klausdorfer Weg 2-24
D-2300 Kiel 14
Germany

ABSTRACT. The influence of air bubble induced sound attenuation on surface generated noise has been investigated experimentally by long term statistics and by a wind jump from breeze to storm. The attenuation due to suspended air, increasing by the n'th power of the wind speed (n between 3 1/2 and 4 1/2) and by about the first power of the frequency, rises to higher values than by any other mechanism, eventually screening the high frequency noise beyond about 10 kHz under storm conditions. However, at frequencies up to several kHz, the bubble suspensions remain transparent enough to leave the second power relation between noise intensity and wind speed up to the hurricane regime unimpaired. The strong high frequency loss transient induced by a rain/hail shower may be also due to air bubbles. The measurements which include propagation conditions of lowest loss confirm that the surface generated shallow water noise depends on the local agitation only, unaffected by remote irradiation, at least beyond about 0.5 kHz. Even at an approaching storm front, a preceding noise increase at frequencies of lowest loss could not be observed; instead the noise increase of the lowest octave measured (25-50 Hz) appears after the passage of the front overhead of the receiver as well as the high frequency noise screening.

1. INTRODUCTION

Three aspects of the influence of attenuation on noise, both generated by sea surface agitation are investigated for the shallow water case, namely the relation of local and remote noise contributions to the gross irradiation, the chronological features of noise and loss during a transient "switch on wind" event of a meteorological front, and passive attenuation properties of air bubble formations, possibly relevant to their active features as noise sources.

The role of the site dependence of the received noise which is essentially a propagation path dependence appears still controversal for the shallow water case [1-4]. A comparison of measured data of different origin and their reliable modelling suffers from uncertainties of the environmental conditions concerning the transmission path

B R. Kerman (ed.), Sea Surface Sound, 295-308.
© 1988 by Kluwer Academic Publishers.

as well as the wind field and the sea surface. Even calibration problems can not be excluded in all cases. Particularly noteworthy are differences of the actual wind stress at the sea surface due to different wind profile types between anemometer height and surface. They are often consistant through monthly averages and within a large wind speed regime [3] and lead to noise spectrum level differences of the same order as attributed to a site dependence.

A decisive experiment to bypass these difficulties should compare the received noise at a fixed position within the entire range of transmission loss occuring in shallow water, simultaneously measured with the same setup. The required transmission loss variability can be achieved without involving a varying sound speed stratification which is often strongly range and time dependent [5]. Measurements of the sea surface induced attenuation at the fixed acoustic shallow water range NORDSEE have revealed a steep loss increase by the wind [6] attributed to bubble clouds, providing a large transmission loss interval at a narrow wind class such as used for parameterisation of surface generated noise.

The behavior of noise and loss developing during a meteorological front event as well as the properties of the bubble induced attenuation determined by independent methods may further contribute to discussions on a possibly varying predominance of bubble suspensions to generate and to dissipate noise.

2. SPECIFICATIONS OF THE EXPERIMENT

The noise and attenuation measuring system (fig. 1) utilises the same hydrophones for both purposes. Source and receiver are positioned below the depth range of bubble clouds to avoid complications due to changing impedance. The configuration yields loss figures directly comparable

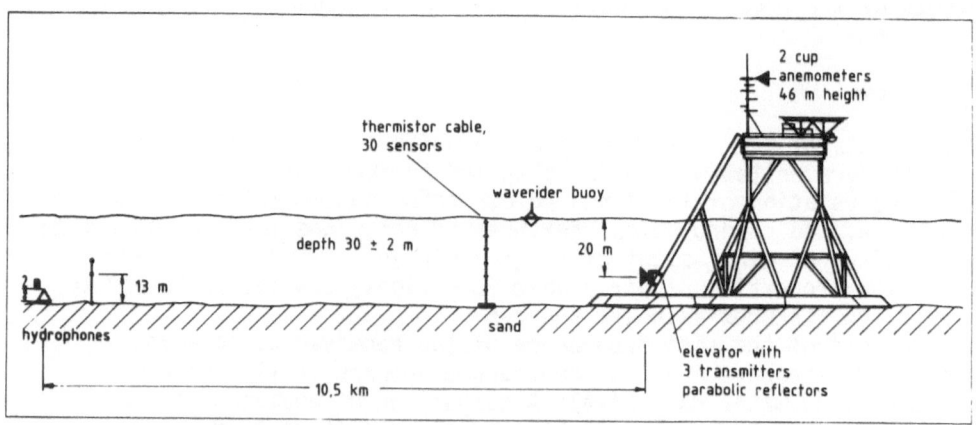

Fig. 1   Simplified sketch of the NORDSEE platform shallow water range for acoustic and environmental measurements.

to the usually measured shallow water transmission loss, produced by multiple bottom and surface bounces, such as the noise emission under-goes when propagating from a distance.

The attenuation within the bubble clouds has been estimated from the range measurements through model inversion [7] for a substitute homogeneous subsurface bubble layer and is compared in sect. 5 to local loss excess measurements close to the surface at the platform tower [8]. The minimum attenuation (isothermal water, nearly calm sea) as well as the geometrical spreading loss was determined separately by explosives along the range.

Components of measuring device
- Noise: 1 hydrophone, omnidirectional, tripod mounted, 1.8 m above sea bottom. 1 hydrophone, omnidirectional, cable mounted, 13 m above sea bottom; buoyant spheroid 5 m on top of measuring hydrophone. Both receivers are cable connected to the platform
- Transmission loss: 2 separate hydrophones, same as for noise measure-ments. 3 transmitters with parabolic reflectors, beamwidths between 3 dB points: 35° at 1 kHz, 13° at 3 kHz, 6° at 8 kHz. Source system elevator 20 m below surface, sources looking towards receiver.
  - Signals: Third octave PR noise centered at 1, 3.2 and 8 kHz.
  - Time schedule of signals: 10 min adjacent intervals of transmission loss and noise, recorded as 32 s average samples.
- Environment monitoring (winter conditions, isothermal water):
  - Wind speed and direction by 2 cup anemometers 46 m above sea surface
  - Wave height by wave rider buoy
  - Temperature profile by vertical thermistor cable, 30 elements.

3.   TRANSMISSION LOSS DEPENDENCE OF SURFACE GENERATED NOISE

Examples of the transmission loss variability from calm to storm are depicted in figs. 2 and 3 [6]. At low wind speeds the loss increase can be attributed to rough boundary scattering [7]. The bubble induced at-tenuation dominates after a threshold-like transition, eventually lea-ding to loss values higher than known for shallow water areas of poor bottom reflectivity [9]. The entire spread of the transmission loss within a single wind class - indicated by the bars of figures 2 and 3 - is a sufficient basis to prove the significance of distant noise contri-butions. The particular wind classes correspond to noise level steps of 2 dB, as far as the second power relation between noise intensity and wind speed is valid [3]. The transmission loss variability of the range (fig. 4) includes the lowest loss measured elsewhere in shallow waters of the North Sea and therefore the largest noise intercept area [11]. The low minimum transmission loss near the platform is due to the sand bottom in connection with the shallow depth with early transition to cylindrical spread.

The refractive propagation of the Baltic Sea underbids the attenua-tion coefficients of the regular reflective North Sea type propagation (fig. 5), it is true, but only for sources at channel axis. The only re-maining situation, where the entire transmission loss for surface

sources may be lower than the variability of fig. 4 is in the Baltic
winter at calm sea which is a refracted-surface reflected case. Then
the transmission loss at 10 km may be lower by 2 and 5 dB at 3 kHz and
8 kHz respectively but not lower at 1 kHz.

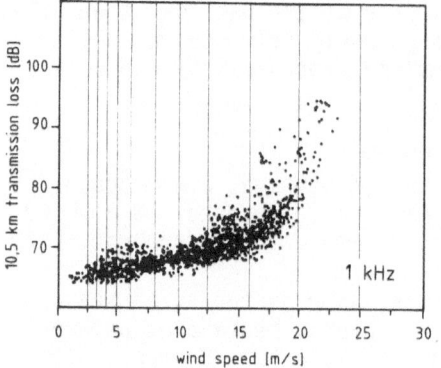 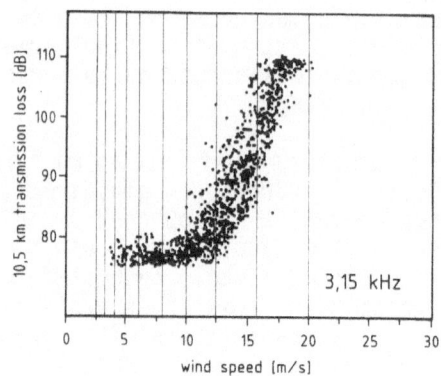

Fig. 2   10.5 km range transmission loss at 1 kHz versus
wind speed. Each dot represents
a 10 min-average.

Fig. 3   Transmission loss at
3.15 kHz versus wind speed.
(Light clipping at lowest loss;
noise limited at highest loss).

The relation between the received noise at a particular wind class
and the simultaneously measured transmission loss is depicted in
figures 6 and 7 for three frequency bands, 2 receiver depths and selected
wind classes. To achieve the large loss spread, more recent measurements than those of figs. 2 and 3 are incorporated.

Obviously the noise spectrum level does not decrease at increasing
loss but exhibits at most a positive trend. This result, relying on
several thousand 10 min-average levels confirms the surface generated
shallow water noise to be determined by the local agitation only, essentially unaffected by contributions propagating from a distant, at least
beyond 1 kHz. This is in accordance with the findings of [1,3].

The reason for the rising noise level of about 2 dB is apparently
the finite class width of the wind speed as shown by fig. 7 where the
class is subdivided into four intervals of equal width: the lowest
noise levels occur more frequently on the left side, the largest values
predominate on the right side. Part of the data, in particular of fig. 6
are contaminated by traffic noise.

However, there is a site dependence not covered by the measurements.
This is due to different bottom reflectivity enhancing the gross noise
level by multiple vertical surface and bottom reflections, amounting to
a theoretical difference of 1-1.5 dB between sand and mud [3].

It remains remarkable that the large variability of the loss which
is also an indicator of the average bubble population density (sect. 5)
has such a little effect on the received noise. If the bubbles still

acted as sound sources in their latest stage as cloud formations, then the received gross noise can only be kept constant if the additional noise production by an increasing cloud density is just compensated by an equivalent absorption. This appears rather unlikely since the direct irradiation from overhead is essentially unaffected up to high wind speeds and moderate frequencies (sect. 5).

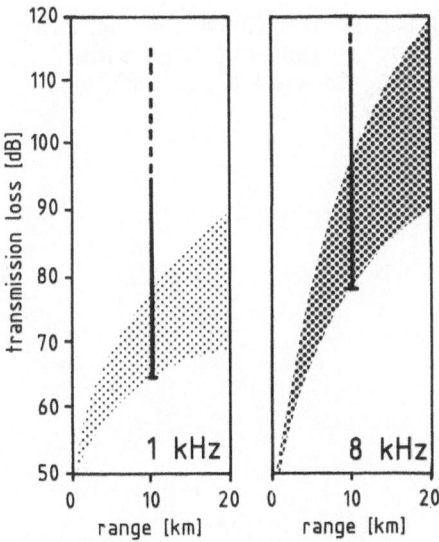

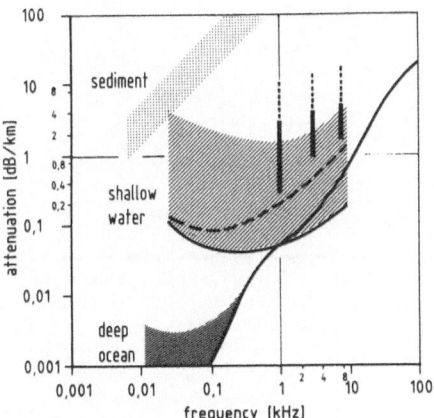

Fig. 4 Transmission loss variability, according to shipborn measurements in shallow waters of the North Sea (shaded area). Bars: Platform range measurements up to high wind speeds. Minimum loss at 3 kHz: 72 dB

Fig. 5 Variability of three attenuation coefficients at the platform range (bars) in comparison with other shallow waters. Below broken line: Baltic Sea.

4. NOISE AND LOSS DEVELOPMENT AT A WIND JUMP

The wind history of a steep meteorological front selected as a "switch on wind" event is depicted together with the transmission loss within the adjacent hours in fig. 8. The critical section is shown at enlarged time scale (resolution: 32 s averaging time) in fig. 9. The wind jump from 6 m/s to 20 m/s, preceded by a reversal of the wind direction arrives at the platform at about 6 h, having passed the receiving hydrophones 10-15 min before. The front propagates opposite to the direction of sound propagation along the range (meteorological analysis of the front direction and speed by Deutsches Seewetteramt, G. Rosenhagen).

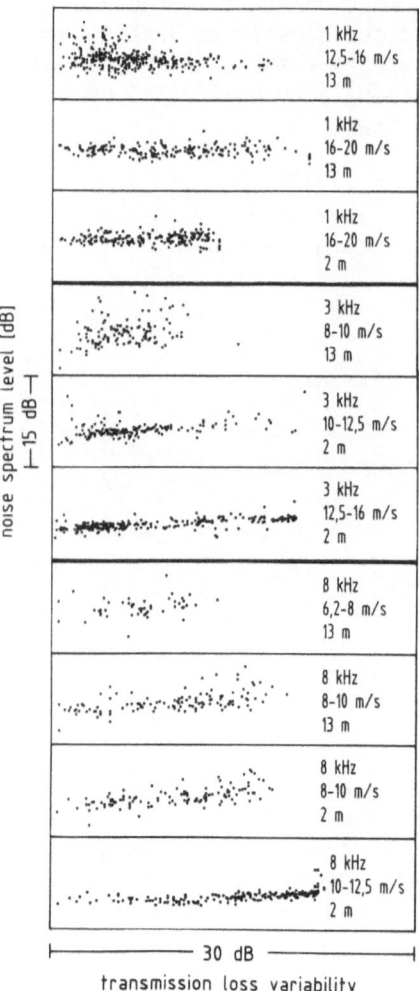

noise spectrum level [dB]

— 15 dB —

| 1 kHz 12,5-16 m/s 13 m |
| 1 kHz 16-20 m/s 13 m |
| 1 kHz 16-20 m/s 2 m |
| 3 kHz 8-10 m/s 13 m |
| 3 kHz 10-12,5 m/s 2 m |
| 3 kHz 12,5-16 m/s 2 m |
| 8 kHz 6,2-8 m/s 13 m |
| 8 kHz 8-10 m/s 13 m |
| 8 kHz 8-10 m/s 2 m |
| 8 kHz 10-12,5 m/s 2 m |

⊢——— 30 dB ———⊣

transmission loss variability

Fig. 6 Ambient noise spectrum level variability versus transmission loss at three reference frequencies and selected wind speed classes measured 2 m and 13 m above the bottom. Each dot represents 10 min average. The depictions are adjusted to the corresponding minimum loss at the left hand side according to wind class and frequency.

Fig. 7 (bottom) Same as fig. 6 but 3 kHz, wind class 12.5-15 m/s, measured 2 m above the bottom. Wind class divided in four subclasses.

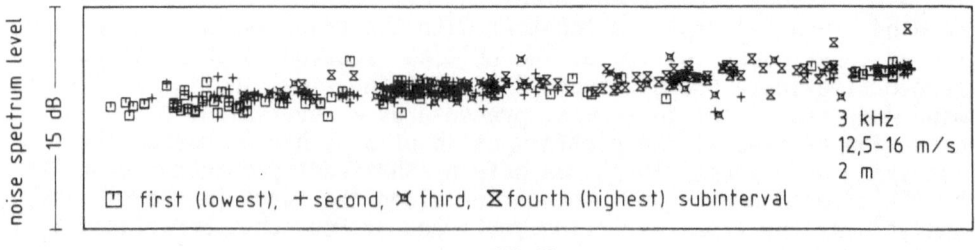

noise spectrum level

— 15 dB —

3 kHz
12,5-16 m/s
2 m

□ first (lowest), + second, ✕ third, ✗ fourth (highest) subinterval

⊢——— 30 dB ———⊣

transmission loss variability

The transmission loss indicated by the signal level increases suddenly by some 30 dB, before the sea state has reacted in terms of the wave height. The signals recover again during the next hours together with the gradually decreasing wind speed despite a still growing wave height and follow superimposed wind maxima (indicated by arrows).

The following acoustic features connected with the front passage appear noteworthy.

(1) The noise increase starts at all frequency components between 0.4 and 20 kHz at the same instant (5:30 h) and at the same rate (the frequency band between .1 and .3 kHz is contaminated by traffic noise). In view of the time delay between the wind record at the tower and the event at the hydrophone position, the noise development appears at most 10 min behind the wind [11]. There is obviously no noise harbinger of the approaching storm at frequencies of lowest loss (which are close to .3 kHz at the range as known from separate measurements). This finding is another independent evidence that the noise irradiation from a distance is only contributing weakly within the frequency regime covered by the measurements at least in comparison with the ever present variability of local origin.

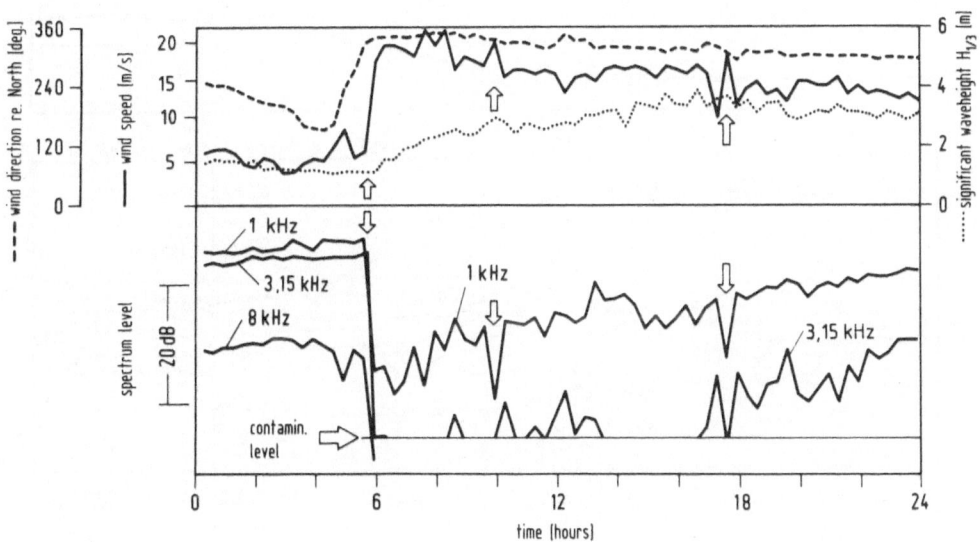

Fig. 8   Wind jump of a meteorological front (top) passing the hydrophone and resulting range transmission loss at three third octave frequency bands (bottom). The arrows indicate corresponding intermediate wind and loss maxima. The front has passed the receiving hydrophone 10-15 min before its arrival at the tower anemometer.

302

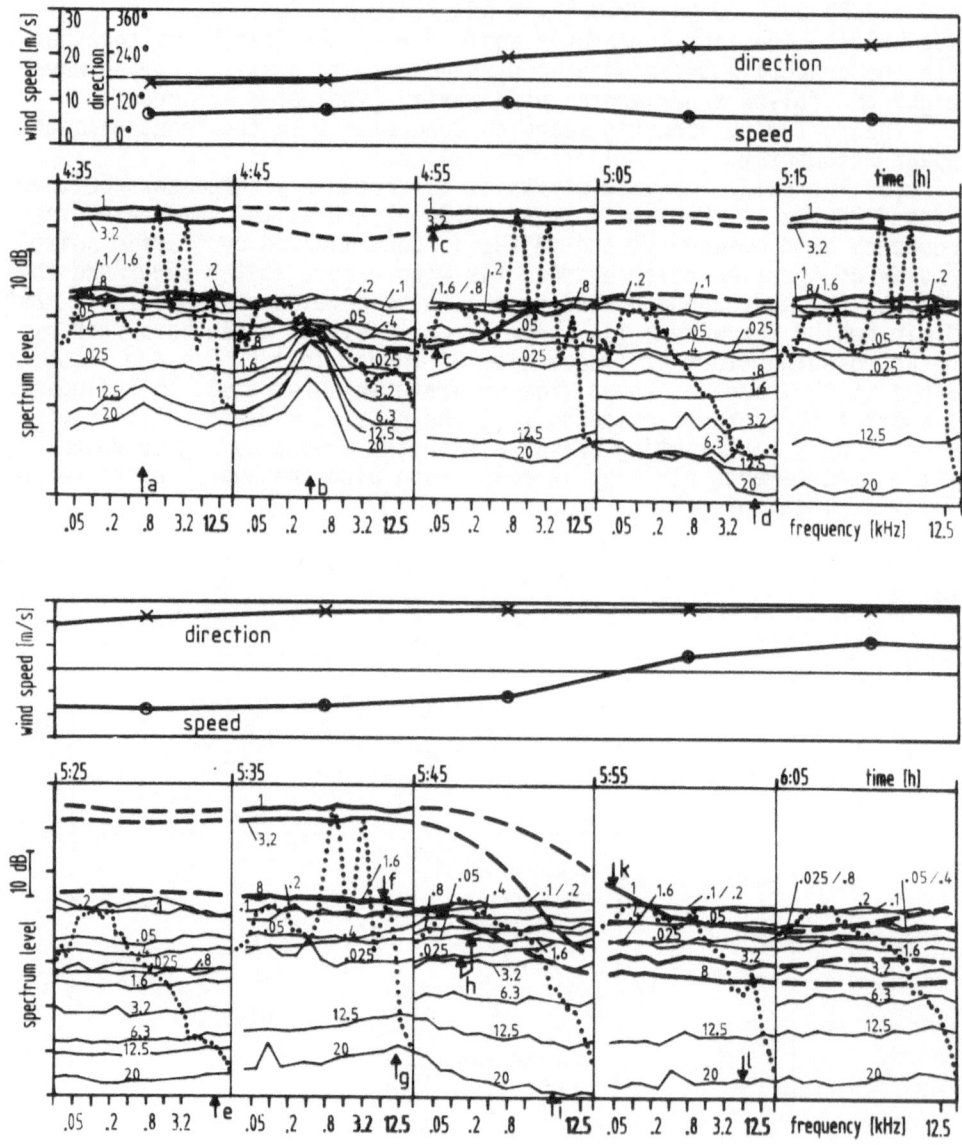

Fig. 9   10 x 10 min sections of fig. 8, enlarged time scale. Synoptic
history of the w i n d vector at the platform sound source and alternating 10 min blocks of n o i s e spectrum level (thin lines: third
octaves at .025, .05, .1, .2, .4, .8, 1.6, 3.2, 6.3, 12.6, 20 kHz)
and received s i g n a l level (strong lines: third octaves at 1, 3.2,
8 kHz. Strong dashed lines: interpolation during signal transmission
interrupt). Dotted line: 10 min average spectrum (lower scale). The
signal transmission (3 peak spectrum) contaminates the noise between
.4 and 12.5 kHz. The events indicated by a-1 are described in the text.

(2) The noise level has reached its maximum value within .4 - 20 kHz before the onset of screening (g at 5:44:30 h) which needs about 10 min to develop to a maximum of 10 dB (i) at 20 kHz. This result suggests that at least an essential portion of the noise in the kHz-regime is produced before bubble clouds sufficient for high frequency screening are developed.

(3) The increase of attenuation starts first at 8 kHz (f) and is finished last at 1 kHz (k). The onset of the noise increase appears to precede the loss increase of the range by several minutes, since 15 min after the onset of the noise growth there is still no indication of increasing loss at 1 and 3 kHz. Regarding the entire range loss excess of about 30 dB, a small fraction of the range length would already produce a measureable loss increase when populated with bubble suspensions. This may be considered another indication that the bubbles appear to end their active phase before they are aged to passive attenuating bubble clouds.

(4) The noise at the lowest frequency bands measured (25-50 Hz) develops latest (h) when the entire noise between 0.4 and 20 kHz has already reached its final level - including the screening effect. This delay appears to indicate a different mechanism of the low frequency noise production.

(5) The front is preceded by rain showers (a, b) obvious through high frequency noise transients which appear to vanish at (d). It is remarkable that a strong shower (b) enhancing the noise level between .4 kHz and at least 20 kHz up to 25 dB is accompanied by an increase of the transmission loss: the 8 kHz-signal is reduced by 12 dB at the beginning of the adjacent 10 min signal transmission section (c). The 3 kHz-signal however is reduced by only 3 dB, the 1 kHz-signal remains unchanged. The recovery time of the signal transmission is less than 10 minutes. This transmission loss transient may be due to air suspensions from a squall (see also fig. 8), recorded at 5:00 h accompanying the rain shower, rather than the result of bubbles from rain drop impact. It should be mentioned that 10 min after (b) a shower of rain and hail was recorded at the platform at a maximum precipitation rate of .2 mm/min as a precursor of the approaching front.

5. PASSIVE FEATURES OF BUBBLE CLOUDS

The description of the dissipative properties of an entire bubble cloud requires the space time distribution of the frequency dependent complex sound speed which is not available up to now. The present investigation is necessarily restricted to characteristics of the attenuation estimated from three independent data sources: The transmission loss of the fixed range, the noise screening and the local subsurface loss excess [8].

(1) Wind speed dependence of the attenuation: From range propagation loss data such as figures 2 and 3 an asymptotic increase of the attenuation coefficient by the n'th power of the wind speed (n between 3 1/2 and 4 1/2) is derived (fig. 12)[6]. Similar wind dependencies are given in a compilation of empirical data [12] for bubble concentrations at a depth of 1.5 m (n ≈ 4.5) and the visible whitecapping (n ≈ 3.75).

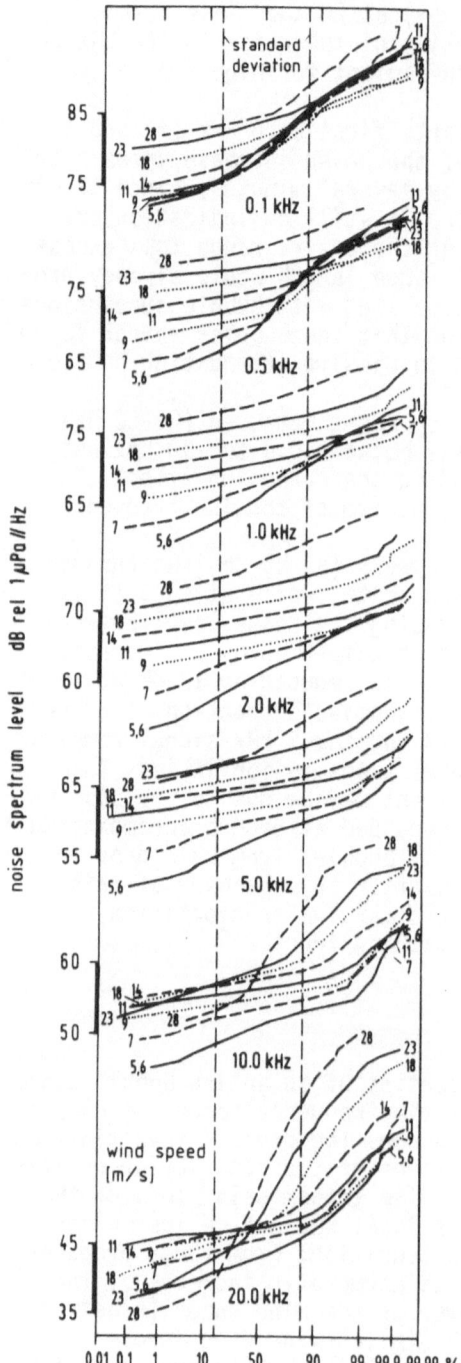

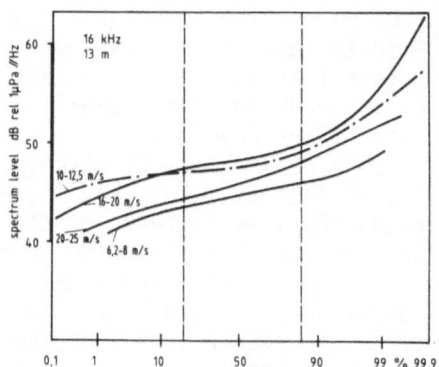

Fig. 11 (above) Same as fig.10 for 16 kHz and 13 m above sea bottom, coinciding with measurements at 2 m bottom distance.

Fig. 10 (left) Cumulative distribution of 30 s averages of noise spectrum level, class width 1 dB. Platform area, December/January. Measured at < 1 m above sea bottom. Parameters: Third octave frequency bands with 8 wind classes each.

Fig. 12 (below) Average attenuation coefficient α and range transmission loss versus wind speed u (log-log-scale) for increasing and decreasing wind. The slope at high wind speeds represents the exponent n of $\alpha \sim u^n$.

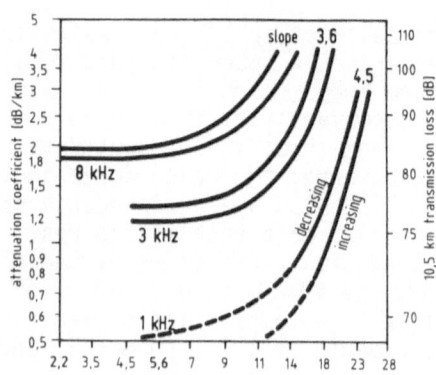

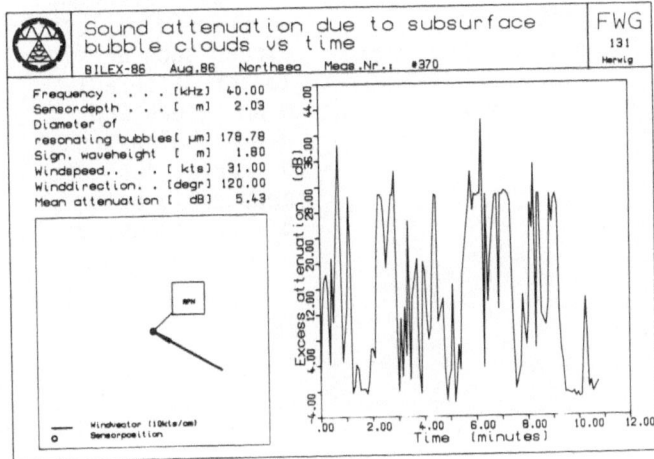

Fig. 13 Excess attenuation of 2 m transmission path, 2 m below wave trough (Herwig, FWG, 86). Level of received sound signal at constant electric source signal. Measuring device lowered by platform crane at a position undisturbed by the platform construction in relation to the wind direction.

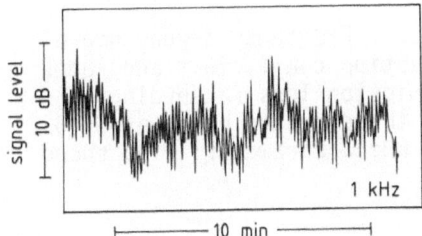

Fig. 14 Short time level fluctuations of transmitted signal at 1 kHz. Platform area, September, 15 m/s wind speed, 3 m significant wave height. Short time periodicity corresponds to main wave period.

The horizontal loss excess measurements of Herwig [8] by a source-receiver pair of 2 m distance, lowered to variable depths below trough yield a rough estimate of n ≈ 3...4.5 at 10 and 20 kHz.

The noise screening [3,11] (fig. 10) show an approximately constant level decrease of 5 dB per windclass at 20 kHz and 10 kHz rather than a power relation of the attenuation coefficient which would strongly vary with depth anyway. The screening loss is determined from the uncontaminated low level side of the distributions of fig. 10 under the assumption of a still valid second power relation between noise spectrum level and wind speed. The screening is partly contaminated by an extreme skewness of the high frequency/high wind speed distributions, possibly due to impact noise from wave induced sediment blast. When the receiver distance to the bottom is enlarged to 2 or 13 m (fig.11) the skewness is reduced and the screening is apparent through the main part of the distribution.

(2) Absolute figures of the attenuation: Based on the platform range measurements the attenuation of a homogeneous substitute subsurface bubble layer has been determined by model inversion taking the bottom loss and the surface scattering loss into account but neglecting the sound speed change [7]: the coefficients range from .07 dB/m at 15 m/s and .3 dB/m at 25 m/s (1 kHz) up to .22 dB/m at 15 m/s and .9 dB/m at 25 m/s (8 kHz). Applying the empirical linear relationship of

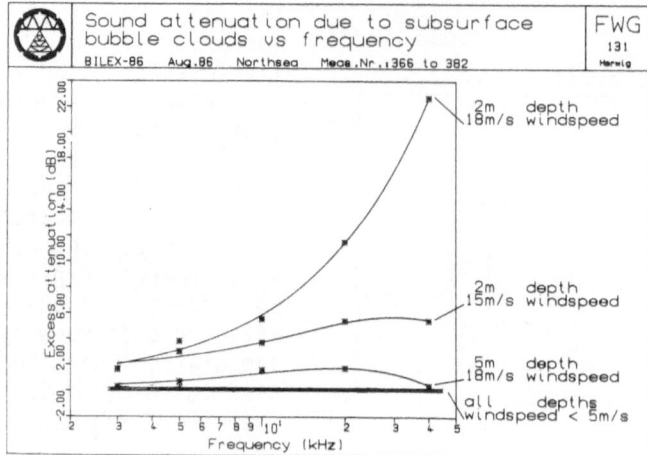

Fig. 15 Frequency dependence of loss excess such as fig. 13 for different wind speeds and depths below trough. No excess loss was measurable below 5 m/s wind speed.

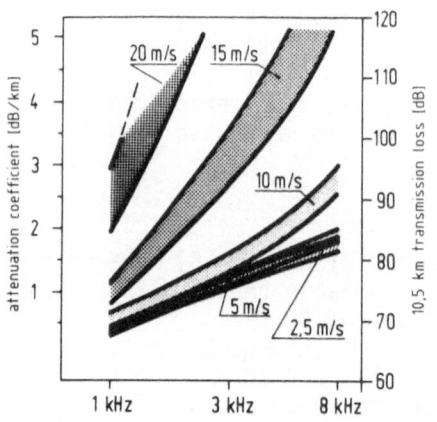

Fig. 16 Frequency dependence of attenuation coefficient and range transmission loss (according to three third octave bands at 1, 3 and 8 kHz). Parameter: wind speed.

attenuation coefficient and frequency (subsect. 3) the substitute layer would attenuate a vertically penetrating noise by 2.25 dB at 25 m/s and 20 kHz against the measured screening of ≈ 10 dB (fig. 9). This discrepancy demonstrates that a more realistic modelling of the bubble cloud attenuation needs information on the impedance structure.

The evaluation of the screening at 20 kHz (fig. 10) yields 4 dB at 14 m/s, 8 dB at 18 m/s, 14 dB at 23 m/s and 19 dB at 28 m/s. Comparable figures have been found by Farmer [14].

The subsurface excess attenuation data of fig. 15 are comparable to the screening values at 20 kHz, giving rise to the assumption that the subsurface measurements cover already an essential portion of the entire loss induced by a bubble cloud.

(3) Frequency dependence of the attenuation; The range measurements (fig 16) as well as the subsurface measurements (fig. 15) exhibit an approximately linear relationship between attenuation coefficient and frequency found at high wind speed within 1 kHz to nearly 40 kHz.

(4) Temporal variability: the measurements of Herwig such as fig. 13 demonstrate the residence time of bubble clouds within the transducer set up window achieves the order of 1 min whereby the loss excess may vanish between adjacent cloud events up to wind speeds of at least 15 m/s. The averaging effect of the 10 km range reduces the fluctuation (fig. 14) but instead exhibits the main period of the sea waves modulating the dissipative structure of the propagation path. This loss fluctuation is one of the essential features of the stochastic nature of shallow water sound propagation limiting the achievable accuracy of loss prediction [6].

6. CONCLUSION

The synoptic measurements of the gross noise and attenuation confirm the surface generated shallow water noise to depend on the local agitation only, at least beyond frequencies about .5 kHz. Further, the measurements provide indications that air bubbles in their latest stage as cloud formations act essentially passive though their internal loss up to several kHz is only small. Due to the empirical description of the bubble induced loss dependence on wind speed and frequency, the second power relation between wind speed and spectrum noise level which has been found to provide the best fit under regular conditions remains unimpaired up to hurricane wind speeds at moderate frequencies. However, a convincing evidence of the acoustic mechanisms of air suspensions needs a synoptic aquisition of the entire life time of breaking wave events together with the relevant environment. It appears necessary to supply the knowledge of the gross "macroscopic" features by high resolution "microscopic" investigations in the domain of space, time and frequency. Such synoptic measurements are under consideration, utilising the research platform NORDSEE to provide fixed positions of the instrumentation set up in close vicinity of and within the breaking wave events.

ACKNOWLEDGEMENT

We are greatly indepted to Dr. L. Ginzkey, B. Scholz, Dr. E. Schunk and G. Wendel for the establishment of the range and the automatic measuring device. Dr. S. Stolte and P. Lobemeier have given valuable support and suggestions in the field of air sea interaction.
Dr. H. Herwig kindly provided preprint material of his subsurface loss measurements.

308

7. REFERENCES

[1] W.A. Kuperman; M.C. Ferla
J.Acoust.Soc.Am. $\underline{77}$ (6), pp. 2067-2073, 1985

[2] S.N. Wolf; F. Ingenito
in: Underwater Ambient Noise, Rep. CP-32 (SACLANT ASW Research
Centre, La Spezia, Italy), 1982

[3] P. Wille; D. Geyer
J.Acoust.Soc.Am. $\underline{75}$ (1), pp. 173-185, 1984

[4] R.M. Hamson
NATO Advanced Res. Workshop on Natural Mechanisms of Surface
Generated Noise in the Ocean, Lerici, Italy, 1987

[5] J. Sellschopp
FLEX 76 Zusammenstellung der Messungen mit dem Thermistor-Schlepp-
kabel, FWG-rpt. 1982-15, 1982

[6] P. Wille; D. Geyer; L. Ginzkey; E. Schunk
12th ICA Associated Symposium on Underwater Acoustics,
16-18. July, Halifax, Canada, 1986

[7] H.G. Schneider
12th ICA Associated Symposium on Underwater Acoustics,
16.-18. July, Halifax, Canada, 1986

[8] H. Herwig
FWG-rpt. in prep.

[9] R. Thiele
in: Acoustics and the Sea Bed, Bath, UK, Bath University Press
pp. 207-213, 1983

[10] P. Wille
Landolt-Börnstein, New Series, Oceanography, Vol. I, chapter 3.2,
pp. 265-382, 1986

[11] P. Wille
NATO Advanced Study Institute on Adaptive Methods in Underwater
Acoustics, Dordrecht, Reidel Publ. Co., 1985

[12] J. Wu
J.Geophys. Res. $\underline{86}$, pp. 457-563, 1981

[13] S.A. Thorpe; A.R. Stubbs
Nature $\underline{279}$, pp. 403-405, 1979

[14] D.M. Farmer; D.D. Lemon
J.Phys.Oceanog. $\underline{14}$, pp. 1762-1778, 1984

MECHANISMS OF SOUND GENERATION AT THE OCEAN SURFACE

J.E. Ffowcs Williams and Y.P. Guo
Department of Engineering
University of Cambridge
Cambridge CB2 1PZ
England

ABSTRACT. Fluid motions at the ocean surface provide a wide variety of mechanisms that might be the cause of underwater sound. This paper is devoted to help understand better these mechanisms, and to help identify those sources most responsible for the received noise in the ocean. Flows at the sea surface, as well as the sound they are generating, are governed by equations of mass and momentum conservation. They can be cast into an acoustic analogy which interprets the sound generation by regarding some terms in the conservation laws as sources that drive the sound field. Though the equations describing the conservation laws are precise, the choice of the 'source' terms is not at all unique; what should, or should not, be regarded as sources of sound is largely a point of view. The arbitrariness of choosing the source terms for an acoustic analogy in aero- and hydro-acoustics is a reflection of the well-known ambiguity that an infinite number of different source descriptions can produce an identical wave field. However, for the problem of underwater sound generation, as indeed for most other flow noise problems, we only have a very crude knowledge of the source flow, so that it is crucially important, in casting an acoustic analogy from the exact equations of motion, to specify the source terms as precisely as possible in order to reveal correctly the dominant sound-producing processes. In doing so, the effects of the ocean surface that separates the air and water must be taken into account with great care; this density interface plays a key role in the problem. Due to its presence, powerful acoustic sources may be induced, while equally important, sources which are efficient in other situations may become negligible. We start with an account of the basic principles of flow noise generation, with particular reference to the ocean sound problem. This enables us to examine a selection of flow processes which are commonly believed to be most likely responsible for the surface motion associated oceanic noise. We will demonstrate that breaking surface waves in the ocean provide the dominant sources of underwater sound by causing rapid momentum variations at the sea surface. These dominant sources arise directly from the splashing of water sprays, and are much more efficient than other processes; entrainment of air bubbles and their subsequent unsteady motions in water are negligible because they suffer from fatal

B. R. Kerman (ed.), Sea Surface Sound, 309–324.
© 1988 by Kluwer Academic Publishers.

destructive interference from their negative images at the sea surface. The interactions of surface waves may also be irrelevant to the ocean sound problem because they can only provide very weak acoustic sources, in comparison, for most of frequencies of interest, with the sources from breaking waves, and for very low frequencies (below 10 Hz, say), with those induced directly by the turbulent airflow.

1. INTRODUCTION

Sound generated by flow processes in the ocean surface layer, such as turbulent airflows and surface wave motions, has long been of interest; it is known to be the major component of the observed ambient noise in the ocean (see, e.g. Wenz 1962). This oceanic noise, originating probably from a variety of flow processes and covering a wide frequency range from infrasonic to tens of kilohertz, is difficult to analyse, and the various existing theories usually deal only with some of the possible mechanisms, and therefore, usually only have limited applications. The aim of this paper is to provide a coherent analysis, within the framework of the Lighthill (1952) theory of flow noise generation, for tackling those surface processes most commonly believed to be the cause of underwater sound.

Since Lighthill's (1952) celebrated paper, our understanding of the mechanisms of flow noise generation has grown steadily (for a review, see Ffowcs Williams 1984). The original theory deals mainly with sound generated by flows of uniform mean density, but it has later been extended, to meet the demand of predicting the noise from hot jet flows, to take account of the effects of non-uniform density (see, for example, Ffowcs Williams 1984; 1986). This extension makes it possible to examine the sound generation by flow processes in the ocean surface layer, where the densities of the two fluids (air and water) differ considerably. Both the jet noise and ocean surface sound problems are characterized by an unstable density interface.

The ocean surface may reduce the radiation efficiency of some flow processes. For example, a bubble which is a simple monopole is silenced when close to a pressure-release plane surface of reflection coefficint -1. The radiation efficiency of such a source is dramatically reduced by the destructive interference of its image in the free surface. It has been shown by Guo (1987c) that the acoustic effects of otherwise noisy bubble motions near the ocean surface can actually be neglected.

The motion of the ocean surface under a turbulent airflow is inevitably nonlinear; wind produced surface waves are intrinsically unstable and they are bound to break (Phillips 1977). Wave breaking is one of the noisiest processes occurring at the sea surface. When waves are breaking, water sprays splash into the airflow, and these cause a rapid variation of the density distribution. As a result powerful acoustic dipole sources are induced by the large variation of momentum on the ocean surface. This acoustic process resembles very much the production of noise by a hot turbulent gas jet where it is known that the dipoles due to mean density inhomogeneity are the dominant sources (Ffowcs Williams 1986). Guo (1987c) has examined this mechanism with particular reference to breaking ocean waves and found that it overpowers other surface layer source processes.

Nonlinear motions of the ocean surface also provide acoustic sources by the wave-wave interaction mechanism which has long been an element of ocean acoustics (Brekhovskikh 1966; Hughes 1976). When two subsonic surface waves, of equal frequency and travelling in almost opposite directions, interact nonlinearly with each other, surface motions of supersonic phase speed may be formed, and these radiate sound. Though this mechanism is physically possible, its relevance to ocean sound generation is arguable. It is evident that the major process that characterizes the energy transfer of ocean surface waves in an active wind field is the breaking; wave-wave interactions is important in transfering energy among different components of the wave field only at a very early stage of its evolution. Wave-wave interaction is a negligible acoustic process in comparison with that of wave breaking for most of the frequency range of interest. Only for sound of very low frequencies (below about 10 Hz, say) is the interaction mechanism a contributor, but then, the direct radiation from the turbulent airflow that produced the surface waves in the first place is more important and the wave-wave interaction is negligible.

2. UNSTEADY FLOWS AS ACOUSTIC SOURCES

Sound is a very weak disturbance about a state of rest and the disturbance, the density fluctuation $\rho'=\rho-\rho_w$ for example, ρ_w being the mean density in water, must comform with the wave equation

$$[\;]^2\rho' = \frac{\partial^2\rho'}{\partial t^2} - c_w^2\nabla^2\rho' = 0, \tag{2.1}$$

c_w being the constant speed of sound in water. The only solution to this equation that complies with the radiation condition in an unbounded space is absolute silence, that is, $\rho'=0$. If we take the view that there is no sound because there is no source, it is then natural to define the quantity $[\;]^2\rho'$ as the source of sound, the vanishing of which guarantees the absolute silence. This definition of sound source, although somewhat arbitrary, is a convenient choice for problems of flow noise generation. When an unsteady flow exists, the density fluctuation ρ' does not satisfy the homogeneous wave equation (2.1); instead, the right hand side is the source distribution derived by Lighthill (1952) from the exact equations of motion.

$$[\;]^2\rho' = \frac{\partial^2 T_{ij}}{\partial x_i \partial x_j}, \tag{2.2}$$

where $$T_{ij} = \rho u_i u_j + p_{ij} - c_w^2\rho'\delta_{ij}.$$

u_i and p_{ij} are respectively the velocity and the compressive stress tensor. The unsteady flow is equivalent to a distribution of acoustic quadrupoles, and that is the essence of the Lighthill (1952) theory.

In the ocean, air bubbles are inevitably present in the surface layer, and it is appropriate to examine their acoustic effects to see if they induce any appreciable sound. Similarly to (2.2), unsteady motions of a bubbly two-phase flow can be regarded as sound sources by casting the equations of motion into (Crighton & Ffowcs Williams 1969)

$$[\,]^2\rho' = \frac{\partial^2 T_{ij}}{\partial x_i \partial x_j} - \frac{\partial S_i}{\partial x_i} + \frac{\partial Q}{\partial t}, \qquad (2.3)$$

where
$$S_i = \frac{\partial}{\partial t}(\rho u_i) + \frac{\partial}{\partial x_j}(\rho u_i u_j + p_{ij})$$

is the force between gas and liquid phases, and $Q = -\rho D\ln(1-\beta)/Dt$ is the effect of gas volume changes due to compressibility. In this, β is the fraction of unit volume that is occupied by air bubbles and the density and velocity are respectively those of the water phase. The non-vanishing of the second and the third term in (2.3) depends on the non-vanishing of the bubble concentration β. If β is zero, ρ and u_i respectively reduce to the density and velocity in pure water. In that case, Q is obviously zero and S_i also vanishes according to the momentum equation. Thus (2.3) reduces to the Lighthill equation (2.2). The definition (2.3) makes explicit the effects of bubbles in an unsteady flow; they induce additional acoustic sources of monopole and dipole form.

In general, the medium we are concerned with may only occupy some parts of the whole space; boundaries are usually present. Boundary surfaces also induce sources of sound, sources at the sea surface being our primary concern here. Boundary surfaces support dipole and monopole sources. This can be shown through the method developed by Ffowcs Williams & Hawkings (1969). To do this, we introduce the Heaviside function $H[f(\underline{x},t)]$ that equals unity for positive arguments and vanishes for negative arguments. Here $f(\underline{x},t)$ is a function whose zero locates the boundary surface, $f=0$ specifying the ocean surface that separates the air and water. We define $f(\underline{x},t)$ to be positive in the water and negative in the air, so that $H(f)\rho'$ is identical to ρ' in the water and vanishes otherwise. By multiplying (2.2) by $H(f)$ and rearranging terms, the Ffowcs Williams-Hawkings equation is derived,

$$[\,]^2 H\rho' = \frac{\partial^2 H T_{ij}}{\partial x_i \partial x_j} - \frac{\partial}{\partial x_i}\left(P_{ij}\frac{\partial H}{\partial x_j}\right) + \frac{\partial}{\partial t}\left(\rho_w u_i \frac{\partial H}{\partial x_i}\right). \qquad (2.4)$$

This equation brings the acoustic effects of the boundary surface into the definition of a sound source in the form of (2.2); the gradient of the Heaviside function, the Dirac delta function, is non-zero only on the boundary surface $f=0$ where it is singular.

Definitions (2.2), (2.3) and (2.4) are all derived from the exact conservation laws of fluid flow, and they all exactly describe both the source flow and the sound field it is generating. In fact, infinitely many choices of definitions similar to (2.2) to (2.4) are possible, depending on the particular situation and depending also on the special features one wishes to emphasize. Of course, if the source flow is known exactly, any one of the infinite number of choices will lead to the same unique result for the sound. However, this is never the case for the problem of sound generation at the ocean surface, nor indeed for most flow noise problems, because we only have a very limited knowledge of the flow in the ocean surface layer. A wide variety of intricate processes are in play and each is a possible mechanism of sound production. Hence, in order to estimate the sound, it is necessary to resort to dimensional analysis and approximations, no matter what defining equation is used. With certain necessary approximations imposed, it is then very likely that one particular choice of sound source is best suitable for any particular mechanism; that is, it may give the best estimate of the sound from a particular approximate description of the flow. This will become apparent in the following sections where (2.2), (2.3) and (2.4) will respectively be used to examine those flow processes commonly regarded as responsible for underwater sound. We also analyse the relations among, and the relative importance of, these mechanisms in a more unified theory which has not been available previously.

3. SOUND SOURCES NEAR A FREE SURFACE

The simplest ideal sources are monopoles which account for the creation of mass. In hydrodynamics, creation of mass is hardly physical, but some flow processes result in the same acoustic effects; unsteady bubble motions are known to be monopole-like acoustic radiators. The sound associated with higher mode oscillation of a bubble is negligible so that only simple volume pulsation (zeroth mode) is acoustically important. That volume pulsation has precisely the same acoustic effects as an ideal monopole source. For a cloud of air bubbles in turbulent water, a situation more likely to occur in the ocean than an isolated bubble, Crighton & Ffowcs Williams (1969) have also shown the monopole structure. This is evident from (2.3); the monopole-like third term arises directly from the compressible motion of gas bubbles. That equation is most suitable for examining the sound from a turbulent flow containing bubbles, and the most important feature is that the bubbles are much more efficient in radiating sound than the quadrupoles of pure turbulence. This can be seen by solving (2.3) in an unbounded space. For flows of low Mach number $M=u/c_w \ll 1$, the acoustic radiation efficiency of the turbulence quadrupoles and the bubble induced monopoles (the first and the third term) are respectively of the order

$$\frac{1}{4\pi} \frac{\Delta}{\ell} M^5 \quad \text{and} \quad \frac{1}{4\pi} \frac{\Delta}{\ell} M^5 \left(\frac{c_w}{c_m}\right)^4 . \tag{3.1}$$

These results are obtained by solving (2.3) for ρ' in terms of the retarded potential solution, from which the radiated sound power, and hence, the radiation efficiency, can be estimated through dimensional analysis. Here ℓ is a typical length scale of the turbulence eddies and

Δ is that of the overall source region. c_m denotes the sound speed in the bubbly flow and is very much smaller than c_w. The second term of (3.1) clearly dominates the first; the acoustic efficiency is increased by the bubbles by the factor $(c_w/c_m)^4$ which is of the order 10^5 (50 dB) if β, the volume fraction of air bubbles, is as small as 0.01 and increases to 10^7 as β reaches 0.1.

These conclusions are derived for an unbounded medium. In the problem of sound generation at the ocean surface, the bubbles, mainly resulting from the entrainment in the breaking of surface waves, are all close to the density interface between air and water. Such an interface has a significant destructive effect on the bubble-radiated sound. Monopole radiation is known to be severely weakened by a negative image field when the real source is close to a pressure-release surface. This weakening is usually so severe that the monopole losses its advantage of effectiveness over other higher order multipoles. Due to the vanishingly small mean density ratio of air to water, the ocean surface can approximately be regarded as a pressure-release surface, and hence, the destructive interference between the bubbles and their negative images can be expected to feature. Indeed, it can be shown by solving (2.3) again with the ocean surface conditions imposed, through the method of exact Green functions for example, that the acoustic radiation efficiency of the bubble induced sources is reduced from the value given by the second term of (3.1), either by the square of the typical Mach number or by the square of the impedance ratio of air to water $(\rho_a c_a/\rho_w c_w)^2$, ρ_a and c_a being respectively the mean density and sound speed in air. Either way, the reduction factor usually offsets the increase factor $(c_w/c_m)^4$ over the pure turbulence radiation, and hence, the bubble-induced monopoles actually degrade to insignificance because of the free surface. This therefore gives grounds to believe that bubble motions in the ocean surface layer may be less important in radiating sound than other processes which induce acoustic dipoles.

Acoustic dipoles are equivalent to unsteady forces, or momentum fluctuations, which are processes taking place most often in the real flow. The unsteady forces caused by turbulent airflows moving across the ocean surface and the momentum fluctuations due to nonlinear interactions of surface waves are typical examples which will be analysed in section 5. Another obvious cause of momentum fluctuations lies in the mean density inhomogeneities which are inevitable when breaking waves exist on the ocean surface. Wind produced surface waves are intrinsically unstable; they are bound to break (Phillips 1977). When this occurs, water sprays splash into the turbulent airflow, which causes a flow in the ocean surface layer that has a rapidly varying mean density. Accelerated density inhomogeneities are known to be important sources of sound (Ffowcs Williams 1984; 1986, for example). This can be demonstrated from equation (2.2). Changes in sound speed in the source region arise mainly from changes of elasticity of the flow due to the entrainment of air bubbles. Because this has a negligible acoustic effect, we assume that the source flow and the water share the same sound speed but have different densities. We denote by ρ_0 the density in the source flow, which is a function of both space and time because the density at a fixed point can be either ρ_w or ρ_a, depending on whether that point is occupied by water elements or aerial elements at the observation time. Since ρ_0 is different from ρ_w, it is more convenient to choose to work with the pressure p, instead of the density fluctuation ρ'. In the ocean sound problem, the pressure is the dominant term in the compressive stress tensor p_{ij} and has a uniform mean value. Noticing that the pressure pertubations are approximately equal to $c_w^2(\rho-\rho_w)$ in the deep water and $c_w^2(\rho-\rho_0)$ in the source flow, (2.2) can be rewritten as

$$\frac{1}{c_w^2}\frac{\partial^2 p}{\partial t^2} - \nabla^2 p = \frac{\partial \rho u_i u_j}{\partial x_i \partial x_j} - \frac{\partial^2 \rho_0}{\partial t^2}. \tag{3.2}$$

Considering that fluid particles maintain a virtually constant density in low Mach number flows where $\partial u_i/\partial x_i = 0$, we have

$$\frac{\partial \rho_0}{\partial t} = -u_i \frac{\partial \rho_0}{\partial x_i} = -\frac{\partial \rho_0 u_i}{\partial x_i},$$

which yields

$$\frac{\partial^2 \rho_0}{\partial t^2} = -\frac{\partial}{\partial x_i}\left(\rho_0 \frac{Du_i}{Dt}\right) + \frac{\partial^2 \rho_0 u_i u_j}{\partial x_i \partial x_j}.$$

Hence (3.2) becomes

$$\frac{1}{c_w^2}\frac{\partial^2 p}{\partial t^2} - \nabla^2 p = \frac{\partial^2 \rho' u_i u_j}{\partial x_i \partial x_j} + \frac{\partial}{\partial x_i}\left(\rho_0 \frac{Du_i}{Dt}\right). \tag{3.3}$$

It is clear from this that the mean density variations induce dipole sources that in general overpower the Reynolds stress quadrupoles at low Mach numbers; the dipoles are more efficient by the inverse of the Mach number.

Before concluding the dominance of the dipole sources, it is necessary to check the reflective effects of the ocean surface; it might be that the density interface degrades the dipoles to higher order multipoles as it does the bubble-induced monopoles. In that case, the domonant sources would still be the Reynolds stresses. The dipole term in (3.3) includes those whose axes are paralell to the mean position of the ocean surface; destructive interference of their images makes them negligible for the same reason as bubble-induced sources are negligible. However, the reflective effects of the ocean surface on those dipoles with axes perpendicular to its mean position are essentially constructive. Hence the dominant source is the x_3-component of the second term of (3.3). This will be confirmed in section 5 and a physical interpretation will be given for the fact that the vertical motion induced dipoles are more powerful than those due to horizontal motions.

Since breaking waves on the ocean surface induce dipole sources that are much more efficient in radiating sound than quadrupoles, the radiation from pure turbulence is of negligible significance. This has long been observed in the ocean, as summarized by Wenz (1962); the predicted noise level from pure turbulence sources is order of magnitude

316

lower than that actually observed in the real ocean. The ineffectiveness
of turbulence quadrupoles in radiating sound, in comparison with dipoles
of the same time and space structure, is guaranteed by the fact that
the typical velocity of the source flow for the ocean sound problem is
always much smaller than the sound speed in either water or air, the
length scale over which the sources vary appreciably being much smaller
than that of the sound generated. Sources of this kind are known as
compact. The acoustic radiation efficiencies of compact multipoles
decrease as their order increases.

4. LINEAR AND NONLINEAR SOURCES

     Though the source definition (2.2) is derived from the exact
conservation laws of fluid motions, it is only a formal statement; we
never know the detailed structure of the source flow in the ocean.
Hence, it is necessary to resort to dimensional analysis and
approximations in order to estimate the sound. In doing so, a difficult
task is to deal with linear terms in the source function. Linear terms
are necessarily present because the acoustic analogy equations describe
exactly both the source flow and its wave field, and even linear fluid
motions in the ocean surface layer are not acoustic. These linear terms
may cause great error because they usually, when the source region is
relatively extensive as in the case of the ocean, represent parts of
the propagating sound field generated; they are not sources at all.
This can be illustrated, by following Ffowcs Williams (1986), through
an example in which two different ways of dealing with the linear terms
in the source definition give two results that differ by an order of
magnitude, one of which, of course, must be in error.

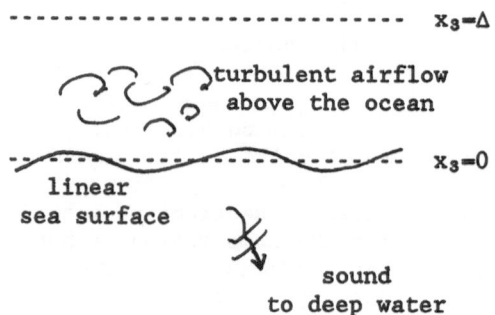

Figure 1. The geometry of the linear ocean surface problem.

     For the purpose of demonstration, we assume that sound is generated
by a turbulent airflow above the ocean, which also linearly deforms the
air-water interface. For this example we neglect the difference in the
sound speed across the interface. Since the mean density changes at the
interface, it is natural to choose its mean position, $x_3=0$ say, as a

control surface (see figure 1). The turbulence sources are then all in the region above the control surface, and the source definition becomes

$$[\,]^2\rho' = \rho_a \frac{\partial^2 u_i u_j}{\partial x_i \partial x_j}, \qquad \text{for } x_3 \geq 0,$$

and

$$[\,]^2\rho' = 0, \qquad \text{for } x_3 \leq 0,$$

where we have replaced T_{ij} by $\rho_a u_i u_j$ because p_{ij} can be approximately regarded as $c^2\rho'$ for the low Mach number flow in the ocean (see, Lighthill 1952). These equations can be solved to find the sound field in the deep water as

$$4\pi c^2 \rho' = \frac{2\rho_a \rho_w}{\rho_w + \rho_a} \frac{x_i x_j}{|\underline{x}|^3} \frac{\partial^2}{\partial t^2} \int_V [u_i u_j]\, d^3\underline{y}, \qquad (4.1)$$

where V is the space occupied by the turbulent airflow and the square brackets imply the retarded time convention. If the typical velocity of the flow is of the order u and the length scale of a turbulence eddy is ℓ, $\partial/\partial t$, the typical frequency scale, should be of the order u/ℓ, and therefore, (4.1) can be estimated, by dimensional analysis (which is the only way applicable because of our very limited knowledge of the Reynolds stresses), as

$$4\pi\rho' \sim \rho_a \frac{\ell}{|\underline{x}|} M^4. \qquad (4.2)$$

This is the correct result which is consistent with the Lighthill theory that predicts the sound of a turbulent flow to be proportional to the fourth power of the typical flow velocity. Note that the constant density in (4.2) is ρ_a, instead of ρ_w, which indicates that most of the sound radiated from the flow is reflected into the air by the ocean surface, and only a small proportion can be recieved in the water, the ratio of the latter to the former being of the order ρ_a/ρ_w. We have directly derived this correct estimate because we have dealt correctly with the linear terms in T_{ij} by choosing the natural ocean surface as the control surface, so that the linear terms vanish and the dominant terms in the source function arise from the nonlinear Reynolds stress tensors.

If on the other hand we chose a linear surface in the air above the turbulence sources, $x_3 = \Delta$ for example, as the control surface, Δ being the height of the turbulence source layer (figure 1), the situation would be quite different. In that case, all the sources are below the control surface and we have

$$[\,]^2\rho' = 0, \qquad \text{for } x_3 \geq \Delta,$$

and

$$[\,]^2\rho' = \frac{\partial^2 T_{ij}}{\partial x_i \partial x_j}, \qquad \text{for } x_3 \leq \Delta.$$

Here T_{ij} cannot be approximated by the Reynolds stresses, as it could previously, because the linear terms are non-vanishing, mainly due to the fact that the mean density in the turbulence source flow is different from that in the water. Both the source flow of mean density ρ_a and the water of ρ_w are now in the region $x_3 \leq \Delta$ where we have a single wave equation with a source function. This source function must then involve terms proportional to the difference of the two mean densities, which is clear from the solutions given by Ffowcs Williams (1986),

$$4\pi c^2 \rho' = \frac{1}{|\underline{x}|} \left(T_+ + \frac{\rho_a - \rho_w}{\rho_a + \rho_w} T_- \right), \qquad (4.3)$$

where $\quad T_+ = \frac{x_i x_j}{|\underline{x}|^2 c^2} \int_V \left[\rho_w \frac{D}{Dt} \left(\frac{1}{1-M_r} \frac{D}{Dt} \left\{ u_i u_j - \delta_{ij} \frac{c^2 (\rho_a - \rho_w)}{\rho(1-M_r)} \right\} \right) \right] d^3\underline{y}, \qquad (4.4)$

and T_- is simply T_+ with x_3 replaced by $-x_3$, and $M_r = x_i u_i / c|\underline{x}|$. In the integrand, the non-constant part associated with the density terms are obviously larger than the Reynolds stress terms by the inverse of the typical Mach number. Hence if we casually apply dimensional analysis to this solution, the Reynolds stresses are neglected and the sound is approximately estimated from the density terms; in that case,

$$4\pi \rho' \sim \rho_w \frac{\ell}{|\underline{x}|} M^3. \qquad (4.5)$$

This is different from (4.2) and is an overestimated result that is higher than the real value by the enormous factor $(\rho_w/\rho_a)/M$. The reason for this is that the linear terms in the integrand of (4.3) are not sources at all. They are parts of the sound field generated elsewhere by the nonlinear Reynolds stresses, and they appear in the source function only because we have chosen a silly position for the control surface. It is clear in this case where the best control surface is; it is not nearly so clear when the surface is highly deformed as it is in a breaking ocean wave system. By carefully transfering the linear 'source' terms in (4.3) into the wave field which they actually represent, it can be shown that (4.3) is actually equal to (4.1). Letting $\underline{x}'$ denote the point $(x_1, x_2, -x_3)$, it is evident that

$$T_- = 4\pi c^2 |\underline{x}| \frac{\rho_w + \rho_a}{2\rho_a} \rho'(\underline{x}', t),$$

and the density terms in T_+ can be expressed as

$$-4\pi c^2 |\underline{x}| \frac{\rho_a - \rho_w}{2} \left(\frac{\rho'(\underline{x}', t)}{\rho_a} + \frac{\rho'(\underline{x}, t)}{\rho_w} \right).$$

On substituting these results into (4.3), it then becomes clear immediately that (4.3) and (4.1) are consistent.

5. SOUND GENERATION AT THE SEA SURFACE

In ocean sound studies, there exist various theories that examine flow processes which are commonly regarded as possible sources of oceanic sound. Being respectively applicable to one particular mechanism, while neglecting the others, these theories give results that are usually very likely to be inconsistent, or, in some situations, even contradictory. Of course, it is impossible to establish the respective relevance to the real ocean of the various theories, to any unambiguous degree, by the limited measurements available from the ocean, because the measured sound cannot be distinguished unambiguously as coming exclusively from one particular process. Hence a unified analysis is much needed to examine the relations among the different theories and to help determine the relative importance of different processes in generating sound. In what follows, we summarize a theory, involving only straightforward manipulations of the basic equations of fluid motions and obvious dimensional analysis, which has been developed by Guo (1987c). We quote his result for the sound in bottomless water, as

$$
\rho' = \frac{-\cos\theta T(\theta)}{2\pi c_w^3 |\underline{x}|} \frac{\partial}{\partial t} \int_{V_0} \left[\rho_0 \frac{Du_3}{D\tau} + \rho_m' \frac{Du_3}{D\tau} + \frac{\rho_a c_a / \rho c_m^2}{\sqrt{1 - \sin^2\theta c_a^2 / c_w^2}} \frac{Dp}{D\tau} \right] d^3\underline{y} + Q,
$$

(5.1)

where $\cos\theta = x_3 / |\underline{x}|$ and $T(\theta)$ is a factor of order one that represents the transmission coefficient of a plane air-water interface for sound propagating from air to water. The integration is over the space V_0 which contains all sources in the surface layer and is bounded by two linear horizontal surfaces $x_3 = \Delta$ and $x_3 = 0$, as shown in figure 2. In (5.1) we have split the density of a fluid particle at $\underline{x}$ into its mean value $\rho_0(\underline{x})$, plus the small deviation $\rho_m'(\underline{x}, t)$ due to compressibility; $\rho(\underline{x}, t) = \rho_0(\underline{x}) + \rho_m'(\underline{x}, t)$. Q is used to denote all contributions to the sound from sources of the same order as, or less efficient than, the pure turbulence quadrupole sources.

Equation (5.1) clearly reveals the source mechanisms of the sound generation. Apart from the turbulence radiation Q, the ocean surface induces two kinds of additional sources. The first, given by the first term in the integrand, is similar to that discussed by Ffowcs Williams (1986) in sound production by mixed jet flow, that is, the unsteady momentum fluctuation in the source flow. The sound of these sources is of dipole type, the axes of the dipoles being perpendicular to the mean position of the ocean surface, and is noisier than that from the pure turbulence radiation Q by the inverse of the typical Mach number of the flow. These vertical dipole sources arise directly from the non-uniform mean density ρ_0, as already shown in section 3; Ffowcs Williams (1986) has shown that, when ρ_0 is uniform, the sound from this term reduces to insignificance. In the ocean, the breaking waves that result in the non-uniform mean density are produced by the gravitational effects (Phillips 1977). The ocean surface agitates much more violently in the vertical direction, in which the gravitational forces act, than in the horizontal directions. Thus the vertical motion induced sources are much more powerful than those due to horizontal motions, a physical explanation to the results derived in section 3. The other kind of sources induced by the surface motions is represented by the second and the third term in the integrand of (5.1), and is associated with compressible motions of the source flow. Compressible motions in the ocean surface layer mainly result from the motions of the entrained air bubbles in the breaking, their volumetrical response to the turbulent pressure fluctuation for example. Motions of this kind has a negligible acoustical effect, which has been clear in section 3 from straightforward physical arguments and will be confirmed shortly by comparing the terms in (5.1).

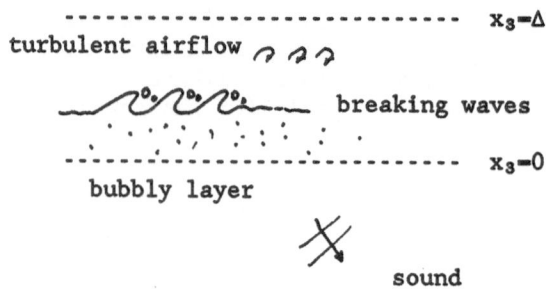

Figure 2. The geometry of the ocean sound problem.

The expression (5.1) can be reduced to that corresponding to the various existing theories when the respective approximations are made. The Lighthill quadrupole radiation Q is the major component of the sound when the ocean surface is only linearly deformed, in which case the integration in (5.1) that is the sound due to nonlinear surface motions vanishes (Ffowcs Williams 1986; Guo 1987c). The water turbulence-induced sound is, of course, of the order

$$Q \sim \rho_w \frac{\ell}{|\underline{x}|} M^4. \tag{5.2}$$

A first order approximation to the turbulent airflow when modelling its sound generation is to specify an unsteady force on the ocean surface and neglect the flow elsewhere. In this case, the water is bounded clearly by a continuous surface above which the pressure remains constant, and the terms in (5.1) all vanish, except for that proportional to $\rho_0 Du_3/D\tau$ which is equal to $-\partial p/\partial y_3$ from the momentum equation. Substituting this into (5.1), we can carry out the y_3-integral by noticing that the upper bound of this integral can be chosen arbitrarily since its contribution is always negligible, the pressure there being no greater than the sound field in air. A detailed proof this has been given by Ffowcs Williams (1986). Hence (5.1) simplifies to Powell's

(1960) form of Curle's (1955) equation for the sound of pressures on a flat surface.

$$\rho' = \frac{-x_3}{2\pi c_w^3 |\mathbf{x}|^2} \frac{\partial}{\partial t} \int_{y_\alpha} \left[p(y_\alpha, \tau) \right] d^2 y_\alpha. \tag{5.3}$$

This has been used respectively by Isakvich & Kur'yanov (1970) and by Guo (1987a) in examining the sound from surface pressures caused by local storms. In a storm, the airflow exerts an unsteady force which moves across the ocean surface. The unsteady force can be considered as composed of Fourier components of phase speeds ranging from subsonic to supersonic values. Evidently, those subsonic components with identical phase speeds to that of free surface waves produce surface waves by resonance (Phillips 1977), while those components of supersonic phase speeds radiate sound. This sound has been found to be an important contributor to the underwater noise below 50 Hz (see, e.g. Isakvich & Kur'yanov 1970; Guo 1987a).

The Brekhovskikh (1966) theory of ocean sound generation by weakly nonlinear interactions of surface waves can also be derived from (5.1). Similarly to the derivations of (5.3), the only nonzero term in (5.1) in this case is also that proportional to $\rho_0 Du_3/D\tau$, because the Brekhovskikh theory also neglects the effects of compressibility and Reynolds stresses in the source region. This time, the integration is performed in the water with its upper boundary set by the surface elevation, $\zeta(y_\alpha, \tau)$, say; ρ_0 is then equal to ρ_w. With this substitution, the total sound is different from Q only if the surface elevation ζ is not exactly linear (Ffowcs Williams 1986). This can be assumed to be the case when examining the Brekhovskikh mechanism since that theory accounts for the acoustic effects of weakly nonlinear surface motions. Thus the y_3-integral can effectively be performed from the mean position of the sea surface $y_3=0$ to $y_3=\zeta$. Expanding this integral in a power series in terms of ζ, retaining only the leading term in the expansion and noticing that $u_3=\partial\zeta/\partial\tau$ on the surface (for details, see Guo 1987c), we can derive

$$\rho' = \frac{-x_3}{4\pi c_w^3 |\mathbf{x}|^2} \frac{\partial^3}{\partial t^3} \int_{y_\alpha} [\zeta^2] d^2 y_\alpha. \tag{5.4}$$

It is easy to identify that this is the Brekhovskikh (1966) sound derived by Hughes (1976) through the method of small parameter perturbation, and by Guo (1987b) from the Ffowcs Williams-Hawkings equation (2.4). Since the specification of the ocean surface, f=0 in equation (2.4), is usually unknown and complicated, particularly in the case of breaking waves, the Ffowcs Williams-Hawkings equation may not be generally applicable to the modelling of ocean sound generation. However, it can conveniently be used to examine the Brekhovskikh theory of sound generation by weakly nonlinear interactions of surface waves. This is a possible physical mechanism because two surface waves of subsonic phase speeds, as is the case for ocean surface waves, can interact nonlinearly with each other to form surface motions of supersonic phase speeds, if the two share the same frequency and traval in almost opposite directions. The supersonic phase velocity elements do, of course, radiate sound.

Theory (5.3) and (5.4) both deal with low frequency sound in the ocean and they respectively model two different processes in the ocean surface layer, namely, the turbulent air flow and the wave-wave interactions. From the limited observations available from the ocean, these two theories both give, to some degree, reasonable interpretations to some measurements, and this makes it difficult to decide experimentally which mechanism is more important than the other, or whether they are equally important. It is then natural to examine these two mechanisms theoretically in an attempt to determine the dominant one. Such a theoretical modelling has been given by Guo (1987b). He has shown that, though the wave-wave interaction mechanism is physically a possible source of sound, the Brekhovskikh theory is very likely to be irrelevant to the sound generation in the real ocean. In the case of low frequency and small wind, where the ocean surface is only weakly deformed and remains intact, the sound from interactions of surface waves is weak in comparison with the direct radiation from the airflow that produced the surface waves. An appreciable sound from wave-wave interactions necessarily implies the violent surface agitations usually associated with breaking. In that case, both (5.3) and (5.4) are negligible compared with the density fluctuation arising from the unsteady momentum effects in the wet air above the water. This therefore leads to the conclusion that the wave-wave interaction mechanism may never be an important source of sound in the real ocean.

When waves are breaking, foaming and splashing are two major features of the surface flow. Foaming entains air bubbles into water, which induces, according to (2.3) that is derived by Crighton & Ffowcs Williams (1969), additional monopoles and dipoles, and splashing causes rapid momentum flux which correspond to powerful dipole sources, as demonstrated in section 3 from the Lighthill equation (2.2). These features also manifest themselves in the general solution (5.1), which is now dominated by the first term in the integrand. Both the second and third term in the integrand result from compressible motions of the bubbles and are monopole-like sources, whilst the first, representing the sources due to momentum fluctuations, arise partly from the water sprays above the ocean and partly from the displacement of water by air bubbles.

The dominant sound source, when breaking surface waves are present, arises from the splashing of water sprays. The sound field from this mechanism is larger than the pure turbulence radiation Q by the inverse of the typical Mach number of the flow which is vanishingly small for the ocean. The bubble induced sound is negligible because it is at most of the same order as Q. This can be shown by estimating the sounds from the second and third term in the integrand. If we still denote by u and ℓ respectively the typical velocity and length scale, the pressure p within the flow should be scaled on $\rho_w uU$, U being the uniform wind speed, and $D/D\tau \sim \partial/\partial t$ is of the order u/ℓ. Hence the second and the third term in the integrand of (5.1) give the sounds as being of the order

$$\rho_w \, \frac{\ell}{|\underline{x}|} \, M^4 \, \frac{c_w^2}{c_m^2} \, M \quad \text{and} \quad \rho_w \, \frac{\ell}{|\underline{x}|} \, M^4 \, \frac{c_w^2}{c_m^2} \, \frac{\rho_a c_a}{\rho_w c_w}. \tag{5.5}$$

These result can be compared with (5.2), the sound from pure turbulence radiation. It is clear then that (5.2) and (5.5) are of the same order, because both the Mach number M and the impedence ratio $\rho_a c_a / \rho_w c_w$ always offset the factor $(c_w/c_m)^2$. Therefore, both the bubbles and the pure turbulence are of negligible significance in radiating sound.

The ineffectiveness of the bubble motions in the ocean surface layer in radiating sound can also be demonstrated by directly comparing their sound with that from water sprays, that is, the sound from the first term in the integrand of (5.1). By scaling p as $\rho_w u U$ and ρ'_m as $\rho_w u U/c_m^2$ by definition, the ratio of the first to the second and the third term are, respectively,

$$\frac{c_m^2}{uU} \quad \text{and} \quad \frac{\rho_w}{\rho_a} \, \frac{c_m^2}{c_a^2} \, \frac{c_a}{U}. \tag{5.6}$$

The first ratio is very much bigger than unity, because c_m, the sound speed in the bubbly flow, is always much bigger than u even at its minimum value that is about 20 m/s and is of the same order as the wind speed U, while the second ratio is evidently at least of the same order as the inverse of the wind Mach number. Hence the bubble sounds are clearly negligible.

6. CONCLUSIONS

Previously, weakly non-linear interactions of surface waves have been regarded as the main contributor to the wave-associated sound in the ocean. However we have shown that this is not the case. The failure in correctly accounting for the ocean sound generation of the previous theories results from two aspects, namely, the ignoring of the turbulent airflow in the low frequency regime and the neglecting of the fully non-linear processes of the surface motions in the high frequency regime. Ocean surface waves may generate underwater sound of very low frequencies by non-linear interactions, but this sound is not necessarily the dominant component because the turbulent airflow that produced the waves can also directly radiate sound.

As frequency increases, the ocean surface moves more and more violently so that interactions of surface waves seem to become important. However the ocean surface cannot be violently deformed without breaking, which is a fully non-linear phenomenon and is obviously, whenever present, the dominant process that generates sound. Any theory that fails to take this into account cannot be expected to correctly interpret the generation of ocean sound. It has been shown that the occurence of breaking on the surface actually prevents the interactions of surface waves from becoming the dominant generation mechanism of the underwater sound; the aerial turbulence radiation remains dominant over the interaction sound until the fully non-linear breaking takes over the leading role in producing sound.

We have examined in detail the sound from surface wave breaking. The breaking of ocean waves is a complex phenomenon involving various flow processes. By examining their relevant acoustic radiation efficiencies, we have identified the splashing of water sprays as the main sources of sound. It is the momentum fluctuations caused by the splashing on the surface that is responsible for the generation of the underwater sound. Unsteady bubble motions, which may be efficient acoustic sources in some situations, have been shown to be negligible for the wave-induced underwater sound in comparison with the splashing.

REFERENCES

Brekhovskikh, L.M., 1966: 'Underwater sound waves generated by surface waves in the ocean.' Izv. Atmospheric and Oceanic Phys. v.2, No.9, 970-980.

Crighton, D.G. and Ffowcs Williams, J.E., 1969: 'Sound generation by turbulent two-phase flow.' J. Fluid Mech., 36(3), 588-603.

Curle, N., 1955: 'The influence of solid boundaries upon aerodynamic sound.' Proc. Roy. Soc. Lon., A231, 412.

Ffowcs Williams, J.E., 1984: 'The acoustic analogy — thirty years on.' IMA J. Applied Math., 32, 113-124.

Ffowcs Williams, J.E., 1986: 'Waves in turbulent mixing layers.' Recent advances in aerodynamics and aeroacoustics. Springer verlag.

Ffowcs Williams, J.E. and Hawkings, D.L., 1969: 'Sound generation by turbulence and surfaces in arbitrary motion.' Phil. Trans. Roy. Soc. Lon., A264, 312-342.

Guo, Y.P., 1987a: 'Waves induced by sources near the ocean surface.' J. Fluid Mech., 181, 293-310.

Guo, Y.P., 1987b: 'On sound generation by weakly nonlinear interactions of surface gravity waves.' J. Fluid Mech., 181, 311-328.

Guo, Y.P., 1987c: 'Sound generation in the ocean by breaking surface waves.' J. Fluid Mech., 181, 329-347.

Hughes, B., 1976: 'Estimates of underwater sound produced by nonlinearly interacting ocean waves.' J. Acoust. Soc. Am., 60(5), 1032-1039.

Isakvich, M.A. and Kur'yanov, B.F., 1970: 'The theory of low-frequency noise in the ocean.' Soviet Phys. Acoust., 16(1), 49-58.

Lighthill, J.M., 1952: 'On sound generated aerodynamically: 1. general theory.' Proc. Roy. Soc. Lon., A221, 564-587.

Phillips, O.M., 1977: The dynamics of the upper ocean. CUP.

Powell, A., 1960: 'Aerodynamic noise and the plane boundary.' J. Acoust. Soc. Am., 32, 8, 982.

Wenz, G.M., 1962: 'Acoustic ambient noise in the ocean: spectra and sources.' J. Acoust. Soc. Am., 34, 1936-1956.

AMBIENT NOISE RADIATION BY "SOLITON" SURFACE WAVES

R. H. Mellen and D. Middleton
Planning Systems Incorporated
95 Trumbull Street
New London, CT 06320
U.S.A.

ABSTRACT.   Acoustic backscattering from surface waves is generally much greater than that predicted by gravity-capillary models. A "soliton" model has been proposed that accounts for both the cross-sections and the Doppler spectra observed in backscattering experiments under known bubble-free conditions. The existence of soliton-like surface waves is supported by wave-gauge experiments. Nonlinear acoustic radiation by wave-wave interaction is also a potential mechanism for ambient noise. Estimates of noise levels for gravity-capillary waves are much too low at frequencies above about 10 Hz, whereas those for random ensembles of solitons show more reasonable agreement with ocean measurements.

1.   INTRODUCTION

Acoustic backscattering from ocean waves is generally much greater than that predicted by conventional gravity-capillary wave models. For small grazing angles and moderate wind speeds, the excesses can be as much as 15-20 dB at the higher frequencies (F>10 kHz). CW Doppler measurements also show no evidence of the predicted dispersion. We have proposed a "soliton" model that accounts for both the cross-sections and Doppler spectra observed in backscattering experiments under known bubble-free conditions [1,2].

The existence of small-scale waves with soliton-like properties is supported by wave-gauge data from wind-wave flume experiments. The saturation effect, common to the gravity régime, is not observed at the higher wavenumbers and spectral levels increase with wind-speed. Phase speed measurements also show no evidence of dispersion except for the lower gravity régime.

Wave-wave interaction is one of many possible sources of underwater ambient noise. Acoustic radiation from a free surface is second-order. Calculations based on conventional gravity-capillary wave spectra have been found to be much too low at frequencies above about 10 Hz. In this paper, we examine soliton-soliton collision as an alternative mechanism that appears to be more consistent with experiment.

B. R. Kerman (ed.), Sea Surface Sound, 325–335.
© 1988 by Kluwer Academic Publishers.

2. WAVE GAUGE EXPERIMENTS

Figure 1 shows point-frequency spectra of elevation measured in a wind-wave flume by Mitsuyasu and Honda [3] for 8.25 m fetch and mean wind speeds of 5, 10 and 15 m/s. The dashed line is the Pierson-Moskowitz [4] spectrum for the fully-developed equilibrium sea:

$$G^2(f) = 5 \times 10^{-6} \ g^2 \ \exp[-5 \times 10^{-4} \ (g/fU_\infty)^4]/f^5 \qquad (1)$$

where g is the gravity constant, f is wave frequency in Hz and U_∞ is the mean wind-speed. The P&M spectrum is saturated in the frequency range shown, the peaks falling below 1 Hz (f(peak)=0.14 Hz for U_∞ =10 m/s). The peaks of the data decrease in frequency with increasing wind speeds but the fetch is far too small to approximate the fully-developed sea. At the higher frequencies, the data fall above the P&M asymptote and show f^{-3} rather than f^{-5} dependence. The spectrum is unsaturated and both amplitude and propagation speed evidently increase with wind speed.

Measurements of wave phase-speed were made in a wind-wave flume by Ramamonjiarisoa et. al. [5] for wind speeds 5 and 8 m/s. Phase speed vs. frequency was determined by correlating the signals from two wave gauges. Separation distances of 3 and 10 cm in the downwind direction were used. Figure 2 compares the data to linear theory (solid curve). Dispersion is evident only below 3 Hz. The f^{-3} asymptote in the frequency spectrum of Figure 1 corresponds to K^{-4} for the wavenumber spectrum if the speed of propagation is non-dispersive; i.e. $K=2\pi f/C$ where C depends only on the wind speed.

Absence of dispersion suggests that nonlinear effects produce wave packets that propagate at nearly constant speeds. Shemdin [6] proposed surface drift as the likely mechanism. The drift is an exponentially decaying current with surface speed about 4% of the wind speed and 1/e thickness of the order of 1 cm. Similar to shallow-water gravity waves, ripples on a drift layer steepen with increasing amplitude due to the drag effect of the deeper water.

Frequency spectra of periodic steep-fronted waves are composed of harmonics of the fundamental and harmonic growth involves a cascade of energy to higher wavenumbers only. Instability produced by wave-wind interaction can destroy periodicity, causing cascade of energy to lower wavenumbers. Under these conditions, unstable wave-packets should tend to decompose into soliton components.

We have proposed a qualitative 4-stage model for wave growth [1,2].
a. Initiation of wind-generated ripples (catspaws) at 16 Hz, near the frequency of minimum phase-speed.
b. Development of surface drift with steepening of the wavefronts. Energy cascade is toward higher wavenumbers.
c. Intermodulation between waves and wind with rapid destruction of periodicity. Cascade of energy reverses to lower wavenumbers.
d. Equilibrium stage where incoming energy is balanced by dissipation. Unstable smaller-scale disturbances should then decompose into soliton components, which are continually created and destroyed at "boundaries."

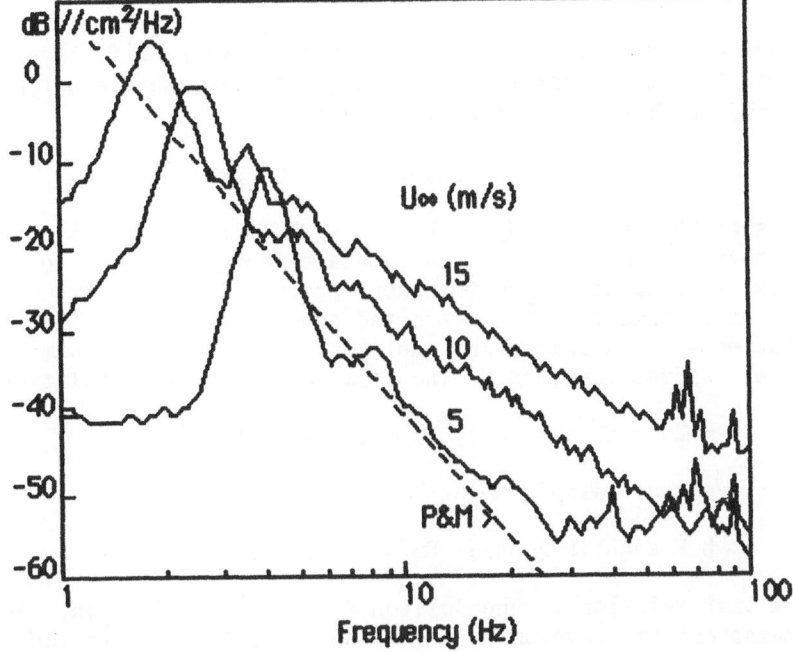

Figure 1. Point-frequency elevation spectra of surface waves.

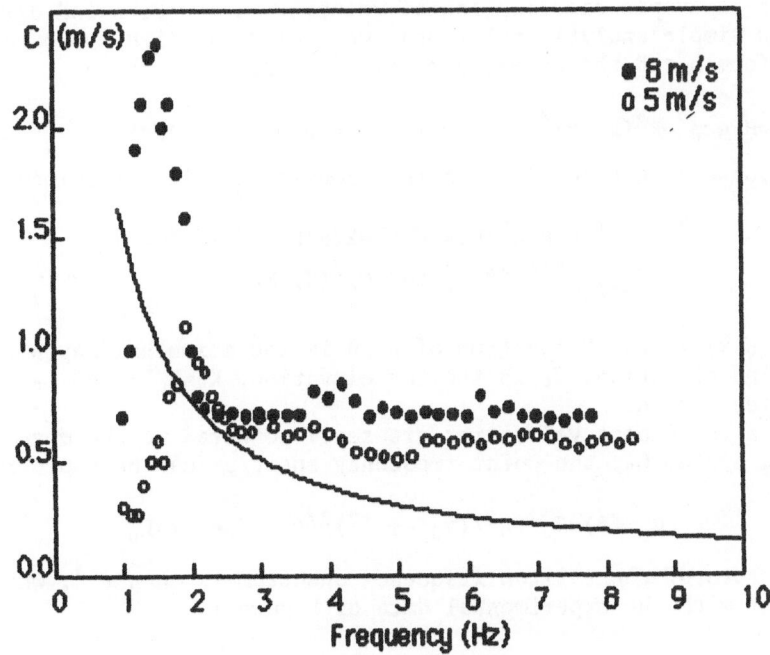

Figure 2. Phase speed of surface waves.

3. SOLITON THEORY

Propagation of the smaller-scale waves on the moving drift-layer will be approximated analytically by the nonlinear plane-wave equation of Korteweg and de Vries (see Lighthill [7]):

$$v_t+(C+v)v_x+(C\ D^2/6)v_{xxx}=0 \qquad C=(gD)^{1/2} \tag{2}$$

where D is an "effective" layer thickness, C is the speed associated with depth-limited waves, v is the excess velocity relative to C and subscripts x/t indicate space/time derivation. Solutions are periodic "cnoidal" waves that resemble distorted sinusoidal waves. The wave period increases with increasing amplitude until the waves act like independent "hydraulic-bumps." The asymptotic "soliton" solution is:

$$v=3V\ \text{sech}^2[k(x-ct)] \quad c=C+V \quad k=[3V/2C]^{1/2}/D \tag{3}$$

In terms of vertical displacement h:

$$h=H\ \text{sech}^2[k(x-ct)] \quad H=2DV/C \quad kH>(kD)^3 \tag{4}$$

where the last relation is the soliton "existence" criterion. Amplitude H and characteristic wavenumber k are thus specifically related and the scaling relation is $H=H_0(k/K_0)^2$.

Since solitons are not distorted by collision, linear superposition can be assumed for the ensemble, which is taken to be a Poisson process with semi-isotropic distribution of velocity in the downwind direction. To obtain simple analytic solutions, we employ Gaussian approximations. The waveform h and the wavenumber pdf W(k) are taken as:

$$h=H\ \exp[-k^2(x-ct)^2] \quad W(k)=2K_0^6\ \exp(-K_0^2/2k^2)/\pi k^8 \tag{5}$$

The soliton-ensemble spectrum is then consistent with experiment; i.e.:

$$G_s^2(K)=_0\!\int^\infty kdk\ H^2(K,k)W(k)=2(H_0K_0)^2/\pi[K_0^2+K^2]^2$$
$$-\pi/2\!\int^{\pi/2} d\theta\ _0\!\int^\infty KdK\ G_s^2(K)=H_0^2 \tag{6}$$

where $H^2(K,k)$ is the K-spectrum of h, θ is the azimuthal angle relative to the wind direction, H_0 is the rms elevation, $K_0=2/L_0$ and L_0 is the correlation length.

Since the excess velocities are small compared to the mean speed in the fixed system C_0, the point frequency spectrum can be approximated:

$$G_s^2(f)\approx H_0^2f_0^2/(f_0^2 + f^2)^{3/2} \quad f_0=C_0/\pi L_0 \tag{7}$$

The asymptote of the soliton frequency spectrum is then f^{-3}, which is consistent with the experimental data of Figure 1.

4. BACKSCATTER STRENGTH

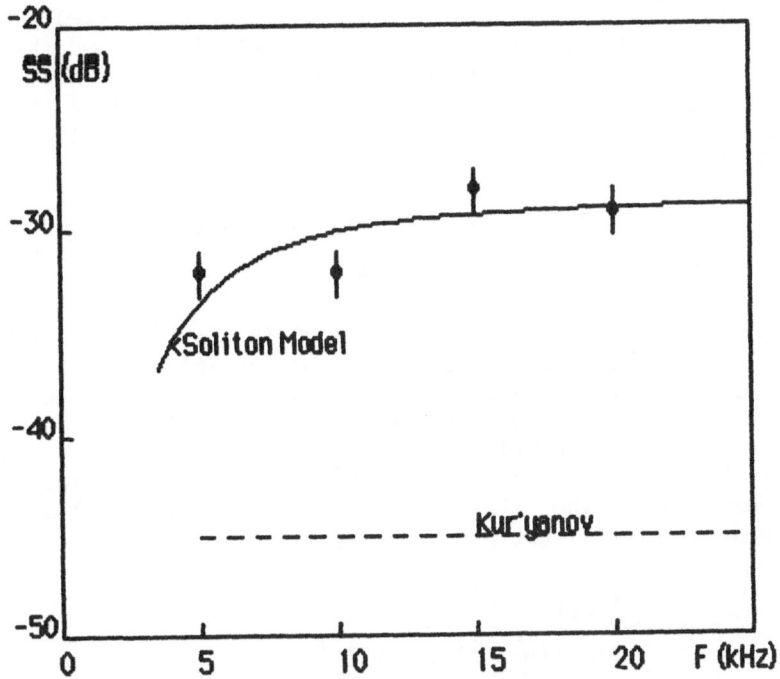

Figure 3. Backscatter strength vs. frequency.

In the Kirchhoff approximation, the backscatter strength SS is:

$$SS=N(\phi,\theta) \; k_a^4 \; G^2(K) \quad K=2k_a \cos\phi \quad k_a=2\pi F/C_a \tag{8}$$

where k_a is acoustic wavenumber, F is acoustic frequency, C_a is sound speed and $K=2k_a \cos\phi$ is the Bragg-scattering condition. The slope factor is $N(\phi,\theta)$, where ϕ and θ are the grazing and azimuthal ray-angles, respectively. Slope effects of isotropic gravity waves on the smaller-scale scatterers can be approximated as [1]:

$$N(\phi)\approx3 \; \sigma^4 \cos^4\phi + 6 \; \sigma^2 \cos^2\phi + \sin^4\phi \tag{9}$$

where σ^2 is the "effective" variance of slopes. Estimates based on photographic surface-glitter experiments [8] show: $\sigma^2\approx(3+5.12U_\infty)\times10^{-3}$ where U_∞ is in m/s.

Figure 3 compares estimates based on Kuryanov's composite-surface model [9] and the "soliton" model with experimental CW data [10] for the mean wind speed 10 m/s and 9° grazing angle ($\phi = 9°$). The soliton spectrum was simply added to the normal gravity-wave spectrum. The values of the statistical parameters for the soliton spectrum $H_0=2cm$ and $L_0=5cm$ were selected for "best" fit.

5. BACKSCATTER DOPPLER

Doppler spectra were obtained by FFT analysis of backscattered signals.

Doppler shift was measured between the spectral peak and the carrier frequency. Calculations for "classical" and "soliton" wave-models are compared with experimental data in Figure 4. The theoretical expressions for mean Doppler-shift for the Bass and Fuks [11] and the soliton models are given by [1]:

$$\Delta F_{B\&F} = (Fg \cos\phi /\pi C_a)^{1/2} + (2F \cos\phi /C_a) U_c \cos\theta_c$$

$$\Delta F_s = (2F \cos\phi /C_a)(U_c \cos\theta_c + C_0 \cos\theta) \tag{10}$$

where C_0 is the speed of the soliton waves, θ is the azimuthal angle of observation, U_c is the speed of the water current and θ_c is the current azimuthal angle, all with respect to the wind direction. Values for the experiment are $\theta=26°$, $U_c=26$ cm/s and $\theta_c=28°$. We estimate $C_0\approx30$ cm/s.

Calculations of the Doppler spread for the two models are compared with data in Figure 5. The spread is measured at the 1/e points of the spectra. Formulae for the "classical" B&F model (solid line) and the "soliton" model (curve 1) are given by [1]:

$$\underline{\Delta F}_{B\&F} \approx 1\times10^{-3} F.$$

$$\underline{\Delta F}_s \approx [(A^2+B^2) F^2 +D^2]^{1/2} \tag{11}$$

where F is in Hz. The term A accounts for phase modulation by large-scale gravity waves, B accounts for the variations in soliton velocity and D is a frequency-independent random amplitude-modulation correction arising from variations in slope due to large-scale gravity waves. The data curve-fit values are: $A=3.3\times10^{-4}$, $B=7.8\times10^{-4}$ and $D=15$. The dashed lines 2 and 3 show the effects for $D=0$ and $B=D=0$, respectively. It is evident that, for small grazing angles and low frequencies, the soliton model ascribes most of the Doppler spread to B and D. At the larger angles, the phase-modulation term A becomes dominant and the two theories yield nearly the same values.

From analysis of backscattering data, it is clear that the soliton model gives good account of both the cross-sections and Doppler spectra in some detail. It is also clear that it is also a possible candidate as an ambient-noise mechanism. Radiation by wave-wave interaction at a "free" surface is a nonlinear function of particle velocity. In the second-order (Bernoulli) approximation, the radiated pressure is assumed proportional to the square of the vertical velocity. Since solitons propagate only in the downwind direction, the radiation mechanism depends on the overtaking effect inherent in a random ensemble. Only the cross-terms in velocity of soliton-pairs will be considered.

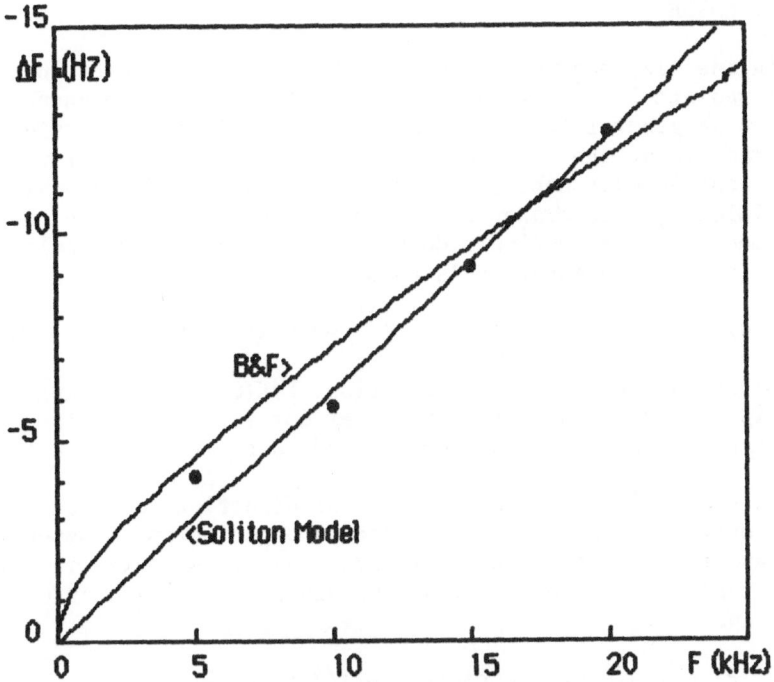

Figure 4. Backscatter Doppler-shift vs frequency.

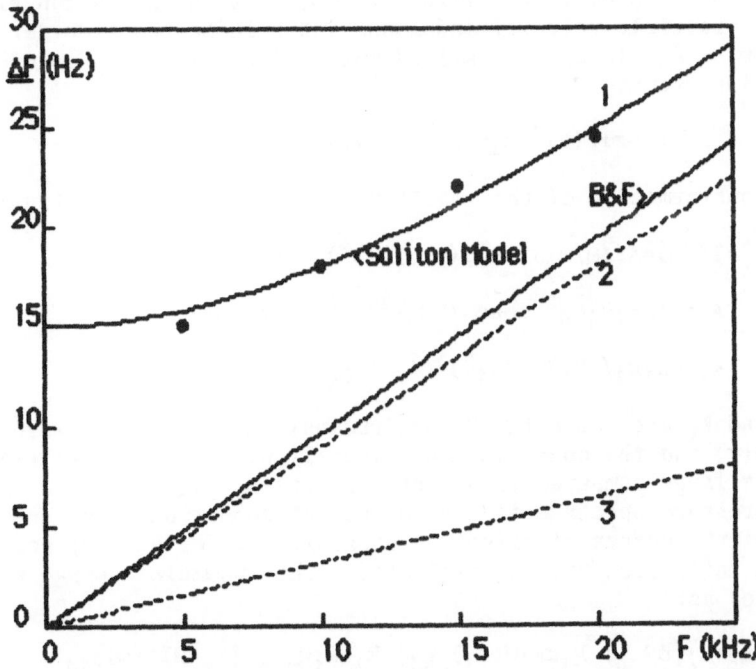

Figure 5. Backscatter Doppler-spread vs frequency (see text).

6. AMBIENT NOISE

Longuet-Higgins [12] showed that monochromatic gravity waves in opposite directions radiate in a vertical direction at twice the frequency. It can be shown in general that any two waves of arbitrary direction and frequency radiate at the sum frequency only if the wavenumber vector sum is a real acoustic wavenumber. For random ensembles of dispersive waves, pressure fields will be evanescent except for nearly-opposite components. Wave frequencies are effectively doubled and radiation is dipole. The spectrum of acoustic pressure P is approximately [13]:

$$P^2(2\omega) \simeq 8\pi\omega^6 [\rho G^2(K)/C_a]^2 K K_\omega \int_0^{2\pi} d\theta\, G^2(\theta)\, G^2(\theta+\pi) \tag{12}$$

where $G^2(\theta)$ is the wave direction-function, $K=K(\omega)$ is the dispersion relationship, $\omega=2\pi F$, subscript ω indicates derivation and ρ is density. Predictions based on Eq. 11 have been found to be much too low for frequencies above about 10 Hz.

Although traveling in the same general direction, solitons should radiate by overtaking collisions because propagation speed depends on amplitude and interacting wavenumbers are "nearly-opposite" in a moving frame. We take the farfield part of the dipole Green's-function for the ij^{th} collision as:

$$P \simeq \rho \iint dS\, [U_i U_j]_t\, \cos\phi\delta(t-R/C_a)/2\pi R C_a \qquad U \simeq -C_0 h_x \tag{13}$$

where ϕ is the normal angle to a field point at range R, S is the surface area, C_0 is mean speed in the fixed system, δ is the delta function, U is vertical particle-velocity and subscript t indicates time derivation. Integration of Eq. 13 can be carried out analytically for "point" sources in the moving frame:

$$x'=x+t(k_i^2 c_i + k_j^2 c_j)/(k_i^2 + k_j^2) \tag{14}$$

Fourier transformation of the result yields the impulse spectrum:

$$I^2(\omega)=A_1(\omega/\omega_0)^6\, \exp(-\omega^2/2\omega_0^2) \qquad \omega_0=K_\lambda(c_i-c_j)$$

$$A_1= (\rho H_i H_j C_0^2 \cos\phi /RC_a)^2\, K_\lambda^4/4\pi\, K_\tau^2(k_i^2 + k_j^2) \tag{15}$$

$$K_\lambda =k_i k_j/(k_i^2 + k_j^2)^{1/2} \qquad K_\tau=(k_{i\tau}^2 + k_{j\tau}^2)^{1/2}$$

where K_λ and K_τ are "effective" longitudinal and transverse wavenumbers, respectively, and the condition for "locally-plane" wave with transverse wavenumbers large compared to acoustic is $K_\lambda \gg K_\tau \gg k_a$.

The pressure spectrum is the incoherent summation of an ensemble of dipole "point" sources of effective area $\pi/K_\lambda K_\tau$. Effects of refraction and bottom reflection will be neglected. The ensemble average will be approximated as:

$$P^2(\omega) \simeq \iint dS\, _0\!\int^\infty k_j dk_j W(k_j)\, _{kj}\!\int^\infty k_i dk_i W(k_i)\, [K_\tau^2 \Omega I^2(\omega)]_{ij} \tag{16}$$

This comes from normalization by dividing by effective area, taking the effective range of collision angles $\Delta\theta \simeq K_\lambda/K_T$ and collision frequency $\Omega \simeq K_0(c_i-c_j)$, where $\Omega>0$ is the overtaking criterion.

After making appropriate substitutions in Eq. 16 and completing the surface integration, we have:

$$P^2(\omega) \simeq A_2 \, \omega^6 \, A_3^5 \int_0^\infty dk_j \, k_j \int^\infty dk_i \, \exp(-F_1)/F_2^5$$

$$F_1 = \{K_0^2 + [\omega A_3/(k_i^2 - k_j^2)]^2\}/2K_\lambda^2 \qquad F_2 = k_i k_j (k_i^2 - k_j^2) \qquad (17)$$

$$A_2 = 2(\rho/\pi C_a)^2 (K_0 H_0 C_0)^4 \qquad\qquad A_3 = K_0^2/V_0$$

where V_0 is the mean excess velocity. Note that K_T vanishes.

Eq. 17 can be solved asymptotically by omitting the low-frequency rolloff term K_0^2 in F_1 and converting to the polar coordinate system $k_i = \beta \sin\alpha$, $k_j = \beta \cos\alpha$, $\beta^2 = k_i^2 + k_j^2$, $\alpha = \arctan(k_i/k_j)$. The result is:

$$P^2(\omega) \simeq A_2 V_0/3K_0 \qquad\qquad (18)$$

The asymptotic spectrum is seen to be independent of frequency, the levels increasing linearly with mean excess velocity. Of course, high-frequency rolloff and other effects are to be expected as the result of surface tension, viscosity and perhaps other factors not yet considered in the model.

Assessment of the soliton model is limited as much by availability of suitable experimental data as by the roughness of the approximations and the lack of specific details on the statistical parameters of the surface. Least contaminated by sources other than local wind appear to be data obtained in the Pacific Ocean with hydrophones at depths up to nearly 5 km [14] (critical depth is approximately 4 km).

Backscatter cross-sections show only moderate wind-speed dependence in the range 5-15 m/s and saturation at speeds greater than about 15 m/s. Rms elevations and correlation lengths are adjusted accordingly.

Taking mean soliton speed on the drift layer as 4% of wind speed is also consistent with backscatter Doppler-shift. To achieve reasonable fit with the noise data, we arbitrarily take the mean excess speed V_0 as 10% of C_0.

The results are shown in Figure 6. Experimental data are shown as points. The corresponding solid curves are numerical calculations with Eq. 17. The gravity-wave component from Eq. 12 is added. The dashed curve shows a typical overall result of Eq. 12 for comparison.

The predicted trends appear fairly consistent with the data. The large excess in the 5 m/s data below 100 Hz is ascribed to the residual long-range propagation effects, which become dominant at all wind speeds at depths much less than critical. The 15 m/s data agree with predicted slope but fall well below predicted level, which suggests that noise may saturate in this range as well as backscattering cross-section. Bubble generation by breaking waves may easily be responsible.

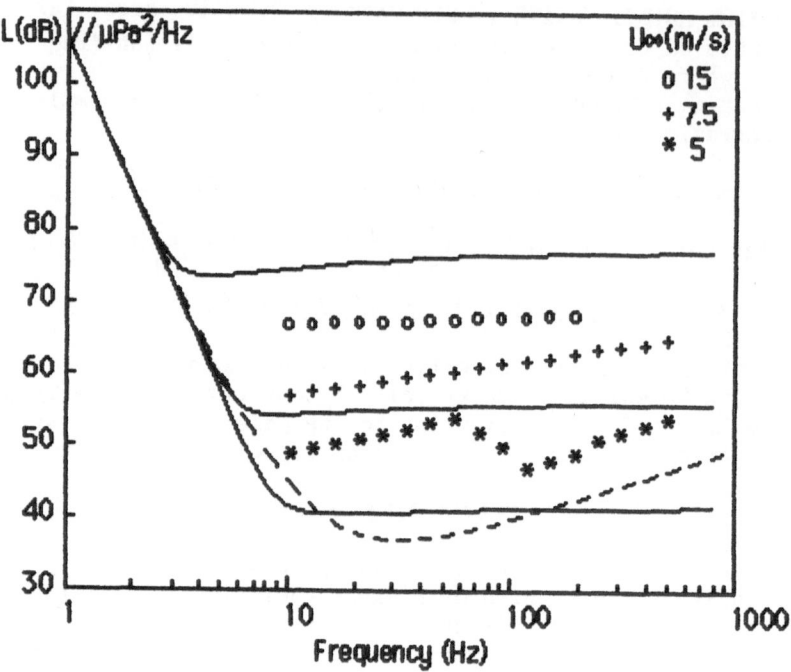

Figure 6. Ambient noise spectra (see text).

7. CONCLUSION

General consistency between acoustic noise and backscatter predictions
give support for a soliton-like process as the basic mechanism for both
phenomena. However, a major problem to be resolved is the wind-speed
dependence of ambient noise. Levels for the soliton model increase as
a very high power of wind-speed and decrease to the fifth power when
the surface K-spectrum is saturated. The turbulent-pressure model of
Isokovitch and Kuryanov [15] has fourth-power dependence. Experimental
data appear to indicate only the second-power or less [16].

It should also be emphasized that the proposed "soliton" model is
heuristic and we lack confirming evidence that the KdV equation really
describes the surface dynamics of wind-driven surfaces. Other processes
could be involved instead of the balance between nonlinearity and
dispersion, as in KdV. However, only minor changes may be required in
the statistical model. For example, Stokes wave-packets have been
considered by Mollo-Christensen and Ramamonjiarisoa [17], employing
similar Gaussian approximations of the waveforms.

Clearly, the surface statistics must be known in much more detail
and more attention given to the hydrodynamical aspects before an acceptable quantitative model can be developed.

REFERENCES

1. D. Middleton and R. H. Mellen, 'Wind-generated solitons: a potential significant mechanism in ocean surface wave generation and surface scattering', IEEE J. Ocean. Eng. 10 471-476 (1985)

2. R. H. Mellen, D. Middleton and J. W. Fitzgerald, 'Sea-Surface back-scattering and the soliton mechanism', NUSC Tech. Doc. 7583, Feb. 1986

3. H. Mitsuyasu and T. Honda, 'The high-frequency spectrum of wind-generated waves', J. Oceanographic Soc. Japan, 30, 185-198 (1974)

4. W. L. Pierson Jr. and L. Moskowitz, 'A proposed spectral form for fully-developed seas based on the similarity theory of S. A. Kitaigorodsky', J. Geophysical Res. 69 5180 (1964)

5. A. Ramamonjiarisoa, S. Baldy and I. Choi, 'Laboratory studies on windwave generation, amplification and evolution', in Turbulent Fluxes Through The Sea Surface, editors: A. Favre and K. Hasselmann (Plenum Press, New York 1978) pp. 403-420

6. O. H. Shemdin, 'Wind-generated current and phase speed of water waves', J. Phys. Oceanography, 2 411-419 (1972)

7. J. Lighthill, Waves in Fluids, (Cambridge University Press, 1979), Epilogue, Part 2, pp. 463-467

8. C. S. Cox and W. H. Munk, 'Measurement of the roughness of the sea surface from photographs of the sun's glitter', J. Opt. Soc. Am., 44 838-850 (1954)

9. B. F. Kuryanov, 'The scattering of sound at a rough surface with two types of irregularity', Sov. Phys. Acoust. 8 252-257 (1963)

10. W. I. Roderick, J. B. Chester and R. K. Dullea, 'High-frequency acoustic backscatter from the sea surface', NUSC Tech. Doc. 7183, July 1984

11. F. G. Bass and I. M. Fuks, Wave Scattering From Statistically Rough Surfaces, (Pergamon Press, New York 1979)

12. M. S. Longuet-Higgins, 'A theory of the origin of microseisms', Trans. Royal Soc. A243 1-35 (1950)

13. B. Hughes, 'Estimates of underwater sound (and infrasound) produced by nonlinearly interacting ocean waves', J. Acoust. Soc. Am. 60 1032-1039 (1976)

14. J. H. Wilson, 'Wind-generated noise modeling', J. Acoust. Soc. Am. 73 211-216 (1983)

15. M. A. Isokovitch and B. F. Kuryanov, 'Theory of low frequency noise in the ocean', Soc. Phys. Acoust. 16 49-58 (1970)

16. A. S. Burgess and D. J. Kewley, 'Wind-generated surface noise source levels in deep water east of Australia', J. Acoust. Soc. Am. 73 201-210 (1983)

17. E. Mollo-Christensen and A. Ramamonjiarisoa, 'Modeling the presence of wave groups in a random wave field', J. Geophys. Res. 83 4117-4122 (1978)

OCEAN NOISE SPECTRUM BELOW 10 HZ - MECHANISMS AND MEASUREMENTS

A.C. Kibblewhite
Department of Physics
University of Auckland
Private Bag
Auckland
New Zealand

ABSTRACT.   Underwater ambient noise is known to be wind dependent
but below 10 Hz spectral levels reach values many orders of
magnitude above those experienced in other parts of the spectrum.
The mechanisms responsible for the transfer of energy from
the wind to the noise field at low frequencies are reviewed.
Nonlinear interactions in the ocean wave field are identified
as the source of both this infrasonic component of the ocean-noise
field and wave-induced microseisms.   The characteristics of
the source, its dependence on sea state and wind speed, and
the nature of the relevant transfer functions are examined.

1.   INTRODUCTION

Spectral levels in all regions of the ambient sea-noise spectrum
are affected by local wind conditions. In his paper Wenz [1]
presented a generalised wind dependent noise spectrum, which
was supported by data from many experiments at frequencies
above 200 Hz, but below 200 Hz it was apparent that the sea-noise
level in general is controlled by non-wind related sources
such as biological and shipping activity.   However later measurements
indicated that the wind also has a considerable effect at frequencies
below 10 Hz, spectrum levels rising several orders of magnitude
above those in other parts of the spectrum [2, 3].
    Several mechanisms have been proposed to account for
the high spectral levels at low frequencies [4-12], but because
reliable ambient-noise data have been sparse at frequencies
below 10 Hz, it has been difficult to assess the agreement
between theory and experiment.   In a recent review Nichols
[13] presented new data and made comparisons with both theory
and other experimental results.   While the lack of detailed
environmental data did not permit definite identification,
his summarised evidence suggested that nonlinear interactions
between ocean gravity waves and/or turbulence were the most
likely noise generating mechanisms.

337

B. R. Kerman (ed.), Sea Surface Sound, 337–359.
© 1988 by Kluwer Academic Publishers.

Nonlinear wave interactions between ocean waves have
also been of interest to seismologists and oceanographers.
In particular they have been cited as a possible source of
wave-induced microseisms [14, 15].   Closely related studies
have thus been pursued in two separate branches of geophysics,
but until recently these studies have been conducted in effective
isolation.   In the last decade however the growing interest
of underwater acousticians in infrasonics has brought the two
communities more closely into contact.

It is the object of this paper to examine the field of
infrasonics from the acousticians stand-point, but in view
of the much longer interest seismologists have had in this
field it seems appropriate to introduce the subject through
a brief review of its impact on seismology.

2.   MICROSEISMS - HISTORICAL SURVEY

In the absence of transient seismic signals such as earthquakes,
the Earth is still found to be subject to minute tremors.
These continuous motions at periods less than 10 seconds (frequencies
greater than 0.1 Hz), are traditionally called microseisms
and define a noise threshold against which all other seismic
signals must be detected. Microseism activity is all pervasive,
being observed on land and on the deep-ocean floor.

Although microseisms have been observed for over 100 years
a clear understanding of their properties and origin has been
slow to appear.   Useful bibliographies are provided by Darbyshire
[16] and Melton [17], but even more instructive are those available
in the doctoral dissertations of Ewans [18], Adair [19], and
Webb [20], all of whom have recently contributed significantly
to the subject.   The value of these sources to this review
is warmly acknowledged.

Because of the sensitivity and passband limitations of
the early instrumentation, observations were generally restricted
to high-amplitude activity.   It was nevertheless quickly recognised
that a close connection existed between microseism activity
and disturbances in the marine environment, and the action
of surf on steep coasts was suggested as a prime source [21].
However evidence in favour of a mechanism involving atmospheric
pressure fluctuations [22] and confusion over deep and shallow
water sources [23] kept the issue in doubt.   Additional confusion
arose with Banerji's report [23] of microseisms characterised
by a spectral peak at double the frequency of those first reported.
Other reports of this property, and further observations of
deep water sources intensified the debate [24].

Understanding developed when Longuet-Higgins [14] recognised
the significance of a theoretical study by Miche [25].   Miche
had evaluated the hydro-dynamical equations governing standing
waves to second order and identified a pressure term which
was proportional to the square of both the amplitude and the

frequency of the ocean surface waves, had twice their frequency, and was unattenuated with depth. Longuet-Higgins recognised this pressure term as possibly being of relevance to the generation of microseisms and proposed a number of ocean disturbances, such as fast moving storms and coastal reflection, as mechanisms capable of producing the necessary standing waves. His original theory has since been augmented by several investigators - see Section 4.

Over the next decade effort was expended to detect and track storms at sea through the microseism energy associated with them [16]. However the method proved unreliable. Throughout this period reservations persisted as to the precise mechanism by which standing waves could be excited [26].

In the 1960's the appearance of improved instrumentation brought renewed interest in the subject. Much of this emphasis stemmed from the concerted effort to discriminate nuclear explosions from natural seismic events. In this period Haubrich et al. [27] identified seismic components at both the same (primary) frequency (PF) and double the frequency (DF) of the active swell. They demonstrated moreover that the DF components occurred at roughly 100 times the intensity of the PF counterpart and could indeed be ascribed to the reflection of swell from the coast.

An explanation for the PF component followed from the theoretical work of Hasselmann in 1963 [15]. In a subsequent development Darbyshire and Okeke [28] introduced damping effects into the elastic system. They derived a relation for the ratio of the microseism and ocean-wave spectral levels but achieved only an order of magnitude agreement with observations.

The first reliable microseism measurements at the ocean bottom were made during this period, following the development of ocean bottom seismometers (OBS). Objectives were two fold.

The first was concerned with the question whether microseism generation did occur in deep water well away from shore. In the majority of cases microseism activity was found to correlate best with wave activity when storms approached a coastal region, but an isolated observation of a source 200 km offshore had left deep-sea sources as a possibility [29, 30]. Observations based on large aperture seismic arrays [31] provided some clarification. Overall it appeared that deep-sea generation could occur if a storm system moved fast enough for opposing wave interactions to be set up in its wake. However a series of later papers again emphasised the lack of a comprehensive explanation for the effects observed, at least in the opinion of some people [32].

The second objective was to explore the lower noise levels it was believed would exist at the ocean bottom, because such sites were well away from man-made noise sources and the influence of surface wind. Noise levels were, however, found to be comparable to average continental levels, and in view of the expense and difficulty of instrument deployment in deep water the use of

OBS packages was discontinued for a period.   Interest in their
use was renewed later, to test whether microseism levels might
be diminished within a submarine borehole.   A program, related
to the Deep Sea Drilling Project (DSDP), has been active since
1979 [33, 34].

Even after a hundred years of research into microseisms
and related phenomena, in the early 1980's many of their characteristics still required clarification.   While the theories of
Longuet-Higgins [14] and Hasselmann [15] appeared to provide
an explanation for the effects observed, no comprehensive and
definitive experimental program capable of resolving the many
issues involved had been reported.

Clarification was to come in the main from the parallel
studies of infrasonic ocean acoustics (frequencies below 20
Hz), prompted by the growing interest of acousticians and geophysicists
in waterborne sound-seabed interactions.   Convincing evidence
now exists, from direct measurements of the gravity-wave spectrum,
that nonlinear wave interaction is the dominant source of short
period microseismic and infrasonic noise generation below 5
Hz [35, 36].   In the following sections we review our current
understanding of the infrasonic noise-field as provided by
recent experiments.

3.   ACOUSTIC AMBIENT NOISE SPECTRUM - HISTORICAL PERSPECTIVE

3.1. Shallow Water Measurements

All communication and detection systems must operate against
a background noise field.   In the case of active sonar this
noise field is usually associated with the system and its mobile
platform [37] but in the application of passive acoustic systems
it is the ambient noise of the sea which is important.

The earliest measurements of sea-noise were made at shallow
water sites.   There were both practical and instrumental reasons
for this.   Before the advent of solid state technology hydrophones
had to operate without the benefit of underwater amplification,
and with the exception of some sophisticated operational systems
about which little is published, cable loss controlled the
distance sensors could be placed off-shore.

Extensive measurements in the Northern Hemisphere slowly
established the general properties of the ocean-noise field
and the nature of the dominant noise sources.   The influence
of the wind in particular was soon apparent, but the importance
of shipping and other sources related to human activity was
also identified.   Much of this research was reviewed by Wenz
in 1962 [1].

Parallel measurements in the South Pacific Ocean were
to emphasise regional differences, particularly the lower ambient
levels resulting from lower shipping density.   These measurements
were also to highlight the importance of geophysical and biological

sources [38-42].

In essentially all cases shallow water spectral data were
characterised by octave or third octave resolution, confined
to frequencies above 12 Hz, and at low sea states often impaired
by mains induced hum in the region of 50/60 Hz.   These restrictions
and the hum problem were to be largely removed in the early
1960's when the introduction of solid state technology provided
the means to more readily detect, record and analyse low-frequency
geophysical data.

3.2.   Deep Water Measurements

The ability to work at lower frequencies and in deeper water
was not however to mean that making acoustical or seismic measurements at sea would become a routine matter.   New technical
problems replaced the old.

First, improved propagation conditions called for recording
systems that could be operated free from the noise (and support)
of a surface vessel. This meant recording periods had to be
restricted in most cases, a distinct disadvantage when dealing
with complex phenomena.

Secondly, such measurement systems are susceptible to
problems which increase the self-noise level at infrasonic
frequencies. Discussions of these sources of contamination
have been given by various authors [43-46].

Inspite of these difficulties a number of experimental
systems were devised and deployed to record ocean-noise in
deep water [34, 47-53].   Other measurements have been based
on fixed operational systems, the characteristics of which
are less well known.   Very few programs have however been
successful in producing reliable noise data below 10 Hz.
Even when successful the recording periods have invariably
been too short to cover a wide range of environmental conditions,
and the supporting environmental data so critical to a quantitative
investigation of the phenomena involved has not been adequate.

Nichols' review of the literature in 1981 [13] highlighted
the limitations of the infrasonic measurements made up to that
time.   Given the instrumental problems involved and the lack
of resolution inherent in the one-third and one octave bandwidths
used in many of the early experiments, it is perhaps not surprising
that it has proved so difficult to measure and characterise
this part of the sea-noise spectrum.

In the following sections we examine infrasonic noise
generation in the ocean on the basis of measurements which
post date Nichols' review, and demonstrate that nonlinear wave
interaction appears to be the dominant source of short-period
microseismic and infrasonic noise generation below 10 Hz.
It is perhaps appropriate that the answers have been provided
by a synthesis of results from seismic and acoustic investigations.
In presenting the evidence it will be instructive to draw heavily
on the New Zealand experience [18, 35, 54-56] but this is in

no way intended to discount other significant contributions
which have provided equally compelling evidence [36, 57-61].

3.3. The Ambient Noise Spectrum

In general shallow-water noise spectra recorded in New Zealand
waters follow the Kundsen curves, although below 100 Hz the
slope is significantly smaller than that observed in the Northern
Hemisphere because of the lower shipping density involved [38].
This basic spectral form can however be distorted by several
interesting phenomena [39, 41].

One of the first "deep" water measurements made in New
Zealand waters was carried out in association with Project
VELA mentioned earlier.   For the particular experiment, CHASE
V [47], solid state technology was available locally for the
first time and underwater amplification was incorporated with
the hydrophones to extend the low frequency response of the
whole electronic package down to 1 Hz.

The receiving system had not been deployed more than a
few hours when a severe storm hit the area.   As the storm
intensified the electronics became so seriously overloaded
as to be nonfunctional.   It was quickly established that the
anomalous noise originated in the sea and was confined essentially
to frequencies below 5 Hz.   The system was hurriedly recovered,
the low frequency roll off adjusted to remove this low frequency
contamination, and happily redeployed in time to record the
1000 ton explosion which took place.

This experience is recounted to indicate the limitations
in knowledge in the early 1960's.   Given the depth at which
the hydrophones were deployed this behaviour was unexpected,
although a greater familiarity with ocean-seismicity would
have revealed similar behaviour had been observed in the VELA
program [47].

A program mounted to explore the phenomenon active below
4 Hz was not successful.   Limitations in the seismic and wave
sensors then available quickly became apparent and the project
was abandoned until the late 1970's when another opportunity
to study the problem arose.   Before describing the next events
it is first appropriate to review the main theoretical proposals
which have been put forward to explain the effects at low frequencies.

4.   NONLINEAR WAVE INTERACTIONS - BASIC THEORETICAL CONSIDERATIONS

As first reported by Miche [25] consideration of second order
effects in the hydrodynamic equations leads to terms representing
the generation of low frequency pressure fluctuations by the
nonlinear interaction of oppositely travelling ocean waves.
In contrast to the progressive waves producing them, the distinctive
features of these second order waves are that the pressure
signals they produce occur at twice the frequency of the interacting

surface waves, are proportional to the amplitude product of
these waves, and do not decrease with depth. Longuet-Higgins'
development of Miche's theory and its further expansion by
Hasselmann were mentioned earlier.

Similar theoretical analyses in the context of underwater
acoustics were carried out by Brekhovskikh [8] and Hughes [10].
When minor errors are corrected, the derived pressure spectrum
can be shown to be essentially the same in all treatments [11].

While dealing with the same geophysical phenomena, the
various theoretical treatments are not readily reconcilable.
For the purposes of the analysis to be described later, we
have found it helpful to re-examine and extend the theoretical
basis of the effects involved. Our concern has been to resolve
more readily the effects of the spreading function associated
with the ocean-wave field and the properties of the geoacoustic
environment. This analysis will be presented elsewhere [54,
55], and only the results essential to the discussion to follow
are presented here.

A simple geoacoustic model is assumed consisting of a
water layer of constant depth, H, overlying an elastic half
space. In common with other treatments the analysis demonstrates
that the spectrum of the source pressure field, induced by
wave action and acting at the mean surface of the sea, is given
by (see Eq 17 of [54]):

$$F_p(2f_w) = \frac{32\pi^4 \rho_1^2 g^2 F_a^2 (f_w) \cdot f_w^3 \cdot I}{\alpha_1^2} \tag{1}$$

where $f_w = f/2$ denotes the frequency of the ocean surface wave;
f is the frequency of the wave-induced source pressure field
and its seismic equivalent, $\omega = 2\pi f$; ρ_1 is the density of sea
water and α_1 the sound velocity in seawater; and $F_a(f_w)$ is
the surface-wave spectral function.

The spectrum of the corresponding underwater noise field,
$F_N(f)$, and that of the microseisms (the displacement of the
seabed), $F_M(f)$, are then given by:

$$F_N(f) = F_p(f) \cdot T_{pN}(f) \tag{2}$$

and

$$F_M(f) = F_p(f) \cdot T_{pM}(f) \tag{3}$$

where

$$T_{pN}(f) = 2 \int_0^1 \frac{|J(\chi^2, \omega^2, z)|^2}{|J(\chi^2, \omega^2)|^2} \chi \, d\chi \tag{4}$$

and

$$T_{pM}(f) = \frac{2}{\rho_1^2 \, \omega^2 \, \alpha_1^2} \int_0^1 \frac{1 - \chi^2}{|J(\chi^2, \, \omega^2)|^2} \, \chi \, d\chi \qquad (5)$$

are the transfer functions relating the source pressure field
on the sea surface to the underwater-noise and seismic fields
respectively. The term $J(\chi^2, \, \omega^2, \, z)$ is a complicated function
of $n_\beta = \alpha_1/\beta_2$, $n = \alpha_1/\alpha_2$, $m = \rho_2/\rho_1$, and $\chi = k\alpha_1/\omega$, while ρ_2,
β_2 and α_2 are respectively the density and velocities of the
shear and compressional waves of the seabed. Further, $J(\chi^2, \, \omega^2) = J(\chi^2, \, \omega^2, \, o)$.

The term I in Eq(1) represents an integral of the spreading
function describing the angular distribution of the surface-wave
field, viz.:

$$I = \int_{-\pi}^{\pi} H(\theta) \, H(\theta + \pi) \, d\theta \qquad (6)$$

where $H(\theta)$ is the normalised spreading function defined as:

$$H(\theta) = G(\theta)/H_o, \quad H_o = \int_{-\pi}^{\pi} G(\theta)d\theta \qquad (7)$$

From Eqs (1) to (5) it is apparent that there should be
a two-to-one frequency relationship between the ambient-noise
and the ocean-wave spectra. Further the variance density
spectral levels of the noise field should be proportional to
the square of the spectral levels of ocean-wave components
producing the exciting pressure field, proportional to the
third power of the frequency, and be dependent on the spreading
function and depth.

5.   THE MAUI EXPERIMENT

5.1. Background

For a number of years the University of Auckland has been engaged
in an environmental study in connection with the development
of an offshore gas field. A major element of this study has
involved an investigation of the wave climate of the Cook Strait
region - see Fig. 1 of [35]. The weather of this region is
complex but because of the influence of a mountain chain the
area experiences well defined changes in the wind field, with
the wind vector often swinging rapidly through 180° from northwest
to southeast. The region therefore provides an ideal environment
for the study of ocean-wave phenomena.

On engineering grounds the general wave-climate study
had of necessity to involve long term recording of the ocean-wave

field and related environmental parameters.   This fact, coupled
with the unique orographic properties of the region, provided
an opportunity to extend the basic study and include investigations
of other wave related processes.   In particular plans were
made to examine wave-induced microseims.

Reliable wave data were provided by a Datawell Waverider
buoy moored in approximately 110 m of water close to the Maui-A
platform, which provided the base for the relevant meteorological
instrumentation.   It was not feasible to deploy a hydrophone
or ocean bottom seismometer for the microseism study, and as
a compromise a long period seismometer was installed ashore
at Oaonui - see for example Fig. 1 of Ref. [35].

Recordings of wave and seismic signals of 20 minute duration
were initiated automatically every four hours.   Wave and seismic
spectra were produced from the analogue records and various
presentations of the data developed as required.   Details
of the instrumentation and analysis procedures used are available
elsewhere [18, 35].

5.2. Microseism Generation-General Observations

Sea-wave/microseism correlations gave unequivocal evidence
for the marine generation of microseisms in the frequency range
0.05 - 1.0 Hz. A comparison of any sea spectrum with its seismic
equivalent would identify activity in the microseism spectrum,
at or close to twice the frequency of that in the wave spectrum.
Figure 2 of Ref. [35], and Fig. 2 below provide good examples
of the development of a local sea and the associated seismic
response.

A general account of the observed wave/microseism behaviour
is available elsewhere [35, 56]. In this review we restrict
discussion to a brief summary of the significant results covered
in these earlier papers, before proceeding to an account of
subsequent investigations.

5.2.1. Microseism excitation.

The joint behaviour of the
ocean-wave and seismic spectra could be interpreted in terms
of wave-wave interactions.   It was demonstrated clearly however
that coastal reflection was not the dominant mechanism by which
opposing wave trains were generated, seismic activity being
just as energetic for offshore as for onshore winds [56].
The most energetic seismic activity was identified with two
opposing seas, although high levels were also observed when
only a single sea was active.   Seismic generation within a
single sea as a natural consequence of spreading of the wave
energy is discussed in Section 6.

5.2.2. The 2:1 frequency relationship.

While a fetch dependence
in wave generation over the active area introduces some ambiguity,
it was clearly demonstrated that, as required by Eq(1), the
microseism energy occurs at essentially twice the ocean-wave

frequency.

5.2.3. Microseism level dependence on ocean-wave amplitude.
One of the spectral parameters routinely calculated was the
significant height (derived from the zeroth moment) of the
microseism and ocean-wave spectra. A linear regression analysis
of these parameters confirmed the second power relationship
predicted by theory and embodied in Eq(1).

5.2.4. Microseism level dependence on ocean-wave frequency.
The dependence of the microseism level on ocean-wave frequency
could not be established directly but an analysis based on
nondimensionalised spectra provided strong evidence in support
of the third power dependence embodied in Eq(1). This conclusion
was reinforced by a subsequent analysis [55].

5.3. Microseism Generation - Quantitative Considerations

While an interdependence between the ocean-wave field and the
microseismic response was established, at the time the results
of Ref. [35] were published uncertainties in respect of the
transfer and spreading functions described in Section 4, precluded
establishing an unambiguous quantitative relationship between
the two wave fields. As a first approximation therefore,
the associated pressure field was derived from the seismic
spectra following the simple procedure proposed by Urick [62].
Figure 1, which is reproduced from [35], provides an example
of such spectra derived from ground displacements recorded
during a particular wave generation event.
    These derived pressure spectra were compared with other
ambient-noise measurements recorded by hydrophones - see Section
IV of Ref. [35]. While some uncertainty (of around 10-20
dB) in absolute levels was apparent, the overall agreement
between the two sets of results was striking - see Fig. 8 of
Ref. [35].
    When the acoustic pressure spectra derived from the ground
displacement were compared with the theoretical spectra calculated
using the observed wave-spectral levels, $F_a(f_w)$, and Eq(1),
it was found that at frequencies close to the spectral peak
the values agreed to within a few decibels. At higher frequencies
(0.4 < f < 1.0 Hz) the theoretical levels calculated on the
basis of Eq(1) were typically 10-15 dB below the experimental
values deduced from the seismic data.
    The significance of these and other small anomalies were
discussed in [35] - Section V. It was concluded that while
the results confirmed the essential features of the formalism
governing nonlinear interactions, further refinement was dependent
on a more complete understanding of the spreading function
I (in Eq(1)), and the transfer functions implicit in Eqs(4)
and (5).
    An initial attempt to resolve these issues was reported

in 1986 [63]. The next section describes the progress made
since then.

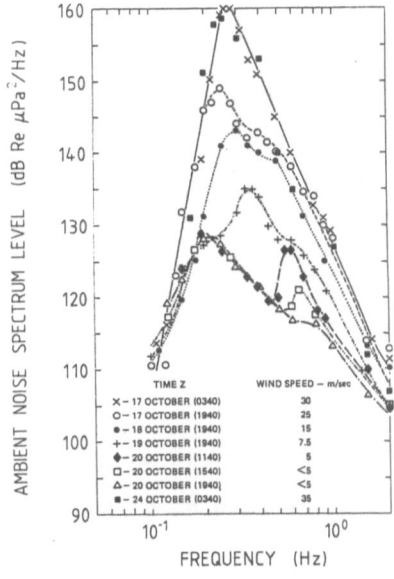

Fig. 1
Ambient noise levels
derived from seismic
spectra as a function
of wind speed.

6. A RE-EXAMINATION OF NONLINEAR INTERACTIONS AND THE INFRASONIC NOISE FIELD

6.1. Average Spectra vs Wind Speed

In the Maui region the wave-field is essentially fetch limited
at the recording site for winds from the southeasterly quarter.
Average ocean wave and seismic spectra (in 2.5 ms^{-1} intervals
of the wind speed at height 10 m, U_{10}) are shown in Fig. 2.
These spectra are associated with SE winds only and strict
criteria were imposed in their selection [18]. The characteristic
growth of the wave peak and its migration to lower frequencies
with increasing wind speed is mirrored in the seismic spectra.
The contribution of an everpresent SW swell is evident in both
at low wind speeds.

6.2. An Analytical Form of the MAUI Wave Spectrum

To establish the acoustic pressure field from Eq(1) it is helpful
to calculate an analytical form of the ocean-wave spectrum.
Discounting the persistent swell from the southwest, the JONSWAP
function, with parameters appropriate to the Maui region, has
been found to describe the MAUI spectrum very well. The JONSWAP
function is traditionally calculated in terms of the peak frequency,

348

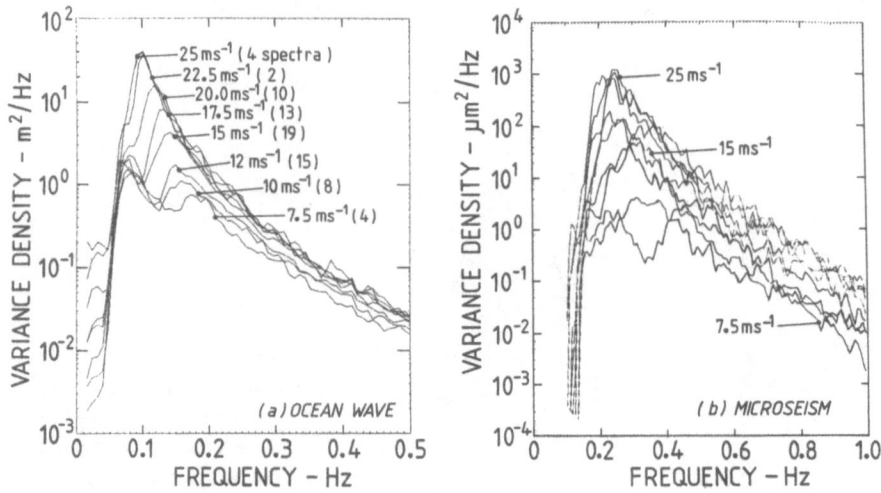

Fig. 2
Ocean wave and associated seismic spectra as a function of wind speed.

but for our purposes a representation in terms of wind speed is more helpful [18, 55]. The resulting one-dimensional MAUI spectra for wind speeds between 2.5 and 30 ms^{-1} are shown in Fig. 3a.

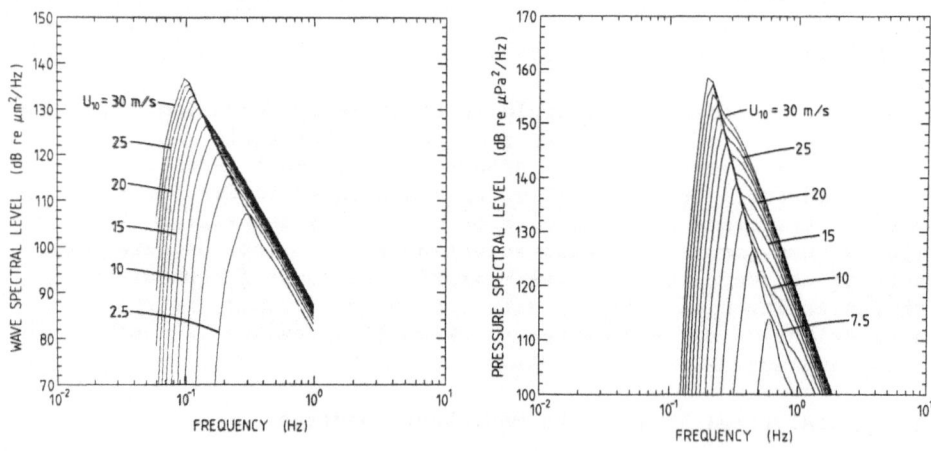

Fig. 3
(a) *Analytical form of the MAUI wave spectrum as a function of wind speed.*

(b) *Source pressure field arising from wave interactions, based on the wave spectra of Fig. 3a.*

6.3 The Spreading Coefficient and the Integral I

In the case of a single sea (ie a sea generated by a steady
wind from a fixed direction blowing across an initially quiescent
surface) a widely accepted form of the spreading function,
$G(\theta)$, is that proposed by Tyler et al [64]:

$$G(\theta) = \cos^{2s}(\theta/2) \qquad -\pi \leqslant \theta \leqslant \pi \qquad (8)$$

where the spreading coefficient, s, is dependent on both wind
speed and frequency.   Several forms of the coefficient s
have been suggested but for ease of comparison with experimental
data we have found it convenient to introduce an adjustable
parameter, δ, and write [55]:

$$
\begin{aligned}
s &= 2/(f/f_0 - 1) & f \geqslant f_0 (1 + \delta) \\
&= 2/\delta & f < f_0 (1 + \delta)
\end{aligned}
\qquad (9)
$$

where $f_0 = g/2\pi U_{10}$ is the "resonant frequency" of the surface
wave corresponding to the wind speed U_{10}.
    In [55] we justify the choice of $\delta = 1$ as a reasonable
approximation. With δ defined the integral I, defined by Eq(6),
can be readily calculated.

6.4 The Source Pressure Field Arising from Wave-Wave Interactions

With $F_a(f_w)$ defined as the Maui version of the JONSWAP function
(see Section 6.2), and using Eqs(6) and (9) to calculate the
integral I, we can derive from Eq(1) the set of theoretical
spectra for the source pressure field of a single sea shown
in Fig. 3b. We note:

  (i)   That for the maximum wind speed the spectral
        levels at the peak and 1.0 Hz are around 160
        and 120 dB re 1 μPa^2/Hz respectively and

  (ii) That the inflexion in the spectral slope on
        the high frequency side of the peak mirrors
        that in Fig. 1.

6.5. The Acoustic Pressure Field Derived from the Seismic Spectra

In our earlier analysis [35] the microseisms were attributed
solely to vertical ground displacement, any shear-wave component
being ignored. Further, use of the simple transfer function
to calculate the incident pressure field from the seismic spectra
meant that the effect of any bottom reflection was also neglected.
While the ambient-noise spectra so derived were strikingly
similar to others recorded by conventional means, certain discrepancies
were identifiable (see Section 5.3). In a subsequent
re-evaluation the implication of both bottom reflection and

shear-wave generation have been investigated.

It is shown elsewhere [54, 55] that when only compressional wave excitation is considered, the transfer functions given by Eqs(4) and (5) take the approximate form:

$$T_{NM} = T_{pM} \simeq a/\rho_1^2 \; \omega^2 \; \alpha_1^2 \tag{10}$$

where $a = (1 - 2n^2 + 2n^4)/2m^2$ and n and m are as defined in Section 4. Equation (10) was evaluated for four reasonable geoacoustic models. The results show that, over the frequency range of interest (0.1 - 1.0 Hz), the expression adequately describes the frequency dependence of the transfer function, the difference from the curves based on the full expression being less than 1 dB. To this extent, therefore, the application of Eq(10) with a = 1 in the earlier analysis [35] is shown to be justified. However the calculation also shows, as expected, that the influence of "a" on the magnitude of the transfer function cannot be ignored. This question was discussed by Kibblewhite and Ewans [35, Section VA], but at that time the influence of "a" could not be established.

As stated earlier, the spectra in Fig. 1 were derived from the corresponding seismic spectra, $F_M(f)$, by applying the approximate form of the transfer function, $1/\rho_1^2 \; \omega^2 \; \alpha_1^2$, where $\omega = 2\pi f$. In Ref. [35] these experimental pressure spectra were compared with ambient-noise data recorded in other ocean environments by conventional methods and attention was drawn to the striking similarities which existed. Attention was also drawn to the anomalous low frequency component in the experimental spectra which became obvious at low wind speeds, and to an inflexion (or subsidiary peak) around 0.5 Hz visible in the spectra for wind speeds of 7.5, 15 and 25 ms^{-1}.

The first of these two effects was shown to arise from the persistent southwesterly swell which is always present in the recording area. Possible explanations were offered for the second feature but a definitive statement could not be given at that time because it was not possible to resolve the relative influences of the spreading and transfer functions.

Later analysis has provided a clearer picture of the spreading function. Based on this analysis the theoretical pressure spectra arising from nonlinear interactions within a single sea assuming $\delta = 1$, are those given by Fig. 3b - see Sections 6.3 and 6.4.

A comparison with those spectra for the same windspeeds but derived from the experimental seismic data (see Fig. 1) shows that:

      (i)   The large low frequency component apparent in the experimental data is not present in the theoretical spectra.

     (ii)   A small but significant difference in the peak frequencies of the two sets of spectra is apparent at high wind speeds.

    (iii)   Both sets of spectra display an inflexion
            around 0.5 Hz at wind speeds above 7.5 ms^{-1}.

    (iv)   While the peak spectral levels are in good
            agreement, the levels at frequencies well
            above the peak are noticeably higher in the
            experimental spectra.

As mentioned earlier, the low frequency peak described in (i) is associated with the southwesterly swell which is always present in the region - see Fig. 2. The JONSWAP form of the Maui spectra presented in Fig. 3a deliberately omitted the swell in the interest of simplicity. In contrast to the explanation given in [35], the differences described in (ii) are now reinterpreted as meaning that early in the event (0340 on 17 October; "x" in Fig. 1) nonlinear interactions were occurring not only between components of the growing southeasterly sea, but also between the new sea and residual components of old northwesterly sea - see Ref. [35]. Further, the inflexion referred to in (iii) is now identified with the spreading function, rather than the transfer function as was suggested in [35] and [63].

7.   NONLINEAR INTERACTIONS BETWEEN MULTIPLE SEAS

To test the validity of the above interpretations an expression was developed for the spectrum of the source pressure field resulting from the combined interactions (during the event of 17 October) of the southwesterly swell, the decaying northwesterly sea and the developing sea from the southeast. In deriving this formula certain reasonable assumptions had to be made regarding the directivity and spectral forms of the interacting wave regimes. Details are given in [55].

    A calculation based on the expression for the multiple seas led to a set of theoretical pressure spectra, see Fig. 4, in much closer agreement with the experimental spectra of Fig. 1, derived from the seismic spectra. In particular we note:

    (i)   The match in spectral levels and peak frequency
        near the spectral peak is much better at high
        wind speeds than in the case for a single
        sea.

    (ii)   The theoretical spectra exhibit a behaviour
        related to the southwesterly swell, which
        mirrors that in the experimental spectra reasonably
        well.

    (iii)   The high frequency spectral values, particularly
        at top wind speeds, are still in general greater
        than those predicted theoretically.

    At this point we recall that the transfer function used in deriving the experimental spectra from the seismic data was the approximate form, $T_{Mp} = 1/T_{pM} = \rho_1^2 \, \omega^2 \, \alpha_1^2$. Use of the

352

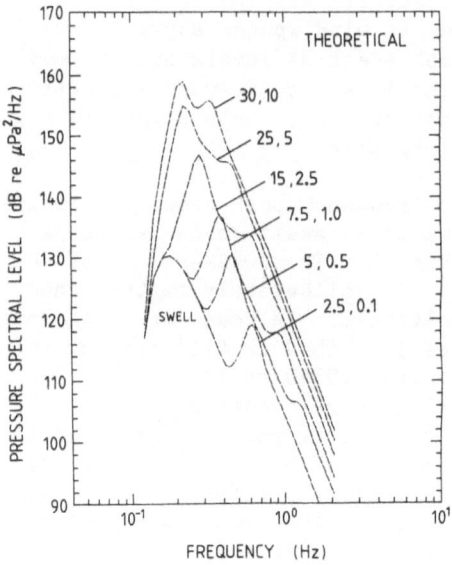

Fig. 4
Theoretical pressure
spectra based on the
interaction of
multiple seas.

complete function increases all the seismically derived pressure
values in Fig. 1 by the value 10 log a. For the geoacoustic
models examined the increase ranges from 6-14 dB depending
on the model chosen.   Such an adjustment obviously increases
the anomalies, particularly at high wind speeds.   Adjustment
for reflection loss alone does not therefore appear to improve
the agreement between theory and experiment.   We next examine
the influence of taking shear wave excitation into account.

8.   THE EFFECT OF SHEAR WAVE EXCITATION

In Ref. [54] it is shown that when the consequences of shear
wave excitation are taken into account, the transfer function
takes the form:

$$T_{Mp} = \rho_1^2 \; \omega^2 \; \alpha_1^2 / a[1 + \eta_T(\omega)] \tag{11}$$

Since $\eta_T(\omega)$ is proportional to ω use of Eq(11) would appear
to bring the experimental data into closer overall agreement
with the theoretical curves, but a test of this clearly depends
on the geoacoustic properties of the Maui region.
      Geophysical data on the area are available from various
sources [55]. Using this information shows that the pressure
spectral values in Fig. 1 should be decreased by a factor ranging
from -1.0 dB at f = 0.1 Hz to 10.5 dB at 2 Hz. Applying these
corrections to the data of Fig. 1 gives the modified ambient
noise spectra of Fig. 5. It can be seen that allowing for
a shear wave component in the recorded seismic spectra leads

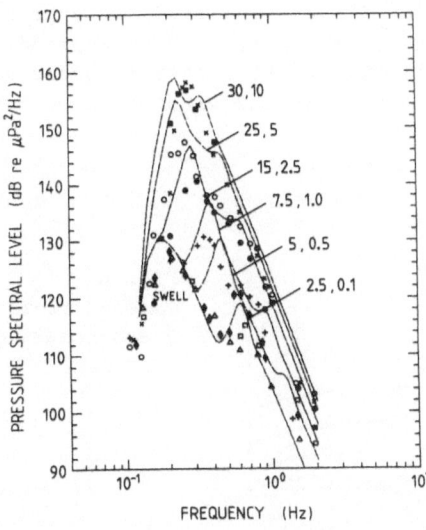

Fig. 5
Comparison of the theoretical pressure spectra with the experimental spectra derived from the seismic data, when bottom reflectivity and shear wave excitation are included.

to little change in the derived pressure spectra around 0.1 Hz, but to significant reductions at 1 Hz.

Bearing in mind the complexities involved, the overall agreement in shape and magnitude of the theoretical curves and experimental data in Fig. 5 is considered good. The fit could be improved by adjusting the levels of the residual swell and component seas, but in the absence of directional wave data this seems an unnecessary refinement. Nevertheless the fact that for both sets of spectra the levels at 1 Hz are close to 120 dB re 1 μ Pa^2/Hz (a value commonly reported for deep-sea ambient noise levels at moderate wind speeds [53, 65]), that the peak values are in close agreement, and that the spectral shape and its wind dependence are similar to those reported in other recent observations [36, 57], gives overall confidence in the arguments presented.

9. THE AMBIENT NOISE SPECTRUM

9.1. Above 0.1 Hz

On the basis of the above analysis it is possible to provide an idealised form of the ambient-noise spectrum - see Fig. 6.

At high wind speeds spectral levels resulting from wave interactions reach maximum values of approximately 160 dB re 1 μ Pa^2/Hz at frequencies around 0.25 Hz. At higher frequencies levels decrease at roughly 18 dB per octave to values around 120 dB at 1 Hz. At low wind speeds the shape of the spectrum below 5 Hz will depend upon the level of the swell component.

In most ocean areas a contribution from this source can be
expected and the lowest recorded levels are likely to approximate
those shown for the New Zealand east coast.   In regions of
no swell the spectral peak will migrate to higher frequencies
in step with that of the wave field, as the wind speed decreases.

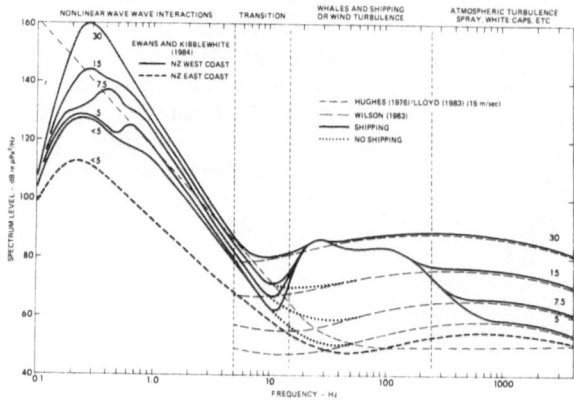

Fig. 6
General form of the
ocean-noise spectrum.

　　　The shape of the spectrum in the transition zone will
be controlled by the level of the shipping and biological component.
Where shipping is significant the spectrum can be expected
to display a minimum of the sort shown, and often observed
[53, 65].   Far away from shipping lanes or at receiving sites
below critical depth the minimum will be less evident.
　　　Between 10 and 200 Hz the levels are controlled by shipping.
In the quietest areas of the ocean, levels at frequencies above
the wave-interaction peak will fall some 100 dB and then be
controlled by the wind through mechanisms yet to be fully understood.
　　　As expected, the earlier theoretical predictions of Hughes
[10] and Lloyd [11] relating to wave interactions are seen
to be in keeping with the curves of Fig. 6.   Without any comment
on the debate surrounding the role of turbulence [66], Wilson's
1983 estimates [67] of the Isakovich/Kur'yanov mechanism [4]
above 5 Hz, are also included for comparison.   These will
be analysed in more detail in a subsequent paper.

9.2　Below 0.1 Hz

Below the spectral peak noise levels fall away in keeping with
the decrease in wave spectral energy at these frequencies.   On
the basis of the seismic spectra there is evidence of another
minimum around 0.05 Hz, noise levels rising again toward even
lower frequencies in a manner controlled by the seismic characteristics of the site [35, 36].
　　　On occasions a subsidiary peak is observed in the New
Zealand spectra below 0.1 Hz, on the low frequency side of
the wave-interaction peak.   This is always associated with

the rapid passage of an energetic cold front across the recording
area.   During such events significant perturbations in tidal
records have been observed, which indicate the excitation of
low frequency shelf waves.   The subsidiary peak in the ambient-noise
record below 0.1 Hz is apparently associated with water movements
arising from such events.   Some of the spectra reported by
others [13, 36] also show a rise in noise levels at frequencies
below 0.1 Hz, no doubt originating from the same atmospheric
cause.

10.   CONCLUSIONS AND FUTURE WORK

Several recent contributions have confirmed the role of wave-wave
interactions in ocean-acoustic phenomena at frequencies below
5 Hz.   The essential elements of the theoretical background
to nonlinear processes were confirmed in [35], but several
aspects required further clarification.   This review reports
in particular on supplementary work which has more clearly
established the relative influence of the spreading and transfer
functions on the shape and magnitude of the related spectra.

An important development has been the examination of the
case of nonlinear interactions between multiple seas.   While
even better agreement between theory and experiment has been
established the present comparison suffers from an inadequate
description of the directional properties of the ocean-wave
field.   Measurements using a directional wave recorder are
currently in progress and will be reported in due course.

Other aspects which will receive attention in the future
include extending the analysis to the case of a multilayered
half space, a study of the directional properties of the nonlinear
component of the noise field, and the measurement of the shear-wave
component of the wave-induced ground motion.

REFERENCES

1.   G.M. Wenz, 'Acoustic Ambient Noise in the Ocean: Spectra
and Sources,' *J. Acoust. Soc. Am.* **34**, 1936-1955
(1962).

2.   A.C. Kibblewhite and R.N. Denham, 'Hydroacoustic Signals
from the CHASE V Explosion,' *J. Acoust. Soc. Am.*
45, 944-956 (1969).

3.   A.J. Perrone, 'Infrasonic and Low-Frequency Ambient Noise
Measurements on the Grand Banks,' *J. Acoust. Soc.
Am.* **55**, 754-758 (1974).

4.   M.A. Isakovich and B.F. Kur'yanov, 'Theory of Low Frequency
Noise in the Ocean,' *Sov. Phys. Acoust.* **16**, 49-58
(1970).

5.   W. Strawderman, 'Turbulent Air Flow Induced Sea Noise,'
NUSC Technical Document 12-190-74, Naval Underwater

356

Systems Center, New London, CT, 28 June 1974.

6. N. Yen and A.J. Perrone, 'Mechanisms and Modelling of Wind-Induced Low Frequency Ambient Sea Noise,' *NUSC Tech. Rep. 5833,* Naval Underwater Systems Center, New London, CT (13 February 1979).

7. J.H. Wilson, 'Very Low-Frequency (VLF) Wind-Generated Noise Produced by Turbulent Pressure Fluctuations in the Atmosphere Near the Ocean Surface,' *J. Acoust. Soc. Am.* **66**, 1499-1507 (1979).

8. L.M. Brekhovskikh, 'Underwater Sound Waves Generated by Surface Waves in the Ocean,' *Izv. Atmos. Ocean. Phys.* **2**, 582-587 (1966) (translated by J. Gollob).

9. E.Y. Harper and P.G. Simpkins, 'On the Generation of Sound in the Ocean by Surface Waves,' *J. Sound Vib.* **37**, 185-193 (1974).

10. B. Hughes, 'Estimates of Underwater Sound (and Infrasound) Produced by Nonlinearly Interacting Ocean Waves,' *J. Acoust. Soc. Am.* **60**, 1032-1039 (1976).

11. S.P. Lloyd, 'Underwater Sound from Surface Waves According to the Lighthill-Ribner Theory,' *J. Acoust. Soc. Am.* **69**, 425-435 (1981).

12. V.V. Goncharov, 'Sound Generation in the Ocean by the Interaction of Surface Waves and Turbulence,' *Izv. Atmos. Ocean. Phys.* **6**, 1189-1196 (1970) (translated by F. Goodspeed).

13. R.H. Nichols, 'Infrasonic Ambient Noise Measurements: Eleuthera,' *J. Acoust, Soc. Am.* **69**, 974-981 (1981).

14. M.S. Longuet-Higgins, 'A Theory of the Origin of Microseisms,' *Philos. Trans. R. Soc. London Ser. A* **243**, 1-35 (1950).

15. K. Hasselmann, 'A Statistical Analysis of the Generation of Microseisms,' *Rev. Geophys.* **1**, 177-210 (1963).

16. J. Darbyshire, 'Microseisms in the Sea, Vol. 1, edited by M.N. Hill pp 700-719, Wiley-Interscience, New York (1962).

17. B.S. Melton, 'The Sensitivity and Dynamic Range of Inertial Seismographs,' *Rev. Geophys. Space Phys.* **14**, 93-116 (1976).

18. K.C. Ewans, 'Ocean Waves, Microseisms and their Interactions,' Doctoral dissertation, University of Auckland, New Zealand (1984).

19. R.G. Adair, 'Microseisms in the Deep Ocean: Observations and Theory,' Doctoral dissertation, University of California, at San Diego, La Jolla, CA (1985).

20. S.C. Webb, 'Observations of Seafloor Pressure and Electric Field Fluctuations,' Doctoral dissertation, University of California, at San Diego, La Jolla, CA (1984).

21. B. Gutenberg, 'Microseisms in North America,' *Bull Seismol. Soc. Am.* **21**, No. 1, 1 (1931).

22. E. Gherzi, 'Cyclones and Microseisms,' *Bietrage Zur Geophysik* **36**, 20 (1932).

23. S.K. Banerji, 'Microseisms associated with Disturbed

Weather in the Indian Seas,' *Phil. Trans. Roy. Soc. Lond.*, Series A, **229**, 287-328 (1930).

24. G.E.R. Deacon, 'Recent Studies of Waves and Swell,' *Ann. N.Y. Acad. Scie.* **51**(3), 475-482 (1949).

25. M. Miche, 'Movements Ondulatoires de la mer en profondeur constante ou decroissante,' *Ann. Ponts Chaussees*, **114**, 25-87 (1944).

26. J.N. Nanda, 'The Origin of Microseisms,' *J. Geophys. Res.* **65**(6), 1815-1820 (1960).

27. R.A. Haubrich, W.H. Monk, F.E. Snodgrass, 'Comparative Spectra of Microseisms and Swell,' *Bull. Seismol. Soc. Am.* **53**(1), 27-37 (1963).

28. J. Darbyshire and E.O. Okeke, 'A Study of Primary and Secondary Microseisms Recorded in Anglesay,' *J. Roy. Astro. Soc.* **17**, 63-92 (1969).

29. G.V. Latham and G.H. Sutton, 'Seismic Measurements on the Ocean Floor, 1. Bermuda Area,' *J. Geophys. Res.* **71**, 2545-2573 (1966).

30. G.V. Latham and A.A. Nowroozi, 'Waves, Weather and Ocean Bottom Microseisms,' *J. Geophys. Res.* **73**, 3945-3956 (1968).

31. R.A. Haubrich and K. McCamy, 'Microseisms: Coastal and Pelagic Sources,' *Rev. Geophys.* **7**, 539-571 (1969).

32. F.I. Monakhov, V.M. Zhak and V.A. Nesterov, 'Mechanisms of Generation of Microseisms by a Storm during the Period October 10-11 1976, near Island Shikotam,' *Isvestia Earth Phys.* **14**(4), 302-303 (1978).

33. J.A. Carter, F.K. Duennebier and D.M. Hussong, 'A Comparison between a Downhole Seismometer and a Seismometer on the Seafloor,' *Bull Seismol. Soc. Am.* **74**, 763-772 (1984).

34. R.G. Adair, J.A. Orcutt and T.H. Jordan, 'Low Frequency Noise Observations in the Deep Ocean,' *J. Acoust. Soc. Am.* **80**(2), 633-645 (1986).

35. A.C. Kibblewhite and K.C. Ewans, 'Wave-wave Interactions, Microseisms and Infrasonic Ambient Noise in the Ocean,' *J. Acoust. Soc. Am.* **78**, 981-994 (1985).

36. S.C. Webb and C.S. Cox, 'Observations and Modelling of Seafloor Microseisms,' *J. Geophys. Res.* **91**, 7343-7358 (1986).

37. D. Ross, 'Mechanics of Underwater Noise,' *Pergamon Press*, New York (1976).

38. R.W. Bannister, R.N. Denham, K.M. Guthrie and D.G. Browning, 'Ambient Sea-Noise Measurements Near New Zealand,' *J. Acoust. Soc. Am.* **55**, 418(A) (1974).

39. A.C. Kibblewhite, 'The Acoustic Detection and Location of an Underwater Volcano,' *N.Z. J. Sci.* **9**, 178-199 (1966).

40. R.I. Tait, 'The Evening Chorus - A Biological Noise Investigation,' *Naval Research Laboratory Report No. 26*, New Zealand (1962).

41. A.C. Kibblewhite, R.N. Denham and D.J. Barnes, 'Unusual Low Frequency Signals Observed in New Zealand Waters,' *J. Acoust. Soc. Am.* **41**, 644-655 (1967).

42. R.A. Norris and R.H. Johnson, 'Submarine Volcanic Eruptions Recently Located in the Pacific by SOFAR Hydrophones,' *J. Geophys. Res.* **74**, 650-664 (1969).

43. J.R. McGrath, 'Infrasonic Sea-Noise Measurements and Experimental Problems,' *Acustica* **39**, 324-327 (1977).

44. M. Strasberg, 'Nonacoustic Noise Interference in Measurements of Infrasonic Ambient Noise,' *J. Acoust. Soc. Am.* **66**, 1487-1493 (1979).

45. F.K. Duennebier, G. Blackinton and G.H. Sutton, 'Current Generated Noise Recorded on Ocean Bottom Seismometers,' *Mar. Geophys. Researches* **5**, 109-115 (1981).

46. J.H. Filloux, 'Pressure Fluctuations on the Open Ocean Floor Over a Broad Frequency Range: New Program and Early Results,' *J. Phys. Ocean* **10**, 1959-1971 (1980).

47. A.C. Kibblewhite and R.N. Denham, 'The CHASE V Explosion - Submarine Topographic Reflections from the Vicinity of Pitcairn Island,' *Deep-Sea Res.* **18**, 905-911 (1971).

48. D.S. Herman, 'The ACODAC Ambient Noise System,' Proceedings of IEEE International Conference on Engineering in the Ocean Environment, Halifax, Nova Scotia (IEEE, New York) Vol. 1, pp 15-21 (1974).

49. A.C. Kibblewhite, J.A. Shooter and S.L. Watkins, 'Examination of Attenuation at Very Low Frequencies using the Deep Water Ambient Noise Field,' *J. Acoust. Soc. Am.* **60**, 1040-1047 (1976).

50. G.B. Morris, 'Depth Dependence of Ambient Noise in the Northeast Pacific Ocean,' *J. Acoust. Soc. Am.* **64**, 581-588 (1978).

51. R.W. Bannister, R.N. Denham, K.M. Guthrie, D.G. Browning, A.J. Perrone, 'Variability of Low Frequency Ambient Sea Noise,' *J. Acoust. Soc. Am.* **65**, 1156-1163 (1979).

52. A.S. Burgess and D.J. Kewley, 'Wind-generated Surface Noise Sources Levels in Deep Water East of Australia,' *J. Acoust. Soc. Am.* **73**, 201-210 (1983).

53. F.D. Cotaras, H.M. Merklinger and I.A. Frazer, 'Ocean Ambient Noise at Very Low Frequencies,' *J. Acoust. Soc. Am.* **74**, Supplement 1, S122 (1983).

54. A.C. Kibblewhite and C.Y. Wu, 'The Generation of Microseisms Infrasonic Ambient Noise in the Ocean by Nonlinear Interactions of Ocean Surface Waves' (submitted).

55. A.C. Kibblewhite and C.Y. Wu, 'A Re-examination of the Role of Wave-Wave Interactions in Ocean Noise and Microseism Generation' (submitted).

56. K.C. Ewans and A.C. Kibblewhite, 'A Study of Ocean Induced Microseisms' (submitted).

57. S.C. Webb and C.S. Cox, 'Pressure and Electric Fluctuations on the Deep Seafloor: Background Noise for Seismic

Detection,' *Geophys. Res. Letters* **11**, 967-970 (1984).

58.   S.C. Webb and C.S. Constable, 'Microseism Propagation
        between Two Sites on the Deep Seafloor,' *Bull. Seismol.
        Soc. Am.* **76**, 1433-1455 (1986).

59.   S.C. Webb, 'Coherent Pressure Fluctuations Observed at
        Two Sites on the Seafloor,' *Geophys. Res. Letters*
        13, 141-144 (1986).

60.   G.K. Duennebier, R.K. Cessaro and P. Anderson, 'Geo-acoustic
        Noise Levels in a Deep Ocean Borehole,' *Ocean Seismo-
        Acoustics*, Nato Conference Series IV Vol 16, Eds
        T. Akal and J.M. Berkson, p743-751, Plenum Press
        (1986).

61.   T. Akal, A. Barbagdata, G. Guide and M. Snoek, 'Time
        Dependence of Infrasonic Seafloor Noise on a Continental
        Shelf,' *Ocean-Seismo Acoustics*, Nato Series IV Vol
        16, p767-778, Eds T. Akal and J.M. Berkson, Plenum
        Press (1986).

62.   R.J. Urick, 'Seabed Motion as a Source of Ambient Noise
        Background in the Sea,' *J. Acoust. Soc. Am.* **56**,
        1010-1011 (1974).

63.   A.C. Kibblewhite and K.C. Ewans, 'A Study of Ocean and
        Seismic Noise at Infrasonic Frequencies,' *Ocean-Seismo
        Acoustics* Series IV Vol 16, pp731-741, Eds. T. Akal
        and J.M. Berkson, Plenum Press (1986).

64.   G.L. Tyler, C.C. Teague, R.H. Stewart, A.M. Peterson,
        W.H. Munk, and J.W. Joy, 'Wave Directional Spectra
        from Synthetic Aperture Observations of Radio Scatter,'
        Deep Sea Res. **21**, 989-1016 (1974).

65.   T.E. Tapley and R.D. Worley, 'Infrasonic Ambient Noise
        Measurements in Deep Atlantic Water,' *J. Acoust.
        Soc. Am.* **17**, 621-622 (1984).

66.   R.G. Adair, 'Comments on the Infrasonic Noise Theory
        of Isakovich and Kur'yanov and its Modification
        by Wilson,' *J. Acoust. Soc. Am.* **81**, 1192-1195 (1987).

67.   J.H. Wilson, 'Noise Generated Noise Modelling,' *J. Acoust.
        Soc. Am.* **73**, 211-216 (1983).

LOW FREQUENCY OCEAN AMBIENT NOISE: MEASUREMENTS AND THEORY

William M. Carey and David Browning
Naval Underwater Systems Center
New London, CT, USA 06320

ABSTRACT. Low frequency ocean ambient noise data are reviewed and summarized. The experimental data, both omnidirectional and directional,
when not dominated by shipping noise, are shown to suggest wind dependent noise at the low frequencies (<500 Hz). Candidate mechanisms are
examined with the result that wave-turbulence interaction at low sea
states and collective bubble oscillations at high sea states are identified as possible sources of this sound. A description of the sonic
properties of bubbly water is presented for low void fractions consistent with those observed in bubble clouds and plumes produced by breaking waves. A description of the collective bubble-water mixture as the
resonant oscillation of a flexible volume with a sonic speed determined
by the properties of the mixture is presented.

INTRODUCTION

The interaction of the wind with the ocean surface has long been recognized as a major source of acoustic noise (Knudsen (1948), Wenz (1962)).
Measurements of the omnidirectional noise at the higher frequencies
(>200 Hz) have been found to exhibit wind-dependent characteristics;
and, when not dominated by shipping noise, the most likely mechanisms
are related to bubbles, spray, and splashes associated with white caps,
as well as capillary wave/wave interactions (Urick (1984)). Furduev
(1966) has proposed that the characteristic broad maxima in the ocean
ambient noise spectrum between 0.2 kHz and 1 kHz be attributed to cavitating bubbles. Kerman (1984) discusses these mechanisms in detail
(also see Fitzpatrick (1959)), but stresses the noise generated by the
non-resonant oscillation of entrained gas bubbles which result from wave
breaking and which are forced by intense velocity of the gravity-capillary waves. For wind speeds with a friction velocity greater than this
critical velocity, Kerman concludes that sound is produced with a velocity to the 3/2 power, frequency to the -2 power, and intensity proportional to the number of bubbles. However, in the absence of white caps,
since noise persists, capillary wave/wave or non-linear wave interactions may be important (Mellen (1987), Kuo (1968)).

At the other extreme of the spectrum (<2-5 Hz), ambient noise associated with ocean microseisms dominates. Recently, this noise has been

B. R. Kerman (ed.), Sea Surface Sound, 361–376.

shown by Nichols (1981) and by Kibblewhite and Evans (1984) to be due to wave/wave interaction. The microseismic effect was postulated by Longuet-Higgins (1950) and confirmed by several authors, including Latham and Nowroozi (1968). Several authors have studied the generation of sound through the second-order pressure effect (Brekhovskikh (1967), Goncharov (1970), Hughes (1976), Lloyd (1981)). Kibblewhite and Evans concludes with theoretical arguments and measurements that the dominant noise source in the 0.1 to 5 Hz range is the non-linear wave interaction. Although difficulties were found in predicting absolute levels, both data and theory showed a frequency dependence to the -6 power.

In the very low frequency (VLF, 2-20 Hz) and low frequency (LF, 20-200 Hz), signals from surface shipping are a significant contributor to the measured noise and have been observed to extend to 500 Hz. In this region, noise contributors can be a great distance from the observation point, and consequently the noise field exhibits the effects of sound propagation in both the horizontal and vertical directions (Carey (1986), Von Winkle (1985)). Wagstaff (1981) showed that, if one knows the locations and types of ships, then one can describe the characteristic of the horizontal noise field. Although the vertical noise distribution, including the broad horizontal maxima, could be qualititatively explained several discrepancies were observed. Wind-driven noise could explain these differences, and the sources of this noise are the subject of this paper.

EXPERIMENTAL EVIDENCE

Omnidirectional noise data at low frequencies which are free from flow and flow-induced vibrations (Strasberg (1984)) are very difficult to obtain. Several investigators (figure 1) measured the spectrum between 2 Hz and 2000 Hz in the deep sound channel or near the bottom. However, most of this data from the relatively heavily trafficked northern hemisphere reflect distant shipping noise in the 2 to 200 Hz range and, consequently, little local wind speed dependence is observed such as shown in figure 1. VLF/LF ambient noise experiments must be carefully examined to ensure that the results are either from distant or local sources.

Wittenborn (1976) (figure 2) performed an experiment with hydrophones that spanned the water column. Hydrophones within the sound channel showed little dependence on local wind speed between 10 Hz and 200 Hz. However, the hydrophone below critical depth showed an inferred local wind speed dependence (10 to 500 Hz) for wind speeds between 5 and 15 kns with levels of 47 dB and 56 dB re 1μPa @10 Hz. The 15 kn spectra showed a slowly varying broad band characteristic between 56 dB at 10 Hz and 65 dB at 500 Hz ($f^{1/2}$). Wittenborn cites an earlier experiment with noise levels of 69 dB for 300 Hz at 30 kns, compared to the 300 Hz levels of 44 dB at 5 kns, 51 dB at 10 kns, and 63 dB at 15 kns. These results suggest two wind noise mechanisms for the cases of low and high sea states with the intensity having a squared velocity dependence (U^2). The abrupt transition between 10 and 15 kns ($U^{6.8}$, based on the levels at 10 and 15 kns and not considered a true velocity dependence), as shown in figure 2, may be a threshold characteristic associated with

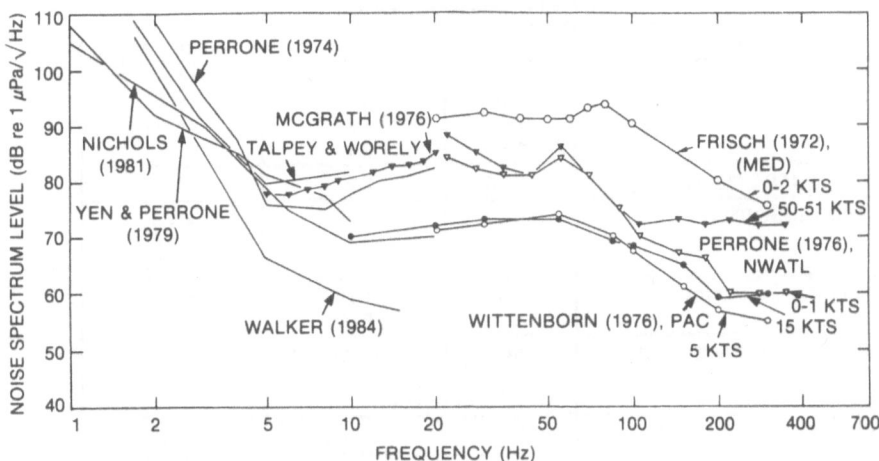

Figure 1. Selected Low Frequency Ambient Noise Measurements. The region less than 5 Hz is dominated by wave/wave interaction. The measurements between 5 Hz and 300 Hz show little local wind speed dependence, but, rather, the effects of distant shipping and other distant noise sources.

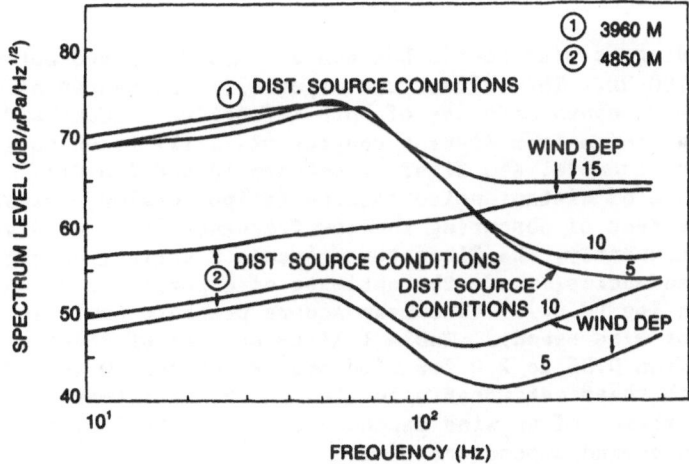

Figure 2. Ambient Noise Level vs. Frequency for the Wittenborn Experiment. The 4850 m deep hydrophone shows the local wind speed dependence (≈200 Hz) with the influence of distant noise sources less than 100 Hz. The 3960 m deep hydrophone is dominated by distant noise sources less 100 Hz.

breaking waves. (These results agree with the observations of Worley (1982), insofar as his data show a threshold-type behavior between the

$$NL = NL_1 + 20nLOG(W.S.)$$

INVESTIGATOR	f	n	w.s.(m/sec.)
PIGGOTT JASA 36(11)	13 13 141 < 50	2.1 0 1.53 2.1	10-20 1-5 3.5-20
PAYNE JASA (1967)	50-100	2.2	
WHITTENBORN (1976)	200-300	1.65 3.4 1	2.5-5 5-7.5 7.5-15
CROUCH JASA 3(2)-72	11 28	1.73 2.09	5-20
SHOOTER JASA 73(6)-81	141 150 177	1-1.39 1.1-1.32 1.36/1.57/.81	5-10 10-15
WORLEY JASA 71(4)-82	100 200	.85-1.5 1.65-2.0	10-15 5-10
WILSON JASA 73(1)-83	10	2.07	5-10
BURGESS JASA 73(1)-83	37	1.66	5-15

FACTORS AFFECTING WIND SPEED DEPENDENCE ARE: 1. DISTANT SOURCES; 2. OVERLAPPING WIND SPEED REGIONS; 3. SOUND PROPAGATION FACTORS.

Table I.   Low Frequency Ambient Noise Wind Speed Dependence

data corresponding to wind speeds between 2.5 and 5 kns and between 5 and 10 kns at 200 Hz. This effect was especially pronounced at 400 Hz.)

Although Wittenborn made use of both refractive effects and bathymetric blockage, noise from distant sources was still found to influence his results (for example, see figure 2 between 10 and 100 Hz). The corrupting influence of distant noise sources (ships, whales, volcanoes, etc.) has the effect of obscuring the low-frequency local wind speed dependence. Consequently, the literature reveals a variety of estimated wind speed dependencies; i.e., the estimate of a parameter n, where $NL = NL_1 + 20 n \log (W.S.)$. (The mean square pressure would increase with 2n power of wind speed.) Table I lists several of these estimates of n, ranging from 0.85 to 2.0 for wind speeds between 10 and 20 m/sec. The problem with these estimates also lies in the fact that the data clearly show a region of no wind dependence, a threshold-type behavior, and region with a wind dependence of n≈2.0.

Figure 3 illustrates this trend with the data of Piggott (1964). One observes the frequency dependent cross-over between the low wind speed and higher winds regions. Furthermore, the lower the frequency, the higher the wind speed will be at which the wind speed dependence point is observed.

Distant noise sources influence vertical noise directionality (Von Winkle (1985), Browning (1982), Bannister (1986)). This influence of the distant source produces a broad maximum in the vertical noise intensity centered on the horizontal. This phenomenon results from the con-

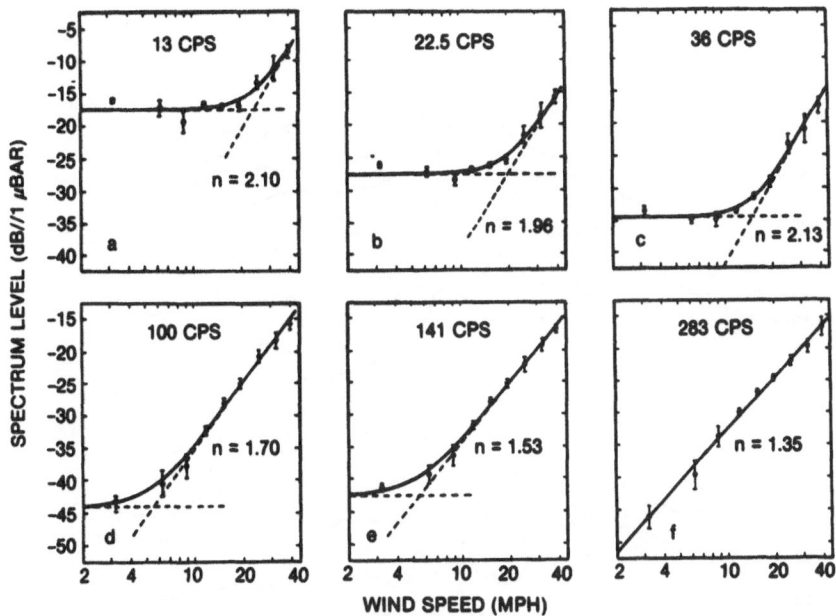

Figure 3. Ambient Noise Spectrum Level vs. Wind Speed, Piggott (1964)

version of higher angle rays to lower angle rays by either reflections from the basin boundaries and seamounts or refractive effects due to shallowing sound channels at the higher latitudes. Wagstaff attributed this effect to surface ships. Since the spectral variation of the horizontal noise is generally smooth, and since ship signatures are narrowband in this frequency range, wind-produced noise over seamounts, slopes, and at high latitudes was speculated to be an important contributor. The broad maximum along the horizontal has been observed in varied geographical locations, such as the sparsely shipped Southern Hemisphere waters of the South Fiji Basin (shown in figure 4). At the lower frequency the data clearly show a broad maximum. At 105 Hz one observes the influence of a single ship. These results are similar to data obtained in the North Pacific and the North Atlantic (Carey (1986)).

The experimental data were examined to obtain measured levels useful in the estimation of the source level of wind-produced noise at the sea surface. These results are shown in Table II, primarily at 50 Hz. The estimated levels based in the Wittenborn data are shown in the table to be between 43 dB at 5 kns and 51 dB at 15 kns, consistent with the estimates by Wilson and Kewley using the same data. Vertical noise cannot be used for local wind-driven noise; however, estimates for a cylindrical basin with sloping sides yields levels in the 50 dB range. Kewley has carefully estimated source levels, and his curves are shown in figure 5.

In summary, we have presented data which indicate the presence of a wind-driven noise in the 10 to 200 Hz region of the spectrum. The low wind speed range ($<$8-10 m/sec) appears to have a weak dependence on the wind speed, $0 < n < 1$; the high wind speed region ($>$7.5 to 15 m/sec) appears to have a dependence of $0.85 < n < 2$. These estimates point to

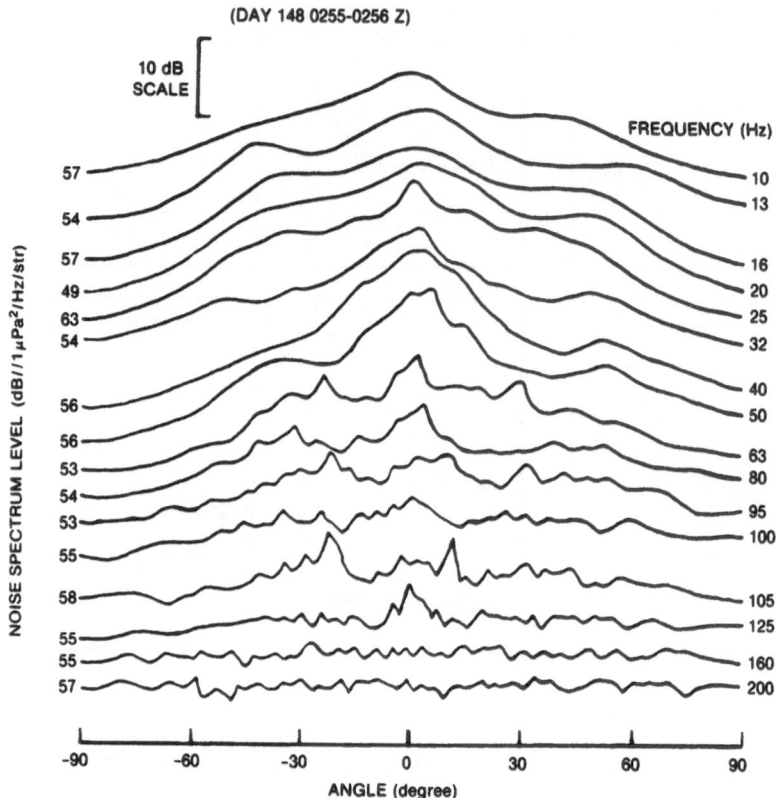

Figure 4. Vertical Noise Spectrum Level versus Angle from the Horizontal, Browning (1986)

the uncertainty in our knowledge of wind speed dependence and spectral characteristics.

POSSIBLE MECHANISMS

The fundamental mechanisms for the production of sound in turbulent regions may be derived from first principles. The basic procedure can be found in several treatments on hydrodynamic noise, most notably Lighthill (1979), Ffowcs Williams (1969), Dowling (1983), and Ross (1976). We have rederived the inhomogeneous wave equation with source terms in appendix A for the purpose of ranking the various mechanisms capable of the production of sound at the surface of the sea in the 10 to 200 Hz range. The basic approach is to write the equations governing the conservation of mass and momentum with source terms. The equation of state is specified, fluctuation quantities assumed, linearization employed, and the inhomogeneous wave equation is formed. The integral solutions to this equation are formulated by use of the Kirchoff Method (Stratton (1941), Jackson (1962)) and of the divergence theorem. The derivation

● OMNIDIRECTIONAL MEASUREMENTS WITH HYDROPHONE BELOW CRITICAL DEPTH:

INVESTIGATOR/FREQ.	10 Hz	50 Hz	100 Hz	w.s.(kns)	
WHITTENBORNE (1982)	48 dB*	50	44	5	N.E. PACIFIC
	50	51	47	10	
	57	58	60	15	
MORRIS (1978)	—	70	63	10	N.E. PACIFIC

● VERTICAL NOISE MEASUREMENTS ALONG THE HORIZONTAL:

INVESTIGATOR/FREQ.	50 Hz	70-80 Hz	100 Hz	
BROWNING (1982)	67 dB	70	65	FIJI BASIN
WAGSTAFF (1981)	—	—	69-66	N.E. PACIFIC
WALES (1981)	95	87	82	N.W. ATLANTIC
AXELROD (1965)	—	—	65-69	N.W. ATLANTIC
FOX (1964)	—	—	60-65	N.W. ATLANTIC
FISHER (1986)	74	—	60	E. PACIFIC
ANDERSON (1979)	—	—	65-69	N. PACIFIC

● SEMI-EMPIRICAL ESTIMATES: * dB//(μPa^2/Hz)

WILSON (1983)   NL=50dB+30LOG(W.S.) @ 50 Hz
BURGESS & KEWLEY (1983)   NL=37.3 +15LOG(W.S.) @ 50 Hz

● SOURCE LEVEL ESTIMATES BASED ON OMNI MEASUREMENTS:

f = 10 Hz   41 (dB//uPa**2/Hz @ 1m)@ 5 kns; 50 dB @ 15 kns
   50 Hz   43 dB @ 5 kns   51 dB @ 15 kns
  100 Hz   37 dB @ 5 kns   56 dB @ 15 kns

● VERTICAL LEVELS YIELD FOR A MEAN WIND SPEED OF 10kns, 50dB@50Hz AND 54-56dB@100Hz

Table II. Ambient Noise Source Levels

in appendix A is similar to those of Huon-Li (1981) and Yen (1979), and the basic result is the following:

$$4\pi C_0^2(\rho-\rho_0) = 4\pi P = \int [\partial q/\partial t]dV/R - \partial/\partial x_i \int [F_i] \, dV/R + \frac{\partial^2}{\partial x_i\partial x_j} \int [T_{ij}] \, dV/R - $$

$$- \int l_i[\partial\rho U_i/\partial t] \, dS + \partial/\partial x_i \int l_i[2\rho_0 U_i u_j + \rho_0 u_i u_j + P_{ij}] \frac{d_i S}{R}$$

The first term, $\int [\partial q/\partial t] \, dV/R$, represents a monopole term. q represents mass addition rate per unit volume. The second term represents an external force acting on the volume and has a dipole character. These two terms could be important in the incorporation of entrained bubble oscillation and translation. The third term is the Lighthill turbulence stress tensor and is known to represent an acoustic quadrupole. The term $\int l_i[\partial\rho U_i/\partial t] \, dS$ involves the motion of the boundary and can act as a monopole. The final integral involves the turbulent and compressive stresses acting on the boundary and is seen to have a dipole character. In particular the term $\partial/\partial x_i \int l_i[2\rho_0 U_i u_j] \, dS/R$ represents the wave turbulence interaction and is dominant since it represents a product of a first order U_i ; and second order term u_j .

Noise generation by the interaction of surface waves and turbulence near the surface was suggested by Goncharov (1970). He calculated levels of 80 dB at 10 Hz and 40 dB at 100 Hz by assuming a Pierson-Moskowitz surface wave spectrum and Kolmogorov's similarity hypothesis. His expression can be shown to be equivalent to the above integral. However, instead of using velocities, he employs the displacement spectrum for the surface wave and turbulence. His expression is $\rho(\omega) \sim 40\pi^2/\omega^2$.

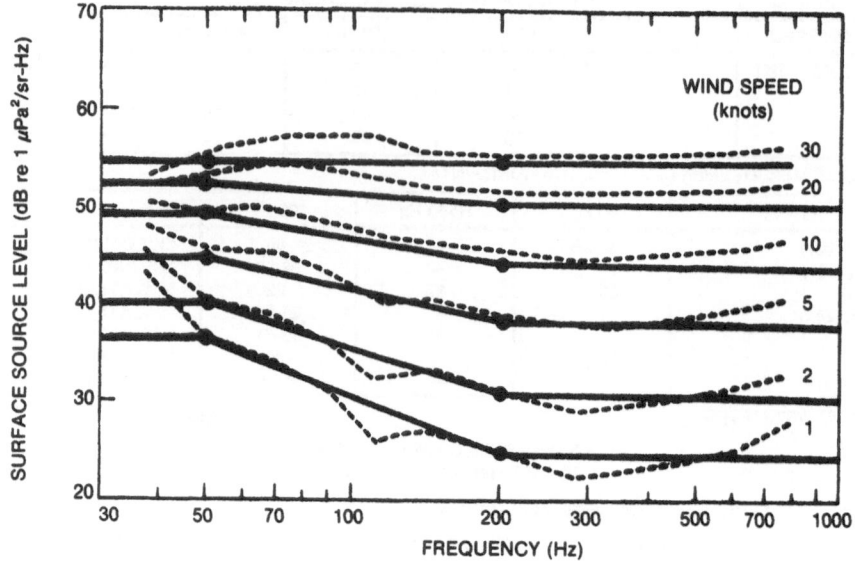

Figure 5. Source Level vs. Frequency and Wind Speed, Kewley (1986)

Yen and Perrone (1979) derived expressions yielding the frequency-dependent radiation characteristics for the wave/wave, wind/turbulence, and
wave/turbulence interaction mechanisms. Their results for the wave/tur-
bulence interaction (70 dB at 10 Hz and 50 dB at 100 Hz) show a linear
dependence on surface wave velocity (U) and an inverse square dependence
on frequency (ω^{-2})

$$P(\omega,k) \simeq 2 \cdot 10^{-2} \, \rho^2 \cos^4\theta \cdot U/\omega^2$$

The Yen & Perrone result contains three interesting factors. The
linear dependence on surface wave velocity is consistent with the previously discussed experimental results prior to wave breaking. The ω^{-2}
dependence is also consistent with the observed behavior at low frequencies; i.e., an overlap region composed of the interaction of the low and
higher frequency roll-offs. However, of particular note is the $\cos^4\theta$ dependence. This sharply peaked angular dependence would accentuate the
role of the ocean bottom and basin boundaries with respect to the vertical noise directionality. Thus, wave/turbulance interaction could be a
source of noise in the 2 to 200 Hz region for those sea states low
enough that breaking waves do not occur, due to the fact that it appears
as a physically realizable mechanism (considering the uncertainty of the
turbulence spectrum).

Kerman (1984) shows that, above a critical wind speed of approximately 10 m/s, small (micrometer (μm)) bubbles are produced and can be a
source of higher frequency sound. Thorpe (1986, 1982) has performed interesting experiments which demonstrate the existence of bubble plumes
and layers composed of μm-size bubbles (mean bubble size approximately
50 μm with densities between 10^4 to 10^6 bubbles/m^3) extending several

meters below the surface. M.Y. Su (1984) has shown that fresh water
breaking waves produced in wave tanks produce bubble plumes which penetrate to depths on the order of signicant wave height, with bubbles of
centimeter diameters due to coalesence (which is absent in salt water).
Several reviews (see bubble references) have been written on the existence and densities of bubbles produced by breaking waves. At high sea
states a residual layer is formed of micro-bubbles with a density that
decreases exponentially with depth, and bubble plumes which are convected to several meters depth by the vorticity beneath the wave. Even
though individual bubble oscillations with these micron-size bubbles
could not produce VLF/LF noise, collective oscillations of the bubbly
mixture driven by the hydrodynamic pressure field could produce sound.

It is well known (see appendix B) that a small amount of bubbles in
water significantly changes the bulk compressibility while not drastically changing the density. Wood has shown that the sonic velocity (C_m)
can be described by the following relationship between void friction
(x), density (ρ_m), and bulk compressibilites (K):

$$C_m^{-2} = [(1-x) \rho_l + x\rho_g][(1-x) K_l + xK_g]$$

The consequence of this result is shown in the figures of appendix B.
Small volume fractions result in large changes in the sonic speed when
the mixture can be treated as a continuum. For example, the sonic velocity of the bubble mixture with a 0.2% volume fraction is approximately 225 m/s. Ffowcs Williams (1969) describes the efficiency of the
radiation from a cloud of bubbly turbulent flow:

$$4\pi C_0^2(\rho-\rho_0) = 4\pi P = \int [\partial q/\partial t] \, dV/R$$

For a compact source with a small gas volume fraction

$$4\pi C_0^2(\rho-\rho_0) = 4\pi P \doteq \frac{1}{R} \frac{\partial}{\partial t} \int q \, dV$$

Ffowcs Williams estimates that

$$q = -\rho \frac{D}{Dt} \ln(1-x) \rightarrow -C_m^{-2} DP/Dt$$

$$dp = C_m^2 \Delta(1-x)\rho = -\rho C_m^2 \Delta x \rightarrow \rho-\rho_0 \sim (\rho/R) \, m^4 \, (C_0/C_m)^2, \, m = u/C_0$$

Thus he concludes that "a cloud of bubbly flow radiates very much more
efficiently than turbulence alone;" that is, the radiation from such a
flow would be $(C/C_m)^4 \approx$ 1975-times larger than the radiation from turbulent flow. However, one must account for the presence of a pressure
release surface.

An alternative approach is to consider the bubble cloud as a flexible sphere of radius a with composite mixture properties and to assume
it is compact with respect to the acoustic wave length and the vorticity
and turbulence scales. Then the forced oscillation of the bubble cloud
in absence of a boundary is $t' = t-R/C_0$, $Q = \int q \, dV = \frac{d}{dt} (\rho_0 v)$

$$P(R,t) = \dot{Q}(t')/4\pi R = \rho_0 \frac{\ddot{V}(t)}{4\pi R} => \frac{3\omega^2 a^3 \rho}{(1-x)} \, m^2(C_0^2/C_m^2) \, f(\omega)/R$$

where f(w) represents the simple harmonic oscillator transfer function.
This forced oscillation of a bubble cloud can have a resonant behavior

(Fitzgerald and Mellen (1982)). The bubble cloud is simply a monopole source and the presence of the boundary can be approximately taken into account via the surface image interference effect. Thus, we find the following for a cloud of micro-bubbles below the pressure release surface driven in forced oscillation by the hydrodynamic forces:

$$|\rho(R,t)|^2 \cong \left[\frac{3\omega^2 a^3 \rho}{(1-x)} m^2 (C_0/C_m)^2 f(\omega) \right]^2 \frac{4}{R^2} \left(\frac{2\pi z}{\lambda} \right)^2 \sin^2\theta$$

This expression shows a frequency-dependent efficiency approximately $(z/\lambda)^2$ at a given bubble cloud depth. This term indicates that, as seas pick up, the deeper the plume, the more efficient the radiation at longer λ. Furthermore, we note that this monopole has an m^2 improvement over the non-compact bubble cloud.

Thus, the low-frequency noise could be caused by wave turbulence prior to wave-breaking and, thereafter, by aggregate bubble (bubble cloud) oscillations exhibiting a threshold-type behavior and velocity squared dependence for the mean radiated pressure.

APPENDIX A: DERIVATION OF THE SOURCE INTEGRALS

The purpose of this appendix is to briefly outline the derivation of source terms important to the production of sound near the surface of the sea.

Conservation of mass: $\partial\rho'/\partial t + \partial\rho'v_i/\partial x_i = q$

Conservation of momentum: $\dfrac{\partial\rho'v_i'}{\partial t} + \dfrac{\partial\rho'v_iv_j}{\partial x_i} = -\dfrac{\partial P_{ij}'}{\partial x_j} + F_{ei}$

where $P_{ij}' = -\rho'\delta_{ij} + \mu D_{ij} + \mu_i\Theta\delta_{ij}$ (ref. Hinze, p. 17)

$\mu_i = 2/3\mu$, $D_{ij} = \partial U_i/\partial x_j + \partial U_j/\partial x_i$ and $\Theta = 1/2 D_{ii} = \partial U_k/\partial x_k$

Taking $\partial/\partial t$ of the continuity equation and $\partial/\partial x_i$ of the momentum equation yields upon subtraction:

$$\partial^2\rho'/\partial t^2 = \frac{\partial^2\rho'v_iv_j}{\partial x_i \partial x_j} + \frac{\partial^2 P_{ij}'}{\partial x_i \partial x_j} - \frac{\partial F_{ei}}{\partial x_i} + \frac{\partial q}{\partial t}$$

$$v = U + u, \ (\partial U/\partial x_i = 0), \ \rho' = \rho_0 + \rho, \ \frac{\partial\rho_0}{\partial t} = \frac{\partial\rho_0}{\partial x_i} = 0$$

$$\partial^2\rho/\partial t^2 - C_0^2 \partial^2\rho/\partial x_i^2 = U_iU_j \partial^2\rho/\partial x_i \partial x_j +$$

$$+ \frac{\partial^2(\rho_0+\rho)u_iu_j}{\partial x_i \partial x_j} + \frac{2\partial^2(\rho_0+\rho)U_iu_j}{\partial x_i \partial x_j} -$$

$$- \frac{\partial C_0^2\rho\delta_{ij}}{\partial x_i \partial x_j} + \frac{\partial P_{ij}}{\partial x_i \partial x_j} - \frac{\partial F_{ei}}{\partial x_i} + \frac{\partial q}{\partial t}$$

For the case of incompressible, invisid flow with no sources or sinks:

$$\partial^2 \rho_0 u_i u_j / \partial x_i \, \partial x_j + 2 \, \frac{\partial^2 \rho_0 \, U_i u_j}{\partial x_i \, \partial x_j} + \partial^2 \rho / \partial x_i^2 = 0$$

Compressible fluid:

$$\rho C_0^2 = P$$

$$1/C^2 \, \partial^2 P / \partial t^2 - \partial^2 P / \partial x_i^2 = \partial q / \partial t - \partial f_{ei} / \partial x_i + \partial^2 T_{ij} / \partial x_i \, \partial x_j$$

$$T_{ij} = 2\rho_0 \, U_i u_j + \rho_0 u_i u_j + P_{ij} - C_0^2 \rho \delta_{ij}$$

(in most instances $P_{ij} - C_0^2 \rho \delta_{ij} = 0$, (Lighthill))

Since $P = \rho_0 \, \partial \psi / \partial t = -i\omega \, \rho_0 \, \psi$, we finally have the wave equation with the source terms:

$$\frac{\partial^2 \psi}{\partial x_i^2} - \frac{1}{C_0^2} \, \frac{\partial^2 \psi}{\partial t^2} = \frac{1}{i\omega\rho_0} \left\{ \frac{\partial q}{\partial t} - \frac{\partial f_{ei}}{\partial x_i} + \frac{\partial^2 T_{ij}}{\partial x_i \, \partial x_j} \right\} = -4\pi \, f(x,t)$$

This inhomogeneous wave equation can be integrated by use of the Kirchoff method (Stratton (1941) and Jackson (1962)) to yield:

$$\psi(x,t) = \frac{\int dV \, [f]}{|x - x'|} + \frac{1}{4\pi} \int dS \, [1/R \, \partial \psi / \partial n - \partial / \partial n (1/R) \psi + 1/C_0 R \, \partial R / \partial n \, \partial \psi / \partial t]$$

This solution, when applied to our specific problem with the properties of $\partial [\;] / \partial x_i$ and $\partial [\;] / \partial y_1$, as well as the divergence theorem, yields the desired results:

$$4\pi C_0^2 (\rho - \rho_0) = 4\pi P = \int [\partial q / \partial t_i] \, dV/R - \partial / \partial x_i \int [f_{ei}] \, dV/R + \partial^2 / \partial x_i \partial x_j \int [T_{ij}] \, dV/R -$$

$$- \int l_i \, [\partial \rho u_i / \partial t] \, dS + \partial / \partial x_i \int l_i \, [2\rho_0 U_i u_j + \rho_0 u_i u_j + P_{ij}] \, dS/R$$

APPENDIX B: MIXTURE THEORY

A.B.Wood (1932) showed that the sonic speed could be calculated for an air-bubble/water mixture by use of the mixture density (ρm) and the mean compressibility (km). The mixture can be treated as a continuous medium when the bubble diameter (d) and spacing between the bubbles (D) are much less than the wavelength of sound. In the case of low frequencies, for the mixture with a volume fraction (X) of gas we can calculate the mean density and compressibility as follows:

$$\rho_m = (1 - x)\rho_l + x\rho_g$$

$$K_m = \frac{-dv_m}{v_m \, dP} \equiv - \frac{dv_l}{v_l \, dP} \, \frac{v_l}{v_m} + \frac{dv_g}{v_g dP} \, \frac{v_g}{v_m} = (1 - x)K_l + x \, K_g$$

This imples that a state of equilibrium prevails and the mixture mass is conserved, and the pressure, P, is uniform throughout the mixture (a low frequency assumption). Since the sonic speed is

$$C^2 \equiv dP/d\rho = (\rho K)^{-1}$$

we have

$$C_m^{-2} = C_{mlf}^{-2} = [(1 - x)\rho_l + x\rho_g] \, [(1 - x)K_l + xK_g]$$

$$C_m^{-2} = (1 - x)/C_l^2 + x^2/C_g^2 + (x)(1 - x) \, \frac{\rho_g^2 \, C_g^2 + \rho_l^2 \, C_l^2}{\rho_l \, \rho_g \, C_l^2 \, C_g^2}$$

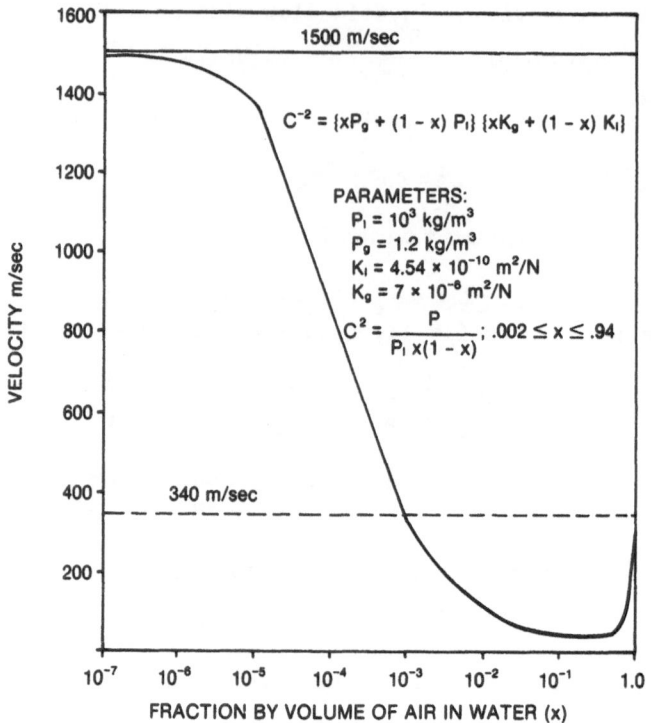

Figure B-1. Low Frequency Approximation for the Mixture Speed

The expression for the sonic speed poses the question of whether the gas compressibility is described by an isothermal or adiabatic process, especially since the single phase sonic speed is known to be adiabatic. However, in the case of an air-bubble/water mixture, the controlling physical factor is the transfer of the heat generated in bubble compression to the surrounding liquid. If the transfer is rapid then the bubble oscillation is isothermal, $(\partial v/\partial P = -v/P, K_g = 1/P)$, as compared to the adiabatic condition $(\partial v/\partial P = -v/\gamma P, K_g = 1/\gamma P)$. Thus, in use of the above equations one must use either for the adiabatic or isothermal case, $C_{gi} = C_{ga}/\sqrt{\gamma}$. Isothermal conditions are most likely to prevail for air-bubble/water mixtures due to the large thermal capacity of water. Examination of the above expressions shows that as $x \to 0$, $C_m^{-2} \to C_l^{-2}$, and as $x \to 1$, $C_m^{-2} \to C_g^{-2}$; as one would expect. The striking characteristic revealed by these equations (shown in figure B-1) is the sharp reduction in the sonic velocity at small volume fractions; i.e., $X=0.002 \to C_m=225$ m/sec. These equations may be approximated for the air/water mixture:

$$C_m^2 = \frac{\gamma P}{\rho_l x (1 - x)} \xrightarrow{\gamma = 1} \frac{P}{\rho_l x (1 - x)}$$

$$C_m (x = 0.5) = 20 \text{ m/sec.}$$

Karplus (1958) used an acoustic tube to determine the standing wave pattern as a function of air volume fraction. His results are shown in figure B-2. Close agreement was found between the inferred sonic speeds

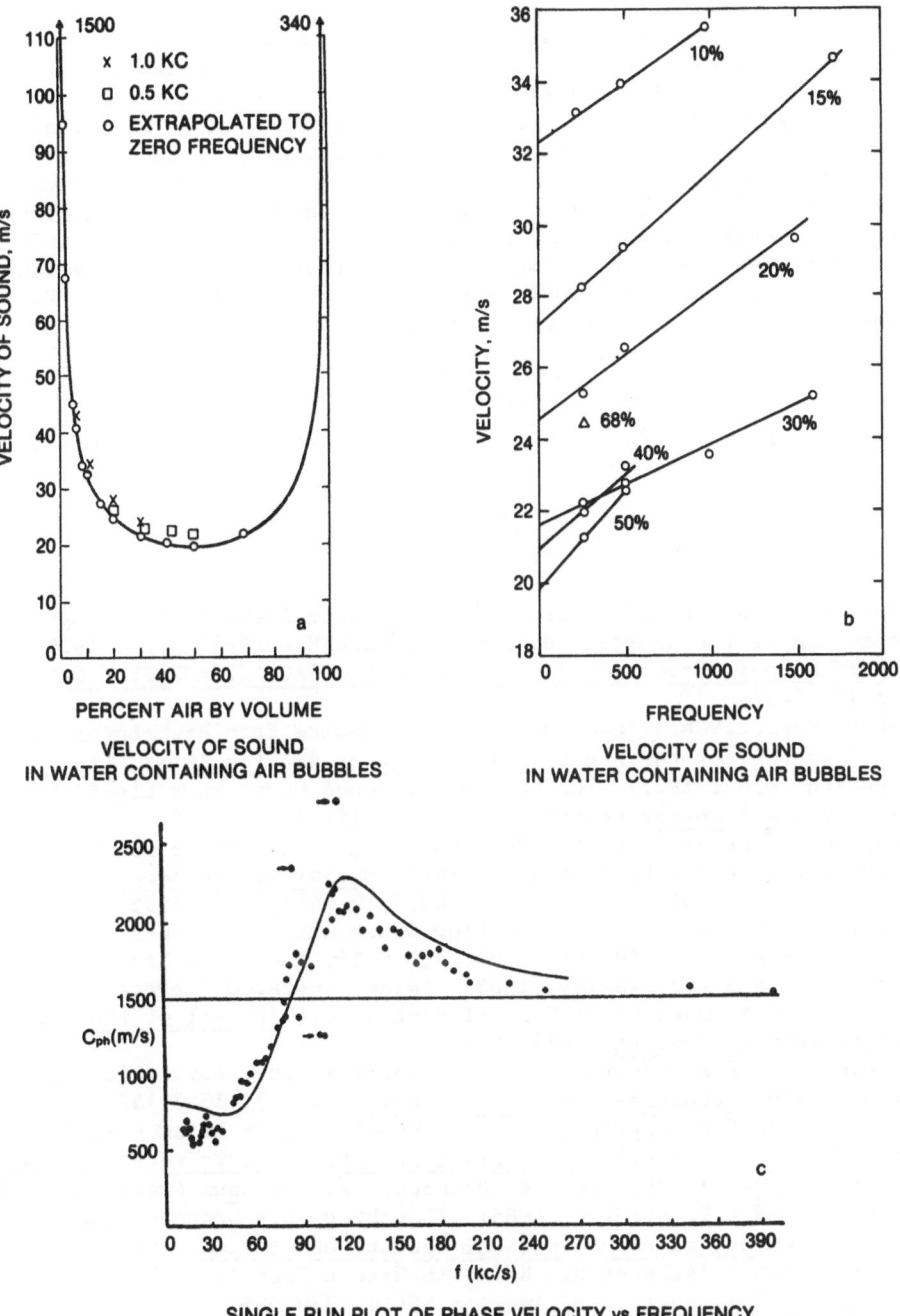

Figure B-2. Measured and Computed Mixture Sonic Speeds. a) Measured Mixture Speed vs. Volume Fraction at 500 and 1000 Hz, Karplus (1958); b) Measured Dispersive Character of Mixture Speed Below 2000 Hz, Karplus (1958); c) Measured and Computed Mixture Sonic Speed Showing the Behavior Below, At, and Above Resonance, Fox, et al. (1955)

374

and results calculated with Wood's expressions. Similar results have also been observed at the low frequencies by Campbell and Pitcher (1955). These results are also observed at the higher frequencies above and below resonance. Several studies and texts have been written on this subject and are listed in the references. An example of the agreement between theory and measurement near the vicinity of bubble resonance is shown in figure B-2c. It is important to note that most calculations performed at these higher frequencies use $Km = K_1 + Kg$, rather than the Wood approach $Km = (1 - x)K_1 + xKg$. This difference is unimportant near resonance and for small volume fraction, but is important as one approaches the low frequencies of interest to this paper. One can show that the correct expression is:

$$\frac{1}{C_{mhf}^2} = \frac{(1 - x)^2}{C_1^2} + \frac{1}{C_{mlf}^2\left((1 - \omega^2/\omega_0^2) + 2i\delta\,\omega/\omega_0\right)}$$

when h.f. and l.f. are the high frequency and low frequency values of the sonic speed.

REFERENCES

Bachmann, W., and R.B. Williams, 1975: 'Oceanic Acoustic Modeling: Proceedings of a Conference Held at SACLANTCEN on 8-11 Sept. 1975.' SACLANTCEN Conf. Proc. 17(2) -- Bubbles NATO SACLANT ASW RES. CENTRE, LaSpezia, Italy

Bannister, R.W., 1986: 'Deep Sound Channel Noise from High-Latitude Winds,' Journal of the Acoustical Society of America 79(1), 41-48

Brekhovskikh, L.M., 1967: 'Generation of Sound Waves in a Liquid by Surface Waves,' Soviet Physics, Acoustics 12(3), 323-350

Browning, D.G., et al., 1981: 'Vertical Directionality of Low Frequency Ambient Noise in the South Fiji Basin,' Journal of the Acoustical Society of America 70(S1), S66a (also NUSC TD 6611, Jan. 1982)

Browning, D.G., et al., 1982: 'Shallow Water Ambient Noise: Offshore Measurements at 20-10,000 Hz.' NUSC TD 6825, NUSC, New London, CT

Burgess, A.S., and D.J. Kewley, 1983: 'Wind-Generated Surface Noise Source Levels in Deep Water East of Australia,' Journal of the Acoustical Society of America 73(1), 201-210

Carey, W.M., and R.A. Wagstaff, 1986: 'Low-Frequency Noise Fields,' Journal of the Acoustical Society of America 80(5), 1523-1526

Carey, W.M., and R.A. Wagstaff, 1985: 'Low-Frequency Noise Fields and Signal Characteristics,' Ocean Seismo-Acoustics: Low-Frequency Underwater Acoustics, T. Aka and J.M. Berkson, eds., Plenum Press, 753-766

Carey, W.M., and M.P. Bradley, 1985: 'Low-Frequency Ocean Surface Noise Sources,' Journal of the Acoustical Society of America, 78(S1)

Cato, D.H., 1976: 'Ambient Sea Noise in Waters Near Australia,' Journal of the Acoustical Society of America 60(2), 320-328

Dowling, A.P., J.E. Ffowcs Williams, 1983: Sound and Sources of Sound, John Wiley & Sons, NY

Ffowcs Williams, J.E., 1969: 'Hydrodynamic Noise.' Annual Review of Fluid Mechanics, W.R. Sears and M. Van Dyke, eds. Annual Reviews, Inc. Palo Alto, CA, 197-222

Fitzgerald, J.W., and R.H. Mellen, 1982: NUSC cont. N00140-82-M-ST20

Fitzpatrick, H.M., and M. Strasberg, 1959: 'Hydrodynamic Sources of Sound.' DTMB Rept 1269, NSRDC, Carderock, MD

Fox, F.E., et al., 1955: 'Ambient-Noise Directivity Measurements.' Journal of the Acoustical Society of America 27(3), 534-539

Furduev, A.V., 1966: 'Undersurface Cavitation as a Source of Noise in the Ocean,' Atmospheric and Oceanic Physics, 2(235), 314-320

Goncharov, V.V., 1970: 'Sound Generation in the Ocean by the Interaction of Surface Waves and Turbulence.' Izv., Atmospheric and Oceanic Physics 6(11), 1189-1196

Hinze, J.O., 1959: Turbulence, McGraw-Hill Book Co., NY

Hughes, B., 1976: 'Estimates of Underwater Sound (and Infrasound) Produced by Nonlinearly Interacting Ocean Waves,' Journal of the Acoustical Society of America 60(5), 1032-1039

Huon-Li, 1981: 'On Wind-Induced Underwater Ambient Noise.' NORDA TN 89, NORDA, NSTL, MS

Jackson, J.D., 1962: Classical Electrodynamics, John Wiley & Sons, NY

Kerman, B.R., 1984: 'Underwater Sound Generation by Breaking Waves.' Journal of the Acoustical Society of America 75(1), 149-165

Karplus, H.B., 1958: 'The Velocity of Sound in a Liquid Containing Gas Bubbles,' Armour Research Foundation, University of Illinois, Proj. c00-248, TID-4500

Kibblewhite, A.C., and K.C. Ewans, 1985: 'Wave-Wave Interactions, Microseisms, and Infrasonic Ambient Noise in the Ocean,' Journal of the Acoustical Society of America 78(3), 981-994

Knudsen, V.O., R.S. Alford and J.W. Emling, 1948: 'Underwater Ambient Noise,' Journal of Marine Research 7, 410-429

Kuo, E.Y.T., 1968: 'Deep-Sea Noise Due to Surface Motion,' Journal of the Acoustical Society of America 43(5), 1017-1024

Latham, G..V., and A.A. Nowroozi, 1968: 'Waves, Weather and Ocean Bottom Microseisms,' Journal of Geophysical Research 73(12), 3945-3956

Lighthill J., 1979: Waves In Fluids, Cambridge University Press, Cambridge, UK

Lloyd, S.P., 1981: 'Underwater Sound from Surface Waves According to the Lighthill-Ribner Theory,' Journal of the Acoustical Society of America 69(2), 425-435

Martin, R.L., 1985: 'Effects of the Ocean Boundaries and Bottom Topography on Acoustic Ambient Noise Fields in the Ocean,' NORDA TN 323, NORDA, NSTL, Mississippi

McDaniel, S.T., 1987: 'Subsurface Bubble Densities from Acoustic Backscatter Data,' ARL/TM-87-57, Applied Research Lab, Penn. State Univ., State College, PA 16804

Mellen, R.H., 1987: private communication.

Minnaert, M., 1933: 'Musical Air-Bubbles and Sounds of Running Water.' Philosophical Magazine 10, 235

Nichols, R.H., 1981: 'Infrasonic Ambient Noise Measurements: Eleuthera,' Journal of the Acoustical Society of America 69(4), 974-981

376

Ross, D., 1976: Mechanics of Underwater Sound, Pergamon Press, NY

Shang, E.C., and V.C. Anderson, 1986: 'Surface-Generated Noise Under Low Wind Speed at Kilohertz Frequencies,' Journal of the Acoustical Society of America 79(4), 964-971

Strasberg, M., 1984: 'Hydrodynamic Flow Noise in Hydrophones,' Adaptive Methods in Underwater Acoustics, H. Urban (ed.), Reidel, 125-141

Stratton, J.A., 1941: Electromagnetic Theory, McGraw-Hill Book Co., NY

Su, M.Y., A.W. Green and M.T. Bergin, 1984: 'Experimental Studies of Surface Wave Breaking and Air Entrainment,' Gas Transference at Water Surfaces, W. Brutsaert and G. Jirka (eds.), Reidel Press, pp 211-219

Thorpe, S.A., 1986: 'Measurements with an Automatically Recording Inverted Echo Sounder: Aries and Bubble Clouds,' Journal of Physical Oceanography 16, 1462-1478

Thorpe, S.A., 1986: 'Bubble Clouds: A Review of Their Detection by Sonar, of Related Models, and of How Kv May be Determined.' Oceanic Whitecaps and Their Role in Air-Sea Exchange Processes, Reidel in assoc. with Galway Univ. Press, 57-68

Thorpe, S.A., 1982: 'On the Clouds of Bubbles Formed by Breaking Wind-Waves in Deep Water, and Their Role in Air-Sea Gas Transfer.' Philosophical Transactions of the Royal Society of London, A304, 155-210

Urick, R.J., 1984: Ambient Noise in the Sea. U.S.G.P.O.: 1984 - 456-963, Washington, D.C.

Von Winkle, W.A., and D.G. Browning, 1985: 'Vertical Noise Directionality in the Deep Ocean: A Review,' Journal of the Acoustical Society of America 78(S1) (also NUSC TD 7561)

Wagstaff, R.A., 1981: 'Low-Frequency Ambient Noise in the Deep Sound Channel--The Missing Component,' Journal of the Acoustical Society of America 69(4), 1009-1014

Wenz, G.M., 1962: 'Acoustic Ambient Noise in the Ocean: Spectra and Sources,' Journal of the Acoustical Society of America 34, 1936-1956

Wilson, J.H., 1983: 'Wind-Generated Noise Modeling,' Journal of the Acoustical Society of America 73(1), 211-216

Wittenborn, A.F., 1976: 'Ambient Noise and Associated Propagation Factors as a Function of Depth Wind Speed in the Deep Ocean,' Tracor Rept. T76RV5060, DTIC(AD 006902), Alexandria, VA

Wood, A.B., 1932/1955: A Textbook of Sound, G. Bell and Sons Ltd., London, 360-364

Worley, R.D., and R.A. Walker, 1982: 'Low-Frequency Ambient Ocean Noise and Sound Transmission over a Thinly Sedimented Rock Bottom,' Journal of the Acoustical Society of America 71(4), 863-870

Wu, Jin, 1981: 'Bubble Populations and Spectra in the Near-Surface Ocean: Summary and Review of Field Measurements,' Journal of Geophysical Research 86 (C1), 457-463

Yen, Nai-Chyuan and A.J. Perrone, 1979: 'Mechanisms and Modeling of Wind-Induced Low-Frequency Ambient Sea Noise,' NUSC TR 5833, Feb. 1979

ESTIMATION OF SOURCE CHARACTERISTICS FROM UNDERWATER NOISEFIELD MEASUREMENTS

R.W. Bannister
Defence Scientific Establishment,
Naval Base Post Office, Devonport, Auckland,
New Zealand
A.S. Burgess and D.J. Kewley
Weapons Systems Research Laboratory,Adelaide,
Australia

ABSTRACT.   Underwater ambient noise measurements in the region surrounding New Zealand are dominated by wind-dependent processes over the frequency band 10-500 Hz.   These data are used to infer surface source levels by removing the relevant ocean transfer function. Resulting directional source levels are presented as a function of windspeed at 40 Hz and 300 Hz.   The angular shape of these curves is consistent with a volumetric source having an active depth equal to half a wavelength.   Monopole source levels are also inferred from these data and depend on the assumed noise processes.   Two cases are analysed - the fixed $\lambda/2$ active depth of the source volume and a depth which tracks the depth of observed bubbles near the surface.

1.   INTRODUCTION

Much of the ocean area surrounding New Zealand is remote from ship traffic.   Underwater noise measured in the region exhibits windspeed dependence over the frequency band 10 to 500 Hz when ship contamination is removed.   Some effort has been put into interpreting these measurements by constructing an ambient noise model which shows that the received ambient noise level consists of various components each of which represents its source function modified by an ocean transfer function.

This paper is an attempt to reverse such normal modelling processes by selecting appropriate noise components and applying the inverse of the ocean transfer function to measured noise and obtaining estimates of the surface source function.   An ocean transfer function for a surface distributed source is developed in the appendix and provides a simple transformation between vertical noise measurements and the vertical angle source characteristics of surface noise.   This relationship is explored first.   The paper also considers the possibilities of using omnidirectional and horizontal noisefield measurements to infer surface source levels.

377

B. R. Kerman (ed.), Sea Surface Sound, 377–389.
© 1988 by Kluwer Academic Publishers.

2. VERTICAL NOISE DIRECTIONALITY MEASUREMENTS

A fundamental requirement in the process of measuring and interpreting wind-dependent noise, is to select data which is dominated by local sources for which ocean surface conditions are both uniform and known.   The vertical noisefield has advantages in this respect since energy from distant sources, which is carried either by convergence zone or deep sound channel paths, will appear in near-horizontal angles and can be ignored.   Noise received at angles greater than about 10 degrees is generated relatively near to a receiver and arrives either by bottom bounce paths or directly from the surface.   It is these paths that are modelled by the transfer function in the appendix and which will be used to infer surface source characteristics.

2.1 The Ocean Transfer Function

The appendix provides a development of the simple transfer function which relates directional surface source levels (per square metre) to the received noise level (per solid angle).   It is interesting to note that, within acceptable limits, this transfer function depends only on the value of bottom loss and a geometrical factor (equation A2).   Since it is independent of experimental factors such as receiver depth and water depth it allows source level estimates, inferred from various investigations, to be superimposed. The proper relationship between angles at the surface, receiver and bottom needs to be preserved by the use of Snell's Law for the particular soundspeed profile.

Care needs to be taken, however, that the region over which surface noise is collected does not grow larger than the scale of uniform ocean-surface conditions.   This range is determined largely by bottom loss (equation A2) and will vary with geographic location, frequency and received angle.   In the worst case of low frequency (<50 Hz), low angle (10 degrees) and low bottom loss (.1 dB per bounce), a noise "catchment area" may have a radius of up to 700 km. At higher frequency and received angle, the catchment area is confined to within a few tens of kilometres from the receiver.

To ensure that only local conditions are sampled, however, it is clear that noise from near-vertical beams of a vertical array should be used.   This was the approach adopted by Burgess and Kewley[3], who used their equivalent of equations A3 and A4 with θ = 90 degrees to infer source level (S.a) from measured noise data (I).   It is the purpose of this section to report upon some preliminary attempts to extend this technique to lower angles and hence to obtain more information on the directional source function.

Figure 1 shows a typical transfer function between measured noise and source levels obtained from equations A3 and A4.   Values from figure 1, when added to measurements of vertical noise (per steradian), will give the directional surface source level (per unit area).   The two major components of the transfer function are separated to show their relative contribution.   At 20 Hz (figure 1a), bottom bounce

effects are significant and so an accurate estimate of bottom loss
(b) is important.   Since the upward and downward components of the
measured noisefield differ by a factor of b (equation A4) this can, in
principle, be used to estimate bottom loss at a given vertical angle.
The geometrical component indicated in figure 1 is simply the
$1/\sin \theta_s$ term in equation A3 which expresses the increased surface
area intersected by a given cone of rays as the average angle reduces.

Figure 1b represents the situation for a higher frequency
(200 Hz).   In this case, the bottom loss contribution is small
because bottom loss is high and relatively little energy is
transferred by these paths.   The transfer function is dominated by
the geometrical effect.

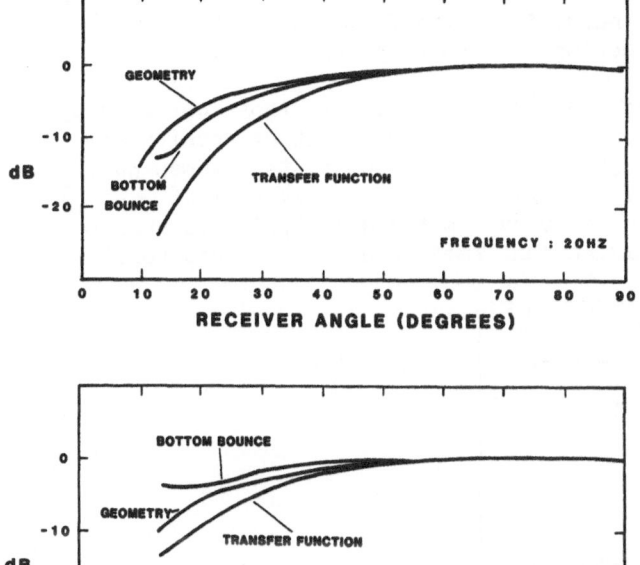

Fig. 1

*The transfer function
between measured
vertical noise
directionality and
the source function
showing important
components.*

It would appear from figure 1, that the transfer function can
be defined but its accuracy is dependent upon the estimation of bottom
loss.   This becomes a problem at low frequency and low angle when
the loss is small.   The technique of determining bottom loss from
measurements of upward and downward going energy has been found to
work well above 100 Hz, where good agreement with independent
estimates has been achieved.   At lower frequencies the estimate
is progressively degraded since the asymmetry of the vertical
noisefield caused by bottom loss is small and unstable.   In this
case, independent estimates of bottom loss have been used.

2.2 Angular Characteristics of the Source

The transfer function of figure 1, when added to vertical noise directionality measurements, will produce source level estimates as a function of angle. Figure 2 gives the result of this process, mapped into the surface grazing angle. The figure shows preliminary information only, since limited data have been used. The measured vertical noise data were all collected in the South Pacific region and were selected so as to exclude contamination by local ships.

The directional source estimates shown in figure 2a (40 Hz) are obtained using an independent estimate of bottom loss in the transfer function which gives rise to different estimates of the source function from up and downgoing noise measurements (equations A3, A4). The figure represents an average of these two estimates. Figure 2b (318 Hz), on the other hand, was obtained by estimating bottom loss from the asymmetry of the noisefield itself, yielding the same estimate of the source function from both up and downgoing measurements.

The results in figure 2 can be compared with those obtained by Burgess and Kewley at 90 degrees. The present 40 Hz results are about 3 dB higher and the 300 Hz data 3 dB lower, than those published by Burgess and Kewley.[3] This difference may simply reflect variability resulting from the very limited amount of data used to produce figure 2 and should be confirmed with further measurements.

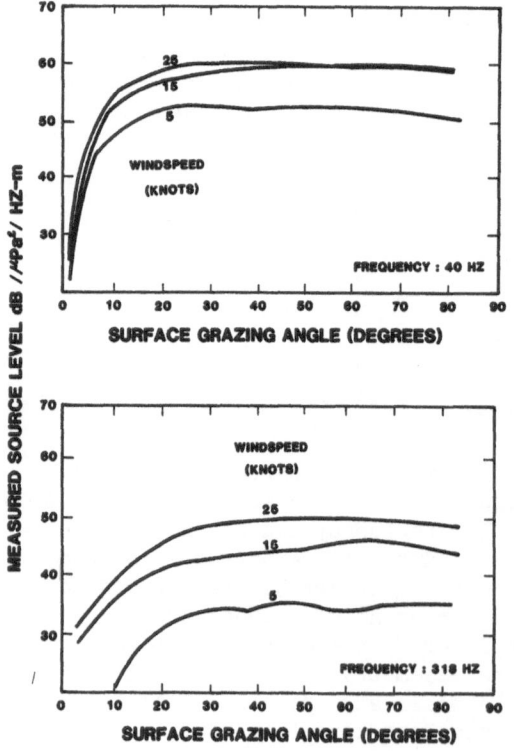

Fig. 2

Source level directionality inferred from measurements of the vertical noise field.

The relative shape of the source level curves as a function of angle, however, represents new information. The reduced source response at low angles is consistent with the presence of a surface interference process.

2.3 The Surface Interference Model

The curves in figure 2 represent the product of the monopole source level (S) and the source directivity function ($a(\theta_s)$), which presumably is the result of surface interference. In order to further interpret these results it is necessary to have some model of the physical processes in mind. To this end, two models are considered - a plane of simple monopoles situated at a single depth below the ocean surface and a set of monopole planes extending vertically to a given depth. This latter model is a volumetric source. The source directivity in each case can be shown to be:

$$a_m (\theta_s) = 4 \sin^2 \psi \qquad \text{[simple monopole plane]} \qquad (1)$$

$$a_v (\theta_s) = 2 \left(1 - \frac{\sin 2 \psi}{2 \psi}\right) \quad \text{[volumetric source]} \qquad (2)$$

where

$$\psi \quad = \quad 2\pi \, z \, \sin \theta_s$$

$$z \quad = \quad \text{source depth of monopole plane or depth of active volume (in wavelengths)}$$

The shape of these functions at low surface grazing angles (θ_s) is particularly sensitive to source depth. Measured curves in figure 2 can be matched with source directivity functions described by equations 1 and 2 to determine the effective source depth. If the source is a simple monopole, its source depth (equation 1) must be less than $\lambda/4$ or interference nulls would be visible in the data at high angles. The measured shape at low angles, however, is more consistent with a source depth of $\lambda/2$. This contradiction indicates that the simple monopole shape is not consistent with data in figure 2.

The angular response of a vertical set of incoherent monopole surfaces (equation 2), however, has no interference nulls other than at zero degrees and is thus more in keeping with observations. The shape of the curves in figure 2 implies an active volume depth of $\lambda/2$ at both frequencies and at all windspeeds considered.

The small data sample used to produce figure 2 makes it difficult to say how significant this result is. At face value it implies that the source is volumetric in nature with an active depth of influence

which extends to at least λ/2 at 40 Hz (20m). This appears to be so
at all windspeeds although data reliability is less at low windspeed.
The indication, however, is not consistent with current suggestions of
the source mechanism known to the author. More data having longer
time averaging will be analysed in future to improve the reliability
of these results. It is considered that the technique itself is
promising and should lead to a useful description of the surface
source function.

2.4 The Monopole Source Level

Results shown in figure 2 include surface interference processes.
In order to explore the monopole source function (S) itself, both the
ocean transfer function and the source directivity function (a) in
equation A3 need to be removed from measured data. It is desirable
to produce frequency spectra of the monopole source level at various
windspeeds since this may provide insight into the source mechanism
uncontaminated by surface interference. The base data used for this
procedure were beam output spectra from a vertical line array,
averaged over a known angle range. Such data is shown in figure 3
for two windspeeds. Some work has been done at low angles
(20 degrees such as figure 3) and some at nearly vertical angles.
The resulting estimate of the source spectrum was the same in each
case, as could be expected. Since the estimate from various beam
angles is consistent, it was decided to proceed by using the large
body of 90 degree vertical array beam spectra reported by Burgess
and Kewley.[3] In their analysis, the ocean transfer function
(identical to equations A3 and A4) was removed, but not the
directivity function. A sample of their source level spectra is
given as broken lines in figure 4.

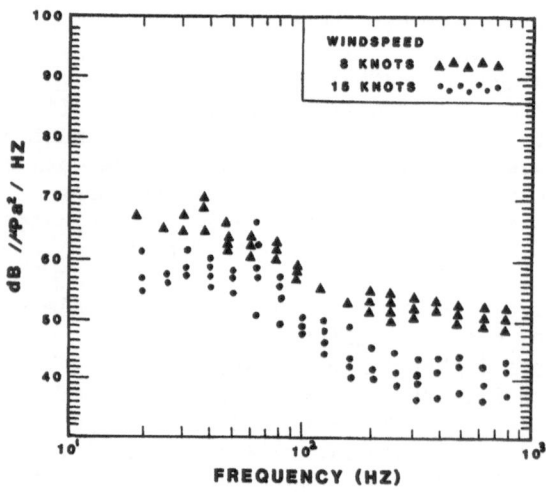

Fig. 3

*Measured beam output
spectra from vertical
line array averaged
over vertical angles
20-25 degrees.*

Since it appears that a simple monopole source is inconsistent with data (section 2.3), this interference process is not pursued further. The matter of obtaining monopole source spectra reduces to removing appropriate values of equation 2 (the volumetric source) from Burgess and Kewley's curves. This immediately raises the question of which volume depth to use.

Earlier indications have been that a depth of $\lambda/2$ is appropriate. This leads to adding 3 dB to Burgess and Kewley's curves shown in (figure 4) at all frequencies and windspeeds.

A further possibility, perhaps more related to physical processes, is based on the notion that bubble plumes and turbulence are observed in the near-surface region. If this physical situation is relevant to noise generation it can be expected that the depth extent of the volumetric source would be coincident with the maximum depth of bubbles. The depth of bubbles is windspeed-dependent and has been measured by Thorpe.[4] Using this description of source depth extent, equation 2 is evaluated and removed from the Burgess and Kewley spectra to produce the solid lines in figure 4. It can be seen that these curves tend to become independent of windspeed at low frequency. The windspeed independence at low frequency could be made more convincing by choosing a windspeed dependence of the source depth other than that indicated from reference 4. This suffices for the present purpose, however, which is to indicate that if there is a volumetric source, whose maximum depth tracks the bubble cloud depth, then there is a frequency region in which there is weak windspeed dependence. Since S is the source level per unit area, the result implies that there is a saturation in the area source function and increased noise is the result of depth increases only. The shape of the solid curves in figure 4 is suggestive of two source mechanisms which apply above and below 100 Hz.

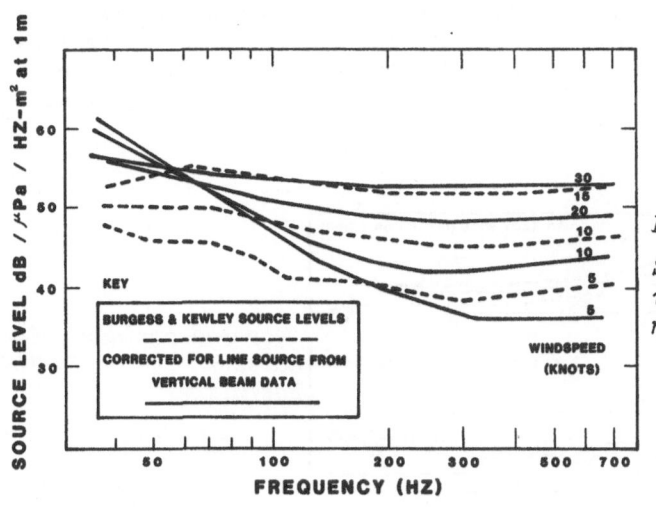

Fig. 4

Source level spectra inferred from noise measurements.

384

2.5 The Use of Source Level Functions

Source level spectra (such as in figure 4) are required by those interested in predicting ambient noise levels. If the transmission model used includes interference effects then curves such as figure 4 are appropriate. It is noted, however, that the same form of source directivity function must be used in a noise model as that employed to infer the monopole source level (eg equation 2). In many cases, the transmission loss model will allow for a simple dipole source such as equation 1. This situation is well served by the Burgess and Kewley curves (Figure 4) plus 3 dB in association with a source depth of $\lambda/2$. This case may be complicated by secondary dipole nulls. If on the other hand, surface interference effects are not included in a noise model, then curves such as those in figure 2 should be used directly.

3. OTHER NOISE MEASUREMENTS

While it is considered that vertical noise measurements are best suited to the task of estimating surface source levels, other data have been used. These are briefly discussed.

3.1 Omnidirectional Noise

Under conditions of high bottom loss, omnidirectional noise data, collected below critical depth have been used to assess surface source levels.[5] Such data were free from the effects of distant (deep sound channel) noise contamination, but still were subject to a bottom bounce contribution as indicated by equation A5. If bottom loss is infinitely high and a $\sin^2\theta$ source directivity function is used, the transfer function reduces to π as shown in equation A6. In general this assumption is not valid (particularly at low frequencies) and equation A5 is preferred. Thus the omnidirectional transfer function can be evaluated by appropriate integration over the various angle-dependent variables if bottom loss is known. Surface source level spectra can thus be inferred from omnidirectional data but no angle dependence can be obtained.

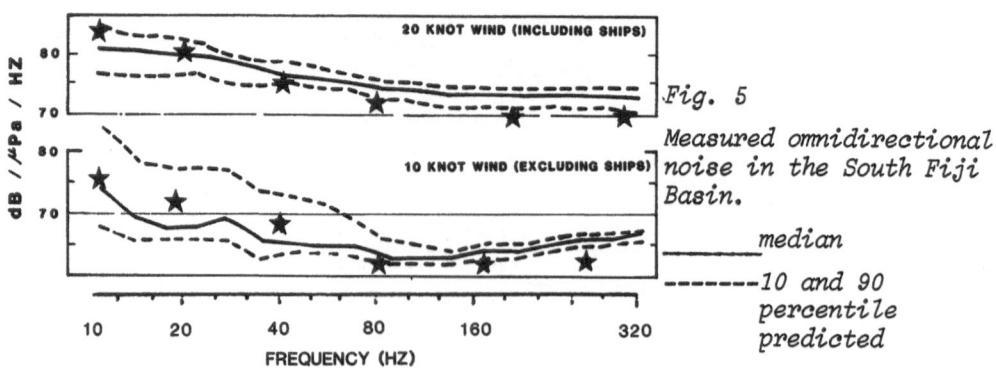

Fig. 5

Measured omnidirectional noise in the South Fiji Basin.

———— *median*

-------- *10 and 90 percentile predicted*

Rather than demonstrate this procedure, figure 5 shows the effect of reversing it to the more normal one of taking the source level from figure 4 and predicting omnidirectional noise levels. Here, omnidirectional measurements from the South Fiji basin are compared with predictions obtained from equation A5. The good agreement indicates reasonable consistency in the processes being discussed.

3.2 Horizontal Noise Data

A recent paper by Ferguson and Wyllie[6] has reviewed methods of inferring source characteristics from beams of a horizontal line array. The work further reveals the interesting fact that data surrounding the beam alias region are very sensitive to source level angular characteristics. Measurements are presented which are consistent with the results above, namely that the source directionality has a form similar to $\sin^2\theta$. The data were collected under extreme sea conditions (30 knots windspeed) where locally generated noise is greater than any distant components.

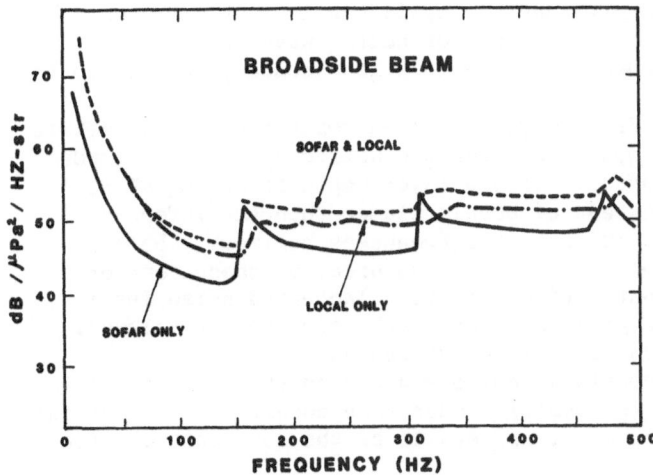

Fig. 6

Modelled broadside beam spectra of horizontal array showing noise conbributions.

To explore the possibilities of this technique at lower windspeeds, the beam outputs of a line array have been modelled in a set of vertical noisefields appropriate to various frequency and windspeed regimes. The main purpose of this exercise is to estimate the degree of contamination caused by the low angle, deep sound channel component arriving from distant sources. All beams of a horizontal array are subject to this interference, particularly at endfire.

Figure 6 summarises the situation for a broadside beam, which is the least contaminated by low angle components. The level of these components is assumed equal to the local noise with 15 knots of wind.

This case, therefore, represents a minimum of contamination from
distant storms at this local windspeed, since more commonly the deep
sound channel component dominates over local noise.   The
contribution of deepsound channel energy and local noise are shown
separately in figure 6.   In the vicinity of the beam alias
frequencies, beam spectra are dramatically different for each
component, reflecting the different vertical noisefield in each case.
It is this feature that Ferguson and Wyllie have described.   However,
the domination of this same frequency region by distant noise sources
is also evident.   The use of this technique would therefore appear to
be limited to high local windspeeds, such as in reference 6, where the
contamination by distant sources is relatively small.

4   SUMMARY AND CONCLUSIONS

The technique of inferring source level characteristics from
underwater noise measurements is explored.   A simple ocean transfer
function is used to map vertical noise directionality data into
estimates of the surface distributed source function.   This method
appears promising and preliminary analysis has indicated that a
volumetric source with maximum depth of half a wavelength is
consistent with measurements.   More work is necessary to confirm this
finding.

While these early data suggest a fixed depth for the volumetric
source, an alternative model in which the active depth is windspeed
dependent is also explored.   If the layer depth is set equal to
the measured vertical extent of bubble plumes then the inferred
monopole source spectra indicate two frequency regimes.   Below
100 Hz, the source level per unit surface area is independent of
windspeed, implying a saturation effect.   Increased noise level
is simply the result of greater layer depth (more source volume).
Above 100 Hz, the source level is windspeed dependent.

The use of omnidirectional noise and horizontal noise to infer
source characteristics is possible under some conditions.   Vertical
noise directionality, however, appears to be the most suitable for
this task.

APPENDIX

The simple ocean transfer function

The received intensity per solid angle from a general cone of rays which intersects a distributed surface source (figure A1) may be estimated as follows:

Downgoing Intensity $\quad i_d = S.A_3 .g.L.a.t/\Omega$

where $\quad\quad\quad\quad\quad$ S $\quad$ = $\quad$ monopole distributed surface source level per m^2

$\quad\quad\quad\quad\quad\quad\quad\quad\quad$ a $\quad$ = $\quad$ the surface source directivity function

$\quad\quad\quad\quad\quad\quad\quad\quad\quad$ A_3 $\quad$ = $\quad$ the surface area intersected

$\quad\quad\quad\quad\quad\quad\quad\quad\quad$ g $\quad$ = $\quad$ the geometrical losses = A_1/A_2

$\quad\quad\quad\quad\quad\quad\quad\quad\quad$ t $\quad$ = $\quad$ the backward transmission loss

$\quad\quad\quad\quad\quad\quad\quad\quad\quad\quad\quad$ = $\quad$ $\cos^2\theta_r /\cos^2\theta_s \doteq 1$

$\quad\quad\quad\quad\quad\quad\quad\quad\quad$ Ω $\quad$ = $\quad$ number of steradians in the cone ($\equiv A_1$).

and $\quad\quad\quad\quad\quad\quad\quad$ L $\quad$ = $\quad$ total absorption loss

Fig. A1.

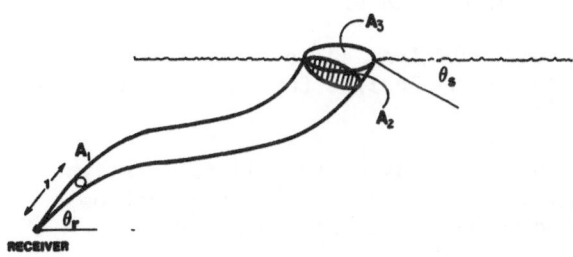

The total loss (L) includes both bottom loss and watercolumn attenuation. It can be shown that, in the present context, watercolumn attenuation is a second order term and will subsequently be ignored for simplicity. Hence L is set equal to the bottom reflection coefficient (b).

Noting that $A_2 = A_3 \sin \theta_s$ it follows that:

$$i_d = S.a.b / \sin\theta_s \qquad \qquad \text{....A1}$$

Consider now n bottom interactions, each contributing a cone such as figure A1. The sum total of the downgoing energy at received angle θ_r is:

$$I_d(\theta_r) = S.a.[1 + b + b^2 + ...b^n]/\sin\theta_s$$

where θ_s and θ_r are connected by Snell's Law, and b is the bottom reflection coefficient (<1).

Hence: $\qquad I_d(\theta_r) = S.a.[(1-b^n)/(1-b)]/\sin\theta_s \qquad \text{....A2}$

$$= S.a/(\sin\theta_s (1-b)) \text{ for } n = \infty \quad \text{....A3}$$

Likewise, upgoing energy is given by:

$$I_u = b.I_d \qquad \qquad \text{....A4}$$

Omnidirectional transfer function

The omnidirectional received energy (I_0) can be obtained by integrating the contribution of an element of solid angle having vertical extent $d\theta$ and covering all horizontal angles. This volume cell contains $2\pi \cos\theta \, d\theta$ steradians. From equations A3 and A4, appropriately weighted by solid angle it follows that:

$$I_0 = 2\pi S.a. \int_{\theta_r=0}^{\pi/2} \frac{1+b(\theta_b)}{1-b(\theta_b)} . \frac{a(\theta_s)}{\sin\theta_s} . \cos\theta_r \, d\theta_r \qquad \text{....A5}$$

where θ_b (ray angle at bottom), θ_s (ray angle at surface) and θ_r (ray angle at receiver) are connected by Snell's Law in a given soundspeed environment.

A simple case of the above used by others[5] is for straight line transmission (all angles the same and $= \theta$), high bottom loss (b=0) and a source directivity function $a = \sin^2\theta$.

In this case:

$$I_0 = 2\pi S.a \int_0^{2\pi} \sin\theta \cos\theta \, d\theta = \pi S.a \qquad \text{....A6}$$

ACKNOWLEDGEMENTS

The authors gratefully acknowledge the continued interest and support provided by members of the underwater acoustics projects group at Defence Scientific Establishment (New Zealand).   In particular, sincere thanks are extended to Drs Ross Chapman, Mike Guthrie and Ralph Marrett for their many valuable and constructive discussions, which have greatly enhanced this study.

REFERENCES

1. Bannister, R.W., Burgess, A.S., Kewley, D.J., "Directional Underwater Noise Estimation", J.Acoust.Soc.Am. Supp. 1 V80 S65 (1986).

2. Wagstaff, R.A., "Low-Frequency Ambient Noise in the Deep Sound Channel - the Missing Component", J.Acoust.Soc.Am. 69(4), 1009-1014 (April 1981).

3. Burgess, A.S. and Kewley, D.J., "Wind-generated Surface Noise Levels in Deep Water East of Australia". J.Acoust.Soc.Am. 73, 201-210 (1983).

4. Thorpe,S.A., "The Effect of Languir Circulation on the Distribution of Submerged Bubbles Caused by Breaking Wind Waves", J.Fluid Mech., 142, 151-170 (1984).

5. Wilson, J.H., "Wind-generated noise modelling", J.Acoust. Soc.Am. 73, (1) 211-216 (January 1983).

6. Ferguson, B.G. and Wyllie, D.V. "Comparison of Observed and Theoretical Responses of a Horizontal Line Array to Wind-induced Noise in the Deep Ocean", priv.comm. (submitted to J.Acoust.Soc.Am.).

NOISE GENERATED BY MOTION OF THE SEA SURFACE - THEORY AND MEASUREMENT

Douglas H. Cato & Ian S.F. Jones
Defence Science and Technology Organisation
Weapons Systems Research Laboratory
Maritime Systems Division
Royal Australian Navy Research Laboratory
P.O. Box 706,
Darlinghurst NSW 2010, AUSTRALIA

ABSTRACT. This paper discusses a model of noise generation by sea surface motion and describes measurements in a controlled experiment to test the predictions of the theory. The model is a development of Lighthill's (1952) theory of noise generation, applied to the case where there is a moving density discontinuity, like the sea surface, within the volume of the fluid. The results show that the sound field is equivalent to that produced by distributions of monopole and dipole sources over surfaces of density discontinuity, in addition to the expected volume distributions of quadrupoles. For a complex sea of linear surface gravity waves, both horizontal and vertical dipoles are shown to dominate the generation of sound at frequencies of 0.1-5 Hz.

1. INTRODUCTION

Of the large number of papers published on sea noise in the last forty years, only a small proportion model theoretically the mechanisms of noise generation by natural processes in the vicinity of the sea surface. Most of these have considered the generation of noise by the motion of the sea surface itself, in the absence of actual wave breaking, and identified the dominant mechanism as a non-linear effect of the interaction of surface waves which are equal in wave length but travelling in opposite directions. Miche (1944) was first to recognise that this effect would result in pressure fluctuations which persisted to the depth of the ocean rather than decaying exponentially with depth like the pressure fluctuations below a passing wave. Longuet-Higgins (1950) presented the equations to determine the production of micro-seisms by this mechanism. Brekhovskikh (1966) applied the theory to determine noise in the ocean to be further developed with somewhat different approaches by Hughes (1976) and Lloyd (1981). The common theme in these works was the development of a perturbation expansion in which the first order solution to the wave equation was the incompressible surface waves and the second, the compressible acoustic effects.

To produce an acoustic field, interacting surface waves must

391

be very close to being equal in wave length and opposite in direction
in order to have phase speeds that couple to the acoustic field.
Although the waves on the ocean surface usually appear as though there
is very little angular spread in the direction of travel, measurements
of directional spread of wave energy (Tyler et al,1974, Mitsuyasu et
al.,1975) have shown that this is true only for the lowest frequency,
longest waves near the spectral peak, and that there is a large spread
at higher frequencies, quite sufficient, as these models have shown, to
produce significant noise.

This paper applies a different acoustical model that starts
from Lighthill's analogy, to determine the noise generation by surface
wave motion in the absence of breaking. It produces similar results to
the other models only when the interaction of waves which are equal and
opposite in wave number are considered. However, we also show that
there is significant contribution from waves which are almost, but not
exactly equal and opposite in wave numbers, and that these sources have
rather different directivities.

2. THE MODEL

The development of an expression for the leading terms in the acoustic
pressure spectrum in the sea, as a result of a complex sea of linear
surface gravity waves, has been presented in Cato and Jones (1987). The
results show that the sound field is equivalent to that produced by
distributions of monopole and dipole sources over surfaces of density
discontinuity, in addition to the volume distributions of quadrupoles.
By assuming that the source distributions are statistically homogeneous
in the horizontal plane we obtain the following expression for the
pressure spectrum

$$P_D(z,\omega) = \int H_i(\omega,\underset{\sim}{k},\underset{\sim}{x}) H_j{}^*(\omega,\underset{\sim}{k},\underset{\sim}{x}) \, \Phi_{ij}(\underset{\sim}{k},\omega) \, d\underset{\sim}{k} \tag{1}$$

where for an infinite ocean and i=1,2

$$H_i(\omega,\underset{\sim}{k},\underset{\sim}{x}) = \frac{i \, k_i e^{-i\underset{\sim}{k}\cdot\underset{\sim}{x}} \, e^{i(\omega z/c)\sqrt{1-(kc/\omega)^2}}}{2 \, k \, \sqrt{(kc/\omega)^{-2} - 1}} \tag{2}$$

while for i=3

$$H_3(\omega,\underset{\sim}{k},\underset{\sim}{x}) = -\tfrac{1}{2} e^{-i\underset{\sim}{k}\cdot\underset{\sim}{x}} \, e^{i(\omega z/c)\sqrt{1-(kc/\omega)^2}} \tag{3}$$

and $\phi_{ij}(\omega,\underset{\sim}{k})$ is the power spectrum of $\Delta\rho u_i u_j$ in the horizontal plane, $\Delta\rho$
being the difference in densities of water and air, and u_i the velocity
at the sea surface. $\underset{\sim}{x}$ is the vector position in the sound field and c
is the speed of sound in water, assumed constant in this model. Also
the ocean bottom is assumed to be perfectly absorbing. $H_i(\omega,\underset{\sim}{k},\underset{\sim}{x})$ are

the coupling factors relating source and received pressure fields. The notation of Cartesian tensors is used.

Equations (2) and (3) show that the sound field is an oscillating function of position (i.e. an acoustic field) when $k < \omega/c$ and exponentially decaying with depth z when $k > \omega/c$. As illustrated in Fig 1, there is a singularity in the expression for $H_i(\omega, \underset{\sim}{k}, z)$ when $i=1,2$ at $k=\omega/c$ which leads to infinite noise in an infinite ocean. To avoid this problem, the values of H_1 and H_2 must be calculated for the size of the ocean basin of interest using the expressions given by Cato and Jones (1982). This leads to a finite maximum in H_1 and H_2 at $k=\omega/c$.

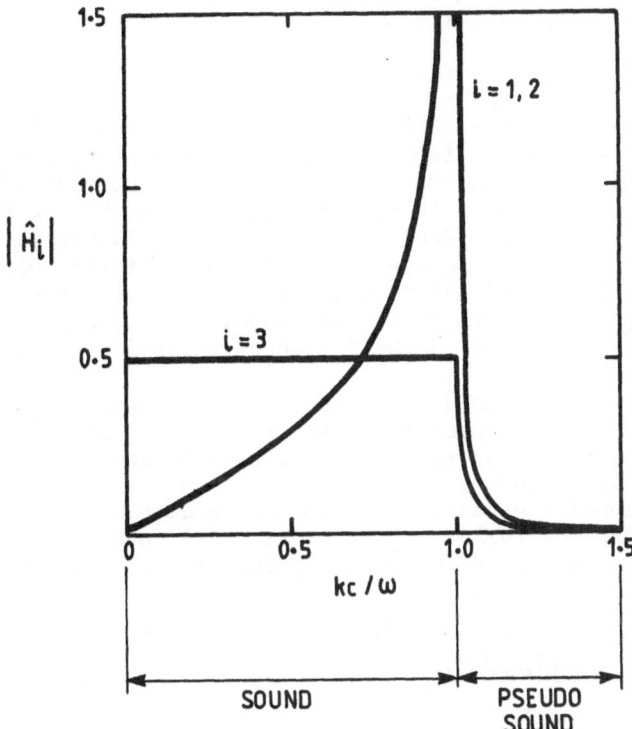

Fig. 1   Coupling factors as a function of the acoustic frequency ω and the sound speed c. The horizontal wave-number magnitude is k.

To evaluate $\phi_{ij}(\omega, \underset{\sim}{k})$ in terms of the spectra $\psi_i(\omega, \underset{\sim}{k})$ of the velocities u_i we have assumed that u_i and u_j are jointly Gaussian and used the convolution

$$\Phi_{ij}(\omega, \underset{\sim}{k}) = \Delta\rho^2 \psi_i(\omega, \underset{\sim}{k}) \bigstar \psi_j(\omega, \underset{\sim}{k})$$

$$= \Delta\rho^2 \iint_{-\infty}^{\infty} \psi_i(\omega', \underset{\sim}{k}')\psi_j(\omega-\omega', \underset{\sim}{k}-\underset{\sim}{k}') \, d\omega' d\underset{\sim}{k}'/(2\pi)^3 \tag{4}$$

We need also to relate $\psi_i(\omega,\underline{k})$ to $\Omega(\omega)$ the spectrum of the surface displacement.

Evaluation of equation (4) is long and somewhat tedious and only the main steps will be highlighted here. Some of these steps, of course, are similar to those used in the other models cited above. Generally we make use of established properties of sea surface waves to simplify and eventually evaluate these integrals. The velocities u_i are, of course, the orbital velocities induced by sea surface wave motion so that the spectra of u_i can be related to the spectrum of the surface displacement. Established theory of linear surface waves in deep water gives a unique relationship between frequency and wave number magnitude which, for frequencies below about 5 Hz (i.e. gravity waves), can be closely approximated as

$$\omega'^2 = gk' \tag{5}$$

where $k' = |\underline{k}'|$ (Phillips 1977). Although some measurements of wind waves have found some departure from this relationship at higher frequencies (Ramamonjiarisoa and Coantic, 1976), it is considered adequate for our purposes. Thus in equation (4), ω' and k' are related by equation (5) as are $\omega'' = \omega-\omega'$ and $|\underline{k}''| = |\underline{k} - \underline{k}'|$. In evaluating the integrals for a particular source spectral component (ie. particular values of ω and $\underline{k}$), ω', $\underline{k}'$, ω'' and $\underline{k}''$ are interdependent. It can also be shown that when $\underline{k}'$ and $\underline{k}''$ are expressed in polar coordinates, their magnitudes are uniquely related to their respective angles α' and α'' for fixed ω and $\underline{k}$.

These constraints, and the assumption that surface wave spectral dependence on k' and α' are separable at the frequencies of interest, allow us to use delta functions to reduce the convolution integrals to one integral with respect to α'.

For the directional dependence of the surface displacement wave number spectrum we use the following empirical model developed from measurements in a bay and in the open sea (Tyler et al, 1974, and Mitsuyasu et al, 1975)

$$G(\alpha') = N|\cos^S\{(\alpha'-\alpha_o')/2\}| \tag{6}$$

where N is a normalising factor. This expression is shown schematically in Fig 2. The dependence of the spectrum on ω' and k' can be reduced to a dependence on ω' alone by virtue of equation (5) so that the final result is given in terms of the frequency spectrum of the surface displacement, which can readily be measured, and the directional spectrum of equation (6).

At this stage the relative contribution of the dipole

distribution over the sea surface and the quadrupole distribution over the body of the ocean can be calculated using the property that orbital

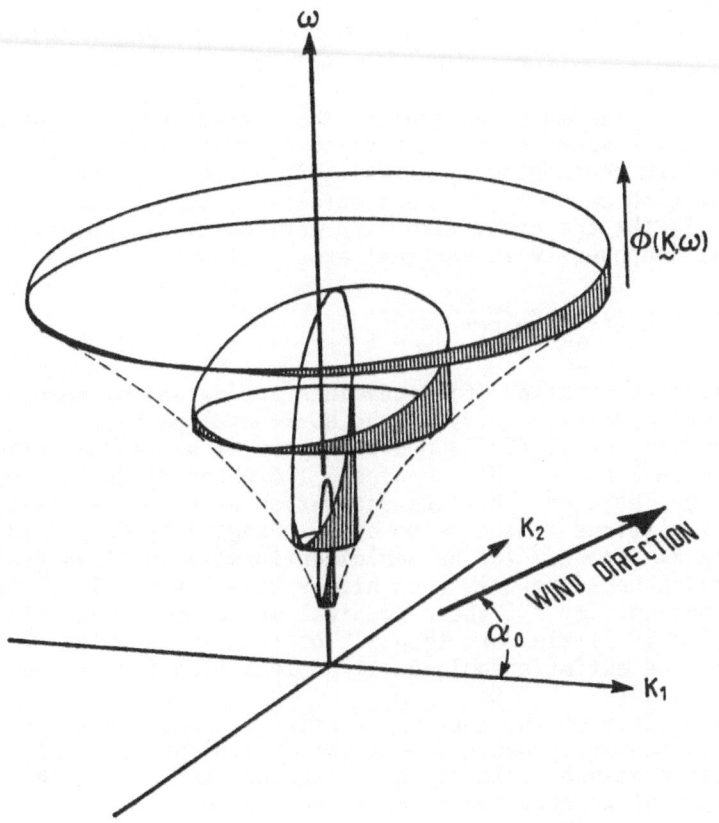

Fig. 2   Schematic of the surface gravity wave spectrum used. The spectra energy lies on the cone $\omega^2 = kg$. For each frequency cut shown the maximum energy is in the wind direction.

motions decay exponentially with depth (Phillips, 1977). We find that in terms of the power spectra of the sound fields at a deep receiver, the quadrupole contribution is of the order M^2 times the dipole contribution, where M is the Mach number of the surface waves. Since M is very small, the contribution from the quadrupoles is negligible compared to that from the dipoles (a factor of order 60 dB for an acoustic frequency of 1 Hz for example).

We eventually obtain an expression for the frequency spectrum of the total pressure field (i.e. acoustic plus non-acoustic) from the dipole sources

$$P(\omega) = \frac{\rho^2 g^2}{4} \omega \Omega^2 (\omega/2) \left\{ (I_1 + I_2) \int_0^\infty H_1 H_1^* \, k \, dk + 2 I_3 \int_0^\infty H_3 H_3^* \, k \, dk \right\} \tag{7}$$

where $\Omega(\omega/2)$ is the frequency spectrum of the surface wave height and $\hat{H}_1 = H_1/\cos\alpha = H_2/\sin\alpha$ where α is the horizontal angle to the wind direction. The term containing I_1 gives the contribution from the dipole component with axes aligned horizontally in the direction of the wind direction, I_2 the component with axes horizontal and normal to the wind, and the I_2 component with vertical axes. Also

$$I_3 = I_1 + I_2 = \frac{\Gamma^2(s/2+1)}{2\pi\Gamma(s+1)} \tag{8}$$

The parameter s, from equation (6) determines the extent of spreading of the wave number spectral energy in the horizontal plane. Measurements by Tyler et al (1974) and Mitsuyasu et al (1975) indicate that s lies between 1 and 2. We choose s = 1.5 which gives $I_3 = 0.1$. It is worth noting, however, that the estimation of noise is relatively insensitive to variations in the value of s, changing by only 2 dB as s varies from 1 to 2. Indeed, for an omnidirectional wave field (s=0) the estimated noise would be only 2 dB higher than for s = 1.5. For a narrower distribution, say s=6 where spectral energy in a direction normal to the wind is little more than 12% of that in the wind direction, the noise estimate would be 11 dB less than for s=1.5.

The behaviour of the coupling factors H_i shows that the pressure field is acoustic when k < ω/c, where c is the speed of sound. Thus the acoustic pressure field may be determined by setting the upper limit of the range of integration in equation (7) to k= ω/c.

For k > ω/c, the pressure field decays exponentially with depth z and increasing k. The depth dependence is of the form

$$\exp\left\{ - z(k^2 - \omega^2/c^2)^{\frac{1}{2}} \right\} \tag{9}$$

$$\sim\exp(-zk) \text{ for large } k$$

Because of the large acoustic wavelengths at the low frequencies at which this orbital motion noise component is dominant, k is small and the non-acoustic pressure field may be significant at depths of hundreds of metres. It may even be dominant for shallow receivers. Since our hydrophones usually respond to all pressure fluctuations, the usual approach of ignoring the non-acoustic component of the field may significantly underestimate the measured value.

As stated above, other models of noise generated by non-linear interaction of surface waves considered only the case where the waves were equal in length but opposite in direction. This is equivalent, in our terminology, to making the approximation $\underset{\sim}{k}' = -\underset{\sim}{k}''$ with the result that $\underset{\sim}{k} = \underset{\sim}{k}' + \underset{\sim}{k}'' = 0$. The effect of this can be seen in Fig 1 which shows coupling factors H_i as a function of k. Note that H_3, the coupling factor for vertical dipoles, shows little variation with k for the acoustic field (k <ω/c). Thus this approximation would be expected to have little effect on the resulting acoustic field from the vertical dipoles. In the case of the horizontal dipoles, however, the coupling factors H_1 and H_2 decrease with decreasing k and approach zero as k→ 0. Thus this approximation which results in k=0, eliminates any contribution from horizontal dipoles. We can therefore make a comparison with previous work by comparing the contribution by the vertical dipoles of equation (7) to the acoustic field. From Fig 1 we see that $H_3 H_3^* \sim 1/4$ in the deep ocean and independent of k for k < ω/c, so that we can evaluate the integral with respect to k. Allowing for the different methods of defining wave height spectra we find our prediction for vertical dipoles is a factor of π/2 (2 dB) larger than the result from equation (33) of Hughes (1976). Lloyd (1981) reported a factor of 2 correction to this equation which then agreed with Lloyd's result. If this is correct, our result for the vertical dipole contribution is a factor of 5 dB larger than these results.

The relative contribution of the horizontal dipoles to the total noise depends on the area of contributing sources and the propagation conditions. In the experiment described below, the water surface dimensions were of the order of a kilometre and the contribution of the horizontal dipoles increased the calculated noise level by 2 dB compared with that from the vertical dipoles alone. In the ocean the relative contribution from horizontally radiating dipoles may be larger because of the much larger area of sources and the preferential propagation offered to the near horizontal rays in the ocean.

3. EXPERIMENT

A carefully controlled experiment was conducted in Woronora Dam, a water supply reservoir near Sydney to measure ambient noise and the results reported in Cato and Jones (1987). Although somewhat irregular in shape, the main water mass is about 1 km across. The hydrophone was placed on the bottom (depth 35m) near the centre of the water mass and connected by cable to a large pontoon about 100m away. Considerable effort was made to minimise the possibility of "flow noise", since this is a likely source of contamination at the low frequencies of interest. The system response was 3 dB down at 0.33 Hz. Surface wave height was recorded using a capacitive wave staff suspended 1.7m from the pontoon on the windward side. Both the noise and wave height data were tape recorded and later replayed in the laboratory to determine the simultaneous Fourier transform of each set of data. Measurements were made during a period when the wind was strong enough to produce a well developed surface wave field and was

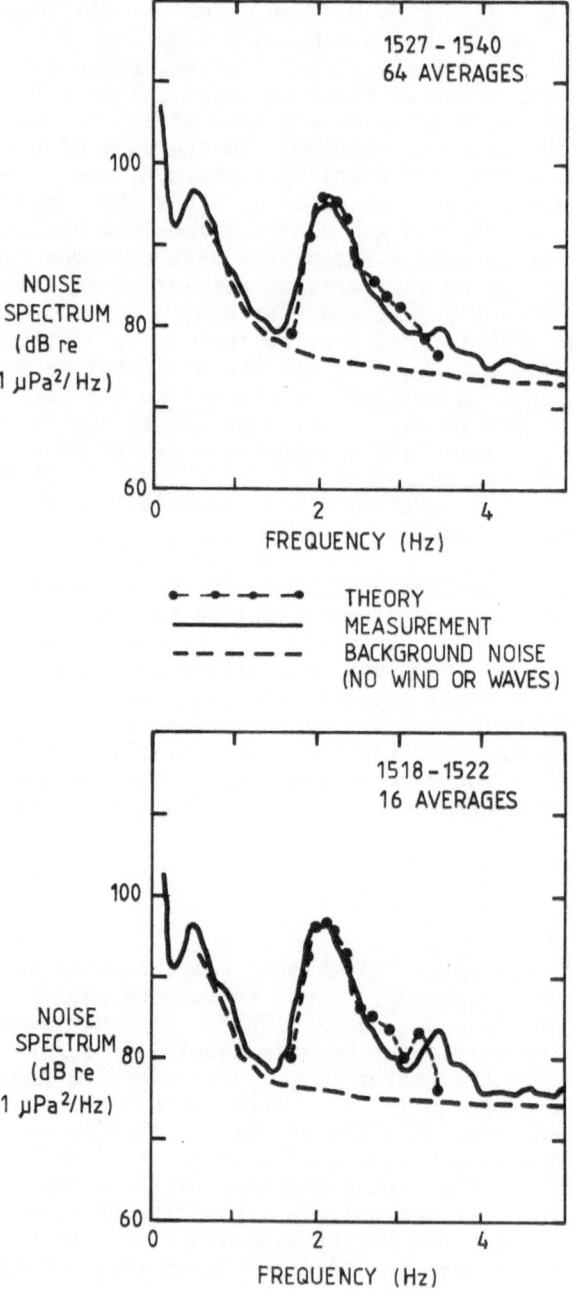

Fig. 3   Comparison between the model and measured results reported in
         Cato & Jones(1987).

steady over the extent of the water mass and over a sufficient period of time to allow adequate averaging in the processing of the data. Measurements were also made at times when there was no wind and the surface was flat to determine the background noise level.

The theoretical noise level prediction applicable to the conditions of measurement was determined by substituting the measured wave height spectrum in equation (7), having estimated the coupling factors for the dimensions of the experiment with the aid of Cato and Jones (1982). Some results are shown in Fig. 3.

Since our model assumes that there is no reflection of acoustic energy from the bottom, we must estimate the contribution from bottom reflections separately. We calculated this using a method given by Brekhovskikh (1960) in which the contributions of successive bottom/surface reflected arrivals are determined using the appropriate reflection losses and phase changes. Using values typical for the rock strata in the area we calculated the contribution of sufficient arrivals to obtain a converging result and found that bottom reflections would have an effect of less than 1 dB. This rather surprising result is due to the fact that the dimensions of the dam are of the order of an acoustic wave length so that there is little phase change between surface and bottom. The phase difference between successive arrivals is dominated by the 180 phase change at the water surface with the tendency for cancellation rather than augmentation.

4. PREDICTION OF OCEAN NOISE

To adequately predict noise at sea we need to extend our model to incorporate the effects of refraction and bottom reflections. Refraction effects were not considered to be important in Woronora Dam because the dimensions are comparable to a wave length. However they would significantly affect the results at sea, especially for the horizontal dipoles. Thus, at this stage we can make only an order of magnitude prediction of noise in the ocean. To do this we used a simple model for the surface wave frequency spectrum that applies to what is termed the equilibrium range of frequencies. While the surface wave spectrum has a peak that is wind dependent, the equilibrium range can be determined by a single empirical constant given by Phillips (1977). While there are other models, such as that of Toba (1978) that appear to fit the data more closely, the essence of our approach is adequately illustrated by Phillips prescription. The result is shown in Fig. 4 and compared with measurements of low frequency noise by Nichols (1981) and by Talpey and Worley, reported by Nichols (1981), and the range of values calculated by Kibblewhite and Ewans (1985) from their measured microseism spectra. The result is shown as a continuous spectrum representing the region above the spectral peak (at a frequency twice the peak of the wave height spectrum). We expect that our predictions will not be substantially different from those of other theories as far as the magnitude is concerned, however, the directionality would be significantly different.

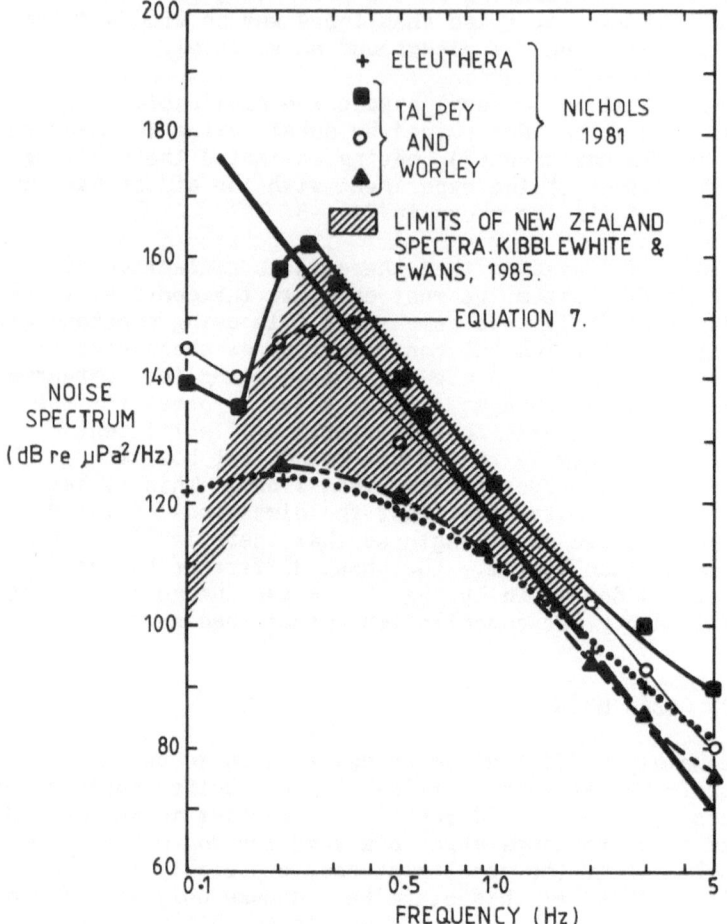

Fig. 4   Comparison of low frequency noise spectra measured by various authors with the prediction of Eq 7. Notice that knowledge about the peak surface wave frequency is needed to predict the acoustic spectral peak.

A comprehensive test of any theory of noise in the deep ocean by this mechanism awaits simultaneous measurements of wave height spectra and noise spectra at sea.

5.   CONCLUSIONS

An explanation of acoustic noise below a field of non-breaking surface gravity waves has been developed based on Lighthill's analogy. If the size of the ocean basin and the frequency spectrum of the surface gravity waves are known, the fluctuating pressure field can be predicted. This field involves both acoustic waves and pseudo-sound and provides a close fit to the experimental results obtained in a freshwater reservoir. The dominating sources are equivalent to distributions of horizontal and vertical dipoles across the sea surface.

Acknowledgements

        We would like to thank Dr. Peter Buchen for helpful simplifications to the theoretical treatment.

References

Brekhovskikh, L.M. (1960) Waves in Layered Media. Academic, New York.

Brekhovskikh, L.M. (1966) 'Underwater sound waves generated by surface waves in the ocean.' Izv. Atmos. Ocean Phys., 2, 582-587.

Cato, D.H. & I.S.F. Jones (1982) 'The sound generated by fluid motion in a horizontally homogeneous ocean.' Marine Studies Centre, University of Sydney, Tech. Rep. 2/82.

Cato, D.H. & I.S.F. Jones (1987) 'Acoustic radiation from surface gravity waves.' (submitted for publication).

Hughes, B. (1976) 'Estimates of underwater sound (and infrasound) produced by nonlinearly interacting ocean waves.' J. Acoust. Soc. Am., 60, 1032-1039.

Kibblewhite, A.C. & K.C. Ewans (1985) 'Wave-wave interactions, microseisms, and infrasonic ambient noise in the ocean.' J. Acoust. Soc. Am., 78, 981-994.

Lighthill, M.J. (1952) 'On sound generated aerodynamically I: General Theory.' Proc. Roy. Soc. (Lon.) Series A, 211, 564-587.

Lloyd, S.P. (1981) 'Underwater sound from surface waves according to the Lighthill-Ribner theory.' J. Acoust. Soc. Am., 62, 425-435.

Longuet-Higgins, M.S. (1950) 'A theory of the origin of microseisms.' Trans. Roy. Soc. (Lon.) A, 243, 1-35.

Miche, M. (1944) 'Mouvements ondulatoires de la mer en profondeur constante ou decroissante.' Ann. Ponts Chauss., 2,1.

Mitsuyasu, H. F. Tasai, T. Suhara, S. Mizuno, M. Ohkusu, T. Honda and

402

      K. Rikiishi (1975) 'Observations of the directional spectrum of ocean waves using a cloverleaf buoy.' J. Phys. Oceanogr., 5, 750-760.

Nichols, R.H. (1981) 'Infrasonic ambient ocean noise measurements: Eleuthera.' J. Acoust. Soc. Am., 69, 974-981.

Phillips, O.M. (1977) The Dynamics of the Upper Ocean. Cambridge University Press, 2nd Edit. pp 336.

Ramamonjiarisoa, A. and M. Coantic (1976) 'Experimental law of dispersion of wind produced waves over a short fetch.' Comptes Rendues de l'Academie des Sciences de Paris (Proc. of the Academy of Sciences of Paris), Series B, 282, 111-114.

Toba, Y. (1978) 'Stochastic form of the growth of wind waves in a single parameter representation with physical implications.' J. Phys. Oceanogr.,8, 494-507.

Tyler, G.L., C.C. Teague, R.H. Stewart, A.M. Peterson, W.H. Munk and J.W. Joy (1974). 'Wave directional spectrum from synthetic aperture observations of radio scatter.' Deep Sea Res., 21, 989-1016.

OBSERVATIONS OF HIGH FREQUENCY AMBIENT SOUND GENERATED BY WIND

D.M. Farmer and S. Vagle
Institute of Ocean Sciences
P.O. Box 6000
9860 West Saanich Road
Sidney, B.C.,V8L 4B2
Canada

Abstract. Observations of ambient sound generated by wind waves provide a basis for studying the structure, distribution and properties of breaking events. The sound detected by a hydrophone represents a spatially filtered average over an area of the ocean surface that increases with increasing measurement depth. Measurements at shallow depths can detect individual breaking events. Observations show great variability in spectral structure with multiple peaks and well defined minima; this structure appears persistent during a particular experimental period (of a few hours) and location but can be quite different at other times and places. The recorded signal fluctuations can be used to infer properties of the distribution of breaking events, in particular their density on the ocean surface.

1. INTRODUCTION

The influence of wind on the ocean is of fundamental importance to dynamical oceanography: to the generation of ocean currents, to the transfer of momentum, heat and gas across the air-sea interface, and to the production of turbulence and mixing within the upper ocean boundary layer. The generation of sound by wind is therefore of special interest, not only from a purely acoustic point of view, but for its potential as a remote sensing signal of processes at the ocean surface. Recent observations have shown how wind generated ambient sound is related to the wind speed, to the distribution of bubble clouds formed by breaking waves, to the temporal and spatial structure of breaking events and to the properties of the wave field itself. In this report we summarise some of the key results, drawing from observations obtained in continental shelf, in-shore and deep ocean experiments.

Our observations were obtained using two different approaches. Longer term measurements were obtained with "WOTAN" ambient sound recorders, moored either on the sea floor, at some intermediate depth or just beneath the surface (Figure 1(a)). These instruments typically record in 3-12 frequency bands between 1 kHz and 30 kHz and can be

403

B. R. Kerman (ed.), Sea Surface Sound, 403–415.
© 1988 by Kluwer Academic Publishers.

deployed for many months at a time. An alternative approach
(Figure 1(b)) makes use of broad band recording between 50 Hz and
22 kHz. Instruments built to measure sound in this way have much shorter
recording periods (8-80h continuous) and have been used in both moored
and free-drifting configurations in conjunction with other sensors,
including active sonar devices.

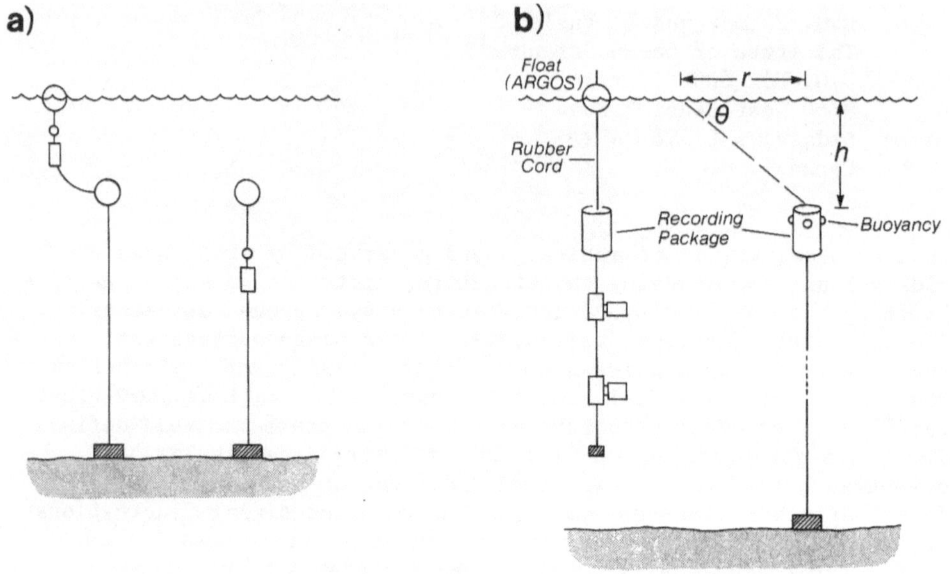

a)

b)

Figure 1(a)   Mooring configurations used for WOTANs.
             Left: instrument at 1m depth for near surface measurements.
             Right: subsurface mooring.
Figure 1(b)   Mooring configuration for broad band instrument.
             Left: free drifting system used in deep ocean.
             Right: subsurface mooring. Notation as used in text.

2.   THE SPATIAL FILTER

A fundamental property of measurements obtained with an omnidirectional
hydrophone is the spatial filter acting on sound sources at the sea
surface. The sound arriving at the hydrophone is determined not only
by the presence of the sources, but also by the effects of spherical
spreading and absorption, and by the source radiation pattern. High
frequency wind generated sound is thought to have a dipole radiation
pattern, although directional measurements are scarce. For a single
dipole source of strength I_o at a horizontal distance r, and detected
at depth h in water of constant sound speed, the received signal I_h
will be

$$I_h = (I_o \sin^2\theta)(h^2+r^2)^{-1}\exp[-\alpha(h^2+r^2)^{\frac{1}{2}}], \tag{1}$$

where α is the absorption coefficient (Farmer & Lemon, 1984). For a
large number of sources, approximating a continuous distribution,
integration over the surface yields the total signal strength I_h:

$$\overline{I}_h = 2\pi \, I_o E_3(\alpha h), \tag{2}$$

where E_3 is the exponential integral function of third order. This
result must be further modified to take account of the presence of
bubbles and of refraction.

The value of the spatial filter function $F(r)$ at radius r, is
found by integration of (1) about an annulus of unit width:

$$F(r) = 2\pi(r \, \sin^2 \theta)(h^2+r^2)^{-1}\exp[-\alpha(h^2+r^2)^{\frac{1}{2}}]. \tag{3}$$

The 'listening radius', r_L, is defined as one half of the second moment
of $F(r)$.

$$r_L = \frac{1}{2}h[\{E_1(\alpha h)-E_3(\alpha h)\}/E_3(\alpha h)]^{\frac{1}{2}}. \tag{4}$$

Figure 2 shows $F(r)$ at 10m and 40m, and r_L at 4.3 kHz and 25 kHz. The
effect of refraction is also shown for the case of a two-layer sound
speed distribution (upper layer sound speed of $1477ms^{-1}$, lower layer
speed of $1481ms^{-1}$ and interface at 10m). The listening radius
increases with depth and decreases with frequency. Typically
refraction effects are modest at a few tens of metres, but can become
important at greater depths.

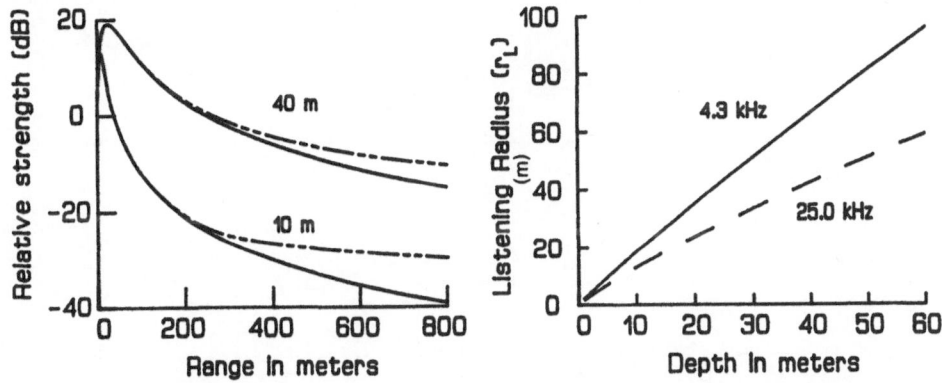

Figure 2(a)   Spatial filter function $F(r)$ for hydrophone at 10m and 40m.
              Dashed lines show effect of refraction as described
              in text.
Figure 2(b)   Listening radius r_L as a function of hydrophone depth, for
              4.3 and 25.0 kHz.

Interpretations of the spatial and temporal variability of high
frequency ambient sound depend on an appropriate representation of this
filter.

3.   SPECTRAL SHAPE

Several studies carried out with WOTAN instruments (Evans & Watts, 1982; Kerman et al., 1983; Farmer & Lemon, 1984; Lemon et al., 1984) have demonstrated general agreement with the spectral shape described by Knudsen et al. (1948). However, this agreement breaks down at higher frequencies and higher wind speeds due to masking effects by the layer of bubbles formed by breaking waves (Farmer & Lemon, 1984). This effect can be strong enough to cause the sound level to decrease with increasing wind speed, at frequencies greater than 10-12 kHz.

The inferred presence of the bubble layer provides an indirect indication of wave breaking. It can be conveniently represented by the spectral slope at higher frequencies, under the assumption that the sound generated by wave breaking has spectral properties independent of the bubble layer. The same assumption can be further exploited to allow derivation of the mean population density of the bubble clouds.

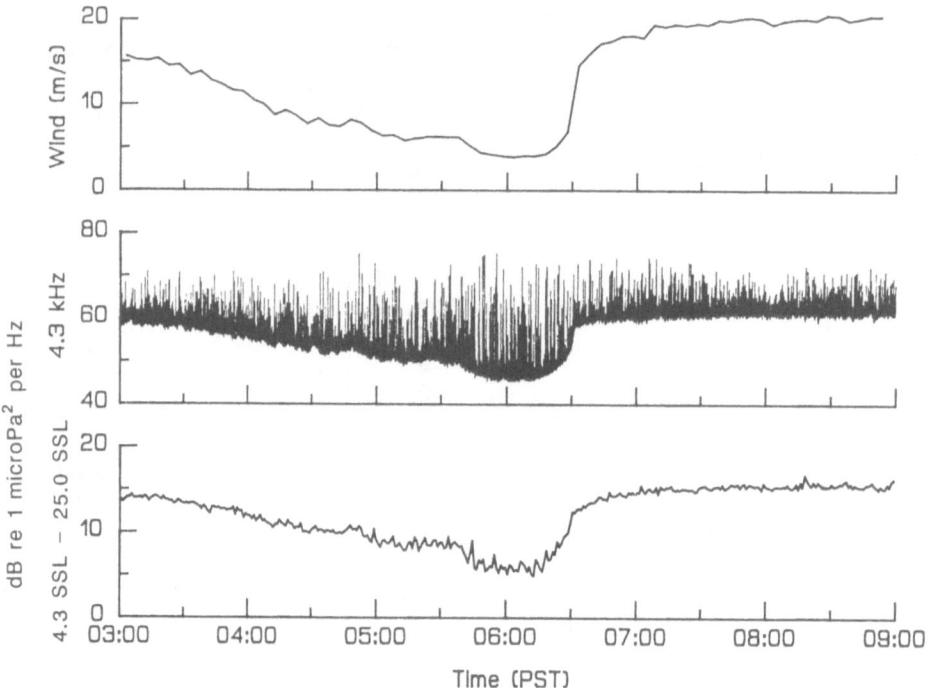

Figure 3   Plot of wind speed, 4.3 kHz sound spectrum level and 4.3-25.0 kHz differenced signal, during passage of a front west of Vancouver Island, B.C. The differenced signal shows the masking effect of bubble clouds with increasing wind speed.

The effect of attenuation by bubbles is apparent in the detailed time series shown in Figure 3 obtained over the continental shelf off Vancouver Island with the hydrophone at a depth of 15m. Passage of a

front across the instrument site produced a steadily decreasing wind
speed, followed by a quite sharp increase. The figure includes
observations of the 4.3 kHz sound spectrum level, sampled at 0.9s, and
also of the difference between the sound spectrum level at 25 kHz and
at 4.3 kHz. The differenced signal, which is low pass filtered at 60s,
provides a measure of sound absorption by the bubble cloud and is seen
to be closely related to the wind speed. We also note here that
fluctuations in the 4.3 kHz sound spectrum level are one-sided; moreover the signal variance is much greater at low wind speeds than at
high wind speeds.

The detailed spectral signature of wave-breaking can be found from
broad-band sound recordings. Individual breaking events can be
detected by simultaneous video observation from the instrument, or just
by listening to the audio signal. Figures 4 and 5 show the temporal
and spectral evolution of one such event recorded at a depth of 15m in
150m of water west of Vancouver Island. The wind speed was $12ms^{-1}$ from
the east and had been from this direction for 5-6h prior to the
measurement. Bursts of sound occurred at intervals of between 12s and
about 200s; that shown in Figures 4 and 5 was the second of a closely
spaced pair.

The temporal development of this event is shown for 5 different
frequencies in Figure 4. Note that the signal peak is reached at
slightly different times at different frequencies; there are two peaks
at 20 kHz. The entire event lasts about 4s. The raw time series are
highly variable and in this figure have been smoothed with a 0.25s
running mean in order to show more clearly the principal features.

Spectra based on 1s averages through the same event are shown in
Figure 5. The time series from which each spectrum are calculated is
centered at 0.5s, 1.5s and 2.5s, as indicated in Figure 4. As the wave
breaks a significant spectral peak appears at about 2.5-3.0 kHz, with a
second peak centered at 1.2 kHz; the spectral gap between these peaks
is especially clear when displayed on frequency-time plots over longer
periods in which the spectral level is colour coded. During the
breaking, significant variability occurs at higher frequencies also.
As the breaking event decays the spectrum gradually reverts to a more
uniform slope which is close to that identified by Knudsen et al
(1948) and shown as a dashed line in the figure.

The same instrument was also used in the FASINEX experiment south-
west of Bermuda in March 1986. The instrument was freely drifting in
this case (Figure 1(b)) with the hydrophone at a depth of 24m; the wind
speed was $12ms^{-1}$. In this experiment the sound recordings were
obtained simultaneously with backscatter measurements of bubble clouds.
The time series was divided into segments between each sonar transmission, and the uncontaminated portions Fourier transformed to
generate the time sequences shown in Figure 6. The duration of the
breaking event is similar to that described previously, but it is much
louder, despite the greater depth of the measurement. The sound
spectrum (Figure 7) is also very different. In contrast to the signal
recorded off Vancouver Island, the FASINEX spectrum has a strong peak,
centered at approximately 1 kHz, with a sharp minimum at 500 Hz.
Notable differences in conditions between the two experiments, include

a much higher spatial density of breaking events in the FASINEX data set. The Vancouver Island experiment was subject to the effects of a strong swell.

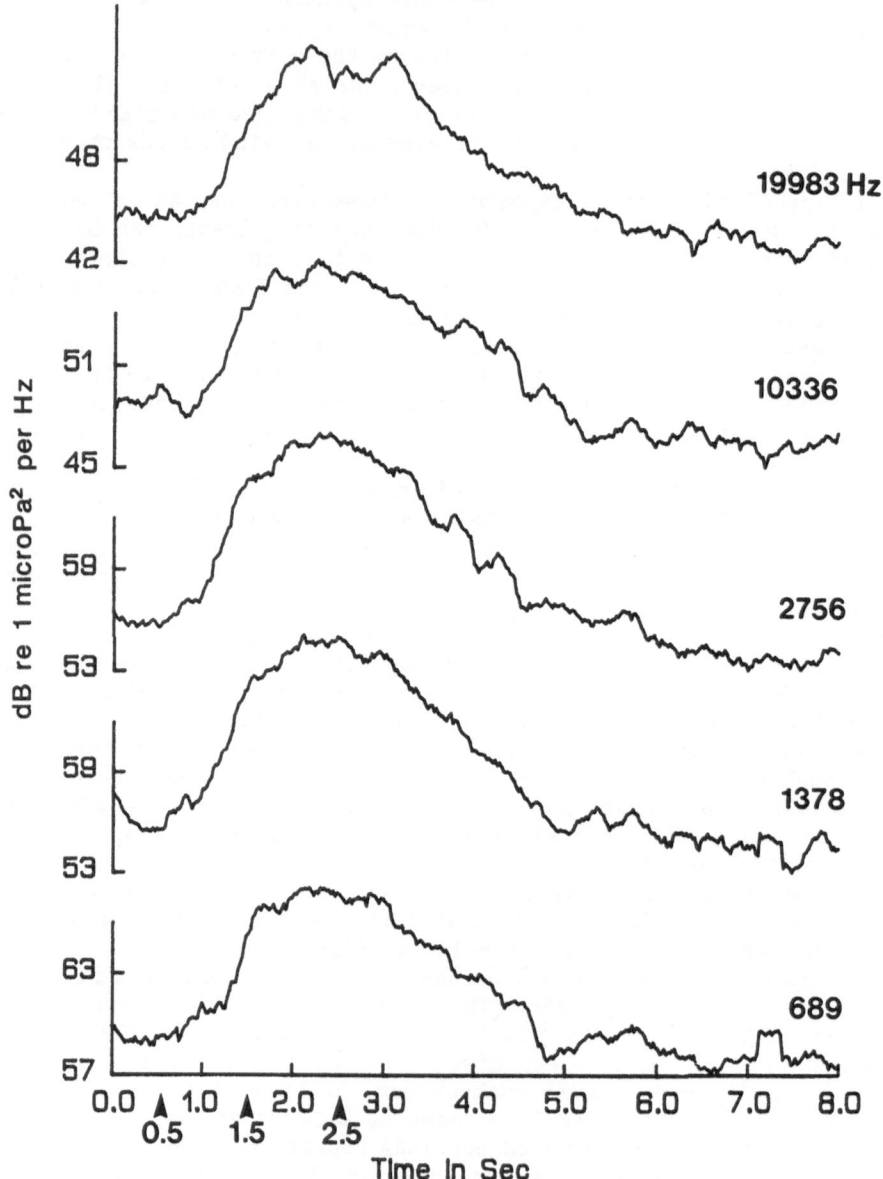

Figure 4   Time series measurements of sound spectrum level is at different frequencies as a wave breaks 15m above a hydrophone (November 1985, La Perouse Bank, B.C.).

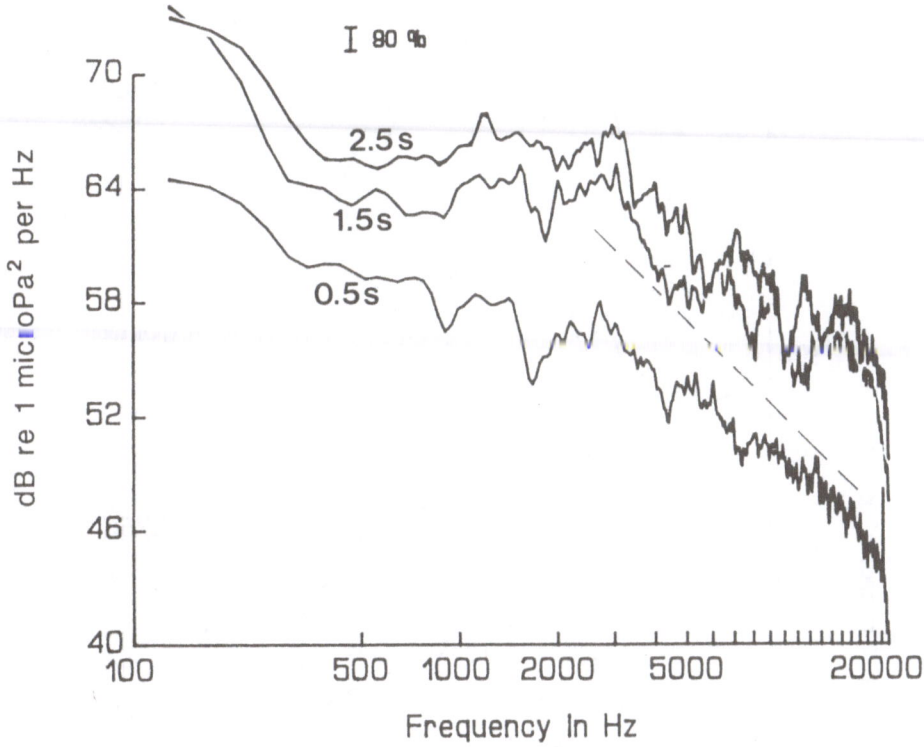

Figure 5   Spectral signature based on 1s averages for the signal shown
in Figure 4. Dashed line has a slope of -17 dB/octave.

Previous measurements of spectra (i.e. Perrone, 1969) have also
revealed a spectral peak, usually at significantly lower frequencies
(300-500 Hz). Kerman (1984) suggests that the location of this peak
depends upon u_*, but the difference in shape between the spectra shown
in Figures 5 and 7, both obtained at similar wind speeds, emphasize
that other factors must be included in a comprehensive model. A
distinguishing feature of these observations is the temporal evolution
of the sound field that they reveal. Shifts in the spectral shape at
different stages of the breaking event, especially in Figure 7, are
presumably associated with changing contributions from different sound
generation mechanisms. For example the sound of spray impact may be
limited to a relatively short portion of the event, with a contribution
to the sound spectrum that differs from that of the breaking and
spilling process. Preliminary analysis also indicates that portions of
the finer structure in the spectra can remain coherent throughout the
breaking event.

410

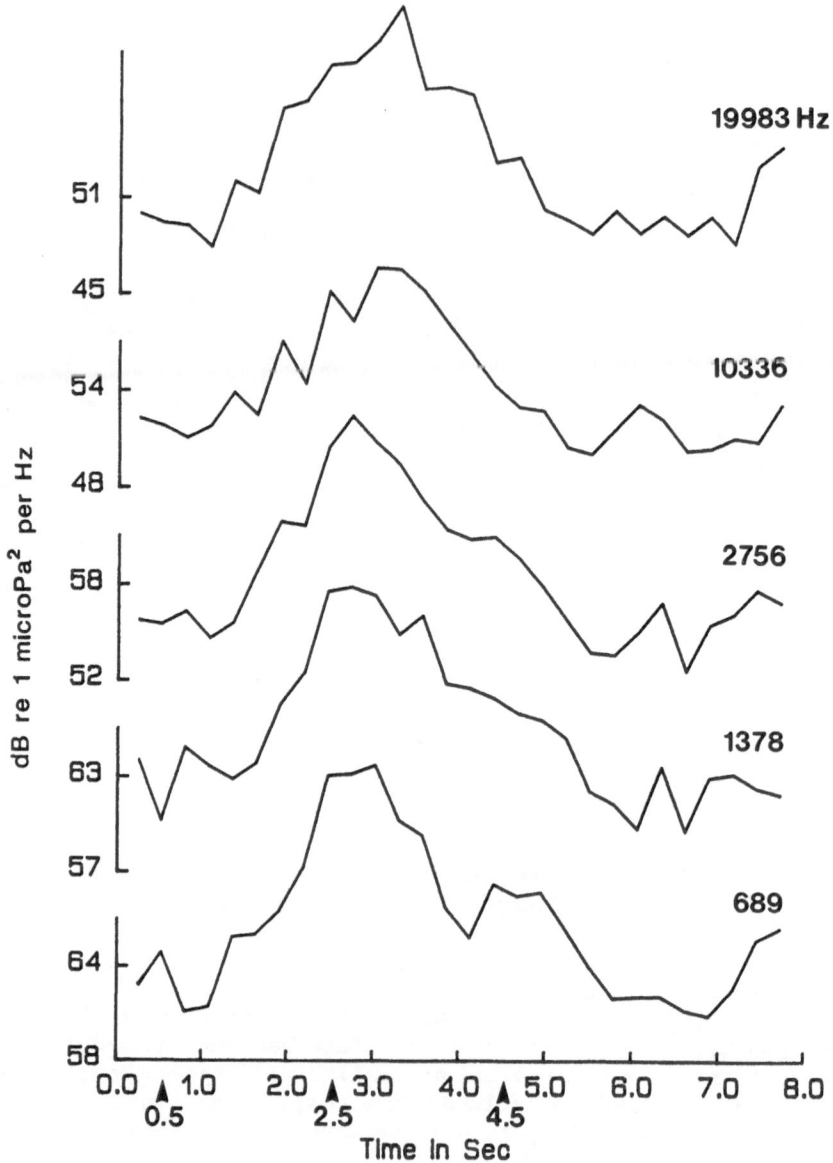

Figure 6 Time series sound spectrum level for a breaking wave in the deep ocean (FASINEX experiment, March 1986). Data is discontinuous to avoid interference effects from simultaneous sonar transmissions.

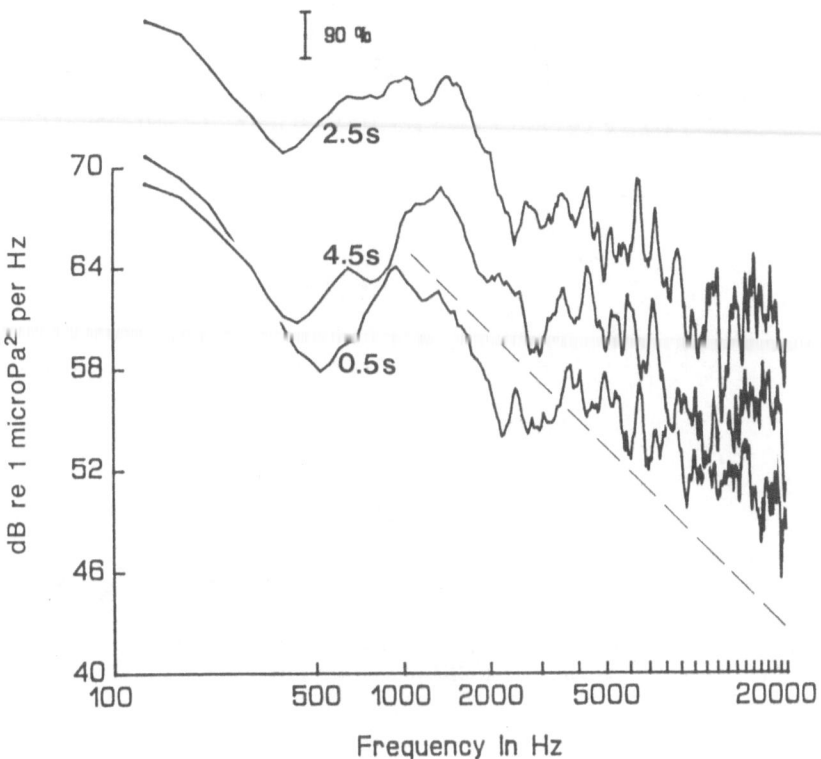

Figure 7   Spectra corresponding to Figure 6.

4.   SIGNAL VARIABILITY AND ITS RELATIONSHIP TO WAVE PROPERTIES

The time series shown in Figure 3 illustrate an important feature of
measurements of wind generated sound. At moderate wind speeds the
signal variance is very much greater than it is at higher wind speeds.
The variance also decreases with hydrophone depth. For example,
Figure 8 shows simultaneous time series measurements at 1m, 10m and 40m
obtained in Georgia Strait, B.C., at a wind speed of $10ms^{-1}$.
        These effects may be explained by the distribution of breaking
waves on the ocean surface, and by the depth dependence of the spatial
filter discussed above. It is clear from Figure 3 that individual
breaking events can be nearly as loud at moderate wind speeds as they
are at high wind speeds. The important difference is that at high wind
speeds there are many more such events within the "listening area";
that is why the overall signal level increases with wind speed. (We
restrict attention here to the 4.3 kHz signal which is free of the
effects of masking by bubbles.)

412

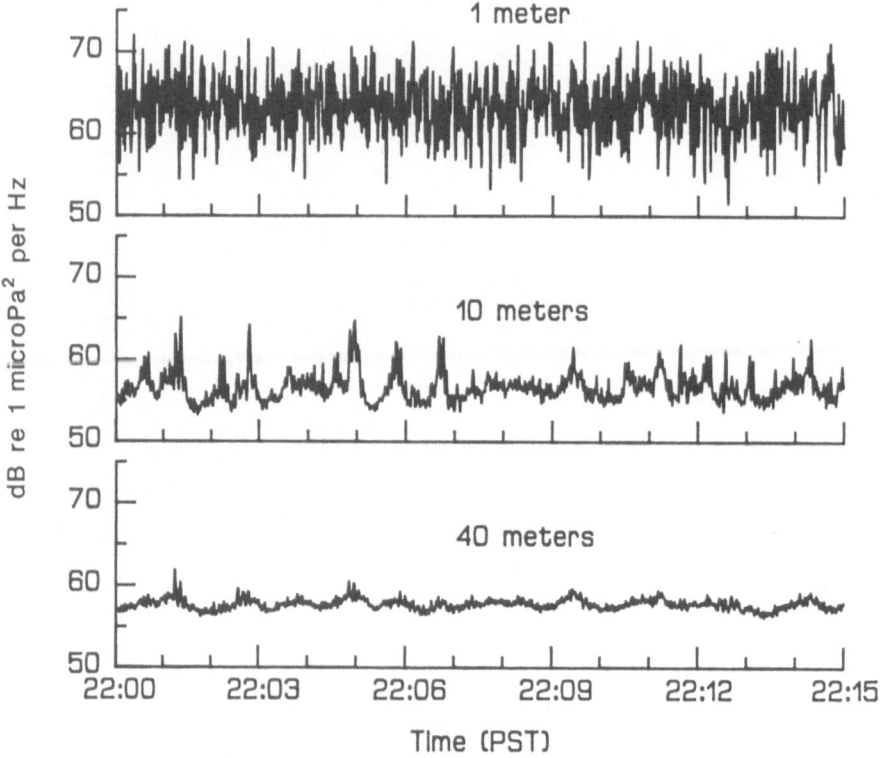

Figure 8 Time series records of sound spectrum level at a frequency of
4.3 kHz at 1, 10, 40m in Georgia Strait during a 10ms^{-1} wind.

This concept has recently been explored with a model of wind
generated sound in which breaking events are randomly distributed on
the ocean surface (Farmer & Vagle, 1987). Taking due account of the
structure of the spatial filter function it was found that the standard
deviation of the signal varies as the inverse square root of the
breaker density:

$$\sigma = 2.5\ \rho^{\frac{1}{2}} r_L^{-1},\qquad\qquad (5)$$

where ρ is the spatial density of breakers and r_L is the listening
radius defined in (4). This model has been applied to observations of
a fetch limited wave field in Georgia Strait. The signal recorded at
40m was used to infer the spatial density of breaking events. The
result is shown in Figure 9, along with the local wind speed; wind
direction was constant and from the south-east for this study. During
the experiment, fluctuations in wind speed produced corresponding
changes in the breaker density. In general, high wind speeds tend to
generate high breaker densities, but there are often deviations from
this rule due either to an enhanced tendency to break when the sea is

adjusting to a changed wind-stress, or to a lag in the response associated with the time required for the wind-wave field to develop.

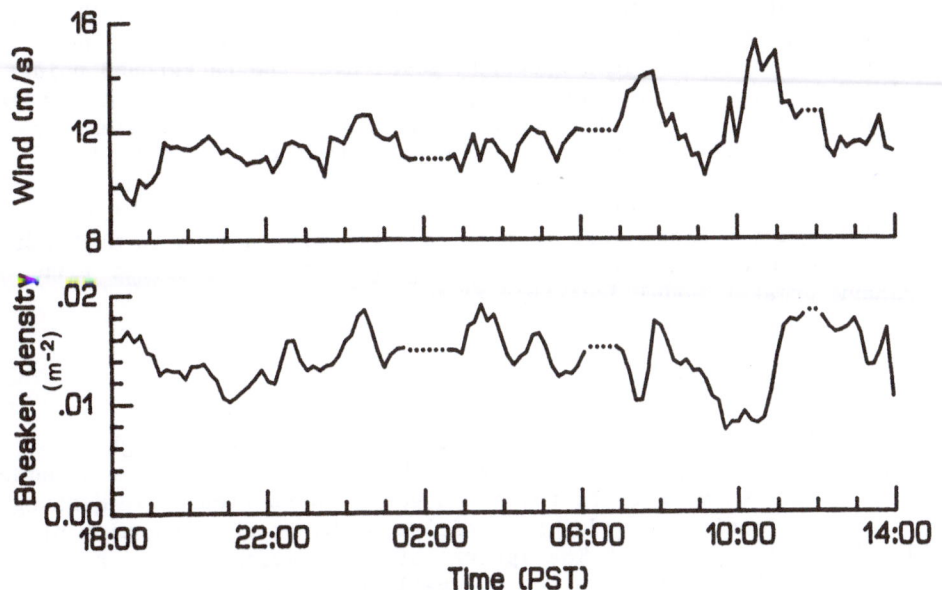

Figure 9.   Plot of wind speed and spatial density of breaking events inferred from observations in Georgia Strait, using a statistical model of breaking waves.

    Figure 8 illustrates another feature of the ambient sound that relates to the surface wave field. There is a longer period variability, especially clear at 10m, but also apparent at 40m, with a frequency of approximately one cycle every 60s. We attribute these variations to the effect of wave groups (see Longuet-Higgins, 1986). Enhanced breaking occurs where the amplitude, and thus wave steepness, is greatest. The wave period in this fetch limited example was approximately 3.5s, implying that amplitude maxima passed above the hydrophone approximately once every 17 wave periods.
    The influence of wave group structure on breaking events has another effect. As pointed out by Donelan, Turner & Longuet-Higgins, 1979, the effect of wave groups is to increase the probability that breaking will be repeated, with the repetition period being twice the wave period and the breaking event occurring one wave-length downwave of the previous breaker. This tendency for breaking to occur as a result of repetition will then appear as an increased contribution to the signal variance at one half the dominant wave frequency. Analysis of the time series obtained in Georgia Strait shows that such a contribution does indeed occur: spectra show a peak rising 5-10 dB above the background signal at a period of about 7s. The shape and size of this spectral peak varies with wind and wave conditions.

Comparison of the time series in Figures 3 and 8 emphasizes differences between the temporal variability in protected waters such as Georgia Strait, and in the open conditions off the west coast of Vancouver Island. The time series in Figure 3 is one sided in the sense that 'spikes' in the data associated with breaking events only represent increases in sound spectrum level over the background noise. Wave breaking is much more sparsely distributed in space and time under these conditions, partly as a consequence of the greater wavelength. In the fetch limited conditions of Georgia Strait the wave field is less coherent, the breaking events more densely distributed and the sound intensity of individual breakers is much lower. The 1m time series in Figure 8 shows that in this case the time series is much more evenly distributed than in Figure 3. This result is consistent with the random breaker model described above.

5. CONCLUDING REMARKS

These examples provide an indication of the variability in high frequency acoustic signals generated by wind over the ocean and of some of the insights they offer to the processes of wave-breaking. Masking of higher frequencies provides a sensitive probe of the injection of bubbles by breaking waves. The spectra of sound during breaking is highly variable, exhibiting peaks and valleys that are locally persistent, but differ from one set of observations to another. Evidently the sound generation is caused by a range of processes which contribute differently depending upon meteorological and local wave conditions. Study of these signals has yet to be related to different types of wave breaking events; their explanation presents a worthwhile challenge for future work.

The temporal variability of the signal can provide a basis for studying the distribution and properties of breaking events on the ocean surface. These are not limited to estimates of the breaker density, but include the acoustic strength of average breaking events and also the dominant wave period.

In concluding we note that the dominant signal in most types of wave measurement (wave-staff, wave-buoy, inverted echo-sounder, etc.) is the wave field itself. The ambient sound field at high frequencies appears to be dominated by wave breaking events. It therefore provides a particularly suitable signal for studying the breaking process.

ACKNOWLEDGEMENT

This work was supported by ONR, Contract No. N00014-85-C-0824, and OERD, Contract No. 67129.

REFERENCES

Donelan, M., M.S. Longuet-Higgins, and J.S. Turner, 1972. 'Periodicity in whitecaps', Nature, London, **239**, 255-261.

Evans, D.L. and D.R. Watts, 1982. 'Wind speed and stress at the se surface from ambient noise measurements', Proc. Int. Symp. Acoustics Remote Sensing Atmos. and Oceans., Dept. Physics, U. Calgary, Calgary, Canada, III69-III78.

Farmer, D.M. and D.D. Lemon, 1984. 'The influence of bubbles on ambient noise in the ocean at high wind speeds', J. Phys. Oceanogr., 14(11), 1762-1778.

Farmer, D.M. and S. Vagle, 1987. 'The distribution of breaking surface waves inferred from ambient sound', to appear.

Kerman, B.R., D.L. Evans, D.R. Watts and D. Halpern, 1983. 'Wind dependence of underwater ambient noise', Bound. Layer Meteor., **26**, 105-113.

Kerman, B.R., 1984. 'Underwater sound generation by breaking wind waves', J. Acoust. Soc. Am. **75**(1), 149-165.

Knudsen, V.O., R.S. Alford and J.W. Emling, 1948. 'Underwater ambient noise', J. Mar. Res. **7**, 410-429.

Lemon, D.D., D.M. Farmer and D.R. Watts, 1984. 'Acoustic measurements of wind speed and precipitation over a continental shelf', J. Geophys. Res. **89**, 3462-3472.

Longuet-Higgins, M.S., 1986. 'Wave group statistics', in Oceanic Whitecaps, E.C. Monahan and G. MacNiocaill (eds), 15-35.

Perrone, A.J., 1969. 'Deep-ocean ambient noise spectra in the north-west Atlantic, J. Acoust. Soc. Am., **46**, 762-770.

ON THE SPECTRA OF WIND GENERATED SOUND IN THE OCEAN

David Shonting
Applied Oceanography Group
Naval Underwater Systems Center
Newport, RI 02841-5047, USA

Nancy Taylor
Applied Science Associates
Narragansett, RI 02882, USA

ABSTRACT. Investigation was made of wind and whitecap generation of sound and its associated spectra from 5-40 kHz. Ambient sound was obtained from a low noise hydrophone placed in 7 m depth, 100 m from shore, in Narragansett Bay, RI, while accompanying records were made of wind speed and whitecapping. Band passed signals were obtained with the SCANR system and high resolution spectra were obtained from FFT analyses on a MASSCOMP computer. High correlation was obtained between rms sound pressure and wind speeds from 4-15 m/s, i.e., below and above the point of whitecap formation. The variance and spectra of the sound pressures indicate an acceleration of sound production between 4.5-8.5 m/s with spectral peaks occurring at 20-25 kHz which are not displayed in the traditional uniformly sloped Knudsen curves. High resolution spectra appear to be a useful for studying ambient sound generating mechanisms.

1. INTRODUCTION

Recent measurements of ambient sound associated with surface environmental effects display clear correlation with rms sound pressure and wind speed, rain and, to a lesser extent, waves and whitecaps. The relation of wind speed to sound can be written in the form suggested by Shonting and Middleton (1987) as

$$U = A(Nsp) + B \tag{1}$$

or in logarithmic (dB) notation of Evans and Watts (1981) as

$$\log_{10} U = A'(Nspl) + B' \tag{2}$$

where: U is wind speed,
        A, B, A' and B' are constants depending on the frequency of

417

B. R. Kerman (ed.), Sea Surface Sound, 417–427.
© 1988 by Kluwer Academic Publishers.

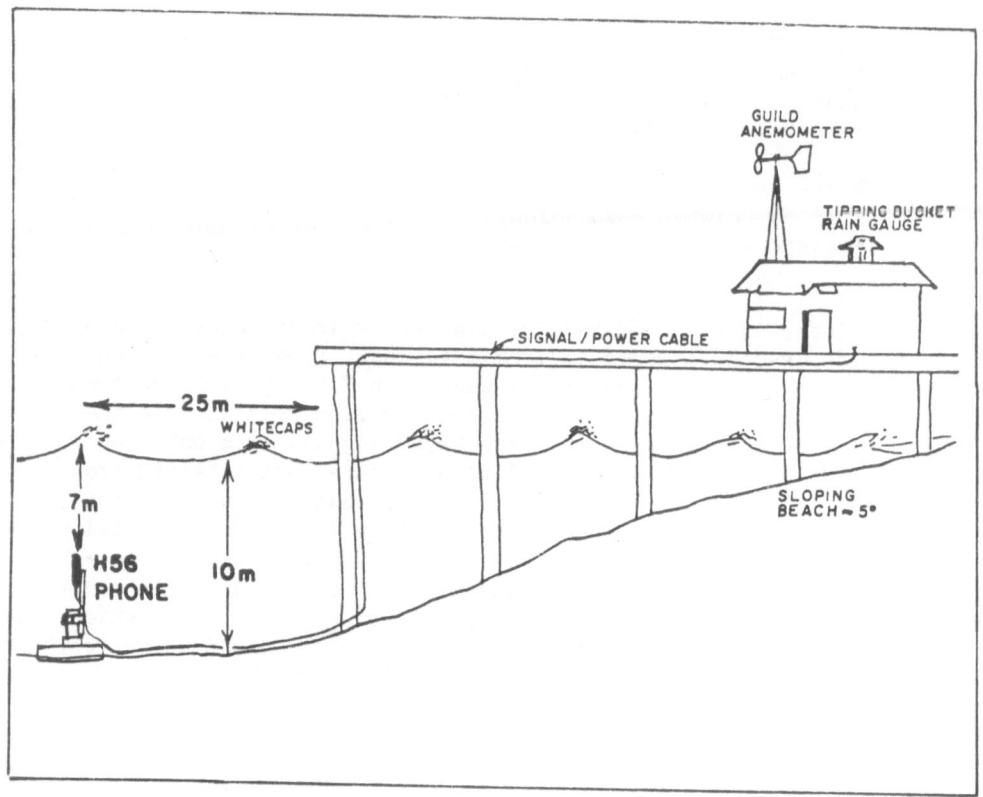

Fig. 1. The Narragansett Bay pier facility for the ambient sound measurements. The H-56 hydrophone is mounted away from the local effects of wave slapping on the pilings and on the beach.

sound generated,
Nsp is sound pressure (rms output from the hydrophone), and
Nspl is the sound pressure level (in dB re 1 μPa).

These empirical relations are produced solely by curve fitting the
acoustic response to observed wind data. Assessment of the true
correlations is hampered by: 1) Difficulty in making environmental
measurements in the same location as the acoustic recording, and 2)
The problem of separating out the wind effects from the other sound
generating phenomena including whitecapping, rain, and coastal
shipping traffic. Moreover, the relations (1) and (2) indicate little
of the physics of the sound generation.

Clearly, the generation of ambient sound is associated with the
production of pressure fluctuations at the sea surface. It appears
that pressure perturbations are most likely caused by three distinct
phenomena: (1) Wind at the water surface producing a boundary layer
which, perhaps by a resonance, creates pressure pulses which radiate
downward, (2) whitecapping which in turn produces spray droplets
impinging on the sea surface, and air bubbles popping through the
surface, and (3) rainfall impacting on the surface which may be
equivalent to the whitecap spray phenomena only more uniformly
occurring.

One expects that resonant frequencies associated with the wind
vis-a-vis droplet impacts should differ, and hence, the frequency
spectra of the sound may hold the key to the definition of, and the
distinguishing among, the ambient sound generation mechanisms.

Studies thus far (e.g., Scrimger et. al., 1987) reveal a unique
sound spectrum from light rain with a peak at about 15 kHz. Recent
work by Nystuen and Farmer (1987) examined the sound spectra
associated with light rain and the alteration of this spectra by light
winds. Unfortunately extrapolation of these results to moderate to
higher wind conditions may be difficult due to the non-linearity and
complexity of the sound producing mechanisms including whitecapping
effects.

Further examination is needed of the spectral character of the
observed ambient sound generated by the different (quantitatively
measured) surface phenomena. As a first step in studying ambient
sound spectra produced by different sea surface effects, we have
examined ambient broadband sound data from Narragansett Bay during
various wind and whitecap conditions. The objective was to explore
characteristics of the ambient sound spectra which may vary with wind
speed and whitecapping intensities while utilizing a new
high-resolution spectral analysis program on the acoustic data.

2.   INSTRUMENTATION AND OBSERVATIONS

The observations were made in Narragansett Bay from an 80 m long pier
(Fig. 1) extending northwesterly from Acquidneck Island with a open
fetch of 12 km (Shonting and Middleton, 1987). The ambient sound
field was monitored using a low noise bottom-mounted hydrophone (Model

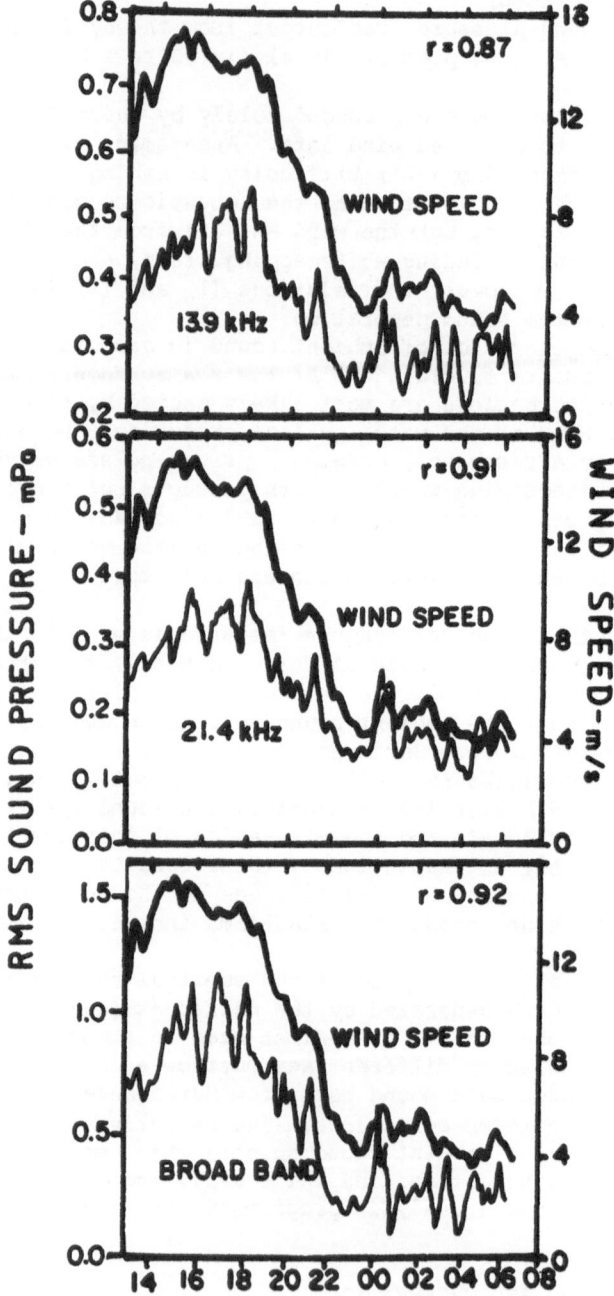

Fig. 2. The SCANR hydrophone output in mPa for the two narrow bands and the broad band as a function of wind speed.

H-56 developed by Henriques, 1972). The unit has a flat response from
1-50 kHz and a sensitivity of 174 dB//volt/1µPa. The hydrophone was
fixed at 7 m depth to a vertical pipe extending from a 320 kg railroad
wheel positioned 25 m out from the pier (Fig. 1). This positioning
reduced interfering effects of the wave slapping on the pilings of the
pier and of waves breaking on the sloping beach some 100 m away. The
signal cable lead to the pier lab shack wherein the data was recorded
on one of two data logging systems.

First, rms sound pressure records made over periods of several
hours or longer were registered on the Self Contained Ambient Noise
Recorder (SCANR) discussed by Shonting and Middleton (1987). This
digitally records, at 1 minute intervals, the rms hydrophone output
over a broad band pass from 5-30 kHz and through 14 and 21 kHz
centered band passes approximately 1000 Hz wide.

In addition, data was obtained to provide energy spectra of the
ambient sound at a variety of wind and wave conditions. For this, the
amplified hydrophone output was directly recorded on an Ampex FR 1300
tape recorder at chosen times representitive of specific sea surface
conditions, e.g., low or high winds, nonwhitecap or whitecap
occurrence of various intensities. To insure that the records were
not contaminated by local ship and boat traffic the records were
monitored with an rms signal conditioner which registered any
anomalous sound pressure levels continuously on a strip chart recorder.

The wind velocity was obtained from the anemometer mounted on the
data shack (Fig. 1) at a height of 6 m above the sea level and about
100 m from the hydrophone. The wind data were recorded on a strip
chart recorder and digitized at 15 minute intervals.

3. DATA AND SPECTRA ANALYSIS

The rms data cassette from the three channels of SCANR were played
back from a MEMODYNE read/write machine into a COMPAQ portable
computer and plotted along with the wind speed.

For the spectral analyses, five to ten minute samples of direct
hydrophone output were recorded on the Ampex tape system in a direct
mode. The ambient sound data were played into a MASSCOMP 500 Series
Computer via a Kronhite Filter which supressed frequencies below 400
Hz and above 40 kHz. Records of 2.048 s were selected and sampled at
a rate of 100 kHz producing 204,800 individual data. For each record
the variance was estimated and Fast Fourier Transform (FFT) analyses
were performed on ten consecutive sets of 2048 data and were then
averaged over the 5-40 kHz band.

4. RESULTS AND DISCUSSION

The acoustic rms response of the SCANR channels to a wide range of
wind speeds is depicted in Fig. 2 as a 21 hr. record which consists of
wind speed plotted with 15 min averages of rms sound pressure at
narrow bands centered at 13.9 and 21.4 kHz and also over the 5-30 kHz

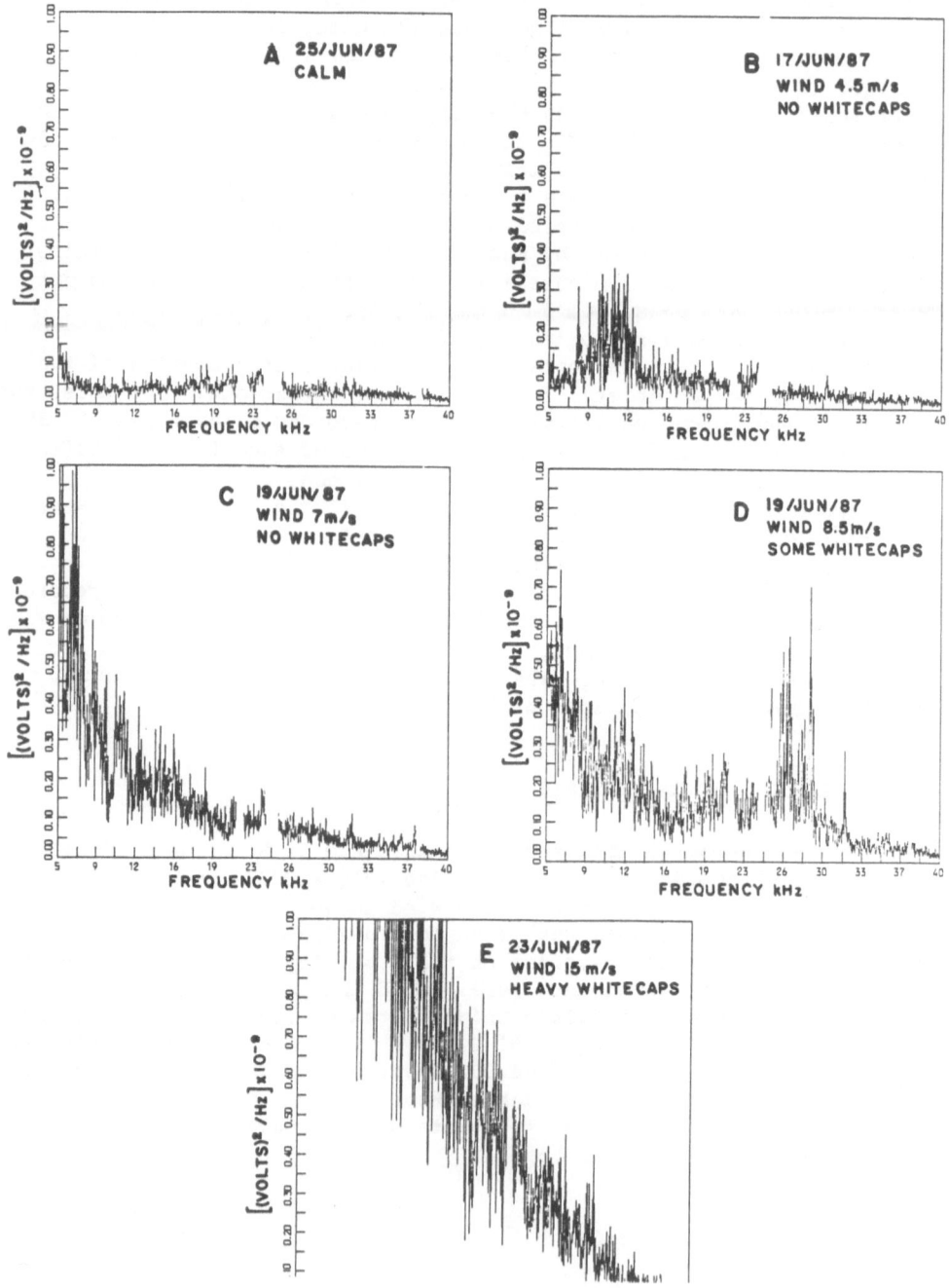

Fig. 3A - 3E. Auto spectra of the hydrophone output for a variety of wind and whitecapping conditions.

broad band.

The similarities of the wind/sound curves and the high correlation coefficients suggest strong coupling throughout the 4-15 m/s range of wind speeds. Even small but rapid fluctuations of wind of 1-2 m/s occurring within 10-20 min (e.g., at 21 hr. and 01 hr.) are reflected by sound pressure changes of 0.1 mPa. This suggests that the observed sound field is very responsive to rapid changes in the local wind. Concomitant changes in the wave or whitecap field altering ambient sound levels would tend to lag the wind speed variations. Interestingly, the sound pressure appears to closely follow the wind fluctuations at lower wind speeds i.e., below 6 m/s where whitecaps are absent.

A sequence of wind/ambient sound data were taken over the period June-July, 1987 and were analyzed on the MASSCOMP computer. Figs. 3A-3E show five spectra, each being an average of the FFT's of ten consecutive time series of the hydrophone voltage output. Each spectrum is representitive of wind conditions ranging from calm to 14 m/s. For intercomparison of the different spectra the ordinates were made identical. Gaps in the spectra at 21 and 24 kHz are excised line spectra introduced by observed self noise in the tape recorder.

The sound spectrum of calm wind conditions (Fig. 3A) shows a flat response from about 7-40 kHz while for winds of 4.5 m/s bursts of energy appear between 7 and 15 kHz. At 7 m/s, while whitecaps have not yet formed, the energy begins to fill at the lower frequencies. By 8.5 m/s as whitecaps were initiated an energy band develops between 19 and 30 kHz. The 15 m/s wind speed presents a full energy spectrum which monotonically decreases to the 40 kHz cutoff. The extrapolated value of the ordinate at 5 kHz attains about 2×10^{-9} (volts)^2/Hz or about 2.6 (uPa)^2.

The total acoustical energy (or variance) registered over the 5-40 kHz bandwidth (equal to the area under the curves 3A-3E) is shown in Fig. 4 to increase monotonically and non linearly with wind speed. It is tempting to suggest that instead of a smooth curve there may be two relatively straight lines intersecting at the wind speed of whitecap formation (e.g., at 6-7 m/s as was suggested for the $\log_{10} U$ relation of eqn. 2 by Evans and Watts (1981). It is thus reasonable to assume that the rate of sound energy (E) production with wind speed (U), i.e., the slope (dE/dU), will increase once whitecapping is introduced, however more data is required to demonstrate this.

The spectra of the sound pressure levels associated with a full range of observed winds speeds from calm to 15 m/s is displayed by the superposition of the smoothed spectral curves converted to dB//(μPa)^2/Hz (Fig. 5). The family of curves display several interesting features. First, the spectral energy level increases with wind speed (an obviously wave and whitecap activity) at all frequency increments from 5-40 kHz, with a general sloping tendency from high to low frequencies. We note however that, acoustical energy appears to be preferentially added faster at 20-35 kHz than at 10-15 kHz. The spectra between 7-25 kHz appears to "fill in" as wind speed increases from 8 m/s to 15 m/s at which point the spectra has a steady downslope with increasing frequency. These effects are suggestive of the well

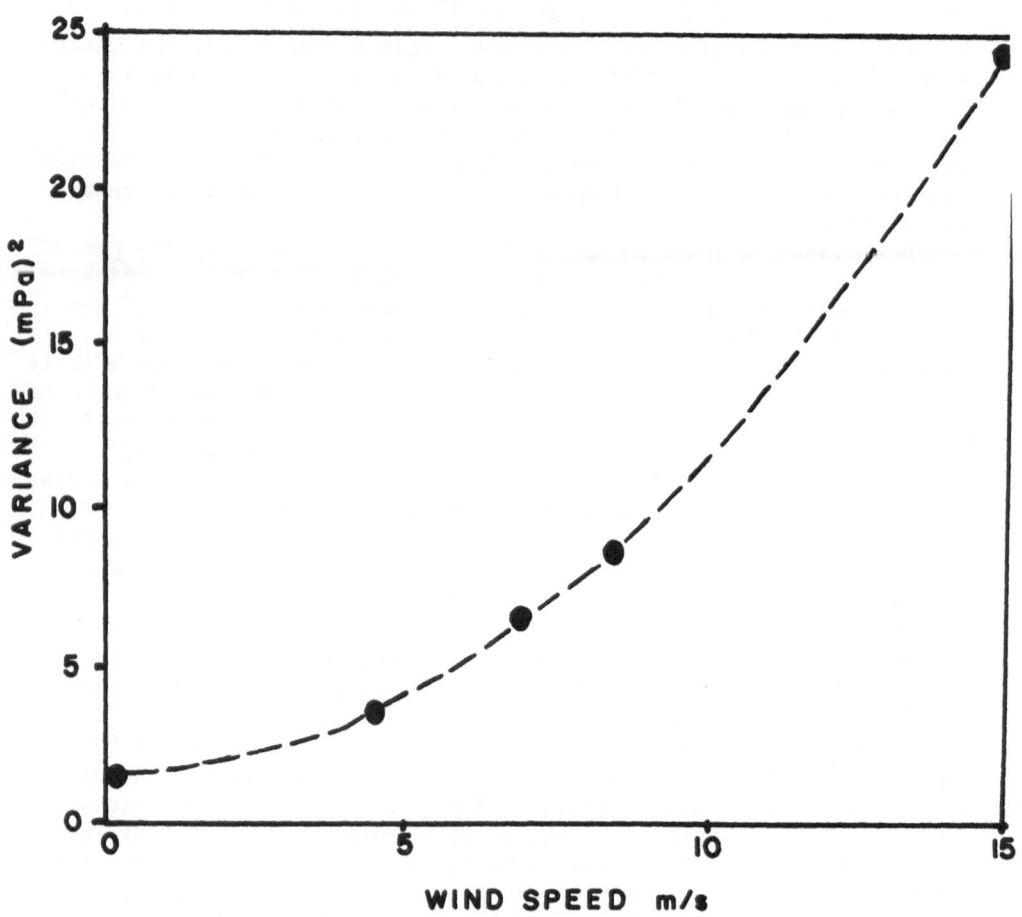

Fig. 4. Variances of records 3A-3E in mPa^2 over the frequency range of 5-40 kHz.

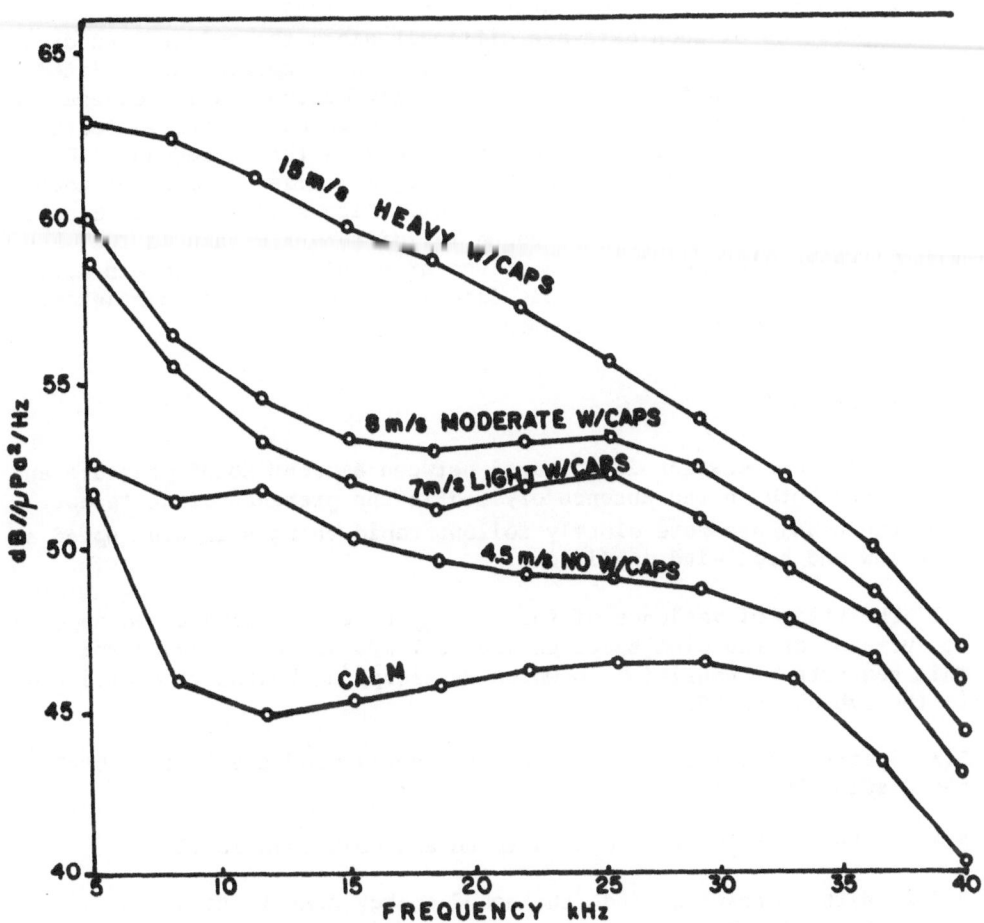

Fig. 5. The spectral density of the ambient sound pressure estimated from the data of Fig. 3.

known free surface wave spectra which, when becoming "fully developed" shift energy to lower and lower frequencies. The anomalous spectral peaking may be attributed to occurrence of additional sound producing mechanisms as whitecaps are formed much as the peak at 15 kHz occurs in the rainfall spectrum (Nystuen and Farmer, 1987).

Comparisons of such data are difficult since the original Knudsen data extended only to 20-25 kHz and some Knudsen spectra are related to sea state (which is, at best, imprecisely defined) while others are compared to wind speed or some estimate of wave height whose accuracy is not known. Moreover, local noise sources in the Narragansett Bay such as industrial and biological activity may bias the ambient sound associated with local wind and whitecaps. Also shallow bay or coastal topography may channel ambient sound far differently than in the open ocean. Certainly, any fair comparison of sound spectrum levels must contain both statistically stable data and be obtained in consistant and accurately measured environmental conditions.

5. CONCLUSIONS

5.1 A high correlation is observed between ambient sound pressure and wind speed both in the absence of, and in the presence of whitecaps. Also the sound pressure closely follows rapid changes in wind speed at both low and high wind conditions.

5.2 The ratio of variance of the sound pressure to wind speed appears to increase as the wind speed exceeds 6-7 m/s where whitecaps form. This suggests a transition in which whitecapping begins to contribute to the ambient sound.

5.3 Spectra of the sound pressure at several wind speeds from 5-40 kHz display the following characteristics:

5.3.1 The spectrum at zero wind speed and calm seas is flat.

5.3.2 With increasing winds the sound energy density tends to increase at all frequencies, the amount of increase varying inversely with frequency.

5.3.3 Peaks and valleys appear in the otherwise smoothly sloped spectra at around 4.5-8.5 m/s wind speeds, suggesting the onset of dynamic pressure-producing mechanism(s) which resonate at certain frequency bands, driven by transient wave turbulence and breaking effects.

5.4 The spectral slopes which vary with both speed and whitecap conditions are in sharp contrast to the the parallel families of curves of the Knudsen plots at lower frequencies. Such spectral slope inflections should be explored further under various sea surface conditions.

5.5 High resolution FFT analyses appear to be a useful tool to examine spectra for possibly identifying and isolating individual ambient sound generating mechanisms.

6. ACKNOWLEDGMENTS

Helpful discussions were held with Prof. Foster Middleton of the Department of Ocean Engineering at URI regarding both the SCANR calibration and the field observations. Paul Cronin of INTEK Engineering cheerfully advised us on the signal cable installation and noise reduction. Jeff Greenberg and Tony Chaves of the Southeastern Massachussetts University assisted in the field measurements and data analyses. Steve Shock from the University of Rhode Island guided our MASCOMP FFT analyses. These studies were supported by the Naval Sea and Air Systems Commands and the Naval Underwater Systems Center, Newport Laboratory.

7. REFERENCES

7.1 Evans, D. and Watts, D.R. 1981. Wind Speed and Stress at the Sea Surface from Ambient Noise Measurements. Proc., of International Symposium of Acoustical Remote Sensing of Atmosphere and Oceans, Univ. of Calgary, Calgary, Canada.

7.2 Henriques, T.A., 1972. An Extended-range Hydrophone for Measuring Ocean Noise, J. Acoustic, Soc. Am., Vol 52/5,1450-1455.

7.3 Knudsen, V.O., R.S. Alford, and J.W. Amling, 1948. Underwater Ambient Noise, J. Mar. Res. Vol. 7, 410.

7.4 Nystuen, T.A., and D.M. Farmer, 1987. The Influence of Wind on the Underwater Sound Generated by Rain. J. Acoustic. Soc. Am., Vol. 82/1, 270-274.

7.5 Scrimger, J.A., D.J. Evans, G.A. McBean, D.M. Farmer, and B.R. Kerman. 1987. Underwater Noise Due to Rain, Hail, and Snow, J. Acoustic. Soc. Am. 81, 79-86.

7.6 Shonting, David, and Middleton, Foster, 1987. Near Surface Observations of Wind and Rain-Related Generation Sound Using the SCANR: An Autonomous Acoustic Recorder. J. Atmospheric and Oceanic Technology. (In Press).

PHYSICAL MECHANISMS OF NOISE GENERATION BY BREAKING WAVES — A LABORATORY STUDY.

Michael L. Banner
School of Mathematics
University of New South Wales
Kensington, N.S.W. 2033.
Australia

Douglas H. Cato
Defence Science and Technology
    Organisation
Weapons Systems Research Laboratory
Maritime Systems Division
RAN Research Laboratory
P.O. Box 706, Darlinghurst, NSW 2010.
Australia

ABSTRACT. Breaking waves are recognised as a significant source of ambient underwater noise in the ocean yet the details of the causal physical mechanisms are poorly understood. This contribution describes the initial findings of a detailed laboratory study aimed at elucidating this process using simultaneous high speed photography and sound measurements under controlled conditions. It was found that the dominant noise occurred in the form of discrete tone bursts which appeared to be associated with the formation of bubbles and coalescing or splitting of bubbles.

1. INTRODUCTION

The earliest studies of noise in the ocean by Knudsen, Alford and Ewling (1948) recognised that breaking waves were a dominant source of noise. Indeed, with a hydrophone close to the water surface it is possible to hear the sound of an individual wave breaking. Wenz (1962) drew on a large amount of data to refine the spectral shape of noise from breaking waves, which he identified as "bubble and spray" noise, as having a broad spectral peak at about 500 Hz. The many experimental studies which have followed have obtained empirical relationships between noise and wind speed (and shown that the correlation was better than between noise and wave height) but there appears to be little experimental work aimed at identifying noise generating mechanisms. Laboratory experiments by Franz (1959) studied the impact noise of water droplets and Strasberg (1956) the noise generated by bubbles formed at a nozzle, in both cases in quiescent water. Theoretical work has also been limited. One of the difficulties is to know which of the many mechanisms to model — impact of water droplets, air entrainment, bubble oscillation, bubble collapse or bursting. Kerman (1984) and Shang and Anderson (1986) have addressed this problem in recent contributions. There is clearly a need for a better under-standing because no existing empirical model is universally applicable

B. R. Kerman (ed.), Sea Surface Sound, 429–436.
© *1988 by Kluwer Academic Publishers.*

except in very broad terms. For example, various observational studies show not only differences in noise levels at any wind speed but also differences in the rate of noise increase with wind speed (Cato, 1978).

This contribution describes the initial results of an investigation aimed at identifying experimentally the processes in wave breaking that contribute to the generation of underwater noise.

2. EXPERIMENTAL CONFIGURATION

The measurements reported here were made in the laboratory wind-wave flume shown schematically in Figure 1.

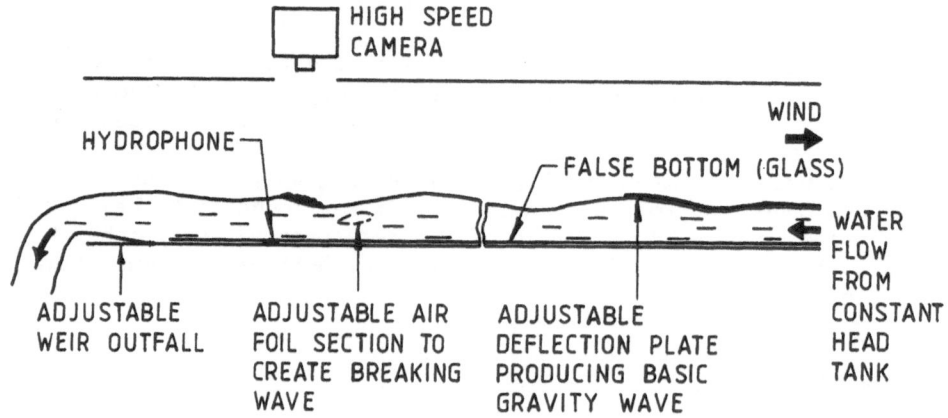

Figure 1. Experimental wave flume configuration: working length, 7.3 m; channel width 0.225 m; water channel depth 0.22 m; air channel depth 0.36 m. The measurement site was halfway along the channel.

The flume could be configured in either of two modes to provide (a) a propagating steep wave train on still water, generated by a constant frequency wavemaker at one end, with an absorbing beach at the other end (b) a quasi-steady wave train, held stationary against the steadily flowing current supplied from a constant head tank and with a quietened outfall. The wavetrain was steepened locally by a subsurface hydrofoil.

For the initial phase of the investigation, it was expedient to utilize the stationary wave configuration. The advantages were lower spurious background noise, and better control over the breaking and hence easier identification of the noise generating events. The applicability of this configuration to transient breaking waves at sea is discussed in some detail by Banner (1987) in a companion paper in this volume. This approach has been used successfully in previous studies relating to other breaking wave phenomena (Banner and Melville, (1976), Banner and Fooks (1985)).

The floor of the flume had a transparent section and a false glass bottom. A hydrophone (LC-10) was installed below the false bottom to shield it from the turbulence in the flow, with negligible acoustic attenuation through the glass. For a given flow velocity, a reference noise level could be established for a very steep unbroken wave induced by the hydrofoil. By very slight adjustments of the shape of the foil, quasi-steady breaking waves of a range of strengths could be produced. Examples of these are shown in the companion paper by Banner (1987) in this volume. In this initial investigation, two types of related observations were made.

The first consisted of analysing noise characteristics of non-breaking and spilling breaking waves. A reference background noise level was determined when the hydrofoil was set to produce a very steep unbroken wave. With slight adjustments to the hydrofoil inclination, the steep unbroken wave transitioned to a quasi-steady spilling breaker. The 'strength' of the breaking, as judged by the downslope extent of the spilling region, could be adjusted via the hydrofoil inclination. For a given breaking wavelength (0.4 m to 0.7 m in this study) the noise was recorded on a taperecorder (NAGRA $\overline{IV}$-SJ) and spectra were determined using a Hewlett Packard 3582A spectrum analyzer.

A second class of observations was made with the aim of correlating events in the breaking wave with the occurrence of noise. A high speed motion camera (Hycam K20S4E) operating at 480 frames per second viewed the working section from above with diffuse illumination from below. This provided a factor of 20 speed reduction to the normal 24 frames per second. Figure 2 provides a typical overhead view of the spilling region showing the instantaneous distribution of entrained air bubbles.

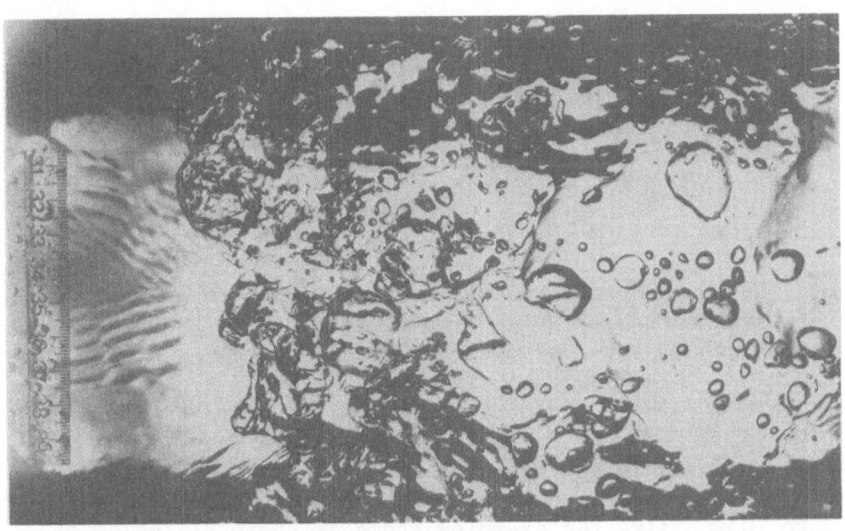

Figure 2. Typical overhead view of spilling region (breaking wave length ~ 0.6 m).

432

The attendant noise was recorded synchronously with the film using a
repeating electronic strobe which illuminated the field of view. The
light flashes were recorded on the film and on a second channel of the
tape recorder as the output pulse from a suitably aimed photodiode.
The film and sound track were viewed subsequently on a flat bed editor
at a speed as low as a few frames per seconds, equivalent to a speed
reduction in excess of 100. The synchronisation was sufficiently
precise (typically within 0.1%). This allowed us to correlate noise
events with physical events occurring during the onset of spilling
breaking induced by controlled (silent) perturbation of a steep
unbroken wave. The wavelength observed was about 0.55 m.

3. RESULTS

For the quasi-steady breaking configuration it was observed that
compared with the unbroken wave, discernible noise was evident from
even slightly breaking waves. However, breaking with air entrainment
was required for the noise to be sufficiently above the background
noise for useful measurements to be made. Figure 3 shows the noise
spectra measured in presence of vigorously breaking and steep unbroken
waves.

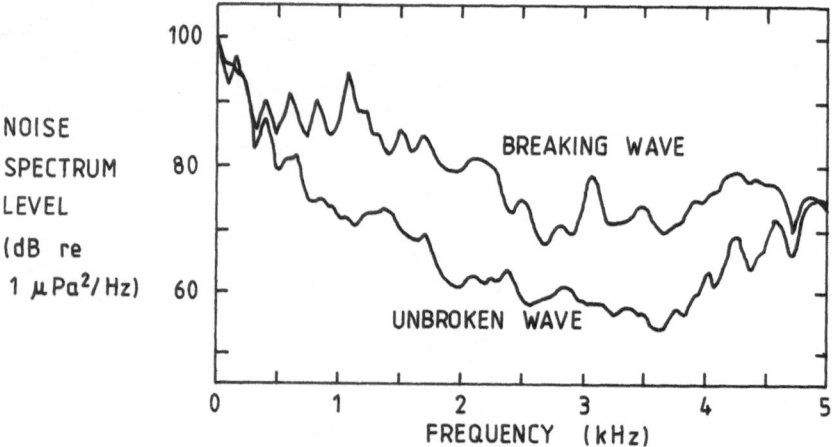

Figure 3. Noise spectra measured in the presence of a breaking and a
steep unbroken stationary wave of wavelength 0.7 m. Background noise
dominates below 300 Hz. Hydrophone distance : 0.2 m.

of comparable wavelength [approx. 0.70 m]. The hydrophone was located
about 0.2 m from the wave crest for both measurements.
Noise observed in the presence of quasi-steadily breaking waves
was dominated by discrete narrow band bursts of noise of duration of
about 3 to 20 ms, like the examples shown in the oscillogram of
Figure 4. These bursts typically showed a very rapid rise and
significantly slower decay.

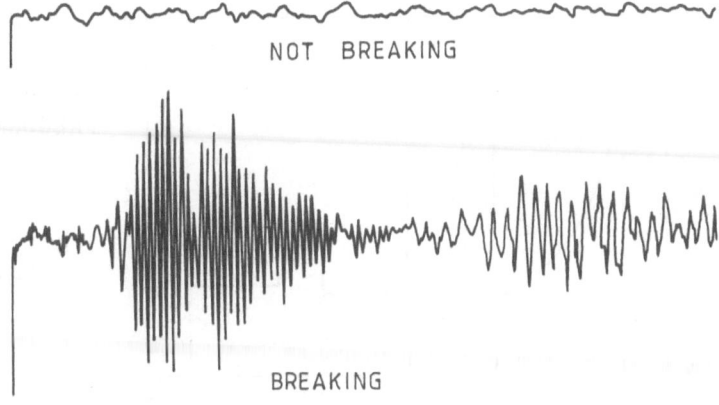

Figure 4. Oscillogram showing the wave form of noise bursts from a breaking wave compared with the noise in the presence of a steep unbroken wave (on the same scale). The record duration was 50 milliseconds.

Based on our analysis of the high speed ciné film and synchronised sound track of the noise events, it was observed that the occurrence of dominant noise bursts appeared to correlate with the generation of bubbles by air entrainment at the leading edge of the spilling region and with the impact and coalescence of bubbles or the splitting of a bubble. Significant noise bursts were not correlated with bursting of bubbles at the air-water interface although there were relatively few examples of these. There were, however, many examples of bubbles undergoing considerable distortion as they were swept back from the breaking edge. As best we could determine from our two-dimensional view, this distortion took the form of alternating elongation of the bubble along orthogonal axes. No noise bursts could be correlated with this motion, however the frequency of oscillation would be less than a few hundred hertz where we were limited by background noise.

Typical distribution of bubble sizes observed in an induced breaking cycle of a 0.55 m spilling breaker are shown in Figure 5, together with the resonant frequency of an air bubble oscillating radially (as a monopole) in water. The range of resonant frequencies corresponding to the observed bubble radii is consistent with the range of frequencies of the observed noise bursts. It should be noted, however, that some bubbles were at the water surface and these would have quite a different resonant frequency.

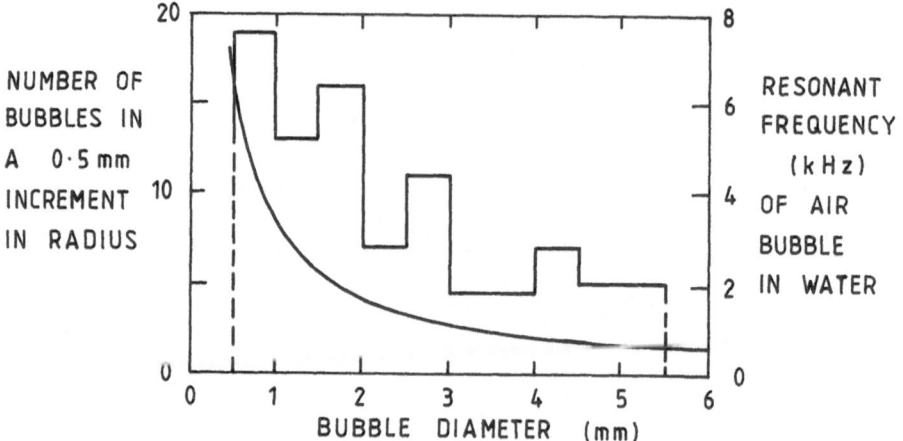

Figure 5. Distribution of the sizes of bubbles entrained by a
stationary breaking wave over a period of about 5 s (histogram).
The curve shows the resonant frequency of an air bubble oscillating
radially in water.

4. DISCUSSION AND CONCLUSIONS

Our initial results suggest that, at least under the conditions
of the experiment, the predominant noise results when bubbles receive
an impulsive impact such as at the point of generation at the breaking
edge or when coalescing or splitting. The character of the noise
bursts is consistent with damped resonant oscillation of air bubbles
in water. Although relatively few bubbles were observed to burst in
the field of view, the bursting did not usually correlate with noise
production. This does not imply that bursting is silent but that the
noise of bursting was more difficult to detect. This may be because
bursting noise levels are lower or it may be that under the conditions
of the experiment the impulsive sound of a bursting bubble is more
difficult to detect than the tone burst of a bubble oscillating at its
resonant frequency. Gross deformation of bubbles as they are swept
back through the wave appears to be less significant as a noise source.
This would correspond to forced oscillation well below the resonant
frequency and the deformation is of the form to suggest a high order
source rather than the more efficient monopole source for the resonant
bubbles. Note that the presence of the water surface, by providing an
out of phase image, may reduce the efficiency of both sources
significantly and would result in noise from oscillating bubbles
appearing as though due to dipoles with axes at or near vertical. This
is consistent with recent measurements by Ferguson and Wyllie (1987) of
oceanic noise (in the frequency range 200 – 450 Hz) at high wind speeds
using a deep horizontal array. Their results are consistent with the
source in the form of a distribution of vertical dipoles across the
sea surface. It is also of interest to recall that a laboratory study
of noise from bubbles formed at a nozzle (Strasberg, 1956) also found
that the dominant noise resulted when bubbles oscillated radially at

the resonant frequency, excited by their formation or at splitting or coalescence.

This study has concentrated on identifying the major sources of noise generation in wave breaking in a controlled laboratory situation. The problem of relating the noise characteristics and levels in our flume with those observed at sea has been deferred to a later stage. Clearly, there are difficulties in achieving this goal. The noise levels in the flume are likely to be enhanced by reverberation (tank modes around 3.5 kHz and 7 kHz are to be expected from the tank width and water depth which were both about 0.2 m.) The noise levels were measured by averaging the contributions from continuously breaking waves, whereas oceanic wave breaking is a transient phenomenon. Also, the generation and dynamics of bubbles in the fresh (reservoir) water used in these experiments are likely to differ from those in sea water because of variations in the relevant properties (Monahan (1969), (1971), Scott (1975)). In the near future, it is planned to measure the reverberation characteristics of the flume and to study the noise characteristics of transient spilling breakers using both fresh and salt water. Preliminary studies have demonstrated the feasibility of this approach.

ACKNOWLEDGEMENT
The assistance of Mr. Steve Preece in the production of the high speed ciné film sequences is gratefully acknowledged.

REFERENCES
M.L. Banner (1987) 'On the mechanics of spilling zones of quasi-steady breaking waves', Proc. NATO ARW on Natural Mechanisms of Surface Generated Noise in the Ocean, Lerici, Italy (this volume).

M.L. Banner and E.H. Fooks (1985) 'On the microwave reflectivity of small-scale breaking water waves' Proc. Roy. Soc A, 399, 93-109.

M.L. Banner and W.K. Melville (1976) 'On the separation of air flow over water waves' J. Fluid Mech. 11, 825-842.

D.H. Cato (1976) 'Ambient sea noise in waters near Australia' J. Acoust. Soc. Am., 60, 320-328.

B.G. Ferguson and D.V. Wyllie (1987) 'Comparison of observed and theoretical responses of a horizontal line array to wind-induced noise in the deep ocean' To appear in J. Acoust. Soc. Am.

G.J. Franz (1959) 'Splashes as sources of sound in liquids' J. Acoust. Soc. Am., 31, 1080-1096.

B.R. Kerman (1984) 'Underwater sound by breaking wind waves' J. Acoust. Soc. Am., 75, 149-165.

V.O. Knudsen, R.S. Alford and J.W. Emling (1948). 'Underwater ambient noise' J. Mar. Res., 7, 410-429.

Monahan, E.C. (1969) *'Laboratory comparisons of freshwater and saltwater whitecaps'* J. Geophys Res. 74, 6961-6966.

Monahan, E.C. (1971) *'Oceanic Whitecaps'* J. Phys. Oceanog. 1, 140-144.

Scott, J.C. (1975) *'The role of salt in whitecap persistence'* Deep Sea Res. 22, 653-657.

E.C. Shang and V.C. Anderson (1986) *'Surface-generated noise under low wind speed at kilohertz frequencies '* J. Acoust. Soc. Am., 79, 964-971.

M. Strasberg (1956) *'Gas bubbles as sources of sound in liquids'* J. Acoust. Soc. Am., 28, 20-26.

G.M. Wenz (1962) *'Acoustic ambient noise in the ocean: spectra and sources'* J. acoust. Soc. Am., 34, 1936-1956.

AUDIO SIGNATURE OF A LABORATORY BREAKING WAVE

Bryan R. Kerman
4905 Dufferin Street
Atmospheric Environment Service
Downsview, Ontario M3H 5T4 Canada

ABSTRACT. Results of an experiment to characterize the sound field
radiated from a breaking wave, both into the air and into the underlying water, are described. The total energy of the sound transient
demonstrates a rapid buildup to a crash and then a steady run-down.
There is sound released preferentially to the air in the late stages
during run-down, probably as a result of the collapse of bubbles in
foam. A simultaneous audio and video dissection of the record reveals
that the sound builds monotonically with the depth of the bubble cloud
on the face of the wave. The audio spectra, averaged logarithmically
demonstrate a similarity structure virtually identical to the overall
ambient noise field observed in the ocean, for frequencies above 4
kHz. There appears to be the possibility of a break in the time evolution between energy at frequencies less than about 2 kHz and those
above, possibly related to forced and free oscillations of a bubble
field. Reverberation effects appear to be limited to less than about
1 kHz. An intermittent, spikey microspectral structure implies that
random sinusoids may be the source of the sound field. Implications of
the source rate of bubbles on detection and counting are discussed.

1. INTRODUCTION

Although breaking waves are thought to be a source of higher frequency
ambient noise, it remains to determine what mechanism within them
actually generates the sound. The sound of a breaking wave, at least
one occurring on a beach, is easily recognizable and characterized by
anyone who has been exposed to the phenomenon. Apparently the first
observations of the underwater sound of breaking waves date to the
Second World War (Knudsen et al., 1948). Surprising little seems to
have occurred in the interim to examine this source of sound. Accordingly this study seeks to investigate the audio characteristics of a
breaking wave, especially in terms of its physical evaluation and to
undertake a preliminary examination of some observed properties to
probe for evidence of the source mechanism. The study is confined to a
laboratory simulation because of questions of repeatability and control

B. R. Kerman (ed.), Sea Surface Sound, 437–448.
© 1988 by Kluwer Academic Publishers.

in a field study.

Various mechanisms including splash due to droplets (Wilson, 1980), resonant bubbles (Crowther, 1988), as well as non-resonant bubbles forced by the turbulent flow (Kerman, 1984), cavitation (Furduev, 1966) and more recently entire bubble cloud oscillation (Prosperetti, personal communication, 1985) have been suggested without collaborative proof. While it is not clear as yet how one will distinguish each mechanism's associated properties, the premise adopted here is that there may be intricate, micro-properties which may indicate one mechanism over another. For example a resonant bubble has a well defined micro-signature - a damped pressure sinusoid where the damping is a well-known function of the frequency which is in turn related to the bubble size (Crowther, 1988). Splash noise also has a well-defined micro-signature (Pumphrey and Crum, 1988) as does a cavitating bubble (Flynn, 1964). On the other hand bubbles forced by turbulence will have a spectrum related to the turbulent (acceleration) spectrum (Kerman, 1984). Ideally it may be possible to dissect an audio record of a breaking wave to identify such micro-structure, and, as a reward, count the number of micro events.

2. EXPERIMENT

The study reported here was conducted in the wave tank of the Canada Centre for Inland Waters in Burlington, Ontario in 1985. The facility is about 100m long, 4.5m wide and 1.5m deep with a programmable wave board which makes exceptionally reproducible breaking waves on demand. The technique of causing breaking is based on concentrating energy locally by having longer, faster waves catch up to shorter, slower waves in a wave pocket (Neumann and Pierson, 1966, p.281). Each individually generated event breaks within centimeters of other such events, after travelling 50m. A general purpose omni-directional hydrophone (ITC 8084A) with a frequency bandwidth over 40 kHz was placed at a depth of 1.2m under the location of maximum wave height during a breaking event. A video camera was placed on the tunnel wall at this location imaging the event from above using forward scattered and reflected light primarily. The audio record was taped using an instrumentation recorder rated to 60 kHz at maximum speed, as well as a video cassette recorder (VCR) with an audio response rated to less than 10 kHz.

The audio record was played back at its recorded speed, digitized at 40 kHz and analyzed using signal processing software on a computer. The video was processed using a special sound-studio cueing device to analyze individual video frames with the (low passed) audio from the VCR. The images were referenced to position markers placed on the tunnel wall so that the breaking wave's evolution could be related to the audio history. The phase velocity of the breaking wave event estimated by this method was 1.9 m/sec. A microphone was placed about 1m above the surface at the location of maximum breaking wave extent and directly above the hydrophone.

A rudimentary reverberation test was attempted by using the

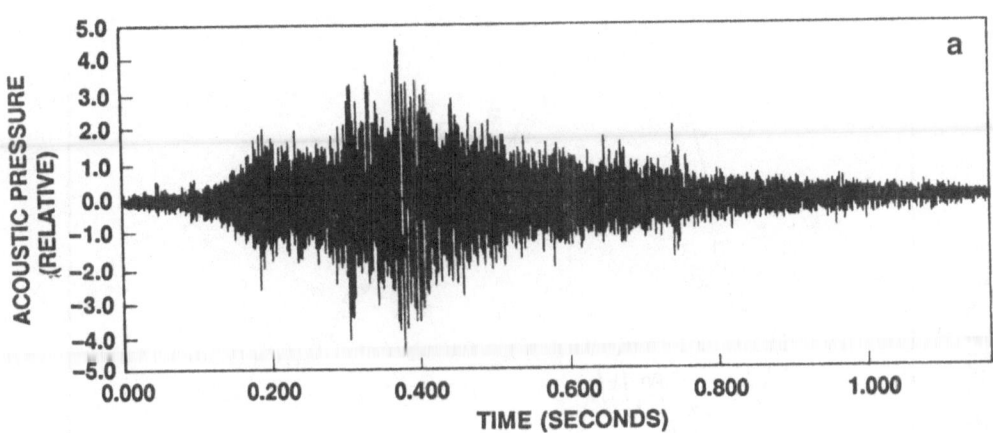

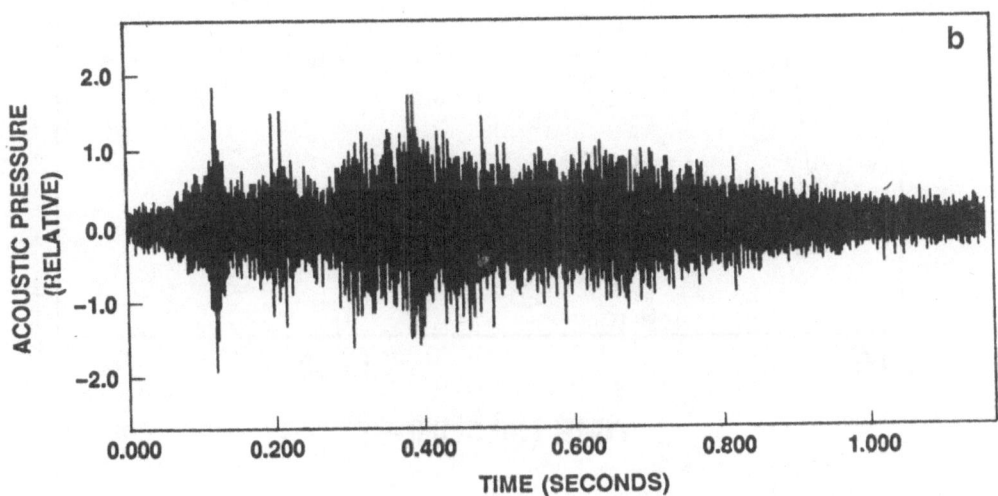

Figure 1 Time evolution of audio record from a breaking wave measured
a) underneath in water, b) above in air. Records not aligned in time.

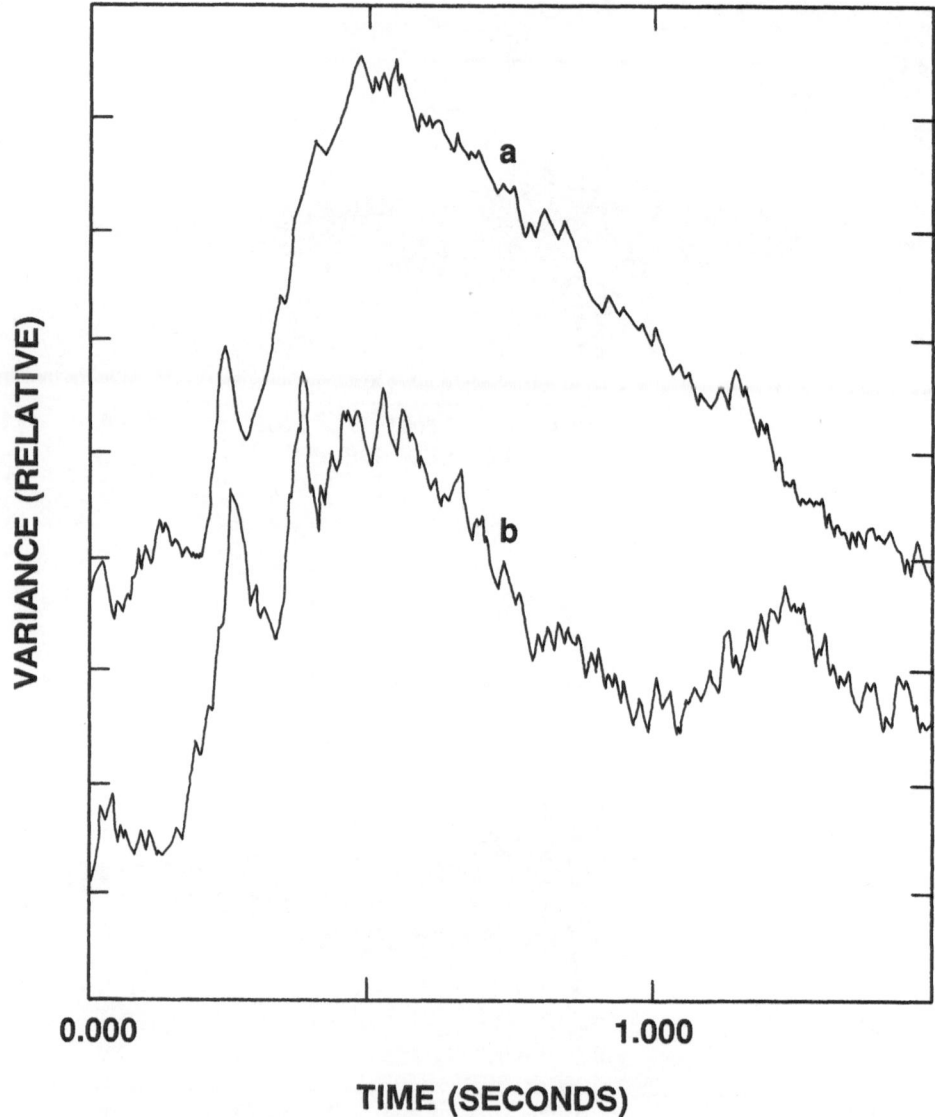

Figure 2 Time evolution of variance of audio records measured a) with a hydrophone and b) with a microphone.

impulse noise from a clip-board held near the breaking location. A more precise impulse source is required to determine the loss rate of impulse energy.

3. ANALYSIS

The time history of the hydrophone and microphone for a typical wave (Figure 1) lasts for only about 1 sec. or about 1.2 wave periods. The buildup to a maximum, (the 'crash') occurs in about 0.3 sec. with the predominance of the sound generation occurring in a regular monotonically decreasing stage (the 'run-out') as shown in the variance history (Figure 2). Amplitude modulations are evident in the original time records leading up to the maximum which last less than 30 msecs; they are regular and repeatable features of the breaking sequence. Another notable feature of the variance history is the secondary peak in the microphone record late in the death throes of the breaker. This is ascribed to foam left behind almost directly below the microphone and above the hydrophone. Apparently the sound generated by the collapse of these bubbles is directed more efficiently into the air. At least in the presence of the retreating breaking wave, the foam generates less acoustic energy underwater than the wave itself.

If the time record is played back at reduced (1/16) speed, prior to the crash, there are several distinct sounds resembling a water splash and a general gurgling prior to the massive onrushing. A similar result for isolated ringing has been reported by Banner and Cato (1988) for a steady breaker of considerably less wave energy. No allowance in this study was made in the time histories for the distance of the breaking wave to the hydrophone and microphone, estimated to be no more than a factor of 1.5 compared to the overhead maximum distance.

The video record indicates that the breaking wave has a short initial plunging stage. However prior to this the wave demonstrates a bent-over, cusp shape with jetting fingers along the leading edge as reported by Worthington (1908) many years ago. These fingers develop into streaks which appear to lengthen and ultimately collapse into a morass of bubbling along the leading edge. The first detected sound occurs with the appearance of an air-water cloud pushed ahead by the wave. As this cloud builds so does the sound until the mixture begins to be left behind by the wave. Within this death stage of the breaker, several foam lines are produced in succession, each perhaps associated with a burst of sound as seen in Figure 2. As these foam lines and the turbulent pool laid down behind the wave die out, the sound level gradually decreases. The sequence of these events is outlined in Figure 3 in terms of an artificially pre-whitened spectrum to emphasis higher frequency content.

The time records were subjected to various spectral analyses to examine the distribution of audio energy. In the first, for the hydrophone, a sequence of relatively long time records (81 m sec) were computed, plotted and then smoothed by eye. They are replotted in Figure 4, using an arbitrary start time of minimum energy. The general trend is to monotonically decreasing energy above 1 to 2 kHz,

442

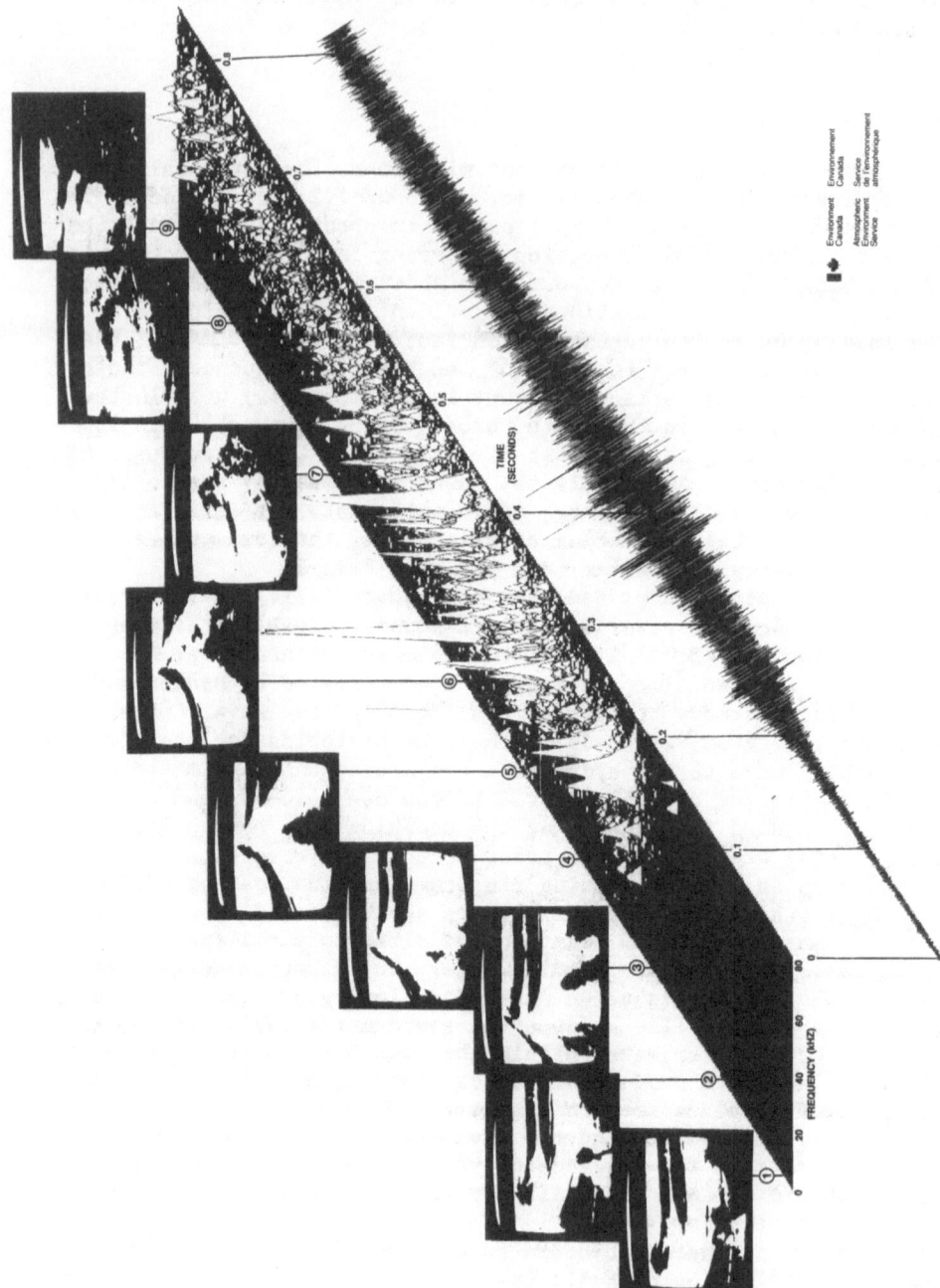

Figure 3 Simultaneous video spectral and observed time evolution of hydrophone record of a breaking wave. The record has been pre-whitened and over emphasizes high frequency content.

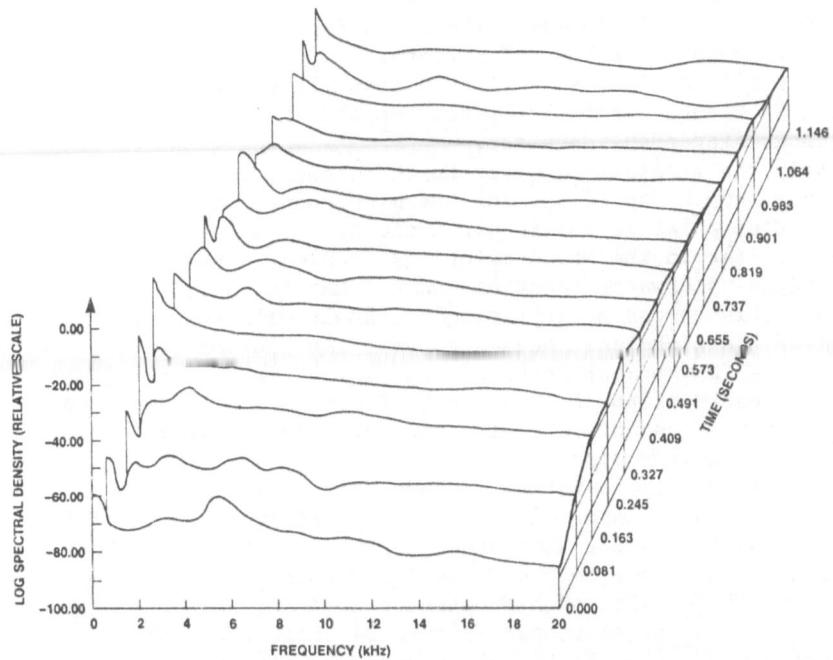

Figure 4 Time evolution of smoothed spectra of hydrophone record of a breaking wave.

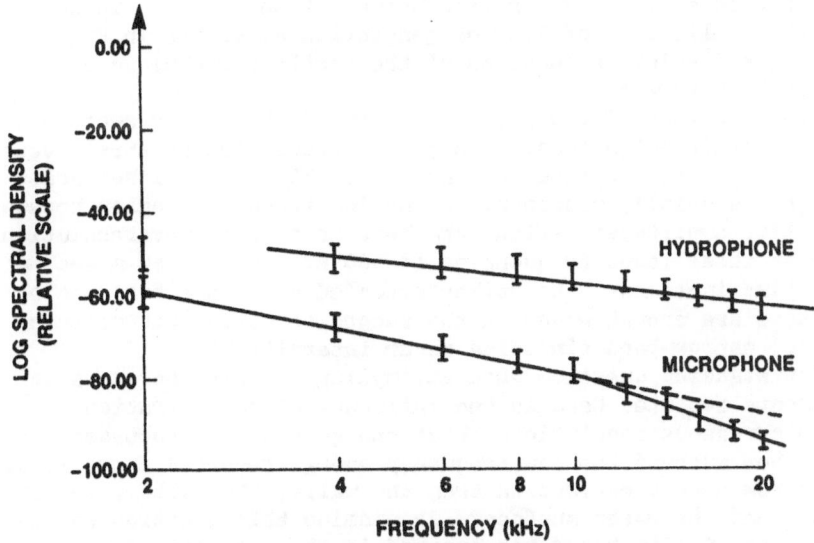

Figure 5 Logarithmically smoothed similarity structure of a) underwater and b) in air audio spectra through a breaking wave.

as well as a rapid increase in the general level, then a monotonic
run-down, as seen in the gridded slice at 20 kHz. These spectra were
then shifted in amplitude logarithmically for maximum alignment in the
frequency range above 4 kHz and averaged. The outrider values at a
sequence of frequencies, as well as a best fit through these points
are given in Figure 5 for the hydrophone and the similarly processed
simultaneous microphone data set.

Clearly the straight line fit to the underwater data is adequate
above 4 kHz, and up to 10 kHz for the microphone. It is believed that
the inflection point in the in-air audio spectrum is imposed by the
arbitrary filter in the sound meter ('A' scale). The slope of the
time averaged underwater spectrum above 4 kHz is -5.4 dB/octave or -18
dB/decade. This compares surprisingly closely with the ambient noise
spectrum of the classical Knudsen curves. Implications of this result
are discussed in the last section. The microphone spectrum demonstrates a steeper slope of about -8.6 dB/octave. The associated
coefficients in frequency are -1.9 and -2.7 respectively for the
underwater and in-air spectra.

Also enlightening in terms of acoustic energy distribution is the
time evolution on a linear plot, the so-called 'waterfall' plot.
Figure 6 demonstrates the dominate acoustic energy in the 250 Hz to 1
kHz region as well as a secondary maximum above 2 kHz. The time
evolution of the unfiltered hydrophone record (Figure 6) is similar to
that of the pre-whitened record (Figure 2). Reference to Figure 2
allows for a connection to the video record of the breaking wave's
development. Of interest to those involved with low frequency ambient
noise is the marked increase in audio energy below 500 Hz at $t-t_0$
$\simeq 0.65$ seconds which coincides with the interval of maximum variance
in the 'crash' stage. Also there is an orderly rundown of the bandwidth of the lower (250 - 1000 Hz) frequency energy. There appears to
be a preferred scale of generation between 2 and 3 kHz. To some
extent the implication of lack of generation elsewhere in Figure 6 is
misleading and solely a function of the plotting technique as
confirmed by Figure 4.

Another feature of the time evolution of the linear spectra is its
spikey, intermittent nature. Finer time resolution spectra revealed
spikes decorrelated in time in a matter of 10's of milliseconds. The
spikes are especially prominent in the low frequency region because of
the relative amplitudes. Although there is a frequency resolution
problem at these lower frequencies (< 500 Hz) for the 50 m sec
segment time increment, the spikes extended to higher frequencies
where there are enough waves in the record to state that certain
energetic, narrow-band sinusoids occur intermittently.

An outstanding question when attempting to make the acoustic
measurements reported here is the influence of reverberation and
possible erroneous conclusions about the generative processes. For
example, how much of the low frequency energy reported above is direct
path and how much a reflection from the walls, the bottom, the end of
the tank, and the water surface? To examine this question initially,
a spring-loaded clip-board was snapped in the vicinity of the
measurement location and its acoustic response recorded and analyzed.

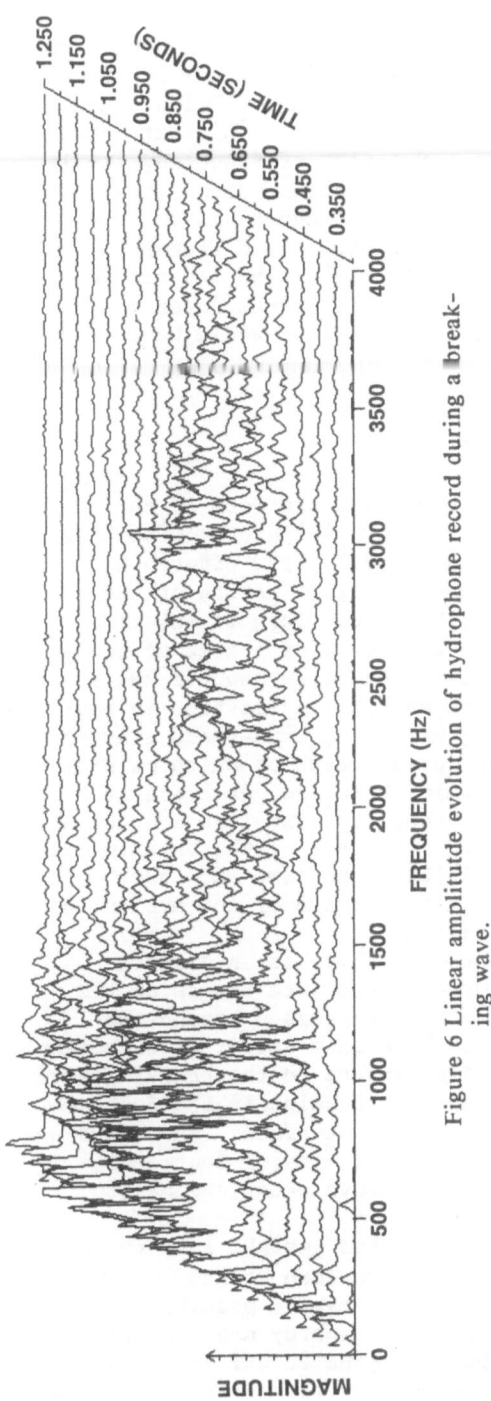

Figure 6 Linear amplitude evolution of hydrophone record during a break-ing wave.

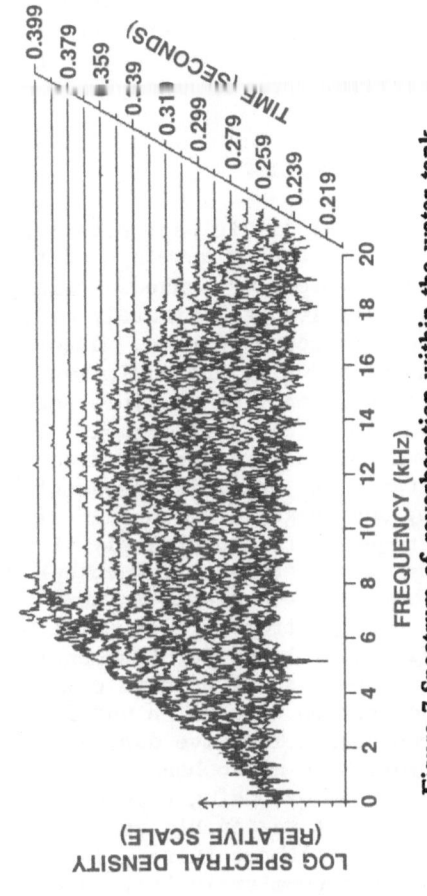

Figure 7 Spectrum of reverberation within the water tank.

A time-evolution, logarithmic spectral density plot is provided
(Figure 7) to demonstrate the relative decay of energy at various
frequencies. While there exists a well-defined edge extending from 20
kHz to about 6 kHz for time from .26 to .4 seconds, there is an
anomalous more rapid decrease from about 1 to 6 kHz. Below 1 kHz
there is a pronounced and continuing reverberation. Without actually
computing the 1/e decoupling time as a function of frequency, it is
felt that reverberation is possibly only a problem below 1 kHz on the
basis that no anomalous peaks appear between 4 and 8 kHz in the
breaking wave spectra as a result of the distinctly different
reverberation at these frequencies. Obviously if the 1/e time was
comparable to the persistence time (comparable to the duration above
the plotting threshold in Figure 7), reverberation would be a major
problem.

4. DISCUSSION

The first significant result of this study was the form of the
acoustic energy envelope of the breaking wave (Figure 2) with time
which resembles a skewed triangle. The orderly form offers the
possibility of ultimately modelling the transient nature of the
oceanic sound sources. Associated with the envelope is the time
development of the audio spectrum. A further useful area of study is
the existence of possible forced oscillation below about 2 kHz and the
possible free oscillations above. The overall implication is that
ambient noise generated by a breaking wave is a time-space average of
transient, primarily decreasing sound events each of which maintains,
by and large, its own self-similar audio structure, at least at higher
frequencies. Accordingly it is conceivable to visualize a model which
has a characteristic envelope radiated from each of the breaking wave
events. These events in turn are distributed in time and space - the
audio spectrum of each event being a mirror image of the ambient noise
field, at least for frequencies above 4 kHz.
    The results provide some evidence that there is an airborne
generation without an underwater source in the late stages of the
run-down. The wave during this event has progressed beyond the
hydrophone-microphone about 1.5 m and there is a carpet of turbulence
and foam extending from about the audio sensors to the breaking wave
crest. A reasonable assumption is that the bubbles in the foam are
popping without introducing sound below the interface. An extension
of this premise is that bubble popping is not a major source of
underwater sound elsewhere in the breaking wave event during its
entire lifetime.
    While this study has provided only limited insight into the sound
generation mechanism during the crash and run-down stages, the spikey
nature of the audio spectrum suggest that it may be possible to isolate
narrow-band micro-structure. According to Pumphrey and Crum (1988) the
bulk of splash noise occurs as a bubble is pinched off from a columnar
jet underneath the impact cavity. Accordingly the controversy between
splash noise and bubble free oscillations is probably best considered

as a problem in determining how the bubble sound is generated – by splashing or by closure or splitting of air cavities. To prove that the discussion is indeed one of the mechanism of bubble oscillation, it is first necessary to prove that the narrow band transients sufficiently resemble bubble free oscillations.

The characteristic form of bubble oscillations is discussed by Crowther (1988) as well as Clay and Medwin (1977). Considerations of the energy exchange of such oscillation leads to the well-known relationship linking the bubble size to frequency. The band-width of such as oscillation is determined primarily by the loss of heat by conductivity during the compression stage of the bubble for frequencies between 2 and 100 kHz. At resonance, the damping constant is a weak function of frequency and depth. One aspect of the characteristic bubble oscillation signature which does not seem to have been represented is the initial buildup stage lasting for 1 to 6 cycles of the ringing, depending on frequency. This initial stage would presumably correspond to the adjustment to sphericity of initial cavities as well as possibly weak finite amplitude effects.

To isolate these characteristic rings within the time record it is useful to construct an inverse filter which, when convolved with the original record, produces a series of spikes where the ringing occurs. The filter is constructed as the impulse response function of a spectrum which is the inverse of the bubble oscillation impulse function, i.e. the characteristic signature. To be effective, the rate of production of bubbles must be such that the average time between generation at a given scale (frequency) exceeds the time resolution of the detector, i.e. inverse filter. This condition can be expressed as

$$\int s_0(r) \, dA < \frac{1}{\beta T(r)} \tag{1}$$

where s_0 is the surface source rate of bubbles (Kerman, 1988), βT is the discrimination time based on the 1/e ring time, T, of a bubble and the integral extends over a discrete surface source, i.e. a white-cap. Initial indications from an application of the filter to this particular wave is that the entrainment (equivalent to the surface source rate) is too large. It is planned to re-run the experiment with a reduced rate of bubbling.

In conclusion, the initial attempt to quantify a breaking wave's audio signature has provided some very interesting and useful insights into the generative process. It remains to systematize the study in terms of wave dynamics and micro-audio structure.

ACKNOWLEDGEMENTS

The author would like to thank Dr. M. Donelan of CCIW for his assistance in arranging for the experiment. As well the technical assistance of Mr. N. Koshyk and Mr. G. Duncan, and the typing help of Mrs. E. Mathis is thankfully acknowledged.

REFERENCES

Banner, M.L. and D.H. Cato, 1988: 'Physical mechanisms of noise gener-
ation by breaking waves - A laboratory study'. Proc. NATO ARW on
Natural Mechanisms of Surface Generated Noise in the Ocean, Lerici,
Italy (this volume).

Clay, C.S. and H. Medwin, 1977: 'Acoustical Oceanography', Wiley-
Interscience, New York.

Crowther, P.A., 1988: 'Bubble noise creation mechanisms'. Proc. NATO
ARW on Natural Mechanisms of Surface Generated Noise in the Ocean,
Lerici, Italy (this volume).

Flynn, H.G., 1964: 'Physical Acoustics', Vol. 1B, (edited by W.P.
Mason), Academic, New York.

Furduev, A.V., 1966: 'Underwater surface cavitation as a source of
noise in the ocean'. Atmos. Oceanic Phys., 2, 314-320.

Kerman, B.R., 1984: 'Underwater sound generation by breaking wind
waves'. J. Acoust. Soc. Am., 75, 149-165.

Kerman, B.R., 1988: 'On the distribution of bubbles near the ocean
surface'. Proc. NATO ARW on Natural Mechanisms of Surface Generated
Noise in the Ocean, Lerici, Italy (this volume).

Knudsen, V.O., R.S. Alford and J.W. Emling, 1948: 'Underwater ambient
noise'. J. Mar. Res., 7, 410-429.

Neumann, G. and W.J. Pierson, 1966: 'Principles of Physical Oceanography', Prentice-Hall.

Pumphrey, H.C. and L.A. Crum, 1987: 'Acoustic emissions associated with
drop impact'. Proc. NATO ARW on Natural Mechanisms of Surface
Generated Noise in the Ocean, Lerici, Italy (this volume).

Wilson, J.H., 1980: 'Low Frequency wind-generated noise produced by the
impact of spray with the ocean's surface'. J. Acoust. Soc. Am.,
68, 952-956.

Worthington, A.M., 1908: 'A Study of Splashes'. Longmans, Green & Co.,
London.

Noise generation by bubbles formed in breaking waves

Reginald D. Hollett and Richard M. Heitmeyer
SACLANT ASW Research Centre
I-19026 La Spezia, Italy

Abstract. A model is presented that describes the high-frequency noise cross-spectrum resulting from the vibrations of bubbles generated in breaking waves. The bubble vibrations are assumed to satisfy a linear model with the excitation occurring at the moment the bubble is formed. In the absence of a clear understanding of precisely how the bubbles are formed, and hence how the vibrations are excited, we postulate two excitation mechanisms. The noise spectra corresponding to each of these excitation mechanisms are evaluated under a simplified propagation assumption using estimates of the physical parameters for a particular wind speed. Although different, both noise spectra have levels comparable to the Wenz spectrum for the corresponding wind speed over the frequency range 0.66 to 10 kHz.

1. Introduction

In his classical paper, Wenz [1] speculated that one possible source of high-frequency, wind-related noise is the vibration of the bubbles located near the ocean surface. Kerman [2] has proposed a model for the noise produced by the bubbles generated in a breaking wave under the assumption that non-linear bubble vibrations are excited by the turbulence generated in the breaking wave. In this paper we present a model for the noise resulting from the linear vibrations of the bubbles generated in a breaking wave, where the vibrations are assumed to be excited at the moment the bubble is formed. A general model is presented for the cross-spectra of the noise produced by individual breaking waves and the noise produced by the aggregate of the breaking waves. We then present simplified expressions for the noise spectra obtained under an idealized propagation assumption. Both the general and the simplified models are applicable for an arbitrary bubble excitation mechanism. To obtain specific results, we postulate that the bubble vibrations are excited either by a sudden change in the pressure

B. R. Kerman (ed.), Sea Surface Sound, 449–461.

on the bubble wall at the moment of formation or by an initial rate of change of the bubble volume. The noise spectra for each of these excitation mechanisms are evaluated over the frequency range 0.66 to 10 kHz using the simplified model and the results are compared to a Wenz spectrum.

2. The general model

The sound produced by the linear vibrations of bubbles is well understood. The pressure waveform is determined from the second time derivative of the volume vibrations which in turn are determined as the solution of a second order linear differential equation with constant coefficients (see Blake [3]). Accordingly, the bubble pressure spectrum $P_b(\omega)$ can be written as the product $H(\omega)F(\omega)$, where $H(\omega)$ is the pressure transfer spectrum and $F(\omega)$ is the pressure excitation spectrum. The pressure transfer spectrum is equal to the bubble pressure spectrum for an impulse excitation of unit strength. This spectrum is given by

$$H(\omega) = -(R/x_0)\omega^2/[-(\omega^2 - \omega_0^2) + i\delta\omega_0\omega], \tag{1}$$

where R is the equilibrium radius of the bubble, $\omega_0 = 2\pi(3.3 \text{ m s}^{-1})/R$, is the natural frequency of the monopole bubble vibrations, δ is the damping constant and x_0 is some reference distance. For the bubble radii of interest here, the pressure transfer spectrum is dominated by a large peak that occurs for $\omega \approx \omega_0$ with a value approximately given by $R/\delta x_0$. The pressure excitation spectrum describes both the applied pressure on the bubble wall and the departure from the equilibrium state at the time at which the bubble is caused to vibrate. This spectrum is given by

$$F(\omega) = (\rho/4\pi R)[i\omega\Delta V(0) + \dot{V}(0)] - \Delta P(\omega), \tag{2}$$

where $\Delta V(0) = V(0) - V_0$, is the initial volume displacement, $V_0 = (4/3)\pi R^3$, is the equilibrium volume of the bubble, $\dot{V}(0)$ is the initial volume rate, and $\Delta P(\omega)$ is the spectrum of the pressure excess acting on the bubble wall. The first term has a significant spectral component at the natural bubble frequency and hence gives rise to high frequency sound. This term will be non-zero whenever a bubble is formed in a non-equilibrium state. The second term will have significant energy at the natural bubble frequency only if the excess pressure waveform changes significantly during a time period of duration $2\pi/\omega_0$. The spectrum of the far-field pressure due to the vibrations of the bubble can be written as the product of the spectrum $4\pi x_0 P_b(\omega)$ and a Green function.

A physical picture for the formation of the bubbles that produce the sound is provided by the description of a spilling breaker by Longuet-Higgins and Turner [4]. According to these authors, the whitecap formed on a spilling breaker can be regarded as a turbulent, air-water mixture that is both accelerated down the forward slope of the wave by gravity and retarded by the entrainment of up-slope momentum from the flow in the wave below. This air-water mixture is lighter than the water in the wave below, owing to the entrainment of air bubbles, and remains distinct from the rest of the wave. Bubbles are generated at the front of the whitecap by the over-running of a layer of air as the front advances and they are generated along the surface of the whitecap by the trapping of air under the turbulent eddies which break out of the surface of the whitecap. We consider only those bubbles that are generated at the whitecap front.

The breaking-wave cross-spectrum is the cross-spectrum of the pressure waveforms generated by a breaking wave observed on two hydrophones. The model for this cross-spectrum is based on two fundamental assumptions. First, neither the sound produced by an individual bubble nor the propagation of that sound to the hydrophones is influenced by the presence of the other bubbles in the breaking wave. Stated in other terms, the bubbles in the breaking wave do not vibrate collectively to produce the sound and the propagation of the sound from an individual bubble is not absorbed or scattered by the other bubbles in the breaking wave. Second, the number of bubbles that are generated in the breaking wave and the positions and times at which those bubbles are generated are described by a Poisson process. Furthermore, the equilibrium radii of those bubbles are statistically independent with a probability density that depends only on the bubble occurrence position.

The setting for the model is illustrated in Fig. 1. The breaking wave occurs at position r and time τ and the sound produced by that wave is observed on hydrophones located at positions z_1 and z_2. The bubbles are generated in a region located at the front of the whitecap that advances with the whitecap front down the leading face of the breaking wave. The total volume swept out by this bubble generation region during the lifetime of the breaking wave is the source region V. The total number of bubbles that are generated in the source region is N_b and the positions and times at which they are generated are $\{(y_m, \nu_m), m = 1, \ldots, N_b\}$, where ν_m is measured relative to the breaking-wave occurrence time τ. By the first assumption, the pressure spectrum observed at each hydrophone from all of the bubbles generated in the breaking wave is the superposition of the received pressure spectra from each of the individual bubbles. Since the pressure radiated by each bubble is described by the linear model, this spectrum can be written in the form

$$P_w(\omega; \mathbf{r}, \mathbf{z}_i) = \sum_{m=1}^{N_b} P_b(\omega; \mathbf{y}_m, R_m) G(\omega, \mathbf{r} + \mathbf{y}_m, \mathbf{z}_i) e^{i\omega\nu_m}, \tag{3}$$

where $P_b(\omega; \mathbf{y}, R)$ is the bubble pressure spectrum for a bubble generated at position $\mathbf{y}$ with equilibrium radius R and $G(\omega, \mathbf{r}+\mathbf{y}, \mathbf{z}_i)/4\pi x_0$ is the Green function describing the propagation from the bubble position $\mathbf{r} + \mathbf{y}$ to the hydrophone positions $\mathbf{z}_i, i = 1, 2$. For fixed values of $\mathbf{y}$, ν, and R, the Green function and the pressure transfer spectrum $H(\omega; \mathbf{y}, R)$ are assumed to be deterministic; the pressure excitation spectrum $F(\omega; \mathbf{y}, R)$ is, in general, stochastic with energy spectrum $\langle |F(\omega; \mathbf{y}, R)|^2 \rangle$, where the average $\langle \cdot \rangle$ is taken for fixed values of $\mathbf{y}$ and R. The

452

breaking-wave cross-spectrum $\langle P_w^*(\omega; \mathbf{r}, \mathbf{z}_1) P_w(\omega; \mathbf{r}, \mathbf{z}_2) \rangle$ represents the mean cross-spectrum over all breaking waves that occur at the position $\mathbf{r}$.

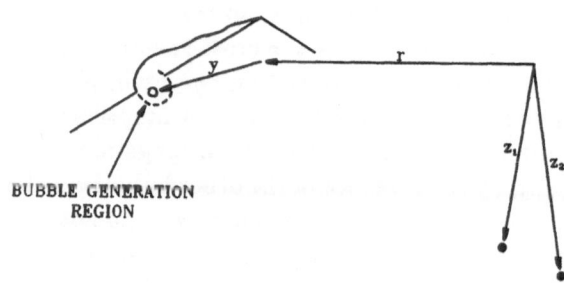

Fig. 1: The model setting.

The Poisson process describes the generation of bubbles by the ensemble of breaking waves that occur at the position $\mathbf{r}$. This process is specified by a bubble generation rate $\mu(\mathbf{y}, \nu)$ that represents the mean number of bubbles generated per unit volume per unit time in the bubble generation region during the breaking-wave lifetime. The mean number generated per unit volume at the point $\mathbf{y}$, $\mu(\mathbf{y})$ is obtained as the integral of $\mu(\mathbf{y}, \nu)$ with respect to the bubble occurrence time ν. The mean number of bubbles generated during the breaking-wave lifetime, M_b, is the integral of $\mu(\mathbf{y})$ over the source region. Finally, the equilibrium radii of the bubbles generated at the position $\mathbf{y}$ are described by the bubble-radius probability density $p(R \mid \mathbf{y})$.

From these assumptions, it is shown in [5] that the breaking-wave cross-spectrum can be written as the sum of two spectra. The first represents the incoherent addition of the sound radiated by the individual bubbles; the second represents the coherent addition of the radiated sound. At the high frequencies of interest here, only the incoherent spectrum contains significant energy. Neglecting the coherent spectrum, the breaking-wave cross-spectrum can be written in the form

$$C_w(\omega; \mathbf{r}, \mathbf{z}_1, \mathbf{z}_2) = M_b S_{bm}(\omega) C_p(\omega; \mathbf{r}, \mathbf{z}_1, \mathbf{z}_2), \tag{4a}$$

where $S_{bm}(\omega)$ is the mean bubble spectrum and $C_p(\omega; \mathbf{r}, \mathbf{z}_1, \mathbf{z}_2)$ is the propagation cross-spectrum. The mean bubble spectrum is given by

$$S_{bm}(\omega) = M_b^{-1} \int_V \mu(\mathbf{y}) S_{br}(\omega; \mathbf{y}) d^3\mathbf{y}, \tag{4b}$$

where

$$S_{br}(\omega; \mathbf{y}) = \int_0^\infty |H(\omega; \mathbf{y}, R)|^2 \langle |F(\omega; \mathbf{y}, R)|^2 \rangle p(R \mid \mathbf{y}) dR, \qquad (4c)$$

is the radius-averaged bubble spectrum and $d^3\mathbf{y}$ is the volume element. The propagation cross-spectrum is given by

$$C_p(\omega; \mathbf{r}, \mathbf{z}_1, \mathbf{z}_2) = \int_V \hat{S}_{br}(\omega; \mathbf{y}) G^*(\omega, \mathbf{r} + \mathbf{y}, \mathbf{z}_1) G(\omega, \mathbf{r} + \mathbf{y}, \mathbf{z}_2) d^3\mathbf{y}, \qquad (4d)$$

where

$$\hat{S}_{br}(\omega; \mathbf{y}) = \mu(\mathbf{y}) S_{br}(\omega; \mathbf{y}) / \int_V \mu(\mathbf{y}) S_{br}(\omega; \mathbf{y}) d^3\mathbf{y}, \qquad (4e)$$

is the normalized radius-averaged spectrum. The derivation of the complete equation for the breaking-wave cross-spectrum, along with the conditions under which the coherent term can be neglected, can be found in [5].

The noise cross-spectrum is defined as the Fourier transform of the cross-correlation function of the pressure due to the aggregate of the breaking waves on the ocean surface. The expression for the noise cross-spectrum is obtained under the assumption that the pressure waveforms received at the hydrophones from different breaking waves are statistically independent and that the occurrence positions and occurrence times of the breaking waves $\{(\mathbf{r}_n, \tau_n)\}$ are described by a Poisson process that is independent of the bubble generation process. The Poisson process is determined by a breaking-wave occurrence rate $\lambda(\mathbf{r})$ that represents the mean number of breaking-wave occurrences per unit area of the ocean surface per unit time at the position $\mathbf{r}$. From this assumption, it is shown in [5] that for the frequency range of interest here, the noise cross-spectrum is well approximated by

$$C(\omega; \mathbf{z}_1, \mathbf{z}_2) = M_b S_{bm}(\omega) \int_{A_s} \lambda(\mathbf{r}) C_p(\omega; \mathbf{r}, \mathbf{z}_1, \mathbf{z}_2) d^2\mathbf{r}, \qquad (5)$$

where $d^2\mathbf{r}$ is the surface area element and A_s is the total area of the ocean surface over which the breaking waves occur. According to (5), the noise cross-spectrum is obtained as the integral of the spectrum $\lambda(\mathbf{r}) M_b S_{bm}(\omega)$ and the propagation cross-spectrum $C_p(\omega; \mathbf{r}, \mathbf{z}_1, \mathbf{z}_2)$ over the surface region A_s. The spectrum $M_b S_{bm}(\omega)$ is the mean energy spectrum from breaking waves that occur at the position $\mathbf{r}$. Consequently, the spectrum $\lambda(\mathbf{r}) M_b S_{bm}(\omega)$ represents a source spectrum per unit surface area from the succession of breaking waves that occur at $\mathbf{r}$ and the product of this spectrum with $C_p(\omega; \mathbf{r}, \mathbf{z}_1, \mathbf{z}_2)$ represents the cross-spectrum observed at $\mathbf{z}_1$ and $\mathbf{z}_2$ for these breaking waves. Thus, the integral of this product over A_s represents the total noise cross-spectrum. The breaking-wave spectrum and the noise spectrum for a single hydrophone located at z are obtained by setting $z_1 = z_2 = z$ in the propagation cross-spectrum of (4) and (5).

3. The simplified model

For a single hydrophone, the shape of both the breaking-wave spectrum and the noise spectrum is determined in part by that of the propagation spectrum $C_p(\omega; \mathbf{r}, \mathbf{z}, \mathbf{z})$. This spectrum describes the propagation of the sound energy from all points in the breaking-wave source region to the position $\mathbf{z}$. A considerable simplification in the noise model results by assuming that the acoustic propagation is approximately that determined for a homogeneous, semi-infinite medium with the source region located below a flat, pressure-release ocean surface.

In the simplified model, the source region has constant depth D and lies directly below a flat ocean surface. The hydrophone is located vertically below the origin of the coordinate system at a depth $z = |\mathbf{z}|$ which is large enough for the hydrophone to be in the far-field region of the breaking waves. The wavenumber vector $\mathbf{k}$ points in the direction of $\mathbf{z} - \mathbf{r}$, at an angle θ relative to the vertical and has magnitude $k = \omega/c$. For these conditions, the Green function is that for a vertical dipole centred at the surface. Using this Green function in (4d) and (4e), the propagation spectrum can be written

$$C_p(\omega; \mathbf{r}, \mathbf{z}, \mathbf{z}) = R(\omega, \theta)T(\omega, |\mathbf{z} - \mathbf{r}|), \tag{6a}$$

where $T(\omega, |\mathbf{z} - \mathbf{r}|)$ is the transmission loss from $\mathbf{r}$ to $\mathbf{z}$, and

$$R(\omega, \theta) = \int_0^D \hat{S}_{br}(\omega; y'')\{2\sin[(\omega/c)\cos(\theta)y'']\}^2 \, dy'', \tag{6b}$$

is the breaking-wave radiation pattern. The normalized radius-averaged spectrum $\hat{S}_{br}(\omega; y'')$ is obtained as the integral of $\hat{S}_{br}(\omega; \mathbf{y})$ over the horizontal dimensions of the source region (y'' is the vertical coordinate in the source region). The breaking-wave spectrum is given by (4) with $C_p(\omega; \mathbf{r}, \mathbf{z}, \mathbf{z})$ given by (6). Assuming that λ is independent of $\mathbf{r}$, the noise spectrum is

$$S(\omega; z) = \lambda M_b S_{bm}(\omega) S_{pi}(\omega; z), \tag{7a}$$

where

$$S_{pi}(\omega; z) = 2\pi \int_0^{\pi/2} R(\omega, \theta) e^{-\alpha(\omega)z/\cos(\theta)} \tan(\theta) d\theta, \tag{7b}$$

is the integrated propagation spectrum and $\alpha(\omega)$ is an absorption coefficient.

In the simplified model, the breaking-wave spectrum is determined as if the energy were radiated from a directional source with a radiation pattern $R(\omega, \theta)$

and an energy source spectrum $M_b S_{bm}(\omega)$. The noise spectrum is determined as if the ocean surface were replaced by a sheet of directional sources with the same radiation pattern $R(\omega, \theta)$ and a source spectrum per unit surface area $\lambda M_b S_{bm}(\omega)$. The radiation pattern is that of a dipole averaged over the depth of the source region with respect to $\hat{S}_{br}(\omega; y'')$. For low frequencies, the dipole radiation pattern is approximately that of an infinitesimal dipole for all depths in the source region. Thus, for low frequencies $R(\omega, \theta)$ has the angular dependence $\cos^2(\theta)$, and both $R(\omega, \theta)$ and the integrated propagation spectrum $S_{pi}(\omega; z)$ have the frequency dependence $[2(\omega/c)D'(\omega)]^2$, where $D'(\omega)$ is the rms depth of the source region with respect to $\hat{S}_{br}(\omega; y'')$. For high frequencies, multiple lobes that occur in the dipole radiation pattern at one depth have maxima in the directions of the nulls that occur in the radiation pattern for another depth. The effect of the depth integration in (6b) is to smooth the angular variation in the pattern over all angles for which the multiple lobes occur. Thus, for high frequencies, the radiation pattern approaches that of a frequency-independent monopole and the frequency dependence of the integrated propagation spectrum is determined only by that of the absorption loss.

There are two problems with the simplified model that result from the idealized propagation assumption used to obtain the model. First, there are propagation effects resulting from the spatial variation in the sound speed and the interaction of the propagating sound with the ocean bottom. These effects are neglected by the assumption of an infinite half-space and a constant sound speed. Second, a problem of potentially greater importance results from the fact that the bubbles that produce the sound are located well within the breaking wave, and over the frequency band considered, the wavelength of the radiated sound from each bubble is comparable to or smaller than the dimensions of the breaking wave. Consequently, one might expect a significantly different radiation pattern than that obtained under the flat surface assumption — especially at the higher frequencies.

4. The excitation postulates

For linear bubble vibrations excited at the moment of formation, high frequency sound will be radiated only if the bubble is formed in a non-equilibrium state or if the bubble is subjected to a sudden change in excess pressure at the moment of formation. We propose two different mechanisms by which the bubble vibrations are excited. The first corresponds to a pure pressure excitation of the bubble. The physical picture underlying this postulate is that the bubbles entrained by overrunning at the front of the breaking wave find themselves instantly underwater and subjected to a sudden pressure change exerted by the weight of the water above them. It is assumed that the bubble is formed in an equilibrium state ($\Delta V(0) = \dot{V}(0) = 0$), and that the excess pressure waveform can be approximated by a step function. The excitation spectrum is then given by

$$F(\omega; \mathbf{y}, R) = -i\rho g d(\mathbf{y})/\omega, \qquad (8a)$$

and the excitation energy spectrum is

$$\langle |F(\omega; \mathbf{y}, R)|^2 \rangle = [\rho g d(\mathbf{y})/\omega]^2, \qquad (8b)$$

where $d(\mathbf{y})$ is the depth of the bubble formed at the position $\mathbf{y}$.

The second postulate corresponds to a pure volume-rate excitation. Here the physical picture is that the bubble is formed by the closure of a cavity and it is the velocity of the surrounding fluid at the moment of closure that causes the bubble to vibrate. It is assumed that the bubble is formed with an equilibrium volume ($\Delta V(0) = 0$), and that there is no sudden increase in the excess pressure at the moment of formation ($\Delta P(\omega) = 0$). To determine the initial volume rate, we use the fact that for the small bubble vibrations for which the linear model is valid, $\dot{V}(0) \approx 4\pi R^2 \dot{R}(0)$, where $\dot{R}(0)$ is the initial radial velocity. Furthermore, by continuity, $\dot{R}(0)$ must be equal to the fluid velocity at the moment of closure, U. Identifying $\dot{R}(0)$ with U and using (2) yields

$$F(\omega; \mathbf{y}, R) = \rho R U. \qquad (9a)$$

To obtain the excitation energy spectrum, we make the additional assumption that the fluid velocity at closure and the equilibrium bubble radius are statistically independent. The spectrum is then given by

$$\langle |F(\omega; \mathbf{y}, R)|^2 \rangle = (\rho R)^2 \langle U^2 \rangle. \qquad (9b)$$

The mean bubble spectrum for either excitation postulate is determined from (4b) with the radius-averaged bubble spectrum determined as an average of the single-bubble spectrum $|H(\omega; \mathbf{y}, R)|^2 \langle |F(\omega; \mathbf{y}, R)|^2 \rangle$ with respect to the bubble-radius density (see (4c)). As might be expected, the two excitation postulates give rise to different mean bubble spectra. This difference is illustrated in the plots of Fig. 2. Both spectra were computed for $\mu(\mathbf{y})$, $H(\omega; \mathbf{y}, R)$ and $p(R \mid \mathbf{y})$ independent of $\mathbf{y}$ over a source region of constant depth $D = 20$ cm and for bubble radii distributed between 0.3 mm and 5 mm with a probability density proportional to R^{-3}. An rms fluid velocity $\langle U^2 \rangle^{1/2}$ of 7 cm s^{-1} was used for the volume-rate excitation. As seen in Fig. 2, both spectra have significant energy in the frequency band 0.66 to 10 kHz, that corresponds to the range of bubble radii used in the computations. For these frequencies, however, the slopes of the two spectra are fundamentally different. The slope for the pressure-jump excitation is about -6 dB/octave over the full band, whereas the slope for the volume-rate excitation increases from about -6 dB/octave at the lower frequencies to about -2 dB/octave at the higher frequencies.

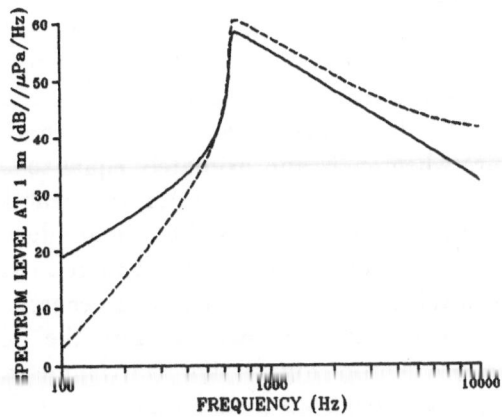

Fig. 2: Mean bubble spectra for 20 cm source region depth: —, pressure-jump excitation; - - -, volume-rate excitation for $\langle U^2 \rangle^{1/2} = 7$ cm s^{-1}.

To explain this, we first compare the single-bubble spectra for the two excitation postulates. The frequency dependence of either of these spectra can be seen through the plots of Fig. 3. These plots show the energy transfer spectrum $|H|^2$ and the excitation energy spectra for the two postulates for a 3.3 mm radius bubble ($f_0 = 1$ kHz), at the rms source depth; the source depth and rms fluid velocity at closure being the same as for the mean bubble spectra of Fig. 2. The frequency dependence of either single-bubble spectrum is seen by adding the corresponding excitation spectrum to the transfer spectrum. For both excitation postulates this results in a single-bubble spectrum with a sharp peak at the natural bubble frequency f_0. However, for the pressure-jump excitation, the single-bubble spectrum approaches zero for frequencies on either side of f_0 due to the −6 dB/octave slope in the excitation spectrum. On the other hand, for the volume-rate excitation, the single-bubble spectrum approaches zero only for frequencies less than f_0 since the excitation spectrum is independent of frequency. The radius dependence of either single-bubble spectrum is determined by that of the energy transfer spectrum and the corresponding excitation energy spectrum ((8b) or (9b)). Here we note that for both excitation postulates, the peak level of the single-bubble spectrum is proportional to $R^4 \delta^{-2}$. To see this, recall that the peak level of the energy transfer spectrum varies as $(R/\delta)^2$. For the volume-rate excitation, $\langle |F(\omega; y, R)|^2 \rangle \propto R^2$, so that the peak level of the single-bubble spectrum is proportional to $R^4 \delta^{-2}$. For the pressure-jump excitation, $\langle |F(\omega; y, R)|^2 \rangle \propto \omega^{-2}$, so that the peak level is proportional to $|H(\omega_0; y, R)|^2 \langle |F(\omega; y, R)|^2 \rangle$ which is proportional to $(R/\delta)^2 \omega_0^{-2}$ or to $R^4 \delta^{-2}$. For natural bubble frequencies less than 10 kHz ($R > 0.33$ mm), δ is only weakly dependent on R (see Strasburg [6]), and

458

thus the peak source level for both excitations is approximately proportional to R^4.

Having made these comparisons, the shapes of the mean bubble spectra for the two excitation postulates are explained as follows. For the pressure-jump excitation, the single-bubble spectrum goes to zero for frequencies on either side of the sharp peak at ω_0, so that for a fixed frequency ω, the major contribution to the integral of (4c) occurs when ω_0 is approximately equal to ω. This occurs for bubble radii $R \approx 2\pi(3.3 \text{ m s}^{-1})/\omega_0$. For these radii, the peak level of the single-bubble spectrum is proportional to R^4 and the bubble-radius density is proportional to R^{-3}. Consequently, the integrand is proportional to R so that the integral is proportional to R^2 or, equivalently, to ω_0^{-2}. Since this occurs only when $\omega \approx \omega_0$, the integral has an ω^{-2} dependence in accordance with the -6 dB/octave slope seen in the mean bubble spectrum of Fig. 2. For the volume-rate excitation, this argument does not apply, even though the peak level of the single-bubble spectrum has the same radius dependence as that of the pressure-jump excitation. This is because the single-bubble spectrum does not go to zero for frequencies $\omega > \omega_0$. As a result, the contributions to the integral of (4c) at a frequency ω from the single-bubble spectra for $\omega_0 \ll \omega$ cannot be neglected. In fact, it is the integration of these contributions that results in the increase in the slope with frequency in the mean bubble spectrum of Fig. 2.

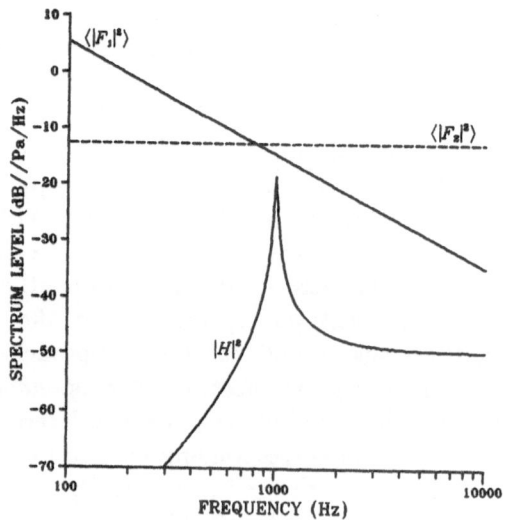

Fig. 3: Single-bubble spectrum components for $R = 3.3$ mm: energy transfer spectrum $|H|^2$; pressure-jump excitation spectrum $\langle|F_1|^2\rangle$; volume-rate excitation spectrum $\langle|F_2|^2\rangle$.

5. Noise spectrum comparisons

The noise spectra for the two excitation postulates were computed using the simplified model for a wind speed of 10 m s^{-1}. The parameters for the mean-bubble spectrum were the same as those of Fig. 2. The breaking-wave occurrence rate λ was inferred from a measured value of the fraction of whitecap coverage w and an assumed mean area A_m and mean lifetime T_m of an individual whitecap event. The fraction of whitecap coverage was taken from the measurements of Ross and Cardone [7], where it is defined as the fraction of the surface covered by 'actively forming whitecaps and large new foam patches'. Using the measured value $w = 0.01$, in the expression $\lambda \approx w/A_m T_m$, with the assumed values $A_m = 10$ m^2 and $T_m = 10$ s, yielded $\lambda = 10^{-4}$ breaking-waves per square metre per second.

The estimates of the mean number of bubbles M_b and the bubble-radius density $p(R)$ were based on a model proposed by Crowther for the generation and distribution of bubbles at the ocean surface [8]. In this model, Crowther uses a dimensional argument to conclude that the surface generation rate $s(R)$ is proportional to $(W/R)^3$, where $s(R)$ is the mean number of bubbles per unit area per unit time of radius R that are generated at the ocean surface, and W is the wind speed. Furthermore, by fitting the bubble density distribution determined by $s(R)$ to backscatter data, he concludes that $s(R) \approx 2.3 \times 10^{-10} W^3 R^{-3}$. To estimate M_b and $p(R)$, we assumed that the only significant source of surface generated bubbles is breaking waves. From the Poisson assumption, it follows that λM_b is the mean number of bubbles generated per unit area of the surface per unit time. Thus, if M_b represents the number of bubbles with radii between R_{min} and R_{max}, then from the definition of $s(R)$,

$$\lambda M_b \approx 2.3 \times 10^{-10} W^3 \int_{R_{min}}^{R_{max}} R^{-3} dR. \tag{10}$$

It follows from (10) and the definition of the bubble-radius density, that $p(R) \propto R^{-3}$ for R between R_{min} and R_{max}. To obtain M_b, we used (10) with $W = 10$ m s^{-1}, $\lambda = 10^{-4}$, $R_{min} = 0.3$ mm, and $R_{max} = 5$ mm. For these values, $M_b = 1.27 \times 10^4$ bubble generations. It is noted that for bubble radii in this range, an R^{-3} bubble density distribution is consistent with that obtained for simulated breaking waves by Monahan and Zietlow [9].

The noise spectra for both excitation postulates are shown in Fig. 4. Also shown is the Wenz spectrum for the 10 m s^{-1} wind speed taken from [1]. Both model spectra show a near constant level at the lower frequencies of the band 0.66 to 10 kHz, followed by a decrease in level at the higher frequencies. At the lower frequencies, the near constant level in both spectra results from the cancellation of the ω^{-2} dependence in the single-bubble spectrum by the ω^2 dependence in the integrated propagation spectrum (see Sect. 3). At the higher frequencies, the slopes of the two model spectra are approximately those of the corresponding mean bubble spectra. This results from the fact that at high frequencies the integrated propagation spectrum is essentially independent of frequency. (For frequencies greater than about 7 kHz, the integrated propagation spectrum decreases with frequency because of the absorption loss, and hence the model noise spectra fall off more rapidly than the corresponding single-bubble spectra.)

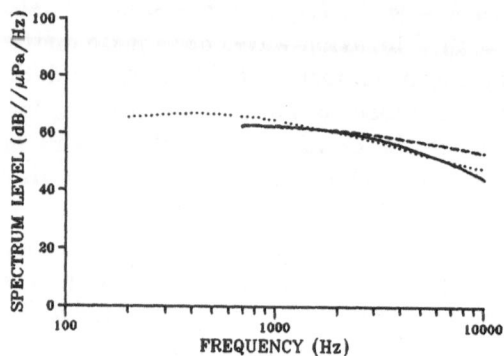

Fig. 4: Noise spectra for 10 m s^{-1} wind speed with $\lambda = 10^{-4}$ breaking waves m^{-2} s^{-1}, $M_b = 1.27 \times 10^4$ bubbles, $p(R) \propto R^{-3}$, R between 0.3 and 5 mm: $\cdots$, Wenz spectrum; —, pressure-jump postulate; - - -, volume-rate postulate.

The noise spectrum for the pressure-jump excitation shows better agreement with the Wenz spectrum than that for the volume-rate excitation at the high frequencies. At these frequencies, the slope of the pressure-jump spectrum is only slightly more negative than the -5 dB/octave slope of the Wenz spectrum, whereas the slope of the volume-rate spectrum is at least 2 dB more positive than that of the Wenz spectrum.

6. Discussion

Both the noise spectra obtained from the simplified model show levels of the same order of magnitude as the Wenz spectrum. Although the values of the physical parameters used to obtain these spectra are physically plausible, there are large uncertainties in these values and hence large uncertainties in the spectra themselves. Nevertheless, the rough agreement between the model spectra and the Wenz spectrum supports the hypothesis that linear vibrations of bubbles generated in breaking waves are a primary source of wind-related noise. The comparison of the high frequency slopes of the model spectra with that of the Wenz spectrum suggests that the pressure-jump excitation is more consistent with measured noise spectra than is the volume-rate excitation. It is important to recall, however, that the idealized propagation assumption used in the simplified model neglects the shape of the breaking wave. If this shape were accounted for, there would be a different frequency dependence in the radiation pattern which could significantly change the shape of the modelled spectra at the higher frequencies.

References

[1] G. M. Wenz, 'Acoustic ambient noise in the ocean: spectra and sources', *J. Acoust. Soc. Am.* **34**, 1936–1956 (1962).

[2] B. R. Kerman, 'Underwater sound generation by breaking wind waves', *J. Acoust. Soc. Am.* **75**, 149–165 (1984).

[3] W. K. Blake, *Mechanics of flow-induced sound and vibration*, vol. 1, 52–53, 371–377, vol. 2, 476–480, Academic Press Inc. (1986).

[4] M. S. Longuet-Higgins and J. S. Turner, 'An entraining plume model of a spilling breaker', *J. Fluid Mech.* **63**, 1–20 (1974).

[5] R. M. Heitmeyer and R. D. Hollett, 'A model for the generation of noise by bubbles formed in breaking waves', SACLANT ASW Research Centre, La Spezia, Italy, Report in preparation.

[6] M. Strasberg, 'Gas bubbles as sources of sound in liquids', *J. Acoust. Soc. Am.* **28**, 20–26 (1956).

[7] D. B. Ross and V. Cardone, 'Observations of oceanic whitecaps and their relation to remote measurements of surface wind speed', *J. Geophys. Res.* **79**, 444–452 (1974).

[8] P. A. Crowther, 'Acoustical scattering from near-surface bubble layers' in *Cavitation and inhomogeneities in underwater acoustics*, ed. W. Lauterborn, 194–204, Springer-Verlag (1980).

[9] E. C. Monahan and C. R. Zietlow, 'Laboratory comparisons of fresh-water and salt-water whitecaps', *J. Geophys. Res.* **74**, 6961–6966 (1969).

ACOUSTIC EMISSIONS ASSOCIATED WITH DROP IMPACTS

H.C. Pumphrey and L.A. Crum
National Center for Physical Acoustics
The University of Mississippi
Oxford, MS 38677
USA

ABSTRACT.   When a liquid drop falling through the air strikes a   flat
horizontal surface there are acoustic emissions associated with this
impact in both the air and the liquid.   Under certain conditions, a
small air bubble can be entrained in the liquid by the impacting drop
and stimulated into volume pulsations that radiate quite strongly.   When
several drops impact the surface in a short time interval, as in
rainfall, the acoustic emissions associated with the drops can be a
major contribution to the total acoustic emission over a considerable
bandwidth.   This paper demonstrates that the major portion of the
acoustic emission associated with rainfall is probably not due to the
direct impact of the drop with the surface but is primarily due to
volume pulsations of gas bubbles entrained in the liquid by the
impacting drop.

1. INTRODUCTION

The   classic   study of the acoustic emissions associated with   drop
impact was performed by Franz [1] in the late fifties and has been   the
standard reference on this subject for the last thirty years.   Franz was
interested in the noise associated with rainfall and concluded that
although gas bubbles are often entrained by drop impacts, their
contribution to the total noise output is small compared to that
associated with the drop impact itself.   Recently Nystuen [2] has
reexamined this problem and concluded that although Franz was correct in
attributing the noise associated with rainfall to be due primarily to
drop impact rather than gas bubble entrainment, there are mechanisms
other than those considered by Franz that influence sound production.
On the other hand we shall present evidence in this paper that the
entrained gas bubble is most likely the principal contributor to the
noise associated with rainfall, and that the noise associated with the
drop impact itself plays only a relatively minor role.

2. EXPERIMENTAL TECHNIQUE

We   have   examined the acoustic signals produced by   individual   as

B. R. Kerman (ed.), Sea Surface Sound, 463–483.
© 1988 by Kluwer Academic Publishers.

well as multiple drop impacts by simple experiments within our laboratory. In our studies of single impacts, we have produced individual droplets of water by gravity-fed hypodermic needles of various sizes. The drops impact the surface of a tank nearly filled with water that has the approximate dimensions of 1m x 1m x 1m. Immersed in the water near the impact site is a Bruel and Kjaer 8103 hydrophone connected to a type 2635 charge amplifier. In the air near the impact site is placed a microphone. The outputs from the microphone and hydrophone are supplied directly to a LeCroy Model 9100 Digital storage oscilloscope. This oscilloscope has many important features which include a 150 Mhz bandwidth, a 32 kbyte internal memory, and a variety of functions that permit signal averaging, fast Fourier transforms and several other mathematical operations.

We have also studied multiple drop impacts by pumping water through a long plastic tube in which are drilled holes of varying sizes. Distrometer measurements of the rainfall drop sizes and distributions demonstrate that our artificial rainfall is quite similar to that produced naturally.

Finally, we have examined both single and multiple drop impacts with high speed cinematography. Utilizing a Photec IV rotating prism camera with frame rates approaching 5000 fps, we can photograph simultaneously the impact of drops on a surface and the output of the immersed hydrophone as displayed on an oscilloscope screen.

3. RESULTS AND DISCUSSION

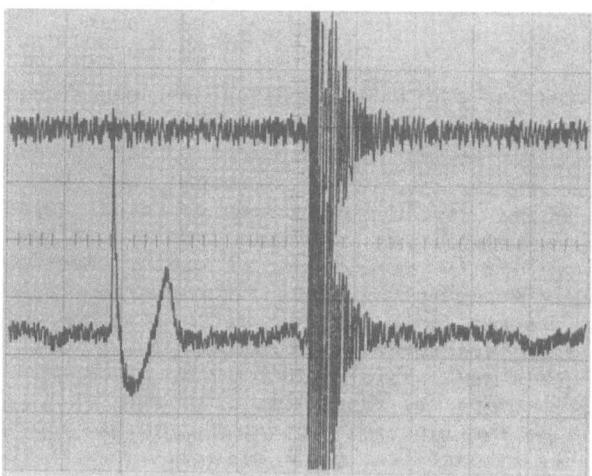

Fig. 1 Sound in air (top trace) and in water, produced by a drop of 3 mm radius falling from a height of 60 cm. Vertical scale, sound pressure level in arbitrary units; horizontal scale, time in 50 ms/div.

Figure 1 shows the microphone (top trace) and hydrophone (bottom trace) outputs when a drop of approximately 3 mm in radius is released from a height of 60 cm. The hydrophone output shows that there is a sharp spike associated with the initial drop impact, a low frequency oscillation associated with hydrodynamic rearrangement of the liquid and a subsequent high amplitude oscillation of a gas bubble that was entrained during the impact. The microphone shows no indications of acoustic emissions during the drop impact and the hydrodynamic rearrangement phase, but a significant acoustic signal is observed in the air due to the pulsating gas bubble. We now examine each of these features individually.

Figure 2 shows an expanded trace of the signal associated with the pulsating gas bubble. Note that the bubble pulsates at a relatively fixed frequency for several cycles before it eventually decays into the noise. The frequency of this bubble oscillation is given by the famous

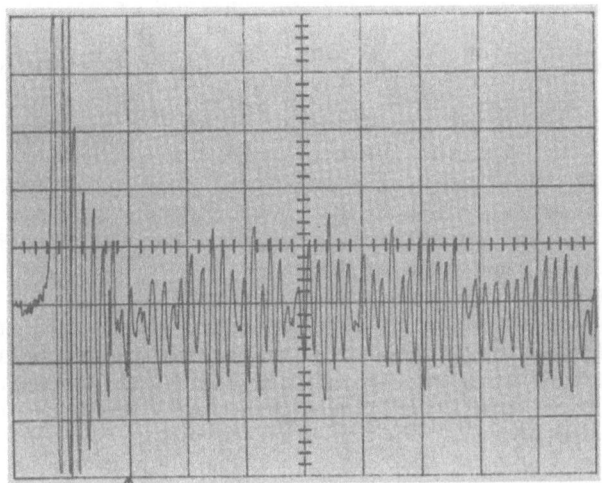

Fig. 2 Sound produced by the oscillations of an entrained air bubble. Vertical scale sound pressure level in arbitrary units; horizontal scale, time in 1 ms/div.

Minnaert [3] formula $f = \frac{1}{2\pi R}\left(\frac{3\gamma P_o}{\rho}\right)^{\frac{1}{2}}$

where R is the bubble radius, Po the ambient pressure, ρ the liquid density, γ = the ratio of specific heats for air. Using an experimental approach similar to the one described here, Pumphrey and Walton [4] were able to demonstrate that the frequencies radiated by entrained gas bubbles associated with drop impact were in close agreement with Eq. (1). In our examination of hundreds of such bubbles, we have observed that although the signals emitted are often complex, whenever simple traces occur with measurable decay times, the damping constant associated with the decay is in agreement with theoretical predictions [5]

It is of interest to examine the bubble production process simultaneously with the acoustic signals emitted from the bubble in order to ascertain how the bubble is forced into oscillation. We have examined this bubble production process with high speed cinematography and these photographs are quite helpful in understanding the phenomenon. Consider Fig. 3 which shows the sequence of bubble production and acoustic emission in detail. For this case, a drop of approximately 1.9 mm radius in size was released from a height of approximately 12 cm. The white vertical line to the right in each figure is the oscilloscope trace of the output of the hydrophone, shown as the circular object in the lower right hand corner of the frame. Note that for this drop release height, little if any acoustic emissions are associated with the impact of the drop with the surface. However, when the air bubble is entrained within the liquid, a significant acoustic emission is observed on the oscilloscope. Since the bubble production is performed so rapidly, it is not possible to determine the exact moment of bubble pulsation at this frame rate of 1000 fps.

A second similar sequence is shown in Fig. 4. For this case, it again appears that no sound is emitted until the bubble is released from the air-filled cavity. These figures appear to demonstrate, however, that the production of a bubble is an acoustically efficient event, with considerably more sound radiation than that associated with the drop impacting the surface.

Let us now consider the effect of the drop impacting the surface and examine the acoustic emissions associated with this event. Recall that in Fig. 1 it is seen that as the drop strikes the surface there is a sharp positive pressure spike followed by a broad low frequency oscillation. This low frequency oscillation falls off rapidly with distance from the impact site and appears to be due principally to liquid forced to flow past the hydrophone by the entering drop. This Bernoulli-type response is only observable when the hydrophone is near the impact site and most probably contributes little to sound radiated to the far field.

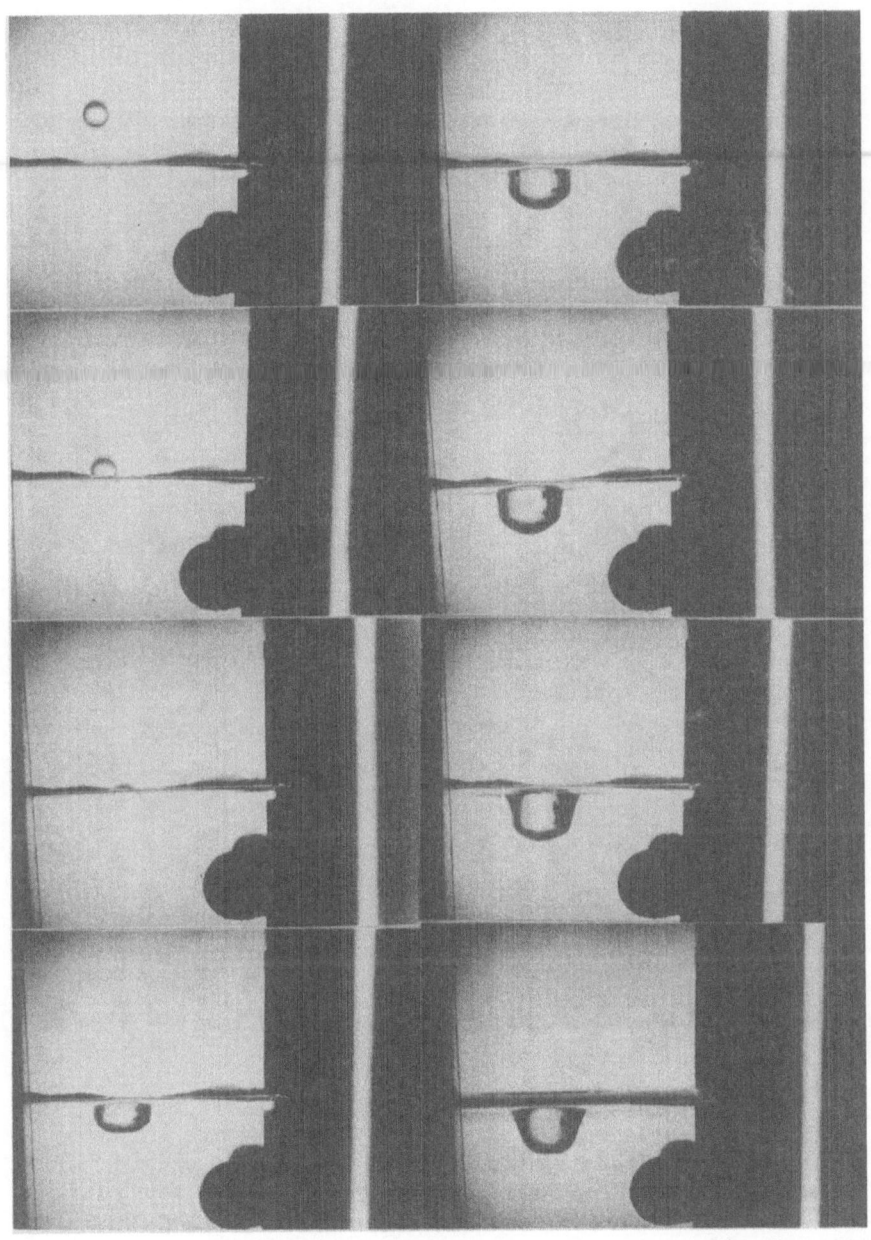

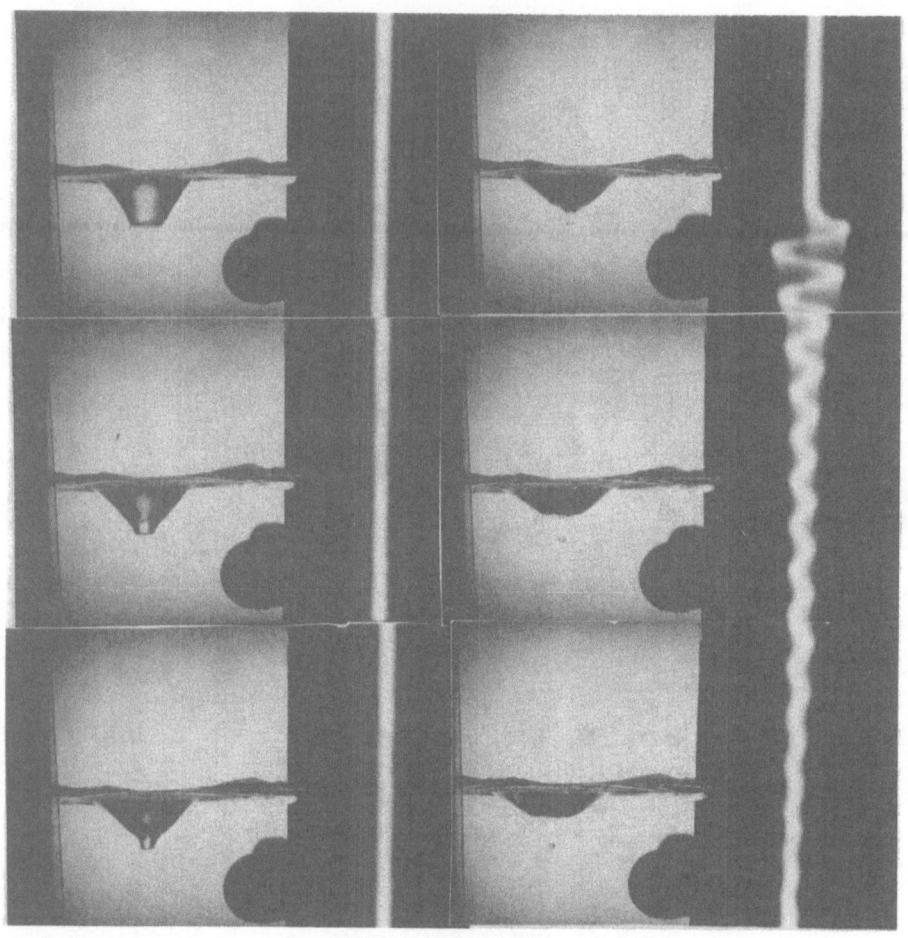

Fig. 3 Frames from a high-speed film, showing impact of 1.9 mm radius drops falling from a height of 12 cm. The oscilloscope trace to the right of each frame provides a record, simultaneous with the picture, of the sound being produced. Frame rate, 1000 fps; the frames are sequential but not necessarily consecutive. Read from top to bottom.

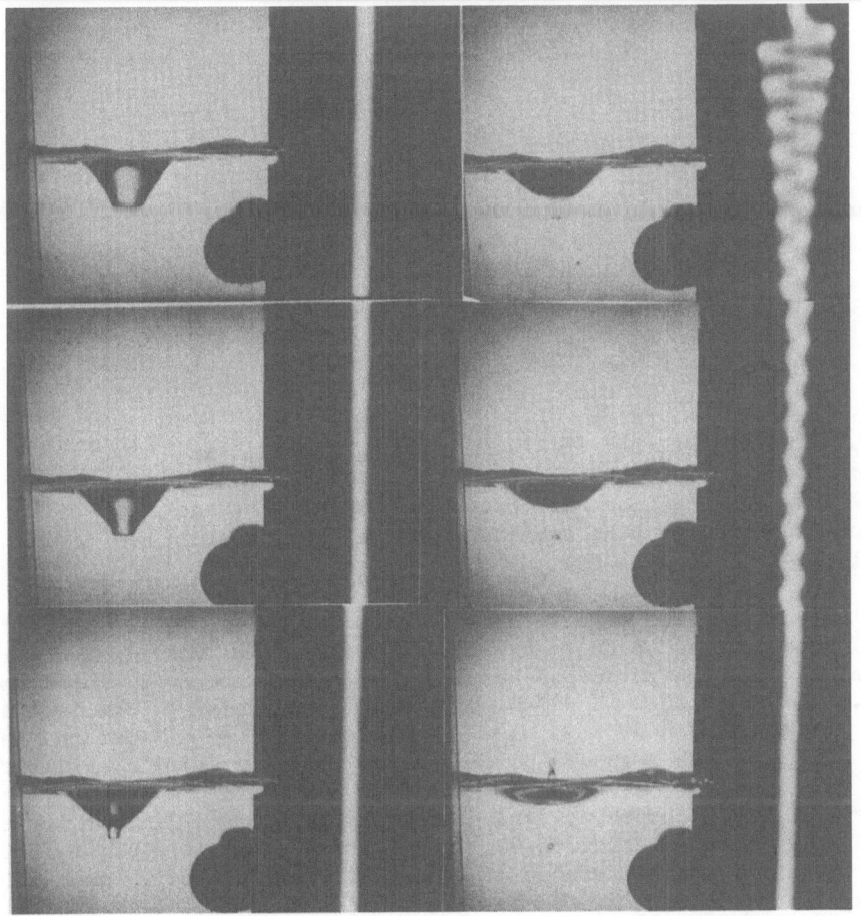

Fig. 4   Frames from a high-speed film, showing
impact of 1.9 mm radius drops falling from a   height
of 12   cm.   The oscilloscope trace to the right   of
each frame provides a record, simultaneous with the
picture, of the sound being produced.   Frame rate,
1000   fps;   the frames are sequential but   not
necessarily consecutive. Read from top to bottom.

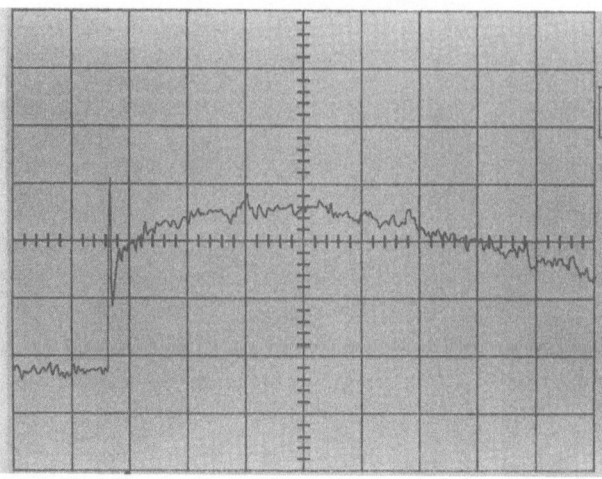

Fig. 5 Expanded view of Fig. 1 showing drop falling
from 60 cm. Note the sharp 'spike' followed by the
slowly varying hydrodynamic pressure. Vertical scale
sound pressure level in arbitrary units; horizontal
scale, time in 0.5 ms/div.

The pressure spike associated with drop impact is of considerable
interest however, as both Franz [1] and Nystuen [2] considered this
component to be the principal contributor to the sound produced by
rainfall. Consider Fig. 5 which shows a 100x expansion of the initial
pressure spike region of Fig. 1. Even with this magnification the
leading edge of the pressure pulse appears to be a simple pressure
spike.

We examine this component more closely by studying a similar impact
with the oscilloscope digitizing 50 times faster. This is shown in Fig.
6, on a frequency scale expanded 10 times from that of Fig. 5. This
final expansion shows that the original impact spike is actually
composed of a reasonably well-defined oscillation on the order of 100
kHz. We have examined hundreds of drop impacts and the vast majority of
them have this high frequency oscillation as the principal component of
the pressure spike. Thus, we see that the drop impact itself (sometimes
called the flow-establishment phase) is composed of a low frequency
nonradiating component and a high frequency strongly radiating
component.

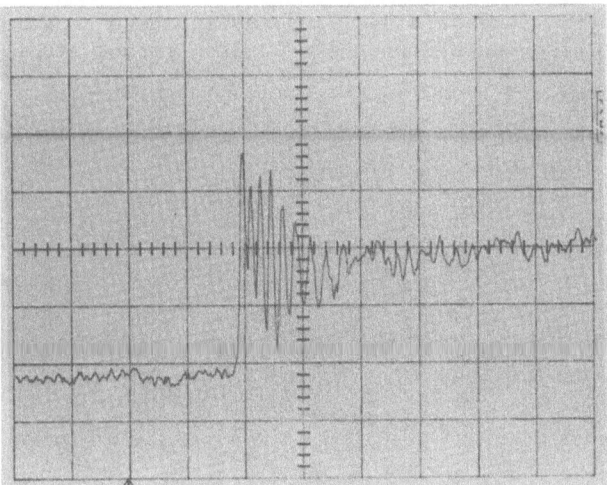

Fig. 6 The same process as Fig. 5 on an expanded timescale. Note the high frequency oscillations following impact. Vertical scale sound pressure level in units; horizontal scale, time in 0.05 ms/div.

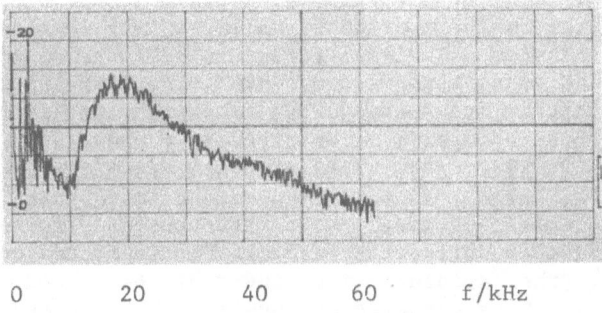

Fig. 7 Acoustic power spectrum of artificial rainfall. The 'spikes' at the low frequency end are resonances of the tank. Vertical scale, sound intensity in with arbitrary reference; horizontal scale, frequency with 10 kHz/div.

472

It is now in order to examine the acoustic emissions associated with the multiple drop impacts of rainfall to see if these emissions can be related to what has been learned from single drop impact studies.

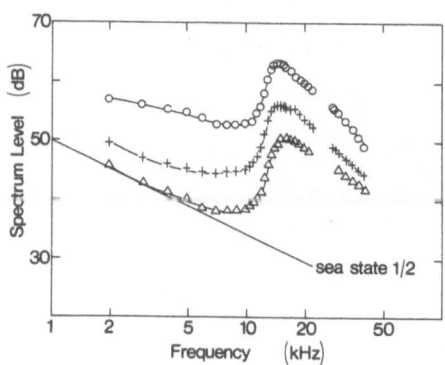

Fig. 8 Spectrum of natural rainfall (after Scrimger et al. [6]). The three curves are for different rain rates. The straight line is the Knudsen curve for Sea State 1/2

Using a simple system wherein we pump water through a plastic tube containing small holes, we have produced multiple drop impacts on the surface of our tank. Again using the LeCroy oscilloscope we can store the acoustic emissions associated with a period of artificial rainfall, take its FFT power spectrum and then average many such spectra to obtain a power spectrum of the rainfall. This result is shown in Fig. 7. Shown for comparison in Fig. 8 is a similar noise spectrum obtained by Scrimger et al. [6] of actual rainfall measurements made in a freshwater lake in British Columbia, Canada. Note that the data in Fig. 7 is plotted on a linear scale while that in Fig. 8 is plotted on a log scale. Furthermore, the absolute magnitudes of the sound pressure levels obtained should not be compared as they are taken under different conditions. Finally, the measurements taken by us in Fig. 7 were made in a reverberant tank and show spectral lines below 8 kHz that have been identified as tank resonances. Ignoring the differences, however, it is seen that the shape of the artificial rainfall spectrum that we obtained is remarkably similar to the natural rainfall one obtained by Scrimger et al. [6]. This good comparison permits us to study "rainfall" in our laboratory.

If one analyses the frequencies associated with the initial drop impact – on the order of 100 kHz and above – and those associated with

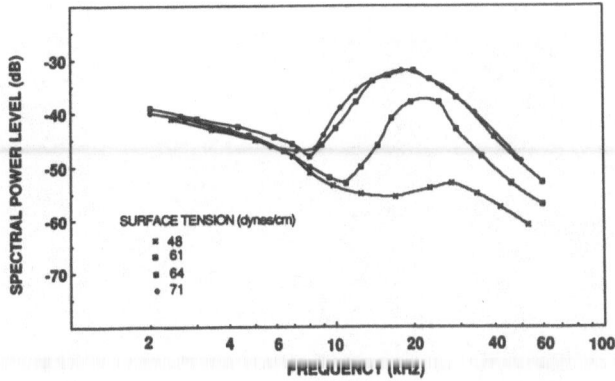

Fig. 9 Acoustic power spectra of artificial rain at different surface tensions.

air bubble oscillation - on the order of 10 kHz - it seems reasonable to assume that the broad maximum near 15 kHz observed by Scrimger et al.[6] is primarily due to bubble oscillations rather than "flow establishment" due to drop impact. A simple experiment can be performed in the laboratory that should critically test which of these acoustic sources is the principal contributor to rainfall noise. If a surface tension reduction agent is added to the water, then the drop impact dynamics should be the same while the gas bubble entrainment - which depends strongly on the surface tension - should be quite different.

Figure 9 shows the results of adding small amounts (parts per million) of a concentrated surfactant, Photoflo, to the liquid. The noise spectrum is drastically altered in a systematic way. The systematic change in the noise spectrum of artificial rainfall with surface tension reduction shows rather conclusively that gas bubbles are intimately involved in the noise production mechanism.

It is helpful to next examine the physical mechanism whereby a reduction in surface tension greatly affects the bubble entrainment process. We can utilize high speed cinematography to assist us in this task. Consider Fig. 10, which corresponds to conditions similar to those in Figs. 3 and 4, except that now we have added sufficient Photoflo to reduce the surface tension from 72 dyn/cm in Figs. 3 and 4 to 30 dyn/cm for this case. Note that the sequence of cavity formation is nearly identical to that seen earlier, except for the

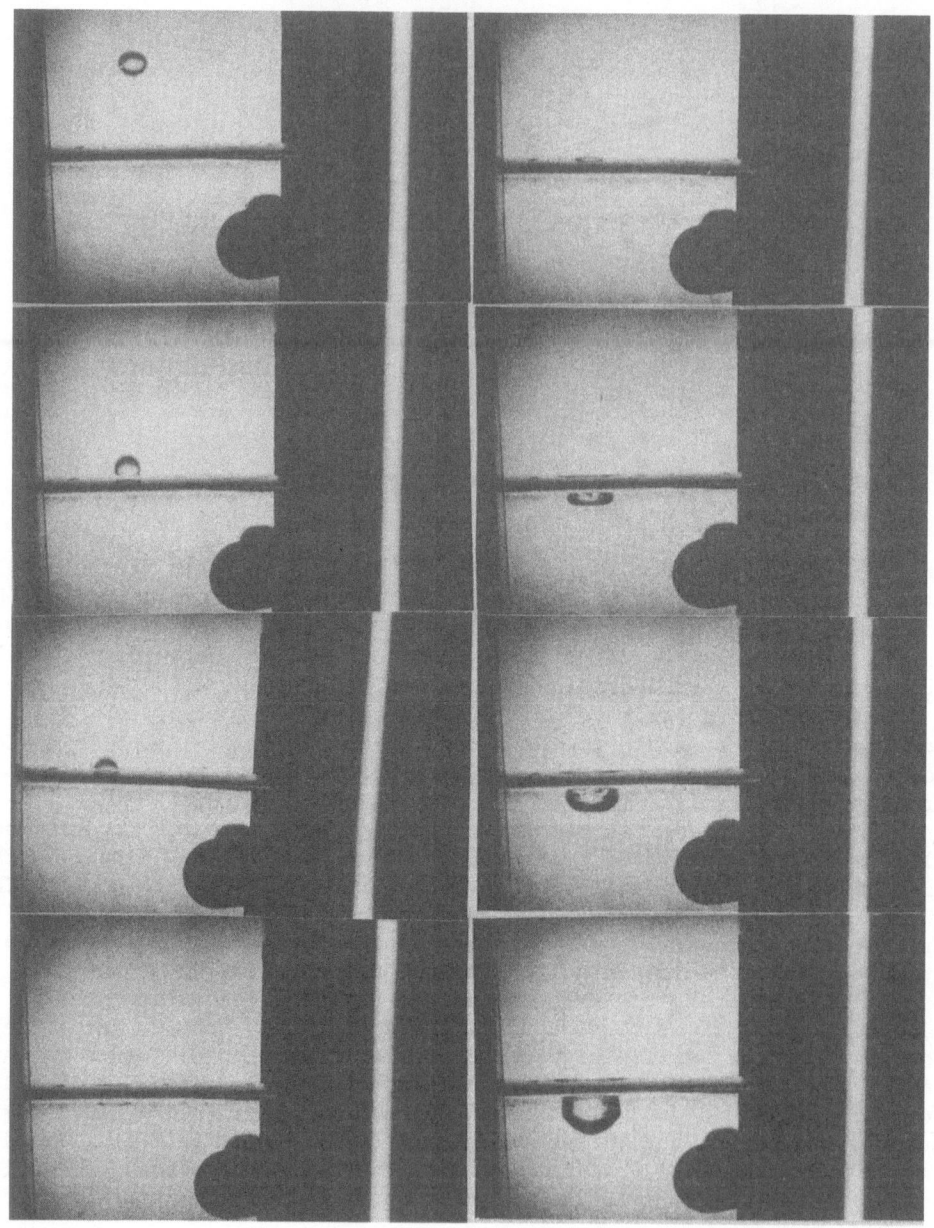

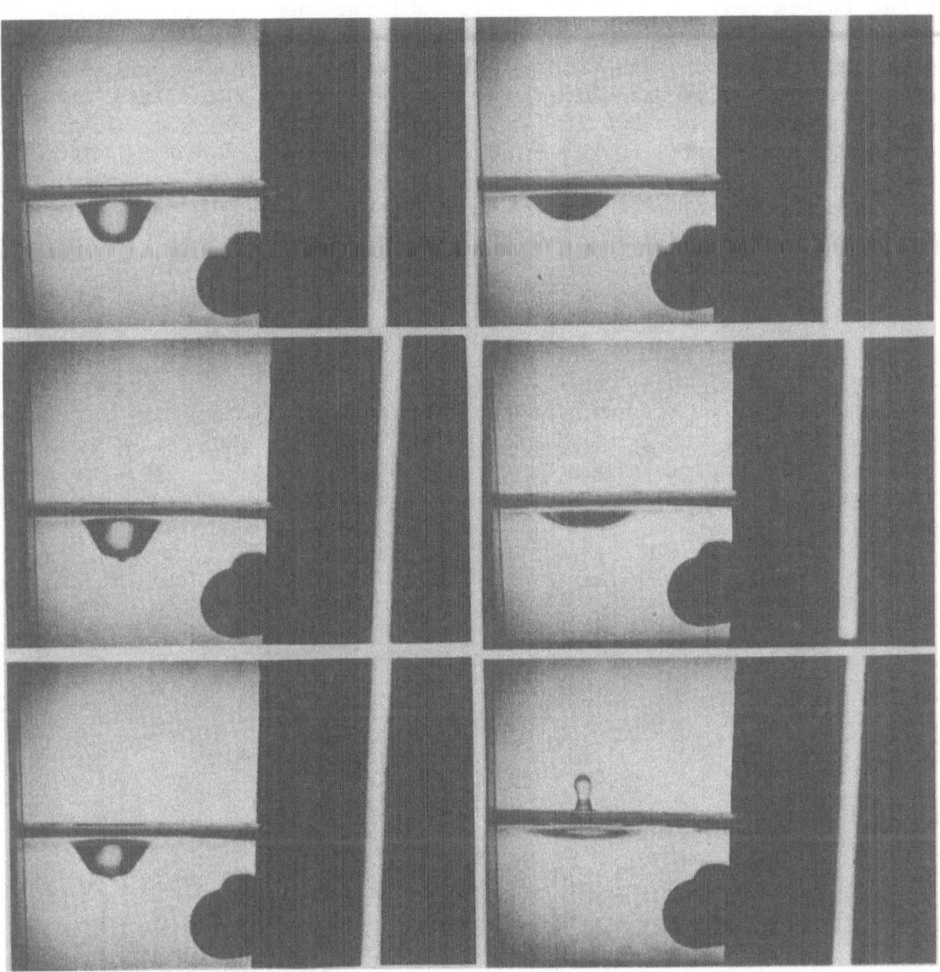

Fig. 10 High speed film of a situation identical to that of Fig. 3, except that the surface tension has been lowered from about 70 dyn/cm to 30 dyn/cm Frame rate 1000 fps. The frames are sequential but not necessarily consecutive.

critical stages where gas bubble entrainment is involved. In the high surface tension case, Figs. 3 and 4, the bottom of the cavity has sharp contours and is practically flat. In the low surface tension case, Fig. 10, the bottom of the cavity stays well-rounded throughout its sequence. In a simple argument, one would say that in Figs. 3 and 4 the surface tension forces are nearly vertical, pulling the cavity upwards so rapidly that gas is left behind. In Fig. 10, the surface tension forces remain primarily horizontal and can't accelerate the fluid rapidly enough to entrain a gas bubble.

An important aspect of Franz's argument that bubbles were not involved in sound production during rainfall was that only small drops induced bubble formation and that rough surfaces due to high rainfall rates discouraged bubble entrainment. We have examined drop impacts for large sizes and high terminal velocities for both high and low surface tensions. High speed cinematographic sequences for the high surface tension case is shown in Fig. 11; the low surface tension case is shown in Fig. 12. It is seen that when the drop impact is rather violent, as in Fig. 11, the pressure spike is now observable at the oscilloscope scale setting used previously. Note also that one can see the beginnings of the low frequency Benoulli-type oscillation. Attention is also called to the complex behavior of the splash - the famous crown - and the multiple surface waves in the air-filled cavity. Finally, when a bubble is produced, it is very small and has a high frequency of oscillation.

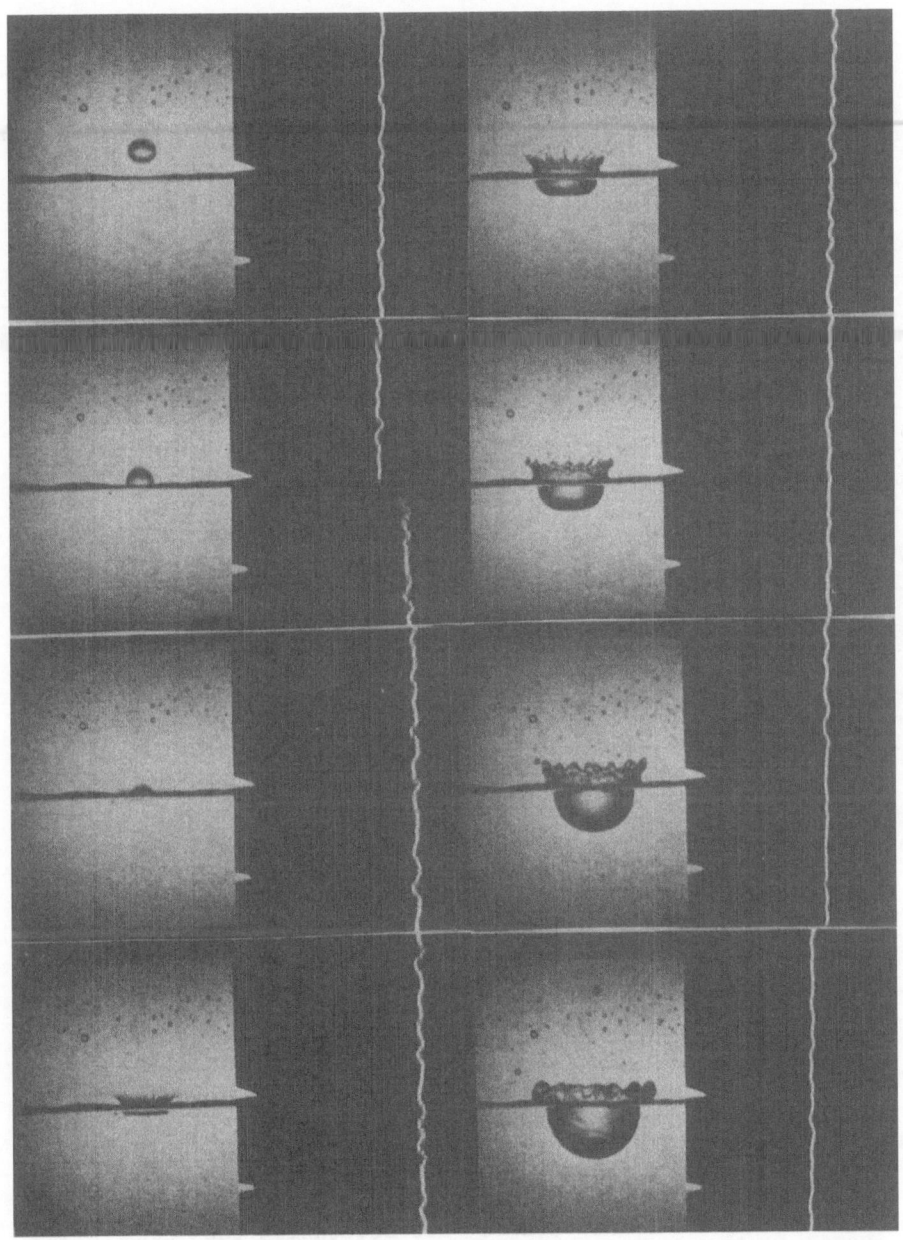

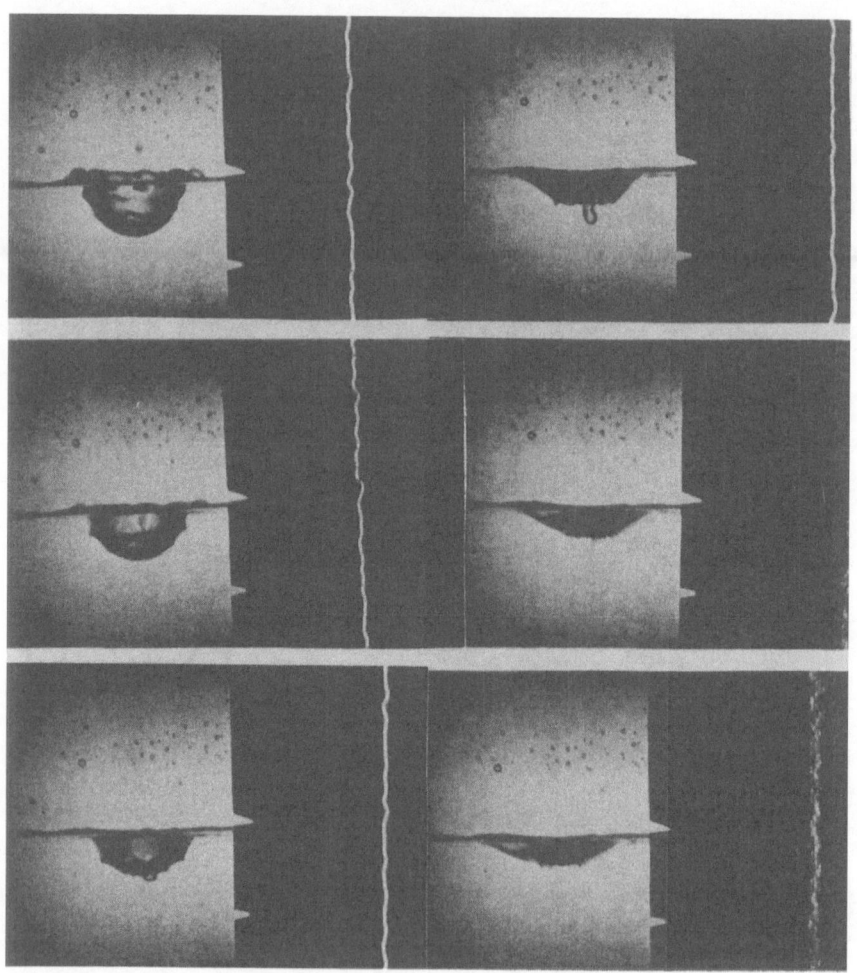

Fig. 11   Frames from a high speed film of a 3mm drop falling from 30 cm.   The film was taken at 1000 frames/second.   The frames are sequential but not necessarily consecutive.

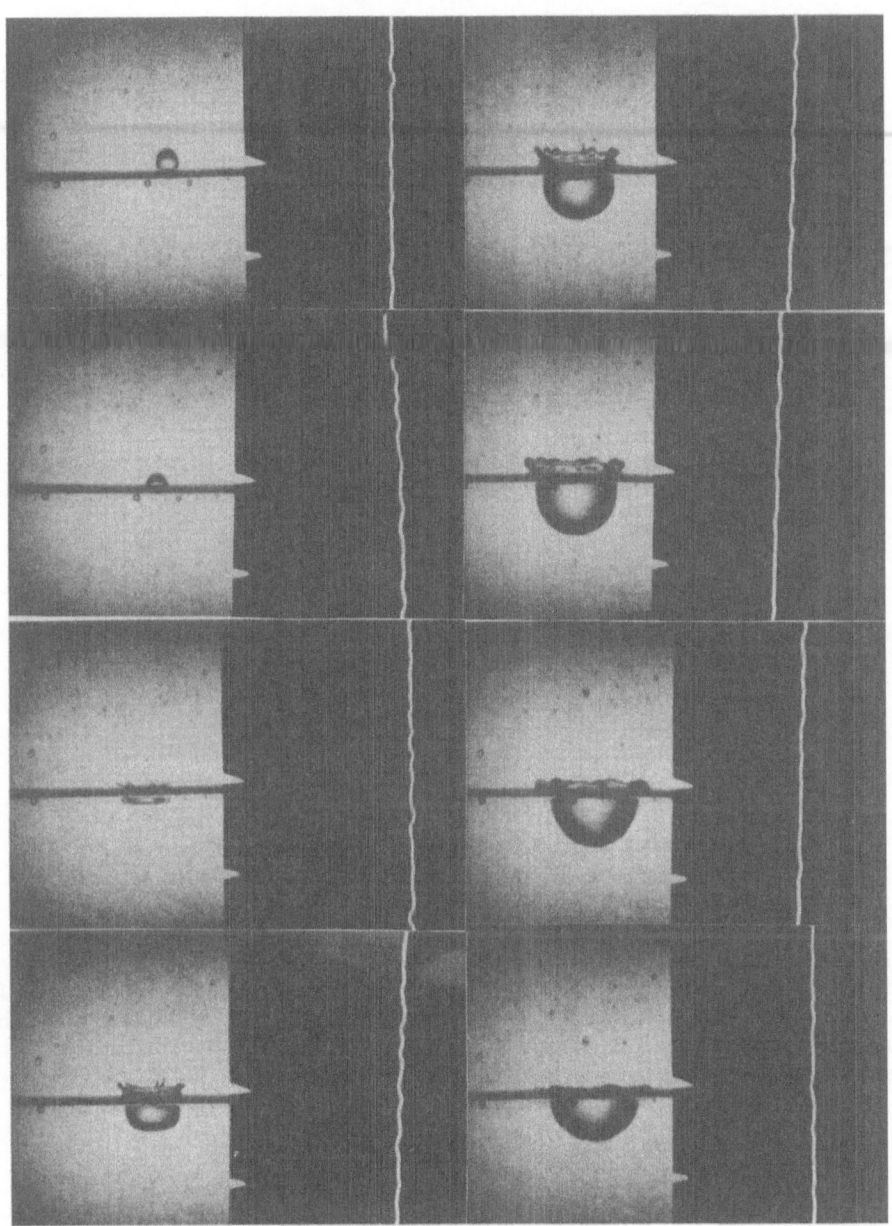

480

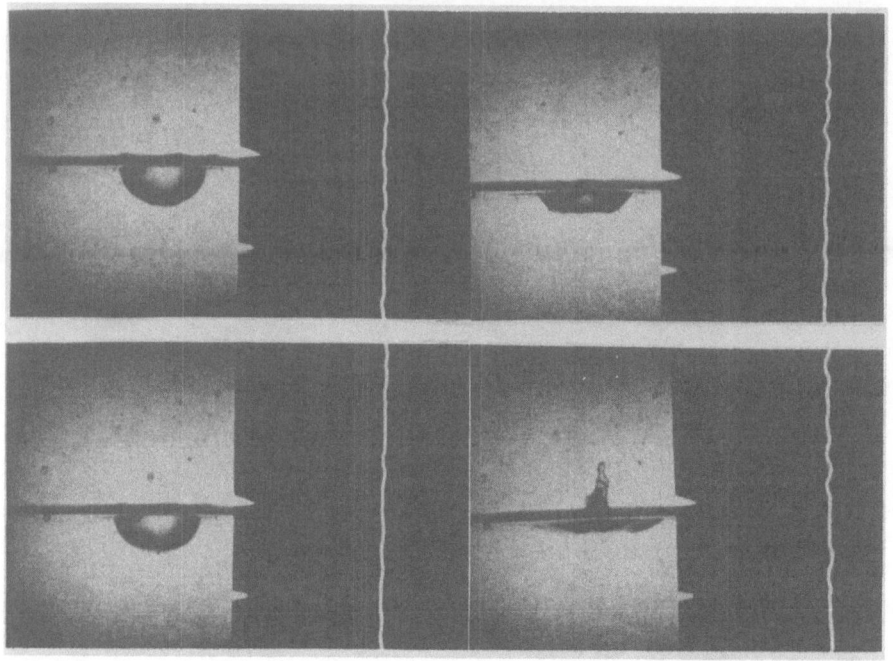

Fig. 12 High speed film of a situation identical to
Fig. 11 except that the surcface tension has been
reduced from 70 to 30 dyn/cm.

Contrast this sequence with that of Fig. 12 in which the surface
tension has been reduced to 30 dyn/cm. In this case the crown hardly
forms, there are few, if any, surface waves in the air-filled cavity and
no bubble is produced. Clearly there is a large difference between the
two sequences and a strong dependence on the surface tension.

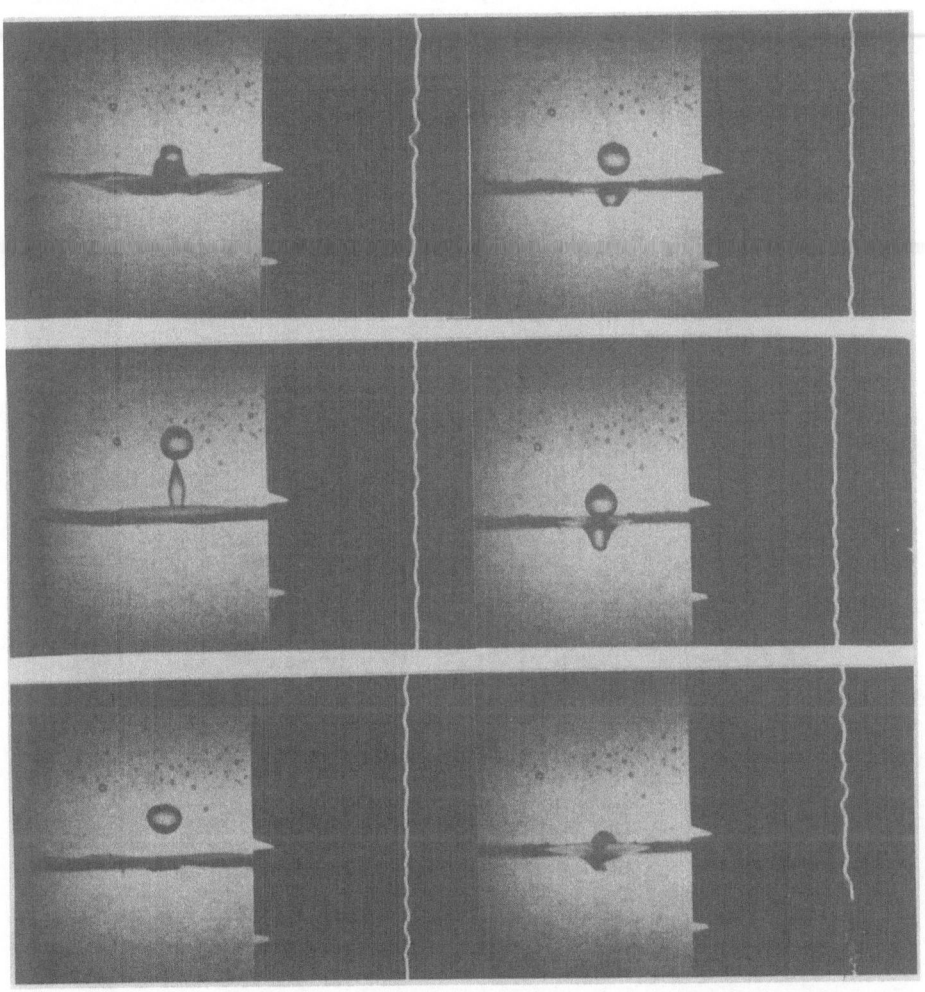

Fig. 13   This figure is a continuation of Fig. 11.
The drop seen is one thrown up by the impact seen in
Fig. 11.

482

When drops impact the surface from a large height, the resulting collapsing air cavity often ejects drops back into the air. This ejection process normally results in a smaller drop that has a better chance of entraining a gas bubble. We have captured this event on film and such a sequence is shown in Fig. 13. Note that when the falling drop strikes the air cavity formed from the recoil of the jet, there is sound produced. Obviously, the surface oscillations during rainfall can be very complex and not merely a combination of single drop impacts.

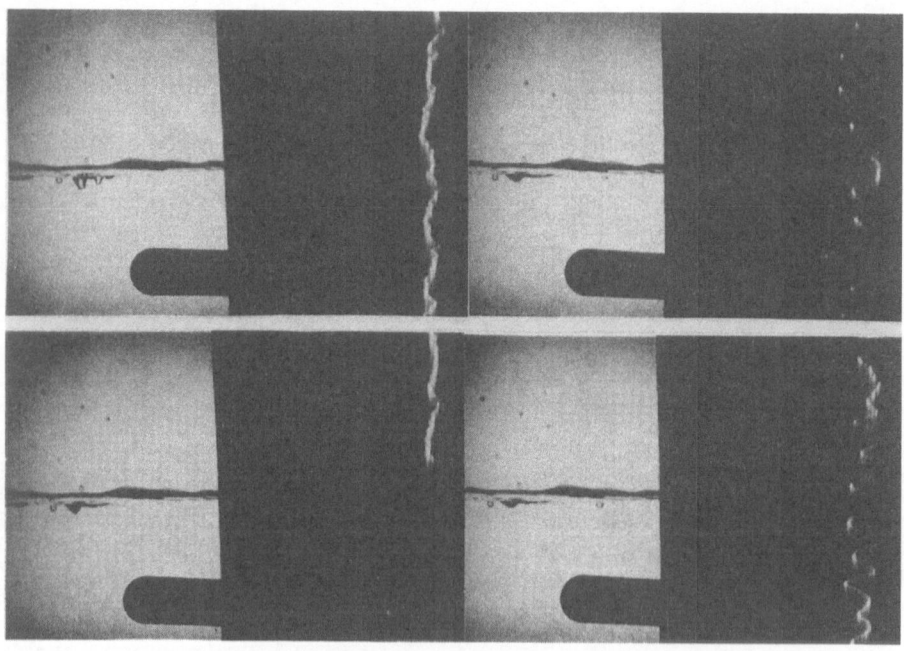

Fig. 14   High speed film of a shower of small drops striking a water surface. These are consecutive frames, taken at 3000 frames/sec.

Finally, we show in Fig. 14 a sequence of frames that show the radiated sound during artificial rainfall. Note that the sound, when produced, consists of bubble-like oscillations and in many cases these can be associated with the production of a bubble. While only a fraction of drops produce bubbles, these still account for most of the sound emitted in the 8-40 kHz region.

4. CONCLUSIONS

Although there are acoustic emissions associated with the collision of a drop with a liquid surface, the frequency of these emissions are in the hundreds of kilohertz bandwidth and thus probably not a principal contributor to the commonly observed spectrum of rainfall. Measurements of the noise spectrum of artificial rainfall demonstrate a strong dependence on the surface tension, indicating that the volume pulsations of gas bubbles play a major role in the acoustic emissions associated with rainfall. High speed cinematography demonstrates that reduction of the surface tension tends to prevent air bubble entrainment and thus considerably reduces the acoustic energy radiated.

5. ACKNOWLEDGEMENT

We gratefully acknowledge many helpful discussions with Andrea Prosperetti and the financial support of the Office of Naval Research.

6. REFERENCES

[1]. G.J. Franz, 'Splashes as sources of sound in liquids', J. Acoust. Soc. Am. 31, 1080-1096 (1959).

[2]. J.A. Nystuen, 'Rainfall measurements using underwater ambient noise', J. Acoust. Soc. Am, 79, 972-982 (1986).

[3]. M. Minnaert, 'On musical air bubbles and the sounds of running water', Phil. Mag. 16, 235-248 (1933).

[4]. H.C. Pumphrey and A.J. Walton, 'Sounds produced by water-drop water-surface impacts', Eur. J. Phys. (submitted for publication).

[5]. C. Devin, Jr., 'Survey of thermal, radiation and viscous damping of pulsating air bubbles in water', J. Acoust. Soc. Am. 31, 1654-1667 (1959).

[6]. J.A. Scrimger, D.J. Evans, G.A. McBean, D.M. Farmer, and B.R. Kerman, 'Underwater noise due to rain, hail, and snow', J. Acoust. Soc. Am. 81, 79-81 (1987).

THE SOUND GENERATED BY PRECIPITATION STRIKING THE OCEAN SURFACE

J. A. Nystuen
Department of Oceanography
Naval Postgraduate School
Monterey, CA 93943

D. M. Farmer
Institute of Ocean Sciences
Sidney, B.C. V8L 4B2
Canada

ABSTRACT. Precipitation striking the ocean surface produces sound underwater. The mechanisms by which raindrops produce sound underwater are the initial impact and entrained bubbles. The influences that wind and waves have on these mechanisms are examined. The implications that these results have on the possibility of using the level of sound generated by rainfall as a means of measuring rainfall in oceanic regions is discussed.

1.   INTRODUCTION

As raindrop formation represents latent heat release into the atmosphere, it is an important driving force for atmosphere and ocean circulation and ultimately affects local, regional and global climate. For this reason accurate measurements of the distribution and intensity of rainfall are needed on all spatial scales. Unfortunately this information is generally unavailable, especially in oceanic regions. At sea traditional methods for rainfall measurements, e.g. rain gauges, are generally unavailable (shipboard rain gauges are known to be unreliable). Satellite techniques are proposed and even in use, but suffer from the lack of adequate surface validation measurements. New surface measurement techniques are needed. One potential technique is the use of underwater ambient noise to monitor rainfall rate. Precipitation, and in particular rain, makes a lot of underwater noise as it strikes the ocean surface. If the sound generated can be related to rainfall rate then this method would be a very attractive way to measure rainfall. The principle sensor, a hydrophone, is relatively inexpensive and durable and could be deployed in a wide variety of locations. Since the deployment is below the surface there are no platform problems (the main problem with ship measurements). Finally the depth of the hydrophone determines the surface area monitored and thus there is a natural spatial averaging which is very important for rainfall measurements as rain is spatially inhomogenous (on all scales).

B. R. Kerman (ed.), Sea Surface Sound, 485–499.
© *1988 by Kluwer Academic Publishers.*

There are many underwater sound sources. For this method to work, rain generated sound must be distinguishable from other sound sources. Fortunately rain generated sound does have a distinct spectral shape and furthermore there appears to be a quantifiable relationship between rainfall rate and spectral energy level (Nystuen, 1986). If rainfall is to be inferred from underwater ambient sound then the physics of sound generation by rain must be well understood. In addition to the question of how a raindrop makes sound underwater (Franz, 1959; Crowther, 1981; Nystuen, 1986, 1987), it is necessary to consider the influence of wind (Nystuen and Farmer, 1987) and the influence that the drop size distribution in the rain has on the underwater sound generated.

2. THE SOUND FROM RAINDROPS

A drop splash produces underwater sound through two different mechanisms (Franz, 1959). The first is the initial impact of the drop onto the surface and the second mechanism is bubble entrainment when it occurs. The observed pressure series from such a splash is shown in Figure 1. The impact pulse is a single cycle, the duration of which

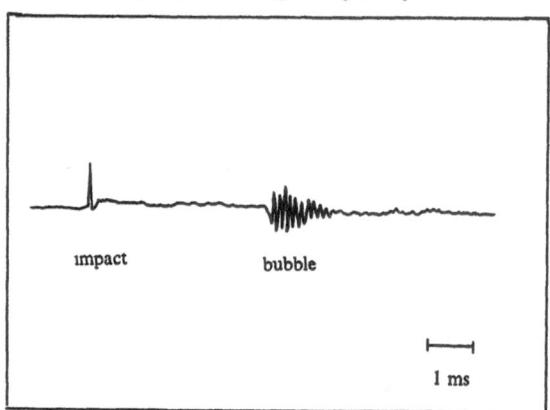

Figure 1. The acoustic signature of a drop splash consists of two parts. The first part is a single cycle pulse from the initial impact of the drop. The second part is a resonant ring from an entrained bubble.

depends on drop shape, size and velocity of impact. The bubble sound is a resonant ring of the newly entrained bubble, the frequency of which is given by

$$f_r = (1/2\pi a) \ (3\gamma P_a/\rho)^{1/2} \tag{1}$$

where a is the bubble radius, P_a is the ambient pressure, ρ is the water density and γ is the ratio of specific heats ($\gamma = 1.4$ in water). While it is apparent for this example that the bubble produced as much or more acoustic energy than the impact, the theories for explaining the sound from rain (Franz 1959; Nystuen 1986, 1987) have concentrated on the sound produced by the impacts alone.

    This decision is based on several different observations. Franz
noted that every drop produces a consistent sound from the impact
mechanism but only an occasional drop splash entrains a bubble.
Furthermore the bubble size varies widely and thus the frequency at
which sound is generated varies widely (Eq. 1). He argued that for an
ensemble of drop splashes (rain) the impact mechanism would dominate
since each drop impact contributed to the sound spectrum in a uniform
manner while the bubble mechanism is sporadic in occurance and spread
over a wide frequency range. Nystuen (1986) accepted this argument,
further noting that the dominant feature of the rain generated sound
spectrum, the broad spectral peak near 15 kHz, is present when the
rainfall rate is very light. Figure 2 presents several ambient sound
spectra recorded during different rain conditions. The spectrum labeled

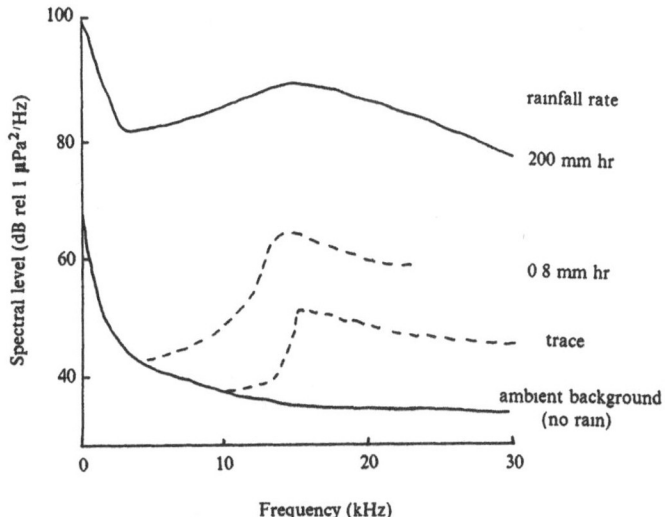

Figure 2. Three underwater sound spectra observed during different rain
conditions. The 200 mm/hr sound spectrum was recorded at Clinton Lake,
IL in Oct 1982. This convective rain contained many large drops (up to
5 mm in diameter). The spectral peak is present but it is more apparent
for light rain (0.8 mm/hr) recorded at Cowichan Lake, BC in Mar 1985.
Few if any raindrops larger than 1.5 mm in diameter were present during
this rain. The very light rain example (trace) was observed in an
outdoor tank at the Scripps Institution of Oceanography (Mar 1984). In
this case no large drops were present and yet the spectral peak is
present. The Clinton Lake and SIO data was collected by Nystuen (1986)
and the Cowichan Lake data was collected by Scrimger (1985; 1987).

"trace" was recorded when the rainfall rate was so light the raingages
couldn't measure it and the splashes on the surface were detected by
outgoing capillary waves only. In this situation, no bubbles were being
trapped underwater, the rain was too gentle, and yet the spectral peak
was there. The acoustic physics for the spectral peak needs to be
explained using the impact mechanism alone.

3. A NUMERICAL SIMULATION OF A DROP IMPACT

A finite difference primitive equation numerical code, the SOLA-VOF code, (Nichols et al. 1980) is used to simulate a drop impact. This code permits multiple free surfaces and includes the influence of surface tension, viscosity and sound speed. Figure 3 shows the shape of the free surface during a calculation where a 3.0 mm drop with impact velocity 8.0 m/s has been placed in contact with a quiescent pool of water. The drop has been flattened to mimic the shape of a real 3.0 mm drop. Raindrops less than 1.2 mm in diameter are nearly spherical but

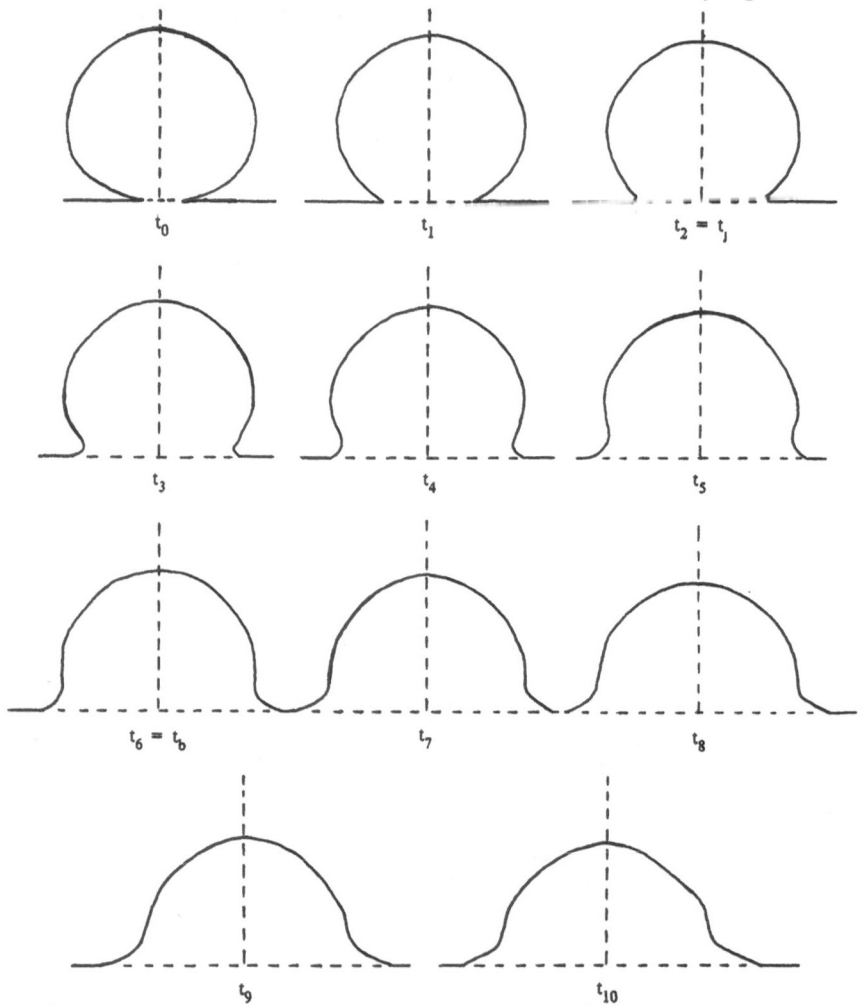

Figure 3. Ten snapshots of the vertical cross-section of a 3.0 mm drop impacting a flat surface at 8.0 m/s. This calculation is done with the SOLA-VOF numerical code (Nichols et al., 1980). The drop shape is realistically flattened. The time step between frames is .01 ms. t_0 is the time of initial contact. A jet appears at the base of the drop at $t_2 = t_j$. The impact ceases to be an acoustic source at $t_6 = t_b$.

as the raindrop size increases, air drag flattens the leading edge of the drop (Pruppacher and Pitter, 1971). There are three times during the drop impact that are important to the acoustics of the impact. These are the time of initial contact, t_0 (frame 1), the time when a lateral jet of water appears at the base of the drop, t_j (frame 3), and the time that the free surface of the drop ceases to show any concavity with the underlying water, i.e. when the drop loses its "drop" shape and appears to simply be a lump on the water surface, t_b (frame 7). After t_b the drop impact ceases to be an acoustic source. Figure 4 attempts to illustrate what is happening. Consider A, the circle of contact with radius b, between a sphere and a plane as the sphere passes through the plane. Initially the circle of contact is a point at the base of the drop. As the sphere passes through the plane, b increases, reaching its

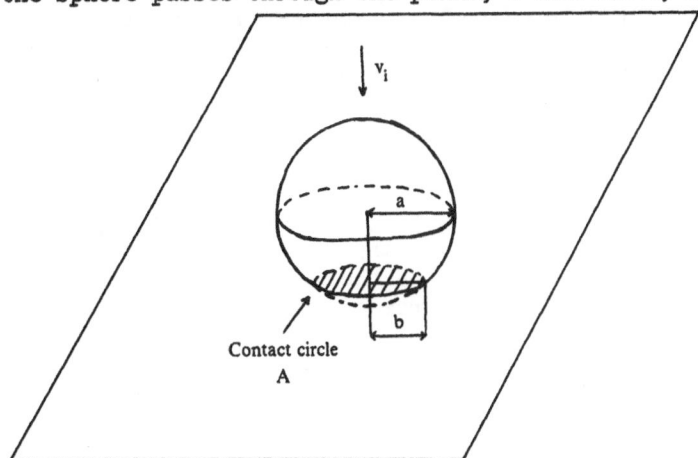

Figure 4. The geometry of a sphere, radius a and velocity v_i, passing through a plane is shown. The intersection between the sphere and the plane defines Contact circle A, which has radius b.

maximum value a, the sphere radius, when the sphere has passed half way through the plane. The radial speed that the contact circle grows at is given by

$$u_a = db/dt = (a-v_i t)/(2at/v_i-t^2)^{1/2} \qquad (2)$$

where a is the sphere radius, v_i is the impact velocity and t is the time after contact.

Now consider a raindrop. As we saw in Figure 3, it deforms during an impact so the model of a rigid sphere passing through the surface is not entirely appropriate. However initially the fluid at the base of the drop has no radial (horizontal) velocity, u_f, whereas u_a is infinite, i.e., $u_a >> u_f$. The contact circle between the raindrop and the underlying water moves radially outward faster than raindrop fluid can accelerate horizontally. Conservation of mass forces the fluid at the base of the drop to compress. This is an acoustic water hammer (Simon, 1904), a sound source arising from the conversion of fluid kinetic

490

energy into acoustic energy, where the pressure increases is given by

$$P = \varepsilon \rho v_i c \qquad (3)$$

where ρ is the initial fluid density, v_i is the impact velocity, c is the speed of sound in water and ε is a number between zero and one. For a classical water hammer $\varepsilon = 1$ but for a drop impact (water onto water) ε is less than one as some of the pressure increase accelerates fluid under the drop downward and fluid at the base of the drop radially. Because of the radial pressure gradient, u_f begins to increase. Once $u_f \geq u_a$, a lateral jet appears at the base of the drop (at time t_j; Fig. 3, frame 3) and the jet acts as a pressure release which causes ε to become smaller and thus reduces the water hammer pressure.

The impact continues to be an acoustic source until t_b, when the contact circle radius, b, equals the drop radius, a. The time t_b is not associated with the acoustic resonance of the raindrop. The acoustic resonance of a raindrop is given by

$$f_d = c/4a \qquad (4)$$

This value ranges from 150 kHz to 3 MHz over observed raindrop sizes. The acoustic energy generated at any given instant of contact quickly radiates away; however as long as there is new drop fluid impacting the underlying fluid there is a source of acoustic energy. The time that

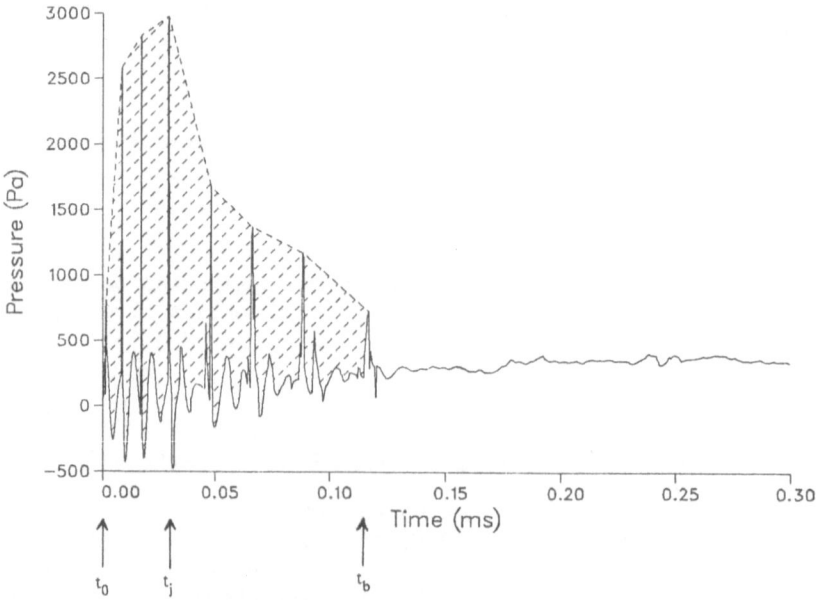

Figure 5. The pressure series recorded at a cell 5.7 mm beneath the impact of a spherical 3.0 mm drop impacting at 4.0 m/s (not a realistic raindrop). The time series shows nine numerical waterhammers, which define the near field acoustic pulse from the impact, superimposed on a low amplitude background. The times t_0, t_j and t_b are marked.

the drop is an acoustic source, t_b, is not a/v_i, as suggested by the example of a sphere passing through a plane. It is shorter since the fluid associated with the lateral jet helps to "fill the gap" under the edge of the raindrop (Fig. 3).

A pressure time series generated by the numerical code is shown in Figure 5. The time series shows a series of high amplitude, short duration numerical water hammers superimposed on a low level background. Each numerical water hammer corresponds to a single grid space increase in b, the radius of the contact circle A. The continuous source is approximated by the shaded area of Fig. 5. Initially there is a very rapid rise in pressure until t_j, the time when the lateral jet appears at the base of the drop. After t_j, the pressure falls steadily until the sound pressure suddenly stops at t_b.

The observed acoustic pulse from a drop impact (Fig. 1) is a single cycle. This is not reproduced by the model (Fig. 5); however, a sound source at a free surface is an acoustic dipole which has a radiation pattern given by

$$P - P_0 = \cos\theta \ \{m/4\pi r^2 + m_t/4\pi rc\} \tag{5}$$

where θ is the vertical angle between the source and the sensor, r is the distance and m is the source strength. The first term in brackets is the acoustic near field term, the second is the far field term. In

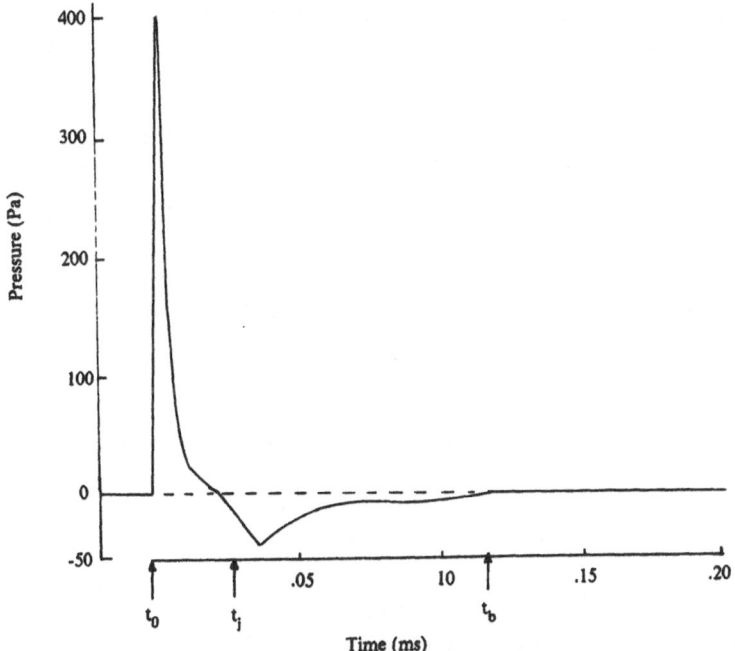

Figure 6. This figure shows the far field acoustic pulse shape obtained by taking a time derivative of the pulse shown in Figure 5. This pulse shape resembles the observed pulse shape of a drop impact (Fig. 1). Its spectral signature is a broad peak centered at $f = 1/t_b$.

the laboratory the sensor is in the far field. The calculated
pressure time series (Fig. 5) is near the base of the drop,
in the near field. The near field pressure is related to the far field
pressure by a time derivative. Figure 6 shows the time derivative of
the pressure pulse from Figure 5. This pulse resembles the observed
acoustic pulse and a Fourier transform of this pulse shape shows a peak
in the power spectrum centered at $f = 1/t_b$.

The time that the impact ceases to be an acoustic source can be
estimated from the numerical simulations. The end of the pulse depends
on impact velocity, drop shape and drop size. For realistic drop sizes
and shapes impacting at their terminal velocities, the pulse duration is
roughly constant and approximately equal to 0.6 ms. Thus natural
raindrops should produce a spectral peak between 15-20 kHz, very near to
that which is observed (Fig. 2).

4   THE ROLE OF THE RAINDROP SIZE DISTRIBUTION

Raindrop size distributions vary depending on environmental
conditions; however, most raindrops are small and only when the rainfall
rate is large are big drops likely to be present. While the small drops
are most numerous they do not contain most of the drop energy in the
rain. If the sound generated underwater by a raindrop is proportional
to drop energy then the large drops will dominate the sound generation
by rain. If large drops generate sound by a mechanism that is different
from smaller drops then a sound measurement technique geared towards the
large drops will be more effective for measuring rainfall rate since

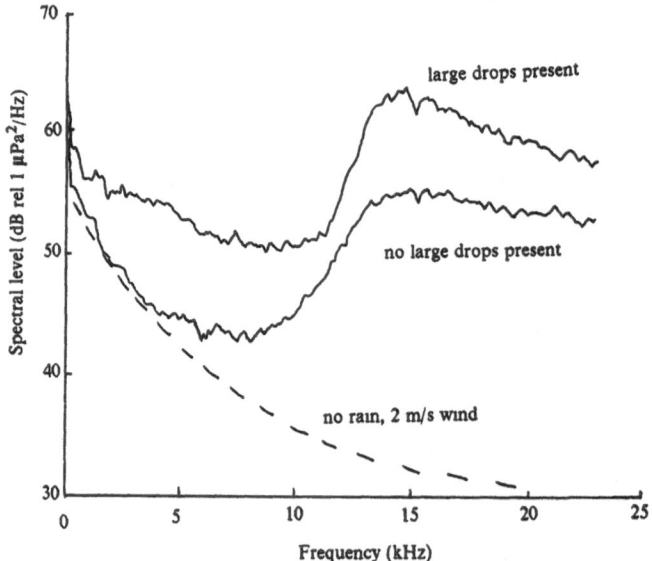

Figure 7. Two ambient noise spectra recorded during light rain at
Cowichan Lake in Mar 1985. In each case the rainfall rate was 0.8 mm/hr
and the wind speed was 1.8 m/s. The difference in the two cases was in
the rain drop size distributions. The sound spectrum from 2 m/s wind is
also shown.

large drops also contain most of the water volume in rain.

There is evidence to suggest that large drops are more effective producers of underwater sound. Figure 7 shows two observed sound spectra recorded under similar wind conditions and rainfall rate. In one case there are few if any drops larger than 1.5 mm in diameter in the rain. The spectral peak is readily observed and the spectral levels and shape below 5 kHz are similar to the ambient sound levels in the absence of rain (the 2 m/s wind generated sound spectrum is also shown). In the other case larger drops are present (~2 mm in diameter). The peak level is much higher suggesting that the large drops are more effective producers of sound via the impact mechanism. This observation was predicted by Lokken and Bom (1972) and a similar observation was noted by Nystuen (1986). In addition to a higher peak level the spectral levels are higher at all frequencies monitored. Since the impact mechanism does not predict an increase in spectral level at low frequencies (below 5 kHz) another sound producing mechanism is needed for large drop splashes.

While bubbles may not be part of the explanation for the spectral peak, newly entrained bubbles are such strong sound sources that when present they are likely to affect the sound spectrum. Underwater bubble formation has been observed during rain (Thorpe and Hall, 1983). Crowther's (1981) laboratory study of splashes showed that the resonant frequencies of the bubbles generated from splashes ranged from 500 Hz to 30 kHz. This wide frequency range suggests that bubbles may be the explanation for the rise in spectral level observed at frequencies less than the peak frequency when large drops are present in the rain.

Sound radiation from an ensemble of bubble sizes would produce a broad band increase in spectral level which would be hard to attribute conclusively to bubbles. On the other hand the acoustic signature of a bubble in the time domain (Figure 1) is very clear. In natural rain the area of water surface monitored acoustically is often so large that individual splashes cannot be identified; however, using an artificial spray the time domain signature of bubbles can be observed. Figure 8 shows the results of an experiment using artificial spray generated by a garden hose with an adjustable nozzle. Two time series are shown, one from the fine spray (part (a)) and the other from coarse spray (part (b)). The character of the two time series are different. The time series from the fine spray shows no bubble signatures, while numerous bubble signatures are present in the time series from the coarse spray, in addition to higher amplitude impact pulses. In both cases the individual drop impacts are not clearly resolved. Part (c) shows the sound spectra calculated from the two time series. For the fine spray the spectrum looks very similar to the spectra observed during light rain (no large drops present). The spectrum generated by the coarse spray shows sound energy generated over a wide frequency range (500 Hz to the limit of the measuring system, 20 kHz). Note that the location of the spectral peak has shifted to 8 kHz. The peak location is proportional to impact velocity and the impact velocities for the large drops in the spray are well below their terminal velocities (large drops need almost 20 m to reach terminal fall speed (Foote and duToit, 1969)). Furthermore, they are not likely to have an equilibrium shape (which for

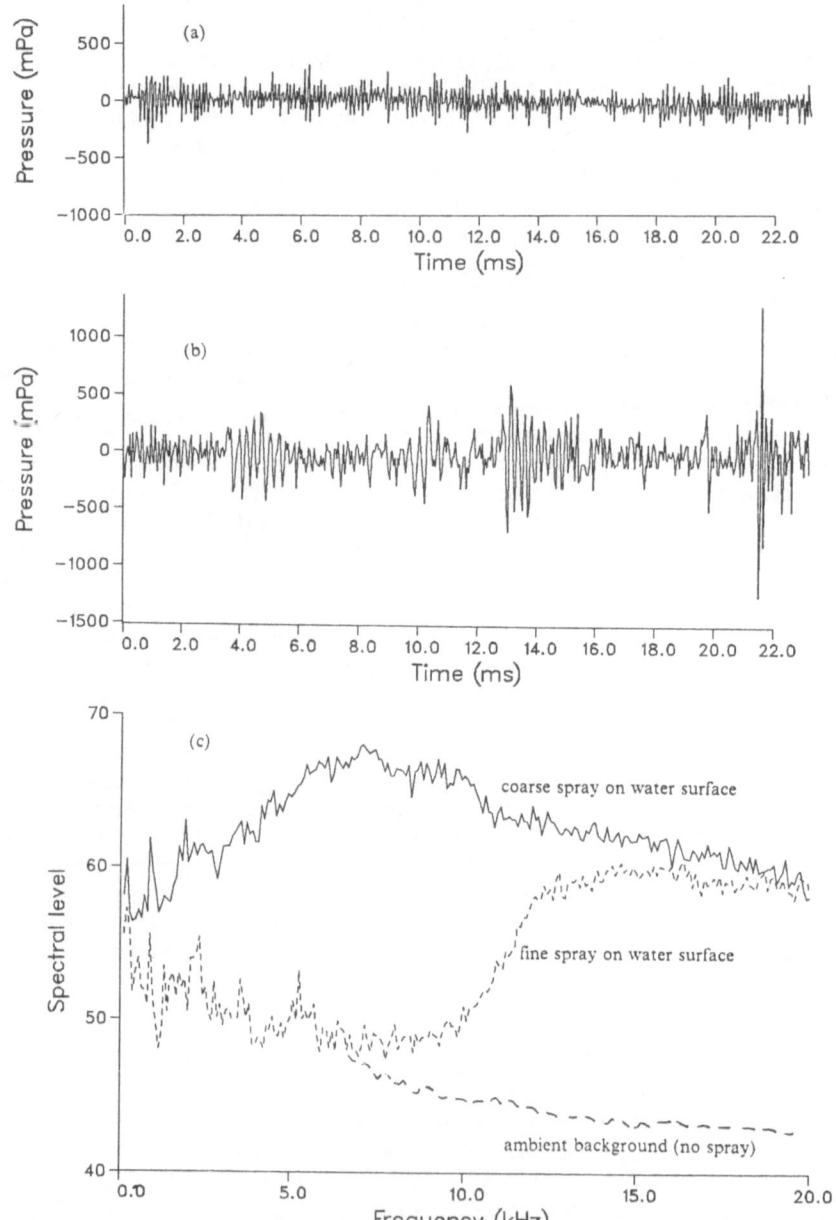

Figure 8. Underwater sound recorded during an artificial spray experiment at the IOS pier in Sidney, B. C. In (a), the pressure time series from the spray of a garden hose with the nozzle set to fine spray shows many impact pulse with no bubble signatures. In (b), a spray with the nozzle set to coarse spray, there are many bubble signatures present. In (c) the resulting sound spectra are shown.

large drops means flattened on the bottom). Since equilibrium shape drops have the shortest acoustic pulses (and thus the highest peak location), non-equilibrium drops will generate a peak at lower frequencies.

While newly entrained bubbles are sound sources they quickly radiate acoustic energy and come into equilibrium with the surrounding water. Now, as ambient bubbles, their role changes. In addition to being effective radiators of acoustic energy, bubbles also absorb acoustic energy at their resonant frequencies. Large bubbles quickly rise to the surface, however smaller bubble remain and can be mixed downward as clouds. These bubble clouds absorb sound, especially above 15 kHz and have been observed to change the shape of the wind generated spectrum (Farmer and Lemon 1984). Such bubble clouds should also affect the shape of rain generated underwater sound.

5.   THE INFLUENCE OF WIND

Wind may also occur when there is rainfall. Nystuen and Farmer (1987) studied the influence of wind on the sound generated by light rain (Figure 9). The rain was light (no large drops present) and so the sound produced by the rain was generated by the impact mechanism only. The influence of wind on the spectral peak is significant. As the wind increases the peak frequency shifts to higher frequency, the peak broadens and the spectral level diminishes. An algorithm for inferring rainfall rate from ambient sound needs to include the influence of wind.

The numerical calculations showed that the duration of the impact pulse from a raindrop was inversely proportional to the impact velocity. A raindrop falling in wind will have a horizontal component of its

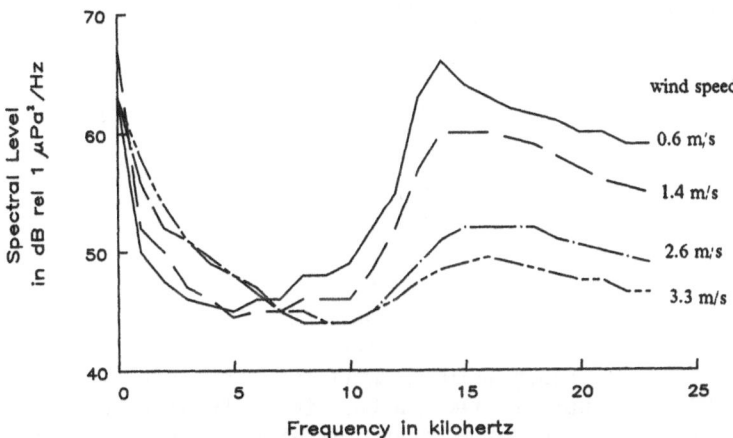

Figure 9. Sound spectra recorded during light rain with variable wind speed (from Nystuen and Farmer, 1987). The rainfall rate for these four spectra was 0.6 mm/hr and the drop size distribution were all very similar. There were few if any larger than 1.5 mm in diameter. As the wind speed increases, the peak shifts to higher frequency, broadens and the spectral level diminishes. The increase in spectral level below 5 kHz as the wind increases is consistant with the increase expected from wind alone and is therefore not likely to be rain generated sound.

velocity equal to the wind speed. This increases the raindrop's net impact velocity and thus shifts the spectral peak to higher frequency. This effect will be most evident in rains containing only small drops because small drops have lower terminal fall velocities than large drops and therefore the added horizontal component from the wind will result in a proportionally larger increase in total velocity. Calculations are shown in Figure 10 along with peak location observations under 58 different rain conditions. The theoretical lines shown in Fig. 10 assume all of the acoustic energy is generated from 0.9 or 2.0 mm drops respectively. In a natural rain there is a distribution of raindrop sizes and since the peak frequency shift varies with drop size the peak energy will be differentially shifted. This will result in a broadened spectral peak with the peak frequency determined by the drop size that dominates the acoustic energy produced by the rain, i.e. large drops.

Wind turbulence in the surface layer will modify both the horizontal and vertical components of the drop velocity. Such turbulence will broaden the distribution of raindrop impact velocities

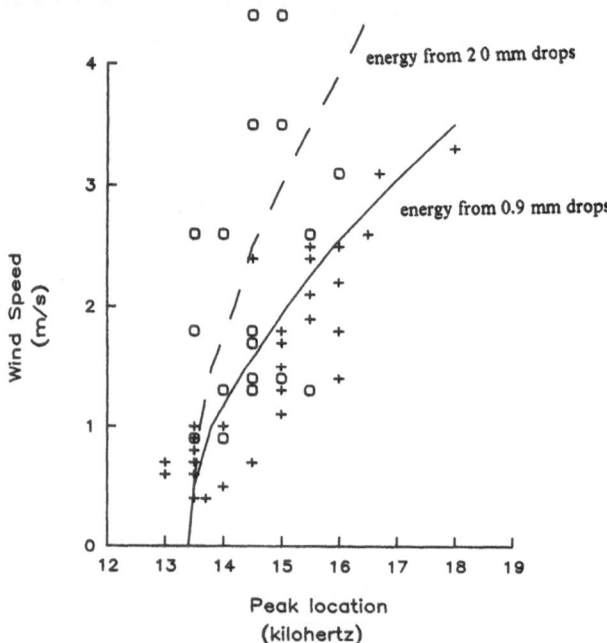

Figure 10. The relationship between wind speed and the observed peak location for 58 sound spectra recorded during rain at Lake Cowichan, B. C. in Mar 1985. The solid curve shows the expected peak location if all the sound generated underwater were due to 0.9 mm drop impacts. The + data points are for spectra recorded during rain with few if any drops larger than 1.5 mm in diameter. For such rain the drop size which contains the most energy is often near 0.9 mm. The dashed line shows the expected peak location if the sound is generated by 2.0 mm drop impacts and the 0 data points show the peak location observed when the rain contained large drops. For this data set the rainfall rates were never greater than 2 mm/hr.

and thus broaden the spectral peak generated by those raindrops. Wind also generates surface waves, and the surface velocities associated with the waves will modify the impact velocities of the raindrops and broaden the spectral peak (Nystuen and Farmer, 1987).

Figure 11 shows an attempt to match this theory with observation. The main difference is in overall spectral level. The current theory does not explain the apparent loss in efficiency of the drop impact as a sound source with increasing wind. This may be due, in part, to ambient bubbles modifying the sound spectrum through nonuniform attenuation as the surface generated sound propagates through the bubbles to the hydrophone, although Farmer and Lemon (1984) do not show any modification of the sound spectrum until the wind speed increases above 10 m/s.

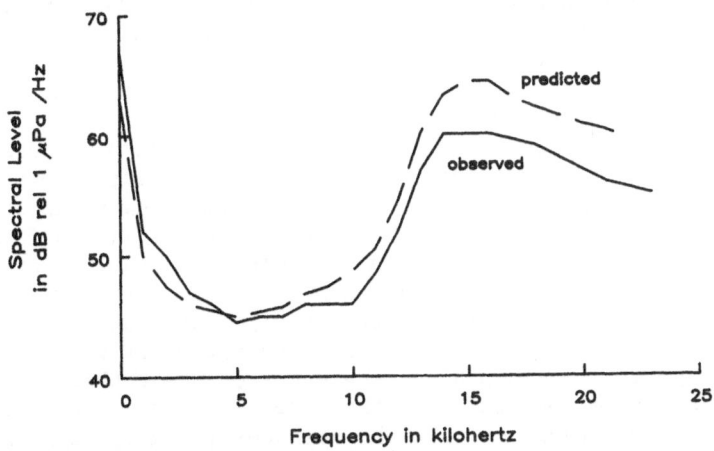

Figure 11. A comparison of theory with observation. The 0.6 m/s spectrum from Fig. 9 has been shifted by a 1.5 m/s wind and the peak broadened by 0.05 m amplitude surface waves. The resulting spectrum is compared to the observed 1.4 m/s spectrum (Fig. 9).

6. CONCLUSIONS

The principle mechanism for underwater sound production by rain is the impact of the raindrops on the ocean surface. The impact has been described as an acoustic water hammer that is modified by the emergence of a jet at the base of the drop during the initial part of the splash. The drop continues to be an acoustic source until the radius of the contact circle between the drop and the underlying water surface reaches the maximum cross sectional radius of the raindrop. In the acoustic far field this water hammer is observed as a single cycle pressure pulse. The duration of the pulse depends on drop size, shape and impact velocity. A remarkable coincidence is that this duration is roughly constant for realistic drop sizes and shapes impacting at their terminal fall velocity. The observed peak in the rain generated sound spectrum coincides with this time interval.

Rain containing large drops is likely to entrain bubbles. Newly entrained bubbles are a known sound source which have to be considered when they are present. While the drop impacts are responsible for the spectral peak characteristic of sound generated by rain, bubbles may be responsible for the increase in spectral level at low frequencies. Bubbles created by drop splashes have been shown to generate sound energy at these frequencies while the impact generated sound is confined to the spectral peak for natural raindrops. The influence of wind on the spectral peak is significant, shifting and broadening the peak as well as reducing the efficiency of the impact mechanism. This influence on the spectral peak is less for rain containing large drops that it is for rain containing only small drops.

These conclusions have a number of implications for our ability to measure rainfall at sea by monitoring the underwater ambient sound. While rainfall generated sound has a distinctive spectral shape, that spectral shape is modified by wind and the raindrop size distribution of the rain. Because of the distinctive spectral peak, rain can be detected when it is present. Quantitative estimation of rainfall rate without knowledge of wind conditions will be difficult; however in the case of light rain (no large drops) the spectrum below 5 kHz is not strongly influenced by the rain. The wind can be estimated from the spectrum in this frequency range and so a wind dependant rainfall rate algorithm can be proposed for light rain. For heavy rain (large drops present) there may also be a wind dependant algorithm, however the data set with which to explore this question does not exist at present. A heavy rain algorithm may depend on the shape of the bubble generated part of the spectrum combined with the spectral level of the peak frequency or it may turn out that the rain generated sound from very heavy rain (large drops present and high rainfall rate) is not very sensitive to the influence of wind.

ACKNOWLEDGMENTS

Financial support was provided by the U. S. Office of Naval Research. Contract No. N00014-85-C-0824, the Office of Energy Research and Development, Project No. 67129 and the Naval Postgraduate School Foundation Research Program.

REFERENCES

Bom, N., 'Effect of rain on underwater noise', J. Acoust. Soc. Am., 45, 150-156 (1968).

Crowther, P. A., 'Near surface bubble excitation and noise in the ocean', in Advances in Underwater Acoustics, Proc. Inst. Acoustics Conference in Portland, Dorset, U. K. (Institute of Acoustics, Edinburgh, Scotland, 1981).

Farmer, D. M. and D. D. Lemon, 'The influence of bubbles on ambient noise in the ocean at high wind speeds', J. Phys. Ocean., 14, 1762-1778 (1984).

Franz, G. J., 'Splashes as sources of sound in liquids', J. Acoust. Soc. Am., 31, 1980-1096 (1959).

Heindsman, T. E., R. H. Smith and A. D. Arneson, 'Effect of rain upon underwater noise levels', J. Acoust. Soc. Am., 27, 378-379 (1955).

Lokken, J. E. and N. Bom, 'Changes in raindrop size inferred from underwater noise', J. Appl. Meteorol., 11, 553-554 (1972).

Nichols, B. C., C. W. Hirt, and S. R. Hotchkiss, 'SOLA-VOF: A solution algorithm for transient fluid flow with multiple free boundaries', Los Alamos Scientific Laboratory Report LA-8355 (1980).

Nystuen, J. A., 'Rainfall measurements using underwater ambient noise', J. Acoust. Soc. Am., 79, 972-982 (1986).

Nystuen, J. A., 'The acoustic physics of a drop impact', part of the final report Development of techniques for measuring rainfall at sea submitted to the Institute of Ocean Sciences (1987). In preparation for J. Acoust. Soc. Am.

Nystuen, J. A. and D. M. Farmer, 'The influence of wind on the underwater sound generated by light rain', submitted to J. Acoust. Soc. Am., 82, 270-274.

Pruppacher, H. R. and R. L. Pitter, 'A semi-empirical determination of the shape of cloud and rain drops', J. Atmos. Sci., 28, 86-94 (1971).

Scrimger, J. A., 'Development of techniques for measuring rainfall at sea', technical data report submitted to the Institute of Ocean Sciences, Sidney, B. C. (1985).

Scrimger, J. A., D. J. Evans, G. A. McBean, D. M. Farmer and B. R. Kerman, 'Underwater noise due to rain, hail and snow', J. Acoust. Soc. Am., 81, 79-86 (1987).

O. Simon, 'Water hammer', Proc. Am. Water Works Assoc., 24, 335-422 (1904).

Thorpe, S. A. and A. J. Hall, 'The characteristic of breaking waves, bubble clouds and near-surface currents observed using side-scan sonar', Cont. Shelf Res., 1, 353-384 (1983).

STUDIES OF MECHANISMS INFLUENCING RAIN NOISE

L. Bjørnø
Industrial Acoustics Laboratory
Technical University of Denmark
Building 425, DK-2800 Lyngby,
Denmark

ABSTRACT. A considerable interest in rain generated underwater ambient noise is reflected internationally in an increasing number of scientific papers published. This paper gives a critical evaluation of more recent publications on the subject and essential deviations between theoretical and experimental results obtained by various authors are pointed out and their prospective reasons are discussed. A detailed exposition of factors influencing rain generated ambient noise in the sea is given for single and multiple raindrops. These factors, which influence the sound generating mechanisms by the raindrop impacts on the water surface are for instance, raindrop diameter and diameter distribution, terminal velocity, raindrop shape, wind profiles, surface tension and temperature, and measurement conditions. Some preliminary experimental results arising from a comprehensive, 2-years research programme on rain generated noise in the sea very recently started at the Technical University of Denmark are given. Finally, the most essential elements and the aim of the 2-years research scheme is discussed and it is concluded that a comprehensive and well-prepared international collaborative research programme is necessary in order to solve the involved scientific questions related to rain generated ambient noise in the sea.

1. INTRODUCTION

Rain falling on the sea surface has turned out to be a considerable contributor to the ambient noise level in the sea. Moreover, the underwater ambient noise spectrum generated by rain has frequently a spectral shape which can be distinguished from the spectral shapes produced by other sources of ambient noise, and the relationship between the spectral levels and the rainfall may be quantified.
For sonar purposes it is necessary to gather information about sources of ambient noise, the spectral composition and the level of the noise, the noise directivity in the horizontal and vertical plane etc.
Rainfall is a climatic factor of great importance and therefore measurement of rainfall has a high priority. However, it has been estimated

B. R. Kerman (ed.), Sea Surface Sound, 501–512.

that 80% of the Earth's precipitation occurs over the ocean where the
smallest number of weather stations are located. Measurements of underwater ambient sound have, therefore, recently been proposed as a way
for determination of the amount of rain falling on the sea surface [1].
A prospective future procedure for rainfall measurements could include
underwater ambient noise measurements at certain geographical locations
combined with the use of satellite measurements and weather radar.
However, a reliable and reproducible relation between the rainfall and
the characteristics of rain produced underwater ambient noise has to be
established first and the reasons for the deviations between rain noise
data produced by various scientists during the past have to be found.
A comprehensive and systematic study of certain features of rain noise
spectra and the mechanisms causing these spectral features will probably call for an extensive international cooperation between underwater
acousticians, hydrodynamicists and meteorologists.

One of the first attempts to describe the underwater noise spectra
arising from rainfall was published by Heindsmann et al [2] who measured
the changes in ambient noise caused by the passage of two heavy rain-
storms over an off-shore bottomed hydrophone system located 1.2 m
above a hard sandy bottom in 36 m of water near the eastern end of
Long Island Sound. They found that an increase in the sound pressure
level of 25 dB in the frequency band of 20 Hz - 20 kHz took place during
the rainfall period, and that during the heaviest rainfall the sound
pressure spectrum level was approximately constant at 77 dB re 1μPa
from below 1 kHz and to above 10 kHz.
A very comprehensive study of the mechanisms of sound production by the
splashes made by gas-to-liquid entry of objects was published by
Franz [3]. He studied the sound spectra produced by single splashes and
by splashes of spray of water droplets in a water tank and he found
that the main sources of underwater sound from a splash are (1) the
impact and passage of the body (water droplet) through the free water
surface leading to the establishment of flow, (2) resonance vibrations
of the body if it has rigidity and (3) volume pulsations of closed
cavities of air in the water. Moreover, he gave some evidence for the
statement that the splashes of water droplets produce underwater sound
like a dipole source with a vertical axis. The two parameters varied
by Franz in his experiments were the impact velocity and the droplet
size, and he found that the conversion of kinetic energy of the droplets to sound radiated into the water could be expressed by $2M^3$ where
M is a Mach number formed by the ratio of the impact velocity of the
droplet and the velocity of sound in water. Thus the radiated sound by
the initial entry of the droplet increases systematically with increase
in droplet size and impact velocity and the half octave spectra of the
impact sound in water show a broad maximum in the frequency range
between 1 kHz and 10 kHz. Franz's results have formed the basis of
several later calculations of underwater sound from large-scale
splashing for instance caused by rain and breaking of waves. Among
these later studies should in particular be emphasized the extensive
work by Wenz [4] who gave an in-depth exposition of the sources and
spectra of acoustic ambient noise in the ocean covering in particular

the frequency range 1 Hz to 20 kHz.

A study of underwater noise due to precipitation on the surface of a small, shallow lake (Lake of Sarzana) was published by Bom [5] who made an analysis in octave bands over the interval 300 Hz to 9.6 kHz estimating the average noise level versus rain rate. Bom found that the noise level in dB versus the logarithm of the rain rate in the interval 1 - 25 mm/h can be represented by a straight line. A comparison with Franz's predictions shows a significant difference in absolute level, in particular at low frequencies. Moreover, Bom's underwater noise spectrum levels seems to peak at 3 - 4 kHz for several rain rates. Nystuen [1] found that the main feature of rain generated noise spectra in the sea is a broadband peak around 15 kHz which in particular was produced by small (<2 mm) water droplets. Large droplets which produce sound energy below 5 kHz contain most of the energy in typical rain. The kinetic energy increases dramatically ($\propto d^4$) with the droplet diameter d.

Based on a numerical model for the drop splash flow field which allows the study of multiple free surfaces to be performed and which permits variation in surface tension, viscosity and droplet shape to be introduced Nystuen discussed several of Franz's findings which disagreed with Nystuen's own results. In particular attention was paid to sound generation by the initial, short and high amplitude pulse (the impulse pressure) and by the flow establishment phase forming the transition to the non-propagating dynamic pressure which can only be detected in the immediate vicinity of the impact. Nystuen found that the initial impact pressure is proportional to the impact velocity V (and not to V^3 as suggested by Franz) and that the assumption that a droplet can be modeled as a rigid sphere entering the water is not valid. However, the droplet impact as an acoustic dipole was confirmed. The most considerable conclusion is, however, that the rain-generated noise spectra should have a broad peak around 15 kHz as found by Nystuen in some lake experiments, and not around 3 kHz as suggested by Franz. Because of the distinct spectral shape of the rain-generated noise and the high sensitivity of the sound level at 15 kHz to rainfall rate, Nystuen concluded, that rainfall can be detected even when wind noise is also present by measurement of underwater ambient noise.

Scrimger et al [6] measured the spectral characteristics of underwater noise generated by rain falling on the surface of a freshwater lake. The measurements were done using a bottom-mounted hydrophone at a depth of 35 m. For wind speeds less than 1.2 m/s the authors found the rain noise spectra to have a sharp peak at 13.5 kHz with a 9 dB/octave falloff on the high-frequency side and a 60 dB/octave falloff on the low-frequency side. Their frequency range of 2 - 10 kHz showed a positive spectral slope of 0-4 dB/octave. Attempts to correlate the rain noise spectra with raindrop-size distribution suggested that larger drops might be responsible for increasing the spectral level in the frequency range below the 13.5 kHz peak. The shape of the spectral regime between 2 and 10 kHz showed many features in agreement with Bom's data.

An apparent lack of agreement exists between the rain noise spectra measured and calculated by the authors [1 - 6]. This lack of agreement calls for a more systematic and comprehensive study of the mechanisms leading to rain produced noise in the sea and to the correlation between these mechanisms and characteristic features of the underwater noise spectra produced by rain. In order to meet this challenge a comprehensive 2-years research programme on rain produced noise in the sea has been started recently at the Technical University of Denmark. As this research programme has only been active for some months some preliminary results alone shall be discussed in this paper.

2. FACTORS INFLUENCING RAIN GENERATED NOISE

In order to study the factors which may influence the mechanisms leading to rain generated noise in the sea the study of the impact course for single droplets and for multiple droplets (rain showers) will be performed independently.

A. Single droplets

Possible sources of underwater sound caused by single droplets are the transient introduction of the droplet into the water, secondary splashes of water droplets thrown up by the entry and oscillations of air bubbles and of cavities open to the atmosphere. The individual contribution to the underwater noise spectra from these mechanisms are influenced strongly by factors related to the droplet size, shape and movement before it hits the water surface. Among these factors should in particular be emphasized. The equivalent raindrop diameter, the size distribution of the raindrops, the terminal velocity of the raindrops, the shape of the raindrops, the wind velocity and its profile, surface tension and raindrop temperature and the condition under which the measurements take place.

1. Raindrop diameter

Typical rain includes a fairly large range of drop sizes. The size of a raindrop is usually given by its equivalent diameter, i.e. the diameter of a sphere of the same volume as the raindrop. As a raindrop of 0.2 mm in diameter can fall about 150 m in an atmosphere having 50% humidity before it is evaporated a lower equivalent diameter of raindrops of interest is normally 0.2 mm. As stable drops of equivalent diameters of 9 mm have been produced under laboratory conditions an upper diameter limit is set to 9 mm. However, in rain equivalent drop diameters above 5 - 6 mm are rarely found.
Several methods have during the past been used for measurements of the sizes of raindrops. The ingenuity covers filter paper, flour balls, photography, capacity measurements, acoustic measurements and radar measurements.

2. Raindrop diameter distributions

The independent variable by raindrop diameter distribution measurements is the number of raindrops falling within certain intervals of
equivalent raindrop diameter. Measurements are performed using a
Joss-Waldvogel RD-69 distrometer, which is designed to identify raindrop sizes by measurement of the pressure pulse from a drop striking
the top surface of the instrument. Knowing the drop-size distribution
is necessary since the drop-size distribution vary during a rain shower
which affects the shape of the underwater ambient noise spectrum. Also
the rainfall rate used by many authors as a parameter governing the
noise spectrum is changed with changes in drop-size distribution. A
much used and simple drop-size distribution expression was formed by
Marshall & Palmer [7] who found that for drop-sizes above 1.5 mm a
negative exponential expression (1) could be used:

$$N_D = N_o \exp(-\lambda D) \tag{1}$$

where N_o = 8000 drops per m^3 per mm diameter interval, D is the
equivalent drop diameter in mm and λ is a function of the rainfall
rate R (mm/h) given by: $\lambda = 4.1 \, R^{-0.21}$. Expression (1) is being used
in our studies which have shown that the number of large raindrops
increases with increasing rainfall rate which also leads to a strong
decrease in the presence of small drops.

3. The terminal velocity

Several studies of raindrop impact noise have been performed at impact
velocities less than the terminal velocity. As the impact velocity
has a strong influence on the energy distribution in the underwater
noise spectra a determination of the individual drop impact velocity
is necessary for the establishment of relations between the impact
mechanisms and the noise spectra. An uncritical application of the
terminal velocity as the impact velocity may lead to erroneous results.
Several attemts to determine the terminal velocity has been done
[8 - 9] for various drop-sizes and for various atmospheric conditions.

4. The drop shape

The drop shape has a considerable influence on the magnitude of the
pulse produced by the drop impact and on the establishment of the flow
in and around the impact location. Drops less than 1 mm in diameter are
nearly spherical while larger drops in particular when they have
obtained their terminal velocity are strongly deformed by influence of
air drag. By changing the drop shape from spherical to a more realistic
deformed shape a 4.8 mm diameter drop will produce a seven times higher
amplitude in the impact pressure pulse [1].

5. The influence of wind profiles

It is a general observation that raindrops very rarely are falling vertically, but at an angle depending on the wind profile and its variation with time. Several models for drops falling in a wind profile have been worked out and wind tunnel tests have been performed in order to study the trajectories of the falling drops. Experiments seem to support that rain drops before the impact on the water surface are falling at an angle determined by their terminal velocity vector and the velocity vector in the wind profile close to the water surface. Also a separation of the various raindrop sizes takes place in a wind profile frequently leading to the observation that early parts of a shower only contain the larger drops while the smaller drops are falling later. Two showers falling after each other may mix their drop-sizes as small drops from the first shower falls together with large drops from the second shower.

Wind profile effects may, moreover, lead to evaporation of water from the raindrops, to collision and coalescense between raindrops and to breaking-up of larger raindrops into several smaller drops.

6. Surface tension and temperature

Only the smallest drops are expected to be influenced by surface tension. However, even if the smallest drops are most numerous in a rain shower the most significant amount of acoustic energy in the water is produced by the impact of larger raindrops. Numerical calculations [1] have shown that changes of surface tension had no noticeable effect on the splash of larger drops, and it is unlikely that surface tension plays a significant role for the mechanisms leading to rain generated underwater sound.

Temperature and thus viscosity may have some effects on the establishment of the flow after the raindrop impact as higher viscosity will lead to a longer flow establishment phase and thus influence the rain noise spectrum.

7. Measurement conditions

Apart from influence on the pressure-time series measured in water arising from the hydrophones, amplifiers, signal processing and display equipment, an influence from the proximity of the water surface may be found. Theoretical and experimental studies have given evidence to the statement that a drop impact at a free water surface should be a dipole source with a vertical axis, because the free surface acts as a pressure release. In order to avoid nearfield effects of this source a distance of one to more wavelengths below the water surface is necessary in order to avoid the influence of non-propagating pressure fields. Moreover, due to the broad spectrum formed by rain generated noise standing wave patterns must be avoided. This will in particular be of importance by the use of laboratory tank facilities of various sizes and shapes, but also by minor lakes the environmental geometries and materials should be considered. A single hydrophone at a fixed position in a lake may produce erroneous results.

B. Multiple raindrops (showers)

The factors influencing single raindrops will also influence multiple raindrops. Moreover, underwater sound generation by multiple raindrop impacts on a water surface will be influenced by interaction between the individual impacts, which will comprise water drops falling on a not-plane (random shape) water surface, drops falling in and out of phase leading to phase cancellation, resonance etc. This more compli- cated noise generation process by multiple raindrop impacts may strong- ly influence the order, the development and the individual contribu- tions of the mechanisms leading to underwater noise generation by single drop impacts. The noise generation process by multiple raindrop im- pacts is possibly the area of rain noise generation where most research is needed in the years to come. Not only extensive experimental studies have to be performed, but a theoretical (numerical) basis has to be developed. In this context it is a question if the theory of "chaos" now being applied to noise generation by multiple bubbles in water (cavitation) can be applied to the description of noise generation in water by the impacts of multiple raindrops of different size distributions.

3. PRELIMINARY EXPERIMENTS AND RESULTS

The experiments have hitherto been performed in two indoor water tanks. One, a nylon tank, 1 x 1 x 2 m, and the other, a steel tank, 2 x 2 x 3 m. The water depths can be varied in both tanks in order to study the influence of the tank wall materials and the tank dimen- sions, in particular on noise produced by multiple drop impacts.

The electronic instruments used for the noise registration and signal processing consists of a broadband transducer, natural frequency around 250 kHz, produced by the Industrial Acoustics Laboratory. This transducer is based on a thick circular disc of PZT and has a directivity index of 26 dB at the resonance frequency. Moreover, a B&K hydrophone type 8101 is being used connected to a preamplifier B&K type 2650. The signal processing is done using a two channel FFT analyzer, B&K type 2032, which permits the use of several signal- processing function. The time signals and the spectra to 25 kHz are displayed on the screen of the FFT analyzer and are plotted on paper using a HP Think Jet printer.

For production of single drops hypodermic needles of various inner diameters were used thus permitting the production of drops of a broad variety of equivalent diameters. By throttling the water supply to the hypodermic needle using a needle valve it was possible to vary the time between each drop, and a drop frequency ranging from 4 to 1/6 per second is used. The drop-size is measured using a test glass which is weighed before and after having received 10 or 20 drops. The multiple drop impacts are produced by the use of a shower head provided with 20 - 100 hypodermic needles of various diameters.

The shower head diameters range from 70 - 250 mm. The hypodermic needles are interchangeable thus permitting various drop-size distributions in the shower to be produced. The drop-size distribution is controlled using the test glass - balance procedure and using the Joss-Waldvogel distrometer.

The impact velocity, which for some experiments is the terminal velocity, is measured using a device based on a glas tube containing a light beam vertical to the direction of the falling drop. When the drop passes through the light beam a signal from a photo multiplier starts a time counter. When the drop hits a piezoelectric sensor after having fallen over a known distance from the light beam crossing the sensor signal stops the time counter. This device is also produced in the Industrial Acoustics Laboratory.

For the study of details of the drop impacts a high speed camera, Deckman-Whitley, permitting up to 35000 frames per second, is being used.

The following large drop equivalent diameters have been studied so far: 3.0, 3.5, 5.2 mm. Due to lack of stability by the largest drops, the diameters 3.0 and 3.5 mm have been most extensively studied and some impact velocities obtained for various fall distances are given in table 1.

Table 1. Impact velocities.

Fall distance m. / Equiv. diam. mm.	1.60	1.73	3.55	7.16
3.00	4.99	5.23	6.62	8.30
3.50	5.20	5.35	7.06	8.73

A comparison with the theoretical curve [9] for the terminal velocity for various drop diameters shows, that at the longest fall distance all three drop sizes have attained their terminal velocities, see figure 1.

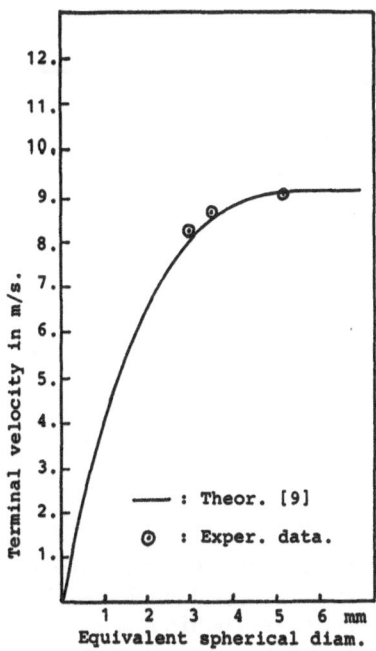

Figure 1

In order to study the pressure-time function, formed by the drop impact on the water surface, at various depths below the water surface, the PZT transducer was positioned at depths ranging from 0 (surface flush mounted with the water surface) to 70 mm below the surface (same depth as used by Franz). Increasing depths are now being studied. In figure 2 are given the pressure-time curves for the impacts measured at a depth of 70 mm below the water surface for a 3.5 mm drop falling over two distances, 1.60 and 7.16 m. The amplitudes are normalized using the individual peak amplitudes measured in the pressure-time courses. Apart from the positive phase at the dimensionless retarded time, 2.5, obtained by Franz which could be due to bubble influences, a good agreement with Franz's pressure-time course is obtained.

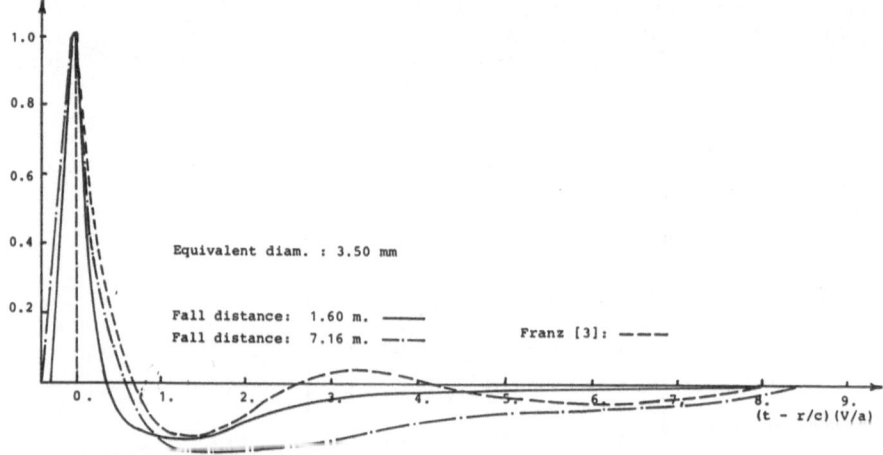

Figure 2

For several impacts, in particular at intermediate velocities, Franz
found a stronger influence from trapped bubbles. The same is the case
for our data. The impact velocities behind the pressure-time curves in
figure 2 were the lowest and the highest (terminal velocity) measured
and no influence from bubble pulsations is present. However,
for several tests, not shown here, secondary splashes caused by droplets thrown up by the entry gave low-frequency contributions to the
underwater sound spectra.
A comparison with Franz's spectra for the underwater sound produced
by the single drop impacts shows a higher spectral level at very low
frequencies, i.e. below 600 Hz, while a somewhat lower level is obtained
in the frequency region 1 - 10 kHz. This higher level at low frequencies
is probably caused by near-field effects measured at the water depths
0 - 70 mm. A considerable reduction in the low-frequency spectral level
is found for increasing depths, which is in agreement with Franz's
finding that the later parts of the pressure-time curves changes with
the depth.
Multiple drop impacts using an equivalent drop diameter of 3.5 mm and
a fall distance of 7.16 m showed, however, an increased spectral level
between 4 and 6 kHz. These parts of the sound spectra are probably
caused by the larger drops, the impacts of which lead to the entrainment
of air bubbles. The levels thus obtained were in agreement with the
measurements by Bom [5], and they are somewhat higher than predicted

using Franz's dimensionless frequency spectra. A more comprehensive
study is now being performed in order to explain these results.
These few, but interesting preliminary results, have inspired further
tests to be done. These tests will comprise a systematic study of the
depth influence on the pressure-time courses and their amplitude spectra
in an attempt to study the transition from the near to the far-field of
the dipole source formed by the single drop impact. This study will
also comprise directivity measurements. The influence of the reservoir
will be studied using various water depth in the tanks and also by the
use of a towing tank, dimensions 6 m deep, 12 m wide and 450 m long,
belonging to the Ship Research Institute. Temperature and viscosity
influences will be studied and simultaneous measurements of pressure-
time functions and high-speed photography of the drop impact course
will be performed. The reproducibility of drop shapes will be studied
using high-speed photography of larger drops in particular. Finally,
rain noise measurements in a lake is being prepared, but a lake with
low ambient noise level arising from trafic and other human activities
have to be found first.

4. PRELIMINARY CONCLUSIONS

In spite of the small number of experimental results obtained so far,
which give information about the underwater ambient noise spectra
caused by the impacts of single and multiple drops of various sizes and
having various impact velocities, there seem to exist a higher low-
frequency noise level in our results than found by others. However, a
more detailed study of the structure of the nearfield may explain
these findings. The sound pressure levels measured in the frequency
range from 1 to 10 kHz for multiple drop impacts seem to be in agreement
with Bom's data. The larger drops, 3.5 mm in equivalent diameter,
studied may lead to excess sound intensity at frequencies in the range
4 - 6 kHz by entrainment of air bubbles.
However, multiple drop impacts did not lead to the broader increased
spectral level around 15 kHz found by Nystuen [1] or the sharp peak
around 13.5 kHz found by Scrimger et al [6]. It is, however, far too
early to draw any final conclusions and much more research has to be
done. Due to the complicated nature of the research which has to be
done, and due to the deviations between earlier data obtained by various
teams of scientists the future studies call for an international
cooperation based on a comprehensive and well prepared research scheme.
It is my hope that the discussions of rain generated ambient noise in
the sea held at this ARW will lead to realistic and detailed plans for a
future international cooperation in rain noise research.

5. REFERENCES

1. Nystuen, J. A., Rainfall measurements using underwater ambient noise. *J. Acoust. Soc. Amer.*, 79 (4), 1986, 972-82.

2. Heindsmann, T. E., Smith, R. H. and Arneson, A. D., Effect of rain upon underwater noise levels. *J. Acoust. Soc. Amer.*, 27, 1955, 378-79.

3. Franz, G. J., Splashes as sources of sound in liquids. *J. Acoust. Soc. Amer.*, 31, (8), 1959, 1080-96.

4. Wenz, G. M., Acoustic ambient noise in the ocean: Spectra and sources. *J. Acoust. Soc. Amer.*, 34, (12), 1962, 1936-56.

5. Bom, N., Effect of rain on underwater noise level. *J. Acoust. Soc. Amer.*, 45, (1), 1969, 150-56.

6. Scrimger, J. A., Evans, D. J., McBean, G. A., Farmer, D. M. and Kerman, B. R., Underwater noise due to rain, hail and snow. *J. Acoust. Soc. Amer.*, 81, (1), 1987, 79-86.

7. Marshall, J. S. and Palmer, W. M., The distribution of raindrops with size. *J. of Meteorology*, 5, 1948, 165-66.

8. Beard, K. V. and Pruppacher, H. R., A determination of the terminal velocity and drag of small water drops by means of a wind tunnel. *J. Atmospheric Sciences*, 26, 1969, 1066-72.

9. Beard, K. V., Terminal velocity and shape of cloud and precipitation drops aloft. *J. Atmospheric Sciences*, 33, 1976, 851-63.

SPECULATIONS ON THE ORIGIN OF LOW FREQUENCY ARCTIC OCEAN NOISE

Ira Dyer
Department of Ocean Engineering
Massachusetts Institute of Technology
Cambridge, MA 02139
USA

ABSTRACT. Arctic ambient noise research has both macrolevel and microlevel aspects. Correlation of average noise with gross environmental parameters, and elucidation of ice source mechanisms from observation of individual acoustic events, are important and mutually supportive. In this review I concentrate on low frequency noise in the central Arctic. I summarize previous papers which show that such noise, on average, correlates highly with environmental moments and forces acting on the ice, and that individual noise events can be aggregated over the entire Arctic basin to yield the average noise. The moments and forces are thus reasoned to relate directly to individual ice source mechanisms. Models of ice motion are proposed as noise sources. Ice cracking and transient unloading at a ridge in response to moments, and ice-block creation and oscillation at an overthrusting site in response to horizontal forces, are shown to be plausible. The models predict the right order of stress in the ice, radiation in the proper frequency range, event signatures such as observed, and the observed source strengths.

An understanding of ambient noise in the Arctic Ocean can be gained at two different levels. At one we would like to know the way in which the average noise in some frequency band varies with gross changes in the natural environment. This can be said to be a study of *macrolevel environmental correlates*. Such studies are of interest not only to those seeking predictive knowledge of ambient noise, but also to those seeking clues to an understanding of ambient noise at a more fundamental physical level.

At this more fundamental level we might think of ambient noise in terms of individual noise generating events, principally because it has come to be believed that such individual events in the aggregate form the average noise condition. This, therefore, can be said to be a study of *microlevel event physics* in which details of the source generating an individual brief event, and the particular ice forces giving rise to it, are to be understood.

B. R. Kerman (ed.), Sea Surface Sound, 513–532.
© *1988 by Kluwer Academic Publishers.*

In this review I provide some overall perspective on both macrolevel and microlevel approaches to the study of Arctic ambient noise (Sections 1 and 2). I then specialize to low frequency central Arctic noise. This is done via discussion in Section 3 on source mechanism clues coming from macrolevel environmental correlates. This continues via review in Section 4 of microlevel event measurements and their aggregation as an acceptable predictor of overall continuous noise. Finally, I complete the microlevel study by speculation in Section 5 on those ice motions that can be taken to be plausible models for noise radiation. Taken as a whole, Sections 3-5 comprise an approach to understanding low frequency central Arctic noise that makes use of, and is consistent with, all available observations.

1. Macrolevel Issues

The study of macrolevel environmental correlates of Arctic ambient noise is complicated by the wide range of ice conditions of potential interest, and of environmental conditions encountered, each of which may be important in some frequency ranges and not in others. Ice conditions might be classified in the gross as those belonging to:

- the *central Arctic*

- the *marginal ice zone*, which resides between central ice and open water, and which can be further delineated by high and low ice concentrations

- the *shear ice zone*, which resides between central ice and ice at the shore line

- the *shorefast ice*, which is ice trapped by the shore's undulating shape.

Each of the foregoing ice types is affected in important ways by its thermal state. This is so because ice morphology depends on long time average freezing or melting conditions, and the ice strength itself depends on long time average temperature.

Since large scale atmospheric patterns and related environmental conditions can change on a time scale of a few days or so, one might also expect environmental correlates to relate on these shorter time scales to the ambient noise. Indeed, environmental conditions and related ice behavior already known to be important, or presumed so, are:

- horizontal stresses in the ice induced by winds and currents

- moments around the horizontal induced, again, by winds and currents

- shear stresses induced by diffusion of the ice field

- tensile stresses in the ice caused by heat flux between the atmosphere and the ocean

- impacts on the ice by snow or ice crystals set in motion by wind

- ice floe bumping associated with differential motion among adjacent ice floes

- flexural stresses induced by surface or internal gravity wave fields

- impacts caused by wave splashing at the water/ice edges

- cavitation and/or rapid ejection of ice particles associated with collapse of voids in melting ice.

Our interest in noise typically resides in the frequency domain from about 3 Hz to 100 kHz, which might be divided into low, mid, and high frequency ranges. I take these ranges to be, respectively, 3-100 Hz, 100-3000 Hz, and 3-100 kHz. All of the foregoing suggest a matrix of some 500 different conditions of ice type, thermal season, environmental forcing function, and frequency range, requiring a tall order of discrimination in researching macrolevel environmental correlates.

Milne's (1972) pioneering work in simultaneously observing air temperature and mid frequency noise under shorefast ice stands as a paradigm of research in macrolevel environmental correlation. More recent work of that genre in the central Arctic by Pritchard (1984) and Makris and Dyer (1986) will be discussed subsequently. While results to date are arguably important, much still needs to be researched in macrolevel environmental correlates.

2. Microlevel Issues

At the micro or more fundamental physical level, some five or more important event signatures have been observed (Townsend-Manning and Dyer, 1987). Each of these presumably can be related to the local ice motion which gives rise to the radiation. These signatures are important since it is likely that certain signature classes belong only to certain types of ice deformation and/or forcing functions. For example, quasi-periodic relaxation oscillations are sometimes observed in the Arctic Ocean and these may relate to stick/slip shear deformation processes.

With the possible exception of snow/ice crystal impacts or wave splashing, noise event mechanisms relate to ice fracture processes. Because they appear to be most responsible for noise, I will restrict attention in this review to fracture processes. Thus, for an event mechanism to be plausible it is necessary that the environmental forces acting on the ice cause local stresses which equal or exceed the ice strength for that particular deformation. Implicit in this is that ice fracture studies, and event physics, can only be pursued properly when one is assured that there is sufficient forcing of the ice to produce fracture. One barrier is that stress/strain laws for ice at low strain rates, such as obtained in the Arctic in response to environmental loads, are still not fully known, leaving some uncertainty on those environments that can lead to fracture. But enough is known of ice rheology and fracture strength to obtain at least an order of magnitude test of fracture plausibility.

Plausible models of ice motion must also explain details of the resulting event signatures. Pressure magnitude, characteristic frequency and bandwidth (or alternatively, time scale) and the directional properties of the event radiation, must flow from knowledge of the localized ice motion and must compare favorably with observations to gain confidence in the physical model.

Finally, the number and type of events should, in the aggregate, predict the continuous noise field.

The most difficult issue in microlevel event physics is experimental input on the detailed ice motion. To date, most of our knowledge on ice motion flows from simple models which are then empirically justified through measurement of the noise itself. This would seem to leave the entire intellectual process open to criticism. But this is no worse than in some other branches of physical modelling which we highly regard (e.g. earthquake seismology) and at least is essential in outlining those ice motion measurements we might care to make and those instruments we must develop to accomplish them.

3. Clues on Source Mechanisms from Correlates of Low-Frequency Central Arctic Noise

An earlier paper (Makris and Dyer, 1986) showed that low frequency central Arctic noise correlates, over time periods of 0(10 days), with stress S and stress moment M, in which both S and M are derived from wind and current shear stresses acting on horizontal surfaces of the ice, and from Coriolis and pressure gradient forces acting on vertical sections of the ice. The correlations were obtained between hourly averages of the noise and S or M but, as I discuss in Section 4, the noise is not a purely continuous process. It is a superposition of a large number of independent noise events, each as brief as 10 ms, and each presumably caused by fracture of the ice.

Pack ice dominates the central Arctic. [It consists of first- and multiyear floes, 2-3m thick, with joints between floes often consisting of heaped and refrozen ice blocks (known as pressure or shear ridges), or of thinner ice sheets resulting from refreezing of earlier ice openings (known as leads).] The pack ice covers the central Arctic nearly completely, which implies that considerable stresses can be accumulated in response to environmental loads, and which further implies that

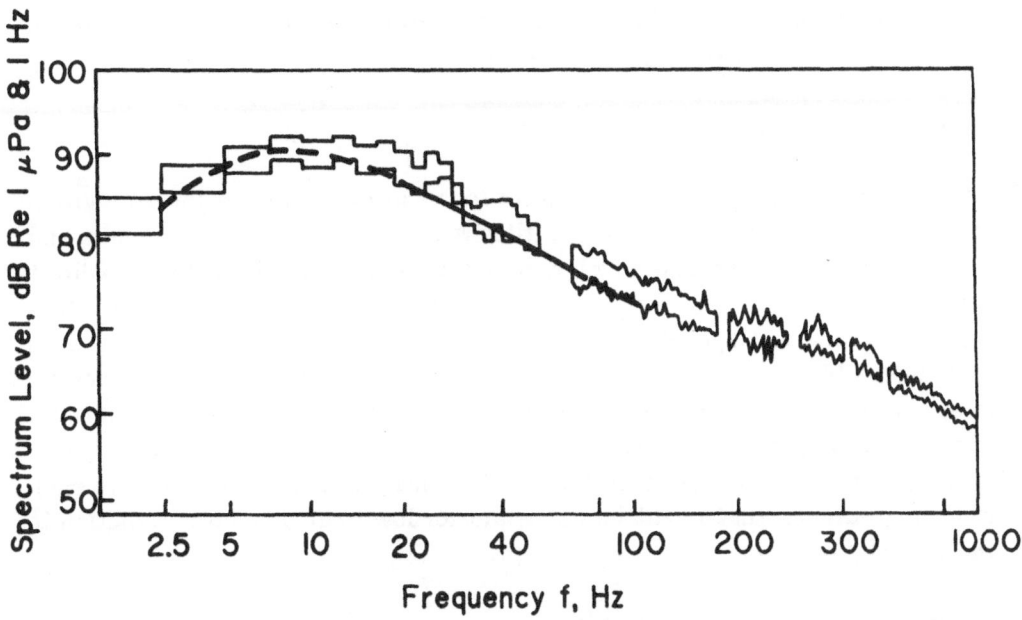

Figure 1: Prediction (shown as a thin line) and data (shown with a spread in level). Prediction based on event data from Fram IV, and on a basin theory (Townsend-Manning and Dyer, 1987), the dashed thin line representing extrapolations from the event data base. Average ambient noise data taken during Fram IV on 21 April 1982, 0808-0821Z, in 2.5 Hz bins.

because of their spatial/temporal infrequency need not be appealed to for a gross understanding of ambient noise. Research on this point, however, could lead to more conclusive analyses and ought to be continued. For example, the shortfall in prediction at some frequencies in Fig 1 might be due to active ridge events, although such shortfalls could be explained more readily by appeal to the range of data inputs used in the aggregate model.

5. Speculations on Ice Source Mechanics

I distinguish between *noise events* and *ice source mechanisms*, the former a group of acoustic observations which are quantitative and individually unambiguous, and the latter a set of ice motions which radiate via processes not yet subject to direct observation, and hence at best known only inferentially and speculatively. It is perhaps curious that speculation might be worthy of formal consideration. I do so with the objective of sorting among the relatively large number of ice processes to seek those that are plausible radiators, and to plan for possible experimental approaches to their verification.

Emphasis is placed in this Section on those ice processes, in response to environmental forcing, that might be important for low frequency noise radiation in the central Arctic.

5.1 Ridge Forcing by Moments

Makris and Dyer (1986) show that the environmental stress moment $\mathbf{M}$ is highly correlated with low frequency ambient noise. Since the thrust of Section 4 is that individual events in aggregate cause the noise, an individual ice fracture process driven by $\mathbf{M}$ is sought.

Ridges are the loci of drag forces induced by winds and currents which, together with normal stress caused by the horizontal ocean pressure gradients, give rise to the stress moment $\mathbf{M}$ acting about the ice sheet's central horizontal plane. This stress moment was measured; $|\mathbf{M}| h/L_M$ was found to have values between about 0.1 and 0.4 Pa-m (Makris and Dyer). These values were observed continuously over a 10 day period, and are implicitly an average over many ridges; h is sheet thickness and L_M is the horizontal scale normal to $\mathbf{M}$ through which the bending moment in the ice can be accumulated.

Figure 2 sketches a natural ridge system and conveys the concept of many individual hummocks or bumps, above and below the ice sheet, making up a ridge. The environmental moment M_e acting on one of these hummocks is, on average

$$M_e = K(|\mathbf{M}| h/L_M) b L_M \tag{2}$$

where K^{-1} is the fractional area of the ice canopy occupied by ridge hummocks

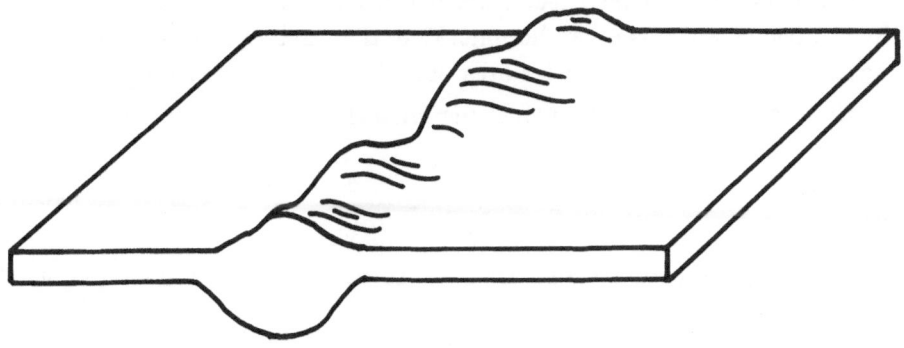

RIDGE LINE WITH HUMMOCKS

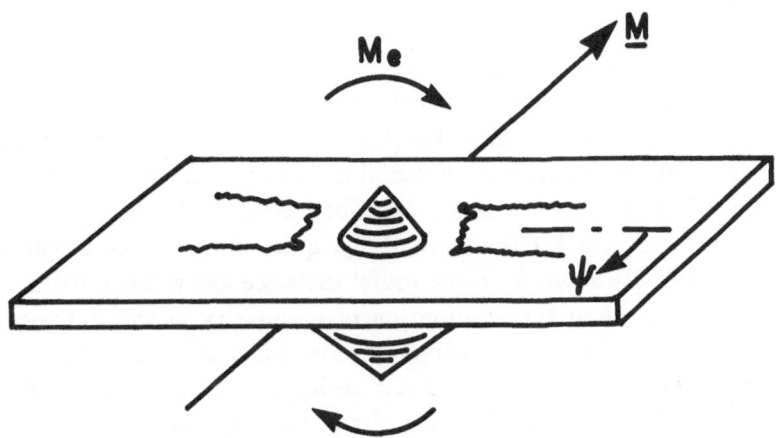

**TWO-CONE MODEL
OF HUMMOCK**

Figure 2: Sketch of a bumpy ridge line. Each bump or hummock is represented by a set of rigid cones which transfers the environmental moment M_e to the ice sheet. Radial cracks, then circumferential cracks, can be generated.

and b is a characteristic ice sheet scale parallel to **M**. Let S_b be the horizontal sectional area of an "average" hummock, d the "average" separation between hummocks, and S_o the smooth ice area enclosed by an intersecting grid of ridge lines. Then the number of hummocks per smooth region is $N_b \approx 2S_o^{\frac{1}{2}}/d$ and

$$K = \frac{S_o}{N_b S_b} \approx \frac{S_o^{\frac{1}{2}} d}{2S_b} \tag{3}$$

Estimates for the foregoing terms are $S_o \sim 10^6 m^2$, $S_b \sim 20m^2$, $d \sim 40m$, so that $K \sim 10^3$.

Because of its greater thickness, a multiyear hummock will have greater strength than the adjacent ice sheet. Thus I assume the hummock to be rigid and M_e to be transferred directly to the sheet. For simplicity, let each hummock be represented by a top and bottom rigid circular cone (Fig 2).

The qualitative nature of the ice deformation can be described. On the downward side of the cones the underside fibers of the ice sheet are in tension; on the upward side the top fibers are also in tension. Based on experience with distributed circular loads on ice (Michel 1978, Frankenstein 1968) and on calculations for localized moments on plates (Chen, pers. comm.), these fibers are apt to rupture first, producing a series of radial cracks and, in effect, forming a radial beam between the hummock and the exterior ice sheet. Then, for large enough loads, the beam will crack circumferentially, quite close to the larger diameter bottom cone, and ultimately at some radial distance away. Such behavior is not crucially dependent upon the assumption of circularity of the hummock; other sectional shapes give rise to qualitatively similar behavior (Michel).

Is M_e large enough for this dramatic series of events? I will not give the criterion for radial cracks, but I do for circumferential cracks which, since they develop later, should provide a suitable basis for acceptance of the entire ice deformation and fracture process.

Consider a hummock that has already developed radial cracks. These cracks form sectors which are expected to subtend quite small angles ($2\psi \lesssim 30°$), given that the stress depends upon $\cos\psi$. A vertical section through one of these sectors is shown in Fig 3. In effect the hummock (represented by the two rigid cones) is connected to the ice sheet by a beam of thickness h and width w.

The vertical displacement η of the ice sheet is a solution of

$$D\nabla^4\eta = -\rho g w \eta \tag{4}$$

Here D is its flexural rigidity given by $D = Eh^3w/12(1 - \nu^2)$, with E as Young's modulus and ν as Poisson's ratio. The product ρg is the foundation modulus per unit width provided by sea water the sheet floats upon (density times gravity acceleration). Equation (4) gives the static, linearly-elastic deformation in response

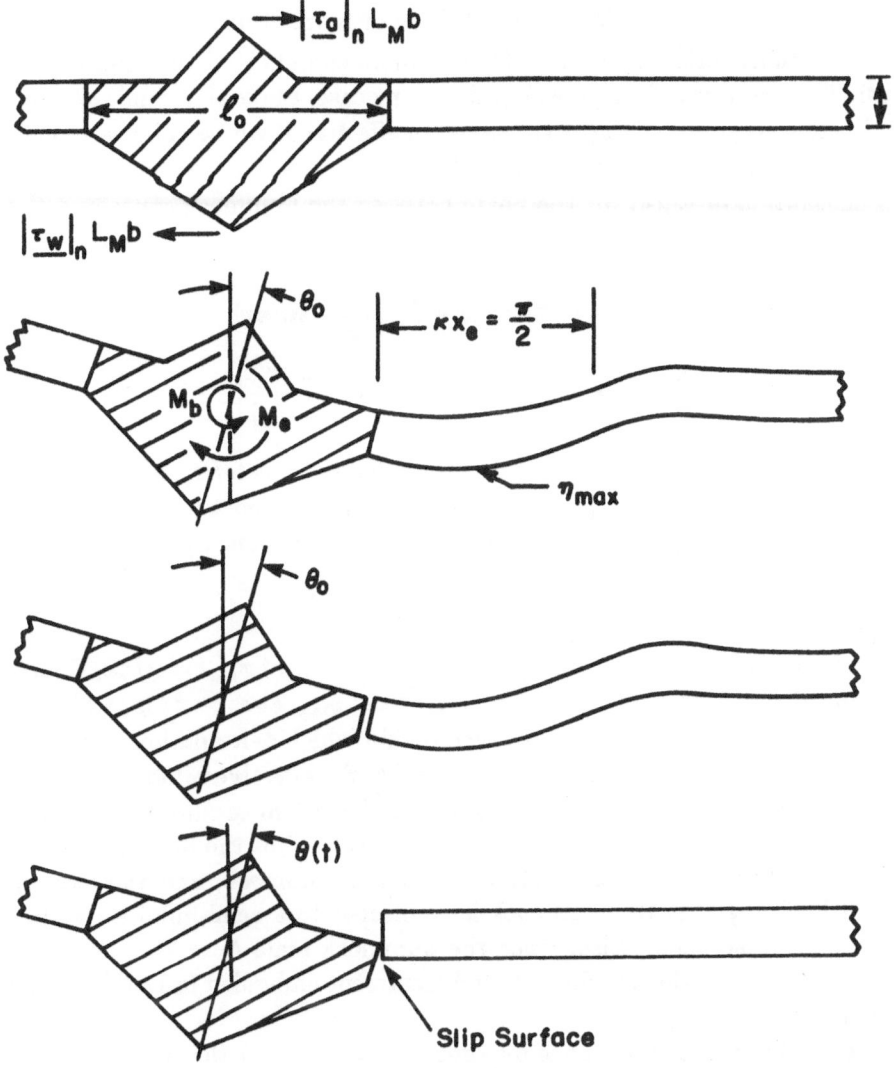

Figure 3: A vertical section through a ridge hummock and the beam between radial cracks. The sequence shows the system at rest, the angular deformation, the fracture at the connection between the rigid cones and the beam, and the return of the beam to its initial position. The environment generates a moment at the hummock, which is indicated symbolically by air and water drag forces in opposition. Use of the shear stress symbols τ emphasizes that these drag forces actually are estimated from measurements of average flow fields and boundary layers and/or shear coefficients (Makris and Dyer, 1986).

to loading, such as from M_e. Linear elasticity is a *dubious* assumption for ice, but other assumptions taking account of one or more of its rheological properties is either equally dubious (Mellor 1983) or more complicated than needed for the plausibility arguments being developed. In any case, I take the standard approach by adjusting solutions of Eq(4) via use of an effective E whose value is a surrogate for some of the rheological properties of ice.

The solution of Eq(4) for the beam between the hummock cones and the full ice sheet is (Hetenyi 1946)

$$\eta = \eta_o e^{-\kappa x}[cos\kappa x + (1 + \frac{2}{\kappa \ell_o})sin\kappa x]$$

$$M = \frac{M_e - M_b}{2}e^{-\kappa x}\left[cos\kappa x - \frac{sin\kappa x}{(1 + \frac{2}{\kappa \ell_o})}\right] \tag{5}$$

where x measures distance from the larger cone of diameter ℓ_o, M is the bending moment, M_b the buoyancy moment applied to the hummock as its rotates, and κ an inverse length measuring the region of major distortion. Specifically

$$\kappa h = [3\rho g h(1 - \nu^2)/E]^{\frac{1}{4}} \tag{6}$$

The beam reaches a maximum deflection $\eta_{max} \approx 1.5\eta_o$ at $x_{max} = \pi/4\kappa$, where η_o is the value at $x = 0$, and is down some 17 dB from η_{max} at $x_e = \pi/2\kappa$, which can be used as a rough measure of the distance within which deformation is important.

For the very small strain rate induced by M, the effective value $E \sim 10^9 Pa$ can be used (Mellor 1983). Then, for h=2.5 m, $\kappa^{-1} \sim 27$ m or about 11 sheet thicknesses, and $x_e \sim 42$m (about 17 sheet thicknesses). Thus the beam's deformation is restricted to distances relatively close to the hummock, and the assumption implicit in Eq(5) that the hummock is connected to a semi-infinite ice sheet on each side is reasonable. Also, since the hummock scale $\ell_o \sim 25$ m is not small compared with κ^{-1}, the one dimensional treatment embodied in Eq(5) is similarly reasonable.

The total bending moment is maximum at $x = 0$, and the buoyancy moment can be shown to be small. Thus $M_{max} = M_e/2$, and the maximum flexural stress in the one-dimensional approximation is:

$$\sigma = \frac{6M_{max}}{h^2 \ell_o} = \frac{3K}{h^2 \ell_o}(|\mathbf{M}| h/L_M)bL_M \tag{7}$$

Take the parenthetical term as observed, i.e. 0.1-0.4 Pa-m, the parallel scale $b \approx 2d/\pi \sim 25$m on the basis that the direction of the applied moment is uniformly distributed in angle with respect to the ridge line, $L_M \sim 2000$ m as the normal scale over which the stress moment accumulates (this is essentially the estimated distance between major cracks), K$\sim 10^3$, h = 2.5m, and $\ell_o \sim 25$ m. Then the maximum flexural stress $\sigma \sim 0.1 - 0.4$ MPa. The flexural strength of sea ice ranges from about 0.2 to 1.0 MPa, dependent upon brine volume (Weeks and Cox, 1984). Within an order-of-magnitude, therefore, the fracture of ice by action of the environmental stress moment M is plausible.

Once broken at its root, the beam can return to its rest position. (See the last two sequences in Fig 3.) If it does so with zero friction at the cracked surfaces, the characteristic frequency of its *dynamic* motion is estimated as

$$f \approx c_g/x_e = 2\kappa c_g/\pi \tag{8}$$

where c_g is the group speed of flexural waves. The latter has been studied by Stein (1986) for an ice sheet floating on water, and it can be noted that for the high strain rate expected in dynamic motion, the assumption of linear elasticity is quite good. Thus c_g is derived from linear elasticity, as is κ, but this assumption is robust for the former and dubious for the latter. Adoption of the parameters used previously gives $f \approx 11$ Hz, a value quite consistent with the range of observation. I expect κ to differ from the value estimated here based on variations in temperature, ice salinity, brine volume, and environmentally induced strain rate, so that a fairer statement of this result is $f \gtrsim O(10\text{Hz})$, the lower bound statement relating to the experience that the plastic nature of ice always leads to higher values of κ than estimated.

The effective dipole force associated with the beam's transient motion is, from the dipole model $F = \ell \partial^2 m/\partial t^2$, where ℓ is vertical separation between accelerated mass m and its image above the water column. Thus, for narrowband events,

$$F_o = (2\pi f)^2 \rho \ell w \int_0^\infty \eta \, dx = \frac{(2\pi f)^2 \rho \ell w \eta_o}{\kappa} \left[1 + \frac{1}{\kappa \ell_o}\right] \tag{9}$$

where

$$\eta_o = \frac{M_e \kappa^3 \ell_o}{2\rho g b}$$

For the assignments used previously I get for the displacement and rotation at the root $\eta_o \sim 13\text{-}52$ mm and $\theta_o \sim 0.6 - 2.4$ deg, values that hardly would be sensed by the casual observer, yet enough to have the potential of ice fracture. Also with $w\sim 5$m from observations on broken ice at ridges (Tucker and Govoni, 1981; Tucker et al, 1985; Welsh 1983), and the vertical dipole separation $\ell \sim h$, Eq(9) yields $F_o \sim 5 - 20$ MN. This is of the proper order in comparison with the event observations discussed in Section 4.

With the assumption that the plate unloads from its deformed to its rest position as a smooth monotonic function of time, such as a Gaussian (Dyer 1984), the radiated acoustic pressure, since it is proportional to the third derivative of displacement, would display a leading and a trailing peak of one sense, and a central peak of the other sense, at the mean frequency governed by Eq(8). Such a signature is most commonly observed at low frequencies, lending further support to the model.

I conclude that moment excitation of pack ice is a plausible cause of ice fracture and consequent event radiation. It develops appropriately high stress, causes dynamic motion in the proper frequency range, and has the proper level of source strength.

A crucial physical result is implicit in the foregoing model. The ice had to be *deformed* to cause fracture; indeed no plausible model is evident to me, for observed values of M, that leads to fracture without local deformation and consequent stress concentration. It is the deformation, when relieved, to which I attribute the acoustic event. Implicitly this motion is taken to be more important than the cracking process itself, and this deserved discussion.

For the crack, defined to be the physical sliding or opening itself, one would expect a characteristic frequency given by (Das et al, 1986)

$$f \approx c_r / \Lambda \tag{10}$$

where c_r is the Rayleigh phase speed and Λ the crack sliding or opening length. With $c_r \approx 1700 m/s$ (Stein 1986) and $\Lambda \sim 5m$ we get $f \sim 340$ Hz, so that such cracking motions might typically be important at frequencies considerably higher than that for deformation unloading. Indeed our experimental work often shows an increase in spectral level around 300 Hz (see Fig 1), which could be attributed to direct crack effects. (Alternatives for explaining this increase are thermal-rather than moment-induced cracking, and stick/slip motions associated with ice shear or diffusion.) But to return to the main point, I favor the unloading model for low frequency radiation, at least because the frequency of the direct crack process is an order-of-magnitude higher.

Langley (1987) has studied radiation from ice sheets containing defect motions such as cracks. While his theory is general, he considers applications to opening (but not sliding) cracks which do not penetrate the entire ice thickness. Although I suggest opening radial cracks and then a sliding root crack, with both types penetrating the entire ice sheet, his findings are nonetheless of interest. For his applications, he cautions that energy in flexural and longitudinal waves carried in the ice sheet from the crack can radiate significantly, especially from discontinuities. The fast crack processes (governed by Eq(10)) then might be unidentifiable as originating from a single locus, but still might contribute importantly to the acoustic spectrum in the 300 Hz region. The slow process governed by Eq(8), which I call unloading, could also be affected by scattering-induced radiation of iceborne waves. Langley's theory implicitly includes such slow processes too, but the method used here to estimate the dipole force, with neglect of iceborne energy, is perhaps a more direct approach. The fact that we can detect a single locus for the slow unloading process lends comfort to the approach, as does the order-of-magnitude agreement between basin aggregation and measured average noise, but his caution is considered in the next paragraph.

The acoustic energy radiated at the source by the dipole can be written as

$$\mathcal{E}' = \beta(ka)^2(k\ell)(fT)\mathcal{E}$$

where the terms multiplying $\mathcal{E}$ form the radiation efficiency, and where $\mathcal{E}$ is the flexural energy stored in the beam before its transient motion. (Clearly I can neglect longitudinal energy for the deformation considered.) The acoustic energy radiated by a scatterer, taken as similar to radiation by the source, that is, caused by a force exerted by the scatterer on the plate, is

$$\mathcal{E}'_1 = \beta(ka_1)^2(k\ell_1)(fT)\mathcal{E}_1$$

where $\mathcal{E}_1$ is the energy of vertical motion at the scatterer. (I neglect angular motion, since its radiation efficiency is smaller, and its corresponding value of $\mathcal{E}_1$ is also smaller.) In the foregoing expressions k is the acoustic wavenumber, a the equivalent source radius, a_1 the equivalent scatterer radius, ℓ_1 the force-image separation, and T the event duration. For a scatterer at range R, I estimate

$$\mathcal{E}_1 \approx \frac{4\ell_1 a_1^2}{\ell a R} e^{-R/R_e} \left(\frac{Z}{2\pi f M}\right)^2 \mathcal{E}$$

where R_e is the e-folding attenuation distance for flexural waves, and where the parenthetical term under the square is the ratio of plate impedance to scatterer impedance, and which I take from classical flexural theory and a simple hummock model to be $c_p^2 h/(10\pi f^2 a_1^3)$, where c_p is the phase speed of flexural waves. This ratio must be 0(1) for validity of $\mathcal{E}_1$. The total energy radiated by the scatterers is

$$\mathcal{E}'_{tot} = \int \mathcal{E}'_1 dN_1 = \frac{2\pi}{KS_b}\int_0^\infty \mathcal{E}'_1 R dR \tag{11}$$

where K is given by Eq(3). For 2.5 m Arctic ice, Stein (1986) estimates $R_e \approx 1500/f^{0.6}$ m and $c_p \approx 80 f^{0.6}$ m/s, with f the frequency in Hz. Then with assignments used previously and $a \sim 3m$, $a_1 \sim 6m$, and $\ell_1 \sim 6m$, I estimate the ratio γ of total scatterer-radiated energy to energy radiated directly from the source to be

$$\gamma = \frac{8\pi}{KS_b}\left(\frac{a_1}{a}\right)^4\left(\frac{a}{R_e}\right)\left(\frac{\ell_1}{\ell}\right)^2\left(\frac{Z}{2\pi f M}\right)^2 \sim 10^{-3}f^{-1} \tag{12}$$

This estimate is valid for $f < 5$ Hz, within which the impedance ratio $Z/2\pi f M$ is 0(1). A better estimate for $f > 20$ Hz is obtained by replacing $Z/2\pi f M$ with 2; the impedance ratio is very small for higher frequencies, telling us that the scatterer is relatively immobile and that the flexural motion adjacent to the scatterer is twice as large as the incoming flexural wave. The adjacent region is then the radiation locus, and

$$\gamma = \frac{32\pi}{KS_b} \left(\frac{a_1}{a}\right)^4 \left(\frac{a}{R_e}\right) \left(\frac{\ell_1}{\ell}\right)^2 \sim 10^{-3} f^{0.6} \tag{13}$$

Both Eq(12) and (13), and a reasonable interpolation between their ranges of validity, demonstrate that the scatterers are unimportant over the entire frequency range of interest. This is so even though there can be many scatterers. From Eq(11) the number of scatterers within R_e is

$$N_e = \frac{\pi R_e^2}{KS_b}$$

which gives $N_e \sim 94$ at 3 Hz to 1.4 at 100 Hz. Evidently, energy associated with an individual scatterer is small enough (due to source-scatterer distance) to render the total scattering negligible. Thus the individual source event dominates.

5.2 Ice Sheet Overthrusting caused by Horizontal Stresses

Makris and Dyer also show that the horizontal stress S is correlated with low frequency noise. Thus I speculate on a model that can explain this observation. Because many of the ideas closely relate to those in Sec. 5.1, my description here will be rather more sketchy.

Figure 4 illustrates a situation in which one ice sheet can be thrust upon another by action of horizontal forces. Two are illustrated; one shows ice being thrust over a ridge rotated by moments, the other shows a thinner sheet thrust over the thicker one.

The solution to Eq(4) in one dimensional approximation is

$$\eta = \frac{P}{2D\kappa^3} e^{-\kappa x} \cos\kappa x$$
$$M = -\frac{P}{\kappa} e^{-\kappa x} \sin\kappa x \tag{14}$$

with P the vertical load at the overthrusting site caused by wedge action in response to the horizontal force. In this circumstance linear elastic theory predicts a maximum fiber stress at $x_b = \pi/4\kappa$, but experience suggests its value is less than that and probably more like $1/3\kappa$. Should the ice break at this position of maximum stress its dynamic frequency for the first free-free mode would be estimated as

$$f \approx \frac{3}{4} \frac{c_p}{x_b} \approx \kappa c_p \tag{15}$$

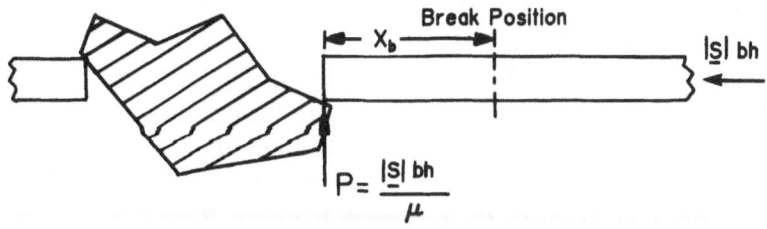

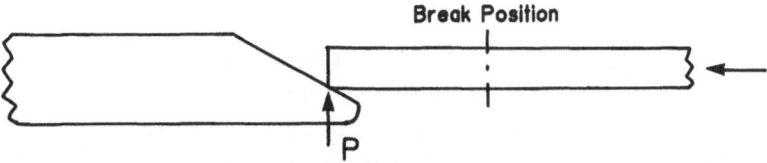

Figure 4: Overthrusting of a sheet on a ridge line (top) and a thinner sheet on a thicker one (bottom). The horizontal force is magnified and redirected to a vertical force by sliding action.

For values used previously $f \gtrsim 0(10)$ Hz, which is in the same range as for moment excitation, and supports the notion that horizontal stress can be important at low frequencies.

Unlike moment-induced deformation and dynamic unloading, the dynamic motion here could be expected to be a damped sinusoid, the broken ice segment serving as a free-free resonant system. Such signatures also are observed at low frequencies, but to a lesser degree than the signature attributed to moment forcing.

Will the ice break? The maximum flexural stress is

$$\sigma = \frac{6M_{max}}{h^2 b} = \frac{1.7(|S|h/L_s)L_s}{\mu \kappa h^2} \tag{16}$$

With use of experimental values for the parenthetical term of 0.02-0.2 Pa, $L_S \sim 100km$ (this is essentially the scale between major leads or openings in the ice), and the coefficient of sliding friction $\mu \sim 0.2$ (Mellor 1983), I get $\sigma \sim 0.08 - 0.8$ MPa. This is to be compared with the range of flexural strength: 0.2-1.0 MPa. Thus, at least at the upper end of this estimate, a 2.5 m ice sheet would be plausibly broken by observed horizontal stresses.

Significantly thinner ice, such as in recently refrozen leads, could break at the lower values of S (see Eq 16). Indeed one would expect all such thin ice to be consumed first and thicker ice, if broken at all, to be consumed in the latter stages of applied horizontal stress.

The source strength of such motion would be comparable to that for moment forcing. The length scales are comparable, as are the strains, since they must lead to the same stress for breaking. Consequently there is no need here to estimate the dipole source strength. Thus one can say that the horizontal stress model is plausible. It has frequencies in the appropriate range, it predicts signatures such as observed, it has the right source strength, and sufficient stress is developed by the observed S for breaking.

6. Summary and Sermon

This review has concentrated on plausible mechanisms for acoustic radiation in central Arctic pack ice at low frequencies. The arguments used here indicate the strong possibility that moments at ridge hummocks and horizontal stresses at overthrusting sites can account for the noise observed in the Arctic, at least for pack ice at low frequencies. The challenge remains to instrument ice either in the laboratory or in the field to test the merit of these speculations.

I make no claim for the validity of these models for higher frequencies at which, indeed, I expect other mechanisms to be of equal or more importance. For example, thermal cracks and other cracking actions are expected to be important. Also ice diffusion is known to occur and thus horizontal shearing motions are expected to contribute. Additionally, in the marginal ice zone, the absence of horizontal stress, and the presence of strong shearing, bumping, and gravity-wave stress, suggest still a different mix.

The program for fundamental ambient noise research in the Arctic must include mechanistic studies that carry with them at least some sense of plausibility. In my experience one can easily be deflected into seemingly reasonable avenues of model development and study. But until such time as assurance comes that ice can actually be broken, sheared, and deformed by the environmental forces nature has provided, the model should be put aside. More strongly, models should be tested not only against the reality of actual environmental loads, but also by comparison with observed frequencies, signatures, and source strengths of acoustic radiation, before they are seriously proposed as possibilities.

References

B. M. Buck and J. H. Wilson, "Nearfield noise measurements from an Arctic pressure ridge," J.Acoust.Soc.Am. 80, 256-264 (1986).

S. Das, J. Boatwright, and C. H. Scholz, Eds., *Earthquake Source Mechanics*, Geophysical Monograph 37, American Geophysical Union, Washington, DC (1986).

I. Dyer, "The song of sea ice and other Arctic Ocean melodies," in *Arctic Technology and Policy*, edited by I. Dyer and C. Chryssostomidis, McGraw Hill, N.Y., (1984), pp 11-37.

G. E. Frankenstein, "Strength of ice sheets", Technical Memorandum 92, pp 79-87, National Research Council, Canada (1968).

V. P. Gavrilo, V.D. Grishchenko, and V.S. Loschilov, "The problem of full scale study of the morphology of hummocks in Arctic ice and the possibilities of modeling hummocking processes", in *Dynamics of Ice Cover*, L. A. Timokhov, Ed. Gidrometeoizdat Publishers, Leningrad (1974). Translated by Amerind Publishing Co., New Delhi, (1984).

M. Hetenyi, *Beams on Elastic Foundation*, U. Michigan Press, Ann Arbor, Michigan (1946).

A. J. Langley, "Acoustic emission from the Arctic ice sheet", submitted to J. Acoust. Soc. Am. (1987).

N.C. Makris and I. Dyer, "Environmental correlates of pack ice noise", J. Acoust. Soc. Am. 79, 1434-1440 (1986).

M. Mellor, "Mechanical behavior of sea ice", Monograph 83-1, Cold Regions Research and Engineering Laboratory, Hanover, NH, (1983).

B. Michel, *Ice Mechanics*, Laval University Press, Quebec (1978).

A. R. Milne, "Thermal tension cracking in sea ice: A source of underwater noise", J.Geophys.Res. 77, 2177-2192 (1972).

R. S. Pritchard, "Arctic Ocean background noise caused by ridging of sea ice", J. Acoust.Soc.Am.75, 419-427 (1984).

P.J. Stein, "Acoustic monopole in a floating ice plate", MIT thesis, (1986).

M. Townsend-Manning and I. Dyer, "Acoustic events radiated by Arctic ice", submitted to J.Acoust.Soc.Am. (1987).

W. B. Tucker III and J. W. Govoni, "Morphological investigations of first-year sea ice pressure ridge sails," Cold Regions Sci. and Tech., 5, 1-12 (1981).

W. B. Tucker III, A. J. Gow and W. F. Weeks, "Pressure ridge morphology and physical properties of sea ice in the Greenland Sea," Proc. Arctic Oceanography Conference and Workshop, NORDA, 214-223, June (1985).

W.F. Weeks and G.F.N. Cox, "The mechanical properties of sea ice: a status report", Ocean Sci. and Eng., *9*, 135-198 (1984).

J. P. Welsh, "Characterization of sea ice", J.Acoust.Soc.Am. *74*, S20 (1983).

Observation of the Sound Radiated by Individual Ice Fracturing Events

Peter J. Stein[*]
Atlantic Applied Research Corporation
4 A Street
Burlington, Massachusetts, 01803
USA

ABSTRACT. This paper discusses some of the observables important to the study of sound radiated into the water by ice fracturing events. Such studies are important in order to understand the mechanisms of Arctic Ocean ambient noise generation. Theoretical work for a homogeneous ice plate indicates that there are three important paths of elastic wave propagation over which energy travels away from an ice event source. These are the flexural wave in the ice, the longitudinal wave in the ice, and the acoustic wave in the water. Each has a distinctive propagation speed, attenuation, and associated pressure field in the water. The experimental work, consisting of the interpretation of four ice events, supports the theoretical work. In each case the event is located and the contributing wave paths identified. In all cases the acoustic wave dominates. The work also demonstrates how event analysis can be used to determine ice properties.

1. INTRODUCTION

The sound radiated by ice fracturing events is the predominant source of Arctic ocean ambient noise. On average, at a given time and place, the noise is a summation of the sound radiated by a great number of ice events. The contributing events occur within an area limited by either sound attenuation in the water or the Arctic Ocean basin edges. However, there will be events strong enough, occurring near enough to the observation point, that their acoustic radiation will dominate over the average background levels. This generally occurs when either the background levels are low or when there is some relatively active nearby reformation of the ice taking place.

This work discusses the observation of such singular events whose acoustic radiation can be distinguished from the general background noise. The reason for studying these events is to develop an understanding of how event statistics (number in time, number in space, strength, directionality, etc.) relate to the environmental conditions. From this we might learn how the environmental loading drives these events and how these events sum to form the background Arctic Ocean ambient noise.

Before we can develop statistics of singular ice event noise, we must think carefully about how an individual event radiates sound and how to interpret its recorded signature. An ice event can transmit energy to a receiver in the water along several paths of elastic wave propagation in the ice-water system. An understanding

[*]Work performed while author at MIT, Dept. of Ocean Engineering, Cambridge, Mass.

533

B. R. Kerman (ed.), Sea Surface Sound, 533–544.

of these paths is required in order to interpret the signature of an event and deduce the source characteristics.

This paper first reviews the theoretical characteristics of the waves which radiate from a localized source in a homogeneous ice plate floating on an infinite half space of water. Using the theoretical work as a guideline, we then give our interpretation of four ice events. In each case we identify the contributing waves and locate the event using the acoustic signatures (inter-sensor arrival times). The first is a lake ice event induced by thermal stresses. The second is an Arctic ice event recorded during a period when the ice was relatively active (a ridge, a formation which results from the crushing together of two ice plates, was forming near by). The third is an ice event recorded during tranquil ice conditions. The last event is not a naturally occurring ice event, but a large explosive charge set off in the water under Arctic pack ice.

2. PATHS OF PROPAGATION FOR A HOMOGENEOUS ICE PLATE

2.1. Overview

As an approximate model we theoretically investigate the case of a homogeneous infinite ice plate floating on an infinite half space of constant sound speed water. We are concerned with the propagation speeds and attenuations of the elastic waves which radiate from a localized source in the ice, then travel in the ice plate and in the water below. This analysis was first performed by Press and Ewing [1] and extended to include absorption in the ice by Stein [2,3]. Greater details of the elastic waves discussed here, and the dependance of their characteristics on the properties of the ice, can be found in these references.

The theoretical work shows that below a frequency-ice thickness product of about 300 Hz-m there are three waves which can radiate from a localized source in the ice. These are the flexural wave in the ice, the longitudinal wave in the ice, and the acoustic wave in the water. Above a frequency-ice thickness product of about 300 Hz-m only the acoustic wave can propagate away from the source with attenuation less than 0.1 dB/m. The higher order modes of propagation in the ice-water system, including surface waves such as the Stoneley or Rayleigh waves, are highly attenuated. Thus we expect that for a given event, at a range of, say, more than 100 m, only the flexural, longitudinal, and acoustic wave can be observed. Let us review the properties of these three waves; their paths of propagation from source to receiver are depicted in Figure 1.

2.2. The Flexural Wave

From a localized source in the ice, the flexural wave radiates cylindrically. For nominal ice conditions the phase speed of the flexural wave is given roughly by

$$c_f = 40 \ (fh)^{0.6} \ \text{m/s} \tag{1}$$

where f is frequency in Hz and h is ice thickness in m. Note the phase speed of the flexural wave is subsonic (slower than the speed of sound in water, roughly 1500 m/s) and dispersive (wave speed is a function of frequency, in this case increasing with frequency).

Since the flexural wave is subsonic its associated pressure field decays exponentially with depth into the water; no energy is radiated away from the ice unless the flexural wave strikes a discontinuity. This exponential decay is depicted by the

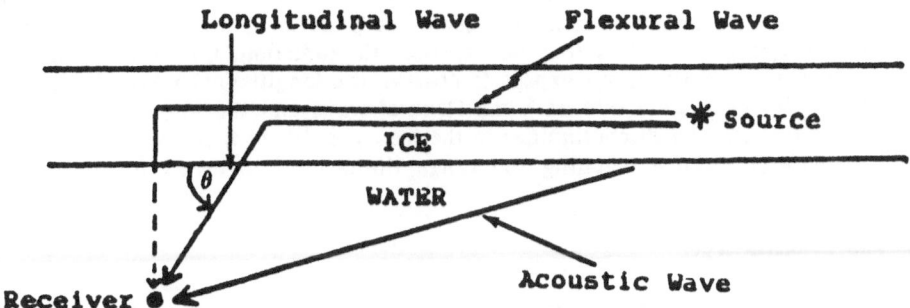

Figure 1. Diagram of the three wave paths over which energy is likely to pass from an ice event source to a receiver in the water. The longitudinal wave travels supersonically in the ice and radiates at angle θ to the receiver. The flexural wave travels subsonically in the ice and has an associated pressure field which decays with depth into the water. The acoustic wave travels directly from the source region to the receiver in the water.

dashed line in Figure 1. Thus we expect to observe the flexural wave only relatively close to the ice.

Since the flexural wave is dispersive its signature has a distinct characteristic with lower frequencies arriving after higher frequencies. Also, the observed speed of propagation is given by the group speed (speed of energy propagation), which in this case is 2.5 times the phase speed.

The flexural wave also suffers moderate attenuation due to absorption in the ice. For ice 3 m thick the attenuation of the flexural wave due to absorption is given roughly by

$$a_f = 0.003 \; f^{0.5} \; dB/m \tag{2}$$

where again f is in Hz. Using the 10 dB down point of absorptive loss as a definition of limiting range, under nominal Arctic conditions, above say 10 Hz, we expect to be able to observe the flexural wave out to about 1 km.

2.3. The Longitudinal Wave

Like the flexural wave, the longitudinal wave also radiates cylindrically away from a localized source in the ice. However, it propagates with frequency independent speed, which for nominal ice conditions is

$$c_l = 3100 \; m/s \tag{3}$$

This wave is supersonic, travelling at roughly twice the sound speed in water. It therefore radiates energy into the water, and we need not be close to the ice to observe the associated pressure disturbance. Acoustic energy from the longitudinal wave radiates away from the ice at an angle θ (given by Snell's law), which is roughly 60 degrees. This angle is shown in Figure 1.

The attenuation due of the longitudinal wave in ice 3 m thick is roughly

$$a_l = 0.0002 \; f \; dB/m. \tag{4}$$

This theoretical attenuation in a homogeneous ice plate is dominated by absorption in the ice which is greater than that due to acoustic radiation. Given this attenuation, above a frequency of 10 Hz we expect to observe the longitudinal wave out to several kilometers. However, note that both the flexural wave and the longitudinal wave can be further attenuated by discontinuities in the ice. It is likely that both of these waves will be scattered before reaching the range limitations governed by absorption in uniform ice.

2.4. The Acoustic Wave

When the source of radiation is localized, such as in an ice event, there is direct radiation into the water from the source region. This radiation spreads spherically in the water at the speed of sound in water. We refer to this wave as the acoustic wave. If all three waves are received, the longitudinal wave arrives first, the acoustic wave second, and the flexural wave third.

Since the energy in the acoustic wave is carried in the water, it suffers relatively little attenuation due to absorption and at low frequencies is relatively unaffected by discontinuities in the ice. For this reason, at far enough range, the acoustic wave will dominate the signature of an ice event.

3. DATA PRESENTATION

3.1. A Lake Ice Event

The first event was recorded in a lake which was approximately 1 km long, 1 km wide, and 6 m deep. The ice was 0.3 m thick and appeared to be relatively homogeneous. A three sensor horizontal hydrophone array was used. The sensors were located 2 m below the ice. The acoustic time history signature of the event is shown in Figure 2. The arrival of all three wave can be clearly seen. The longitudinal wave arrives first, roughly 0.1 s into the trace, and the acoustic wave arrives next, roughly 0.15 s into the trace. The flexural wave arrival follows the acoustic wave arrival and has ever lower frequencies arriving later. The acoustic wave arrival is strongest.

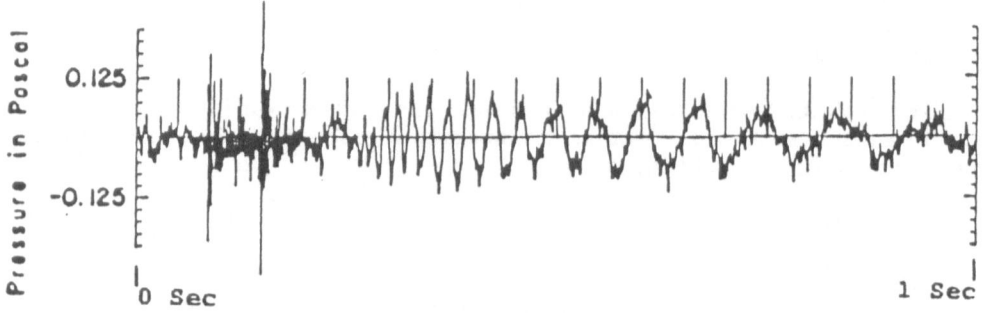

Figure 2. Pressure versus time trace for a lake ice event. Data are bandpass filtered from 10 to 1000 Hz. Trace is 1 s long with 0.05 s tick marks. Depth of hydrophone is 2 m. Location of event determined to be at a horizontal range of 170 m. Event recorded at 05:46:48 EST on 18 January 1984, Canaan Street Lake, Canaan, New Hampshire.

Inter-sensor arrival times were used to determine the longitudinal and acoustic wave speeds. The acoustic wave was found to propagate at 1430±20 m/s and the longitudinal wave was found to propagate at 3350±100 m/s. These values are in good agreement with theoretical values. The longitudinal wave is slightly faster than the theoretical value given in Eq. 3, possibly due to the higher stiffness of fresh water ice. Using the difference in arrival time of the longitudinal wave and the acoustic wave, along with their measured wave speeds, the event was located 170±5 m from the sensors. The flexural wave arrival pattern closely matched that predicted by using an impulsive source, the estimated location, and the theoretical wave speed given in Eq. 1.

3.2. An Arctic Ice Event.

The next event to be interpreted was recorded during the Fram II experiment when an ice ridge was building nearby. This event was recorded on 24 channel horizontal hydrophone array, with an approximate aperture of 1 km, with the sensors at 90 m depth. The event as recorded at one of the hydrophones is shown in Figure 3. The signature appeared roughly the same at all of the hydrophones, except for a decrease in level farther away from the event. If more than one wave type arrived strongly at the hydrophones the event signature would change with range (since the three wave types travel at different speeds). Thus, one of the three possible waves dominates the event shown in Figure 3. The flexural wave can be discounted because of the depth of the hydrophones. Estimation of the speed of propagation can tell whether it is the longitudinal or acoustic wave.

Inter-sensor time delays were estimated using cross-correlations. These time delays were used to locate the event by finding the location which gave the smallest least-squares error in a linear regression fit of range versus time delay [2]. This method also gives an estimate of the wave propagation speed. The results are shown in Figure 4 where arrival time is plotted versus range. The slope of the line in Fig. 4 gives a wave speed of 1450±12 m/s. This indicates that the acoustic wave is the dominant wave arriving at the hydrophones. The event was located 680±25 m from the array apex. Further review of the signature at each hydrophone, using the estimated location and the theoretical longitudinal wave speed, concluded that the longitudinal wave was not detected.

3.3 Another Arctic Ice Event

The next event to be discussed is also a naturally occurring ice event. However, it differs from the last event in that it was recorded during tranquil ice and low ambient noise conditions. This event was recorded during the Fram IV experiment, also on a 24 channel, approximately 1 km aperture, horizontal hydrophone array with the sensors at 90 m depth. Its acoustic signature at 20 of the hydrophones is shown in Figure 5. The signature at a vertical geophone attached to the upper surface of the ice is shown in Figure 6.

Arrival times were estimated for the most energetic part of the event, which was assumed to be the acoustic wave, and used to locate the event via the same method as the last event. These arrival times are indicated by the upward arrows in Figure 5. The result of the location is depicted in Figure 7. The event was located 300±25 m from the array apex. The wave speed for the most energetic arrival was estimated to be 1430±40 m/s. Certainly this arrival is the acoustic wave.

Estimated arrival times for the longitudinal wave, using the above location and a wave speed of 3100 m/s, are given by the downward arrows in Figure 5. Indeed a weaker arrival precedes the most energetic at most of the hydrophones. Note also that

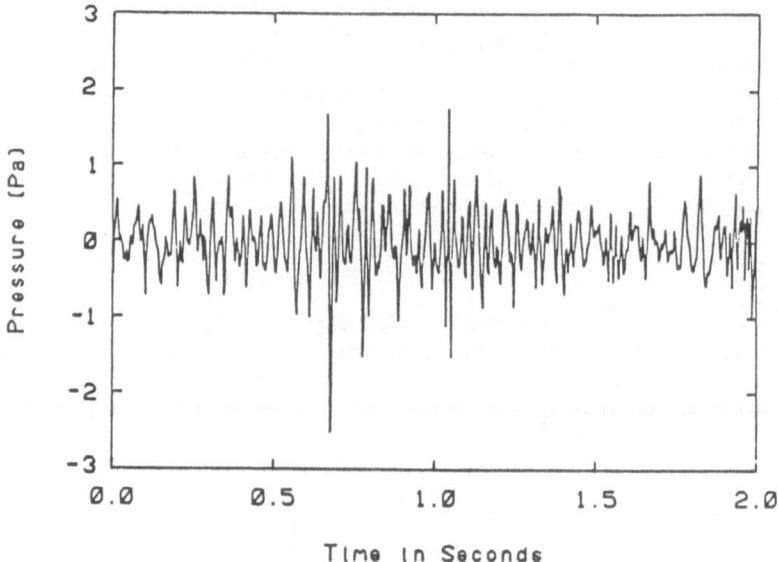

Figure 3. Pressure versus time trace for a Fram II ice event which occurred on 18 April 1980, at approximately 0530 Z. Data are bandpass filtered from 15 to 80 Hz. Event range for this sensor estimated to be 680 m.

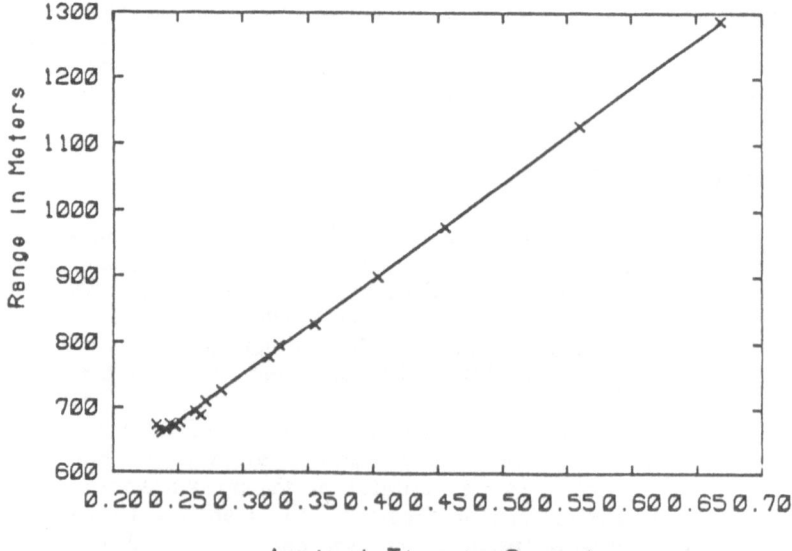

Figure 4. Time delay versus range for the event shown in Figure 3. Time delays estimated by cross-correlation. Slope of best-fit line gives a wave speed of 1450 m/s indicating that the acoustic wave is dominant.

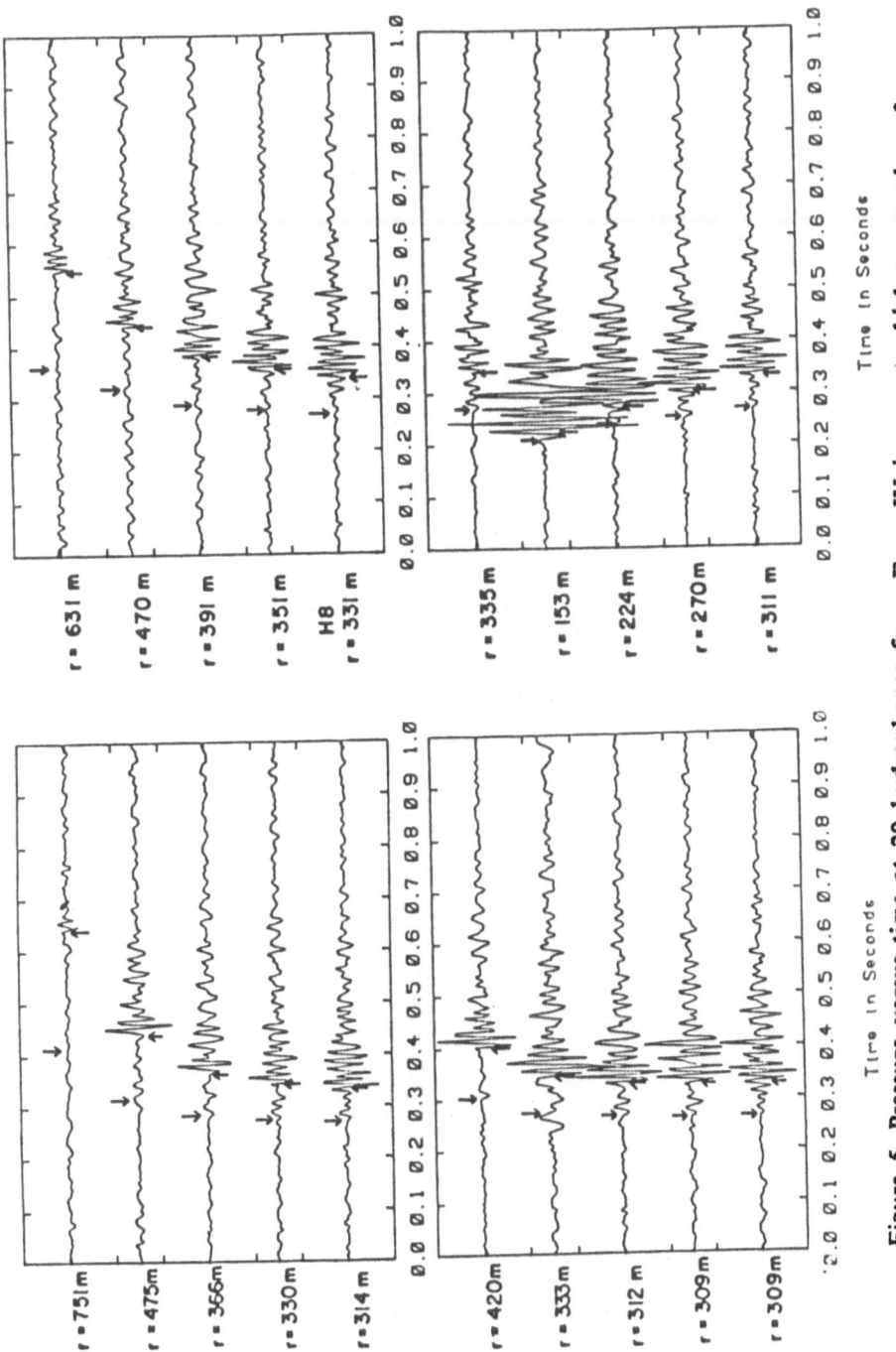

Figure 5. Pressure versus time at 20 hydrophones for a Fram IV ice event which occurred on 3 April 1982, at approximately 0135 Z. Data are bandpass filtered from 15 to 80 Hz. Arrow directions show the estimated arrival times of the longitudinal wave (downward) and acoustic wave (upward). Range shown was determined after location using estimated acoustic wave arrival times.

540

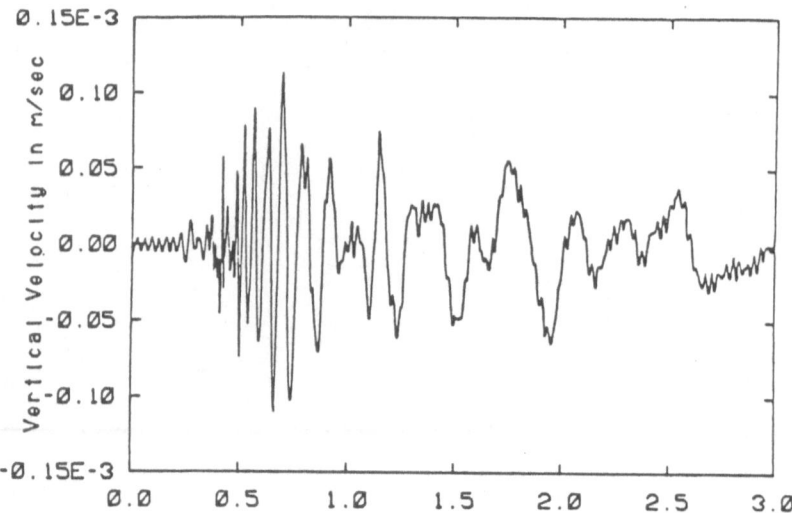

Figure 6. Vertical ice velocity versus time for the event shown in Figure 5. Data are bandpass filtered from 1 to 80 Hz. Measurement taken by a geophone on the upper surface of the ice. Range to the event from this sensor was estimated to be 400 m.

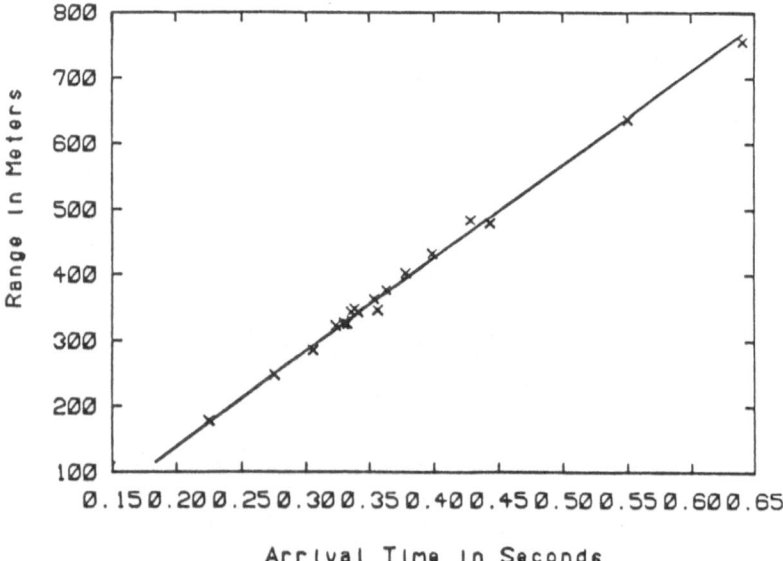

Arrival Time in Seconds

Figure 7. Time delay versus range for the upward arrows shown in Figure 5. Best-fit line gives a wave speed of 1430 m/s indicating the most energetic arrival in the event of Figure 5, as marked by the upward arrows, iş the acoustic wave.

the first small deflection off the 30 Hz noise at the geophone (0.25 s into the trace in Fig. 6) agrees with the approximated arrival time for the longitudinal wave at that sensor.

The flexural wave, as expected due to its decay with depth, is not detected at the hydrophones. It is, however, clearly detected at the geophone. The dispersion pattern of the arrival in Fig. 6 was well predicted using the theoretical flexural wave speed for 3 m thick ice and the estimated location of the event.

3.4. A 55 lb. Charge Detonation

The third event to be interpreted was not a naturally occurring ice event, but a 55 lb. charge of TNT set off at 90 m depth roughly 1 km from the same Fram IV hydrophone array described in 3.3. The acoustic signature at five of the hydrophones, along with their horizontal range from the blast, is shown in Figure 8.

At each phone one sees the longitudinal wave arriving first, and the direct acoustic wave, which saturates the hydrophones, arriving second. Since the longitudinal wave travels faster, the time between longitudinal and acoustic wave arrival at a particular hydrophone increases with range. Note also how rapidly the longitudinal wave amplitude is decaying with range. Clearly the attenuation is faster than that resulting from geometrical spreading, implying that absorption or scattering is playing an important role.

The arrival times versus range for the longitudinal and acoustic waves are shown in Figures 9 and 10. The best-fit lines drawn give a measured acoustic wave speed of 1440±10 m/s and a measured longitudinal wave speed of 3060±80 m/s. This is in very good agreement with the theoretical values. An average pressure level versus range for the longitudinal wave is given in Figure 11. The best-fit line in this drawing leads to a longitudinal wave exponential attenuation of 0.016 dB/m. The estimated peak frequency of this blast is 40 Hz, so the measured attenuation is roughly twice the theoretical value given in Section 2.3.

4. DISCUSSION

The lake ice event discussed in Section 3.1 showed the ideal case when the longitudinal, flexural, and acoustic waves could be detected. Here the event signature could be used to measure the wave speed in ice and, via the flexural wave speed, the ice thickness. Analysis of such ideal lake ice events, recorded at a small fraction of the cost of Arctic ice events, might be a good starting point for source characterization.

The Fram II event displayed in Section 3.2 had a signature in which only the acoustic wave was detected. The absence of the flexural wave was expected since the event was recorded only on hydrophones at 90 m depth. Lack of longitudinal wave detection could be due to high background noise levels combined with weak excitation of the longitudinal wave by the source, or strong scattering of the longitudinal wave by discontinuities in the ice.

In the Fram IV ice event discussed in Section 3.3, all three waves were detected (the flexural wave by use of a geophone) and their measured wave speeds agreed with theoretical values. However, the acoustic wave clearly dominated the signature in the water. As in the Fram II event, either the longitudinal wave was weakly excited by the source or it was scattered by discontinuities in the ice.

The results of Section 3.4, in which a 55 lb. charge of TNT was used to measure longitudinal wave properties, gave a measured attenuation which was on the same order as the theoretical value which accounts only for absorption. This means that the ice at

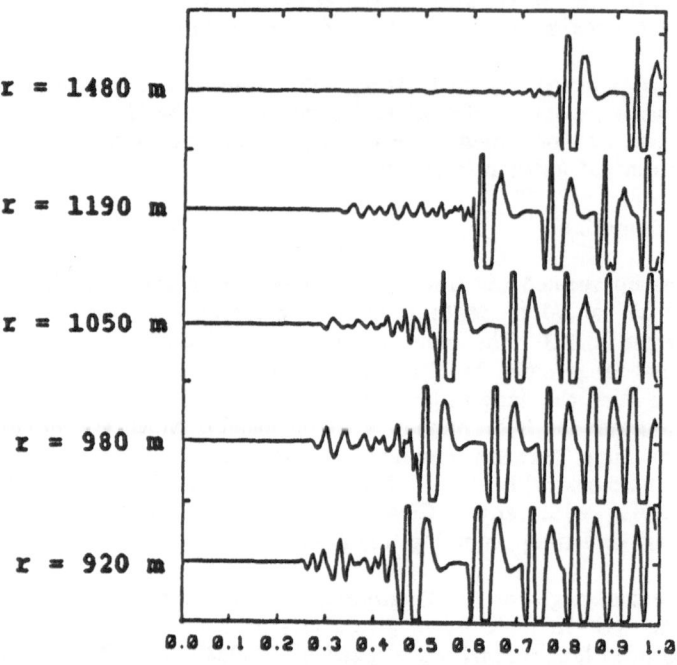

Figure 8. Pressure versus time series at five hydrophones showing the arrival of energy from a 55 lb. charge of TNT set of approximately 1 km from the Fram IV array apex. Range from each of the five hydrophones is as indicated. Data are bandpass filtered from 10 to 80 Hz and are clipped at the maximum levels shown. The experiment was performed on 22 April 1982. The first arrival at each sensor is the longitudinal wave and the second saturating arrival the acoustic wave.

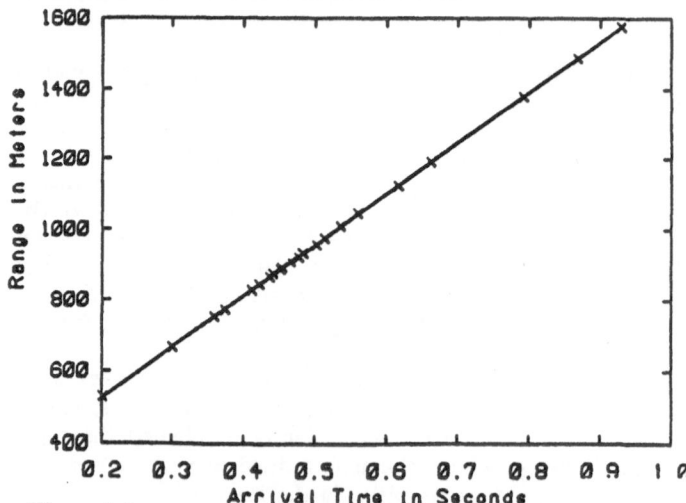

Figure 9. Time delay versus range for the acoustic wave radiating from a 55 lb. charge. Slope of best-fit line gives a measured speed of 1440 m/s for the acoustic wave.

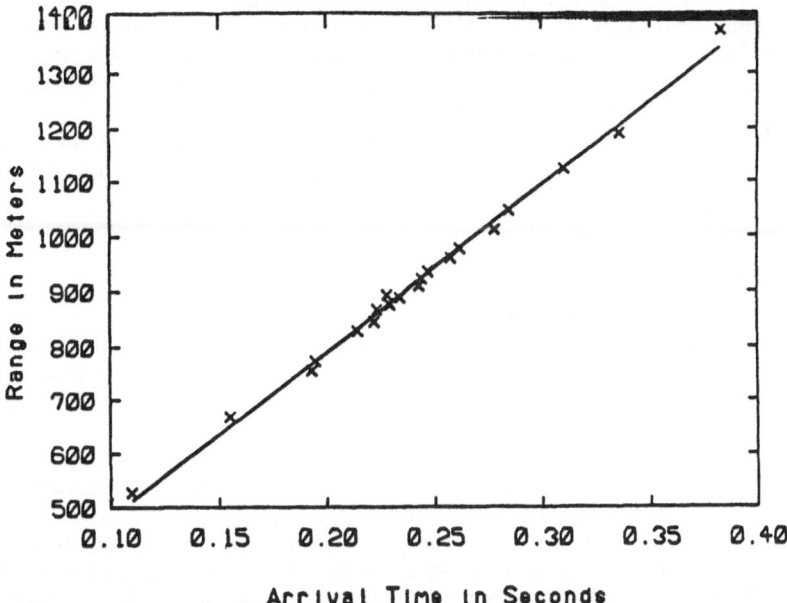

Figure 10. Time delay versus range for the longitudinal wave radiating from a 55 lb. charge. Best-fit lines gives a measured speed of 3060 m/s for the longitudinal wave.

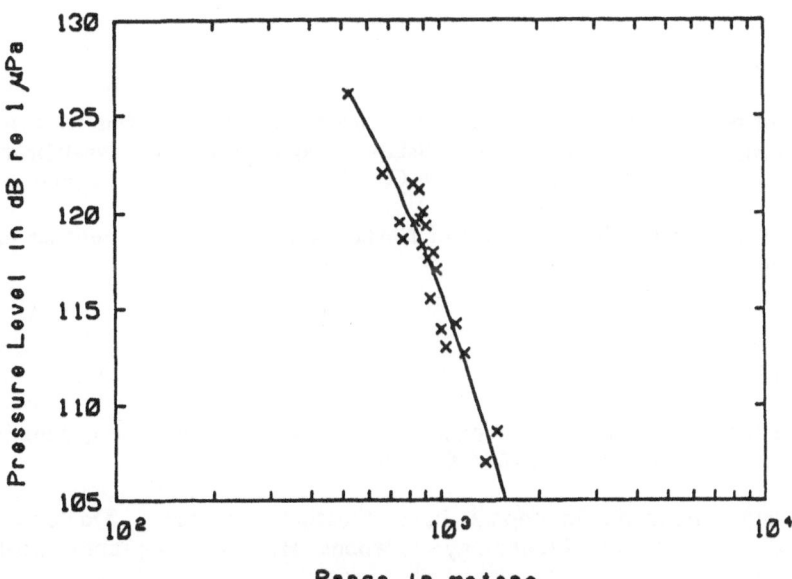

Figure 11. Average pressure level of the longitudinal wave arrival plotted versus horizontal range. Line drawn is the best least-squares fit of the theoretical attenuation curve for the longitudinal wave and gives a longitudinal wave attenuation of 0.016 dB/m.

Fram IV was continuous enough to support the longitudinal wave. Visually the ice at Fram IV changed little over the course of the experiment. This could indicate that in the event of Section 3.3, the weakness of the longitudinal wave was due to poor excitation and not scattering. Still, this is not conclusive since longitudinal wave scattering may be highly variable in time and space.

One last thing to note is that the Fram II ice event was roughly an order of magnitude louder at the same range than the Fram IV ice event. From this we can conclude that ice event transients are not of constant source strength.

5. CONCLUSIONS

This work is intended as a take off point for much more extensive studies of the acoustic radiation from nearby ice events. In terms of propagation paths, we are confident in our understanding of the wave speeds in the ice-water system. However, we are uncertain of the attenuation of these waves.

In all cases of event interpretation the longitudinal wave was either absent or much weaker than the acoustic wave. It is therefore possible that quantifying the acoustic wave directivity and power will be most important to the Arctic Ocean ambient noise problem. However, it is not certain whether the absence of the longitudinal wave in the signatures is due to lack of excitation by the source or scattering. If the longitudinal wave is well excited, and then scattered into the water, it may well be an important contributor to Arctic ocean ambient noise. The same holds true for the flexural wave as its energy can also be scattered into the water by ice discontinuities. To address these issues, and many more involved with ice event source characterization, more events need to be interpreted in conjunction with more measurements of elastic wave attenuation and scattering in ice.

ACKNOWLEDGEMENTS

Many thanks to Prof. Ira Dyer, my thesis supervisor, of which this work is part, for his support, encouragement, and friendship. Also thanks to Dr. Neal Brown and Dr. Steve Africk (Atlantic Applied Research Corp.) for their assistance in preparation of this paper.

This work was funded by the Office of Naval Research under the guidance of Dr. Leonard Johnson and Mr. Bob Obrochta.

REFERENCES

[1] F. Press and W.M. Ewing, "Propagation of Elastic Waves in a Floating Ice Sheet", Trans. Am. Geo. Un., 32(5):673-678, 1951.

[2] P.J. Stein, "Acoustic Monopole in a Floating Ice Plate", Doctoral Thesis, Massachusetts Institute of Technology / Woods Hole Oceanographic Institution, Feb. 1986.

[3] P.J. Stein, "Interpretation of a Few Ice Event Transients", Submitted to the J. Acoust. Soc. Am., 1987.

THE CORRELATION OF MID-FREQUENCY PACK ICE NOISE WITH ENVIRONMENTAL
PARAMETERS

Ruth Eta Keenan
Science Applications International Corporation
346 Gifford Street
Falmouth, Massachusetts, USA   02540-2996

ABSTRACT.   The mechanisms generating the FRAM IV mid-frequency Arctic
ambient noise are examined.   The 100-Hz, 200-Hz and 300-Hz ambient
noise levels are correlated with temperature, temperature rate of
change, heat flux, wind speed, the ice stress moment and the 10 to
20-Hz ambient noise levels.   The time series and spectral character of
several individual noise events are presented.   The events are local,
within 15 km of the receiving array, and peak in the mid-frequencies.
There is no correlation with the measures of thermal stress, a weak
correlation with wind speed and the ice stress moment and a .67
correlation with the 10 to 20-Hz levels.

1.   INTRODUCTION

Other investigators have correlated environmental parameters with
ambient noise to study the generating mechanisms.   Two separate
mechanisms have been identified which generate ambient noise, pressure
ridging and thermal stress.   Based on these results Arctic ambient
noise is modeled as the composite of the pressure ridging and thermal
stress spectra.   The pressure ridging spectrum peaks about 10-20 Hz
and the thermal stress spectrum has a broad peak about 300 Hz, both
spectra decay monotonically about the peak.   Using this model the FRAM
IV mid-frequency data were analyzed.   The data were collected during
the FRAM IV exercise in the spring of 1982 at approximately 83°N, 20°E
on an array of hydrophones placed 91 m below the ice.

Studies at low frequencies[1,2,3] have shown high correlation
between ambient noise and various measures of the large scale ice
stress that builds pressure ridges.   Makris and Dyer[1] have shown that
the rms pressure of the FRAM IV 10 to 20-Hz ambient noise band has a
.87 correlation with the stress moment on the ice given by a composite
of the wind, current, Coriolis and normal stresses on the ice.   Lewis
and Denner[2] found a .83 correlation for the AIDJEX 32-Hz ambient noise
levels with a linear combination of the ice translation, normal and

B R Kerman (ed), Sea Surface Sound, 545–554

shear deformations, vorticity and divergence measures. Pritchard[3] calculated the pressure ridging energy during the AIDJEX exercise and found a weaker .68 correlation with the 32-Hz ambient noise levels. At the mid-frequencies Milne[4] conducted studies in the Canadian Archipelago that showed a striking diurnal variation in the octave band noise centered about 300-Hz that mirrored the diurnal fluctuations in the air temperature.

If this simple noise model is correct then the FRAM IV mid-frequency ambient noise should either be thermally generated or it is the high frequency tail of the pressure ridging spectra. The former case should correlate well with a measure of the thermal ice stress and the latter with a measure of the large scale ice stress that builds pressure ridges.

2. DISCUSSION

To test the Arctic ambient noise model the 100-Hz, 200-Hz and 300-Hz pressure levels were correlated with measures of the thermal stress, temperature, temperature time rate of change, $\Delta T/\Delta t$, and heat flux and the pressure ridging measures, 10 to 20-Hz noise levels, ice stress moment and wind speed. Table 1 lists the correlation results at zero time lag between the various time series. In all cases the maximum correlation coefficient was located at zero time lag.

Table 1. Correlation Results

	100-Hz	200-Hz	300-Hz
100-Hz	--	.88	.89
200-Hz	--	--	.94
Temperature	.05	.18	.07
$\frac{\Delta T}{\Delta t}$	.09	.08	.11
Heat Flux	.15	.13	.18
Wind Speed	.31	.39	.37
Stress Moment	.28	.37	.37
10 to 20-Hz	.67	.62	.62

The 100-Hz, 200-Hz and 300-Hz ambient noise levels are illustrated in Figures 1, 2, and 3, respectively; straight lines are used between data points. The inter-frequency correlation among the mid-frequency levels is very high which suggests the same mechanism dominates in the 100 to 300-Hz region. Spectral analysis of the air temperature, Figure 4, showed a strong diurnal component whereas spectral analysis of the mid-frequency ambient noise time series

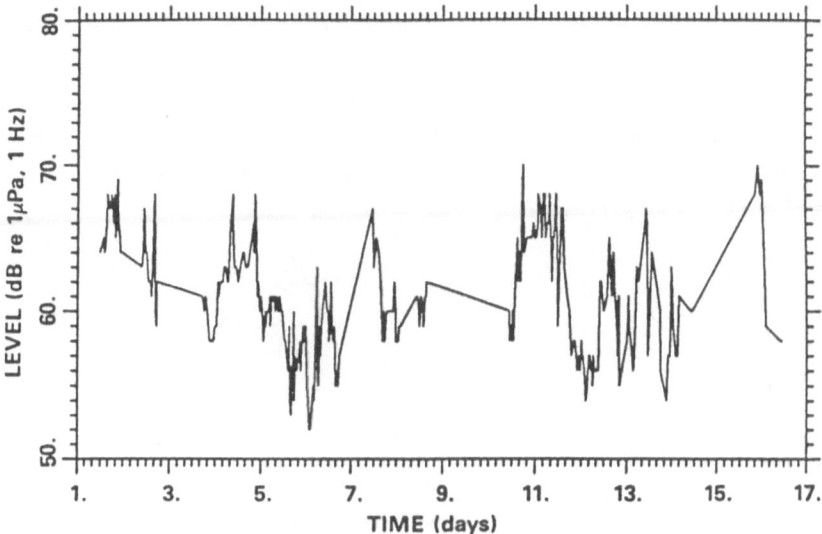

Figure 1. FRAM IV 100-Hz spectral noise levels

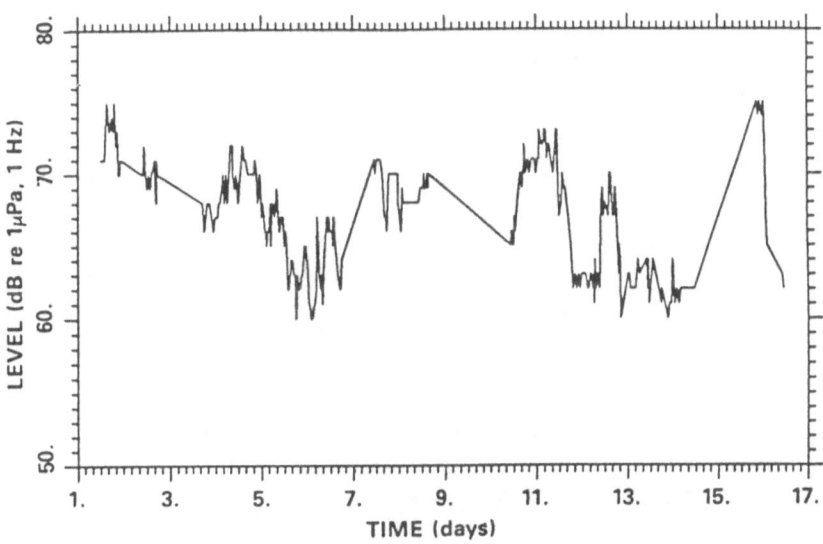

Figure 2. FRAM IV 200-Hz spectral noise levels

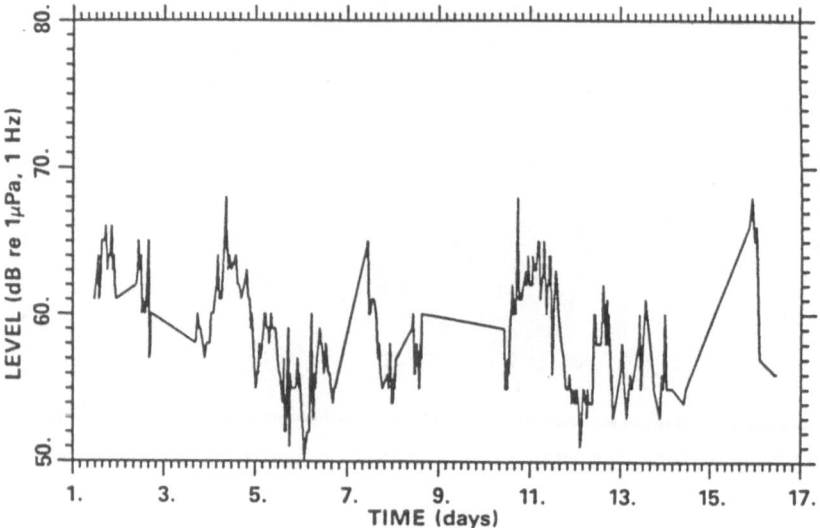

Figure 3. FRAM IV 300-Hz spectral noise levels

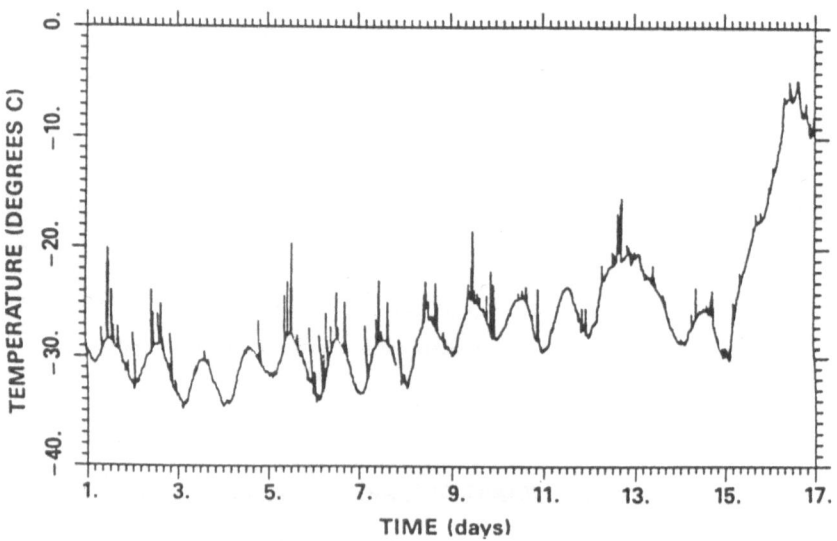

Figure 4. FRAM IV air temperature

showed none. The correlation results between the mid-frequencies and air temperature was less than .2. The other measures of thermal ice stress showed similar poor correlations with the mid-frequency ambient noise. Milne[4] observed that the temperature time rate of change is a better measure of the thermal ice stress than the temperature alone. He found greater ambient noise levels during time periods in which the temperature falls faster. The temperature rate of change was correlated with the mid-frequency noise but showed no better correlation than that with temperature. Based on Lewis'[5] observation that heat flux is an appropriate measure of thermal ice stress the mid-frequency noise was correlated with the measure of temporal variability of the heat flux given by equation 1.

$$HF~(t) = \Big(Air(t) - Ice(t)\Big) * WS(t) \tag{1}$$

$$
\begin{aligned}
HF &= \text{measure of heat flux variation with time} \\
Air &= \text{air temperature} \\
Ice &= \text{ice temperature} \\
WS &= \text{wind speed}
\end{aligned}
$$

The correlation with the heat flux measure was no better than the correlation with temperature. Based on these results we conclude the thermal noise mechanism does not dominate the FRAM IV mid-frequency ambient noise.

The 10 to 20-Hz ambient noise and ice stress moment data were taken from Makris and Dyer.[1] The 10 to 20-Hz ambient noise levels are replotted in dB re 1 µPa, 1 Hz in Figure 5. The mean 10 to 20-Hz

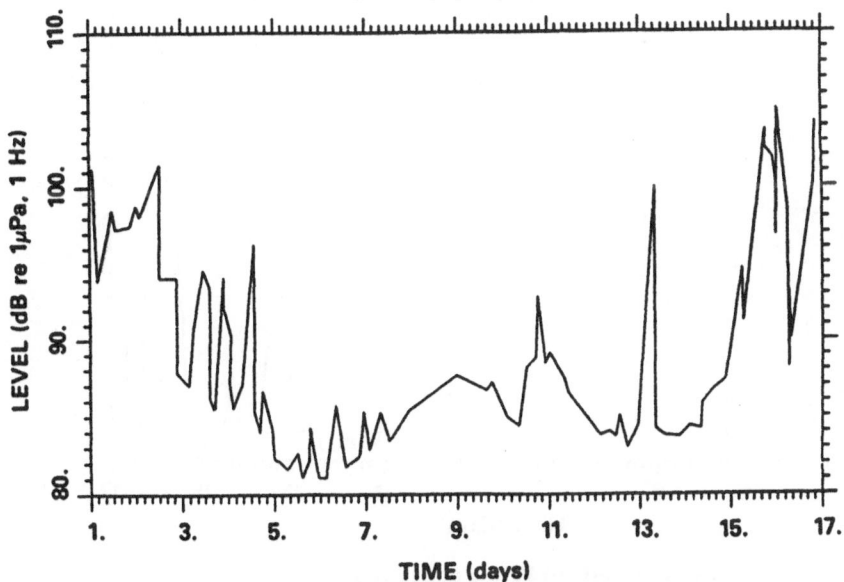

Figure 5. FRAM IV 10-20 HZ spectral noise level corrected to 1-Hz band (Makris and Dyer[1])

ambient noise levels show the same envelope variation as the 100-Hz noise levels. The correlation between the two time series is .67. If the mid-frequency noise is the higher frequency tail of the pressure ridging noise then some deterioration in the cross frequency correlations might be accounted for by propagation effects.[6] For example, consider a pressure ridging event that increases the 10-Hz noise by 3 dB. If this noise source is far enough away so that energy at 100 Hz suffers 3 dB or more loss than the 10-Hz energy, then the mid-frequency noise will not rise with the 10-Hz noise. However, this explanation does not account for the cases where the mid-frequency noise increases without a corresponding increase in the 10-Hz noise. This effect can be observed on day 11 of the noise time series in Figures 1 and 5.

The mid-frequency noise was correlated with Makris and Dyer's[1] ice stress moment, (illustrated in Figure 6) and the wind speed (Figure 7). Both functions show weaker correlations with the mid-frequency ambient noise than the 10 to 20-Hz levels. For the stress moment this may be due to the shorter time series and in the case of the wind speed it may be due to the fact that wind speed is only a partial measure of the forcing function generating the noise.

The time series comparison and correlation results indicate that although the pressure ridging spectrum noise influences the mid-frequency noise levels there are still uncorrelated oscillations between the time series.

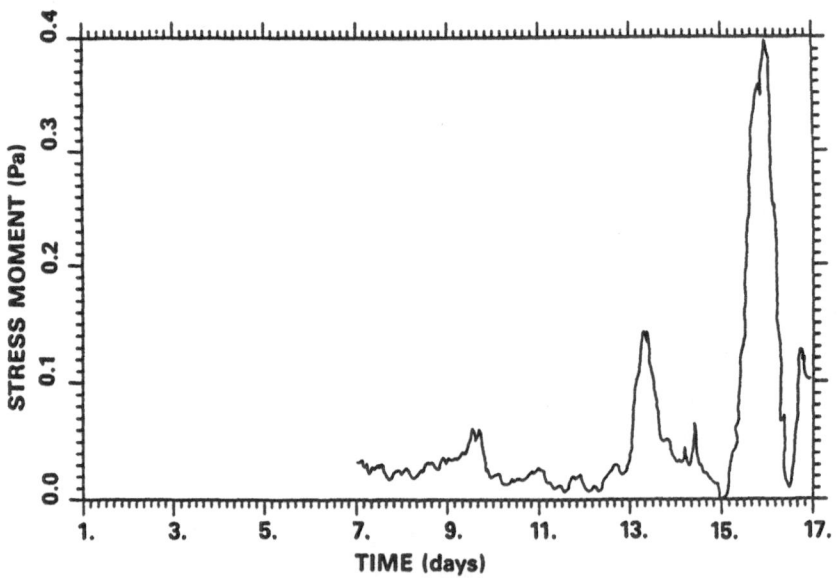

Figure 6. FRAM IV ice stress moment
(Makris and Dyer[1])

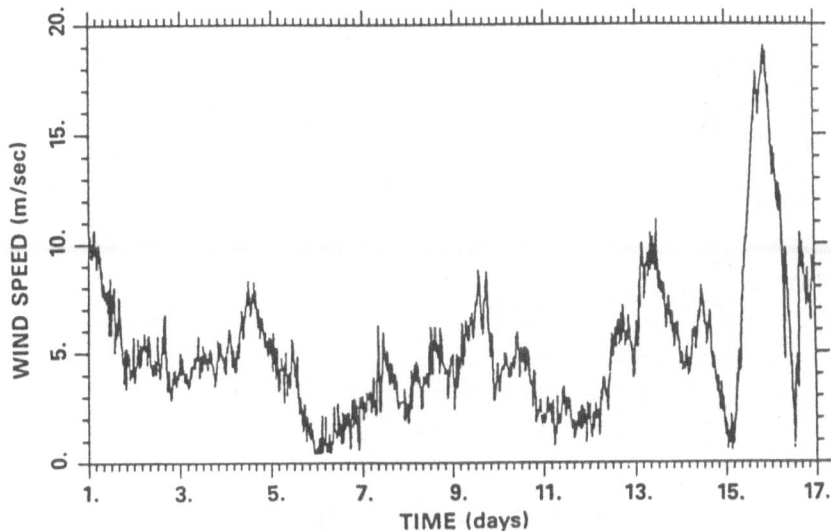

Figure 7. FRAM IV wind speed

Figure 7. FRAM IV wind speed

Several mid-frequency noise events[7] were located in the FRAM IV data; three events, designated a pop, straf and a hoot, are presented here. The popping event time series and associated spectrum is illustrated in Figure 8 using a Kronhite filter below 80 Hz. The event lasted about 20 msec and had a broad spectrum peaked about 400 Hz. The event was localized 15 km from the array apex by eigenrays that reached the hydrophones after one surface bounce. The straf event time series and associated spectrum is illustrated in Figure 9 using a Kronhite filter below 80 Hz. This event lasted 300 msec. The spectra is peaked about 150 Hz. This event was localized 4 km from the array apex by direct path eigenrays. A number of hoots were observed that lasted approximately 25 seconds. Figure 10 illustrates the time signature of an individual event with the spectrum averaged 25.6 seconds over the series of events. The hoot shows a damped sinusoidal character peaked about 117 Hz. These events were localized 500 m from the array apex.

These particular events did not affect spectra averaged more than one minute because their duration was so short; however, averaged spectral shapes did vary from the pressure ridging spectrum shape to spectrum with multiple peaks.

The observation of individual noise events with spectral peaks in the mid-frequencies may indicate the existence of other such events that were not distinguishable but contribute to the variation in spectral shape.

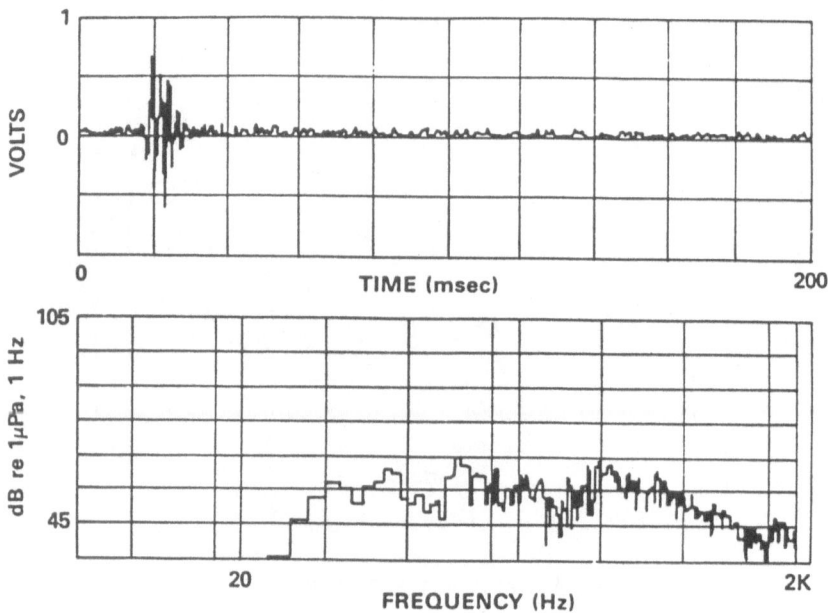

Figure 8. Spectrum analyzer plot of 'pop' noise (Kronhite filter 80 Hz) time series and associated spectrum

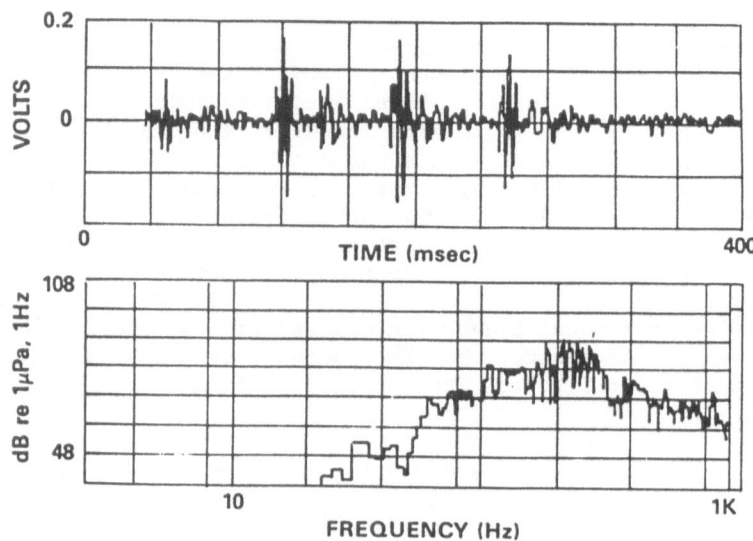

Figure 9. Spectrum analyzer plot of 'straf' noise (Kronhite filter 80 Hz) time series of event with associated spectrum

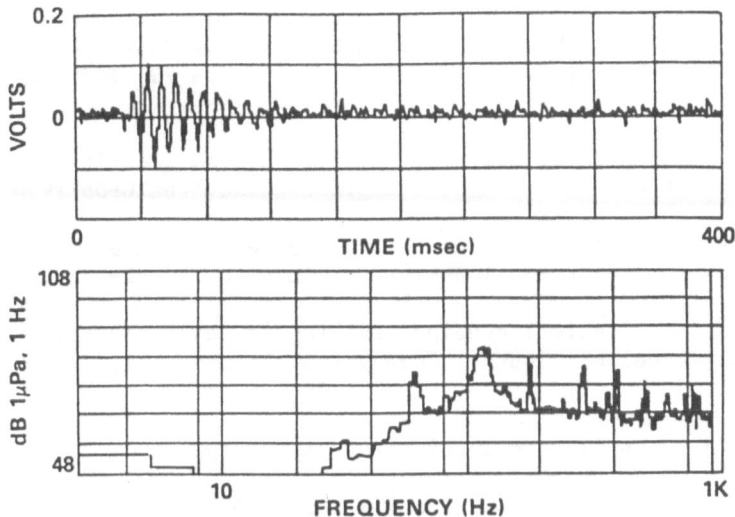

Figure 10. Spectrum analyzer plot of 'hoot' noise (Kronhite filter 80 Hz) time series of single event, 25.6-second averaged spectrum

3. CONCLUSIONS

The composite pressure ridging and thermal stress spectrum ambient noise model is modified to allow for mechanically induced cracking events that have spectra which peak in the mid-frequencies. The mid-frequency FRAM IV data is not thermally induced; it shows a .67 correlation with the 10 to 20-Hz noise and the mean levels of both time series show similar fluctuations with time. The correlation deteriorates because the actual levels do not fluctuate synchronously. The data suggests the mid-frequency ambient noise is a combination of the high frequency tails of the pressure ridging spectrum and the spectrum of smaller scale cracking events with mid-frequency peaks.

4. ACKNOWLEDGEMENTS

This work has been sponsored by the ONR AEAS Arctic program under the auspices of Mr. B.N. Wheatley and Mr. G.A. Gotthardt. The enlightening discussions with Dr. Ira Dyer of MIT provided many insights into the data interpretation.

5. REFERENCES

[1] Makris, N.C., and I. Dyer, 'Environmental correlates of pack ice noise,' Journal Acoustical Society of America 79, 1434-1440, 1986.

[2] Lewis, J.K., and W.W. Denner, 'A study of sea ice kinematics and their relationships to Arctic ambient noise,' SAIC-85/1950, 1986.

[3] Pritchard, R., 'Arctic ocean background noise caused by ridging sea ice,' Journal Acoustical Society of America 75, 419-427, 1984.

[4] Milne, A.R., 'Thermal tension cracking in sea ice: A source of underwater noise,' Journal Geophysical Research 77, 2177-2192, 1972.

[5] Lewis, J.K., private communication, April 1987.

[6] Buck, B.M., private communication, May 1987.

[7] Keenan, R.E., and L. Gainey, 'FRAM IV ambient noise: (100 Hz - 500 Hz) data analysis Volume II,' SAIC-86/1893, 1986.

HIGH FREQUENCY AMBIENT SOUND IN THE ARCTIC

David M. Farmer and Sherman R. Waddell
Institute of Ocean Sciences
P.O. Box 6000
9860 West Saanich Road
Sidney, B.C. V8L 4B2
Canada

ABSTRACT. Sound pressure levels recorded in six frequency bands from 50 Hz to 14,500 Hz provide an acoustic description of the evolving processes of ice cracking and disintegration from 2 April to 7 August, 1986 in Dolphin and Union Strait. In the early spring the dominant sources of sound are thermal stress cracking and wind effects. In early June the sound level drops to very low values. Thereafter, the progressive deterioration of the ice leads to an increase in sound levels and short term variability. These acoustic signatures can be related to satellite and meteorological observations.

INTRODUCTION

The ambient sound field in the Arctic is profoundly different to that found in most of the world's oceans, and reflects the special characteristics and dynamical response of the ice cover. Attention has recently been directed towards the acoustic environment in the Arctic basin and marginal ice zone (Makris and Dyer, 1986; Pritchard, 1984). Here we report results of a relatively long time series measurement in one of the channels of the Canadian Archipelago (Figure 1). In some respects interpretation of the sound field in the ice covered channels is simpler than it is in the mobile ice-pack of the open Arctic. Much of the year the ice is landfast, and the channel boundaries limit the area from which more distant signals can arrive. The sound field can thus be more readily traced through meteorological data and satellite imagery to specific physical processes associated with its generation. The present study follows this approach with ambient sound observations 32 times per hour and averaged over 28s; the data were recorded in 6 different frequency bands between 50 Hz and 14,500 Hz through 2 April to 7 August, 1986 in Dolphin and Union Strait.

B. R. Kerman (ed.), Sea Surface Sound, 555–563.
© 1988 by Kluwer Academic Publishers.

556

Figure 2 shows satellite images from late spring to the final break-up of ice in early August. Disintegration of the ice sheets proceeds eastwards across Amundsen Gulf. During this period various different physical processes contribute to the recorded signal.

SEASONAL VARIATION

The time series observation at 1000 Hz is shown in Figure 3. This provides a general view of the long term change in both signal level and quality. Maximum signals are observed during spring. This is followed by a 2-3 week quiet period in early June. Thereafter the maximum signal level progressively increases until the end of the record.

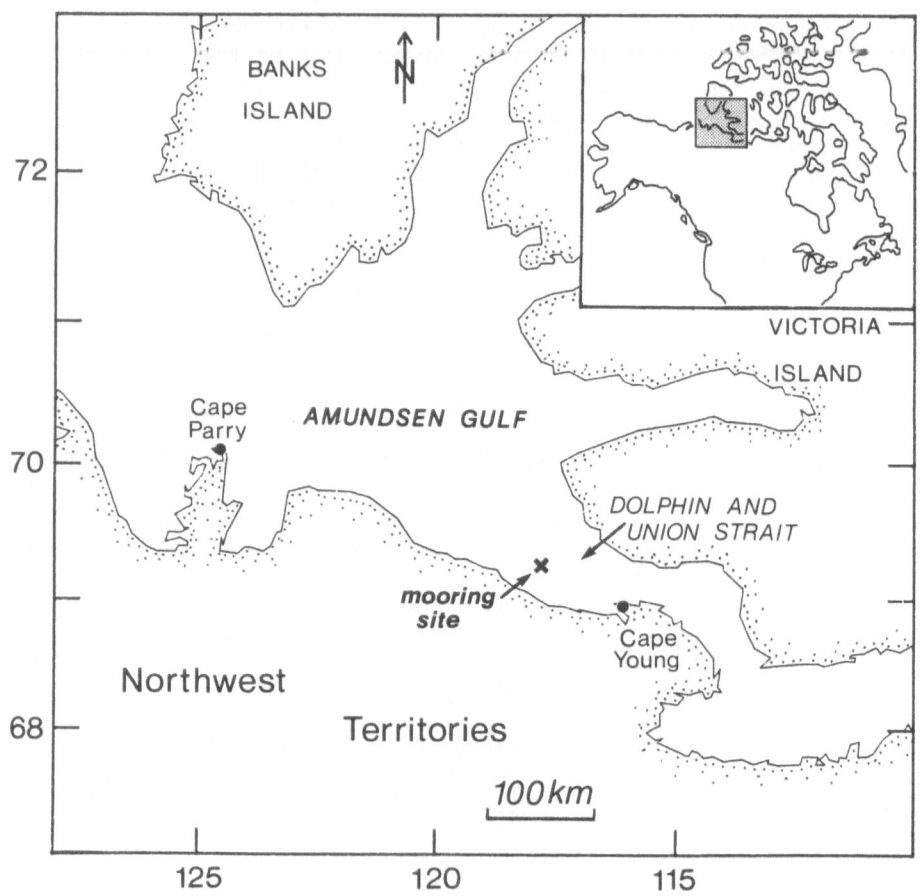

Figure 1. Location of the measurement site.

SPRING

Strong acoustic signals have been previously reported by Ganton and Milne (1965) and Milne (1972) and have been attributed to thermal stress cracking and wind effects. The unique feature of the present record is that it shows the development of these signatures in great detail over an extended period. This allows a more complete distinction to be made

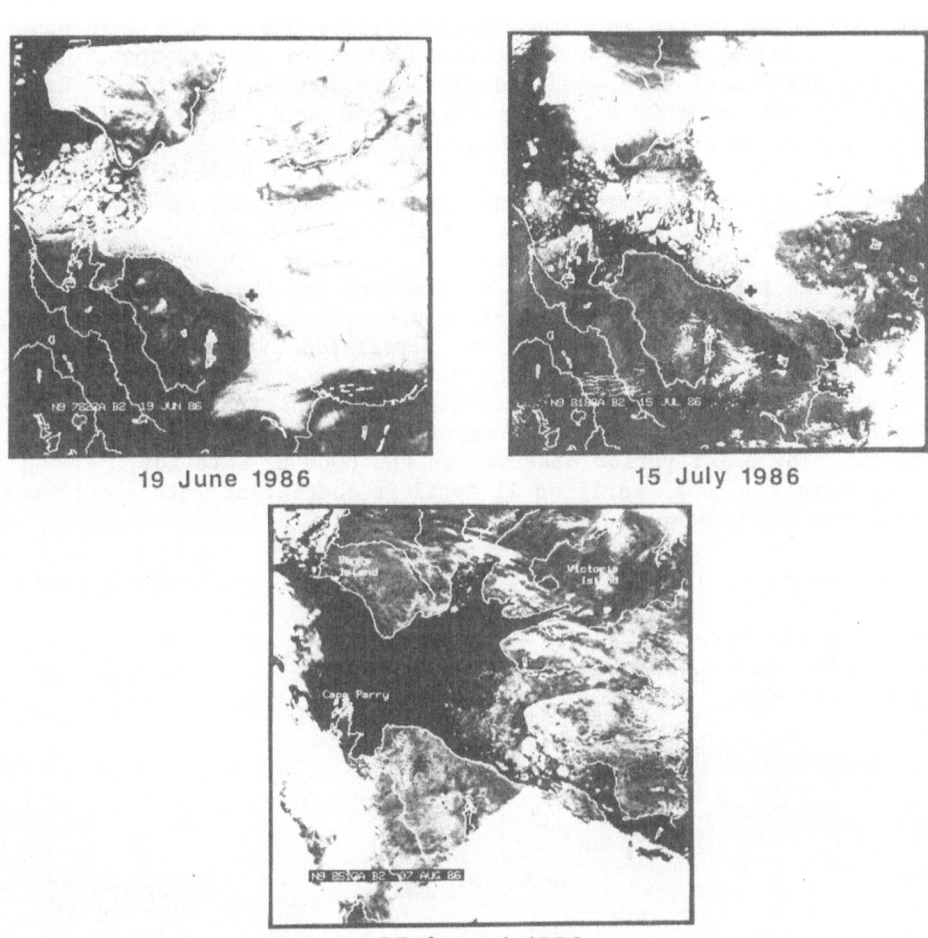

19 June 1986   15 July 1986

07 August 1986

Figure 2. NOAA-9 band 2 (near IR) images of ice conditions in Amundsen Gulf and Dolphin and Union Strait. The mooring site is marked with a thick +. The ice appears bright while open water or very thin ice appears dark. Coastlines have been added to the image for reference.

between the various physical factors responsible and demonstrates the importance of long wave radiation in the ice surface heat balance.

Time series records of acoustic and meteorological data over a 14 day period are shown in Figure 4. Here the sharply diurnal signal of thermal stress cracking is clearly evident. It is centered at local midnight, starts abruptly when the sun sets below 22° (at this time of year), reaches its maximum value in 3-4 hours, and takes somewhat longer to decay at dawn (5-6h). However, this relatively simple structure breaks down from 19-24 April. At this time the sky is obscured by cloud, as shown by the infrared transmissivity plot. This weather change is accompanied by much increased wind for part of the time.

The effects of both wind and cloud are apparent in the acoustic signal at 1000 and 8000 Hz. When cloud is present the thermal stress cracking is suppressed. Thus the thermal stress at the ice surface must be dominated by the net radiation balance. For example on 19 April the thermal stress cracking signal has two separate peaks; at this time cloud cover is building up and is intermittent. Clear skies return 24 April and the diurnal cracking signature returns. The close relationship between thermal stress cracking and long wave radiation implied by these observations differs from previous work (Milne, 1972), in which thermal stress cracking was correlated with air temperature.

Thermal stress cracking has a much stronger contribution at 1000 Hz than at 8000 Hz while wind induced noise has a dominant contribution at high frequencies as shown by the strong correlation with wind speed at 8000 Hz. The quiet period observed in the 8000 Hz data for approximately 7 hours early on 21 April is consistent with a calm period before the wind changed direction.

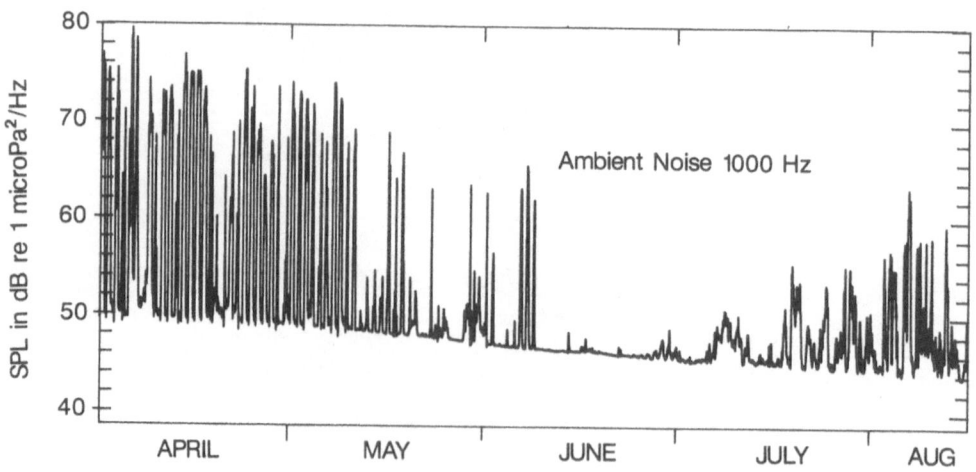

Figure 3. Ambient sound level in dB re 1 microPa^2/Hz measured at 1000 Hz for the duration of the experiment.

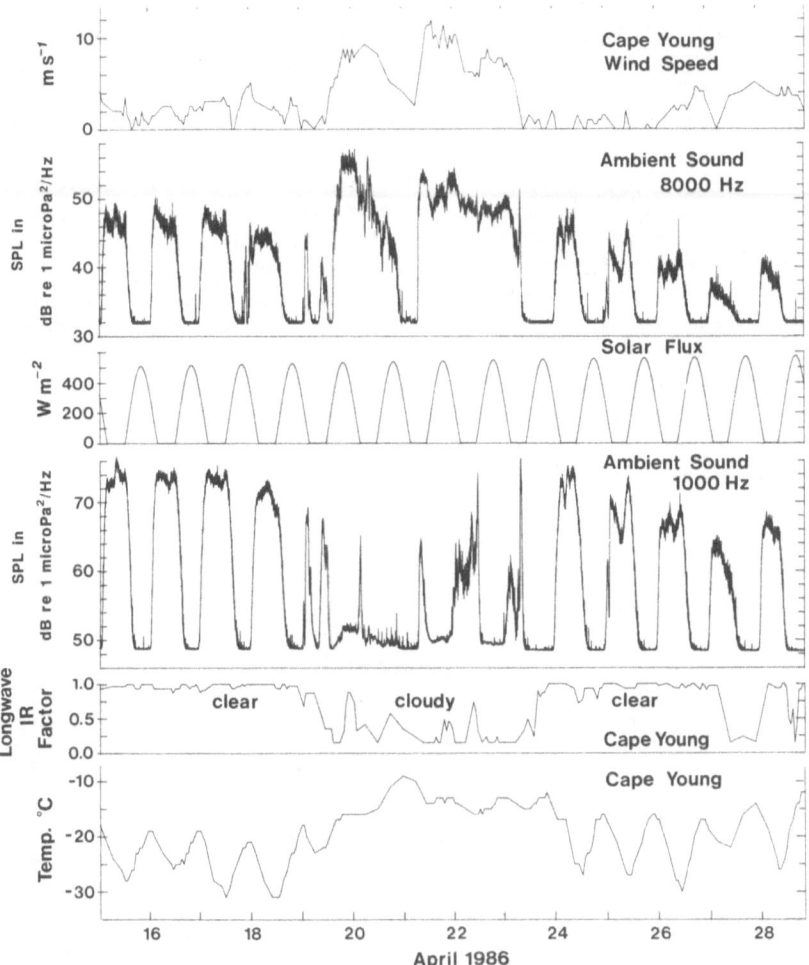

Figure 4. Ambient sound measured at 1000 and 8000 Hz is shown together with wind speed and temperature observed at Cape Young. The Cape Young cloud reports were used to calculate the effect of cloudiness upon longwave outgoing infrared radiation using the cloud correction factor of Laevastu (1967). This factor is in effect a relative transmissivity with 1 representing cloud free conditions and 0 representing opaque conditions. The solar flux is calculated from ephemeris data and includes geometrical corrections for sun altitude. Notice the phase of the solar flux with respect to the diurnal ambient sound signal. Also note the relationship between wind speed, cloud cover and the ambient sound data in each frequency band.

TRANSITION PERIOD AND ICE BREAK-UP

In June the ice is ponding and weakening; the background noise level drops to very low values (Figure 3). Unlike the open ice pack it is land-fast and remains essentially motionless until break-up. Disintegration of the ice cover is a drawn-out process which can be tracked in detail with simultaneous satellite images and the acoustic record. In late June (21-28) while the ice cover remains intact, a clearly formed diurnal signal occurs (see Figure 5). In contrast to the thermally induced cracking in spring (also shown in Figure 5), this signal reaches its maximum in late afternoon and is thought to represent the distant, broken ice response to the local sea-breeze. Simultaneous current measurements near the acoustic instrument show no corresponding diurnal signal.

The sound of distant ice mobilisation picks up again 1 July (see Figure 6). That the signal is indeed from distant ice is inferred from the spectral signature which is dominated by low frequencies. Satellite imagery shows that the mobile ice edge is 250 km distant at this time.

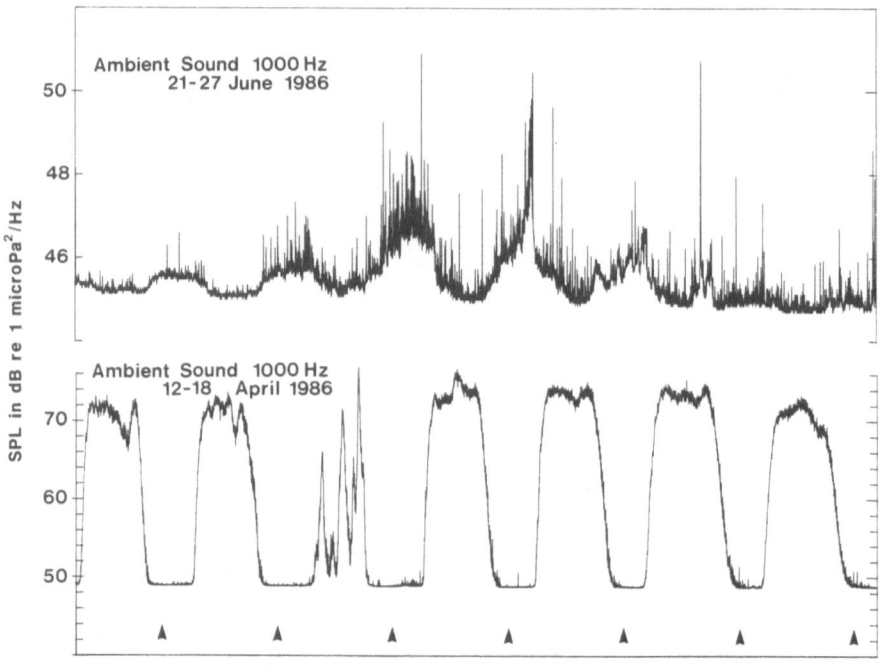

Figure 5. Diurnal sound signal 21 - 27 June compared with diurnal sound signal from early spring. The arrows indicate the time of local solar noon. Note the difference in phase of the two signals and the difference in amplitude scales.

In contrast to this spectral signature, however, is the sharp peak at 14,500 Hz at the start of the time sequence. We attribute this peak to the formation of the crack that reached across the mouth of Dolphin and Union Strait and widened into a lead which became visible on the satellite image of 2 July.

Thereafter the overall signal level, and most notably the variance calculated over a 24h period (Figure 7), increase progressively as the mobile ice limit advances towards the instrument site. As the ice edge approaches, the signal level at higher frequencies, increases. The daily variance at 1000 Hz provides a good indication of ice activity. Localised break-up of the ice begins on 11 July. The interpretation of acoustic signals at this frequency as a signature of ice crushing and rubbing is consistent with the laboratory studies of Bogorodskii et al. (1969).

The overall signal levels increase until the end of July. By this time ice had mobilised over the instrument. In early August numerous floes are present, but the interactions between floes are evidently reduced at this time. The instrument continued to record until 7 August; the transition to ice free conditions had just begun, but presumably could be easily recognised by the appearance of the distinctive spectral signature of wind noise.

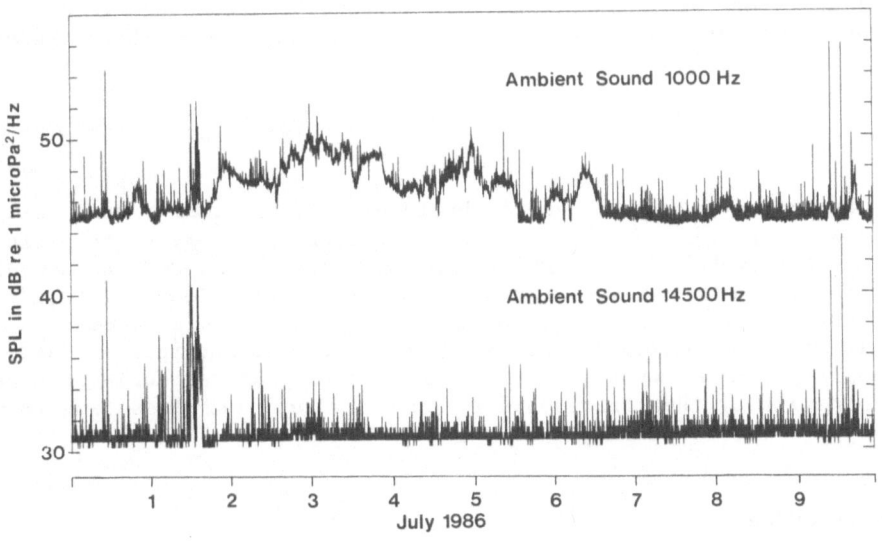

Figure 6. Distant ice breakup measured at 1000 Hz and 14500 Hz. Initial mobilisation of the ice sheet occurs about 1200Z, 1 July 1986 followed by a period of ice sheet disintegration and floe-floe interaction from 2 - 7 July 1986.

562

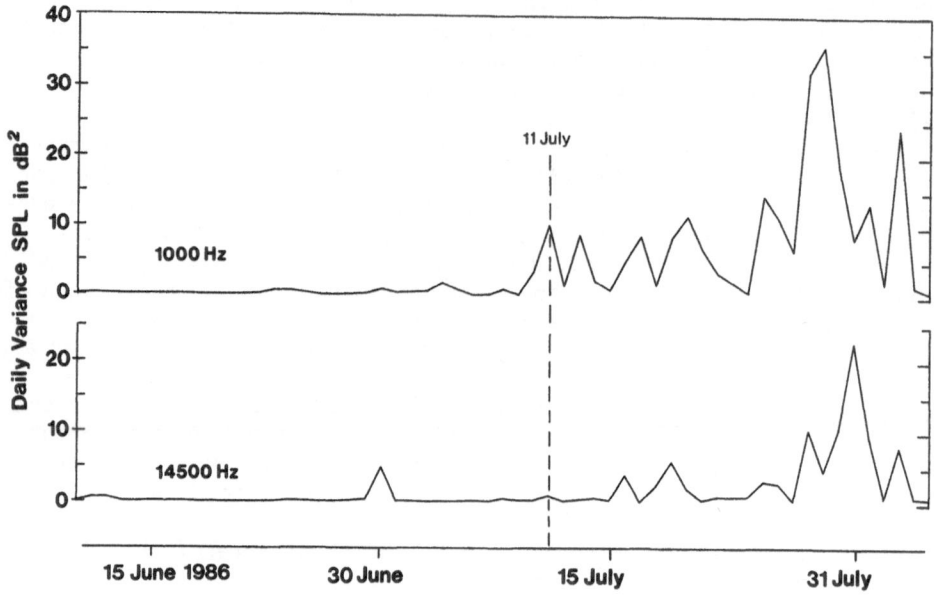

Figure 7. Variance of the sound pressure level at 1000 Hz and 14500 Hz
calculated over intervals of one day. Notice the sudden change of
variance associated with local ice breakup in Dolphin and Union Strait
beginning about 11 July 1986.

CONCLUDING REMARKS

     The type of acoustic record obtained here, provides a much less
detailed description of the frequency structure and very short period
fluctuations (<60s) than broad band measurements. However the densely
sampled long time series can yield a data set that allows separation of
contributions from a range of physical processes. Longer term
measurements are needed to resolve seasonal cycles and inter-annual
variability. There will also be interesting signatures during freeze-up
and formation of the first solid ice cover. Long term measurements of
this type are now underway.

ACKNOWLEDGEMENT

This work was supported by the Panel on Energy Research and Development,
Project No. 67136.

REFERENCES

Bogorodskii,V.V., V.P. Gavrilo, V.S. Grigor'ev and A.V. Gusev, 1969,
          'Sound generation mechanisms in the breaking of ice samples in
          a liquid', Soviet Physics - Acoustics., 15, pp 161-164

Ganton, J.H. and A.R. Milne, 1965, 'Temperature- and wind-dependent
          ambient noise under midwinter pack ice', J. Acoust. Soc. Am.,
          38, pp 406-411.

Laevastu, T., 1967, 'Cloud factor in long-wave radiation formulas',
          J. Geophys. Res., 72(16), p 4277.

Makris, N.C. and I. Dyer, 1986, 'Environmental correlates of pack ice
          noise', J. Acoust. Soc. Am., 79(5), pp 1434-1440.

Milne, A.R., 1972, 'Thermal tension cracking in sea ice: A source of
          under-ice noise', J. Geophys. Res. 77(12), pp 2177-2192.

Pritchard, R.S., 1984, 'Arctic Ocean background noise caused by ridging
          of sea ice', J. Acoust. Soc. Am., 75(2), pp 419-427.

Waddell, S.R. and D.M. Farmer, 1987, 'Ice breakup: observation of the
          acoustic signature', submitted to J. Geophys. Res., May 1987

ARCTIC OCEAN NOISE GENERATION DUE TO PACK ICE KINEMATICS AND HEAT FLUXES

James K. Lewis
Warren W. Denner

Science Applications International Corporation
1304 Deacon
College Station, Texas 77840
United States of America

ABSTRACT. Noise, ice motion, and a heat flux model are used to study the generation of arctic ambient noise under pack ice. During the summer, observations indicate that most under-ice noise is generated by the ice rushing through the water. Differential ice motion appears to be insignificant in producing ambient noise. This situation changes by fall when the total deformation of the pack ice becomes important in producing lower frequency noise (10 and 32 Hz). Correlations indicate that differential motions of other forms (primarily ice convergence) become important only during winter. As for higher frequency noise (1000 Hz) during fall and winter, correlations with ice motion are low and ambiguous. The generation of higher frequency noise by the combination of ice motion and thermal microfracturing is the likely cause of the seemingly sporadic nature of this frequency band. Anomalous observations are presented which are investigated using a variety of environmental data sets. A model of thermal heat flux within a slab of ice is used to demonstrate the character of one set of observations. However, there are distinct deficiencies in that the model fails to reproduce the phase of the thermal microfracturing and certain periods during which microfracturing is absent.

1. INTRODUCTION

Ambient noise in ice-infested waters is significantly different from that in open oceans. The reason for this, of course, is the presence of the thin veneer of sea ice, an often highly brittle, wind-driven substance which can cover large areas. One may ask what specific mechanisms within sea ice result in under-ice noise. This is a difficult question to answer for a number of reasons. First, the extent, thickness, and rigidity of sea ice vary by season and region, thus introducing temporal and spatial variations in the cause/effect relationships. Also, we have yet to identify all the mechanisms within an ice cover which could possibly cause under-ice noise. Moreover, one mechanism could cause noise across a broad band of frequencies, with the result being the masking of other noise variations induced by a different mechanism. Finally, the difficulty of working in the remote and cold regions of the poles has resulted in a relatively limited collection of noise and environmental data with which to study.

Despite these problems, some progress had been made in relating under-ice noise levels to sea ice phenomena. Milne[1] found that higher frequency noise (order of KHz)

B. R. Kerman (ed.), Sea Surface Sound, 565–581.

variations could be related to the thermal cracking of sea ice. In his work, this occurred during the low air temperatures of the polar winter, but only when the air temperature dropped. Greene and Buck[2] correlated atmospheric pressure gradients with ambient noise levels in the Beaufort Sea, obtaining correlations of the order of 0.6. This was a reasonable result, seeing that wind is the primary forcing of sea ice. Pritchard[3] used a ridging model to calculate energy dissipation, and he then correlated the energy levels with noise variations. His results showed that ridging energy could account for up to 64% of the noise intensity at 10 Hz and 32 Hz for several 20 day periods during the 1975-76 winter in the Beaufort Sea. Makris and Dyer[4] used current, wind, and ice motion data to calculate internal ice stresses and bending moments from the steady-state equations for ice motion. Although they worked with a limited time series, they were quite successful in correlating the magnitudes of the stress and moment variations with ambient noise at low frequencies (10-20 Hz). The respective correlations were 0.81 and 0.87, with an approximately zero time lag.

The work we present here is similar to previous research in attempting to relate various sea ice phenomena to under-ice noise variations. The noise data with which we worked were those collected during the Arctic Ice Dynamics Joint Experiment (AIDJEX) in 1975 and 1976. The characteristics of the noise data are discussed in Lewis and Denner[5]. In this paper, we used a statistical model to investigate ice/noise relationships. We used observed ice motion as determined by satellite-tracked drifters on the ice, a method which eliminates the need to formulate how the ice responds to forcing given its thickness, inertia, rheology, etc. Moreover, here we considered the ice motion in terms of its displacement per unit time (translation speed), its spreading rate, rotation rate, and shape change rate.

As pointed out above, some progress has been made in studying lower frequency noise generation. However, higher frequency noise (>300 Hz) generation has not been as thoroughly considered. Higher frequency noise reflects small scale fracturing of the ice pack (fracture scales of the order of 1 to 5 m). For ice, such scales include fracturing induced by thermal stresses. Thus, higher frequency noise variations in the arctic can often be the result of a combination of kinematic as well as thermal processes. As a result, frequencies >300 Hz are more complicated than lower frequency arctic ambient noise and are less well understood. To consider this problem, time series of higher frequency noise were studied in detail along with all available environmental data. In particular, the thermal microfracturing of ice was studied using analytical and numerical models of heat flux through the ice.

Because of the seasonal variations of the characteristics of sea ice in the Arctic Basin, we calculated statistical relationships for three different periods: summer, fall, and winter. These relationships were calculated for arctic ambient noise at 10 Hz, 32 Hz, and 1000 Hz. Our results imply how the increasing compactness of the ice during fall and winter affect the generation of noise. Moreover, the available environmental data indicate a distinct seasonal variation in the characteristics of the thermal microfracturing of sea ice.

2. THE DATA

2.1 ICE POSITION DATA

The data used in this study were from the Arctic Ice Dynamics Joint Experiment which began in March 1975 and ended in May 1976 (Fig. 1). Four sites on the Beaufort Sea ice pack were initially manned, and were tracked by the Navy Navigation Satellite System (rms position accuracy of approximately 60 m). Ice kinematics were determined

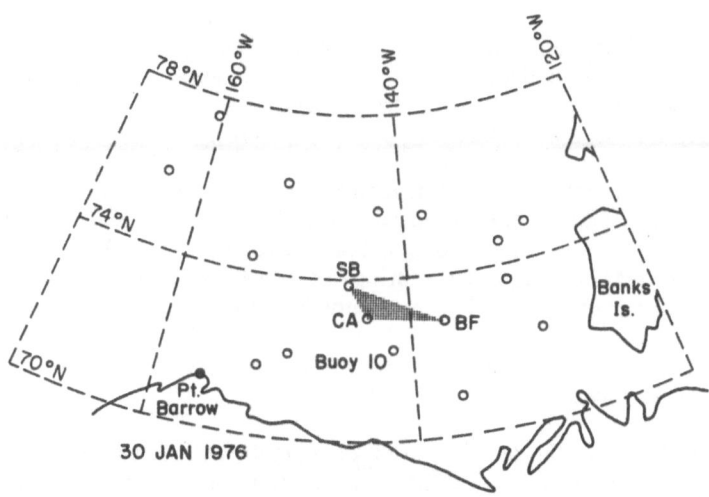

Fig. 1. Locations of a number of AIDJEX data buoys and manned camps (SB, CA, and BF) for 30 January 1976. Station 10 acoustic data discussed in this paper were collected at Buoy 10.

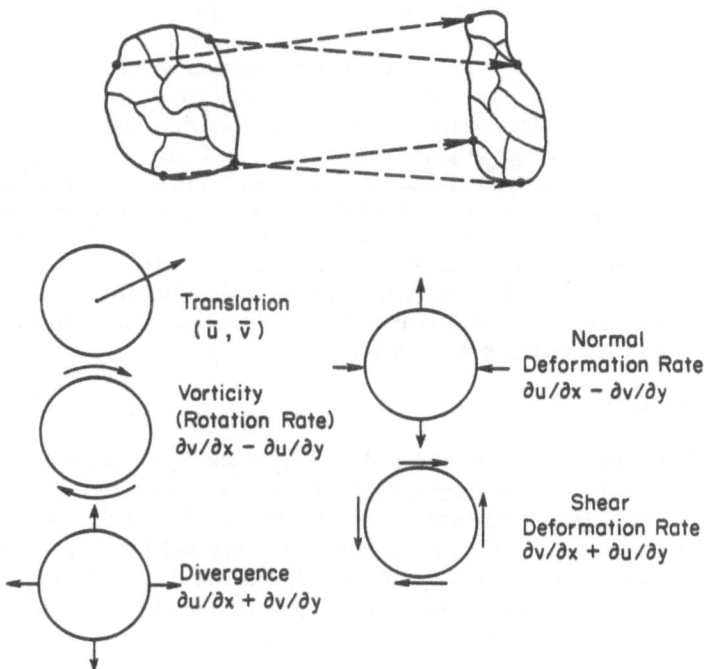

Fig. 2. Mathematical and physical definitions of the five basic components of the motion of a parcel of ice.

for periods during which at least three of the four manned camps were providing position data[6].

The region of ice delineated by the manned camps was of the order of a hundred kilometers. The position data were at three hour intervals, and the coordinate system was centered at the North Pole with the positive x direction in the Beaufort Sea being approximately northeastward and the positive y direction being approximately northwestward.

As opposed to defining ice kinematics based on the motion of each individual manned camp, the stations are considered to delineate a parcel of ice. The kinematics are then defined as the translation of that ice parcel as well as any rotation, area change, or deformation that occurs during the translation (Fig. 2). The motion of the ice parcel is decomposed into five independent components: translation rate (U), vorticity (V) (rotation rate), divergence (D), normal deformation rate (N), and shear deformation rate (S). The last four of these parameters describe relative motion within the ice parcel as it translates[7]:

V - the change in the orientation of the parcel (without a shape or size change),
D - the change in the size of the parcel (without an orientation or shape change),
N - the change in the shape of the ice parcel due to forces acting normal to the sides of the parcel (without a size or orientation change), and
S - the change in the shape of the ice parcel due to forces acting parallel to the sides of the parcel (without a size or orientation change).

We will refer to these above four parameters as the differential kinematic parameters (DKP).

Time histories of U and the DKP were calculated and then correlated with corresponding noise variations. In a previous calculation using the same position data, McPhee[8] considered steady-state conditions by low-pass filtering the data. This effectively removed the energy at periods which included inertial oscillations.

The utility of this type of analysis is that each mode of motion has a specific physical interpretation for sea ice. For example, arctic ambient noise generated by lead and ridge formation is related to the divergence D. Also, U is related to the ice rushing through the water. Finally, shape changes caused by deformation (N and S) can be considered in terms of simply a rearrangement of individual floes within the ice parcel (no area change). Thus, N and S represent the degree to which the floes are moving past one another in the rearrangement process.

2.2 NOISE DATA

Omni-directional noise data were collected by eight buoys deployed around the manned camp array during various periods from the spring of 1975 to the spring of 1976. These data were 1/3 octave bands centered at 3.2 Hz, 10 Hz, 32 Hz, and 1000 Hz. The data were 45 s averages sampled every 3 hours. Greene and Buck[9] indicated that the 3.2 Hz signal was likely contaminated by strum. Therefore, these data are not considered here. Lewis and Denner[5] discussed the remaining noise data in detail, with one of their conclusions being that the summer 10 Hz noise signals were also contaminated by cable strum.

To perform the correlations appropriately, we used noise data from two stations near the manned camp array after converting the noise from decibels to pressure amplitudes relative to 1 uPa. The first station was Buoy 10 (Fig. 1), which was ~100 km south of the manned camp array during the summer and fall of 1975 and the winter of

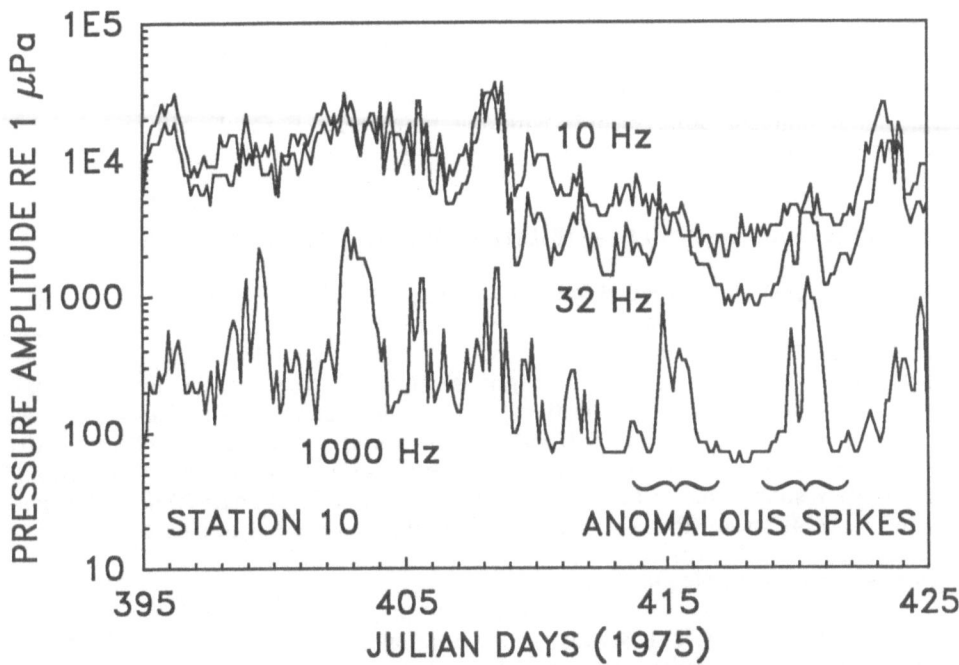

Fig. 3. Ambient noise variations at 10 Hz, 32 Hz, and 1000 Hz at Station 10 during February 1976.

1976. The second station was Buoy 66 which was ~70 km north of the manned camp array only during the summer and fall of 1975.

Time histories of the noise data can be found in Lewis and Denner[5]. Fig. 3 gives an example of the noise time series for the winter of 1976 (February). The data show that the 10 and 32 Hz signals are quite similar and that the short-term fluctuations with the greatest energy were found in the 1000 Hz data. One feature that was noted in all the acoustic stations during the winter was the period of relatively low noise levels during Julian days 412-422 (Fig. 3). This corresponds to a time during which no ice motion was detected at the manned camps. One will note two anomalous spikes in the 1000 Hz noise record during this period. These spikes are of the order of 20 dB, with some signatures at 10 and 32 Hz.

2.3 AIR TEMPERATURE DATA

At each acoustic station, air temperature was also measured at 3 hr intervals. The sensible heat flux between the air and sea ice is proportional to $F = T_a - T_i$, where T_a is the air temperature and T_i is the temperature at the surface of the ice. Microfracturing of the ice is assumed to occur only if $F < 0$. If we also assume that the ice temperature takes about 3 hrs to adjust to the air temperature, then

$$T_i \text{ at } t=0 = T_a \text{ at } t=-3 \text{ hrs}$$

and so

$$F = (T_a \text{ at } t) - (T_a \text{ at } t\text{-}3 \text{ hrs}).$$

Since we assume that a zero or positive F does not result in any thermal fracturing, we correlated our ambient noise records with the time series F for F < 0 and 0 for F greater than or equal to zero.

3. CORRELATIONS BETWEEN AMBIENT NOISE AND ICE KINEMATICS

Ambient noise levels from acoustic stations 10 and 66 were correlated with the variations of the ice kinematic parameters plus the air temperature fluctuations for the summer and fall of 1975. During the winter, only Station 10 was close enough to perform the correlations. Unfortunately, coinciding noise and high resolution ice kinematic data do not exist for the spring, so correlations could not be calculated for that season. We first consider the correlations at 10 Hz and 32 Hz for the three seasons (Tables I-III). Since these two low frequency signals are so similar for most of the year[5], one would expect similar results. Due to the situation of the summer 10 Hz data likely being contaminated by cable strum, the correlations for that data set are not discussed.

3.1 CORRELATIONS - 10 Hz AND 32 Hz

During the summer, Table I shows that the translation speed U of the ice was well correlated with the 32 Hz ambient noise signals for both stations. Of course, this is not a surprising result since, in most cases, the ice pack will try to move as it is put under

Correlations (% Variances)

	10 Hz		32 Hz		1000 Hz	
	Stat. 10	Stat. 66	Stat. 10	Stat. 66	Stat. 10	Stat. 66
U	----	----	0.74(54.6)	0.82(67.3)	0.63(40.3)	----
U^2	0.65(42.4)	0.70(48.9)	0.72(52.3)	0.80(63.6)	0.65(41.7)	0.76(58.1)
D	0.48(23.2)	0.41(16.8)	0.28(8.1)	0.40(15.7)	0.30(8.7)	0.23(5.3)
V	0.47(22.5)	0.31(9.7)	0.41(16.5)	0.37(13.6)	0.44(19.0)	0.27(7.1)
N	-0.04(0.2)	-0.11(1.2)	0.01(0.0)	0.0(0.0)	-0.09(0.8)	0.0(0.0)
S	0.09(0.9)	-0.02(0.0)	0.06(0.3)	0.09(0.8)	0.08(0.6)	-0.07(0.5)
F	0.02(0.0)	0.06(0.4)	0.03(0.1)	0.05(0.2)	0.0(0.0)	0.07(0.5)
$\lvert D \rvert$	0.40(15.9)	0.50(25.1)	0.27(7.2)	0.40(16.3)	0.20(3.9)	0.19(3.8)
$\lvert V \rvert$	0.39(15.2)	0.31(9.5)	0.33(10.8)	0.35(12.5)	0.23(5.5)	0.13(1.7)
$(N^2+S^2)^{\frac{1}{2}}$	0.23(5.3)	0.38(14.3)	0.30(8.7)	0.42(17.4)	0.14(1.9)	0.32(10.1)
Maximum Correlation	0.73	0.74	0.76	0.84	0.69	0.81

Table I. Linear correlation coefficients between the summer AIDJEX ambient noise data sets (Stations 10 and 66) and various ice kinematic parameters (manned camp position data). The percent variances accounted for are in parentheses. The maximum correlation is that obtained by the linear combination of various ice kinematic parameters and air temperature fluctuations.

stress. We tested the hypothesis that differential motion was the true cause of the noise by eliminating U in the correlation calculation. When this was done, the highest correlations between the DKP and F and the noise were only 0.48 for Station 10 and 0.59 for Station 66. These correlations imply that the 32 Hz signal is indeed primarily a response to the ice rushing through the water as opposed to differential motion.

The maximum correlation at Station 10 using various combinations of all the summer environmental parameters was 0.76. For Station 66, the maximum correlation was 0.84. Lewis and Denner[5] calculated an e-folding distance for 32 Hz during the summer of 440 km. Since the noise data were collected on the order of 100 km from where the ice motion data were collected, one might expect that these correlations are the largest obtainable.

We note that there was little difference between the summer correlations for U and U^2. Thus, the summer data does not allow one to interpret whether or not the lower frequency ambient noise is better related to the speed of the ice or to the kinetic energy. The fall data (Table II) seemed to answer this question more clearly, with significantly higher correlations between U and both the 10 Hz and 32 Hz signals. However, during this fall season we see that differential motion became quite important. This is particularly true for total deformation $(N^2+S^2)^{\frac{1}{2}}$. Using linear combinations of the DKP and F could not produce correlations significantly higher than just using $(N^2+S^2)^{\frac{1}{2}}$. However, U and $(N^2+S^2)^{\frac{1}{2}}$ can be combined to obtain correlations that range from 0.72 (10 Hz at Station 10) to 0.83 (32 Hz at Station 66). Thus, the data indicate that low frequency noise during the fall was primarily a result of the ice translating through the water plus the individual ice floes moving past one another as the shape of the ice parcel changed.

During the winter (Table III), the importance of $(N^2+S^2)^{\frac{1}{2}}$ decreased while other forms of differential motion became more important. The best correlations were with U, the speed of the ice. However, at 10 Hz we could combine the DKP and F to obtain

Correlations (% Variances)

	10 Hz		32 Hz		1000 Hz	
	Stat. 10	Stat. 66	Stat. 10	Stat. 66	Stat. 10	Stat. 66
U	0.72(52.5)	0.76(58.5)	0.70(49.2)	0.70(48.7)	0.34(11.5)	0.28(7.7)
U^2	0.61(37.7)	0.74(55.4)	0.59(35.4)	0.62(38.8)	0.28(7.6)	0.20(4.0)
D	0.07(0.5)	0.23(5.5)	0.09(0.8)	0.18(3.2)	0.11(1.3)	0.26(6.6)
V	-0.21(4.4)	-0.48(22.7)	-0.21(4.3)	-0.42(17.6)	0.08(0.6)	-0.20(3.9)
N	-0.06(0.4)	0.01(0.0)	-0.09(0.8)	-0.02(0.1)	0.11(1.3)	0.01(0.0)
S	-0.46(21.0)	-0.40(16.2)	-0.49(24.2)	-0.34(11.8)	-0.32(10.1)	-0.06(0.4)
F	0.06(0.4)	0.10(1.0)	0.01(0.0)	0.07(0.4)	-0.08(0.6)	0.04(0.2)
|D|	0.41(17.0)	0.58(33.4)	0.35(12.4)	0.55(29.9)	0.17(2.9)	0.40(16.3)
|V|	0.46(20.9)	0.67(44.6)	0.44(19.0)	0.59(34.4)	0.16(34.4)	0.28(7.9)
$(N^2+S^2)^{\frac{1}{2}}$	0.63(39.4)	0.72(51.9)	0.57(32.3)	0.65(41.8)	0.11(1.3)	0.24(5.7)
Maximum Correlation	0.79	0.86	0.76	0.78	0.49	0.45

Table II. Linear correlation coefficients between the fall AIDJEX ambient noise data sets (Stations 10 and 66) and various ice kinematic parameters (manned camp position data). The percent variances accounted for are in parentheses. The maximum correlation is that obtained by the linear combination of various ice kinematic parameters and air temperature fluctuations.

a correlation of 0.71. At 32 Hz, a combination of the DKP and F gave a correlation of 0.64. This would indicate that differential motion was quite important at 10 Hz while U and the DKP were both important at 32 Hz.

For the 32 Hz winter signal, D, N, and U were found to give a correlation of 0.74. Once again, D<0 and N>0 tended to produce noise. This and the 10 Hz correlations were not as encouraging as one might have expected considering the large space scales of the 10 Hz and 32 Hz winter data (order of 800 km)^5. This could be an indication that other processes are at work during the winter which produce low frequency noise but have no associated ice movement. For example, winter-time ice may not move under a limited forcing as a result of internal ice stresses. However, one would still expect the generation of noise due to the load put on the ice.

3.2 CORRELATIONS - 1000 Hz

The summer-time correlations between ice motion, F, and the 1000 Hz noise are shown in Table I. We first point out that this higher frequency noise is best correlated with U^2. [Indeed, at Station 66 the correlation calculation between U and 1000 Hz resulted in a negative constant in the least-squares fit. This implies a negative baseline (minimum) noise level, which is physically impossible.] The best correlations that were obtained using various combinations of the DKP and F (U being excluded) were 0.47 for Station 10 and 0.44 for Station 66. Thus, unlike the lower frequencies, the 1000 Hz summer signal appears to be primarily a function of the kinetic energy of the ice parcel.

The fall correlations at 1000 Hz are shown in Table II. One immediately notices that the maximum of these correlations is considerably lower than the correlations for other frequencies and other seasons. The e-folding space scales calculated by Lewis and Denner^5 give us a clue as to the probable reason for the low correlations. They found that the minimum e-folding length at 1000 Hz occurred during the fall and was 170 km.

Correlations (% Variances)

	10 Hz	32 Hz	1000 Hz
U	0.60(35.9)	0.66(43.7)	0.60(35.5)
U^2	0.46(21.3)	0.56(31.1)	0.62(38.8)
D	-0.29(8.4)	-0.38(14.2)	-0.27(7.5)
V	0.05(0.2)	0.12(1.4)	0.01(0.0)
N	0.01(0.0)	-0.04(0.2)	-0.11(1.3)
S	-0.10(1.1)	-0.23(5.3)	-0.31(9.4)
F	0.18(3.0)	0.13(1.5)	-0.07(0.1)
|D|	0.49(24.0)	0.42(18.0)	0.30(8.8)
|V|	0.50(25.4)	0.49(23.9)	0.38(14.1)
$(N^2+S^2)^{\frac{1}{2}}$	0.39(15.4)	0.38(14.5)	0.36(12.8)
Maximum Correlation	0.76	0.78	0.67

Table III. Linear correlation coefficients between the winter AIDJEX ambient noise data sets (Station 10) and various ice kinematic parameters (manned camp position data). The percent variances accounted for are in parentheses. The maximum correlation is that obtained by the linear combination of various ice kinematic parameters and air temperature fluctuations.

Thus, the problem of using noise and ice kinematic data from two different locations is enhanced during the fall. However, suppose we assume that the relative magnitudes of the various correlations give an indication of the relative importance of various parameters. This leads us to some very interesting results.

The fall correlations at the two stations were strikingly different. At Station 10 (in the southern part of the study area), we see that the speed of the ice parcel as well as the shearing deformation were relatively important. However, at Station 66 in the central part of the study area, the absolute value of divergence was the parameter that was best correlated with the 1000 Hz signal. Thus, in the southern region, which would typically have more open water during the fall, the 1000 Hz noise appeared to be generated by translation and by the types of deformation which cause elongation along a northwest/southeast axis. But farther north, where conditions would be more solidly frozen, the opening (D>0), and closing (D<0) of leads and perhaps ridging (D<0 for zero percent water) appear to have been most responsible for generating noise.

These differences between the mechanisms which generated the 1000 Hz ambient noise may be the cause of the low e-folding scale found during the fall. Spatial variations in the frozen state of the ice pack could allow for one mode of response in the south and another in the north. Thus, any given forcing of the ice could result in different noise signals at different locations. This would translate to a spatial incoherence.

The winter correlations (Table III) showed that the 1000 Hz signal went back to being slightly more coherent with U^2 than with U. The highest correlation that can be obtained using all variables except U^2 (or U) is 0.55. Thus, ice parcel translation at Station 10 still appears to be of some importance during winter. As seen in Table III, several of the DKP have correlations of the order of 0.3–0.4. The magnitude of these correlations are relatively small with respect to that of U^2 or U. But since a combination of their effects gives a 0.55 correlation (close to 0.60 for U and 0.62 for U^2), it is difficult to speculate as to whether translation plays a more important role than differential motion in the generation of higher frequency noise.

Of particular interest at the 1000 Hz frequency is the influence of the air temperature fluctuations, F. The correlations indicate that the effect of thermal cracking is either a) insignificant or b) not being properly calculated. It is the latter that we believe to be true.

4. HIGHER FREQUENCY NOISE AND THERMAL MICROFRACTURING

As can be seen, there are several factors concerning higher frequency arctic ambient noise that warrant investigation. Firstly, higher frequency ambient noise during non-summer months tends to have considerably shorter space and time scales than the lower frequency ambient noise. Somewhat shorter space scales can be expected since higher frequency noise is attenuated more as it travels through the water column. However, the major influence for sea ice processes in the arctic is the driving by the atmosphere. Thus, the large differences between higher and lower frequency scales is difficult to explain.

For the summer, the space and time scales of higher frequency noise are as one would expect for noise generated primarily by ice motion. The shorter time scales for non-summer months are taken as a reflection of multiple noise-generating mechanisms at frequencies >300 Hz. These mechanisms may not necessarily be correlated in time with one another, the result being a relatively spurious and variable signal. This would produce the shorter time scales. Altogether, the higher frequency time scales imply that one could expect significant changes in higher frequency noise levels within a day during the summer but within as little as 9-12 hrs during non-summer months.

The higher frequency space scales have a minimum of 170 km in fall and then increase to 300 km in the spring and summer[5]. The increase through spring can be taken as a reflection of the greater compactness and rigidity of the ice field. The summer-time scale reflects microfracturing due to ice motion. The drop from 300 km during summer to 170 km during fall could be the result of spatial variations in the frozen state of the ice pack. But the extremely short space scales during winter (240 km) and spring do not appear to be logical. If the primary forcings are atmospheric, one would expect the scale of such forcings to be of the order of 500 km, even for temporally incoherent processes.

A second factor concerning higher frequency noise is its seemingly poor correlation with environmental parameters. During non-summer months, the horizontal movement of the ice accounts for only a part of higher frequency noise (Tables II and III), while the estimates of sensible heat fluxes have no apparent correlation. This implies that the heat flux through sea ice must be considered more thoroughly and exactly. Moreover, poor correlations may be an indication that we have yet to discover a significant mechanism for the generation of higher frequency noise. For example, we again bring up the noise time histories during the winter, days 412-422 (Fig. 3), and the corresponding ice motion characteristics. As previously noted, there were two 1000 Hz anomalies during that time, each with a 20 to 25 dB signature. A much reduced noise increase (10 dB) was also seen at 32 Hz. These anomalies occurred during a time of no perceivable ice motion. Yet 32 Hz under-ice noise during the winter is motion-induced[5,10]. Thus, it is likely that the 1000 Hz anomalies were also motion-induced.

A review of all the AIDJEX data for the winter period in question showed that the two 1000 Hz anomalies had signatures to some degree at all of the 5 operational acoustic stations. Two-dimensional contour maps (not shown) detail large-scale events covering most of the Beaufort Sea. For example, the first event (days 414-416) can be seen to move into the study area from the north and eventually engulf all but the eastern Beaufort Sea. During day 416, the trailing edge of the phenomena could be seen moving in from the north as the system moved southward out of the study region. We can only speculate as to the cause of these anomalies, but the pattern, scale, and the noise variations at 32 Hz imply some form of atmospheric pressure wave. Such a phenomena could produce vertical motion of sea ice without any horizontal movement. Thus, one would expect mostly microfracturing of the ice, with larger scale cracking also possible.

4.1 HEAT FLUX IN SEA ICE

The vertical flux of heat within sea ice may be written as

$$o_I \, c_I \, (dT/dt) = k_I \, (d^2T/dz^2) \tag{1}$$

where o_I is the ice density (917 kg/m^3), c_I is the specific heat of ice (a function of temperature and salinity), T is the ice temperature, t is time, z is positive upwards from the ice surface, and k_I is the thermal conductivity of ice (2.03 W m^{-1} °C^{-1}). Assume that we have surface temperature fluctuations, the magnitude of which decays exponentially with the depth of the ice. If o_I, c_I, and k_I were constant within the ice, then the solution to (1) is

$$T = (T_s - T_b) \, z \, / \, H \; + \; T_s \; + \; T_s' \; e^{az} \cos(ft + az) \tag{2}$$

where T_s is the mean surface temperature, T_b is the mean temperature at the ice bottom, H is the ice thickness, T_s' is the amplitude of the surface temperature fluctuations,

a^{-1} is the e-folding depth of the temperature fluctuations, f is the frequency of the temperature fluctuations, and

$$a = (o_I \, c_I \, f/2 \, k_I)^{1/2}.$$

For a daily period and typical values of o_I, c_I, and k_I, a^{-1} is of the order of 20 cm. Thus, most of the microfracturing of sea ice can be expected to occur within the top 30 cm of the ice for a diurnal heating cycle. In this case, one would expect arctic ambient noise generated by thermal microfracturing to coincide closely with the drop in temperature of the ice surface.

Examples of such microfracturing have been discerned using the available environmental data from the fall of 1975 of the AIDJEX period. In Fig. 4 we show several noise spikes that occurred during the following conditions: small or decreasing wind and/or ice speeds, falling air temperatures, and an increase from lower atmospheric pressure to higher atmospheric pressure. These are the classical conditions during which one expects to find the advection of a cold air mass into a region. As a low pressure system (and its associated front) passes, cold air is advected into an area, and the winds begin to decrease on the approach of a high pressure system. As the surface is cooled by the colder air, the ice fractures as it contracts. This fracturing continues as long as the air temperature is cooler than the ice and as long as the thermal wave propagating through the ice results in cooling.

Of course, nature is rarely as simple as our analytical solutions. Another of the AIDJEX data sets, this time for the spring of 1976, implies a more complicated form of thermal microfracturing. In Fig. 5 we see diurnal fluctuations of the 1000 Hz noise which appear to be related to air temperature fluctuations. However, in this case the noise tends to occur when the air temperature hits its minimum and, in some cases, even as it is increasing. This indicates that there continued to be cooling of the ice during the entire period of lower air temperatures. Moreover, there are numerous instances in which the daily noise pattern is absent from the record, even though there exists substantial air temperature variations. The periods during which there were strong noise signals occurred primarily when the atmospheric pressure was relatively high and when the atmospheric inversion level was relatively low (typical of cloud-free conditions). It would appear that these spring noise fluctuations are of a considerably different nature than those of the fall.

4.2 MODELING HEAT FLUX IN ICE

We must reconsider our heat flux within the ice. First, the specific heat of sea ice increases by an order of magnitude from the surface to about 2-3 m because of the warmer and saltier ice at the bottom. Secondly, thermal conductivity can vary considerably with temperature and salinity. Finally, the mechanical characteristics of sea ice (volumetric expansion and elastic modulus) are also a function of salt content and temperature. Thus, to consider such variations, one is required to numerically model heat flux in sea ice.

If the ice is divided into n layers, (1) can be approximated in finite difference form as

$$H_n \, o_I \, c_I \, (dT/dt) \tag{3}$$

$$= 2 \, k_I \, ((T_{n+1}-T_n)/(H_{n+1}+H_n) - (T_n-T_{n-1})/(H_n+H_{n-1}))$$

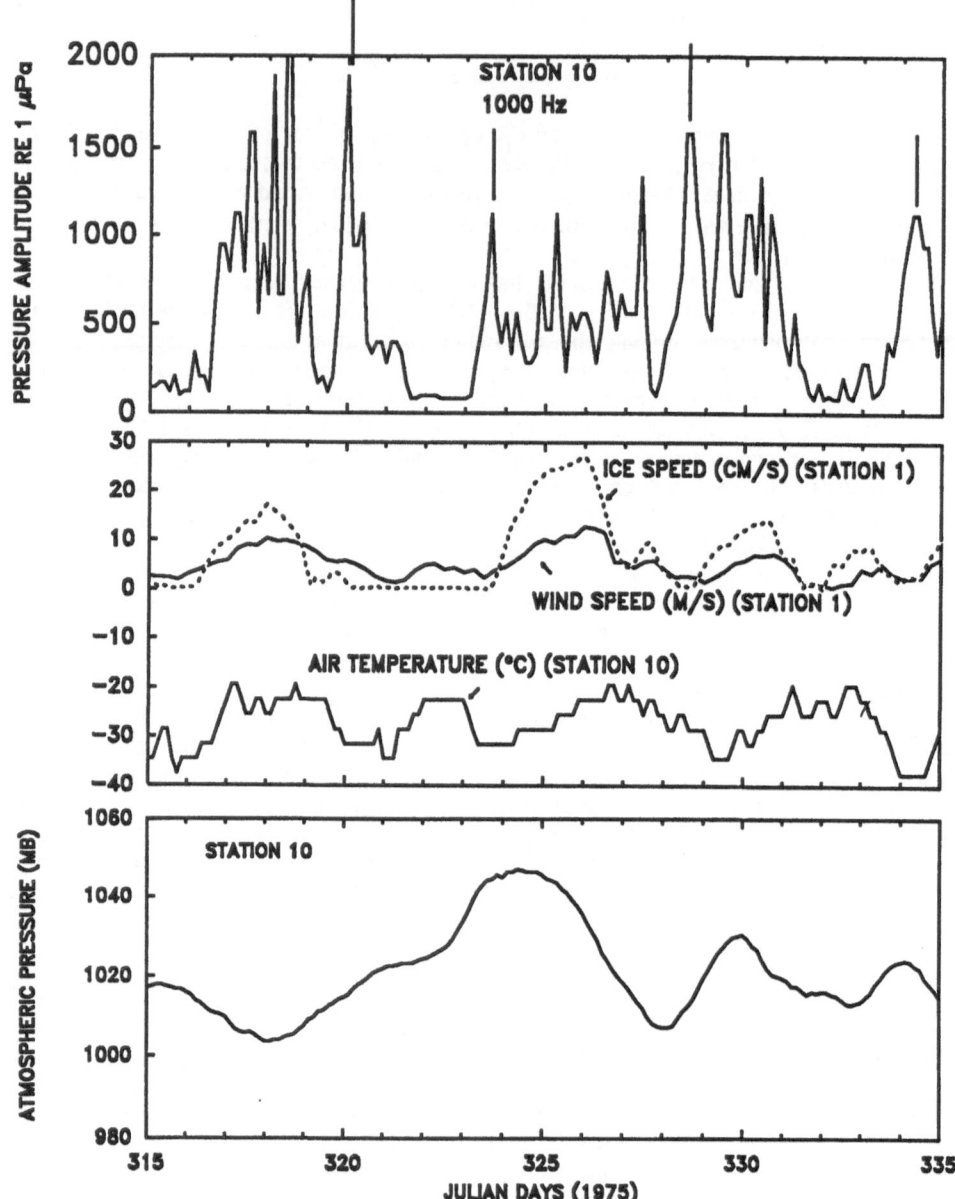

Fig. 4. Time histories of 100 Hz under-ice noise (top), ice speed, wind speed, and air temperature (center), and atmospheric pressure (bottom) in the central Beaufort Sea during November 1975. The vertical lines (top figure) denote times during which the 1000 Hz noise was, at most, weakly associated with ice kinematics and wind forcing.

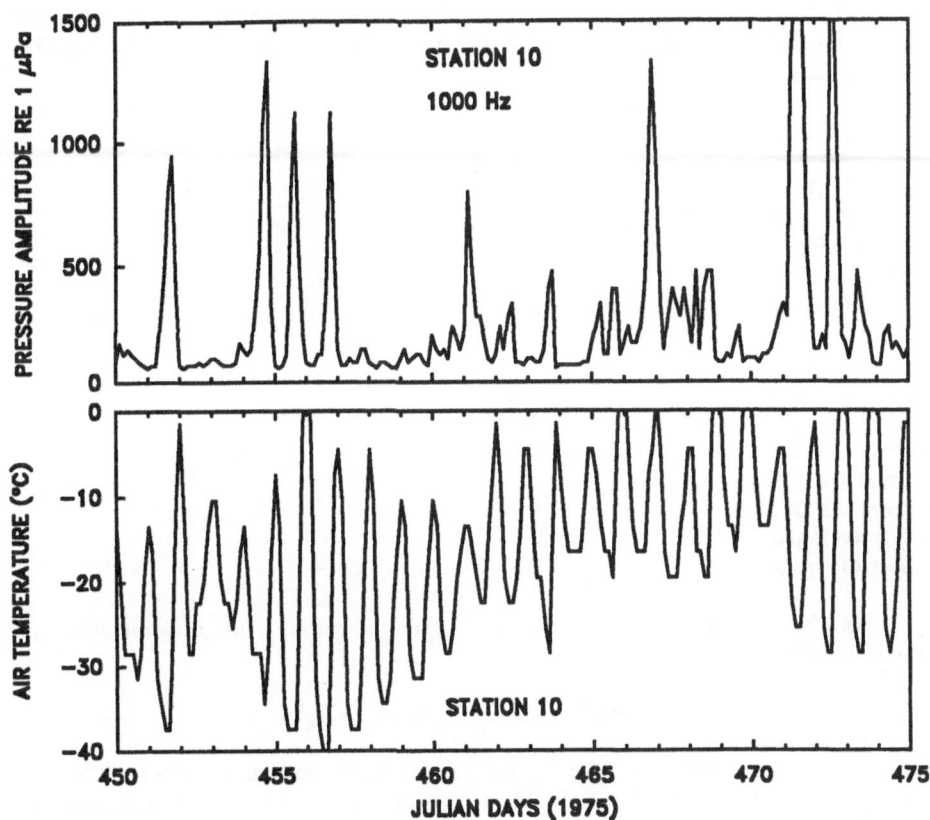

Fig. 5. Fluctuations of 1000 Hz noise and air temperature during March and April 1976 in the central Beaufort Sea.

where n is the layer number (increasing upwards), T represents the average temperature of the nth layer, and H_n is the thickness of the nth layer. For the bottom layer of the ice, a temperature of $-2°$ C is typical of arctic ice conditions, a result of the temperature of the underlying ocean water. Thus, the conductive heat flux from the ocean in the bottom layer n = b modifies (3) such that

$$H_b \; o_I \; c_I \; (dT/dt) \tag{4}$$

$$= 2 \; k_I \; ((T_{b+1}-T_b)/(H_{b+1}+H_b) - (T_b+2°)/H_b)$$

In the top layer of ice, there exist heat exchanges at the ice surface that include sensible and evaporative heat fluxes (QSENS and QEVAP), short wave solar radiation (QSLR), longwave radiation from the ice to the atmosphere (QBI), and longwave back radiation from the atmosphere (QBA). Thus, for the top layer n = t, (3) becomes

$$H_t \; o_I \; c_I \; (dT/dt) = QSLR + QSENS + QEVAP + QBI + QBA -$$

$$- 2 \; k_I \; (T_t-T_{t-1})/(H_t+H_{t-1})) \tag{5}$$

where the Q terms are expressed in W/m^2. The above expression accounts for the flux of heat from the underlying layer of ice and from the ice surface. Comparing (3) and (5), we see that

QSLR +

where T_s is the te
calculate T_s, whic
and evaporative h
   The unknowr
terize the Q terms

QSLR -
QSOLAR and alb
the AIDJEX exp
ables[12]. For typ
maximum for QS(
zon from ~2000
Albedos averaged
for most of the da

QSENS - using the air temperature T_a along with T_s, the sensible heat flux is parameterized by[13]

$$QSENS = o_a \, C_p \, C_s \, U \, (T_a - T_s)$$

where o_a is the air density (~1.395 kg/m^3 for air temperatures of about -20°C), C_p is the specific heat of air (10^3 W s/kg°C for cold air), and C_s is a bulk transfer coefficient (1.2×10^{-3}). Based on the three hourly AIDJEX air temperature data, we specify T_a as a function of time for this model.

QEVAP - is considered negligible. Maykut[13] compared a number of observations and found QEVAP to be considerably smaller than QSENS during spring and fall. Thus, for our purposes, we ignore this term.

QBI - is specified directly as a function of the surface temperature T_s of the ice using

$$QBI = e_L \, s \, (273.16 + T_s)^4$$

where e_L is the effective emissivity for ice (0.97) and s is the Stefan-Boltzmann constant (5.67 $\times 10^{-8}$ W/m^2°K).

QBA - is specified directly as a function of the air temperature T_a and the fractional cloud coverage C using[14]

$$QBA = e^* \, s \, (273.16 + T_a)^4$$

where e^* is the effective emissivity of the atmosphere. Since the increase in atmospheric moisture associated with greater cloud coverage results in an increase in e^*, we specify C in the model and use

$$e^* = 0.7855 \ (1 + 0.2232 \ C^{2.75}).$$

Thus, equations (3) - (6) give us n+1 equations in n+1 unknowns for n levels within the ice. Using our analytical solution as a standard, the numerical model was validated and tested. Preliminary tests with the model have all been with constant coefficients of expansion and elasticity. When sensible heat fluxes were prominent (wind speed of 10 m/s), the estimated microfracturing (noise level) began slightly after 1500 hrs (local). The fracturing increased to a maximum at 0000 hrs, and then fell close to zero by 0600 hrs. Without sensible heat flux, the phase of the microfracturing was shifted by several hours, with maximums at about 0100 to 0200 hrs. However, at no time was a maximum in fracturing seen to occur at about 0600 hrs as was recorded in the AIDJEX and other data (Fig. 5 and reference 1).

The model runs indicate that the primary balance of heat flux for the spring-like conditions are in the radiative heat flux terms. Since the data imply that cloud conditions can effectively stop thermal microfracturing, a number of runs were made varying the percent cloud coverage C. This effectively increased the amount of long-wave radiation from the atmosphere back to the ice. However, it was found that the estimated microfracturing was little reduced when the cloud cover was increased. Thus, the lack of consistent, daily noise spikes may be a result of other factors in addition to varying cloud cover.

5. MONITORING AND PREDICTING ARCTIC AMBIENT NOISE

5.1 MONITORING ARCTIC AMBIENT NOISE

The primary method for monitoring arctic ambient noise is through the use of buoys drifting on the ice. Only a limited amount of arctic ambient noise has been monitored in this way for a number of reasons. Firstly, these buoys and their deployments on the ice pack are considerably more expensive than typical satellite-tracked buoys. The acoustic instrumentation is responsible for some of the additional costs. Also, acoustic buoys cannot be simply dropped by air since holes must be drilled through the ice for the cables and instrumentation. Moreover, the deployment of such buoys throughout the arctic is limited due to policies of various nations. Finally, there are many ice conditions for which the use of such buoys is impossible. For example, shallow water conditions often preclude the hanging of hydrophones from ice since it might become grounded or be crushed.

It appears that indirect methods may hold more promise in being able to monitor arctic ambient noise. In Sections 1 and 3, we discussed a number of parameters that have been found to be correlated with under-ice noise. It seems possible that relatively accurate estimates of lower frequency noise levels (and summer-time higher frequency noise levels) could be produced using observed ice motion (buoys, satellite imagery, etc.). Even more accurate lower frequency estimates might be possible by using internal ice stresses[4]. A number of techniques have been developed to estimate internal ice stresses from observed ice motion and wind fields. Kheysin and Ivchenko[15] developed momentum, energy, and continuity expressions which allow one to take initial conditions and known drift velocities and compute a field of internal stresses. Such techniques use observed forcing and responses to balance the equations of motion and, as such, can provide reasonable stress estimates[16].

Indirect estimates of non-summer higher frequency noise does not seem feasible at this time. As our knowledge of thermal microfracturing grows, this situation will likely change. But scale analyses indicate that a fine grid of observations will be required to produce reliable higher frequency noise predictions. Time scales of variations can be as short as three hours during the winter for higher frequency noise. Thus, the sampling of environmental parameters may have to be of the order of one hour or less. Moreover, space scales for higher frequency noise can be as small as 170 km, implying a sampling network of the order of <100 km in regions of interest.

5.2 UNDER-ICE NOISE PREDICTIONS

Prediction of arctic ambient noise levels relies on the prediction of those processes that cause the noise. Using ice models, it may be feasible to produce lower frequency and summer-time higher frequency noise predictions up to 5 to 10 days. These predictions could be based on the calculated ice motion or internal stresses. Such models may eventually be able to produce accurate estimates of thermal microfracturing within the ice. Taken along the lines of the model in Section 4, it would not be unrealistic to use wind, pressure, and heat forcing to drive an ice model as well as calculate small scale fracturing parameters. This would be the basis of providing predictions for higher frequency noise during non-summer months.

6. SUMMARY AND CONCLUSIONS

The correlations between noise, ice motion, and temperature changes have provided the following conceptual model for the generation of arctic ambient noise at lower frequencies. During the summer, most of the lower frequency noise is generated by the ice rushing through the water. Differential ice motion appears to be insignificant in producing under-ice noise. This is likely the result of the relatively large percent of open water along with the fact that the internal structure of ice weakens and decays during summer. However, this situation changes by fall when the total deformation ($N^2 + S^2)^{1/2}$ as well as the translation of the pack ice becomes important in producing lower frequency noise. The motion implied by the deformation is that of individual ice floes moving past one another during a rearrangement process. Thus, noise would be generated as the bonds between floes are broken and as the floes grind and slip past each other.

It is only during winter that some of the other forms of differential motion become important in producing lower frequency noise. Indeed, at 10 Hz the DKP can be combined to account for a significant amount of the noise. We found that correlations were higher when the divergence D was used as opposed to the absolute value of D. The negative correlation with D would imply that ridging and rafting become important generating processes for lower frequency noise during winter-time conditions.

The correlations using the 1000 Hz data result in a somewhat different scenario for the generation of higher frequency noise in the arctic. First, the summer correlations imply that the kinetic energy related to the ice parcel translation is the primary factor in higher frequency noise generation. This noise/motion relationship is supported by space and time scale analyses[5]. During the fall, all correlations drop considerably. The fall correlations indicate that U and -S are important parameters in the southern Beaufort while the absolute value of D is the important parameter farther north. This implies a high degree of regional variability in the generation of higher frequency ambient noise.

The effect of our parameter reflecting sensible heat flux, F, was small. However, the data indicate distinct thermal microfracturing events. During fall, sensible heat fluxes can be seen to cause spikes in the 1000 Hz data. But spring-time spikes which appear to be the result of daily heat fluxes are not readily understood or modeled. The correct formulation of the thermal microfracturing of sea ice appears to be a critical step in our understanding, monitoring, and forecasting of higher frequency under-ice noise.

ACKNOWLEDGEMENTS

This work was supported by the Office of Naval Research, Arctic Programs, through contracts N00014-85-C-0531 and N00014-87-C-0115.

REFERENCES

1. Milne, A.R., 1972: 'Thermal tension cracking in sea ice: a source of under-ice noise.' *J. Geophys. Res.*, 77, 2177-2192.

2. Greene, C.R., and B.M. Buck, 1977: ' Influence of atmospheric pressure gradient on under-ice ambient noise.' *U.S. Navy J. Underwater Acoustics*, 28 (4).

3. Pritchard, R.S., 1984: 'Arctic Ocean background noise caused by ridging of sea ice.' *J. Acoust. Soc. Am.*, 75 (2), 419-427.

4. Makris, N.C., and I. Dyer, 1985: 'Environmental correlates of pack ice noise.' *J. Acoust. Soc. Am.*, 79 (5), 1434-1440.

5. Lewis, J.K., and W.W. Denner, 1987: 'Arctic ambient noise in the Beaufort Sea: seasonal space and time scales.' In press, *J. Acoust. Soc. Am.*

6. Thorndike, A.S., and J.Y. Cheung, 1977: 'AIDJEX measurements of sea ice motion 11 April 1975 to 14 May 1976.' *AIDJEX Bull.* No. 35, Univ. of Washington, Seattle, 149 pp.

7. Kirwan, A.D., 1975: 'Oceanic velocity gradients.' *J. Phys. Oceangra.*, 5 (4), 729-735.

8. McPhee, M.G., 1978: 'The free drift velocity field across the AIDJEX manned camp array.' *AIDJEX Bull.* No. 38, Univ. of Washington, Seattle, 158-170.

9. Greene, C.R., and B.M. Buck, 1977: 'Arctic noise measurement experiment using Nimbus 6 data buoys.' *U.S. Navy J. Underwater Acoustics*, 27 (4).

10. Dyer, I., 1984: 'The song of sea ice and other Arctic Ocean melodies.' In *Arctic Policy and Technology* (I. Dyer and C. Dhryssostomidis, eds.), Hemisphere Publ. Corp., New York, 11-37.

11. Colony, R., and A.S. Thorndike, 1984: 'An estimate of the mean field of Arctic Sea ice motion.' *J. Geophys. Res.*, 86 (C6), 10623-10629.

12. Pautzke, C. G., and G. F. Hornof, 1978: 'Radiation program during AIDJEX: a data report.' *AIDJEX Bull.* No. 39, Univ. of Washington, Seattle, 165-185.

13. Maykut, G. A, 1987: 'Surface heat and mass balance.' In *Sea Ice Geophysics* (N. Untersteiner, ed.), Chap. 5. NATO Advanced Study Institute, Ser. C, Math. Phys. Sci. D. Reidel, Hingham, Mass.

14. Maykut, G. A., and P. E. Church, 1973: 'Radiation climate of Burrow Alaska, 1962-1966.' *J. Appl. Met.*, 12, 620-628.

15. Kheysin, D. Y., and V. O. Ivchenko, 1976: 'Pressure distribution in consolidated ice.' *Oceanology*, 15, 5, 542-546.

16. Lewis, J. K., R. D. Crissman, and W. W. Denner, 1986: 'Estimating ice thickness and internal pressure and stress forces in pack ice using Lagrangian data.' *J. Geophys. Res.*, 91 (C7), 8537-8541.

ACOUSTIC AMBIENT NOISE IN THE ARCTIC OCEAN BELOW THE MARGINAL ICE ZONE

Michael J. Buckingham* and Chi-fang Chen
Department of Ocean Engineering
Massachusetts Institute of Technology
Cambridge, MA 02139
USA

*On leave of absence from the Royal Aircraft Establishment
Farnborough, Hampshire, GU14 6TD, England

ABSTRACT. Experimental evidence suggests that the ambient acoustic noise in the Arctic Ocean below the marginal ice zone (MIZ) is generated by the clashing of ice floes on the sea surface. The noise is spiky in character, with a spectral density over the frequency range from 50 Hz to 1 kHz which varies with frequency as f^{-n}, where the index n varies slowly with time, showing an average value approximately equal to 2. Over a period of five days or so, n was observed to vary between a low value of 1.0 and a high value of 3.0. A theoretical model of the MIZ ambient noise is presented, based on a floe/floe collision mechanism, which predicts time series and spectra whose features are broadly consistent with those observed in the data.

1. INTRODUCTION

In June 1984, as part of the international Marginal Ice Zone Experiment (MIZEX), recordings were made of the omnidirectional ambient acoustic noise in the Arctic Ocean below the marginal ice zone (MIZ). The measurements were made in the Fram Strait, west of Svalbard, in an area approximately centred on $80.6°N, 8°E$. Hydrophones were suspended sixty metres below a large ice floe, which drifted through the MIZ over a period of several weeks. The noise data are available in the form of time series and power spectra, both of which are discussed in this paper. A theoretical model of the noise, based on an ice floe collision mechanism, is also presented and shown to be consistent with the major features in the data.

2. THE NOISE DATA

Typically, the time series representing the acoustic pressure fluctuations at the

583

B. R. Kerman (ed.), Sea Surface Sound, 583–598.
© 1988 by Kluwer Academic Publishers.

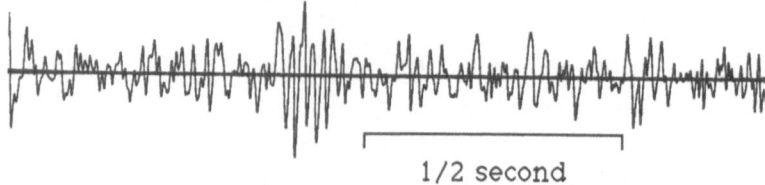

1/2 second

Figure 1: Time series from the MIZEX noise data, after bandpass filtering between 20 Hz and 100 Hz.

hydrophone are rather spiky, in accord with earlier observations of noise in the MIZ by Diachok and Winokur [1]. Individual pulses, standing out more or less prominently from the background noise are a common feature of the time series (Fig. 1). Many of these pulses decay to a level below the background noise within a cycle or two. Presumably, these pulses are the acoustic signatures of energetic events (e.g. floe/floe collisions or flexural ice cracking) occurring on the surface in the vicinity of the hydrophone. The duration of the pulses ranges from milliseconds to fractions of a second, which is qualitatively consistent with a distribution of floe diameters ranging from a few metres to several hundred metres.

The power spectra of the MIZEX noise, averaged over thirty-two or more samples, show considerable variability but generally display a broad maximum centred around 20 Hz, and a decay from about 50 Hz to beyond 1 kHz of the form f^{-n}, where f is frequency and n is a constant for a given spectrum with a value close to two. Below 5 Hz or so, the spectra usually rise rapidly with decreasing frequency, due to spurious non-acoustic effects such as flow past the hydrophone. Fig. 2 shows three examples of spectra observed on different days of MIZEX. Between 50 Hz and 1 kHz all three spectra show a power law dependence on frequency, but with different logarithmic slopes, that is, different n values.

Over an interval of two weeks, between Julian days 167 and 181, n slowly oscillated in an apparently systematic manner between extreme values of 1.0 and 3.0, with a quasi-period of approximately five days. This behaviour is illustrated in Fig. 3, which shows the n values of many spectra plotted as a function of time through the experiment. The gaps in the figure are due to gaps in the data. The reason for the slow time dependence displayed by the n values is not understood, nor is it known whether a systematic variation in n is a universal feature of the MIZ ambient noise. If so, then the n value must presumably correlate with some environmental factor, such as the wind vector or the ice compactness, but so far no compelling evidence is available which favours one mechanism rather than another.

Several mechanisms almost certainly contribute to the ambient noise in the

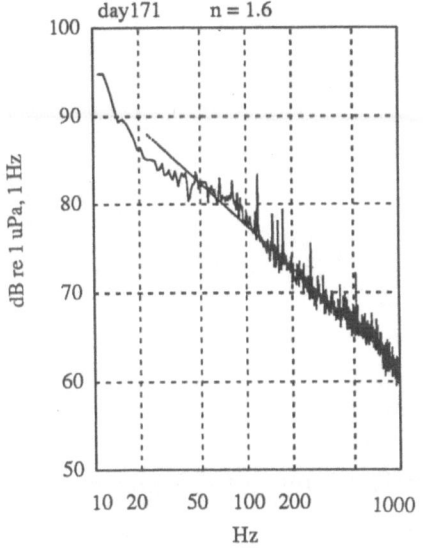

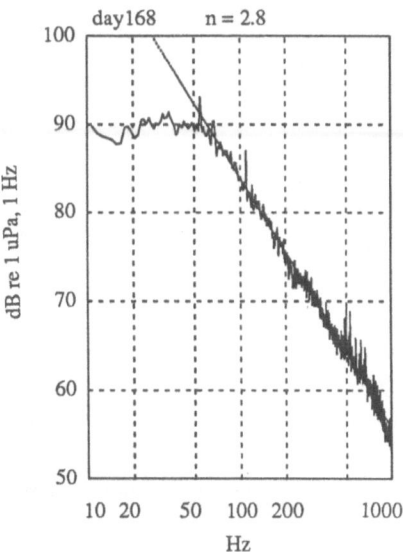

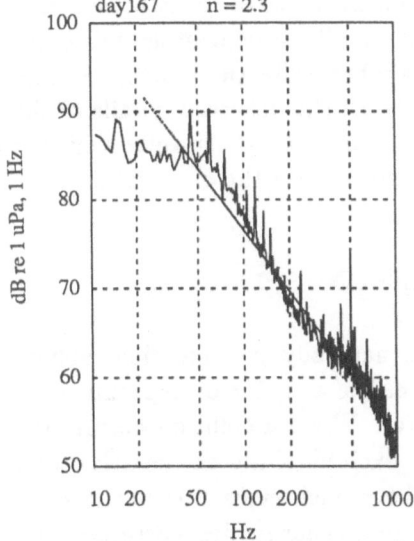

Fig.2 Noise spectra on three different days from the MIZEX data. Note the changing n value.

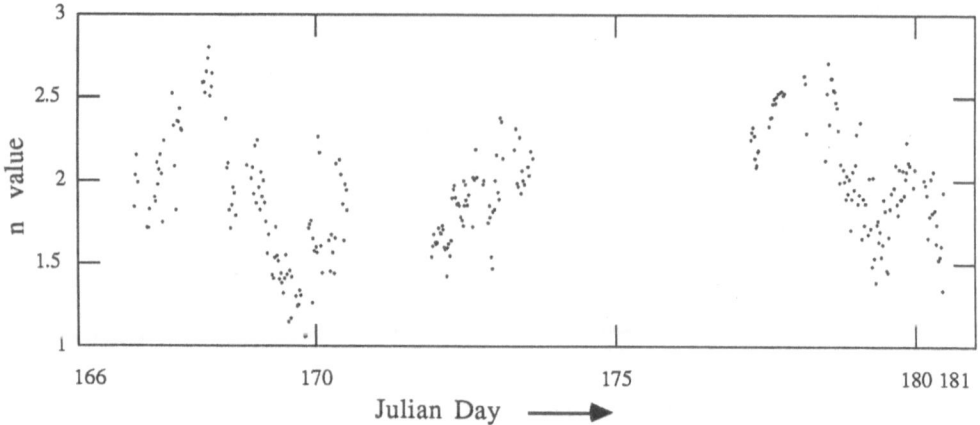

Figure 3: Observed n values as a function of time through the experiment.

MIZ [2]. These include floe/floe collision processes, that is bumping, rubbing and grinding; ice floe cracking events; floe slapping by waves; floe washover; and floe bobbing and flexing. At low frequencies, below 1 kHz but above say 10 Hz, the predominant source mechanism is likely to be floe collisions, since such events are highly energetic with moderately efficient mechanical/acoustic coupling.

In an attempt to model the low frequency ambient noise in the MIZ, we have concentrated on bumping between floes and ignored all other possible noise generating mechanisms. Several simplifications are inherent in the model, the aim at this stage being, not to reproduce theoretically all the observed details of the noise, but to obtain a representation of the time series and the spectra which is both physically plausible and in reasonable agreement with the observations.

3. THE GREEN'S FUNCTIONS FOR A COLLISION

Each ice floe in the model is represented as a neutrally buoyant fluid sphere (i.e. shear is neglected) in which the sound speed is a factor of approximately two higher than that in the surrounding seawater. When a collision occurs, the floe distorts and various modes of vibration are excited. Only the zeroth order, or radial, mode of vibration is considered in the model, since this is the most efficient acoustic radiator to the far field. The higher order modes, corresponding to dipole, quadrupole and higher order multipoles, contribute to the near field but not significantly to the far field.

To ensure that only the radial mode is excited, the impact associated with a collision is represented in the model by a fictitious point source located at the

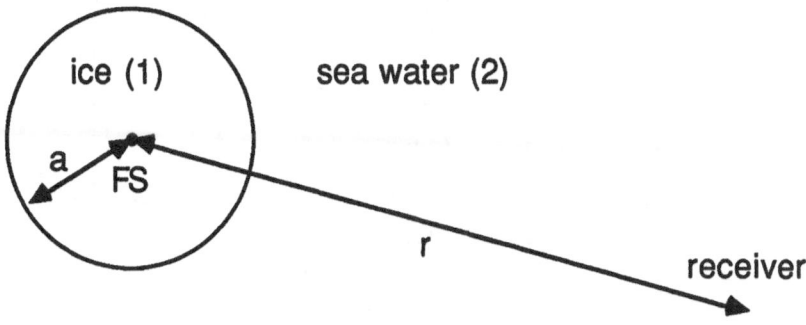

Figure 4: An ice sphere of radius a (region 1) immersed in an infinite ocean (region 2). A fictitious source, FS, is at the centre of the sphere and the receiver is at range r from the source.

centre of the sphere. The time dependence of this source depends on the nature of the floe/floe collision process, as discussed in Section 4. First we are interested in the spectrum of the field generated by an impulsive point source, that is the Green's function of the Helmholtz equation.

The Green's function for the field in the ice sphere and in the surrounding seawater is determined by solving the inhomogeneous Helmholtz equation, subject to the requirements that the pressure and normal component of velocity should be continuous across the spherical ice/seawater boundary. For this part of the problem the ice sphere is assumed to be immersed in an infinite ocean (Fig. 4), the effect of the pressure-release sea surface being taken into account later. The method of solution involves two finite Hankel transforms [3], one taken over the internal region (the ice sphere) and the other over the external region (the ocean). This procedure yields a set of simultaneous equations which may be solved to give the following expression for the Fourier transform (with respect to time) of the velocity potential, $G_2(jw)$, in the seawater due to the (fictitious) impulsive source at the centre of the sphere:

$$G_2(jw) = (4\pi r)^{-1} \{exp[-jk_2(r-a)]\} / [cos(k_1 a) + j\beta sin(k_1 a)]. \tag{1}$$

In this expression, a is the radius of the ice sphere, r is the range from the source point, ω is the angular frequency, k_2 is the wavenumber in the seawater, k_1 is the wavenumber in the ice, $j = (-1)^{\frac{1}{2}}$, and

$$\beta = k_2/k_1 . \tag{2}$$

Details of the analysis leading to the Green's function in equation (1) are given in Ref. 3.

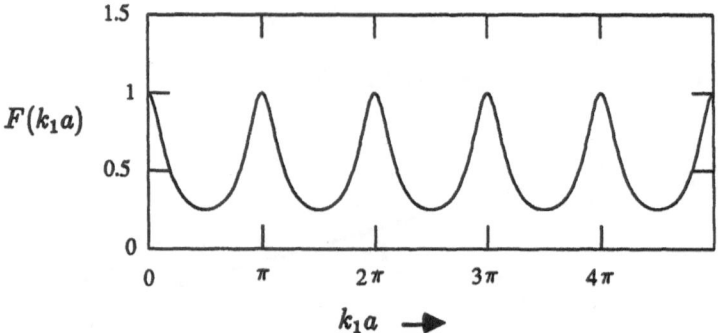

Figure 5: The function $F(k_1 a)$ from equation (4) plotted with $\beta = 2$.

From equation (1) it follows that

$$|G_2(jw)|^2 = (4\pi r)^{-2} F(k_1 a) ,\tag{3}$$

where

$$F(k_1 a) = 1/[cos^2(k_1 a) + \beta^2 sin^2(k_1 a)] .\tag{4}$$

Note that when $\beta = 1$, that is, when there is no acoustic mismatch between the sphere and the surrounding fluid (which is not our case), the function $f(k_1 a)$ is unity and the expression for the field in equation (3) is independent of frequency. Then the field follows a simple spherical spreading law. When $\beta \neq 1$, the magnitude of the field does depend on frequency, as illustrated in Fig. 5, showing $F(k_1 a)$ in equation (4) with $\beta = 2$, a value which is representative of our ice/seawater problem. The peaks in $F(k_1 a)$ occur when $k_1 a$ is a multiple of π, that is, when the diameter of the ice sphere is an integer number of wavelengths. (Incidentally, when $\beta < 1$ a similar plot to that in Fig. 5 is obtained except that the peaks are shifted to values of $k_1 a$ which are odd multiples of $\pi/2$).

On taking the inverse Fourier transform of equation (1), the time dependence of the field in the seawater is obtained. This is the Green's function $g_2(t)$ of the wave equation; it takes the form of a sequence of impulses, as follows:

$$g_2(t) = [2\pi r(1 + \beta)]^{-1} \sum_{n=0}^{\infty} b^n \delta \{t - [(r - a)/c_2] - [(2n + 1)a/c_1]\} ,\tag{5}$$

where c_1 and c_2 are the speeds of sound in the ice and seawater, respectively,

$$b = (\beta - 1)/(\beta + 1)\tag{6}$$

and $\delta\{ \}$ is the Dirac delta function. The parameter b in equation (6) is simply the reflection coefficient for normal incidence at the ice/water interface.

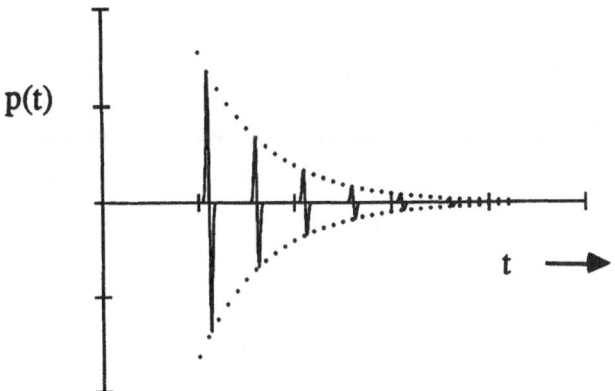

Figure 6: Series of decaying, equispaced pressure doublets observed at the receiver due to the source in the ice sphere. The first doublet is the direct arrival and remainder arise from internal reflections at the ice/seawater boundary.

The $n = 0$ term in equation (5) represents the direct arrival at the receiver from the source at the centre of the sphere. The remaining terms represent successive equispaced arrivals, due to reflections from the ice/seawater boundary. The nth arrival is attenuated by the factor b^n, representing the losses incurred in the n reflections from the interface between the two media. The pressure signature in the water is proportional to the time derivative of $g_2(t)$, that is, the pressure consists of a sequence of decaying equispaced doublets. Such a pulse is illustrated schematically in Fig. 6.

With $\beta = 2$, the attenuation due to a single reflection is, from equation (6), $b = 1/3$. The pulses observed in the time series data are usually not much more than a factor of three above the root-mean-square background noise (e.g. see Fig. 1). If such a pulse corresponds to a direct arrival from a floe collision, then it is unlikely that the successive reflected arrivals would be observable in the data, due to the relative attenuation that they suffer. Thus, the only perceptible contribution to the time series is likely to be the $n = 0$ term in equation (5). This term (and indeed all the terms with $n > 0$) is represented by a delta function in the expression for $g_2(t)$, as may be expected since it is just the retarded potential associated with the impulsive (fictitious) source. We are now interested in establishing a more realistic time dependence for the source function, which means examining the floe collision mechanism.

590

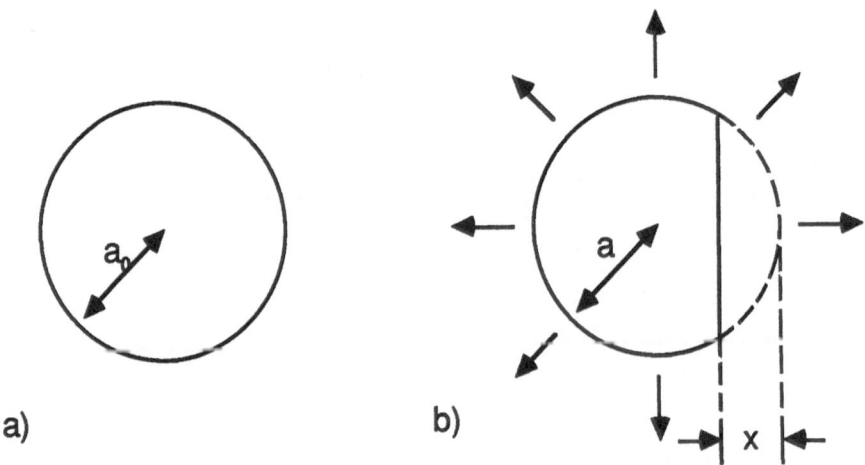

Figure 7: a) Ice sphere in equilibrium and b) undergoing deformation due to a collision.

4. THE MECHANICAL/ACOUSTIC COUPLING

When two ice floes collide, one often undergoes cracking and breaking while the other deforms due to the force of contact. This force is expected to rise to a maximum and then decay away to zero [4]. The noise created by the cracking and breaking processes is likely to be rich in high frequency components, above the range of interest here, but the floe deformation process may give rise to lower frequency acoustic noise, below 2 kHz.

To model the collision, we first assume that the ice sphere representing the floe undergoing deformation flattens out at the point of contact, as illustrated in Fig. 7. We further assume that the floe maintains a constant volume, and deforms instantaneously. As the contact area increases, the radius, a, of the sphere must also increase, that is, a is a function of time during the floe/floe interaction. Thus the sphere resembles a simple acoustic source, radiating sound into the surrounding seawater. The source strength is

$$s(t) = 4\pi a_o^2 da/dt , \qquad (7)$$

where a_o is the equilibrium radius of the sphere. If x(t) is the deformation due to the collision (see Fig. 7), then some simple algebra shows that, for $x \ll a_o$,

$$da/dt = (x/2a_o)dx/dt \qquad (8)$$

and thus the source strength is

$$s(t) = 2\pi a_o x \, dx/dt \, . \tag{9}$$

To determine the time dependence of x we assume a simple linear mass, spring and damper model for the deformation process. With $x = 0$ at $t = 0$, and $dx/dt = V$ at $t = 0$, where V is the closing speed of the two ice floes undergoing the collision, we find that

$$x(t) = (V/\gamma)exp(-\Omega_o t/2Q)sinh(\gamma t) \, . \tag{10}$$

In equation (10), Ω_o is the angular resonance frequency of the floe interaction, Q is the mechanical loss factor, and

$$\gamma = (\Omega_o/2Q)(1 - 4Q^2)^{\frac{1}{2}} \, . \tag{11}$$

For $Q >> \frac{1}{2}$,

$$\gamma \simeq j\Omega_o \tag{12}$$

and

$$x(t) \simeq (V/\Omega_o) \, exp(-\Omega_o t/2Q) \, sin(\Omega_o t). \tag{13}$$

Thus, the source strength from equation (9) is

$$\begin{aligned} s(t) &\simeq (\pi V^2 a_o/\Omega_o) \, exp(-\Omega_o t/Q) \, sin(2\Omega_o t), \, for \, 0 \le t \le \pi/\Omega_o \\ &= 0, \, otherwise. \end{aligned} \tag{14}$$

The reason for the limits on t in this expression is that the floe/floe interaction is assumed to occur over one half of a resonant cycle. The Fourier transform of the source strength in equation (14) is

$$S(jw) \simeq 2\pi V^2 a_o/\omega^2 \, , \tag{15}$$

which is valid provided $\omega >> \Omega_o$. This inequality, that is the requirement that the acoustic frequency be very much higher than the mechanical resonance frequency of the floe interaction, is easily satisfied for the acoustic frequencies of interest here. Note that S in equation (15) is independent of Ω_o.

The velocity potential, $\phi_2(t)$ of the acoustic pulse radiated into the seawater as a result of the floe collision is now given by the convolution of the Green's function $g_2(t)$ in equation (5) with the source strength in equation (14):

$$\phi_2(t) = g_2(t) \otimes s(t) \, . \tag{16}$$

Taking only the direct arrival in equation (5) as being significant, the velocity potential is

$$\phi_2(t) = s(t)/[2\pi r(1+\beta)] \, . \tag{17}$$

592

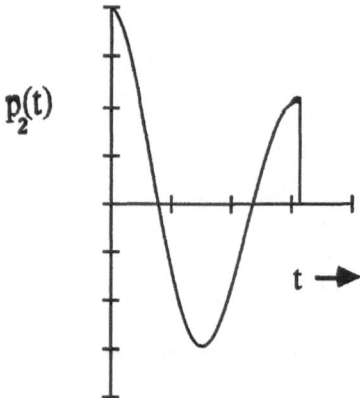

$p_2(t)$

$t \longrightarrow$

Figure 8: Pressure pulse from equation (18) for $Q = 5$.

The associated pressure pulse is proportional to the time derivative of this expression:

$$
\begin{aligned}
p_2(t) &= \rho_2 s'(t)/[2\pi r(1+\beta)] \\
&= [\rho_2 V^2 a_o/\{r(1+\beta)\}]\, exp(-\Omega_o t/Q)\, cos(2\Omega_o t),\ for\ 0 \le t \le \pi/\Omega_o \\
&= 0,\ otherwise,
\end{aligned}
\tag{18}
$$

where ρ_2 is the density of seawater.

The shape of the pressure pulse in equation (18) is shown in Fig. 8 for a value of $Q = 5$. The pulse shows two peaks, the second attenuated with respect to the first by a factor which depends on Q. The frequency Ω_o depends on the mechanical properties of the floe undergoing the deformation when a collision occurs. In particular, Ω_o depends on the size of the floe, which may have a diameter anywhere from a few metres up to several hundred metres. Thus, according to equation (18), pulse durations spanning several orders of magnitude should be expected in the MIZ ambient noise. This range of pulse lengths is indeed observed in the noise data.

The power spectrum of the pressure pulse at the receiver is

$$
\begin{aligned}
|P_2(j\omega)|^2 &= \omega^2 \rho_2^2 |G_2(j\omega)|^2 |S(j\omega)|^2 \\
&= [V^2 a\rho_2/(4r\omega)]^2 F(k_1 a)\,,
\end{aligned}
\tag{19}
$$

where the expressions for G_2 and S have been substituted from equations (3) and (15), and the subscript zero has been dropped from the sphere radius a in the square brackets. Equation (19) states that the pressure spectrum due to a single collision varies as ω^{-2} modulated by the function F shown in Fig. 5. We are now interested in using the result for a single collision in equation (19) as the basis for

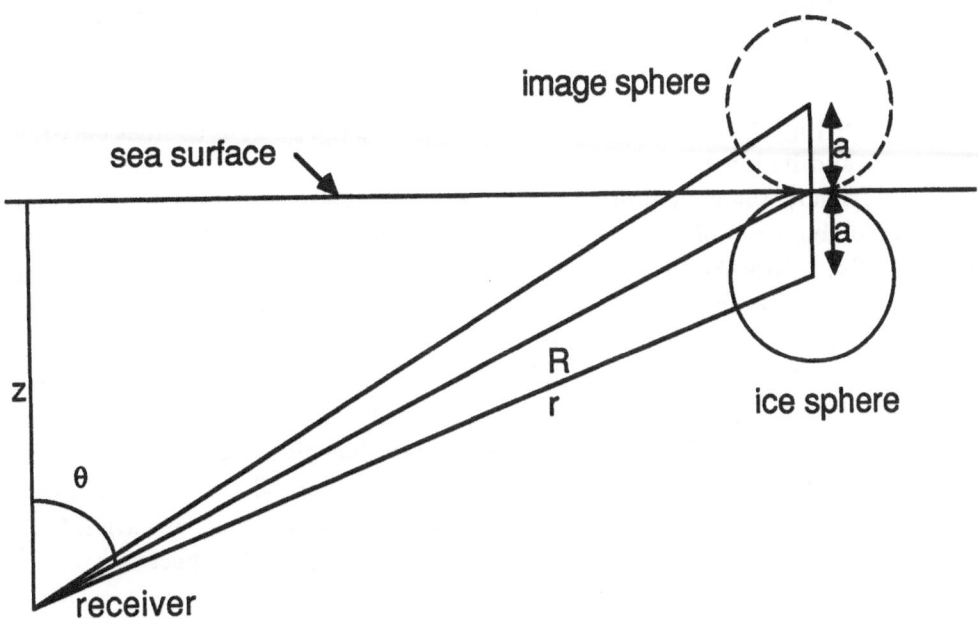

Figure 9: The source and its negative image in the sea surface.

establishing the spectrum of the noise field generated by all the collisions between floes of various sizes distributed across the ocean surface.

5. THE SPECTRUM OF THE TOTAL NOISE FIELD

The result for the pressure spectrum in equation (19) is based on the assumption of an unbounded ocean. To take into account the effect of the pressure-release surface, a negative image of the vibrating ice floe must be included in the calculation (Fig.9). If the floe diameter were much less than the wavelength of interest, the source and its image would appear as a dipole. Equation (19) would then be modified by a multiplicative factor proportional to $\omega^2 cos^2\theta$, where θ is the angle between the vertical and the point of contact of the source floe and its image, as shown in Fig. 9.

However, in our case, the floe diameters are not negligibly small but are comparable with the wavelengths of interest. A straighforward geometrical argument shows that in such a situation equation (19) should be modified as follows:

$$|P_2(j\omega)|^2 = [V^2 a\rho_2 sin(\beta k_1 a\ cos\theta)/(2R\omega)]^2 F(k_1 a) , \qquad (20)$$

where R (see Fig. 9) is the range from the receiver to the point of contact between the source and its image.

Consider now the contributions to the noise at the receiver due to those collisions involving floes of radius a, which we assume to be distributed uniformly (statistically) across the sea surface. Those collisions occurring within an annulus on the surface of radius w and thickness dw will give rise to a pressure field in the form of a random pulse train, in which all the pulses have the same shape. The power spectrum of such a pulse train is given by Carson's theorem [5,6]. This theorem, which is based on a statistical argument, shows that the spectrum of the continuous pulse train has the same frequency dependence as that of a single pulse within the train. Thus, the spectrum arising from all collisions involving floes of radius a within the annulus is proportional to $|P_2(j\omega)|^2$ in equation (20); and this must be integrated over the whole surface to give the spectrum, $\overline{S_a}$, of the noise due to all such collisions:

$$\overline{S_a(\omega)} = 4\pi\nu z^2 \int_0^{\pi/2} |P_2(j\omega)|^2 tan\theta sec^2\theta\ d\theta , \qquad (21)$$

where z is the depth of the receiver below the surface and ν is the mean rate of arrivals per unit area of surface. On substituting for $|P_2(j\omega)|^2$ from equation (20), bearing in mind that $R = z/cos\theta$, equation (21) becomes

$$\overline{S_a(\omega)} = \pi\nu(V^2 a\rho_2/\omega)^2 F(k_1 a) \int_0^{\pi/2} tan\theta\ sin^2(\beta k_1 a\ cos\theta)\ d\theta . \qquad (22)$$

The integral here can be evaluated exactly to give

$$\overline{S_a(\omega)} = \pi\nu(V^2 a\rho_2/\omega)^2 F(k_1 a)[C + 1n(2\beta k_1 a) - ci(2\beta k_1 a)] , \qquad (23)$$

where $C = 0.577215\ldots$ is Euler's constant and ci( ) is the cosine integral. Note that when $k_1 a << 1$, the cosine integral is approximated by

$$ci(2\beta k_1 a) = C + 1n(2\beta k_1 a) - (\beta k_1 a)^2 + \ldots , \qquad (24)$$

and equation (23) reduces to the spectral form appropriate to dipole sources. In this limiting case the frequency dependence displayed by the spectrum $\overline{S_a}$ is that of the function $F(k_1 a)$ given in equation (4).

The spectrum of the noise in equation (23) is a function of the floe radius a. Now, the floe radii in the MIZ are not all equal but are distributed over two or three orders of magnitude. Assuming that the ice fracturing giving rise to the formation of floes occurs in a random fashion, that is, it is Poisson controlled, as discussed by Wadhams [7], then the probability density function of the floe size distribution is a decaying exponential:

$$p(a) = (1/\overline{a}) \, exp(-a/\overline{a}) \,, \tag{25}$$

where $\overline{a}$ is the mean floe radius. Note that the mean-square radius is related to the mean radius as follows:

$$\overline{a^2} = (1/\overline{a}) \int_0^\infty a^2 \, exp(-a/\overline{a}) da$$
$$= 2(\overline{a})^2 \,, \tag{26}$$

which is a result that will be used later.

If a_1 and a_2 are the lower and upper limits on the range of floe radii encountered in the MIZ, then the spectrum of the total noise field at the receiver is

$$\overline{S_T(\omega)} = \int_{a_1}^{a_2} p(a) \, \overline{S_a(\omega)} da \,. \tag{27}$$

Within the integration range here, the function in square brackets in equation (23) for $\overline{S_a(\omega)}$ shows only a weak dependence on the floe radius a, and may be satisfactorily approximated as a constant in which a is replaced by the mean radius $\overline{a}$. Then the integral which has to be evaluated in equation (27) to give the spectrum of the noise is

$$I = \int_0^\infty a^2 \, exp(-a/\overline{a}) \, F(k_1 a) da \,, \tag{28}$$

where little error has been incurred by replacing the lower and upper limits on the integral by zero and infinity, respectively. The function $F(k_1 a)$, given in equation (4), can be written in the form

$$F(k_1 a) = 1/[cos^2 k_1 a + \beta^2 sin^2 k_1 a]$$
$$= [2/(\beta^2 + 1)] \, coth(u) \, [1 + 2 \sum_{n=0}^\infty e^{-nu} cos(2nk_1 a)] \,, \tag{29}$$

where u is defined by the identity

$$coth(u) = (\beta^2 + 1)^{\frac{1}{2}}/(2^{\frac{1}{2}}\beta) \,. \tag{30}$$

It can be shown that the contribution to I from the summation in equation (29) is negligible over the frequency range of interest here, and thus I may be approximated as

$$I \simeq \left(2^{\frac{1}{2}}/[(\beta^2 + 1)^{\frac{1}{2}}\beta]\right) \int_0^\infty a^2 \, exp(-a/\overline{a}) da$$
$$= \left(2^{\frac{1}{2}}/[(\beta^2 + 1)^{\frac{1}{2}}\beta]\right) \overline{a}\,\overline{a^2} \,. \tag{31}$$

On substituting this expression back into equation (27), and making use of the result in equation (26), we find finally that the spectrum of the noise is

$$\overline{S_T(\omega)} \simeq \left(4\pi\overline{a}^2\nu(V^2\rho_2)^2/[2^{\frac{1}{3}}\beta(\beta^2+1)^{\frac{1}{3}}\omega^2]\right)[C + ln(2\beta k_1\overline{a}) - ci(2\beta k_1\overline{a})]. \quad (32)$$

Perhaps the first point to note about equation (32) is that it is dimensionally correct. Secondly, it varies with frequency approximately as ω^{-2}, but modified slightly by the weak frequency dependence displayed by the term in square brackets. For a mean floe radius $\overline{a} = 30m$ the actual n value predicted by equation (32) is $n = 1.8$, which is about in the middle of the range of n values observed in the data. As it stands, equation (32) does not allow for the variation in n values observed in the data (see Fig. 3). However, a number of simplifications are embedded in the predicted spectrum, including the assumptions that the mean rate of collisions per unit area of sea surface and the closing speed between colliding floes are independent of floe size. If assumptions such as these were relaxed, then it is conceivable that, as the ice compactness changed with the prevailing environmental conditions, a modified form of equation (32) could yield a time dependent n value for the spectrum of the noise.

6. CONCLUDING REMARKS

The theoretical model presented here of ambient noise in the MIZ due to floe/floe collisions gives rise to acoustic pulse shapes with a range of durations similar to those observed in the time series data. It also leads to a spectral form over the range 50 Hz to 1 kHz which varies with frequency approximately as ω^{-2}. This is similar to the frequency dependence of the measured spectra.

No attenuation is included in the model, since the MIZ is far from being an "infinite" ocean, and attenuation effects of the spectral shape of the noise are unlikely to be very significant. This is especially true in the light of recent evidence by Yang et al. [8], which suggests that the noise in the MIZ may be predominantly a local phenomenon originating in noise "hot spots" at the ice edge, possibly associated with ice/ocean eddies. If this turns out to be the case, that is, the noise propagation paths in the MIZ are relatively short, then frequency dependent attenuation effects should be negligible.

The essential feature of the noise model discussed in the paper is a fluid sphere, representing an ice floe, which is set into oscillation when a collision occurs. Although this representation of the floe as a noise generator leads to reasonable agreement with observations of the noise in the MIZ, some of the features of the model are difficult to justify on the basis of physical arguments. An alternative model would be a solid ice sphere supporting random vibrations over its surface. Such vibrations could be induced by floe/floe collisions, waves slapping against the floe, or a number of other mechanisms. The origin of the vibrations is largely immaterial, the important point being that they exist and radiate acoustic energy into the surrounding water. By characterizing the surface vibrations in terms of a spatial coherence function, it should be possible to determine the properties of the radiated field. This type of model is currently being examined in connection with ambient noise in the MIZ.

ACKNOWLEDGEMENTS

We are greatly indebted to Professor Ira Dyer of MIT for making the MIZEX ambient noise data freely available to us, and for many valuable discussions on ambient noise processes in the Arctic Ocean. Drs. Orest Diachok and T.C. Yang of NRL very kindly provided a preprint of their recent paper (Ref. 8) on noise hotspots in the MIZ. This work was supported by the Office of Naval Research, contract number N00014-86-K-0325.

REFERENCES

1. O.I. Diachok and R.S. Winokur, 'Spatial variability of underwater ambient noise at the Arctic ice-water boundary,' *J.Acous.Soc.Am.*, **55**, pp. 750-753, (1974).

2. MIZEX Bulletin VIII: Winter MIZEX 87/89 Science Plan, Section 2.5, Acoustics, pp. 21-24, April 1986.

3. M.J. Buckingham, 'The acoustic field of a point source in a fluid sphere immersed in a second fluid medium,' to be submitted to the *J.Acous.Soc.Am.*

4. D.S. Sodhi and C.E. Morris, 'Characteristic frequency of force variations in continuous crushing of sheet against rigid cylindrical structures,' *J.Cold Region Science and Technology*, **12**, pp. 1-12, (1985).

5. M.J. Buckingham, *Noise in Electronic Devices and Systems*, Ellis Horwood (John Wiley), pp. 34-37, (1983).

6. M.J. Buckingham, 'A theoretical model of ambient noise in a low-loss, shallow water channel,' *J.Acous.Soc.Am.*, **67**, pp. 1186-1192, (1980)

7. P. Wadhams, 'The ice cover', Chapter 2 of *The Nordic Seas*, edited by Burton G. Hurdle, Springer-Verlag, New York, p. 72, (1986).

8. T.C. Yang, G.R. Giellis, C.W. Votaw and O.I. Diachok, 'Acoustic properties of ice edge noise in the Greenland Sea,' submitted for publication in *J.Acous.Soc.Am.*

ICE EDDY AMBIENT NOISE

Ola M. Johannessen
Nansen Remote Sensing Center/Geophysical Institute,
University of Bergen, Norway

Susan G. Payne
Applied Research Laboratories,
The University of Texas at Austin,
Austin, Texas, U.S.A

Ken V. Starke and Gerry A. Gotthardt
ASW Environmental Acoustics Support Program,
Office of Naval Research,
Arlington, Virginia, U.S.A.

Ira Dyer
Department of Ocean Engineering ,
Massachusetts Institute of Technology,
Cambridge, Massachusetts, U.S.A.

ABSTRACT.   Mechanisms affecting ice floe interactions in the marginal ice zone (MIZ) are discussed. These mechanisms are assumed to be the major forcing functions driving the ambient noise levels in the MIZ. Results from previous studies are briefly summarized with hypotheses presented concerning the effect of ice eddies and grease ice on ambient noise. Data are presented from an experiment held in winter 1985, in which sonobuoys were planted in the water just off the compact ice-edge and in the adjacent ice-field containing two eddies. The ambient noise levels are found to be relatively high, with a significant variability over the sampled region. A broad noise peak across the eddy/ice-edge region suggests the eddy as a distributed ambient noise source.

INTRODUCTION

A marginal ice zone (MIZ) exists where polar and temperate climate systems interact, resulting in an edge of ice cover with strong horizontal and vertical gradients in the atmosphere and the ocean (MIZEX Group 1986, Johannessen, O.M. et al., 1987). Significant interactions occur between the air, ice and ocean on various time and space scales that are potentially important for acoustic propagation (Mellberg, 1987) and ambient noise. The region is very dynamic with wind, gravity waves, swell and current causing floe-floe interactions, ice breakup, and changing

599

B R. Kerman (ed.), Sea Surface Sound, 599–605.
© 1988 by Kluwer Academic Publishers.

ice concentrations. During an on-ice wind, surface waves will propagate into the ice cover, increasing the floe-floe interactions in the vicinity of the edge. On-ice swell can penetrate up to 100 km into the pack (Wadhams, 1978), impacting floe size distributions and the extent of the MIZ. A consequence of both of these forcings is a compact ice-edge at which Diachok and Winokur (1974) have reported a sharp and large noise peak. With a diffuse ice edge (caused by off-ice winds, or winds blowing parallel to the East Greenland ice edge from the south), these authors reported a more gradual noise increase and a reduced peak level as the edge was approached from the complete ice cover.

In at least one situation, a thin ice cover off the edge or in open water region may reduce ambient noise levels below those that would be expected without the ice. In cold water regions with freezing air temperatures and winds less than 10 m/s, grease ice is formed. This soupy, viscous liquid rides on top of the waves, drastically damping their amplitude and breaking (Johannessen, 1986), and thereby potentially reducing the noise levels.

Another phenomenon that may contribute to the observed ambient noise field was postulated by Yang et al (1985) to be ice-ocean eddies which are consistently observed in the MIZ (Johannessen, O.M. et al, 1983, Johannessen, J.A. et al., 1987). This hypothesis arose from the observation of ambient noise "hot spots" along the ice edge that were the dimensions of eddies that have been observed in the MIZ. However, no synoptic oceanographic data were available to validate the presence of eddies in the vicinity of these measurements. We would anticipate increased noise levels in regions of higher ice concentrations which are associated with horizontal velocity shear caused by eddy circulation or a meandering ice edge.

In the summer of 1984 and the winters of 1985 and 1987, investigations of ambient noise in the Fram Strait, East Greenland and Barents Sea MIZ were carried out as part of the MIZEX program (Mizex Group, 1986). Nine flights were conducted by Norwegian P3 aircraft during which sonobuoys were dropped in the MIZ. In this paper, results are reported from one of the flights during which sonobuoys were laid in ice-edge eddies. Subsequent papers will report the results from the other flights during which acoustic data pertaining to a variety of environmental, meteorological and ice conditions were collected.

OBSERVATIONS.

The set of data reported upon in this paper was obtained on 30 April 1985, between 1100Z and 1900Z in the northern Greenland Sea. The air temperature was -10°C. The depth to the bottom of the ocean in the experiment area was more than 2000m. Strong northerly winds (12-15 m/s), which had been blowing for a week, moderated so that winds on this day were calm, and visibility was good. Aerial reconnaissance of the ice edge between 75°N and 80°N revealed an abundance of eddies with their circulation, in the absence of winds, displayed by the ice-floe distribution mirroring the ocean circulation. Two eddies (A and B, shown in Figure 1) were selected for investigation. Eddy A was a small eddy, 20km in diameter, positioned within the ice-edge; eddy B was situated at the ice-edge. Clearly visible in Figure 1 are the striations and convergence bands in the ice-field arising from the orbital motion of the eddy.

Imagery from the NOAA satellite were used to obtain detailed information about the marginal ice zone, and in particular to locate the ice-edge and the eddies. The characteristics of the ice-field, shown in Figure 2, were drawn up from all available sources - satellite imagery, aerial photography from the P3, and visual characterization by an ice-observer. The ice-concentration of the field was estimated at 90-95% at an average. The edge was generally compact except where the eddy circulation had swept the ice out into the open ocean creating a diffuse arm, indicated by a dotted ice edge between the two eddies.

Data were collected by Squadron 333 of the Royal Norwegian Air Force using a P3B aircraft equipped with an AN/ARR-72 receiver and a 28-track, analog, FM Wideband II tape recorder. All ambient noise data were received on AN/SSQ-57A, 53B and 57XN5 sonobuoys

Figure 1. View of the two ice-edge eddies A and B sampled in this study, (aerial photography by O.M. Johannessen).

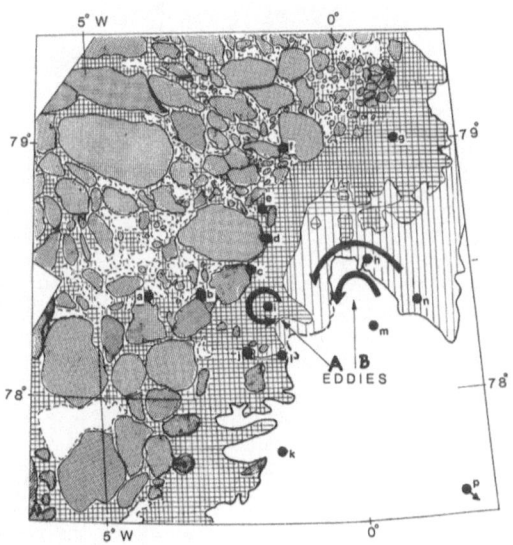

Figure 2. A schematic diagram of the ice-field. The letters a through p indicate the locations of the sonobuoys. stippling: multiyear ice, 3-4m thick, large floes; crosshatching: multiyear ice, small floes, 20-100m size; lines: first year ice, small floes, 1m thick

602

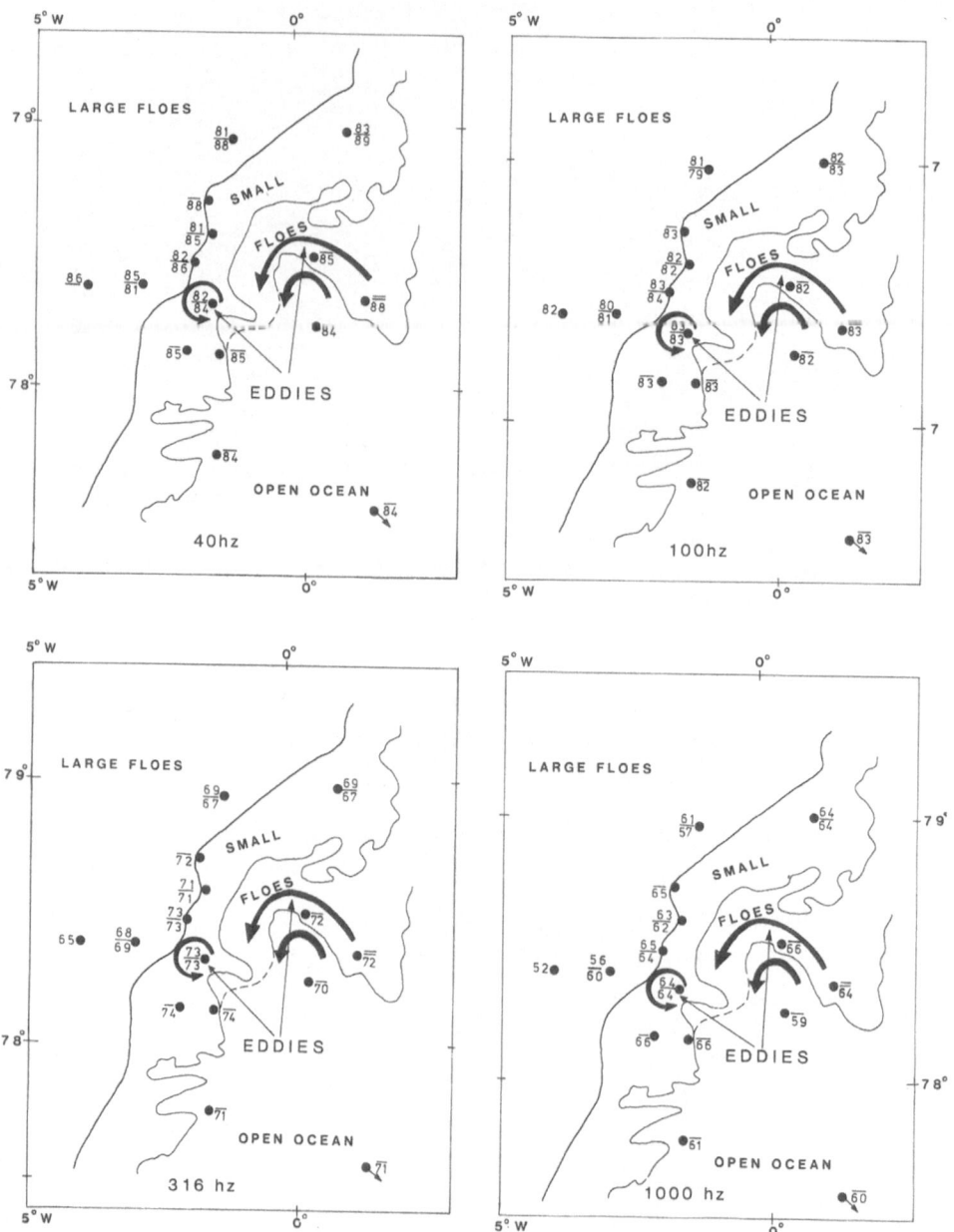

Figure 3. Median ambient noise levels (dB//mPa2/ Hz) in third octave bands centered at 40 Hz. 100 Hz, 316 Hz and 1000 Hz. No bar over the number represents 18m depth, one bar 124m and two bars 305m.

with operating depths of 18, 124, and 305 m. The positions of the sensors are shown in Figure 2, labelled a through p. At some of these locations there are colocated sonobuoys at different depths. Airborne bathythermograph (AXBT) data were also collected, and were used to corroborate the eddy features.

The acoustic data were processed in a bandwidth of 1000 Hz with a bin resolution of 2.5 Hz. Sixteen power spectra (6.4 seconds of data) were averaged together to produce a resultant spectrum every minute. Standard sonobuoy frequency response curves were used in the calibration process, with a published uncertainty of 2 dB. Averages were obtained in one-third octave bands, centered at 40 Hz, 100 Hz, 315 Hz and 1000 Hz. These data were then sorted to produce spectrum levels for each of these frequency bands.

There were no surface ships in the area during the experiment. Biologics were heard intermittently on most of the hydrophones.

RESULTS

The noise levels for each buoy in each frequency band are shown in Figure 3. Examination of the spatial variability of the noise levels leads to several conclusions. First, the levels vary considerably over the 60 X 60 mile area. Simply comparing the minimum and maximum noise levels from Figure 3, it is seen that the variability is least at 100 Hz (5 dB), and increases with frequency (9 dB at 316 Hz, 14 dB at 1000 Hz). The noise is also more variable at the lowest frequency (8 dB difference at 40 Hz).

Second, three noise regions can be outlined - the ocean, the edge/eddies region, and the large floe region. The noise level in edge/eddies region is generally the highest, with the large floe region being the quietest. This is true for all frequencies although the trend at 40 Hz is marginal. The noise appears to have a broad peak across the eddy/edge region, and, despite the edge being compact, does not show the very sharp peak on the edge as was observed by Diachok and Winokur (1974). The size of the peak in the edge region increases with frequency, and the level decreases faster with range into the ice than it does with range into the open ocean.

The noise levels that were observed in the open ocean during this experiment are close to those of Diachok and Winokur (despite lighter wind conditions). However, the noise in the edge region is about 4 dB lower than the Diachok and Winokur 1974 study and the noise in the large floe region, at about 60 km from the edge, is higher by about 10 dB at 100 Hz, 3 dB at 316 Hz and 5 dB at 1000 Hz if we compare the results from similar depth hydrophones. To put the noise levels in perspective, comparisons can be made between typical shipping and wind noise levels. The noise levels in the open water a few kilometers off the ice-edge correspond to a sea state 5 level at 316 Hz and to a sea state 3 level at 1000 Hz, according to the Wenz curves (1962). The low frequency measurements correspond to a heavy shipping density.

DISCUSSION

With the calm winds and slight sea state, ice interactions are mainly caused by the horizontal shear velocity that arises due to the current and the eddy circulation. The broadening of the ice-edge noise peak across the eddy region may be a manifestation of the bands of high ice-concentration. With the presence of an eddy, it is as though the bands of concentrated ice found at the edge is repeated over the eddy diameter. Thus, the eddy appears to act as a distributed ambient noise source. The ice-concentration may not be as great as that caused by strong on-ice winds due to the differing magnitudes of the forcing fields, but the area with the denser ice-concentration is significantly larger in the eddy region than along the ice-edge. The effect on the noise is thus expected to be an area of higher noise level rather than a line of high noise along the ice edge. This argument is lent credence by the "hot spot" results of Yang et al. (1985), discussed above.

604

If the above hypothesis is correct, one might expect higher noise levels within an eddy region in the areas with higher shear velocity and ice-concentration. The horizontal shear velocity would be a result of the strength of the orbital motion of the eddy and the background flow, modified or enhanced by the wind effect, thus causing a variable noise field also within the ice eddy. Unfortunately, this experiment did not sample the eddy region sufficiently densely to determine if this was indeed the case.

In March 1987, acoustic and environmental data including SAR images were collected in ice-eddy fields on two consecutive days during the MIZEX 87 experiment. We succeeded to placing the sonobuoy sensors in an ice eddy much closer together than in the 1985 experiment in order to address the hypothesis described above. The processing and analysis of these data are not complete at this time, but preliminary analyses indicate that the above hypothesis describe above is correct. This new result will be reported in 1988.

Finally we suggested that observations of ambient noise can be used to indicate regions of high internal ice stress. This will be another method to gain new information of this term in the equation for the ice motion, when a proper inverting scheme has been worked out. If this can be done a mesoscale ice-ocean coupled model, including an internal ice term, can also be used in prediction of the ambient noise field. SAR observation of ice concentration, floe size distribution, types of ice and ice kinematics will be key input parameters in such a model.

ACKNOWLEDGEMENT

The P3 - 333 Squadron of the Royal Norwegian Air Force is sincerely acknowledged for outstanding work in dropping sonobuoys in a dense pack ice region. This work was supported by the Royal Norwegian Air Force, the ASW Environmental Acoustics Support Program, the Arctic Program both Office of Naval Research, and the University of Bergen.

REFERENCES

Diachok, O.I. and R.S. Winokur, 'Spatial variability of underwater ambient noise at the Arctic ice-water boundary', *J. Acoust. Soc. Am.* **55**, 750-753, 1974.

Johannessen, J.A., O.M. Johannessen, E. Svendsen, R. Shuchman, T. Manley, W.J. Campbell, E.G. Josberger, S. Sandven, J.C. Gascard, T. Olaussen, K. Davidson and J. Van Leer: 'Mesoscale eddies in the Fram Strait marginal ice zone during the 1983 and 1984 marginal ice zone experiments', *J. Geophys. Res.*, **92**, 6754-6772, 1987.

Johannessen, O. M., J.A. Johannessen, J. Morison, B.A. Farelly and E. Svendsen, 'Oceanographic conditions in the marginal ice zone north of Svalbard in early fall 1979 with an emphasis on mesoscale processes', *J. Geophys. Res.* **88**, 2755-2769, 1983.

Johannessen, O. M., 'A note on the grease ice microlayer effect on remote sensing', in*ONRL Workshop Proceedings - Role of Surfactant Films on the Interfacial Properties of the Sea Surface*, C-11-86, ONR London, 49-59, 1986.

Johannessen, O. M., J.A. Johannessen, E. Svendsen, R.A. Shuchman, W.J. Campbell and E. Josberger, 'Ice-edge eddies in the Fram Strait marginal ice zone', *Science*, **236**, 427-439, 1987.

Mellberg, L., O.M. Johannessen, D.N. Connors, G. Botseas and D. Browning, 'Modeled acoustic propagation through an ice edge eddy in the East Greenland Sea marginal ice zone', *J. Geophys. Res.*, **92**, 6857-6868, 1987.

MIZEX Group, MIZEX East 83/84: 'The summer marginal ice zone program in the Fram Strait/Greenland Sea', *EOS Trans. AGU*, **67**, 513-517, 1986.

Yang, T.C., C.W. Votaw, G.R. Giellis and O.I. Diachok, 'Horizontal directionality of ice edge noise', in *Proceedings of the Arctic Oceanography Conference and Workshop*, Naval

Oceanographic and Research Development Activity, June 11-14, 1985.

Wadhams, P., 'Wave decay in the marginal ice zone measured from submarine'. *Deep Sea Research*, **25**, 23-40, 1978.

Wenz, G.M., 'Acoustic ambient noise in the ocean spectra and sources',*J. Acoust. Soc. Am.* **34**, 1936-1956, 1962.

BREAKING WAVES

Panel: M.L. Banner
        H.E. Huang
        M.S. Longuet-Higgins
        W.K. Melville
        O.M. Phillips (Chairman)

The discussion of the papers on breaking waves was lively and fruitful, with contributions from many participants. There was a good deal of give and take: I will attempt here only to summarize the consensus conclusions.

What is a breaking wave and how do we identify it? This question, at first sight trivial, was quickly seen to involve the threshold used for detection, whether it is a sudden change in elevation in a jump meter, the occurrence of foam patches of a certain size or the detection of sea spikes above a certain threshold in radar returns from the sea surface. Both Longuet-Higgins and Kerman pointed out that the process might usefully be described in terms of fractals, the frequency of occurrence increasing as the detection threshold or resolution scale decreases, and the fractal dimension should be found. New kinds of field measurements are needed to characterize ocean wave breaking – length of breaking fronts per unit area of the sea surface, size distribution and separations as functions of the resolution scale. New parameterizations are needed – instead of expressing whitecap coverage as a function only of u_*, the measurements should be expressed possibly as functions of u_*/c_o, where c_o is the speed of the dominant wave, and/or of u_*/c_m where c_m is the speed of breaking waves at the detection threshold. The detailed statistics are also likely to be dependent on the width of the directional distribution of the wave spectrum and on the presence or absence of swell.

The detailed dynamics of breaking waves seem to be a fertile field

B. R. Kerman (ed.), Sea Surface Sound, 607–610.
© 1988 by Kluwer Academic Publishers.

for investigation, following the pioneering studies of Melville reported here. It was agreed that there is a range of scales over which the dynamics of breaking may be largely self-similar, but scale effects are likely to show up at small scales, when the ratio of breaking wave scale to the capillary-gravity scale $(\gamma/g)^{1/2}$ is not large and also possibly at larger scales with such intense bubble entrainment that the buoyancy forces on the air-water mixture become important on the time scale of the breaking. These questions need experimental examination involving large facilities - potentially useful in this regard are those in Norway and Germany. The first step of such experiments should extend the structural information we already have from the measurements of Melville and of Banner (reported later) and also start on developing an understanding and parameterization of the formation of bubble clouds, their origins, density and dynamical behaviour as well as their influence on the breaking process itself. There are a number of intriguing unsteady processes even in a quasi-steady spilling breaking event - the oscillations noted by Banner and Duncan as well as small scale surface instabilities in plunging breakers.

Measurement techniques are not simple. Breaking at sea is sporadic and transient. Laboratory studies have involved either flow past submerged foils to generate quasi-steady breaking or the use of 'chirped' mechanically-generated waves to produce energy convergence at an observation point. Another possibility suggested by Longuet-Higgins to generate a three-dimensional breaking region is the use of variable depth to provide focusing, say in the lee of a circularly symmetrical bank. Quasi-Lagrangian methods of measurement, with small freely floating meters may be possible for larger scales of breaking.

The actual processes of bubble generation are little understood in detail but likely to be very important acoustically as the Banner-Cato film (shown later in the meeting), indicated clearly. There appear to be a variety of sources - entrapment of small bubbles at the toe of a breaker (which then ring), the break-up of a larger, engulfed bubble, the entrainment at the upper surface by vigorous bubbles and the amplification of perturbations in the high shear zone at the surface of a roller. It is probable that the mix of mechanisms is dependent on scale and experiments should attempt to isolate them in turn if understanding is to be gained, rather than tackling the whole mix at once. The formation of spray may be important acoustically; spindrift is a likely candidate.

There was some discussion about the influence of wind on the breaking process. Under moderate sea conditions, increasing the wind stress increases the density of breaking, but it was argued that the time scale involved in each breaking event is so short that the presence or absence of wind during this breaking is probably irrelevant. Clearly, this is not so if the surface wind shear is so large that Kelvin-Helmholtz instability occurs with spindrift (a sheet of spray detaching from the surface). The acoustical effects of wind on breaking waves is difficult to study experimentally in most laboratory facilities because of fan noise.

Wave breaking is clearly a source of high frequency ocean noise. As discussed previously, ringing following entrainment of isolated bubbles is clearly important, but it is not clear whether a mass of entrained bubbles acts in the same way. Are the frequencies of the small-scale turbulence eddies in the shear zone at the base of a roller sufficiently high to excite resonance? What are the acoustical consequences of splashing in the flow as opposed to that produced by an isolated falling drop in the laboratory? A recent theory by Guo (J. Fluid Mech., 1987) reported by Ffowcs-Williams suggests that splashing is important, but it also predicts that radiation from bubbles is negligible, which is contrary to observations described at this meeting. Are there characteristic acoustic signatures that might be used in analysis of observations that are more revealing than short time spectra? This would involve inverse filtering if the response functions were known adequately.

Bubble clouds also reflect, scatter and absorb incident sound and acoustical measurements were recognized as a critically important adjunct to study of the breaking events themselves, particularly their distribution in space and in scale. Important advances seem possible. Passive measurements will become more important when the sound generation mechanisms are better understood; active acoustic measurements can already delineate bubble clouds and their distributions.

To summarize the summary: it is clear that the interaction between the acoustical and hydrodynamical research communities has provided stimulus on both sides. Increasing awareness of the questions asked, the techniques available, the results already obtained as well as the aspects still not understood, will surely lead to advances in both dynamical oceanography and ocean acoustics.

STUDY OF THE DISTRIBUTION OF BUBBLES AND TURBULENCE IN AND NEAR A BREAKING WAVE

Panel: S.C. Ling
S.O. McConnell
S.T. McDaniel
H. Medwin
M-Y. Su
S.A. Thorpe (Chairman)

1. I don't think I can adequately summarise the discussion we had. It became a riotous assembly within minutes of our commencement, and it was perhaps fortunate that only some 40 minutes were available for discussion, for the level both of sound (with several people on their feet at one time) and the potential for violence, were both tending towards dangerous, even violent, limits by the end.

2. This 'summary' is thus a highly edited version of what was said, as well as what might have been said had the Panel had more time for discussion.

3. How to study turbulence, is rather close to the subject of the Panel chaired by Owen Phillips and appeared not to justify further discussion by this Panel, although I shall return to the subject later. (Incidentally I heard little clamour for a study of turbulent motion in the sea for reasons directly associated with acoustics, and I should like to know what attention it should be given in this respect). It appeared appropriate to ask, 'Exactly what information is needed by acousticians about bubble clouds and turbulence?', in the hope of defining the objectives of the Panel discussion as they bear on the subject of the symposium.

4. We were told that the most important measurement to be made was the volume fraction of bubbles as a function of depth and time,

611

B. R. Kerman (ed.), Sea Surface Sound, 611–615.
© 1988 by Kluwer Academic Publishers.

with changes to be resolved over time scales of milliseconds. This is an objective which, although demanding in time resolution, falls short of the task to which I am more accustomed to hear voiced, that of measuring the number distributions of bubble sizes and its variation with depth. It seems to me that one might learn a lot about the volume fraction by measuring the speed of sound at various frequencies and depths over a short (say 1-2m) horizontal path length and with an instrument raised by buoyancy from the sea bed remote from obstructions. A 'yo-yo' mooring is quite feasible. This however could not be used at levels above the wave troughs and would fail to measure in the region of the breaking wave crest. Perhaps here sensors mounted on a catamaran towed ahead of the ship wake could be useful.

5. It is perhaps cautionary to recall that bubble clouds are very patchy in space and time. Use of the time-space averages in the construction of ray paths could be very misleading indeed. It would be nice to know if there is any interest in the distribution of bubble clouds. David Farmer has suggested that the rows of bubbles found in Langmuir circulation may act as a diffraction grating for intermediate frequency sound, and have a directional effect. Is this known to be important?

6. The discussion of the techniques for measuring bubble size distributions was lead by Herman Medwin. He mentioned two types of measurement, both of which suffer from serious shortcomings. The acoustic methods depend on knowledge of the bubble damping coefficients, about which there is insufficient information. The photographic method used by Koloveyev was criticised by Johnson and Cooke, whilst their own method appears to have poor resolution for low bubble radius. The optical system described by Ming Su suppose that conditional and multiple sampling may be needed to resolve size spectra in breaking waves if sufficient bubbles are

to be measured in these regions to achieve statistical reliability. A technique based on laser holography devised by workers at Cal.Tech. has, we are told, very high resolution (down to 10μm) but appears not to have been used in condition in which the bubbles are predominantly those generated by breaking waves. A further alternative suggested by Dr. Carey involved the detection of γ rays emitted from, perhaps, a cobolt source. I have no information of the accuracy or resolution of this technique. There is also a technique used by Ferren MacIntyre using coloured light which might be tried. (This is referred to in 'Oceanic Whitecaps' a Conference Proceedings, edited by Ed Monahan and Dr. Mac Niocaill and published in 1986 by Reidel).

7. I wondered whether the briefly-mentioned idea of acoustic tomography might not be a way to proceed. It might not be too difficult to provide an array which could either resolve scales of the order of one bubble cloud (with a horizontal array of transducers in positions fixed in relation to each other), or perhaps a larger array to sample many clouds, and to determine the speed of sound variation and hence the volume fraction as a function of position. I should welcome more discussion. This type of measurement, possibly together with optical measurements, might go some way towards satisfying a Recommendation of the Panel that direct, side by side, comparisons of optical and acoustic scattering or attenuation measurements should be made at various wind and sea states (and I think we should try to obtain simultaneous good wave data, including some measurement of breaking waves perhaps using video as described by David Farmer). Measurements are also suggested in coastal and open ocean areas. I am personally less confident about the value of doing this. Whilst, fetch is something we should bear in mind, I have not yet seen a reliable technique for measuring surface tension (or other surface film paramaters) in windy conditions, and if this is a reason for a variation between some coastal measurements and those in deep sea,

we need more reliable environmental measurements. One man's coastal region is another man's deep ocean perhaps? Can we be sure that we know? Comparative measurements in fresh water and salt water may be instructive. I think we may find a difference but will this help us to understand sound in the sea? Maybe.

8. The discussion turned to the possible alternative ways to advance knowledge, through analytical or numerical calculations, via laboratory experiments or in sea-going experiments. I believe that we should not discount any of these techniques, although detailed estimates of the bubbles and turbulence generated by breaking waves do not yet appear feasible except by observational or empirical means. Numerical and analytical estimates of bubble properties based, say, on energy considerations (as Paul Crowther has shown), or estimates of the frequency and distribution of breaking waves (papers by Owen Phillips and Dr. Papadimitrakis) are indeed of great value. But we are asking for more details of the bubbles and turbulence and for tests of these ideas. Laboratory experiments have yet to explore adequately the parameter range of breaking waves, and it would make sound economic and scientific sense to continue and to extend them to examine bubbles and noise, the creation rate of bubbles, and their dispersion. We do not yet understand even the fundamental scales which govern the bubble size distribution. It seems unlikely that conventional ideas of turbulence (e.g. the Kolmogarov scale) will be useful in the anisotopic, highly time-dependent, 2-phase flow in the immediate temporal and spatial vicinity of a breaking wave. Bubble fragmentation processes there appear to be beyond any simple hydrodynamic models. One might look to scaling laws as Bryan Kerman has suggested. This could involve surface tension, viscosity, the speed of the jet injected by the surface wave, fluid density, the acceleration due to gravity, sound speed, and so on. Solubility (don't forget air is a mixture of gases) and diffusivity may be important only later in determining that part of the bubble cloud which is

left after a brief initial period of breaking, probably a silent spectre drifting with the current, an attenuator of sound but not an active generator unless forced to ring as David Farmer suggests.

9. It is the sound generation on which we must again focus. (How easy it is to divert from this subject. You know, there really are very good reasons for knowing more of bubble clouds e.g. gas transfer - and one might hope that a closer collaboration between hydrodynamicists and acousticians will be to the advantage of each party. Both have much to learn from the techniques and data of the other). Dr. Ling advocates closer study of the processes of sound production in the vicinity of the high speed jet of a plunging breaker. Here gas-filled vortex cores are formed, or line vortices. Some work on this was done by Dr. Neil Thomas at Cambridge (he is now at Birmingham University, U.K.) in the pool at the end of a spillway. But much more work is needed, both in the laboratory and at sea, before we have sufficient information to provide a process-based model of turbulence, bubbles, and sound production in a breaking wave.

10. Finally let me mention an experiment planned to look at the distribution of bubbles and turbulence in the sea. The experiment will use a submarine as a platform, with Dr. Tom Osborn's turbulence sensors mounted on a rigid tripod near the bow. David Farmer will use multi-frequency active sonar to examine the bubble clouds immediately before the turbulence in them is measured. We plan to view the formation of the clouds and the breaking waves ahead of the submarine, using side-scan sonar on the submarine itself and cameras mounted on a helicopter hovering above. The experiment includes wave measurements, CTD profiles, and heat flux, perhaps also oxygen saturation levels, and hopefully will provide information on the Panel topic, that is 'the distribution of bubbles and turbulence in and near breaking waves'.

PANEL DISCUSSION REPORT

SOURCES OF SOUND AT THE OCEAN SURFACE; BUBBLES AND OTHER NOISE

Panel: P.A. Crowther
           D.M. Farmer
           J.E. Ffowcs Williams (Chairman)
           A. Prosperetti
           P. Wille

The source processes generating noise at the interface of two fluids with different speeds, densities, temperature and speed of sound have been the central theme of aero-acoustic studies since the noise of jet engines first became a problem. The theoretical foundation of that study is Lighthill's classic (1951) work 'Noise generated aerodynamically', a theory so elegant and persuasive that it tended to bias experimenters to measure and report only those parts of the subject perceived to be in accordance with that theory. A frequency doubling of the sound over the velocity field of the source, the scaling of sound power on the eighth power of source speed and an amplifying effect of velocity shear and speed of sound variability were verified under rather special well defined circumstances. Real sound-generating flows are much richer in structure than that conforming with this crude characterisation; it has taken some thirty years to see clearly the acoustic consequences of density contrast for example. That work can guide now the modelling of the sea surface source region and it leads to some surprising new views on the subject.

The clearest identifying characteristic of an acoustic field is that all its constituent elements have phase speeds equal to or greater than the speed of sound. Gravity waves on the ocean surface rarely have such supersonic elements and the oceanic sound field is therefore only weakly coupled to the surface wave system. An evanescent monochromatic linear surface wave is absolutely uncoupled to sound, and the weak coupling that occurs at second order in wave amplitude, between near-colliding, or near-standing, surface waves and

617

B. R. Kerman (ed.), Sea Surface Sound, 617–620.
© 1988 by Kluwer Academic Publishers.

the sound field has come to be regarded as the probable source of underwater sound at low frequency. Longuet-Higgins (1950) proposed this model of acoustically-induced microseisms, though in acoustics credit for identifying that source mechanism is usually given to Brekhovskikh (1966). Guo's (1987) work takes a much more general approach to the problem and leads to the conclusion that more efficient surface source processes would invariably dominate the 'microseism' mechanism. All linear surface waves are coupled to sound during their growth phase, and Guo has shown that the underwater sound-producing ability of the wind is greater than that of the waves once the wind has stopped generating them. The frequency doubling characteristic between source and sound fields is a common feature of all sources involving non-linear source motions, a non-linearity that is at the heart of Lighthill's acoustic analogy. In fact Guo demonstrated with a particular model of the aerial turbulence that the Brekhovskikh mechanism could never be relevant; when the surface waves are weak the direct air-induced sound is stronger and when they are strong they no longer conform with the weak interaction presumption of that source model.

Guo has gone on to develop the hydro-acoustic analogy to very rough ocean conditions involving spray formation, near surface bubbles and wave breaking effects. Bubbles, though usually a dominant aspect of acoustics whenever they arise, were discounted by Guo on the grounds that their monopole field would be destroyed by that of their cancelling (negative) image in the free surface. Resonant scattering is unlikely to occur because the turbulence induced hydrodynamic pressure field at bubble resonance frequency will have too small a length scale to 'ring' a bubble coherently. This may be so, but this workshop has seen convincing evidence that the formation of bubbles is a process that is both noisy and nice. Banner and Cato's film showed clear evidence of ringing sounds whenever bubbles appeared in a turbulent water flow and there can be no doubt at all that the bubble formation process in the lovely experiments of Crum and Pumphrey is a process that excites bubble resonance and a strong associated sound field. Whether or not bubble formation is a strong oceanic source process remains a question, and I come away with the feeling that it is. Bubble formation seems likely to be the strong acoustic influence that 'bubble presence' would have been were it not for the destructive interference of the nearby free surface. There are many interesting questions surrounding this aspect that need further research, not the least of which is the possibility that rough surface reflection lacks the destructive interference potential of coherent flat pressure-release boundaries.

Oceanic spray and raindrops induce dipole sources with axes normal to the mean surface, a drop of speed U and diamater d converting its energy into waves within a time of order d/U. That kind of process, in which an inhomogeneous rough air/water interface is accelerated, is the one identified by Guo as the most likely surface source of strong underwater sound.

This workshop has seen much more excitement generated in the area of sound production than in sound propagation. Intricate sound propagation codes have been constructed to incorporate fine features of the ocean over a large scale, and it was interesting to note that the view that simple models perform as well as (if not better than) the giant codes was not significantly countered. The long distance propagation problem will probably be much simpler when it is properly understood.

Throughout the workshop there has been a recurring theme that experimental evidence is difficult to reconcile with theory. The ocean problem is very complex, involving many different parameters. Too rarely are all the relevant parameters measured in any one experiment. Workshops such as this will do an enormous service if experimenters can be made aware of all the aspects which feature in the sound generating process and which therefore need to be measured if an experiment is to give conclusive evidence for one effect over another. For this there is a clear need for new instrumentation. Bubble content might be measured with new electric or acoustic or optical probes of some kind. Sound velocity measurements are needed in detail, and the capacitance between a wire and an air/water mixture might, thought Longuet-Higgins, provide a useful new technique of measurement.

It is certain that we need sensors that can resolve both the space and time structure of wave fields if sound and surface wave fields are to be clearly identified and separated. A new facility might be justified for this, and the scale might be large enough to merit international collaboration. There was a strong view that the subject should be kept as broad as possible; the field is not yet mature enough to justify a narrow specialist concentration of resources, and the prospects of real progress coming from a broad frontal attack are better than those from a limited-objective study.

There is likely to be a strong correlation between some aspects of the aerial noise above the rough ocean and that under the waves, and this provides pressure to extend the ocean acoustics experimental programme into the air, a programme whose success depends on the quality of instrument arrays that can be deployed at one time.

PANEL DISCUSSION REPORT

WAVE AND TURBULENCE NOISE

Panel: W. M. Carey
D. H. Cato
A. C. Kibblewhite (Chairman)
R. H. Mellen

1. INTRODUCTION

The Panel was tasked to review our current understanding of low frequency ambient noise generated by the ocean surface, as revealed by this Workshop. The frequency range of interest has been defined as 0.01 to 200 Hz and for the purpose of discussion this has been divided into four subsegments. Specific comments are made on each frequency interval separately, and these are followed by a brief summary of the various noise sources cited at this Workshop, which are or might be relevant to the whole frequency band.

2. REGION 0.1 - 5 HZ

It is observed experimentally that ocean-noise spectral levels rise sharply below 5 Hz, maximum values reaching on the order of 160 dB re 1 Pa^2/Hz at frequencies around 0.2 Hz, some 100 dB above those encountered at higher frequencies. These high levels have long been recognized as being sea state related, but the fact that the effects are observed in both deep and shallow water has ruled out a simple connection with ocean surface waves.

In seeking an explanation for these high noise levels several mechanisms have been proposed. One ascribes the effects observed to nonlinear interactions between components of the wavefield [1], a mechanisms which has also been proposed as the source of marine induced microseisms [2]. Others invoke atmospheric turbulence as the noise source [3,4] or the combined interaction of atmospheric turbulence with the surface wavefield [5].

621

B. R. Kerman (ed.), Sea Surface Sound, 621–627.
© 1988 by Kluwer Academic Publishers.

Because of the difficulties associated with field programs at these frequencies, experimental studies that satisfactorily resolve between the various theoretical proposals have been slow to appear. However, as reviewed at this Workshop, strong evidence now exists to identify wave interaction processes as the dominant mechanism active in this part of the spectrum (6).

The Workshop has, nevertheless, also heard of a new analysis in support of atmospheric turbulence (7). This analysis argues that the depth independent pressure component of the wave-wave interaction field is swamped in deep water by effects which have their origin in atmospheric turbulence. The evidence relating to wave-wave interactions is questioned primarily on the ground that at the frequencies and water depths involved the true acoustic component of the noise field arising from nonlinear interactions is not resolved from the depth dependent component of the pressure field.

While no experimental data exist where the infrasonic noise field is measured as a function of water depth at one site, the results from various programs involving a wide range of water depths (including one very shallow water experiment involving somewhat higher frequencies (8)), have all been consistent with wave interaction processes. It is also relevant that the pressure levels observed experimentally are in good agreement with those predicted by wave interaction theory. If atmospheric turbulence is to dominate the effects resulting from nonlinear wave processes to the extent proposed, then, unless wave interaction theory itself is wrong (and the basic theory is not in dispute), predicted levels will be much higher than those actually observed.

Furthermore, several distinctive features of the observed noise spectra (e.g., the 2:1 frequency relationship with the wave spectra, the high frequency spectral slope, the influence of multiple seas on spectral shape) are readily explained in terms of wave interaction processes. Earlier turbulence theories have not been able to do this and it is not yet demonstrated whether this latest approach will be any more successful.

On the evidence available, the Panel concludes that wave-wave interactions remain the dominant source of the high, wind-dependent noise levels observed below 5 Hz.

3. REGION BELOW 0.1 HZ

At frequencies below the wave related peak in the ocean-noise spectrum, the spectral levels decrease in step with the fall in the spectral levels of the prevailing ocean wavefield. On the basis of seismic evidence noise levels continue to fall with decreasing frequency, reaching a minimum around 0.05 Hz before again increasing to comparatively high levels (6).

Occasionally, a narrow band of acoustic energy appears as a small secondary peak in the spectrum at frequencies around 0.06 Hz. Evidence suggests that such distortions are associated with long period sea-surface oscillations, arising from the passage of an energetic cold front across the area. Spectral levels within this secondary peak are typically two orders of magnitude lower than those in the peak associated with the wave-wave interactions, but two to three orders of magnitude above the levels of the noise minimum which normally prevails at these frequencies.

4. REGION 5 - 20 HZ

The frequency interval 5 - 20 Hz can be considered as a transition band between the region where the mechanisms dominant at extremely low frequencies (< 5 Hz) are replaced by those active at higher frequencies. While reliable experimental data for this region are limited, the leveling of the spectral slope above 5 Hz implies a wind-speed dependence which is not in accordance with the predictions of conventional gravity/capillary wave models, and indicates the dominance of another source such as atmospheric turbulence (4,9).

With respect to the former, the discrepancy may rest in the wave model rather than the wave-wave interaction mechanism per se. Indeed, recent wave-guage data from flume experiments suggest that saturation and dispersion do not occur at high wave numbers. The soliton model described at this meeting (10) is an attempt to formulate these modified properties into acoustic models. Backscattering cross sections and doppler spectra are shown to become consistent with predictions when surface fine structure is modelled heuristically as a random ensemble of solitons or hydraulic bumps. Ambient noise spectra predicted for collisions of such features also appear to give better agreement with experiment than conventional dispersive wave models. These developments confirm the need for a nonlinear wind/wave model that is consistent with experiment in all pertinent detail, so that the mechanism can be properly compared with atmospheric turbulence and other potential sources of noise in this frequency region. At this time the situation is not clear.

5. REGION 20 - 200 HZ

Experimental evidence obtained with both omnidirectional and vertically directive systems suggests that from 20 to 200 Hz the noise field is dependent on the wind in a more direct way. Source levels are found to be typically 50 dB re 1 Pa^2/Hz at frequencies around 50 Hz for low wind speeds (~ 5 ms^{-1}). However, since acoustic propagation is very good at low frequencies (< 200 Hz) these data are often corrupted by distant noise sources. This fact, coupled with the inadequacy of the meteorological data in most measurement programs, means that only a qualitative assessment of the wind dependent component of ocean noise has been established for this part of the spectrum, although at high wind speeds the spectrum level appears to vary as 20 n log (wind speed) with n$\approx$2.

6. SUMMARY OF NOISE SOURCES - FUTURE WORK

Several surface related sources of sound have been proposed at this Workshop as being active or possible active in the frequency range of interest here. These can be listed as:

1. Wave-wave interactions
2. Atmospheric turbulence
3. Wave/atmospheric turbulence interactions
4. Oceanic turbulence
5. Gross motions of the sea surface
6. Density discontinuities at the surface
7. Bubble convection by turbulence
8. Bubble cloud oscillations driven by hydrodynamic forces.
9. Soliton-like disturbances

The first of these is believed to be dominant at frequencies less than 5 Hz and may be important to somewhat higher frequencies. Atmospheric and oceanic turbulence may also be important in the transition range between 5 to 20 Hz, but neither appear to be viable mechanisms above 20 Hz because of their quadrupole nature. Wave/turbulence interaction on the other hand, with its dipole nature and linear velocity dependence, remains a possible mechanism up to frequencies of the order of 100 Hz (1, 11). Long period sea-surface motion has been identified as significant at very low frequencies (< 0.1 Hz), but has not been invoked at higher frequencies. The new mechanisms based on density discontinuities, turbulence induced bubble convection, bubble cloud oscillations due to hydrodynamic forces (12), and solitons (10), have yet to be properly assessed as possible noise-generation mechanisms at low frequencies.

The bubble related mechanisms have focused attention on the role of gas bubbles in changing the effective compressibility of the fluid, and through this the sound speed. Sound speed is known to depend on gas volume fraction and this important fact may have important consequences at frequencies below 200 Hz, for sound generation, and scattering and propagation in the upper mixed layer. Furthermore, bubble layers and bubbly mixtures are known to enhance nonlinear effects and thus may support soliton-like propagation.

Overall the Workshop has shown that substantial progress has been made in identifying sources of surface noise. It is nevertheless clear that further work is required to resolve the various issues raised at this Workshop, and it is recommended that additional experimental work with directional hydrophone systems be carried out to quantify more adequately the characteristics of the wind-dependent noise mechanisms that are active in the VLF/LF region of the spectrum. An essential element of these programs must be the parallel collection of comprehensive environmental data, including directional wave spectra. Too often in the past this essential information has been given inadequate attention.

REFERENCES

1. Lloyd, S.P., "Underwater Sound from Surface Waves According to the Lighthill-Ribner Theory," J. Acoust. Soc. Am. **69**, 425-435 (1981).

2. Longuet-Higgins, M.S., "A Theory of the Origin of Microseisms," Philos. Trans. R. Soc. London Ser. A243, 1-35 (1950).

3. Isakovich, M.A. and B. F. Kuryanov, "Theory of Low Frequency Noise in the Ocean," Sov. Phys. Acoust. 16, 49-58 (1974).

4. Wilson, J.H., "Very Low-Frequency (VLF) Wind Generated Noise Produced by Turbulent Pressure Fluctuations in the Atmosphere Near the Ocean Surface," J. Acoust. Soc. Am. **66**, 1499-1507 (1979).

5. Gorcharov, V.V., "Sound Generation in the Ocean by the Interaction of Surface Waves and Turbulence," Izv. Atmos. Ocean. Phys. 6, 1189-1196 (1970) (translated by F. Goodspeed).

6. Kibblewhite, A.C., "Ocean Noise Spectrum Below 10 Hz - Mechanisms and Measurements," This volume.

7. Flowcs-Williams, J.E., "Mechanisms of Underwater Sound Generation at the Ocean Surface," This volume.

8. Cato, D.H. and I. S. F. Jones, "Noise Generated by Motion of the Sea Surface-Theory and Measurement," This volume.

9. Adair, R.C., "Comments on the Infrasonic Noise Theory of Isakovich and Kuryanov and its Modification by Wilson," J. Acoust. Soc. Am. 81, 1192-1195 (1987).

10. Mellen, R.H. and D. Middleton, "Ambient Noise Radiation by 'Soliton' Surface Waves," This volume.

11. Yen, N.and A. J. Perrone, "Mechanisms and Modelling of Wind-Induced Low Frequency Ambient Sea Noise," NUSC Tech. Report. 5833, Naval Underwater Systems Center, New London, CT (13 February 1970).

12. Various. This volume.

PANEL DISCUSSION REPORT

PRECIPITATION NOISE

Panel: L. Bjorno (Chairman)
L.A. Crum
J. Nystuen

It was generally agreed that much more research is needed in order to
explain the mechanisms leading to rain generated noise in the sea. The
discussion was structured according to the following questions:

1. Which subjects need more research?

2. How shall these subjects be studied, i.e. can we design some
   good experiments?

3. Who shall study the subjects and will it be possible to do
   these studies in an international cooperation between more
   research groups?

1. RESEARCH TOPICS

It was felt that a strong need exists for more data related to the
following subject areas.

* Impact studies and bubble pulsation studies related to much
  smaller drops than studied previously (3 mm in equivalent
  diameter) have to be performed in order to study the influence of
  nearfield effects and in order to study the influence of the
  magnitude (dimensions and shape) of the test tanks used.
  Directivity measurements were recommended.

* Drop impacts from large heights on surfaces of random topography
  should be studied.

B. R. Kerman (ed.), Sea Surface Sound, 629–631.
© 1988 by Kluwer Academic Publishers.

* A more systematic study of the influence of the drop shape including high-speed photography and sound pressure measurements is needed.

* The air motion produced by the falling raindrops should be studied using, for instance, schlieren or shadowgraph techniques.

* The influence of electric fields on charged drops should be studied.

* The influence of shear flow in the water surface on the drop impact and its sound producing mechanisms should be studied.

* A systematic study of influence of the impact velocity using various fall distances for the same drop size should be made.

* A need is also felt for more data on the bubble size distribution from various characteristic rain rates.

* An in-depth study of the influence of surface tension including the influence of pollution, pH-values, sulphuric acid, oil spills etc. is needed.

* The influence of salinity on bubble production still needs much more research.

* The influence of the angle of drop impact with the water surface caused by the wind should be studied as no data are available in this context.

* The concerted interaction between increasing number of random time impacts by drops on the water surface should be studied.

* Oscillations of the drops during their penetration through the water surface need more research.

* The combined effects of wave breaking and raindrop impacts should be studied.

* Also the impact noise from snow flakes and hail needs more research.

* In particular, a need for more field tests was emphasized including the effects of heavy rain and wind.

* There is a need for improvements in the instruments used, for the development of special transducers for studies in various frequency ranges. In particular, the need for careful calibration was emphasized.

2. EXPERIMENTAL DESIGN

The design of good experiments which can give the data needed will demand a high degree of ingenuity of the research groups and it was generally felt that the tests to be set up could not be developed within the time frame allotted to the panel discussion, but should be left to the research groups' own decision.

3. INTERNATIONAL CO-OPERATION

The active participation in the research subjects suggested under item 1. is open to everybody.

Professor Leif Bjorno offered to act as the international coordinator for a common research program involving research in areas suggested under item 1. Proposals for international collaboration related to rain generated noise in the sea can be sent to Professor Bjorno, who in collaboration with the proposer will work out an international collaboration scheme.

ICE NOISE

Panel: I. Dyer (Chairman)
        D. Farmer
        J. Lewis

The number of distinct ice source mechanisms creating Arctic ambient noise is large. In many cases, several mechanisms at times can contribute simultaneously to the noise spectrum in the same frequency range. Thus observational programs aimed at identifying and quantifying relevant mechanisms should:

- include the widest possible description of the natural environment;

- be supported by time series of the presumed relevant environmental variations;

- be designed to observe noise and environmental variations over temporal and spatial scales sufficient to resolve the presumed relationship between noise mechanisms and the environment.

While one can identify perhaps 10 or so important environmental variations as causes of ice fracture, and hence ice noise radiation, heat flux appears quite worthy of near-term emphasis. Milne first provided the firm basis for its importance, but present researchers have identified many fundamental questions needing answers:

- the role of radiative heat flux, along with other flux terms, is apt to be crucial in setting thermally-induced stress in the ice, but has not yet been demonstrated quantitively via field measurements;

B. R. Kerman (ed.), Sea Surface Sound, 633–634.
© 1988 by Kluwer Academic Publishers.

- the influence of ice surface properties in heat transfer to the ice needs to be accounted for;

- thermal gradients in the ice could be directly measured or inferred by heat flux estimates, but the plausibility of the thermal source mechanism needs to be demonstrated via ice stress considerations and crack observations, the latter including acoustic, geophonic and/or visual techniques.

Experiments identifying and quantifying source mechanisms must be heavily oriented towards the natural environment. There is an important role, however, for laboratory or lake tests. Such idealizations are to be approached not as primary research tools, but as secondary ones able to isolate particular phenomena and to test specific hypotheses. Caution especially needs to be given to the largely uncertain elastic-plastic properties of natural sea ice, and its fracture strength measures, in interpreting laboratory or lake experiments. Also morphological properties of natural sea ice are likely difficult to reproduce in the laboratory or in a lake, especially with respect to their stochastic variations. But with these cautions, laboratory and lake experiments can be vital adjuncts to field research.

Because of their central role in determining ice fracture events and hence noise radiation, the elastic-plastic properties and strength measures of natural ice should be researched beyond knowledge already available. Of particular interest is the crack propagation speed in large-scale ice samples, the crack arresting properties of natural ice under strain, and their relationship to environmental parameters such as ice temperature, brine content, crystalline axis alignment, etc. While laboratory studies of such issues no doubt are to be encouraged, it also seems clear that basic knowledge on this topic can come from carefully instrumented field studies. Arctic ambient noise research thus is an inverse as well as a direct arena of study.

SUBJECT INDEX

Acoustic
  absorption, 225, 228, 237
  backscattering, 225, 227, 325
  coherent, incoherent spectra,
    452
  cross-spectrum, 451
  forward scattering, 248
  near surface radiation
  patterns, 454
  pressure excitation spectrum,
    450, 456
  pressure transfer spectrum, 450
Acoustics
  Lighthill's theory, 167, 330,
    391, 617
  backscatter cross-section, 624
  image effects, 618, 619
  in-air audio spectrum, 444
  inverse filtering, 447, 609
  self-similar spectra, 446
  sound speed, 612, 613, 620
  sources, 617
  splash noise, 609
  tomography, 613
  water tank, reverberation, 438
Aerosol
  surface flux, 90
Ambient noise
  Knudsen region, 131
  angular source characteristics,
    380
  capabilities and objectives,
    126, 127
  coupling to ocean bottom, 352
  depth dependence, 284
  effective source depth, 381
  environmental effects, 284
  evaluation experiment, 123, 128
  evidence of wind effect, 362
  forced collective bubble
    oscillations, 369
  generation by laboratory
    breaking wave, 429
  high frequency, wind, 403, 417
  low frequency spectrum, shallow
    water, 340
  low frequency spectrum, deep
    water, 341
  low frequency generation, 361

Ambient noise
  model, ocean transfer function,
    377, 378, 387
  plane and volumetric sources,
    381
  radiation by 'soliton' waves,
    325
  sea surface motion, generation,
    391
  shallow water, 273, 281
  source characteristics from
    noise field, 377
  source correlation function,
    281
  source levels from omni-
    directional sensor, 384, 388
  spatial properties, 265
  spatial variability, 132, 145
  spectrum, low frequency, 337
  spectrum, wind speed
    dependance, 423
  surface interference model, 381
  tone busts in spilling breaker,
    432
  vertical noise directionality,
    364, 378
  wave-turbulence interaction,
    367, 368
  wave-wave interaction, theory,
    338, 391, 397
Aries
  sonar system, 180
Boundary-layer
  air-water coupling, 51
  below wind wave, 51
  similarity structures, 53
Bubble density
  acoustic estimate, 225, 237
  depth integrated, 226, 234
Bubble observations
  filtered water, 203
  glacial lake, 204
  depth-dependence, 243
  near-shore effect, 232
  open ocean, 228, 242, 247
  size-dependence, 244
  temperature dependence, 230
Bubble
  acoustic generation, 608
  acoustic radiation, 159

636

638